Second Supplements to the 2nd Edition of

RODD'S CHEMISTRY OF CARBON COMPOUNDS

ELSEVIER SCIENCE B.V.
Sara Burgerhartstraat 25
P.O. Box 211, 1000 AE Amsterdam, The Netherlands

ISBN: 0-444-81483-3

This book is printed on acid-free paper.

Printed in The Netherlands.

Second Supplements to the 2nd Edition of

RODD'S CHEMISTRY OF CARBON COMPOUNDS

VOLUME I

ALIPHATIC COMPOUNDS
★

VOLUME II

ALICYCLIC COMPOUNDS
★

VOLUME III

AROMATIC COMPOUNDS
★

VOLUME IV

HETEROCYCLIC COMPOUNDS
★

VOLUME V

MISCELLANEOUS

GENERAL INDEX
★

Second Supplements to the 2nd Edition of

RODD'S CHEMISTRY OF CARBON COMPOUNDS

A modern comprehensive treatise

Edited by
MALCOLM SAINSBURY
*School of Chemistry, The University of Bath,
Claverton Down, Bath BA2 7AY, England*

Second Supplement to

VOLUME II ALICYCLIC COMPOUNDS

Part B: Six- and Higher-membered Monocarbocyclic
Compounds (Partial: Chapters 6-8 in this Volume)
Part C: Polycarbocyclic Compounds excluding Steroids
Parts D and E: Steroids

ELSEVIER
Amsterdam – London – New York – Tokyo 1994

Contributors to this Volume

J. CONNOLLY

Department of Chemistry, The University, Glasgow G12 8QQ, Scotland

D. EWING

School of Chemistry, University of Hull, Hull HU6 7RX, England

D.H. GRAYSON

Department of Chemistry, Trinity College, University of Dublin
Dublin 2, Ireland

J.R. HANSON

School of Chemistry and Molecular Sciences, University of Sussex,
Falmer, Brighton BN1 9QJ, England

A. HEWSON

University of Sheffield, Pond Street, Sheffield S1 1WB, England

R. HILL

Department of Chemistry, The University, Glasgow G12 8QQ, Scotland

S.L. JENSEN

Organic Chemistry Laboratories, The Norwegian Institute of Technology,
University of Trondheim, N-7034 Trondhiem, Norway

R. LIVINGSTONE

Department of Chemical and Physical Sciences, The University,
Queensgate, Huddersfield HD1 3DH, England

S.B. MAHATO

Indian Institute of Chemical Biology, 4 Raja S.C. Mullick Road,
Jadavpur, Calcutta-700 032, India

A.P. MARCHAND

Organic Chemical Dynamics, National Science Foundation,
1800 G Street, N.W., Washington, DC 20550, U.S.A.

M. SAINSBURY

School of Chemistry, The University, Claverton Down,
Bath BA2 7AY, England

M. WILLS

School of Chemistry, The University, Claverton Down,
Bath BA2 7AY, England

F.J. ZEELEN

Floraliastraat 2, 5384 GP Heesch, The Netherlands

Preface to volume II B (partial), C, D and E

The theme of this volume is alicyclic chemistry, a very broad canvas indeed, embracing natural products such as terpenes, steroids and carotenoids and an array of synthetic compounds, including both medium-size and large ring monocycles and polycycles.

In order to cope with these diverse and ever expanding subject areas an international team of authors has been recruited and I am extremely grateful to them for their expertise in selecting and presenting the major advances in the this area of chemistry since the first supplement was published in 1974. It will be noted that in the first supplement Chapter 8 dealt with cycloheptanes, cyclooctanes and macrocyclic compounds as a single topic, here the main chapter is subdivided so that macrocycles are covered separately.

Much of this supplement deals with the chemistry of the terpenoids, a subject in which there has been a surge of interest, following the discovery of many new structural types with wide ranging biological activities. It has not been possible to provide a blanket coverage of this huge area, thus Chapter 15 focuses solely on the steroids, Chapter 16 on glycosides and Chapter 17 on triterpenoids. The cardiotoxic glycosides, toad poisons etc. which formed the basis of Chapter 17 in the first supplement have not been selected for detailed attention; similarly steroid saponins and sapogenins, the subjects of Chapter 18 in the previous work, have not received special treatment. These topics are now considered more generally in both Chapters 15 and 16.

Chapter 19 in the first supplement, dealing with the biosynthesis of carotenoids and steroids, is also deleted from this volume, not because there is a lack of interest but because all the important innovations in these areas are to be found in Chapters 7 (carotenoids) and 15 (steroids).

Malcolm Sainsbury
Bath
June 1993

Contents
Volume II B (partial), C, D and E

Alicyclic Compounds; Six- and Higher-membered Monocarbocyclic Compounds (Partial)

Polycarbocyclic Compounds excluding Steroids

Steroids

Chapter 6. Acyclic and Monocyclic Monoterpenoids
by D.H. GRAYSON

Chapter 7. The Carotenoid Group of Natural Products
by S. LIAAEN-JENSEN

Chapter 8a. The Cycloheptanes and Cyclooctanes
by D.F. EWING

Chapter 8b. Large Alicyclic Ring Systems
by D.F. EWING

*Chapter 9. Polycarbocyclic Compounds with Separate Ring Systems,
and Spiro Compounds*
by M. WILLS

Chapter 10. Polycyclic Compounds. Fused or Condensed Cyclic Systems
by M. SAINSBURY

Chapter 11. Polycarbocyclic Bridged Ring Compounds
by A.P. MARCHAND

Chapter 12. Bicarbocyclic Natural Products
by R. LIVINGSTONE

Chapter 16. Glycosides, Saponins and Sapogenins
by S.B. MAHATO

Chapter 17. Triterpenoids
by J.D. CONNOLLY and R.A. HILL

List of Common Abbreviations and Symbols Used

A	acid
Å	Ångström units
Ac	acetyl
a	axial
as, asymm.	asymmetrical
at.	atmosphere
B	base
Bu	butyl
b.p.	boiling point
c, C	concentration
CD	circular dichroism
conc.	concentrated
D	Debye unit, 1×10^{-18} e.s.u.
D	dissociation energy
D	dextro-rotatory; dextro configuration
d	density
dec., decomp	with decomposition
deriv.	derivative
E	energy; extinction; electromeric effect
*E*1, *E*2	uni- and bi-molecular elimination mechanisms
E1cB	unimolecular elimination in conjugate base
ESR	electron spin resonance
Et	ethyl
e	nuclear charge; equatorial
f.p.	freezing point
G	free energy
GLC	gas liquid chromatography
g	spectroscopic splitting factor, 2.0023
H	applied magnetic field; heat content
h	Planck's constant
Hz	hertz
I	spin quantum number; intensity; inductive effect
IR	infrared
J	coupling constant in NMR spectra
J	Joule
K	dissociation constant
k	Boltzmann constant; velocity constant
kcal	kilocalories
M	molecular weight; molar; mesomeric effect
Me	methyl
m	mass; mole; molecule; *meta-*
m.p.	melting point
Ms	mesyl (methanesulphonyl)

[M]	molecular rotation
N	Avogadro number; normal
NMR	nuclear magnetic resonance
NOE	Nuclear Overhauser Effect
n	normal; refractive index; principal quantum number
o	*ortho-*
ORD	optical rotatory dispersion
P	polarisation; probability; orbital state
Pr	propyl
Ph	phenyl
p	*para-*; orbital
PMR	proton magnetic resonance
R	clockwise configuration
S	counterclockwise configuration; entropy; net spin of incompleted electronic shells; orbital state
S_N1, S_N2	uni- and bi-molecular nucleophilic substitution mechanism
S_Ni	internal nucleophilic substitution mechanism
s	symmetrical; orbital
sec	secondary
soln.	solution
symm.	symmetrical
T	absolute temperature
Tosyl	p-toluenesulphonyl
Trityl	triphenylmethyl
t	time
temp.	temperature (in degrees centrigrade)
tert	tertiary
UV	ultraviolet
α	optical rotation (in water unless otherwise stated)
$[\alpha]$	specific optical rotation
ϵ	dielectric constant; extinction coefficient
μ	dipole moment; magnetic moment
μ_B	Bohr magneton
μg	microgram
μm	micrometer
λ	wavelength
ν	frequency; wave number
χ, χ_d, χ_μ	magnetic; diamagnetic and paramagnetic susceptibilities
(+)	dextrorotatory
(−)	laevorotatory
−	negative charge
+	positive charge

Second Supplements to the 2nd Edition of Rodd's Chemistry of Carbon Compounds, Vol. II B(Partial), C, D and E, edited by M. Sainsbury

1

Chapter 6

ACYCLIC AND MONOCYCLIC MONOTERPENOIDS

D.H. GRAYSON

1. Introduction

This Chapter reviews developments in the chemistry of acyclic
and monocyclic monoterpenoids which have taken place between
1973 and 1992. It seeks to provide a representative overview
of these developments rather than a catalogue of each and
every new result which has been reported. Thus, reactions
which are of general rather than specific application have
been given prominence, and particular attention is drawn to
some areas of the subject where there has been notable
activity. These include biotransformation reactions, the conversion of acyclic monoterpenoids into cyclic systems, and
applications of monoterpenoids as chiral reagents and auxiliaries for asymmetric synthesis.

Pressure on space has meant that there can only be a brief
discussion of the novel polyhalogenated monoterpenoids which
have been isolated from marine sources, and no attempt has
been made to survey the large array of novel iridoids and
seco-iridoids which have been discovered in the plant kingdom. In cases such as these the reader is directed to appropriate Review articles which have been published.

A wealth of information on the monoterpenoids is available
from sources such as the "Dictionary of the Terpenoids", Vol.
1, ed. J.D. Connolly and R.A. Hill, Chapman and Hall, London,
1991, J.S. Glasby, "Encyclopaedia of the Terpenoids", Wiley,
New York, 1982, chapters in "Specialist Periodical Reports -
Terpenoids and Steroids", published by The Chemical Society/
Royal Society of Chemistry, London, written by A.F. Thomas
(Vol. 4, 1974 and Vol. 5, 1975), R.B. Yeats (Vol. 6, 1976,
Vol. 7, 1977, Vol. 8, 1978 and Vol. 9, 1979) and by D.V. Ban-
thorpe and S.A. Branch (Vol. 12, 1983), chapters in "Special-

2

ist Periodical Reports - Aliphatic and Related Natural Prod-
uct Chemistry", also published by The Chemical Society/Royal
Society of Chemistry, London, written by D.H. Grayson (Vol. 1,
1979, Vol. 2, 1981 and Vol. 3, 1983), and in D.H. Grayson,
Nat. Prod. Rep., 1984, 1, 319, M.S. Carson and D.H. Grayson,
Nat. Prod. Rep., 1986, 3, 251, and D.H. Grayson, Nat. Prod.
Rep., 1987, 4, 377, 1988, 5, 419, 1990, 7, 327 and 1992, 9,
531.

13-C NMR spectra of many acyclic and monocyclic monoterpenoids have been recorded (F. Bohlmann, R. Zeisberg and E.
Klein, Org. Magn. Resonance, 1975, 7, 462: A.-ur-Rahman and
V. Uddin, "13-C NMR of Natural Products", Vol. 1, "Monoterpenes and Sesquiterpenes", Plenum, New York, 1991) as have
other spectroscopic data (A.A. Swigar and R.M. Silverstein,
"Monoterpenes: Infrared, Mass, Proton and Carbon-13 NMR
Spectra, and Kovats Indexes", Aldrich Chem. Co., Milwaukee,
1981).

The synthesis of monoterpenoids has attracted continued
attention, and routes to various compounds have been critic-
ally discussed (A.F. Thomas and Y. Bessiere in "Total Synthesis of Natural Products", Vol. 7, ed. J. ApSimon, Wiley, New
York, 1988). Industrial-scale routes to economically valuable
monoterpenoids have also been reviewed (J.M. Derfer and M.M.
Derfer in "Kirk-Othmer Encyclopedia of Chemical Technology",
Vol. 2, 3rd edn., ed. M. Grayson and D. Eckroth, Wiley, New
York, 1983).

Three monoterpenoid cyclases have been isolated from the
flavedo of *Citrus limonum* (G. Portilla and M.C. Rojas, Biochem. Int., 1989, 18, 173). Two of these can utilise both
geranyl (1) and neryl diphosphates (2) as substrates, and the
third acts specifically on linalyl diphosphate (3). Two cyclase enzymes which have been isolated from leaves of *Salvia*

(*E*)-1
(*Z*)-2

(3)

officinalis, and for which geranyl diphosphate (2) is a substrate have been carefully studied and their stereochemical

preferences determined (R. Croteau et al., J. Biol. Chem., 1988, 263, 10063). Useful reviews on monoterpenoid biosynthesis include articles on enzymology (O. Cori, Phytochemistry, 1983, 22, 331), on the mechanisms of enzyme-mediated biocyclisation reactions (D.E. Cane, Ann. N.Y. Acad. Sci., 1986, 471) and on the biosynthesis and catabolism of cyclic systems (R. Croteau, Chem. Rev., 1987, 87, 929). Some monoterpenoids such as linalool and geraniol can sometimes be derived from L-leucine rather than from mevalonic acid, and this biosynthetic pathway has been further explored (see, for example, P. Anastasis et al., J. Chem. Soc., Perkin Trans. 1, 1987, 2427).

The desirability of being able to obtain monoterpenoids from plant sources without relying on variable natural growth and harvesting conditions has led to a significant number of studies on their production using plant cell culture systems (review: A.J. Parr, J. Biotechnol., 1989, 10, 1). For example, callus cultures of *Santolina chamaecyparissus* form myrcene (4) and β-phellandrene (5) in yields similar to those obtained from the parent plants (M.A. Baig, D.V. Banthorpe and S.A. Branch, Fitoterapia, 1989, 60, 184).

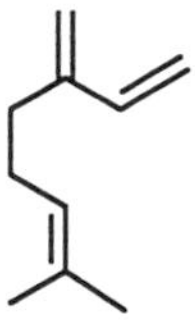

(4) 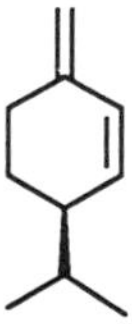   (5) 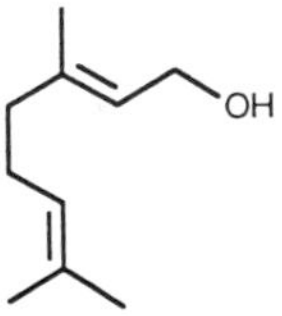   (6)

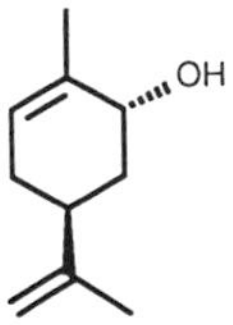

(7) 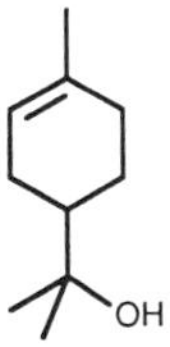   (8)

It has become increasingly apparent that glycosylated monoterpene alcohols are commonly present in plants, and that these are often readily formed. Thus, the injection of geraniol (6) into certain apples is quickly followed by the appear-

ance of its β-D-glucoside (R.B.H. Wills and F.M. Scriven,
Phytochemistry, 1979, 18, 785). The β-D-glucosides of *trans*-
carveol (7) and of α-terpineol (8) have been found in various
Citrus spp. (Y. Matsubara et al., Yukagaku, 1988, 37, 13),
and the aglycon specificities of several β-D-glucosidases
towards a range of monoterpenyl-β-D-glucosides obtained from
Rosa spp. have been determined (I.E. Ackermann et al., J.
Plant Physiol., 1989, 134, 567). High concentrations of mono-
terpenyl glucosides occur in Muscat grapes during growth, the
free terpenols not becoming prevalent until after the fruit
reaches commercial maturity (P.J. Williams et al., Dev. Food
Sci., 1985, 10, 349). Glycosylation may play a part in facilitating storage or transport within the plant cell (*cf.* E.
Stahl-Biskup, Flavour Fragrance J., 1987, 2, 5 and H. Pfander
and H. Stoll, Nat. Prod. Rep., 1991, 8, 69).

The rationale for monoterpenoid production by plants has
received consideration (G. Ourisson, Pure Appl. Chem., 1990,
62, 1401). It seems reasonable to speculate that the allelopathic characteristics exhibited by many plant oil constituents has led to the predominance of one plant species over
another in a given location. For example, monoterpenoids
exuded by the fire-sensitive shrub *Conradina canescens* grow-
ing in Florida inhibit the seed-germination and growth of
some accompanying fire-prone grasses (G.B. Williamson et al.,
J. Chem. Ecol., 1989, 15, 1567). The allelopathic properties
of monoterpenoids have been reviewed (S.D. Elakovich, A.C.S.
Symp. Ser., 1988, 380, 250), and further investigation of
this phenomenon seems likely to provide leads for the development of new, ecologically acceptable agrochemicals.

2. Artemisyl, Santolinyl, Chrysanthemyl and other Irregular Systems

The nor-monoterpenoid (9), specific rotation -71°, has been
isolated from *Artemisia schimperi*, and may arise via oxidative degradation of the corresponding triene, but its absolute

(9)   (−)-(10)   (11)

configuration has not yet been determined (B.M. Abegaz and W. Herz, Phytochemistry, 1991, 30, 1011). (S)-(-)-Lyratrol (10), obtained from *Cyanthocline lyratra*, has been synthesised from chrysanthemic acid (R.G. Gaughan and C.D. Poulter, J. Org. Chem., 1979, 44, 2441), and various lyratryl esters have been found in *Chrysanthemum coronarium* (F. Bohlmann and U. Fritz, Phytochemistry, 1979, 18, 1888). Isolyratrol (11) is a constituent of *Artemisia arbuscula* (W.W. Epstein and L.A. Gaudioso, Phytochemistry, 1984, 23, 2257), and neolyratrol (12) is present in *A. tridentata* var. *rothrockii* (W.W. Epstein, L.A. Gaudioso and G.B. Brewster, J. Org. Chem., 1984, 49, 2748). The same source provides the volatile (-)-rothrockene (13) which has been synthesised (M.R. Barbachyn, C.R. Johnson and M.D. Glick, J. Org. Chem., 1984, 49, 2746).

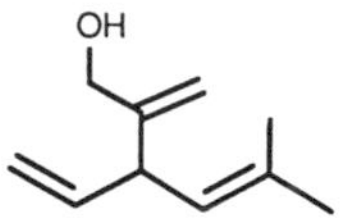

(12)

(-)-(13)

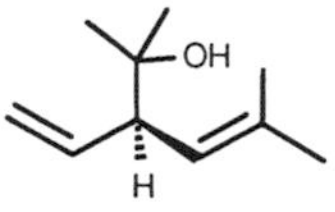

(14)

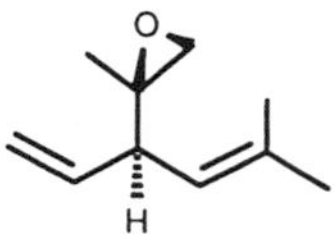

(15)

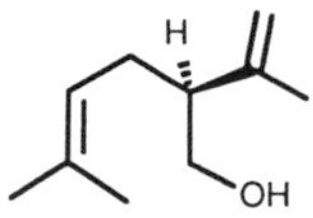

(+)-(16)

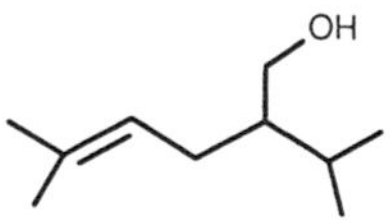

(17)

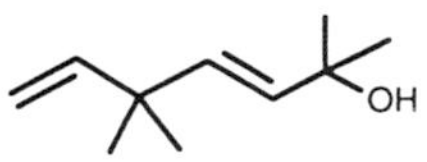

(18)

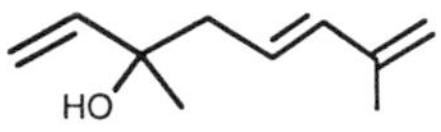

(19)

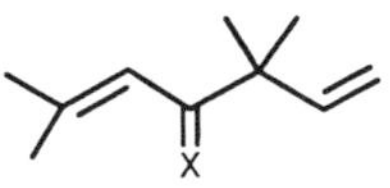

(20) X = H, OH
(21) X = O

Santolina alcohol (14) is predominant in *Ormenis mixta* (B. Toulemonde and D. Beauverd, Perfums, Cosmet., Aromes, 1984,

60 65), and the oxidosantolinatriene (15) has been reported
from *Artemisia tridentata* (T.A. Noble and W.W. Epstein,
Tetrahedron Lett., 1977, 3933).

Several syntheses of lavandulol (16) (e.g., M. Majewski et
al., J. Org. Chem., 1981, 46, 2029), dihydrolavandulol (17)
(T. Fujita et al., J. Chem. Technol. Biotechnol., 1982, 32,
476), yomogi alcohol (18) (e.g., M. Wada et al., Chem.
Letters, 1977, 557) and hotrienol (19) (e.g., M. Guittet and
S. Julia, Synth. Commun., 1979, 9, 317) have been accomplishe
Artemisia alcohol (20), which is the allylic isomer of yomog:
alcohol, has also been synthesised (J.F. Ruppert and J.D.
White, J. Org. Chem., 1976, 41, 550), and many routes to
artemisia ketone (21) have been devised (see, for example,
G. Delaris, J.P. Pillot and R.C. Rayer, Tetrahedron, 1980, 3(
2215). A useful general reaction which provides a single-ste;
synthesis of (21) involves a coupling between the lithium
enolate of the prenoyl synthon (22) and dimethylallylmagnes-
ium chloride (C. Fehr and J. Galindo, J. Org. Chem., 1988, 5:
1828).

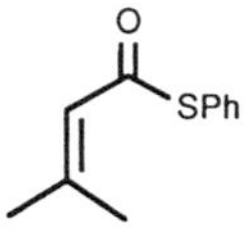

(22) 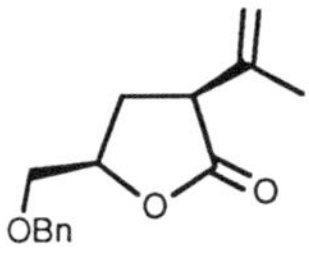   (23)

The butanolide (23), derived from (S)-1-O-benzyl glycerol,
has been exploited as a common intermediate for synthesis of
the four compounds (+)-lavandulol (16), (-)-santolinatriene
(10), (+)-rothrockene (*ent*-13) and the unnatural (-)-*cis*-
chrysanthemol (24) (S. Takano et al., J. Org. Chem., 1985, 50
931). Pyrolysis of the lithium salt of the *p*-toluenesufonyl-
hydrazone (25) yields santolinatriene (10) (T. Sasaki et al.,
J. Org. Chem., 1973, 38, 4095).

(24)    (25)

Rearrangement reactions characteristic of artemisyl, santolinyl and lavandulyl compounds have been reviewed (C.D. Poulter, Agric. Biol. Chem., 1974, 22, 167), as have biosynthetic pathways to the non H-T monoterpenoids (C.D. Poulter, Acc. Chem. Res., 1990, 23, 70). These may (Scheme 1) involve rearrangements of the chrysanthemyl cation (26) which could behave as a common intermediate for both the santolinyl and artemisyl systems (W.W. Epstein and L.A. Gaudioso, J. Org. Chem., 1979, 44, 3113, C-H.R. King and C.D. Poulter, J. Amer. Chem. Soc., 1982, 104, 1413). In accord with this hypothesis,

SCHEME 1

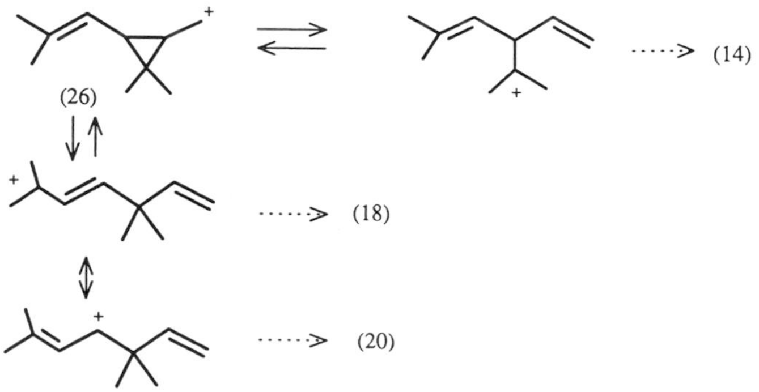

in vitro solvolyses of the chrysanthemyl derivatives (27) afford mainly yomogi (18) and artemisia alcohols (20), together with some santolina alcohol (14) (C.D. Poulter et al., J. Amer. Chem. Soc., 1977, 99, 3816).

(27) R = ODNB, OMs, OP, etc.

Evidence has been obtained for an alternative pathway involving "direct" biosynthesis from dimethylallyl and isopentenyl diphosphates wherein a sulfhydryl enzyme may play a leading part (D.V. Banthorpe, S. Doonan and J.A. Gutowski, Phytochemistry, 1977, 16, 85). A definitive description of the biogenesis of these irregular monoterpenoids, which are found mainly in members of the Compositae, has yet to be provided

Much work on the synthesis of chrysanthemic acid (28) has been carried out, stimulated by its importance as the iso-prenoid moiety of the natural pyrethroid insecticides such as pyrethrin-1 (29). While there have been many syntheses of racemic (28), some of them very ingenious (see, for example, M.J. Bullivant and G. Pattenden, J. Chem. Soc., Perkin Trans. 1, 1976, 256: W.T. Brady, S.J. Norton and J. Ko, Synthesis, 1983, 1002), emphasis has shifted to the development of ster-eocontrolled and enantioselective routes because of the dep-endence of biological activity upon stereochemistry (for reviews see T. Aratami, Kagaku, Zokan (Tokyo), 1985, 133: A. Krief et al., Pure Appl. Chem., 1990, 62, 1311). For example, the reaction of 2,5-dimethylhexa-2,4-diene with ethyl diazo-acetate which is catalysed by the copper(1) complex of the

(+)-(28)

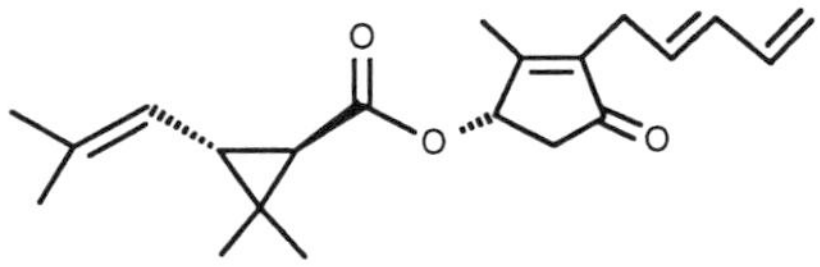

(29)

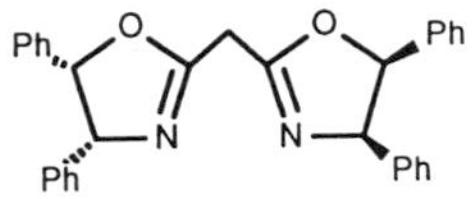

(30)

chiral ligand (30) yields the *trans*-chrysanthemate (28) of 94% e.e. together with traces of the corresponding *cis*-isomer (R.E. Lowenthal and S. Masamune, Tetrahedron Lett., 1991, 32, 7373). Natural tartaric acid has been converted (Scheme 2) into chrysanthemic acid (28) and into the *cis*-dibromo analogue (31) which is a precursor to some synthetic deltamethrin pyrethroids (A. Krief, W. Dumont and P. Pasan, Tetrahedron Lett., 1988, 29, 1079). (+)-*trans*-Chrysanthemic acid (28) has also been obtained by enantioselective hydrol-ysis of the corresponding racemic methyl ester using pig liver esterase (M. Schneider, N. Engel and H. Boensurann, Angew. Chem., 1984, 96, 52). Routes to optically active *cis*-chrysanthemate derivatives from chiral glyceraldehyde have also been described (A. Krief et al., Synthesis, 1990, 275).

SCHEME 2

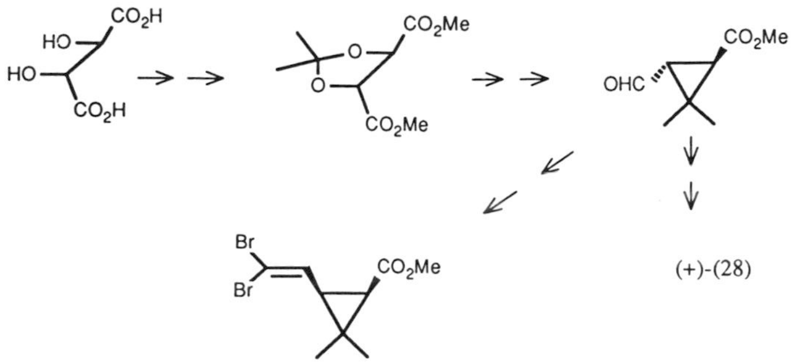

(31)

(+)-(28)

3. 2,6-Dimethyloctanes

The novel hydroperoxides (32) and the cyclic peroxides (33)
have been obtained from *Artemisia aucheri* (A. Rustaiyan et
al., Phytochemistry, 1987, 26, 2307). Other new dimethyloct-
ane derivatives include the linalyl carboxylic acid (34) from
A. santolinifolia (K. Nabeta and H. Sugisawa, Saibo Kogaku,
1985, 4, 382), its geometric isomer (35) from *Kickxia spuria*

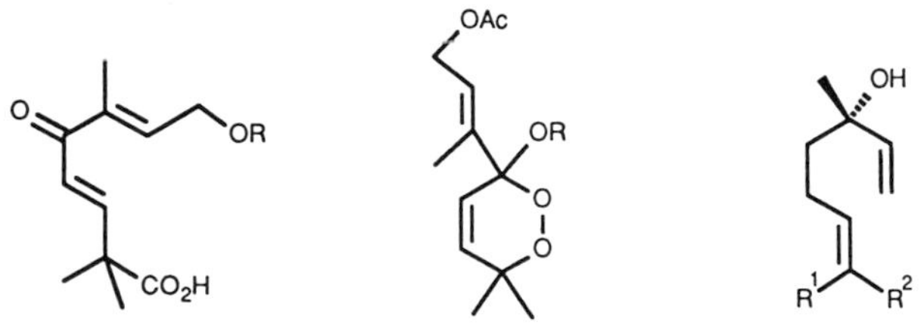

(32) R = H or Ac

(33) R = H or Me

(34) $R^1 = CO_2H$; $R^2 = Me$
(35) $R^1 = Me$; $R^2 = CO_2H$

(M. Nicoletti, L. Tomassini and M. Serafini, Fitoterapia, 1989, <u>60</u>, 252), the dehydroneryl isovalerate (36) from *Anthemis montana* (F. Bohlmann and H. Kapteyn, Tetrahedron Lett., 1973, 2065), and the isogeraniol (37) from a cultivar of *Vitis vinifera* (O. Shoseyov et al., Vitis, 1990, <u>29</u>, 159).

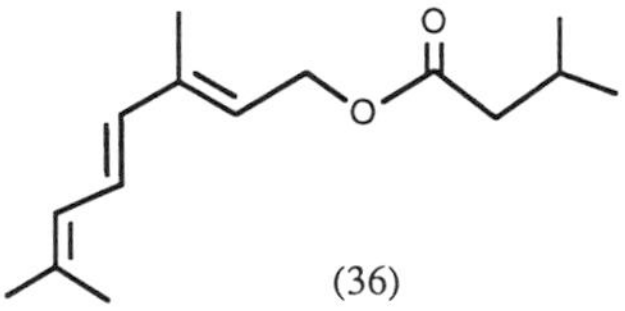

(36)

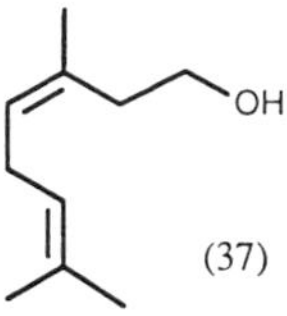

(37)

Geranyl (38) and linalyl (39) glucosides occur in shoots of *Camellia sinensis* (T. Takeo, Phytochemistry, 1981, <u>20</u>, 2145: T.Takeo, Phytochemistry, 1981, <u>20</u>, 2149) and these, together with the corresponding neryl (40) and (*S*)-citronellyl (41) glucosides have been synthesised in good yields by treating the alcohols with acetobromoglucose in the presence of silver silicate followed by acetate hydrolysis (H. Paulsen et al., Liebigs Ann. Chem., 1985, 1513: K.E. Ishag , H. Jork and M. Zeppezauer, Fresenius' Z. Anal. Chem., 1985, <u>321</u>, 331).

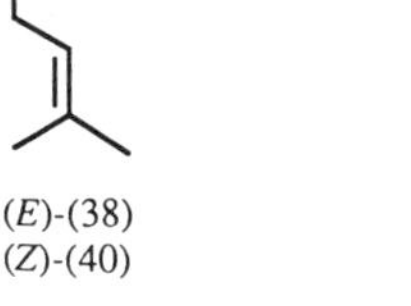

(*E*)-(38)
(Z)-(40)

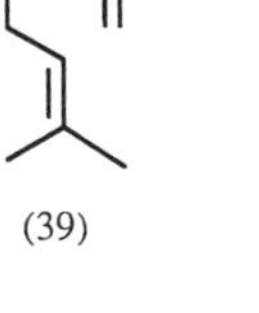

(39)

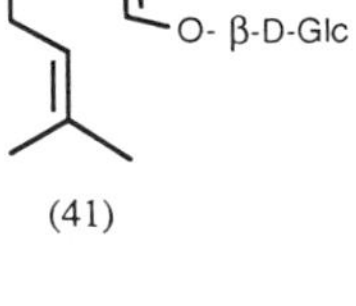

(41)

cis-β-Ocimene (42) is a major component of the oil which is obtained from the Jaffna cultivar of *Mangifera indica* (A.J. MacLeod and N.M. Pieris, Phytochemistry, 1984, 23, 361), whilst trans-β-ocimene (43) is prevalent in *Lindera glauca* (H. Nii et al., Nippon Nogei Kagaku Kaishi, 1983, 57, 433) and in *Parabenzoin trilobum* (*idem.*, *ibid.*, 1983, 57, 663). A synthesis of trans-α-ocimene (44) has been described (O.P. Vig et al., Indian J. Chem., 1977, 15B, 25).

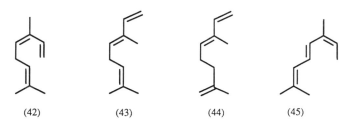

| (42) | (43) | (44) | (45) |

Pyrolysis of allo-ocimene (45) yields products which arise as a result of [1,5]- and [1,7]-hydrogen shifts and [3,3]-electrocyclisations (K.J. Crowley and S.G. Traynor, Tetrahedron, 1978, 34, 2783).

Ipsenol (46) and ipsdienol (47) are pheromones of the male bark beetle *Ips paraconfusus*. Racemic ipsenol has been obtained by pyrolysis of the cyclobutene (48) (S.R. Wilson and L.R. Phillips, Tetrahedron Lett., 1975, 3047), and by an ene reaction between isoprene and 3-methylbutanal which is catalysed by dimethylaluminium chloride (B.B. Snider and D.J. Rodini, Tetrahedron Lett., 1980, 21, 3377). Both ipsenol (46)

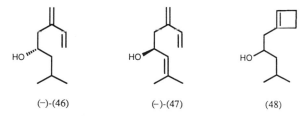

| (−)-(46) | (−)-(47) | (48) |

and ispdienol (47) can be made (Scheme 3) from the useful synthon (49) which is accessible from isoprene (A.G. Martinez and J.L.M. Contelles, Synthesis, 1982, 742). The absolute configuration of (−)-ipsenol (46) has been established by its synthesis from L-leucine (K. Mori, Tetrahedron Lett., 1975, 2187) and from (S)-lactic acid (B.M. Trost and M.S. Rodriguez, Tetrahedron Lett., 1992, 33, 4675), and (−)-ipsdienol (47) has been synthesised from a chiral irontricarbonyl complex

12

(M. Franck-Neumann, D. Martina and M.P. Heitz, Tetrahedron Lett., 1989, 30, 6679).

SCHEME 3

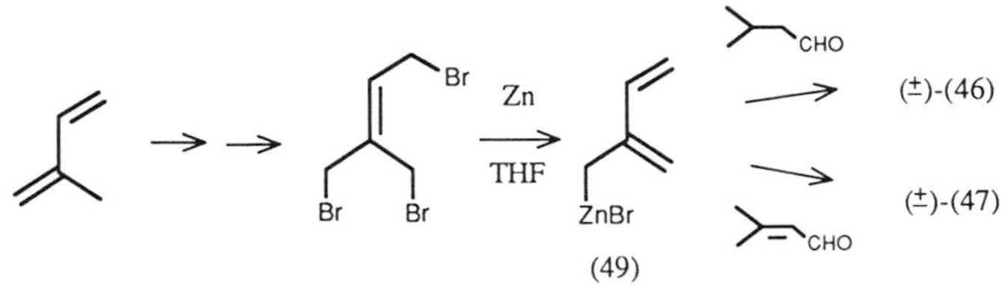

Brown has described (Scheme 4) an elegant enantioselective route to ipsenol and ipsdienol by which they are each formed in 96% e.e. (H.C. Brown and R.S. Randad, Tetrahedron, 1990, 46, 4463). The enantioselective isoprenylation reaction involved is clearly capable of wider application in monoterpenoid synthesis.

SCHEME 4

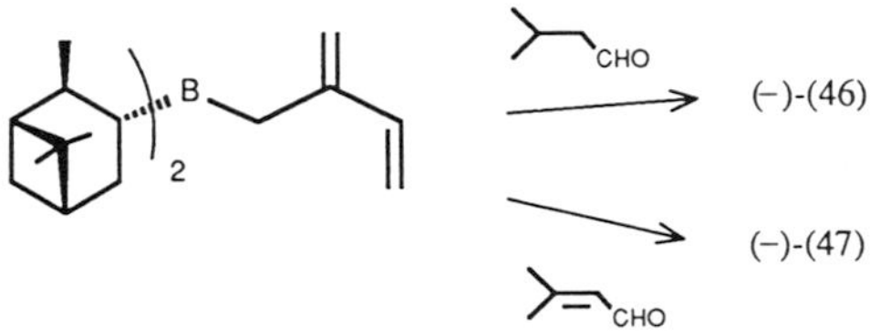

Ipsenone (50) has been found in the plant kingdom (*Lippia multiflora*) for the first time (G. Lamaty et al., Phytochemistry, 1990, 29, 521).

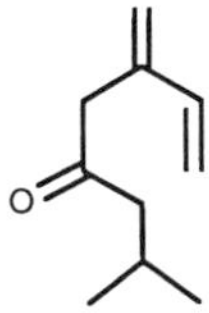

(50)

Myrcene (4) reacts with chloroform - NaOH in the presence of tertiary amines to give the dichlorocarbene adduct (51) regiospecifically (Y. Kimura, K. Isagawa and Y. Otsuji, Chem. Letters, 1977, 951). The same product can be obtained by protecting the conjugated diene system of (4) as its iron-tricarbonyl complex (52), reacting this with dichlorocarbene, and then decomplexing using cupric chloride (G.A. Taylor, J. Chem. Soc., Perkin Trans. 1, 1979, 1716). Tricarbonylmyrcene-iron (52) cyclises to (53) when it is treated with a catalytic amount of fluoroboric acid, thus providing an entry to the ochtodane skeleton (A.J. Pearson, Aust. J. Chem., 1976, 29, 1841).

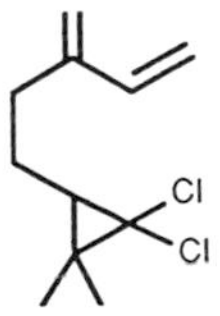

(51) 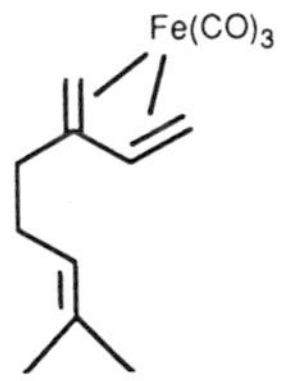 (52) 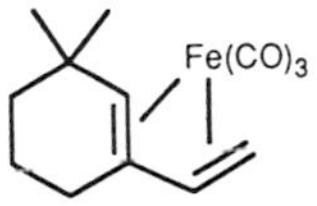 (53)

The dye-sensitised photo-oxidation of myrcene (4) leads initially to the isomeric hydroperoxides (54) and (55) which can then react further to form the peroxy-hydroperoxides (56)

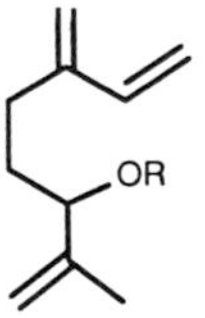

(54) R = OH
(58) R = H

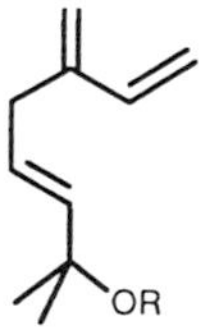

(55) R = OH
(59) R = H

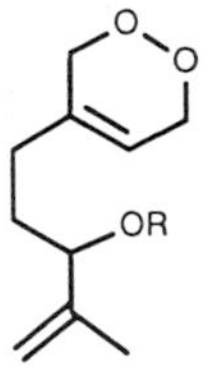

(56) R = OH
(60) R = H

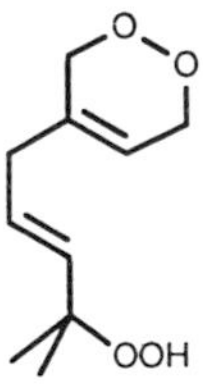

(57) 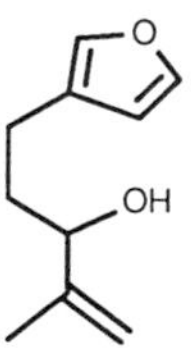 (61) 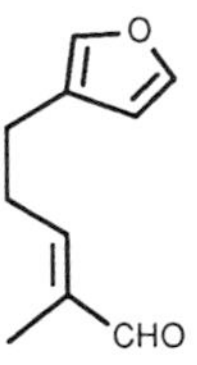 (62)

14

and (57) (M. Matsumoto and K. Kondo, J. Org. Chem., 1975, 40, 2259), results which have been rationalised using frontier molecular orbital theory (L.A. Paquette and D.C. Liotta, Tetrahedron Lett., 1976, 2681). If the photo-oxidation reaction of (4) is carried out in the presence of tetrabutylammonium borohydride then the alcohols (58) and (59) are obtained via *in situ* reduction of the intermediate hydroperoxides (P. Baeckstrom et al., Acta Chem. Scand., Ser. B, 1982, 36, 31). The alcohol (59) is a natural product, having been found in the frass of *Ips paraconfusus* and in the gut of *I. amanitus*. Continued photo-oxidation of (58) leads to the peroxy-alcohol (60). This can be converted by iron(2) salts into the furan (61) which undergoes oxidation with accompanying allylic rearrangement to give perillenal (62), the main volatile component in glands of both sexes of the pine sawfly *Neodiprion sertifer*.

Irradiation of myrcene (4) in the presence of copper(1) triflate leads to the bicyclic olefin (63) (K. Avasthi, S.R. Raychaudri and R.G. Salomon, J. Org. Chem., 1984, 49, 4322). This result contrasts with the direct and sensitised photolyses of (4) which yield bicyclo[3.1.1]heptanes or bicyclo-[2.1.1]hexanes, respectively. Irradiation of myrcene (4) in the presence of, for example, methyl 3-oxobutanoate gives the β-keto ester (64) (H. Takeshita, K. Komiyama and K. Okaishi, Bull. Chem. Soc. Jpn., 1985, 58, 2725).

(63)   (64)

Reaction of myrcene (4) with Mg-THF in the presence of a Lewis acid gives "myrcenemagnesium" which may have the structure (65). This reacts with water to yield the dienes (66) and (67), and with electrophiles such as acetone which gives the carbinols (68) and (69) in the ratio 31 : 9 (S. Akutagawa and S. Otsuka, J. Amer. Chem. Soc., 1976, 98, 7420: R. Baker, R.C. Cookson and A.D. Saunders, J. Chem. Soc., Perkin Trans. 1, 1979, 1809 and 1815).

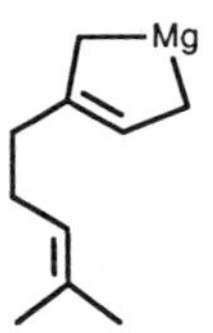

(65)

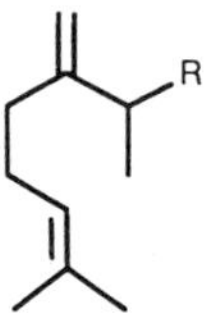

(66) R = H
(68) R = C(OH)Me$_2$

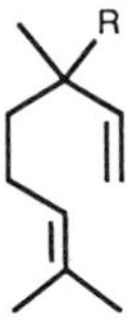

(67) R = H
(69) R = C(OH)Me$_2$

The Pd-catalysed hydrosilylation of myrcene (4) proceeds regioselectively to give only the silane (70), but mixtures are obtained under Rh catalysis (I. Ojima and M. Kumagai, J. Organomet. Chem., 1978, 157, 359). Regiospecific 1,4-hydrocobaltation of myrcene gives (71) which can be oxidised by singlet oxygen to (72), hydrolysis of which gives linalool (73) (A.R. Howell and G. Pattenden, J. Chem. Soc., Perkin Trans. 1, 1990, 2715). When myrcene (4) is reacted with [(MeCN)$_2$PdCl$_2$] in the presence of water the useful complexes

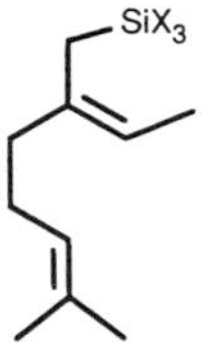

(70)

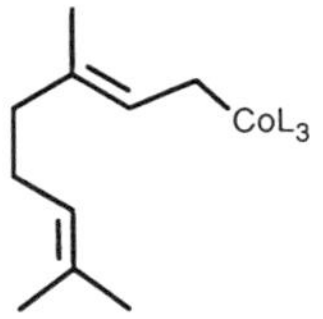

(71)

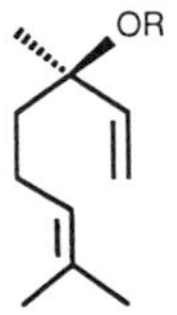

($\pm$)-(72) R = OCoL$_3$
(−)-(73) R = H

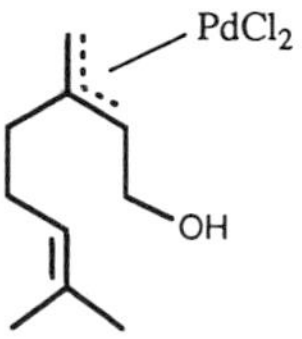

(74)

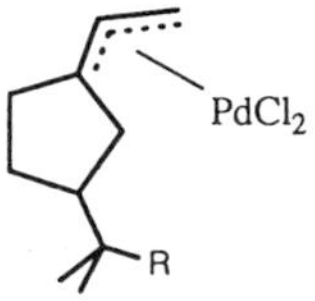

(75) R = OH or Cl

(74) and (75) are formed (M. Takahashi et al., Chem. Letters, 1979, 53).

Isoprene reacts with diethylamine in the presence of $PdCl_2$-Ph_3P - CO_2 to give predominantly the geranyl derivative (76) (M. Hidai et al., Bull. Chem. Soc. Jpn., 1980, 53, 2091). The same amine can be obtained by treatment of myrcene (4) with lithium (or BuLi) and diethylamine (K. Takabe et al., Org. Synth., 1989, 67, 44), and the neryl isomer (77) is formed when isoprene is reacted with BuLi - diethylamine (K. Takabe et al., Org. Synth., 1989, 67, 48). Reaction of isoprene with NH_3 - $Pd(acac)_2$ - $(BuO)_3P$ under controlled conditions gives

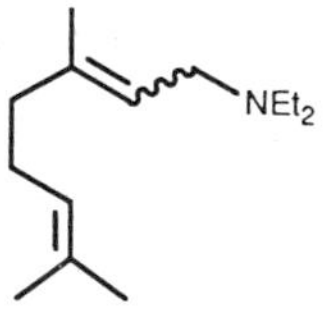

(E)-(76)
(Z)-(77)

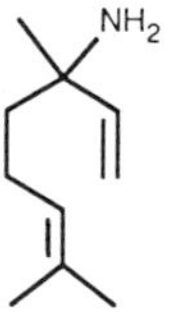

(78)

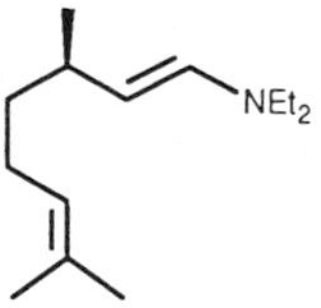

(−)-(79)

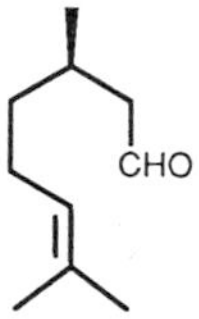

(+)-(80)

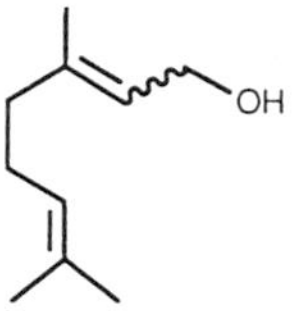

(E)-(81)
(Z)-(89)

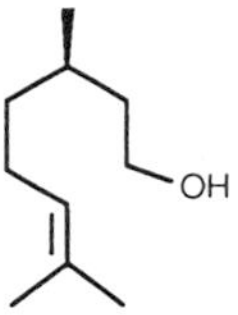

(+)-(82)

mainly linalylamine (78) (W. Keim and M. Roper, J. Org. Chem., 1981, 46, 3702).

When the nerylamine derivative (77) is isomerised using Noyori's $Rh[(+)-\{BINAP\}(cod)]^+$ ClO_4^- catalyst the (−)-citron-ellyl compound (79) is formed in very high chemical and optical yield. The geranylamine (76) affords the corresponding (S)-citronellyl derivative (ent-79) with the same catalyst (K. Tani et al., J. Chem. Soc., Chem. Commun., 1982, 600). Hydrolysis of (−)-(79) leads to (R)-(+)-citronellal (80) of high optical purity (K. Tani et al., Org. Synth., 1989, 67, 33). The mechanisms of these Rh-catalysed isomerisation

reactions have been investigated (S. Inoue et al., J. Amer.
Chem. Soc., 1990, 112, 4897), and the methodology has been
exploited on an industrial scale (see S. Akutagawa, T. Saka-
guchi and H. Kumobayashi, Dev. Food Sci., 1988, 18, 761).

Homogeneous hydrogenation of geraniol (81) using (S)-BINAP
based Ru catalysts leads to (+)-citronellol (82) of up to 98%
e.e. (H. Takaya et al., J. Amer. Chem. Soc., 1987, 109, 1596),
and similar results have been obtained by using the cationic
Rh-BINAP catalyst referred to above (S. Inoue et al., Chem.
Letters, 1985, 1007).

All of these major contributions from Noyori's laboratory
have had a most significant effect upon the availability of
optically active acyclic monoterpenoids.

The geranylamine N-oxide (83) rearranges thermally to give
the hydroxylamine (84) which is cleaved to linalool (73) by
Zn - AcOH (V. Rautensrauch, Helv. Chim. Acta, 1973, 56,
2492). The geranyl trichloroacetimidate (85) undergoes [3,3]-
sigmatropic rearrangement to give the amide (86) which can be
hydrolysed to linalylamine (78) (L.E. Overman, J. Amer. Chem.
Soc., 1976, 98, 2901). The related pseudourea behaves in a
similar manner (S. Tsuboi, P. Stromquist and L.E. Overman,
Tetrahedron Lett., 1976, 1145).

(83) R = N(O)Me$_2$
(85) R = OC(=NH)CCl$_3$
(87) R = OC(=NH)—N

(84)

(86)

Linalool (73) forms solid complexes with both α- and β-
cyclodextrins, with guest-host ratios of 1 : 2 and 1 : 1,
respectively. The photo-oxidisability of (73) is significantly dimished by binding to the cyclodextrins (Y. Ikeda et
al., Yakugaku Zasshi, 1982, 102, 83). (-)-Linalool (73) has
been utilised as chiral starting material for a synthesis of
(+)-frontalin (88), an aggregation pheromone of the southern
pine beetle Dendroctonus frontalis (F. Cooke, G. Roy and P.D.
Magnus, Organometallics, 1982, 1, 893).

(88)

Compressed monolayers of geraniol (81) or of nerol (89)
which are supported on 50% aqueous sulfuric acid have been
found to be relatively inert to chemical change. An expanded
monolayer of geraniol gives mainly acyclic linalool (73) via
allylic rearrangement under these conditions, whereas expanded monolayers of nerol undergo almost complete cyclisation
to α-terpineol (8), presumably through conformers of high
energy and low probability which are disfavoured in the compressed film (J. Ahmad and K.B. Astin, J. Amer. Chem. Soc.,
1986, 108, 7434).

Irradiation of nerol (89) in the presence of octanethiol
and a radical initiator causes it to isomerise to geraniol
(81) (A. Murata et al., Nippon Kagaku Kaishi, 1982, 657). The
same transformation can be carried out thermally (PhSH –
AIBN) but the product is then contaminated with citronellal
(80) (A. Murata, S. Tsuchiya and H. Suzuki, Nippon Kagaku
Kaishi, 1982, 1223).

In a reaction of general application neryl phenyl ether (90)
is converted with retention of geometry into the alkene (91)
when it is treated with $LiBEt_3$ - $Pd(PPh_3)_4$ (R.O. Hutchins and

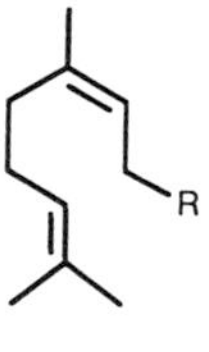

(90) R = OPh
(91) R = H

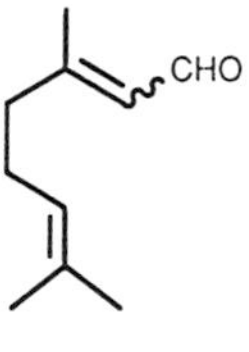

(E)-(92)
(Z)-(93)

K. Learn, J. Org. Chem., 1982, 47, 4380).

It is usually difficult to oxidise sensitive allylic alcohols to the derived aldehydes without some (E)/(Z) isomerisation taking place. Swern oxidation of geraniol (81) yields

(E)-citral (92) exclusively (A.J. Mancuno, S-L. Huang and
D. Swern, J. Org. Chem., 1978, 43, 2480), as does oxidation
of (81) using $Ru_2O.nH_2O - O_2$ (M. Matsumoto and N. Watanabe,
J. Org. Chem., 1984, 49, 3435). The analogous oxidation of
nerol (89) gives only the corresponding (Z)-isomer (93).
Reaction of geraniol (81) with $Me_3SiCl - Na_2CO_3 - CH_2Cl_2$

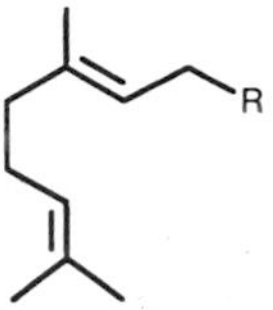

(94) R = Cl   (101) R = Bu
(97) R = Br   (102) R = SO_2Ph
(99) R = OAc   (103) R = CO_2H
(100) R = $OCOCMe_3$ (106) R = OBu

affords geometrically pure geranyl chloride (94) (M. Lissel
and K. Drechsler, Synthesis, 1983, 314). The method is of
general application. The chloride (94) can be converted
efficiently into the trisammonium salt of the biosynthetic
intermediate geranyl diphosphate (1) via reaction with tris-
(tetrabutylammonium) pyrophosphate (A.B. Woodside, Z. Huang
and C.D. Poulter, Org. Synth., 1988, 66, 211). When geranyl
chloride (94) is treated with Rieke barium (made from BaI_2 -
lithium biphenylide) at -78°C in THF it is converted into
geranylbarium chloride (A. Yanagisawa, S. Habaue and H.
Yamamoto, J. Amer. Chem. Soc., 1991, 113, 8955). This can be

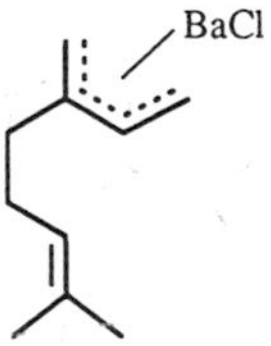

(95)

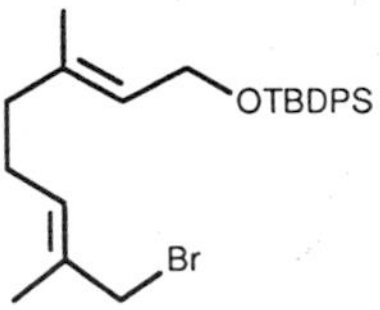

(96)

regiospecifically coupled at C-8 with terpenyl allylic bromides. For example, reaction of (95) with (96), which is derived from geraniol, yields the diterpene geranylgeraniol with
complete retention of all double bond geometry. The procedure
can then be reiterated to provide higher homologues (E.J.
Corey and W-C. Shieh, Tetrahedron Lett., 1992, 33, 6435).
    Geranyl bromide (97) couples to give the H-H dimer (98)
when it is treated with $[(Ph_3P)_3CoCl]$ (D. Momose et al.,

Tetrahedron Lett., 1983, <u>24</u>, 921), and geranyl acetate (99) gives the same product with Zn - (Ph$_3$P)$_4$Pd (S. Sasoaka

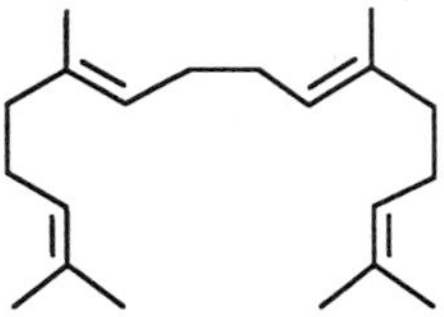

(98)

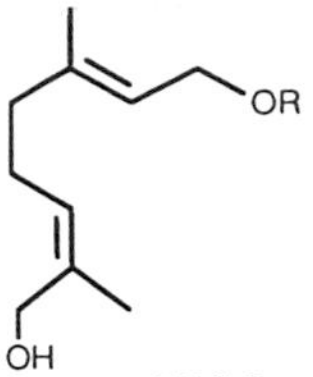

(104) R = Ac
(107) R = Bu

et al., Chem. Letters, 1985, 315). In another coupling procedure, geranyl pivalate (100) reacts with, for instance, BuLi in the presence of copper(1) iodide to give the alkylated product (101). The reaction is general for allylic pivalates (H.J. Liu and L.K. Ho, Can. J. Chem., 1983, <u>61</u>, 632). The lithium salt of geranyl phenyl sulfone (102) can be carboxylated with carbon dioxide and the sulfonyl function then reductively cleaved using Na-Hg to give homogeranic acid (103). The neryl derivative behaves analogously (P. Gosselin, C. Maignan and F. Rouessac, Synthesis, 1984, 876).

When geranyl acetate (99) is oxidised using catalytic amounts of selenium dioxide in the presence of tert-butyl hydroperoxide the major product os the alcohol (104) (M.A. Umbreit and K.B. Sharpless, J. Amer. Chem. Soc., 1977, <u>99</u>, 5526). Reaction of geranyl acetate with stoichiometric SeO$_2$

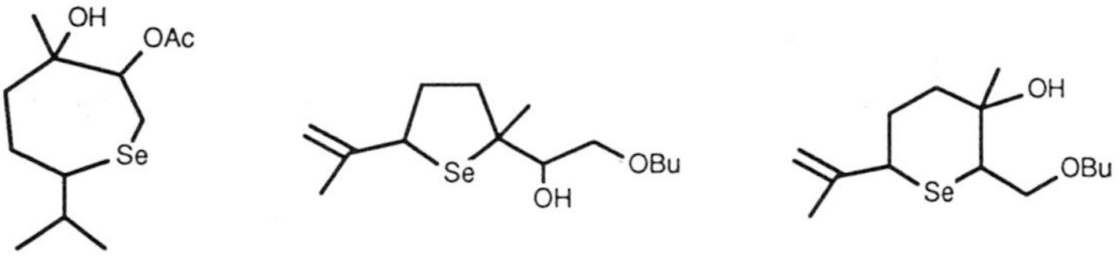

(105)

(108)

(109)

gives the same alcohol (104) together with the cyclic selenide (105). Geranyl butyl ether (106) yields the alcohol (107) and the selenides (108) and (109) under the same conditions (A.M. Moiseenkov, N.Y.Grigoreva and A.V.Lozanova, Dokl. Akad. Nauk S.S.S.R., 1986, <u>289</u>, 114: N.Y. Grigoreva et al., Izv. Akad. Nauk S.S.S.R., Ser. Khim., 1986, 2514).

Oxidation of geranyl acetate (99) with neutral aqueous potassium permanganate yields the tetrahydrofuran (110). This is hydrolysed by one equivalent of potassium hydroxide to give

the remarkable chloroform-soluble potassium complex (111),
m.p. 56-58°C, which can also be prepared from the corresponding triol using aqueous potassium acetate (R.E. Hackler, J.
Org. Chem., 1975, 40, 2978).

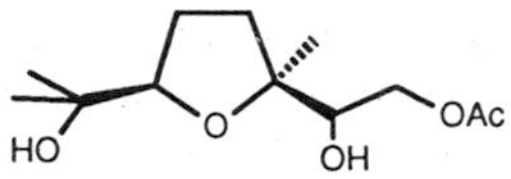

(110)

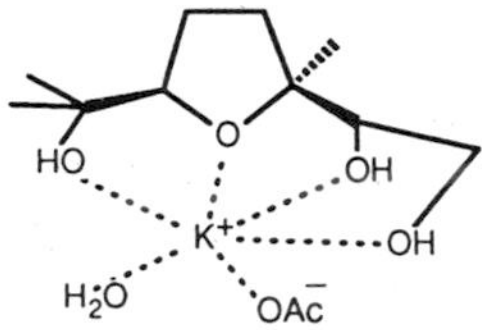

(111)

In an extraordinary but valuable reaction the digeranyl
ethanolamine (112) reacts with sodium nitrite in aqueous
acetic acid to give mainly the alkyne (113) (S.L. Abidi, J.
Org. Chem., 1986, 51, 2687). Citronellol (82) reacts under
the same conditions to give (114) in 95% yield (S.L. Abidi,
Tetrahedron Lett., 1986, 27, 267). Evidence for a reaction
mechanism has been presented (E.J. Corey, W.L. Siebel and J.
C. Kappos, Tetrahedron Lett., 1987, 28, 4921).

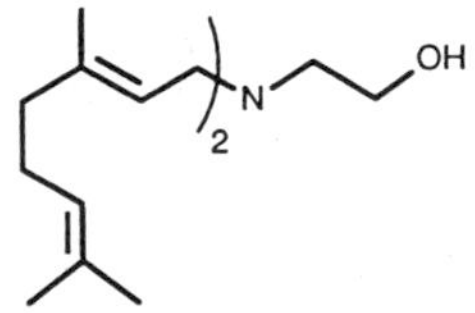

(112)

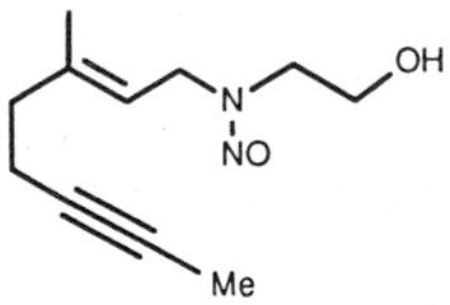

(113)

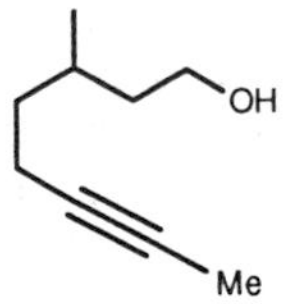

(114)

Phenols can be condensed with citronellal (80) to give
hexahydrocannabinoids (115) by heating the reactants in

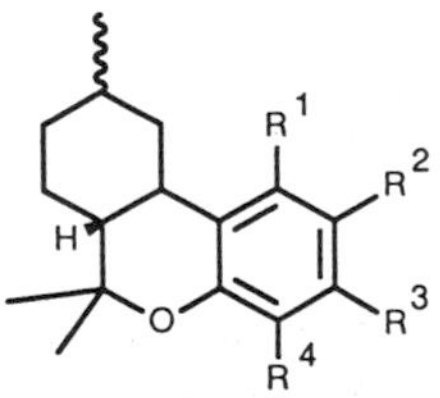

(115)

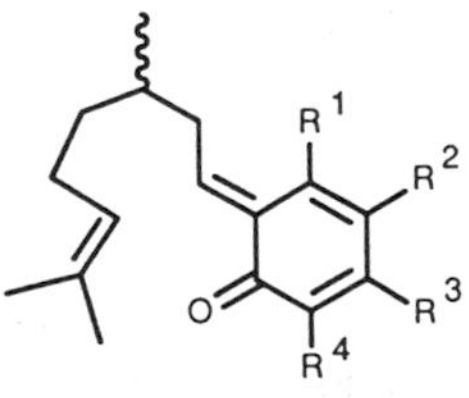

(116)

quinoline solution (W.S. Murphy et al., J. Chem. Soc., Perkin Trans. 1, 1992, 3397). The reaction may proceed via an inter-mediate quinone methide (116). More highly activated phenols give similar products under milder conditions involving treat-ment with phenylboric acid in acetic acid (W.S. Murphy, S.M. Tuladhar and B. Duffy, J. Chem. Soc., Perkin Trans. 1, 1992, 605).

(R)-Citronellic acid (117) has been synthesised by the copper(1) iodide-mediated reaction of homoprenylmagnesium bromide with the (R)-(+)-propiolactone (118) (T. Sato et al., Tetrahedron Lett., 1980, 21, 3377). The process is general for RMgX.

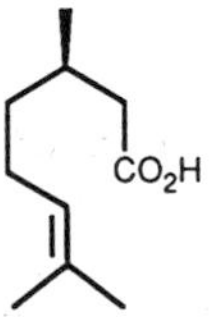

(117) 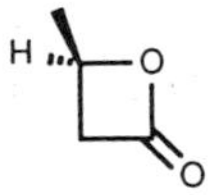   (+)-(118)

Geraniol (81) can be regiospecifically converted into its racemic 2,3-epoxide (±)-(119) by tert-butyl hydroperoxide in the presence of vanadium(5) or molybdenum (6) catalysts. If these are polymer-bound they may be re-used (T. Yokoyama et al., Chem. Letters, 1983, 1703), but the advent of the Sharp-less asymmetric epoxidation reaction has made the synthesis of homochiral (119) an almost trivial procedure. Thus, geran-iol (81) gives the (2S,3S)-epoxide (119) when it is treated

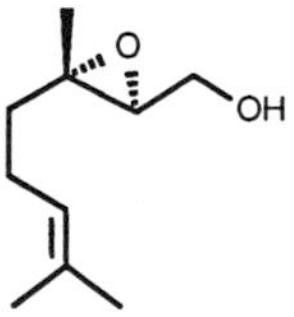

(119)

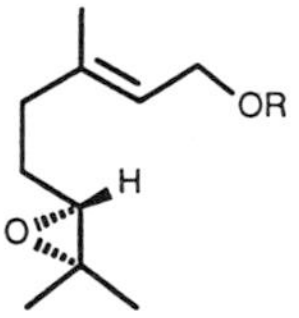

(+)-(120) R = H
(±)-(121) R = Bu

with tert-butyl hydroperoxide - (+)-diethyl tartrate - titan-ium tetraisopropoxide, and the (2R,3R)-analogue when (−)-di-ethyl tartrate is employed (T. Katsuki and K.B. Sharpless, J. Amer. Chem. Soc., 1980, 102, 5974). The p-toluenesulfonate of

(2S,3S)-(119) can be converted into (+)-linalool (ent-73) by treatment with Te - HOCH$_2$SO$_2$Na (rongalite) under alkaline conditions (R.P. Discordia, C.K. Murphy and D.C. Dittmer, Tetrahedron Lett., 1990, 31, 5603: D.C. Dittmer et al., J. Org. Chem., 1993, 58, 718).

Oxidation of geraniol (81) with benzeneseleninic acid gives mainly the racemic 6,7-epoxide (120) (P.A. Grieco et al., J. Org. Chem., 1977, 42, 2034), and the (+)-enantiomer of (120) has been synthesised from L-glutamic acid (S-i. Yamada, N.Oh-Hashi and K. Achiwa, Tetrahedron Lett., 1976, 2557). The oxidation of geranyl butyl ether (101) by NaOCl - tetraphenyl-porphinemanganese(3) under phase-transfer conditions yields the 6,7-epoxy derivative (121) (M. Nali et al., Gazz. Chim. Ital., 1987, 117, 207).

Treatment of the trimethylsilyl ether of 2,3-epoxygeraniol (119) with diethylaluminium tetramethylpiperidide gives trans-β-ocimene (43): the corresponding neryl derivative yields myrcene (4) under the same conditions (S. Tanaka et al., J. Amer. Chem. Soc., 1975, 97, 3252).

Geraniol (81) is cyclopropanated under Simmons-Smith cond-itions to give the 2,3-derivative (122) (W. Sobotka and E. Chojecka-Koryn, Bull. Pol. Acad. Sci., Chem., 1984, 32, 207), and by tri-isobutylaluminium - methylene iodide to give the 6,7-compound (123) (K. Maruoka, Y. Fukutani and H. Yamamoto, J. Org. Chem., 1985, 50, 4412).

(122)   (123)

The conversion of acyclic monoterpenoids into cyclic syst-ems has long been of interest, and there have been several developments in this area. The autoxidation of allo-ocimene (45) in DMSO or DMF solutions leads to the p-menthadienols (124) and (125) (M. Nomura, Y. Fujihara and Y. Matsubara, Nippon Kagaku Kaishi, 1980, 779). Photolysis of citronellyl iodide (126) gives, in part, the trans-p-menthene (127) (P.D. Gokhale et al., Tetrahedron, 1976, 32, 1391). Citronellol (82) is oxidised by buffered pyridinium chlorochromate to citron-ellal (80) but suffers oxidative cyclisation to pulegone

(±)-(128) in acidic media (E.J. Corey and J.W. Suggs, Tetrahedron Lett., 1975, 2647). (+)-Citronellal (80) gives a mixture of isopulegol (129) and neoisopulegol (130) when it is refluxed for 15 hr in hexane solution or heated at 210°C for 1 hr (L.N. Misra, S.K. Srivastava and M.C.Nigam, Riv. Ital. EPPOS, 1980, 62, 365). Isopulegol (129) is also formed from (80) by an ene-like reaction which takes place on silica gel at elevated pressure (W.G. Dauben and R.T. Hendricks, Tetrahedron Lett., 1992, 33, 603).

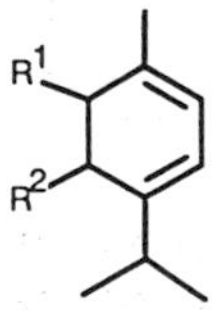

(124) R^1 = OH; R^2 = H
(125) R^1 = H; R^2 = OH

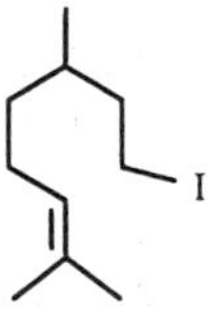

(126)

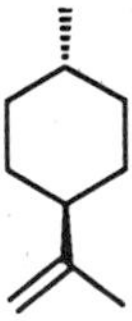

(127)

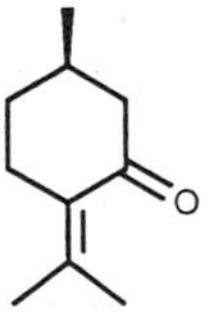

(+)-(128)

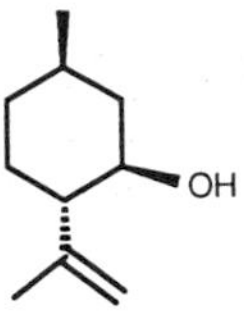

(-)-(129)

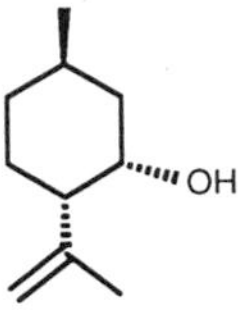

(+)-(130)

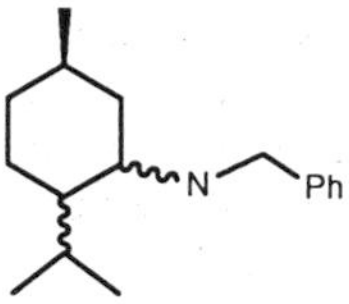

(131)

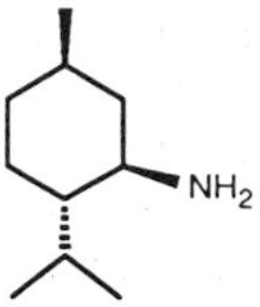

(132)

The benzyl imine of citronellal (80) cyclises with tin(4) chloride to give a mixture of amines (131) which affords menthylamine (132) (73%) on catalytic hydrogenation (G. Demailly and G. Solladie, Tetrahedron Lett., 1977, 1885).

Geranyl acetate (99) has been converted via the derived cyclic borane into the menthanol (133) (R. Murphy and R.H. Prager, Aust. J. Chem., 1976, 29, 617), and dehydrolinalyl acetate (134) cyclises to a mixture of carvenone (135) and 2-acetoxycar-2-ene (136) when it is exposed to zinc chloride (H. Strickler, J.B. Davis and G. Ohloff, Helv. Chim. Acta, 1976, 59, 1328). The cyclic ether (137), derived from nerol (89), rearranges to the isopiperitenol (138) of 25% e.e. when it is treated with lithium bis[(S)-1-phenylethyl]amide (J.A. Marshall and J. Lebreton, J. Org. Chem., 1988, 53, 4108).

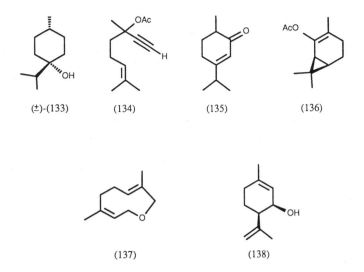

(±)-(133)   (134)   (135)   (136)

(137)   (138)

The ocimenones (139) rearrange to isopiperitenone (140) and filifolone (141) on treatment with aluminium chloride (P. Weyerstahl, W. Zombick and C. Gansav, Liebigs Ann. Chem., 1986, 422).

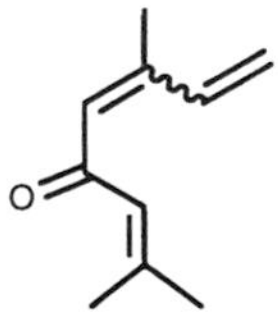

(139) 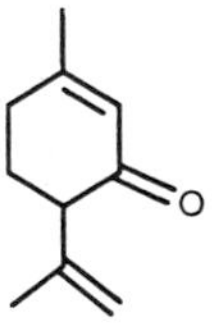   (140) 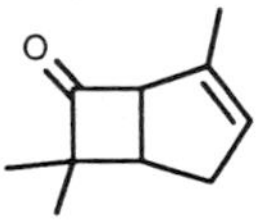   (141)

4. Naturally-occurring Halogenated Monoterpenoids

A number of polyhalogenated monoterpenoids have been isolated, principally from red marine algae. Most of the compounds are formally derivatives of 2,6-dimethyloctatrienes or 2,6-dimethyloctatetraenes, and many of them likely arise from myrcene (4) via peroxidase-mediated halogenation reactions (see S.L. Neidleman and J. Gieger, "Biohalogenation: Principles, Basic

Roles and Applications", Ellis Horwood, Chichester, 1986).
The majority of these halogenated compounds contain bromine
and/or chlorine. In spite of the presence of iodine in sea
water, iodides have not been encountered, but this may simply
reflect their greater lability and susceptibility to solvolytic processes.

The digestive gland of the sea hare *Aplysia californica*
contains (142) and (143) which originate from the red alga
Plocamium coccineum upon which the animal feeds (D.J. Faulkner et al., J. Amer. Chem. Soc., 1973, 95, 3413: M.R. Willcott
et al., Tetrahedron Lett., 1973, 3967). Other halogenated

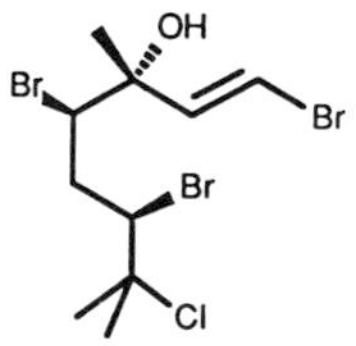

(142) 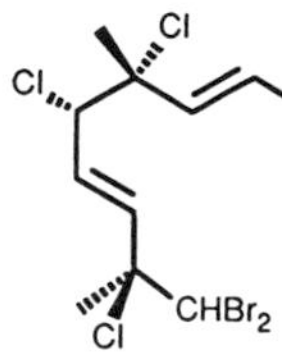   (143) 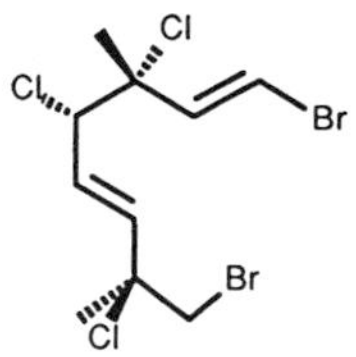   (144)

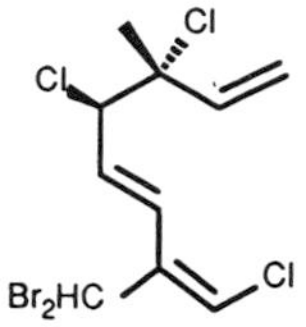

(145) 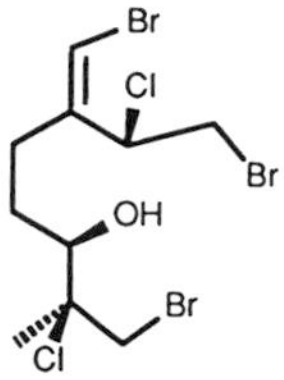   (146)

monoterpenoids found in *Aplysia* spp. include (144) from *A.*
californica (C. Ireland et al., J. Org. Chem., 1976, 41,
2461), (145) from *A. limacina* (F. Imperato, L. Minale and
R. Riccio, Experientia, 1977, 33, 1273), and kurodainol (146)
from *A. kurodai* (A. Katayama et al., Agric. Biol. Chem., 1982,
46, 859). All of these compounds are probably of algal origin.
Additional *Plocamium* metabolites include (147)-(153) from *P.*
cartilagineum (J.S. Mynderse and J.D. Faulkner, Tetrahedron,
1975, 31, 1963), costatol (154), the hemiacetal costatone
(155), and the derived nor-monoterpenoid costatolide (156)
from *P. costatum* (R. Kazlauskas et al., Tetrahedron Lett.,
1976, 4451: D.B. Stierle, R.M. Wing and J.J. Sims, Tetrahedron Lett., 1976, 4455), and the insecticidal compounds (157)

and (158) from *P. telfairiae* (K. Watanabe et al., Pestic. Biochem. Physiol., 1990, <u>37</u>, 275).

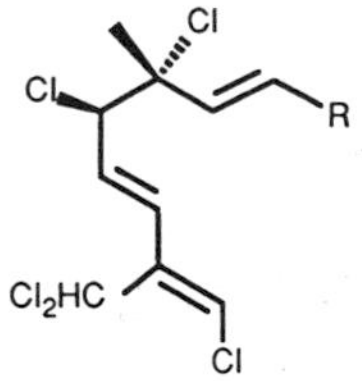

(147) R = H
(148) R = Br

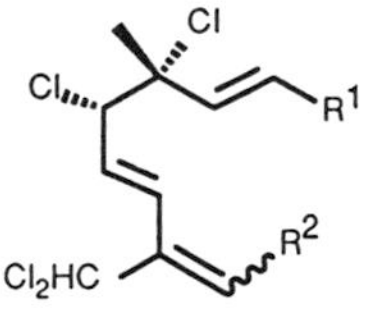

(149) R^1 = R^2 = H
(150) R^1 = H; R^2 = (*E*)-Cl
(151) R^1 = H; R^2 = (*Z*)-Cl
(152) R^1 = Br; R^2 = (*E*)-Cl
(153) R^1 = Br; R^2 = (*Z*)-Cl

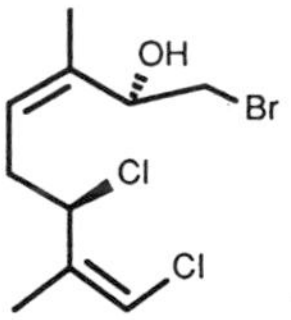

(154) 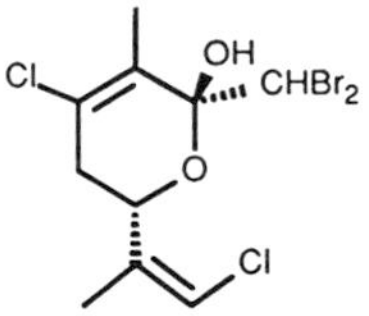   (155) 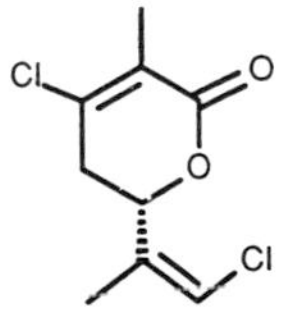   (156)

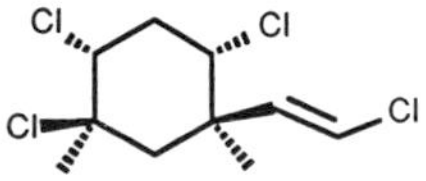

(157) 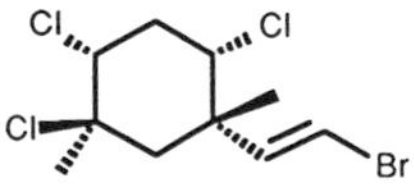   (158)

Another source of halogenated monoterpenoids is *Desmia* (*Chondrococcus*) *hornemanni* which contains the myrcene derivatives (159)-(168) (N. Ichikawa, Y. Naya and S. Enomoto, Chem. Letters, 1974, 1333: B.J. Burreson, F.X. Woolard and R.E. Moore, Tetrahedron Lett., 1975, 2155). *D. hornemanni* also contains the bromide (169) which possesses the ochtodane skeleton, and which has been synthesised via the brominative cyclisation of myrcene (4) (K. Yoshihara and Y. Hirose, Bull. Chem. Soc. Jpn., 1978, <u>51</u>, 653).

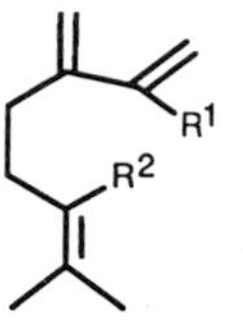

28

(159) R^1 = Cl; R^2 = H
(160) R^1 = Br; R^2 = H
(161) R^1 = R^2 = Cl
(162) R^1 = Cl; R^2 = Br

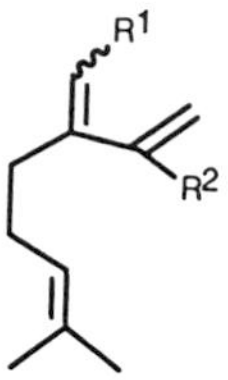

(163) R^1 = (Z)-Br; R^2 = H
(164) R^1 = (E)-Br; R^2 = H
(165) R^1 = (Z)-Br; R^2 = Cl
(166) R^1 = (E)-Br; R^2 = Cl

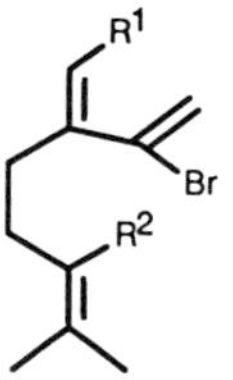

(167) R^1 = Br; R^2 = Cl
(168) R^1 = Cl; R^2 = Br

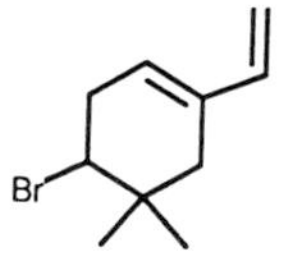

(169)

5. Cyclobutanes

(+)-*cis*-Grandisol (170) is an important pheromone produced by males of the cotton boll weevil *Anthonomus grandis*, and it can be used in the field in mid-season to attract and trap the female insects. There have been a number of syntheses of the racemic material (*cf.* J.A. Katzenellenbogen, Science, 1976, 194, 139: C.A. Henrick, Tetrahedron, 1977, 33, 139), and (+)-(170) has been synthesised from (-)-β-pinene by a route which confirms its absolute configuration (P.D. Hobbs and P.D. Magnus, J. Amer. Chem. Soc., 1976, 98, 4594). An alternative route to (+)-grandisol (170) and to its enant-

iomer which requires the classical resolution of an intermediate has also been published (K. Mori, Tetrahedron, 1978,
34, 915).

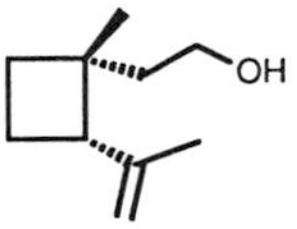

(+)-(170) 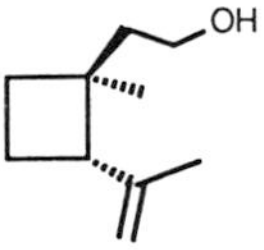   (171)

The related fragranol (171), obtained from *Artemisia frag-
rans*, has also been synthesised (B.M. Trost and D.E. Keeley,
J. Org. Chem., 1975, 40, 2013), as has the pheromone (172) of
the *Citrus* mealybug *Planococcus citri*, which was prepared
from (+)-verbanone (Scheme 5) via a photochemical route (B.A.
Bierl-Leonhardt et al., Tetrahedron Lett., 1981, 22, 389).

SCHEME 5

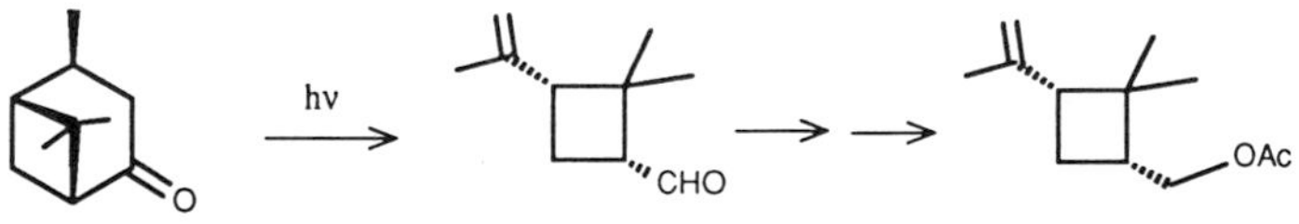

(+)   (+)-(172)

The tricyclic acetal (+)-lineatin (173), a pheromone which
has been obtained from the frass of the Douglas fir beetle
Trypodendron lineatum, and which is attractive to both sexes,

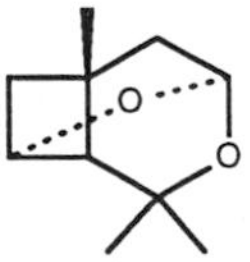

(+)-(173) 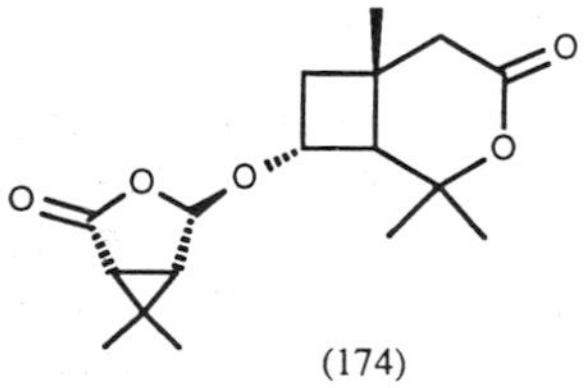   (174)

has been synthesised via the chrysanthemate-derived intermediate (174), the absolute configuration of which was verified
by X-ray analysis (K. Mori et al., Tetrahedron, 1983, 39,
1735).

6. Cyclopentanes

This section discusses monoterpenoids which possess the
iridane skeleton (175) but cannot, because of limitations on
space, deal with the plethora of iridoid derivatives which
are based upon the related framework (176) and which have
been isolated from plant sources. Iridoids reported between
1980 and 1989 have been catalogued together with their
physical and spectroscopic data and a plant source index (C.
A. Boros and F.R. Stermitz, J. Nat. Prod., 1990, 53, 1055).
Other reviews of the subject area include articles on synthesis (L.F. Tietze, Angew. Chem., 1983, 95, 840: P. Junior,
Planta Med., 1990, 56, 1), chemistry (A. Bianco, Stud. Nat.
Prod. Chem., 1990, 7, 439), biosynthesis (Y. Asakawa et al.,
Phytochemistry, 1988, 27, 3861) and applications (F. Stermitz,
A.C.S. Symp. Ser., 1988, 380, 397).

(175)   (176)

Iridanes frequently occur in the insect world where they
commonly perform defensive functions. The repellant iridodial
(177) is found in the secretion from the anal gland of the
Australian cocktail ant *Iridomyrmex nitidiceps* (G.W.K. Cavill
et al., Tetrahedron, 1982, 38, 1891), and has been synthesised
(P. Ritterskamp, M. Demuth and K. Schaffner, J. Org. Chem.,
1984, 49, 1155). *Trans,trans*-Dolichodial (178) is found in *I.
humilis* (G.W.K. Cavill et al., Insect Biochem., 1976, 6, 483),
and also in the plant *Teucrium marum* (C. Beaupin et al.,
Rivista Ital. Essenze-Profumi, Piante Offic., Aromi, Saponi,
Cosmet., Aerosol, 1978, 60, 93). It has been synthesised
several times (e.g., Y. Morizawa, K. Oshima and H. Nozaki,
Tetrahedron Lett., 1982, 23, 2871). Chrysomelidial (179)
forms part of the defensive secretion of *Plagiodera versi-*

colora (J. Meinwald et al., Proc. Nat. Acad. Sci. U.S.A.,
1977, 74, 2189) and has been synthesised from (+)-limonene
(J. Meinwald and T.H. Jones, J. Amer. Chem. Soc., 1987, 100,
1883). The diastereomer of (179) is dehydroiridodial (180)
which is found in *Actinidia polygama*, and which has been synthesised from (-)-limonene (K. Yoshihara, T. Sakai and T.
Sakan, Chem. Letters, 1978, 433).

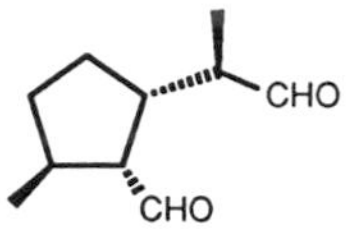

(+)-(177) 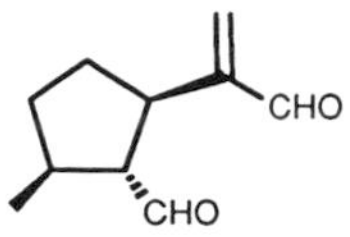 (178) 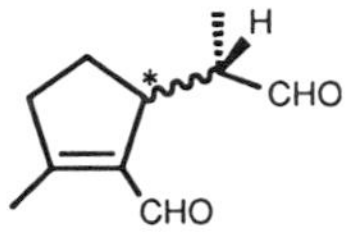 (S)-(179)
(R)-(180)

Cultures of *Rauwolfia serpentina* cells oxidise the hydroxy-
geraniol (181) to a mixture of (182)-(184), all of which are
biosynthetic precursors to iridodial (177) and thus to the
iridoids (S. Uesato et al., Tetrahedron Lett., 1987, 28,
4431). Treatment of the aldehyde (184) with base leads to a
mixture of racemic chrysomelidial (179) and dehydroiridodial
(180) which then undergoes a Cannizarro reaction to give 1,2-
isodehydroiridomyrmecin (185) (F. Bellesia et al., Tetrahedron
Lett., 1986, 27, 381). The same mixture of chrysomelidial and
dehydroiridodial is formed when (184) is treated with 50% formic acid (S. Uesato et al., J. Chem. Soc., Chem. Commun., 1987,
1020). Evidence that photocyclocitral A (186) may be involved
in the biosynthesis of iridoid precursors has been presented
(R. Grandi et al., J. Chem. Res. (S), 1984, 194).

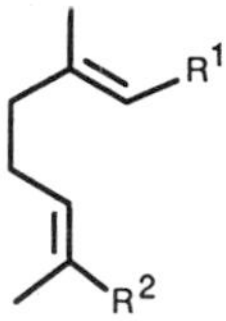

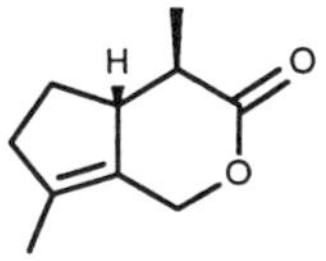

(185)

(186)

(181) $R^1 = R^2 = CH_2OH$
(182) $R^1 = CH_2OH$; $R^2 = CHO$
(183) $R^1 = CHO$; $R^2 = CH_2OH$
(184) $R^1 = R^2 = CHO$

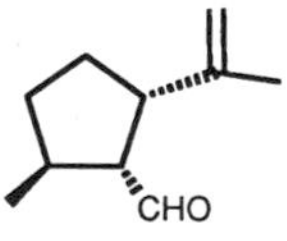

Other iridanes which have been synthesised include (+)-dehy-
droiridodiol (187) and (-)-isodehydroiridodiol (188) (M.
Yamaguchi et al., Tetrahedron Lett., 1986, 27, 959).

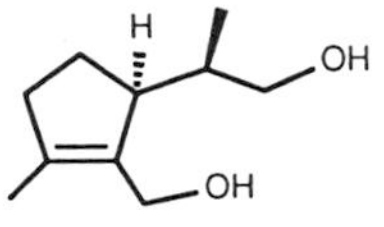

(+)-(187)

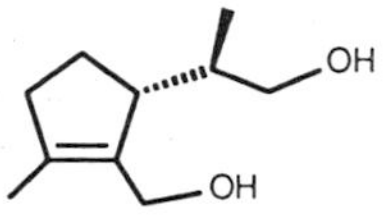

(-)-(188)

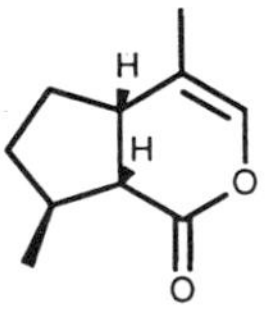

(189)

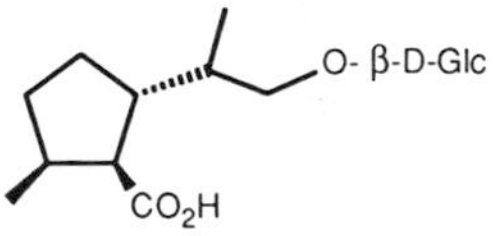

(190)

Some compounds in the iridane group act as feline attract-
ants. Nepetalactone (189) has been synthesised (S.E. Denmark
and J.S. Sternberg, J. Amer. Chem. Soc., 1986, 108, 8277), and
nepetarioside (190) has been isolated from *Nepeta cataria* and
characterised (F. Murai et al., Chem. Pharm. Bull., 1987, 35,
2533).

7. Tetramethylcyclohexanes

The absolute configuration of picrocrocin (191), a bitter substance found in *Crocus sativus*, has been determined (R. Buchecker and C.H. Eugster, Helv. Chim. Acta, 1973, 56, 1121) as,
by interconversion, has that of picrocrocinic acid (192) from
Gardenia jasminoides grandiflora (Y. Takeda et al., Chem.
Pharm. Bull., 1976, 24, 2644). The presence of β-cyclocitral
(193) in lake waters has been correlated with populations of
phytoplankton of the genus *Microcyctis* (F. Juettner, B. Hoefl-
acher and K. Wurster, J. Phycol., 1986, 22, 169). The acid
(194) has been found in Californian crude oil (R. Buchecker
and C.H. Eugster, Helv. Chim. Acta, 1973, 56, 2563), and the
compounds (195) occur in *Piqueria trinervia* (F. Bohlmann and
A. Suwita, Phytochemistry, 1978, 17, 560).

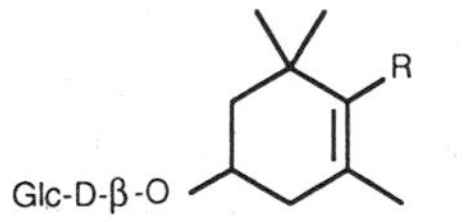

Glc-D-β-O—

(191) R = CHO
(192) R = CO$_2$H

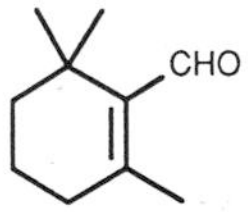

(193)

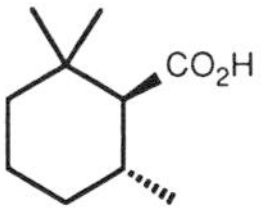

(194)

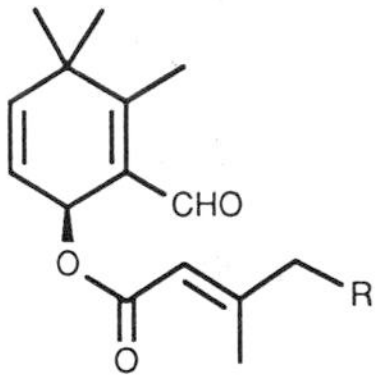

(195) R = H, OH, OAc

β-Cyclocitral (193) can be converted into safranal (196) via bromination using NBS followed by dehydrobromination (W.M.B. Konst, L.M. van der Linde and H. Boelens, Tetrahedron Lett., 1974, 3175), but α-cyclocitral (197) decarbonylates under these conditions. Acid-catalysed cyclisation of the L-proline derived enamine (198) of citral gives (+)-α-cyclocitral (197) (M. Shibasaki, S. Terashima and S. Yamada, Chem. Pharm. Bull., 1975, **23**, 279).

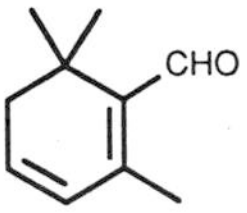

(196) 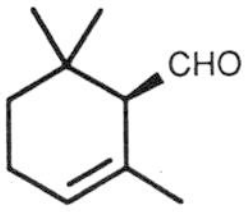   (+)-(197) 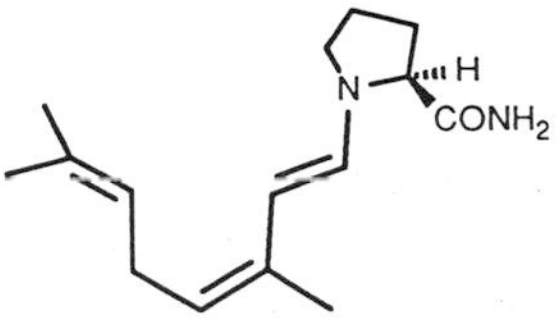   (198)

α-Pyronene (199) rearranges to (200) on brief exposure to acidic conditions, but further reaction leads to a mixture of β-pyronene (201) and γ-pyronene (202) (W. Cocker, K.J. Crowley and S.G. Traynor, J. Chem. Soc., Chem. Commun., 1974, 982).

Sensitised photo-oxidation of α-pyronene (199) yields the peroxide (203) which can be reduced with triphenylphosphine to give the epoxide (204). Oxidation of (199) using peroxybenzoic acid in diethyl ether leads to the isomeric epoxide (205)

34

(W. Cocker, K.J. Crowley and K. Srinivasan, J. Chem. Soc.,
Perkin Trans. 1, 1978, 159). Photo-oxidation of β-pyronene
(201) gives the peroxide (206) which is reducible to the diol
(207) (W. Cocker, K.J. Crowley and K. Srinivasan, J. Chem.
Soc., Perkin Trans. 1, 1973, 2485).

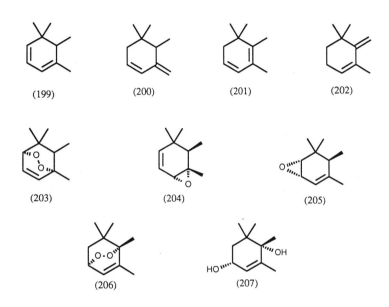

Epoxidation of δ-pyronene (208), obtainable from myrcene
(4), using 3-chloroperoxybenzoic acid gives mainly the mono-
epoxide (209) together with some of the di-epoxide (210). The
alternative mono-epoxide (211) can be accessed from δ-pyronene
via formation of the bromohydrin (212) using aqueous NBS followed by treatment with base. Under appropriate conditions, the
epoxide (209) can be converted into either α-cyclocitral (213)
or β-cyclocitral (193) (D. Serramedan et al., Tetrahedron
Lett., 1992, 33, 4447).

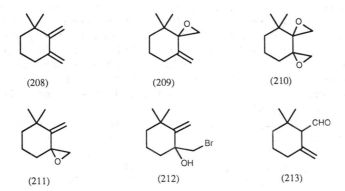

(208)   (209)   (210)

(211)   (212)   (213)

8. Menthanes

(a) o-Menthanes

The phenolic ether (214) and the *o*-menthatriene derivative
(215) have been obtained from *Piqueria trinervia* (F. Bohlmann
and A. Suwita, Phytochemistry, 1978, <u>17</u>, 560). Also found in

(214) R = OMe
(218) R = H

(215)   (216)   (217)

P. trinervia is the potent plant growth inhibitor Piquerol A
(216), the structure of which has been confirmed by X-ray
analysis (M. Scoriano-Garcia et al., Chem. Letters, 1983,
617). The lactone (217) has been isolated from the urine of
the koala bear *Phascolarctos cinereus* (T.A. Southwell,
Tetrahedron Lett., 1975, 1885). A synthesis of (217) has been
reported (N. Lander and R. Mechoulam, J. Chem. Soc., Perkin
Trans. 1, 1976, 484).

Reduction of *o*-cymene (218) using calcium in ammonia yields
a mixture of *o*-menthane derivatives (V.V. Bazylchik, P.I.
Federov and N.M. Ryabushkina, Zh. Org. Khim., 1978, 14, 969),
and the menthenone (219) has been synthesised (Scheme 6) by a
photochemical route (D.K.M. Duc et al., Synthesis, 1981, 139).

SCHEME 6

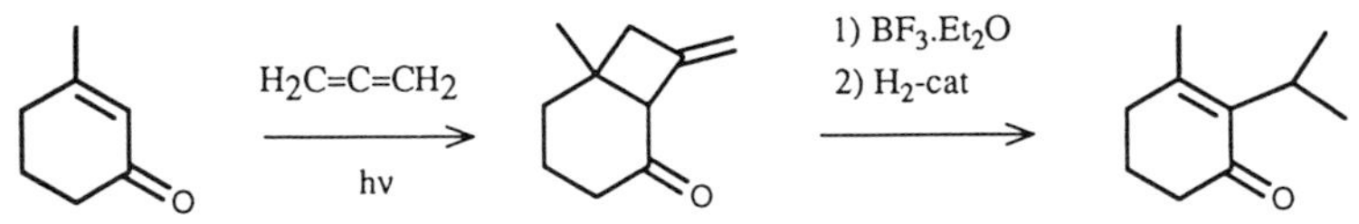

(219)

The Diels-Alder adduct (220) is a precursor for the *o*-menth-
enol (221) and the *o*-menthadiene (222) (V.V. Bazylchik et al.,
Zh. Org. Khim., 1985, 21, 1450), and the (±)-*o*-menthone (223)
has also been synthesised (J. Ficini, A. Eman and A.M. Touzin,
Tetrahedron Lett., 1976, 679).

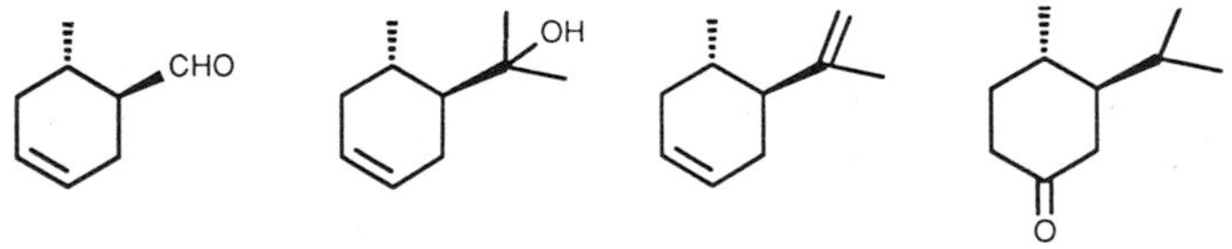

(220)   (221)   (222)   (223)

(b) m-Menthanes

New naturally-occuring *m*-menthane derivatives which have been described include the *m*-menthadienol (224) from *Cannabis sativa* resin (E. Stahl and R. Kunde, Tetrahedron Lett., 1973, 2841) and the *m*-menthadienes (225) and (226) isolated from Russian turpentine (I.I. Bardyshev et al., Chem. Natural Compounds, 1974, 10, 325). A synthesis of (225) has been described (N. Miyaura et al., Chem. Letters, 1974, 1411). Fully aromatised derivatives which have been discovered include the isobutyrate (227) from *Eupatorium* spp. (F.Bohlmann et al., Phytochemistry, 1977, 16, 1973) and from *Senecio* spp. (F. Bohlmann et al., Phytochemistry, 1977, 16, 965), and the resorcinol (228) which occurs in *Macowania hamata* (F. Bohlmann and C. Zdero, Phytochemistry, 1977, 16, 1583).

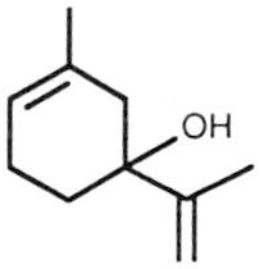

(224) 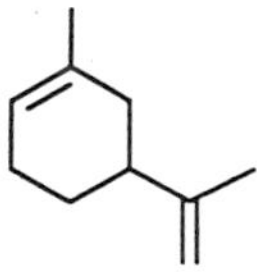   (225) 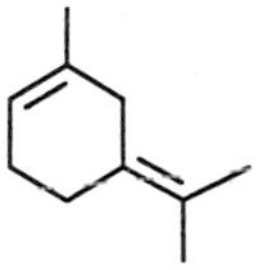   (226)

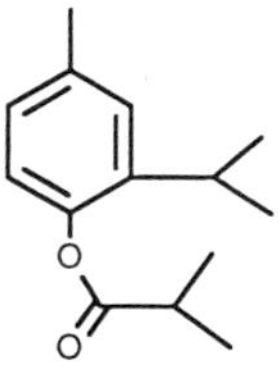

(227) 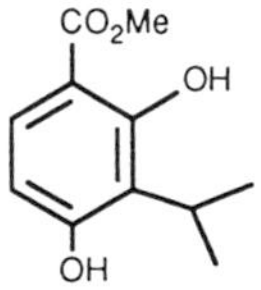   (228)

Many naturally-occuring *m*-menthanes may arise via ring-cleavages of carane systems. Thus, for example, the epoxycaranc (229) yields the *m*-menthenone (230) when it is treated with zinc bromide (B.C. Clark et al., J. Org. Chem., 1978, 43, 519).

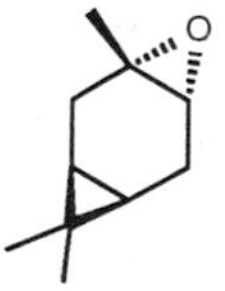

(229) 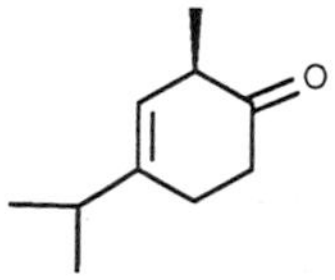   (230)

(c) p-Menthanes

These form the largest group of the monocyclic monoterpenoids
and are of significant economic and chemical importance. The
menthenethiol (231) has been isolated from *Citrus paradisi* and
exhibits a grapefruit aroma at p.p.b. levels (E. Demole, P.
Enggist and G. Ohloff, Helv. Chim. Acta, 1982, <u>65</u>, 1785: *cf.*
Eur. Pat. Appl. EP 54847, 1982). The racemates of the diols
(232) occur in the leaves of *Eucalyptus citriodora* and possess
allelopathic properties (H. Nishimura et al., Agric. Biol.
Chem., 1982, <u>46</u>, 319). The dihydrocarvone derivative (233),
from *Mentha haplocalyx* (D. Ding and H. Sun, Zhiwu Xuebao,
1983, <u>25</u>, 62), exhibits mosquito repellent activity (J. Verghese, PAFAI J., 1983, <u>5</u>, 27). Terpinene-4-ol (234) from
Artemisia vulgaris may be repellent to the yellow fever mosquito *Aedes aegypti* (Y.S. Hwang et al., J. Chem. Ecol., 1985, <u>11</u>,
1297).

| (231) | (232) | (233) | (234) |

The unusual orthoester (235) has been isolated from *Mentha*
piperita and has been synthesised (M. Koepsel et al., Prog.
Essent. Oil Res., Proc. Int. Symp. Essent. Oils, 16th, 1985,
241). The equally unusual peroxy-compound (236) has been obtained from *Satureja gilliessi* (V. Manriquez et al., Acta
Crystallogr., Sect. C: Cryst. Struct. Commun., 1990, <u>46C</u>,
802). The pulegone derivative schizonepetoside C (237) occurs
in *Schizonepeta tenuifolia* (M. Kubo et al., Chem. Pharm. Bull.,
1986, <u>34</u>, 3097).

Two naturally-occuring *p*-menthane dimers have been discovered. These are (238) which is found in *Cymbopogon martinii*
(A.T. Bottini et al., Phytochemistry, 1987, <u>26</u>, 2301), and
(239) from the heartwood of *Callitris macleayana* which is a
Diels-Alder dimer of the menthadienone (240) (R.M. Carman et
al., Aust. J. Chem., 1986, <u>39</u>, 1843). Indeed, oxidation of
carvacrol (241) with lead(4) acetate yields the racemic acet-

ate (242) which is converted into the dimer (239) when it is
heated in ethanolic sulfuric acid (R.M. Carman, S. Owsia and
J.M.A.M. van Dongen, Aust. J. Chem., 1987, 40, 333). A similar
synthesis of (239) can be achieved in one step by treating
carvacrol (241) with sodium periodate (S.K. Paknikar and J.
Patel, Chem. Ind. (London), 1988, 529).

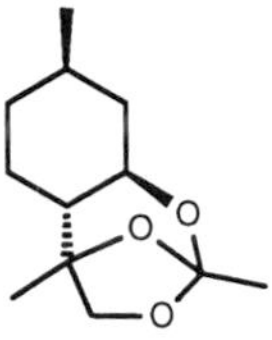

(235)

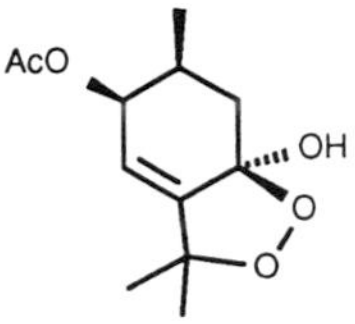

(236)

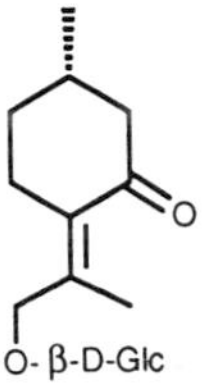

(237)

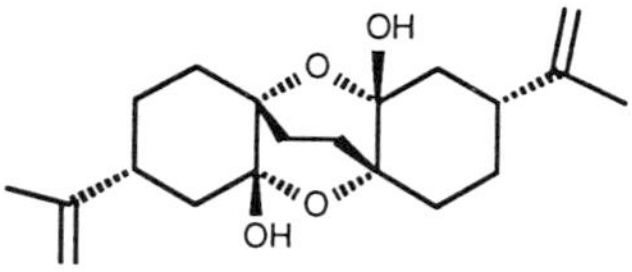

(238)

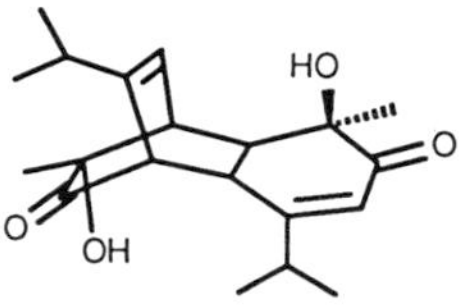

(239)

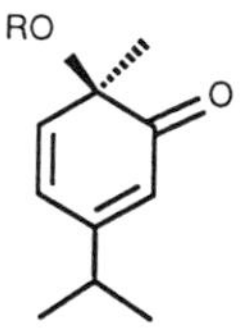

(240) R = H
(242) R = Ac

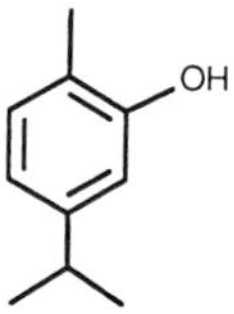

(241)

40

The uroterpenols (243), which may be derived from dietary
limonene (244) and which have been isolated from human urine
as their β-D-glucuronides, have been synthesised (A. Kergomard and H. Veschambre, Tetrahedron Lett., 1975, 835) and the
absolute configurations of both diastereoisomers have been
determined (R.M. Carman, K.L. Greenfield and W.T. Robinson,
Aust. J. Chem., 1986, 39, 21). (+)-Limonene (244) is metab-
olised in human liver to the 7,8-epoxides (245) and (246)
(precursors to the uroterpenols?), and their stereochemistries
have been determined unambiguously (R.M. Carman, J.J. de Voss
and K.L. Greenfield, Aust. J. Chem., 1986, 39, 441).

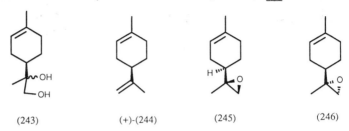

(243)   (+)-(244)   (245)   (246)

(+)-Limonene (244) yields the dienes (247) and (248) when it
is pyrolysed, but these are racemic because of the intervention of the diradical (249) (K.J. Crowley and S.G. Traynor,
Tetrahedron, 1978, 34, 2783).

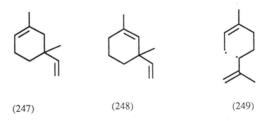

(247)   (248)   (249)

Many useful 7-substituted p-menthenes (250) have been obtained from limonene (244) by treating it with mercury(2)
tetrafluoroborate and a nucleophile and then reducing the
resulting organomercurial using sodium borohydride (M.C.S. de
Mattos et al., Tetrahedron Lett., 1992, 33, 4863). (+)-Limon-
ene (244) yields the vicinal diamine (251) when it is reacted
with HgO - fluoroboric acid - aniline and the initially-
formed 7-amino-8-mercuri adduct is then heated to 80°C (J.A.F.
Barluenga et al., J. Org. Chem., 1991, 56, 2930).

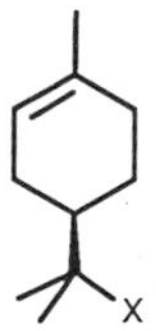

(250) X = OH, OMe, OAc,
NH$_2$, N$_3$, NHCOCH$_3$

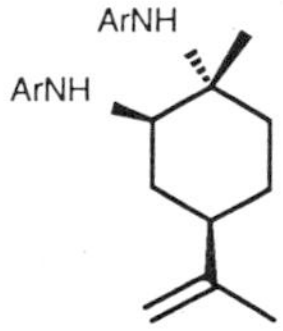

(251)

The allylic oxidation of inexpensive limonene (244) can lead
to valuable products. Reaction of (244) with palladium chloride - cupric chloride - AcOH gives *trans*-carveyl acetate (252)
(A. Heumann, M. Reglier and B. Waegell, Angew. Chem., 1982,
94, 397).

Functionalisation of limonene (244) at C-9 can be acheived
in various ways. Thus, lithiation followed by reaction with an
electrophile leads to compounds (253) (H. Suemune et al., Chem.
Pharm. Bull., 1984, 32, 4632), and Prins elaboration of (244)
yields bergamyl acetate (254) (M. Armenteros, J.P. Zayas and
H. Velez, Rev. Cienc. Quim., 1981, 12, 1). Hydroformylation
of limonene (244) using [dichloroplatinum(bisdiphenylphosphino-
propane)] - tin(2) chloride - CO gives exclusively the exocyclic aldehyde (255) (L. Kollar et al., J. Organomet. Chem.,
1990, 385, 147). Other isopropenyl derivatives behave in a
similar way.

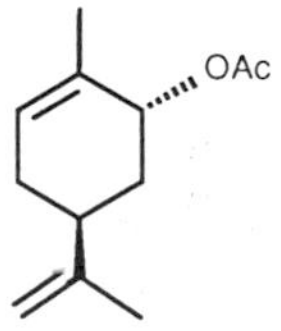

(252)

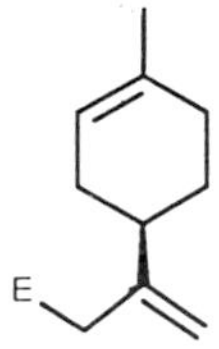

(253)

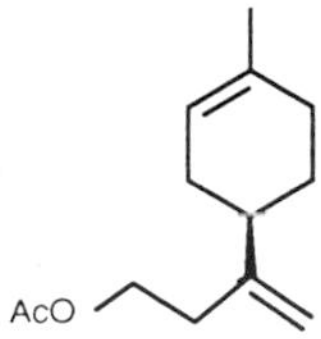

(254)

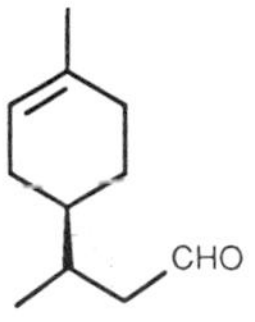

(255)

The epoxylimonene (256) can be rearranged to *trans*-carvomen-
thone (257) using a zeolite catalyst, or to the ring-contrac-
ted aldehyde (258) with molecular sieves coated with potassium
permanganate (T. Kurata, T. Koshujama and H. Kawashima,
Yukagaku, 1987, 36, 206). The naturally-occuring ascaridole
(259) rearranges to the di-epoxide (260) under catalysis by

42

FeCl$_2$(PPh$_3$)$_2$ (M. Suzuki et al., J. Org. Chem., 1989, <u>54</u>, 5292).

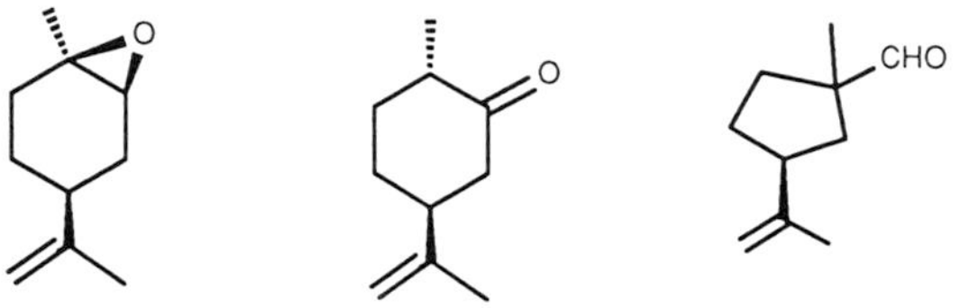

(256)   (-)-(257)   (258)

(259)   (260)

In a reaction which is of general application, (-)-perillyl alcohol (261) is exclusively cyclopropanated at the 7,8-double bond to give (262) when it is treated with tri-isobutylaluminium - diiodomethane (K. Maruoka, S.Sakane and H. Yamamoto, Org. Synth., 1989, <u>67</u>, 176). In another procedure, which is a general method for the 1,3-transposition of allylic

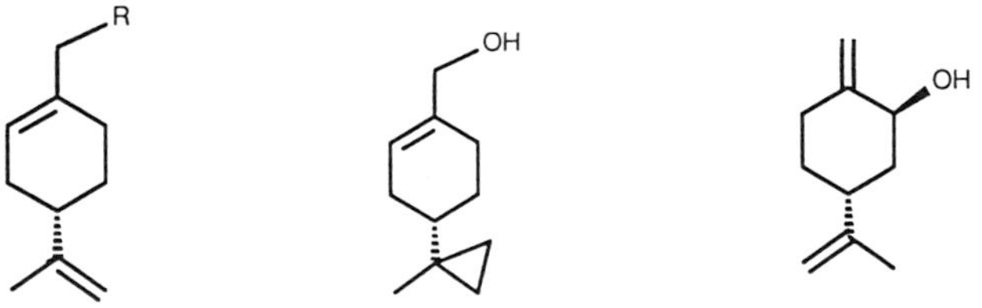

(-)-(261) R=OH   (262)   (264)
(263) R=SnBu$_3$

alcohols, perillyl alcohol (261) is converted via its xanthate into the stannane (263) which yields the sensitive exocyclic dienol (264) on treatment with 3-chloroperoxybenzoic acid

(Y. Ueno, H. Sano and M. Okawara, Synthesis, 1980, 1011). The 1,2-transposition of a carbonyl group such as that of menthone (265) to give tetrahydrocarvone (266) can be achieved (Scheme 7) using organosilicon chemistry (W.E. Fristad, T.R. Bailey and L.A. Paquette, J. Org. Chem., 1980, 45, 3028).

SCHEME 7

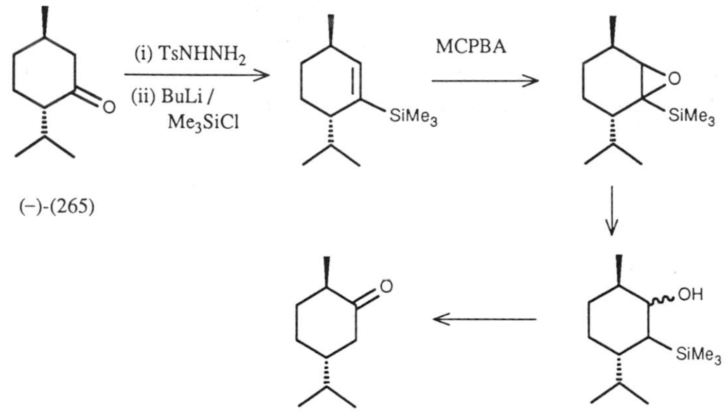

(−)-(265)

(+)-(266)

Menthone (265) can be converted into the Mannich product (267) which gives the diosphenol buchucamphor (268) on sens- itised photo-oxidation (H.H. Wasserman and J.L. Ives, J. Amer. Chem. Soc., 1976, 98, 7868). Buchucamphor (268) can be reduced to piperitone (269) by treatment of the derived dimethylthio- carbamate (270) with LiI − AcOH (A.A. Ponaras et al., J. Org. Chem., 1988, 53, 1110).

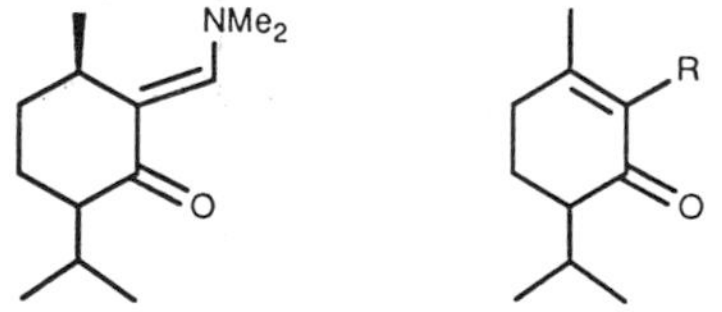

(267)

(268) R = OH
(269) R = H
(270) R = OC(=S)NMe$_2$

Isomerisation of the dihydrocarvone (257) using rhodium(3) chloride leads to a 1 : 7 mixture of 7,8-dihydrocarvone (271) and carvenone (135) (P.A. Grieco et al., J. Amer. Chem. Soc., 1976, <u>98</u>, 7102). Reduction of carvone (272) using Te - sodium borohydride saturates the endocyclic double bond to give *trans*-dihydrocarvone (257) exclusively (M. Yamashita, Y. Kato and R. Suemitsu, Chem. Letters, 1980, 847). This simple methodology must surely replace the inconvenient and sometimes erratic reduction of α,β-unsaturated ketones using alkali metals in liquid ammonia. The exocyclic double bond of carvone (272) is selectively reduced by homogeneous catalytic hydrogenation using Wilkinson's catalyst to give the dihydro derivative (271) (S.G. Levine and B. Gopalkrishnan, Tetrahedron Lett., 1979, 699).

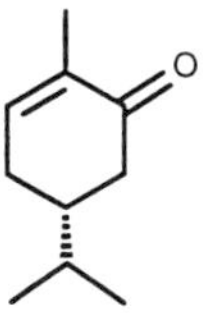

(-)-(271)

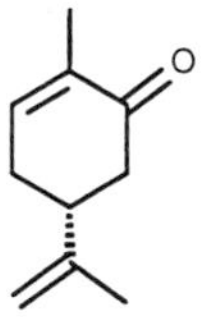

(-)-(272)

Carvone (272) is oxidised to the piperitenone derivative (273) by lead(4) acetate (J. de P. Teresa and I.S. Bellido, Anales de Quim., 1976, <u>72</u>, 76). The Diels-Alder reaction of carvone (272) with isoprene which is catalysed by aluminium chloride yields the adduct (274) which should have application in sesquiterpenoid synthesis (E.C. Angell et al., J. Org. Chem., 1985, <u>50</u>, 4696).

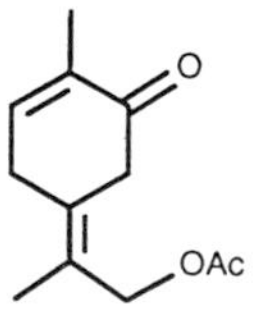

(273)

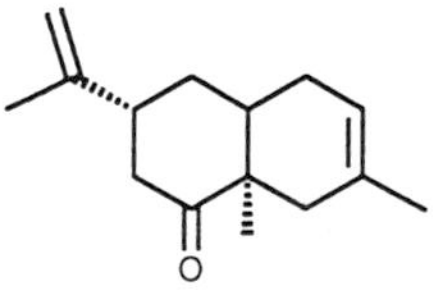

(274)

Biotransformations of monoterpenoids which can be effected by bacteria or by fungi have attracted considerable attention and, often in the context of biosynthetic investigations, so have transformations carried out by plant cell cultures.

(+)-Limonene (244) is oxidised to (+)-perillic acid (275) by *Pseudomonas incognita*, which also biotransforms various 2,6-dimethyloctane derivatives (J.R. Devi and P.K. Bhattach-aryya, Indian J. Biochem. Biophys., 1977, 14, 359). The transformation of (+)-limonene (244) into the diol (276) by *Corynespora cassiicola* DSM 62475 can be carried out on a very large scale (W.R. Abraham, B. Stumpf and K. Kieslich, Appl. Microbiol. Biotechnol., 1986, 24, 24), and the same diol (276) is produced from limonene by *Diplodia gossypina* (W.R. Abraham et al., Eur. Congr. Biotechnol., 3rd, 1984, 1, 245).

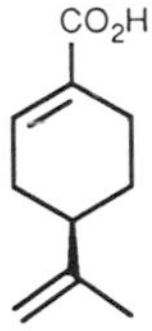

(+)-(275) 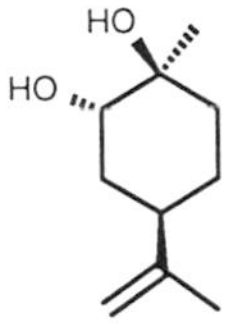   (276)

Another large-scale bioprocess involving monoterpenoids is the enantioselective hydrolysis of (±)-menthyl acetate (277) to give (-)-menthol (278) and (+)-menthyl acetate which is mediated by *Bacillus subtilis* (I.K. Brooks, M.D. Lilly and J.W. Drodz, Enzyme Microb. Technol., 1986, 8, 53). A related

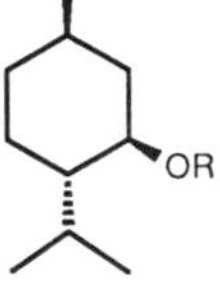

(±)-(277) R = Ac
(−)-(278) R = H
(±)-(279) R = COCH$_2$Cl

enantioselective hydrolysis has been carried out by using *Pseudomonas* sp. NOF-5 to convert (±)-menthyl chloroacetate (279) (800 kg) into (-)-menthol (278) (225 kg) and (+)-menthyl chloroacetate (388 kg) which are separable by fractional dis-tillation (T. Inagaki and H. Ueda, Nippon Nogei Kagaku Kaishi, 1987, 61, 49). Other applications of *Pseudomonas* NOF-5 have been reviewed (T. Inagaki, Bio. Ind., 1989, 6, 811).

Menthenols are also amenable to biotransformation. The fungus *Rhodotorula mucilaginosa* readily hydrolyses (-)-isopulegyl acetate (280) to give the alcohol (129), but the (+)-acetate is very little affected and (+)-neoisopulegyl acetate (281) is inert (A. Monkiewicz and J. Gora, Acta Biotechnol., 1985, 5, 263).

(-)-*cis*-Carveol (282) is converted into the potent germination inhibitor (+)-bottrospicatol (283) by *Streptomyces bottropensis*, and the same organism converts the enantiomer of (282) into (-)-isobottrospicatol (Y. Noma and H. Nishimura, Agric. Biol. Chem., 1987, 51, 1845).

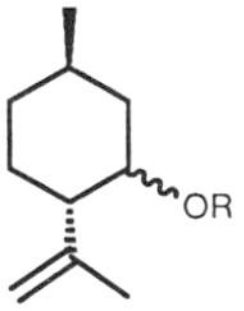

(−)-(280) R = β-OAc
(+)-(281) R = α-OAc

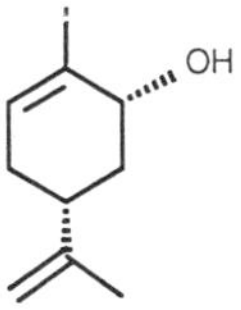

(−)-(282)

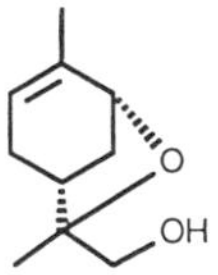

(+)-(283)

Bioreactions of a number of menthones and menthenones have also been investigated. (-)-Menthone (265) is converted into the isomeric keto acids (284) by *Pseudomonas fluorescens* (N. Sawamura, S. Shima and H. Sakai, Agric. Biol. Chem., 1976, 40, 649), and (-)-piperitone (285) yields mainly the ketol (286) with *Aspergillus niger* (E.V. Lassak et al., Aust. J. Chem., 1973, 26, 845). (+)-Pulegone (128) is reduced to (-)-menthone (265) by cultures of *Nicotiana tabacum* cells, and

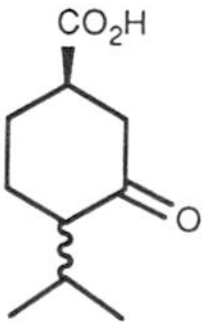

(284)

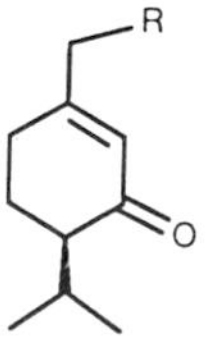

(285) R=H
(286) R=OH

the saturated ketone then undergoes further hydroxylation and reduction reactions (T. Suga et al., Phytochemistry, 1988, 27, 1041). Oxidation of (+)-pulegone (128) by mammalian cyto-chrome P-450 leads to menthofuran (287) (R.H. McClanahan et al., J. Amer. Chem. Soc., 1988, 110, 1979). (-)-Carvotanacet-one (271) is converted into (+)-carvomenthone (266) and then into (-)-carvomenthol (288) by *Pseudomonas ovalis*. The enant-iomer of (271) gives mainly (-)-isocarvomenthone (289) (Y. Noma, S. Nonomura and H. Sakai, Agric. Biol. Chem., 1974, 38, 1637). (-)-Carvone (272) gives (+)-dihydrocarvone (257) and (+)-neodihydrocarveol (290) when it is exposed to a suspen-sion culture of *Medicago sativa* (A. Kergomard et al., Phyto-chemistry, 1988, 27, 407). Further biotransformations of

(287)   (-)-(288)   (-)-(289)   (+)-(290)

carvone (272) have been carried out using *Nicotiana tabacum* (T. Hirata et al., Phytochemistry, 1982, 21, 2209), *Beauveria sulfurescens* (A. Kergomard, M.F. Renard and H. Veschambre, J. Org. Chem., 1982, 47, 792), and *Rhodotorula mucilaginosa* (A. Mironowicz et al., Pol. J. Chem., 1982, 56, 735).

During the last ten years part of the remarkable resurgence of interest in monoterpenoid chemistry has been due to the discovery that many examples which are readily available in good optical purity from the chiral pool can be transformed into valuable chiral reagents and auxiliaries. This is espec-ially true of members of the bicyclic pinane and camphane classes but also applies to the *p*-menthane series.

It is possible to achieve the kinetic enantiodifferentiat-ion of racemic 1,3-alkanediols via their enantioselective acetalisation by (-)-menthone (265) (J. Harada et al., J. Org. Chem., 1992, 57, 1637). The menthyl derivative (291) reacts with elemental fluorine in acetonitrile solution to give an intermediate which can be methanolysed to yield the α-fluoro-β-keto ester (292) of 98% e.e. (T. Iwaoka et al., Tetrahedron: Asymmetry, 1992, 3, 1025). Chloromethyl menthyl ether (293) reacts (Scheme 8) with the racemic tributylstannyl alcohol

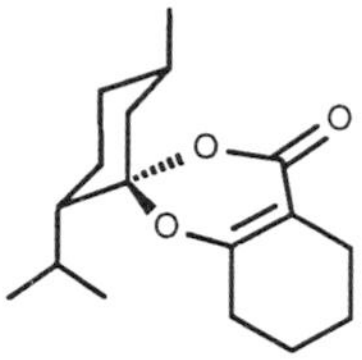

(291) 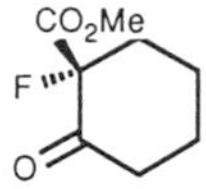 (292) 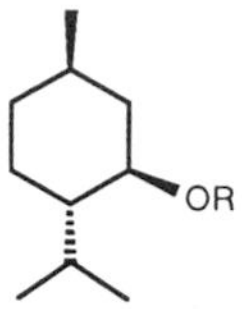 (293) R = CH₂Cl
(298) R = AlCl₂

(294) to give a mixture of the diastereoisomeric acetals
(295) which can be separated by chromatography. Reactions of
(295a) with benzaldehyde followed by further chemistry leads
to the β-hydroxy ester (296) of greater than 90% e.e. The
enantiomer of (296) is similarly obtained from (295b) (V.J.
Jephcote, A.J. Pratt and E.J. Thomas, J. Chem. Soc., Chem.
Commun., 1984, 800).

SCHEME 8

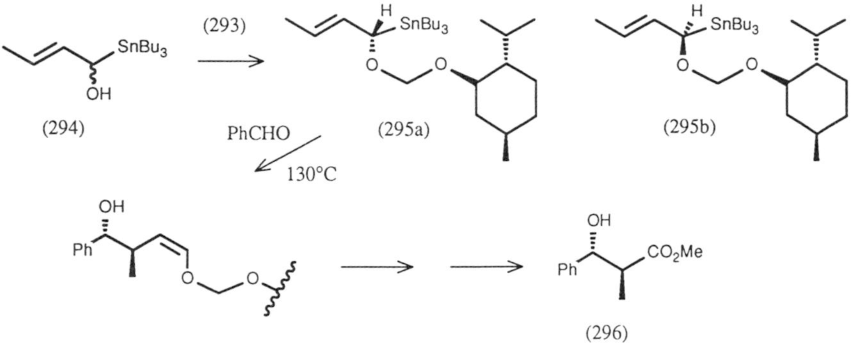

Other applications of the chloro ether (293) have been described (resolution of an oxime: R. Croteau, M. Felton and R.C.
Ronald, Arch. Biochem. Biophys., 1980, 200, 524, resolution of
an organometallic compound: S.G. Davies et al., J. Chem. Soc.,

Chem. Commun., 1986, 307). The menthoxymethyl group, derived
from (293), can be utilised as a chiral protecting group for
alcohol functions when its presence permits % e.e. measurements to be made by NMR at various stages of a reaction sequence (D. Dawkins and P.R. Jenkins, Tetrahedron: Asymmetry,
1992, 3, 833). The europium complex (297) can be used as a
chiral shift reagent for NMR spectroscopy (V.M. Potapov et
al., J. Gen. Chem. (U.S.S.R.), 1975, 45, 2071).

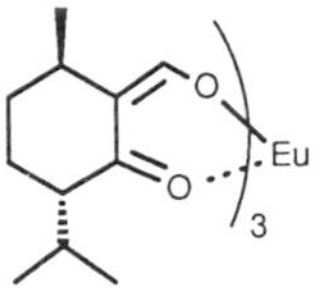

(297)

Menthyloxyaluminium dichloride (298) is a chiral Lewis
acid catalyst (C.J. Northcott and Z. Valenta, Can. J. Chem.,
1987, 65, 1917), and soluble chiral menthyl- and neomenthyl-
phosphine rhodium complexes have been used as enantioselective hydrogenation catalysts (D. Valentine, R.C. Sun and K.
Toth, J. Org. Chem., 1980, 45, 3703). The alane (299), derived from (-)-menthol (278), can reduce the acetylenic ketone (300) to the corresponding (S)-alcohol of 92% e.e. (M.
Falorni, L. Lardicci and G. Giacomelli, J. Org. Chem., 1989,
54, 2383).

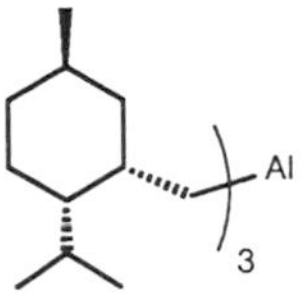

(299)

(300)

The first optically pure, stable selenonium ylid, with (R)-
configuration at Se, has been resolved via the (-)-menthyl
ester (301) (N. Kamigata et al., J. Chem. Soc., Perkin Trans.
1, 1992, 1721).

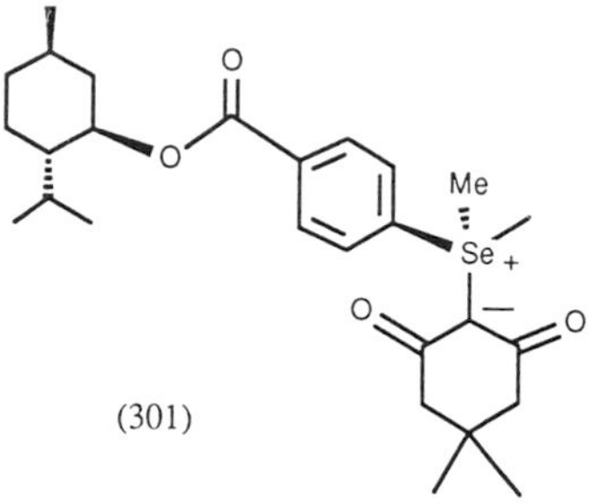

(301)

Menthyl esters have been widely used in asymmetric synthesis. Reaction of the (-)-menthyl phenylglyoxalate (302) with lithium tetra-alkylaluminates (prepared via hydroalumination of vinyl precursors) yields menthyl (R)-mandelates (303) of moderate d.e. selectively (G. Boireau et al., Tetrahedron Lett., 1985, 26, 4181). Nickel(0)-catalysed [3+2]-cycloaddition of (-)-menthyl acrylate (304) with methylenecyclopropane gives the cyclopentene (305) of 25% d.e. The more efficient (-)-8-phenylmenthyl acrylate (306) gives, under the same conditions, the corresponding cyclopentene (307) of 64% d.e. (P. Binger, A. Brinkmann and W.J. Richter, Tetrahedron Lett., 1983, 24, 3599). Further applications of (-)-8-phenylmenthyl derivatives are discussed below.

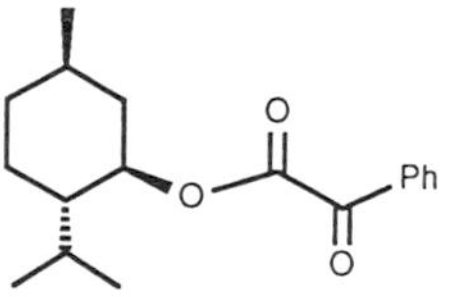

(302)

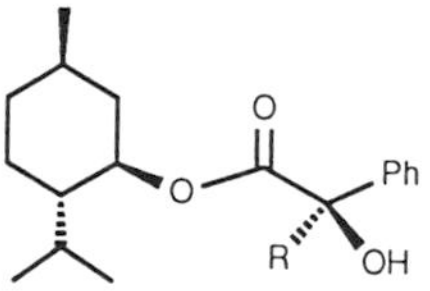

(303)

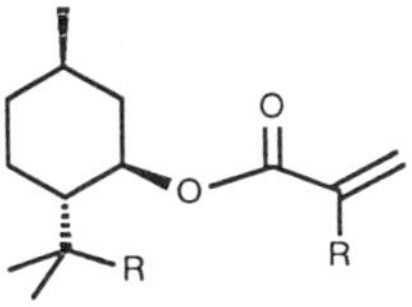

(304) $R^1 = R^2 = H$
(306) $R^1 = Ph; R^2 = H$
(308) $R^1 = H; R^2 = CF_3$

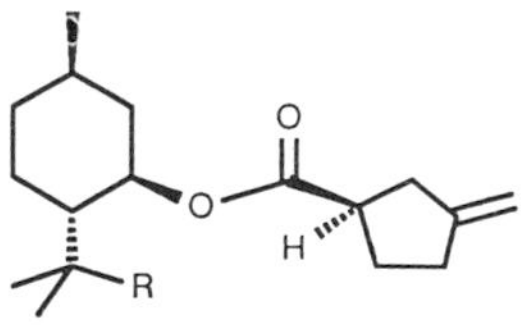

(305) R = H
(307) R = Ph

Addition of RMCu to the trifluoro ester (308) leads to the difluoro compound (309). This can be converted into the trifluoromethyl compound (310) using tetrabutylammonium fluoride - TsOH, but a 1 : 1 mixture of diastereoisomers is produced (T. Kitazume and T. Onogi, Synthesis, 1988, 614). On the other hand, anodic oxidation of the menthyl phenylacetate (311) in the presence of HF - MeCN - triethylamine leads to the α-fluorinated esters (312) of reasonable d.e. (L. Kabore et al., Tetrahedron Lett., 1990, 31, 3137).

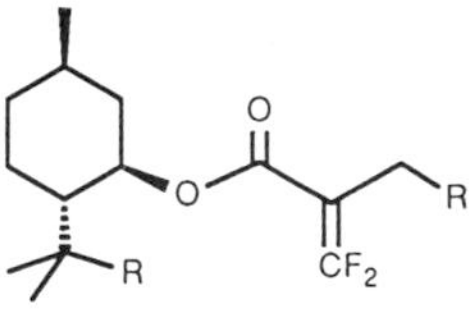

(309)

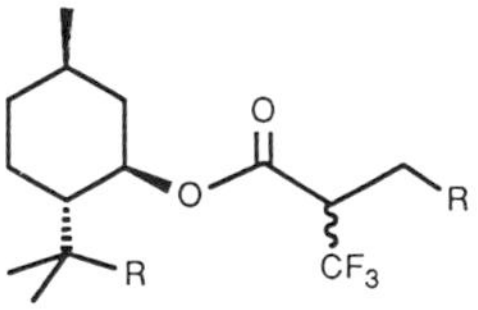

(310)

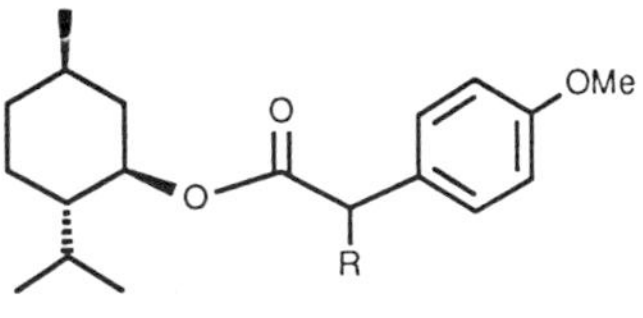

(311) R = H
(312) R = F

The γ-menthyloxybutenolide (313) undergoes Diels-Alder reaction with butadiene to yield an adduct which gives the lactone (314) of 99% e.e. after methanolysis (B.L. Feringa and J.C. de Jongh, J. Org. Chem., 1988, 53, 1125). Similarly, the 1,3-dipolar cycloaddition of benzenenitrile oxide to (313) gives exclusively the adduct (315) (B. de Lange and B.L. Feringa, Tetrahedron Lett., 1988, 29, 5317).

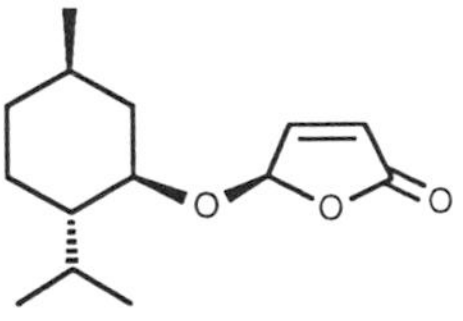

(313) 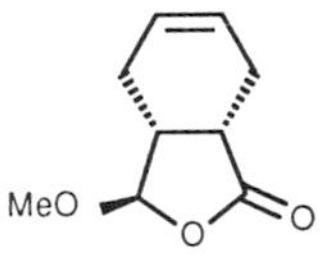   (314) 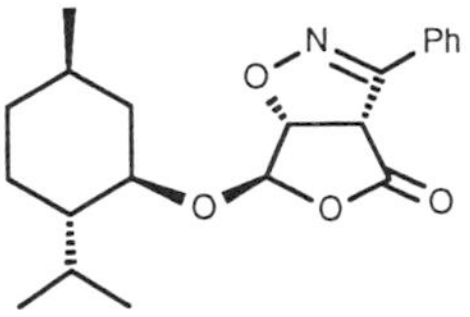   (315)

The irontricarbonyl complex (316) yields the cyclohexenone (317) on reaction with the sodium salt of dimethyl malonate followed by decomplexation and hydrolysis (G.A. Potter and R. McCague, J. Chem. Soc., Chem. Commun., 1990, 1172), and the N-menthyl ketenimine (318) reacts with dimethylsulfoxonium methylide to give optically active methyl ketones after acidic hydrolysis (K. Hiroi and S. Sato, Chem. Pharm. Bull., 1985, 33, 4691).

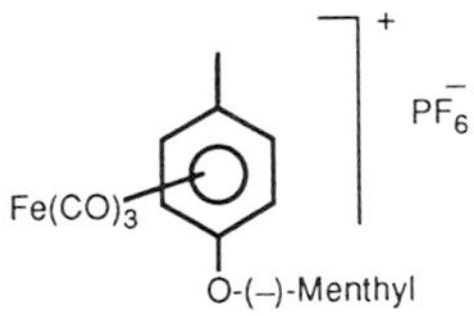

(316) 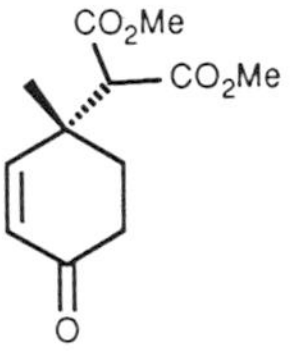   (317) 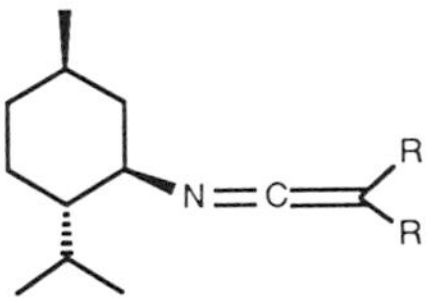   (318)

Whilst menthyl derivatives like those discussed above offer useful levels of asymmetric induction, the related (-)-8-phenylmenthol (319), which is also available as its (+)-enantiomer, is a far superior chiral auxiliary. The preferred solution conformation of (319) and of many of its acylated derivatives is represented by (320) where the effective shielding of one face of the attached group R is apparent (cf. J. Runsink et al., J. Chem. Soc., Perkin Trans. 2, 1988, 49).

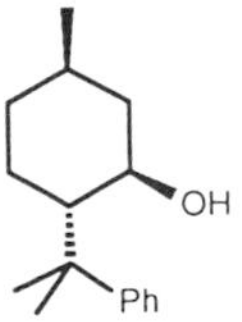

(-)-(319) 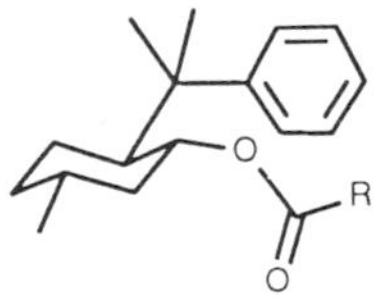   (320) 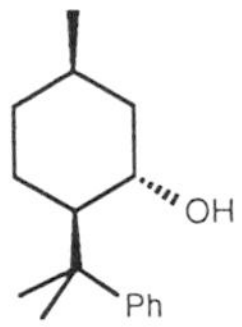   (321)

(-)-8-Phenylmenthol (319) is synthesised from (+)-pulegone
(128) via conjugate addition of PhMgBr followed by reduction
of the intermediate ketone to an 87 : 13 mixture of (319) and
its *epi,ent*-form (321). These can be separated by fractional
crystallisation of the derived chloroacetates (O. Ort, Org.
Synth., 1987, 65, 203) or by preparative HPLC (J.K. Whitesell
et al., J. Org. Chem., 1986, 51, 551). Improved procedures
for obtaining (-)-(319) and some of its 8-aryl analogues
have been described (D. Potin, F. Dumas and J. Maddaluno,
Synth. Commun., 1990, 20, 2805), and (+)-(319) can be prepared from unnatural (-)-pulegone (*ent*-128) which is accessible
via the oxidative cyclisation of (-)-citronellol (*ent*-82) (H.
Buschmann and H.D. Scharf, Synthesis, 1988, 827).

One of the first applications of 8-phenylmenthol in asymmetric synthesis was the diastereoselective formation of
Diels-Alder adducts from its (+)-propenoate (*ent*-306) (E.J.
Corey and H.E. Ensley, J. Amer. Chem. Soc., 1975, 97, 6908).
The reaction of cyclopentadiene with (-)-(306) has been carefully investigated and optimised (W. Oppolzer et al., Tetrahedron Lett., 1981, 22, 2545).

The conjugate addition of cuprates R^3CuBF_3 to various alk-
enoates (322) derived from (-)-8-phenylmenthol (319) yields
products (323) which can be hydrolysed to give recyclable

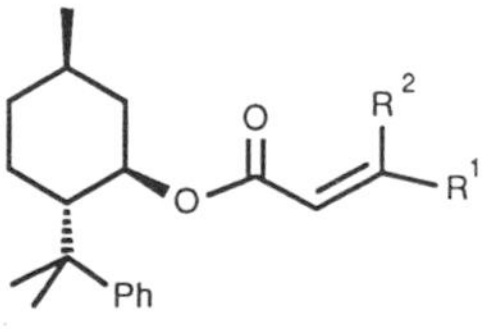

(322) 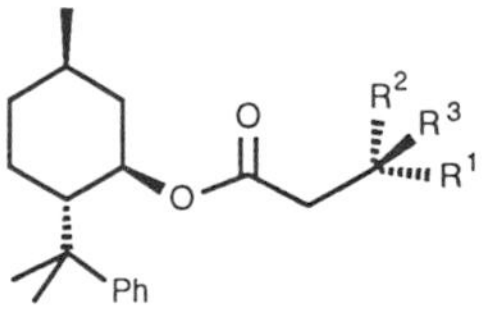   (323)

chiral auxiliary and β-substituted carboxylic acids of high
optical purity (W. Oppolzer and H.J. Loher, Helv. Chim. Acta,
1981, <u>64</u>, 2808).

The lithium enolate (324) undergoes Michael addition to
methyl (E)-but-2-enoate to give mainly the threo-product
(325) (E.J. Corey and R.T. Peterson, Tetrahedron Lett., 1985,
<u>26</u>, 5025).

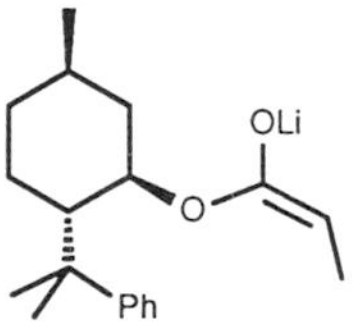

(324) 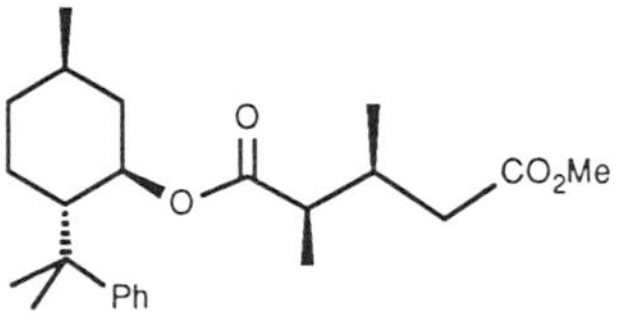   (325)

Glyoxylates (326) derived from (-)-8-phenylmenthol (319)
are extremely versatile tools for the synthesis of a wide
range of optically active products. Thus, (326: R = H) reacts
with Grignard reagents to give the α-hydroxy acid precursors
(327) of 90 - 99.4% d.e. (J.K. Whitesell, A. Bhattacharya and
K. Henke, J. Chem. Soc., Chem. Commun., 1982, 988).

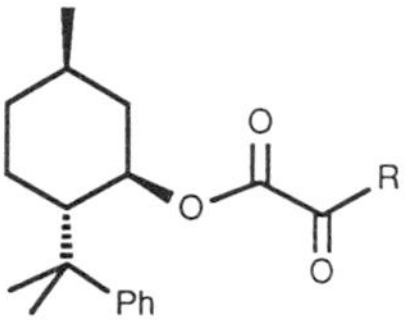

(326) R = H, Me, or Ph 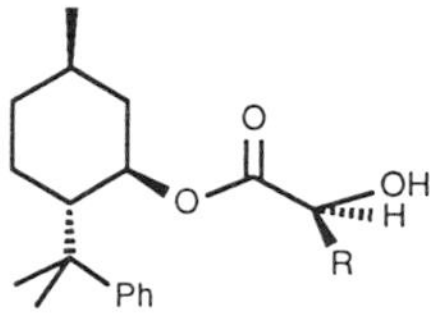   (327)

The same glyoxylate (326: R = H) also undergoes a tin(4)
chloride-catalysed ene reaction with 1-hexene to give (328)
of 97.6% d.e. (J.K. Whitesell et al., J. Chem. Soc., Chem.
Commun., 1982, 989).

Reduction of the pyruvate (326: R = Me) or the phenylglyox-
ylate (326: R = Ph) using potassium tri-isopropoxyborohydride
in THF yields, respectively, lactate or mandelate esters
(329) with d.e. values of 90% (J.K. Whitesell, D. Deyo and
A. Bhattacharya, J. Chem. Soc., Chem. Commun., 1983, 802).
The same paper describes how reaction of the (-)-8-phenylmen-
thyl pyruvate (326: R = Me) with PhMgBr leads to the atrolac-
tic ester (330) while reaction of the (-)-8-phenylmenthyl

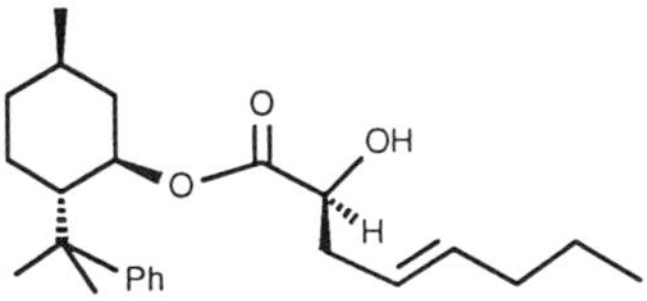

(328)

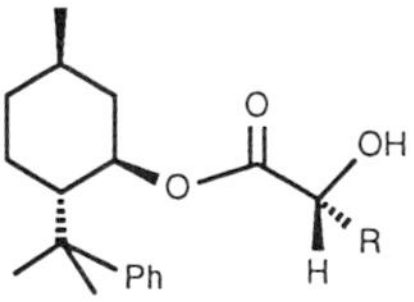

(329)

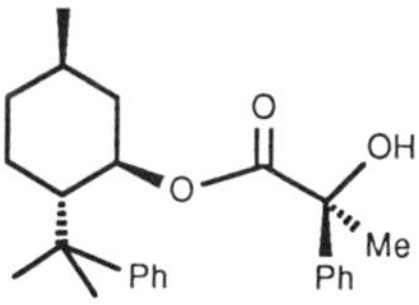

(330)

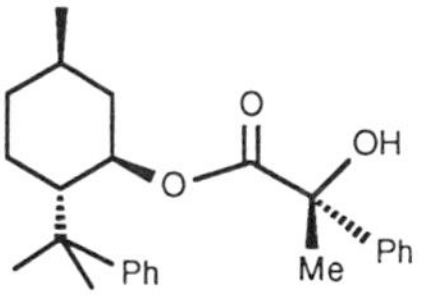

(331)

phenylglyoxalate (326: R = Ph) with MeMgBr gives the diaster-eoisomeric product (331). A review of this and other related work has been published (J.K. Whitesell, Chem. Rev., 1992, 92, 953).

Second Supplements to the 2nd Edition of Rodd's Chemistry
of Carbon Compounds, Vol. II B(Partial), C, D and E, edited by M. Sainsbury
© 1994 Elsevier Science B.V. All rights reserved.

Chapter 7

The Carotenoid Group of Natural Products

SYNNØVE LIAAEN-JENSEN

1. Scope of presentation

In this series the carotenoid group has previously been reviewed in detail by J.B. Davis (2nd Ed. Vol. II B, pp. 231-346; First Supplement Vol. II B, pp. 191-357), appearing in 1967 and 1973 respectively. Both presentations were exhaustive, covering individual carotenoids in an encyclopedic manner and dealing with topics such as nomenclature, occurrence, function, isolation, structure elucidation, chemical, spectroscopic and other physical properties, and synthesis as well as biosynthesis. Also included were the vitamins A and related compounds and the visual process.

Since 1973 a large and interesting development has occurred within the field of carotenoids. The number of known, naturally occurring carotenoids has increased from ca 250 in the 1973 review to around 600 in 1992. A detailed treatment in line with the previous presentation in the series is not possible within a limited frame. The present review will consequently be selective rather then exhaustive, focusing on important trends in carotenoid research during the two last decades. Attempts are made to direct the reader to available, detailed compilations and other review type literature and to avoid repetition. Only true carotenoids by the IUPAC definition (IUPAC-IUB, Pure Appl. Chem. 1975, 41, 405-431) will be considered here, thus excluding Vitamins A and other retinoids as well as smaller metabolic degradation products.

Great progress is currently being made within carotenoid biosynthesis by means of genetic engineering. For instance the gene cluster for carotenoid biosynthesis in procaryotes may now be transferred to carotenoidless bacteria and yeasts (G.A.Armstrong et al. Proc. Natl. Acad. Sci. USA, 1990, 87: 9975; B.S.Hundle et al., Photochem. and Photobiol., 1991, 54: 89). However, carotenoid biosynthesis or metabolism will not be covered here, but treated in detail elsewhere (Carotenoids Vol. 3, Biosynthesis and metabolism, eds. G. Britton, S. Liaaen-Jensen and H. Pfander, Birkhäuser, Basel, 1995).

2. Relevant literature overviews

Progress in the carotenoid field is reviewed triannually via the printed versions of the plenary lectures of the International Carotenoid Symposia, generally sponsored by IUPAC. Reference is made to the third symposium in Cluj, Romania (Pure Appl. Chem., 1973, 35: 1-130), fourth in Berne, Switzerland (Pure Appl. Chem., 1977, 47: 97-243), fifth in Madison, Wisconsin (Pure Appl. Chem., 1979, 51: 435-886), sixth in Liverpool, England (Carotenoid Chemistry & Biochemistry eds. G. Britton and T.W.Goodwin, Pergamon Press, 1982) seventh in München, Germany (Pure Appl. Chem., 1985, 57: 639-821), eight in Boston, USA (Carotenoids Chemistry and Biology, eds. N.I. Krinsky, M.M. Mathews-Roth and R.T. Taylor, Plenum 1989) and ninth in Kyoto, Japan (Pure Appl. Chem., 1991, 63: 1-176).

A useful compilation on naturally occurring carotenoids, first issued in 1971 (O. Straub, Key to Carotenoids, Birkhäuser, Basel), appeared recently in a second expanded edition (ed. H. Pfander, Birkhäuser, Basel, 1987, 296 p). Structures and complete literature references concerning isolation, spectroscopic properties, chemistry, partial and total syntheses are given. New structures from the 1987-1992 period are included as an appendix elsewhere (G. Britton, S. Liaaen-Jensen and H. Pfander eds. Carotenoids, Vol. 1A. Isolation and analysis, Birkhäuser, Basel, 1993).

The monograph "Carotenoids" edited by O. Isler (Birkhäuser, Basel, 1971), covering all aspects on carotenoids has served as a useful reference text, and the 2nd edition of T.W. Goodwin "The Comparative Biochemistry of the Carotenoids" has appeared as two volumes, Vol. I Plants (Chapman & Hall, 1980), and Vol. II Animals (Chapman & Hall, 1984). Whereas the former monograph is no longer available, a new series of six multi-author volumes on carotenoids will be issued annually from 1993 and onwards by the same publisher (Birkhäuser, Basel) with G. Britton, S. Liaaen-Jensen and H. Pfander as editors.

In the meantime several other reviews have appeared on special carotenoid topics: e.g on stereochemistry (S. Liaaen-Jensen in Progr. Chem. Org. Nat. Products 1980, 39: 123-172a, on allenic and acetylenic carotenoids (S. Liaaen-Jensen in Natural Product Chemistry, ed. T.I. Atta-ur-Rahman, 1990, Elsevier, pp. 133-169), on marine carotenoids (S. Liaaen-Jensen, 1978, In Marine Natural Products Chemistry eds. D.J. Faulkner and W.H. Fenical, 1977, Plenum, 239-259; T. Bjørnland and S.

Liaaen-Jensen in <u>The Chromophyte Algae, Problems and Perspective</u>, eds.
J.C. Green, B.S.C. Leadbeater and W.L. Diver, 1989, Clarendon Press, pp.
37-61).

Attention should also be paid to the more practical approaches dealt
with in <u>Methods in Enzymology</u>. General carotenoid methodology is treated
by G. Britton; (<u>Methods Enzymol</u>, 1985, <u>111</u>: 113-149); see also various
other papers which will appear in Vol. <u>213</u>, 1992.

3. <u>Trends in carotenoid chemical research</u>

In retrospect progress in the field during the last two decades has
been greatly determined by improved methodology. Thus it is in principle
possible nowadays to elucidate the detailed structure of a new carotenoid
on the 1 mg scale. Major advances in chromatography include HPLC with a
diode array detector. Sophisticated two-dimensional and other special NMR
techniques allow the identification of each proton and carbon atom in a
carotenoid and close definition of spacial proximity within the molecule.
Whereas NMR may only provide information on relative configuration,
additional CD data, in comparison with carotenoids of known chirality,
will in most cases serve to define absolute configuration. Moreover,
progress in the enantiospecific total synthesis of chiral carotenoids has
resulted in the syntheses of several structurally complex naturally
occurring carotenoids as optically pure isomers, thus providing final
proof of structure. Particularly for the identification of minor
metabolites on the microgram scale, the availability of authentic
synthetic reference carotenoids is of great importance.

The last two decades are characterized by large interest in
sterochemical aspects of carotenoids. Whereas until <u>ca</u> 1970 it was
assumed that a given carotenoid constitution occurred in Nature with one,
preferred chirality, several examples have been demonstrated of the
natural occurrence of enantiomers and diastereomers in different
organisms. Particularly in marine animals mixtures of optical isomers are
encountered. The availability of new NMR techniques, the application of
CD spectra to stereochemical problems and excellent HPLC systems are
responsible for this progress. So far X-ray analyses are not successfully
performed for complex chiral carotenoids, presumably due to the difficulty
in obtaining perfect, regularly packed carotenoid crystals, (see <u>Caroteno-
ids</u> Vol. 1B, <u>Spectroscopy</u>, eds. G. Britton, S. Liaaen-Jensen and H.

Pfander, Birkhäuser, Basel, 1993).

More recently there has been a great revival of the interest in the geometrical isomerism of carotenoids. The early studies were compiled in L. Zechmeister's monograph Cis-trans isomeric Carotenoids - Vitamins A and Arylpolyenes (Springer, Wien, 1962). However, it was the availability of high field ^{1}H NMR spectroscopy in the eighties that enabled unequivocal assignment of E/Z configuration of the polyene chain. Moreover cis-isomeric carotenoids have recently been shown to be much more frequent in Nature than originally thought. Since E/Z isomerization occurs readily in solution during isolation, special precautions must be made for the demonstration of naturally occurring cis-isomers, vide infra. The general awareness of artifact formation in the carotenoid field, e.g under saponification conditions, has increased considerably in recent years.

Examples will be discussed illustrating new structural types that have been discovered. Particularly marine animals and microalgae have offered interesting carotenoid chemistry. However, re-examination of more classical sources such as flowers of higher plants have revealed interesting minor carotenoids. Hitherto no other heteroatoms than oxygen have been encountered in the skeleton of naturally occurring carotenoids, despite a search for analogues containing sulfur or halogen.

A special trend in carotenoid chemical research has been an increased interdisiplinary approach. Structure determination has been important for biosynthetic and metabolic studies. Chemosystematic applications have been useful, particularly for microalgae. New developments are seen concerning the several functional aspects of carotenoids. The more detailed role of carotenoids in photosynthesis, photochemistry and photo-protection are being revealed. Commercial application of carotenoids now includes aquaculture. A large market is opened for astaxanthin as a feed ingredient to salmon and Crustacea. In recent years much interest is centered around medical applications including the potential cancer-preventive effect of β,β-carotene.

4. Improved methodology

As already pointed out improved methodology has been the basis for rapid progress in carotenoid chemical research during the last two decades. The methods largely responsible for more refined results are (a) (a) Isolation and chromatography. Improved isolation techniques are imperative for the isolation of pure carotenoids for spectroscopic characterization. Attention must be paid to the removal of colourless contaminants as well as the separation of carotenoid components, including geometrical isomers.

The state of the art relating to isolation and analysis has recently been dealt with by K. Schiedt and S. Liaaen-Jensen (in Carotenoids. Vol. 1A. Isolation and Analysis, eds. G. Britton, S. Liaaen-Jensen and H. Pfander, Birkhäuser, Basel, 1993) and concerning chromatographic techniques by H. Pfander et al. in the same source.

High performance liquid chromatography has become a routine tool during this period. Diode array detectors allow the recording of visible spectra during the chromatographic run, a very convenient and important device when dealing with sterically unstable carotenoids. An up to data treatment on HPLC is presented by Pfander et al. in Carotenoids. Vol. 1A (eds. G. Britton, S. Liaaen-Jensen and H. Pfander, Birkhäuser, Basel, 1993). From the vaste literature a few selected examples of efficient separations will be given.

Isocratic normal phase separation on nitrile columns has many successful applications. For example the separation of diasteromeric carotenoids with long distance between chiral centers (M. Vecchi et al., Helv. Chim. Acta, 1982, $\underline{65}$: 1050), the separation of diastereomeric carotenoid dicamphanates (M. Vecchi and R.K. Müller, J. High Res, Chrom. Comm. 1979, $\underline{2}$: 195), of geometrical isomers of zeaxanthin and lutein (F. Khachik et al., J. Chem., 1992, in press), and the separation of ten different E/Z isomers of astaxanthin (G. Englert and M. Vecchi, Helv. Chim. Acta, 1980, $\underline{63}$: 1761). However, for the separation of eleven E/Z isomers of β,β-carotene an alumina column was employed (M. Vecchi et al., Helv. Chim. Acta, 1981. $\underline{64}$: 2746) and for E/Z isomers of the C_{50}-tetrol bacteriorube-rin a reversed phase system developed by R. Riesen and H. Pfander (Abstr. 9th Int. Carot. Symp. Kyoto, 1990, p.82) was employed (M. Rønnekleiv and S. Liaaen-Jensen, Acta Chem. Scand., 1992, in press).

Silica columns are successfully employed for the separation of diastereomeric carotenoid carbamates (A. Rüttimann et al, Helv. Chim. Acta, 1983, $\underline{6}$: 212; K. Schiedt et al., Helv. Chim. Acta, 1988, $\underline{71}$: 881) and for the separation of lutein and zeaxanthin (K. Schiedt, dr.techn. Thesis, Univ. Trondheim, 1987). Silicagel coated with H_3PO_4 is efficient for the separation, without tailing, of ketocarotenoids including α-ketols and enolized ß-diketones (M. Vecchi et al., J. High Res, Chrom, & Chrom, Comm. 1987, 10: 349).

C-18 Reversed phase columns have been employed for the separation of carotenol fatty acid esters (F. Khachik and G.R. Becher, J. Chrom. 1988, 449: 119), fruit carotenoids (F. Khachik et al., Agric Food Chem., 1989: 1465) and carotenoids from human plasma (F. Khachik et al. Agric, Food Chem. 1989: 1465). Using a ternary gradient system fast separation of a large variety of phytoplankton carotenoids was accomplished in a somewhat complicated reversed phase system (S. Wright et al., Marine Ecol, Progr, Sec. 1991, 77: 183).

(b) NMR spectroscopy. Great advances in the NMR spectroscopy of carotenoids have occurred during the last two decades due to the introduction of high field instruments ($\geq$ 400 MHz) and the availability of novel two-dimensional and other special NMR techniques. Thus two-dimensional ^{1}H^{1}H and ^{1}H^{13}C NMR spectra are becoming routine in the carotenoid field, and sophisticated methods such as ^{1}H-detected one- and multiple bond hetero-COSY, 1D and 2D ROESY and TOCSY provide additional structural information.

Excellent compilations on the progress in this field have been presented by G. Englert (in Carotenoid Chemistry & Biochemistry, eds. G. Britton and T.W. Goodwin, Pergamon, 1982, pp. 107-134; Pure & Appl. Chem., 1991 63: 59-70). A comprehensive up-to-date chapter on MMR of caro-tenoids is in press (G. Englert in Carotenoids. Vol. 1B. Applied Spectro-scopy, eds. G. Britton, S. Liaaen-Jensen and H. Pfander, Birkhäuser, Basel, 1993).

The assignment of each proton and carbon, as well as coupling constants, are now feasible for even complex carotenoid structures. As an example is mentioned 7',8'-dihydroneoxanthin-20'-al 3'-lactoside (P457, Fig. 1), (G. Englert, Pure & Appl. Chem., 1991, 63: 59).

^{1}H NMR spectroscopy has also made unequivocal assignment of geometrical isomers of carotenoids possible. Of note is the structural assignment of several geometrical isomers of fucoxanthin (Fig. 2), namely all-trans, 9'-cis, 13'-cis, 9',13'-dicis, 13,9'-dicis and 13,13'-dicis (tentative) (J.A. Haugan et al., Acta Chem, Scand., 1992, 46: 389).

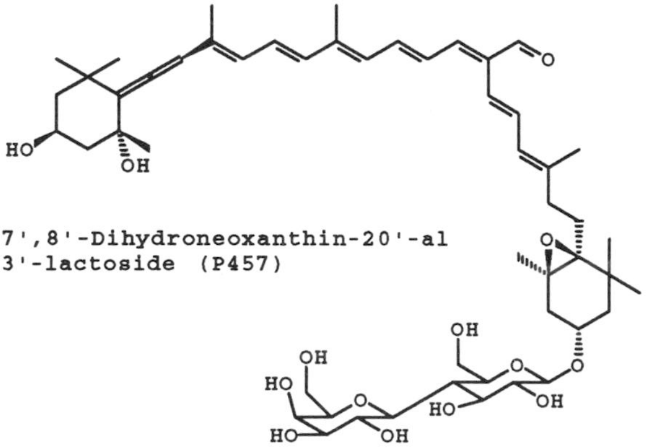

7',8'-Dihydroneoxanthin-20'-al
3'-lactoside (P457)

Fig. 1

For pure carotenoids NMR spectroscopy is currently the most important
single method for structural investigations, even on the microgram scale.
(c) Circular dichroism. Modern studies on the optical properties of carotenoids started with a fundamental paper on the ORD properties of
carotenoids (P.M. Scopes et al., J. Chem. Soc. C, 1969: 2527), introdu-
cing the additivity hypothesis for the contribution of the two chiral end
groups connected to the polyene chromophore.

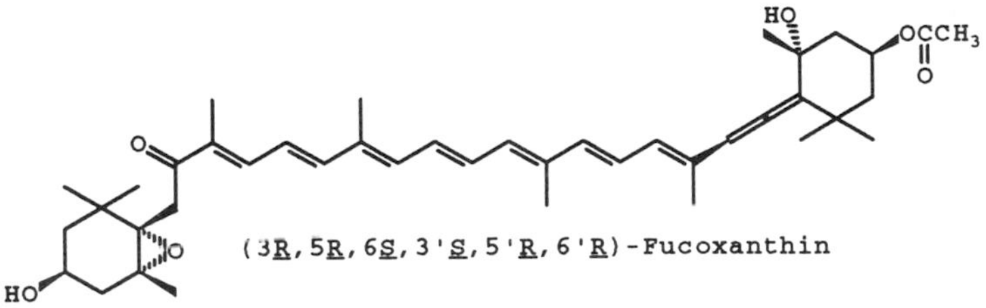

(3R,5R,6S,3'S,5'R,6'R)-Fucoxanthin

Fig. 2

In subsequent work CD spectra, rather than ORD spectra, have been
employed in the carotenoid field. The early progress has been summarized
(S. Liaaen-Jensen, Progr. Chem. Org. Nat. Prod., 1980, 39: 123-127) and an
up-to-date chapter on this topic is being published (K. Noack and R.
Buchecker in Carotenoids. Vol. 1B. Applied spectroscopy. eds. G. Britton,
S. Liaaen-Jensen and H. Pfander, Birkhauser, Basel, 1993). In summary CD
spectra are essential for studies on chiral carotenoids and may now be

measured also in the visible region. Carotenoid CD spectra are classified as conservative, intermediate or non-conservative (V. Sturzenegger et al., Helv. Chim. Acta, 1980, 63: 1074). Conservative CD spectra invert upon introduction of a cis(Z) double bond. Hence only CD spectra of geometrically pure isomers are meaningful. The conformational rule states that the preferred chiral conformation for a chiral β-ring determines the sign of the Cotton effect caused by that end group (see A.G. Andrewes et al., Acta Chem. Scand. 1974, B28: 730). This rule has been extensively used for dicyclic carotenoids with two chiral β-end groups. Temperature effects on conformational equilibria have been investigated (K. Noack and A.J. Thomson, Helv. Chim. Acta, 1979, 62: 1902, K. Noack in Carotenoid Chemistry & Biochemistry, eds. G. Britton and T.W. Goodwin, Pergamon 1981, pp. 135-154). Using the additivity hypothesis several 2'-substituted monocyclic carotenoids have been stereochemically correlated by means of CD spectra (H. Rønneberg et al., Phytochemistry, 1985, 24: 309). CD spectra of carotenoproteins are determined by the protein and not by the carotenoid prosthetic group.

(d) Other spectroscopic techniques. The current understanding of the physical phenomena of electronic absorption spectra (UV/VIS) of carotenoids is being treated by B.E. Kohler (in Carotenoids. Vol. 1B. Applied spectroscopy, eds. G. Britton, S. Liaaen-Jensen and H. Pfander, Birkhäuser Basel, 1993), and practical aspects treated by G. Britton and A. Young in the same source. Visible spectra still represent the simplest method for preliminary chromophore assignment.

Mass spectrometry remains the method of choice for the determination of molecular weight and elementary composition of carotenoids. The fragmentation pattern upon electron impact provides additional structural information. Advances have been treated by H. Budzikiewics (in Carotenoid Chemistry & Biochemistry, eds. G. Britton and T.W. Goodwin, 1982, pp. 155-166), and by C.R. Enzell and I. Wahlberg (in Biochem. Appl. Mass Spectrom., Vol. II, ed. G. Waller, Wiley, New York, 1980 pp. 407-438). A comprehensive up-to-date treatment including recent developments is being published by G. Enzell and S. Bromann (in Carotenoids. Vol. 1B. Applied spectroscopy, eds. G. Britton, S. Liaaen-Jensen and H. Pfander, Birkhäuser, Basel, 1993), including fragmentation patterns and new ionization techniques other than electron impact. Fast atom bombardment, chemical ionization and particle desorption represent useful, less commonly employed methods. Combined HPLC-MS methods is a great perspective for the future.

Infrared spectroscopy is still useful for the confirmation of special functional groups. For instance the detection of carotenoid sulfates was first based on IR-spectroscopy (T. Ramdahl and S. Liaaen-Jensen, _Acta Chem. Scand._, 1980 _B34_: 773). The recent availability of FT-IR instruments means improved quality also of carotenoid IR-spectra, _e.g._ for the detecting the weak -C≡C- stretching vibration of monoacetylenic carotenoids. An up-to-date treatment of IR spectroscopy for carotenoids is described by K. Bernhard (in _Carotenoids. Vol. 1B. Applied spectroscopy_ eds. G. Britton, S. Liaaen-Jensen and H. Pfander, Birkhäuser, 1993).

Resonance Raman represents the most recent spectroscopic method of importance for the carotenoid field. The method is complementary to IR spectroscopy and is used for demonstrating the presence of polyene chromophores of _e.g._ _in situ_ biological samples, for carotenoproteins and particularly for carotenoids bound to photosynthetic units (Y. Koyama, _J. Photochem. Photobiol. B_. Biol., 1991, _9_: 265-280); Y. Koyama and H. Hashimoto in _Carotenoids in Photosynthesis_ eds. A. Young and G. Britton, Springer, _in_ press).

5. Partial and total synthesis

(a) _Partial synthesis_. In this section partial syntheses of carotenoids are generally considered to involve chemical derivatizations maintaining the carotenoid skeleton intact. At a time when spectroscopic techniques were less developed and offered less structural information than at present, such reactions were of great importance for the structural elucidation of carotenoids. The topic was reviewed by the present author in 1971 (in _Carotenoids_, ed. O. Isler, Birkhäuser, Basel, pp. 61-188) and in progress reports written by J.B. Davis (this series, 2nd Ed. Vol. IIB, 1967, pp. 258-265, First Supplement Vol. IIB, 1973, pp. 213-217).

Common derivatizations such as alkaline hydrolysis (saponification) of esters, acetylation or silylation of hydroxy groups, selective oxidation of allylic hydroxyl and complex metal hydride reduction of carbonyl functions are still of importance in conjunction with spectroscopic studies for structural elucidation. These reactions may all be carried out on the micro scale (C.H. Eugster in _Carotenoids_, Vol. 1A. _Isolation and analysis_, ed. G. Britton, S. Liaaen-Jensen and H. Pfander, Birkhäuser,

Basel, 1993).

Of new important derivatization reactions employed during the last two decades the formation of camphanates for α-ketols of mixed configuration should be mentioned. This allows HPLC separation of the resulting diastereomers (M. Vecchi and R.K. Müller J. High Res. Chrom. Chrom. 2: 195). Also of importance is the formation of carbamates of racemized carotenols, again for subsequent HPLC separation of the diastereomeric esters (A. Ruttimann et al. J. High Res. Chrom. Chrom. Comm., 1983, 6: 612). Both reactions serve to provide diasteromers useful in the assessment of the optical purity of chiral sec. carotenols.

(b) Total synthesis. Several chiral carotenoids were prepared with great expertice by total synthesis prior to the seventies (H. Mayer and O. Isler in Carotenoids, ed. O. Isler, Birkhäuser, Basel, 1971, pp. 325-576). However, in general most carotenoids were prepared in the optically inactive forms as mixtures of enantiomers or diastereomers. During the last two decades general progress in enantioselective synthesis has had an impact also on the carotenoid field. Several carotenoids have now been synthesized as pure optical isomers, some with up to eight chiral centers in the carotenoid skeleton. Representative examples (Fig. 3) of C_{40}-carotenoids include: (3R,3'R,6'R)-lutein (H. Mayer and A. Ruttimann, Helv. Chim. Acta, 1980, 63; 1451), all eight optical isomers of ε,ε-carotene-3,3'-diol (H. Mayer in G. Britton and T.W. Goodwin eds. Carotenoid Chemistry and Biochemistry, Pergamon, 1981, p. 55), (3S,5R,6R,3'S, 5'R,6'S)-neoxanthin (A. Baumeler and C.H. Eugster Helv. Chim. Acta 1992, 75: 773) and with eight chiral centers (3R,4S,5R,6R,3'R,4'S,5'R,6'R)-tetrahydroxypirardixanthin (U. Hengartner, Abstr. 9th Int. Symp. Carotenoids, Kyoto, 1990, p. 22, see Fig. 26), nor-carotenoids: (3S,3'S)-actinioerythrol (R.K. Muller et al., Helv. Chim. Acta 1978, 61: 2881), (3S, 5R,6R,3'R,5'R,6'R)-peridinin (M. Ito et al., J. Chem. Soc. Perkin Trans. 1. 1990: 197), C_{50}-carotenoids: (2R,6R,2'R,6'R)-decaprenoxanthin (M. Gerspacher and H. Pfander, Helv. Chim. Acta, 1989, 72: 151, see Fig. 4), (2S,2'S)-bacterioruberin J.P. Wolf and H. Pfander Helv. Chim. Acta 1986, 69: 62).

The syntheses of optically active carotenoids has been reviewed (H. Mayer, Pure Appl. Chem. 1979, 51, pp. 535-565; H. Pfander, Pure Appl. Chem, 1991, 63, pp. 23-34). Chiral centers have been generally introduced by using optically active, naturally occurring compounds such as camphor

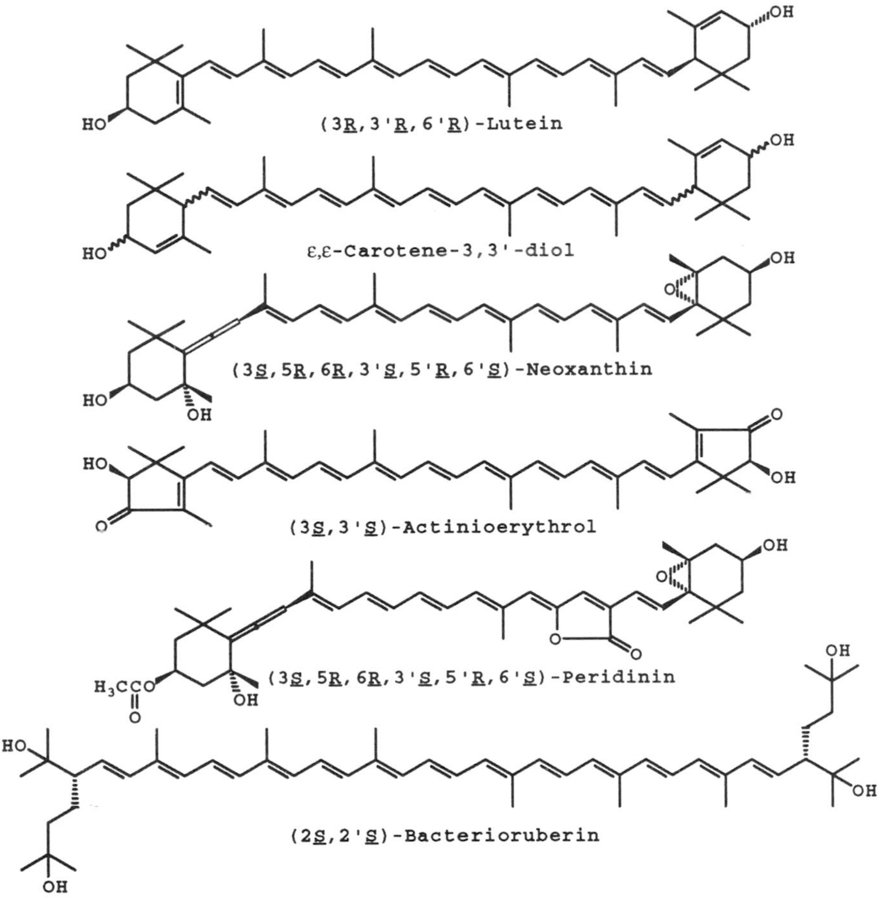

Fig. 3

as starting materials, by selective reduction with baker's yeast or by means of the Sharpless epoxidation reaction. A full volume covering synthesis of carotenoids is in preparation (Carotenoids. Vol. 3. Synthesis, eds. G. Britton, S. Liaaen-Jensen and H. Pfander, Birkhäuser, to be published 1994).

6. New structures

As already pointed out the number of known naturally occurring carotenoids for which structures have been assigned has increased from ca. 250 in the 1973 review (J.B. Davis, First Supplement Vol. IIB, pp. 191-357) to around 600 in 1992. Several of these structures are not unequivocally proved, and in many cases their stereochemistry rests on biosynthetic analogies. The easiest entry to carotenoid structures is via Key to Carotenoids (2nd. Ed., ed. H. Pfander, Birkhäuser, 1987) with a recent supplement (Appendix in Carotenoids Vol. 1A Isolation and analysis eds. G. Britton, S. Liaaen-Jensen and H. Pfander, Birkhäuser, 1993).

Whereas several of the new structures merely represent variations on old themes, emphasis here is made to focus on i) new structural elements and ii) structural types where particular progress has been made in stereochemistry and/or chemistry. Only selected examples are given. The general approach is to treat modifications of the size of the carbon skeleton first : higher carotenoids > C_{40}, nor-carotenoids < C_{40}, C_{30}-carotenoids (diapo) and apocarotenoids < C_{40}. Particular functional group modification are considered next.

(a) Higher carotenoids (C_{45}, C_{50}). The term homocarotenoids (Key to Carotenoids, Birkhäuser, 1971; 1987), is not approved by IUPAC. With reference to the detailed review by J.B. Davis (this series, First Supplement Vol. IIB pp. 309-319) only a few new, major higher carotenoids have been encountered, namely A.g. 471, (Fig. 4), (N. Arpin et al., Acta Chem. Scand., 1975, B29: 921) and glycosidic modifications also of bacterioruberin, (N. Arpin et al., Acta Chem. Scand., 1972, 26: 2526). However, with access to high field ^{1}H NMR some of the previous structures have been revised, namely for C.p., 450 (A.G. Andrewes and S. Liaaen-Jensen, Tetrahedron Lett. 1984, 25: 1191), C.p. 473 (ibid) and sarcinaxanthin (S. Hertzberg and S. Liaaen-Jensen, Acta Chem. Scand., 1977, B31: 215). Studies on the chirality of C_{50}-carotenoids involved total synthesis of model C_{42}-carotenoids (A.G. Andrewes et al., Acta Chem. Scand., 1974, B28: 737; 1976, B30: 214; 1977, B31: 212), ^{1}H NMR and CD correlations. Confirmation was obtained by total syntheses of decaprenoxanthin (A.K. Chopra et al., J. Chem. Soc. Chem. Commun., 1977, 357; J.P. Ferezou and M. Julia, Tetrahedron 1985, 41: 1277), sarcinaxanthin (J.P. Ferezou and M. Julia, Tetrahedron, 1990, 45: 475), tetraanhydrobacterioruberin (J.E. Johansen and S. Liaaen-Jensen, Acta Chem. Scand., 1979, B33: 551), and a C_{45}-model (A.G. Andrewes et al., Acta Chem. Scand., 1984, B38: 871), see also H. Pfander (Pure Appl. Chem., 1991, 63: 23). Structures with full stereochemistry are given in Fig. 4.

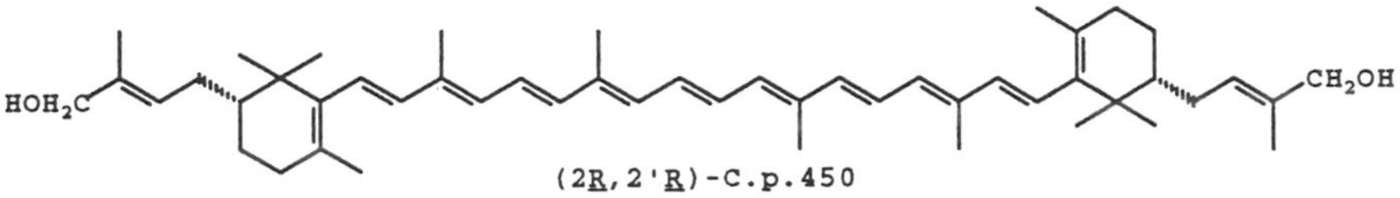

(2R,2'R)-C.p.450

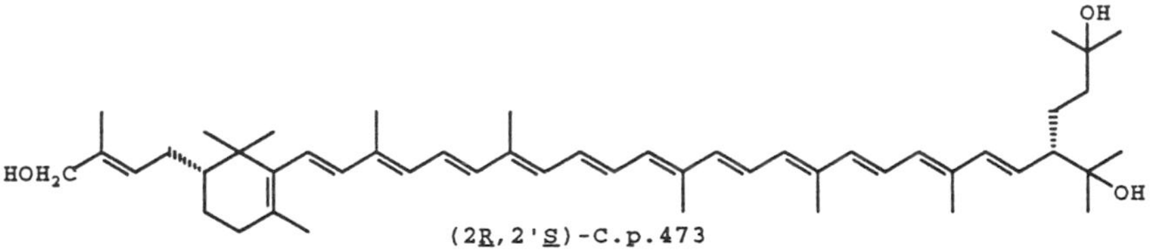

(2R,2'S)-C.p.473

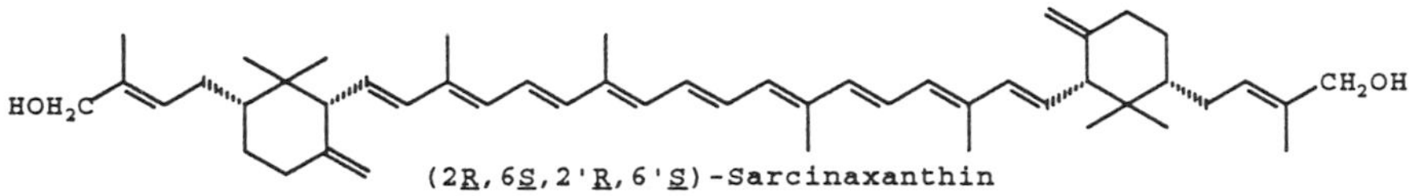

(2R,6S,2'R,6'S)-Sarcinaxanthin

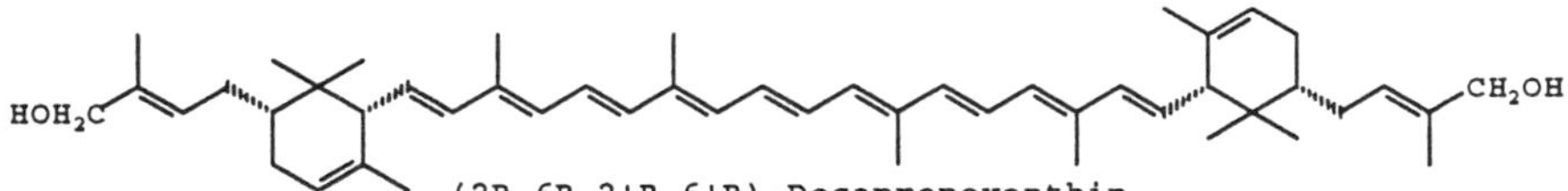

(2R,6R,2'R,6'R)-Decaprenoxanthin

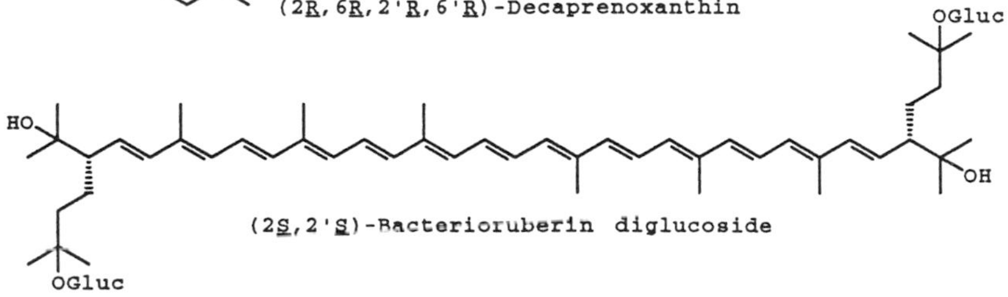

(2S,2'S)-Bacterioruberin diglucoside

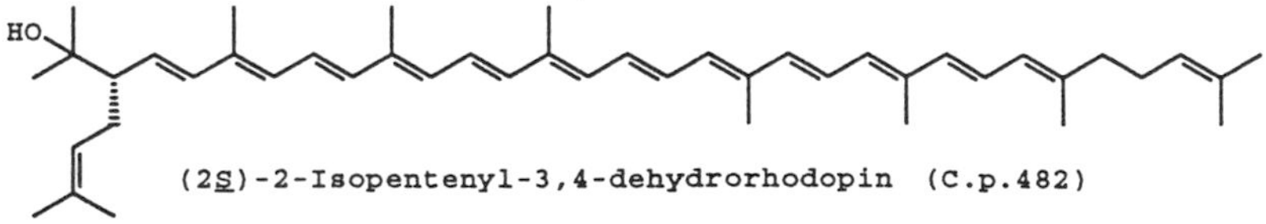

(2S)-2-Isopentenyl-3,4-dehydrorhodopin (C.p.482)

Fig. 4

(b) <u>Nor-carotenoids</u>.   Nor-carotenoids are by the IUPAC definition
carotenoids where formally a CH_3, CH_2 or CH group has been eliminated.
Only some ten carotenoids belong to this class: i) Cyclopentene
carotenoids of the actinioerythrol (Fig. 3) type where C-2 or C-2,2' are
lacking and ii) C_{37}-skeletal carotenoids of the peridinin (Fig. 3) type
where C-10,11,20 are lacking in the polyene chain. Of new representatives
belonging to the former type i) the C_{39}-skeletal 2-nor-astaxanthin diester
is mentioned,which provides roseerythrin upon alkali treatment in the
presence of oxygen (G.W. Francis <u>et</u> <u>al</u>., <u>Acta Chem. Scand</u>., 1972, <u>26</u>
1097) and a natural C_{39}-cyclopentenedione (P. Beyer <u>et</u> <u>al</u>., <u>Z. Natur-
forsch</u>. C: <u>Biosci</u>. 1979, <u>34</u>: 179), Fig. 5.

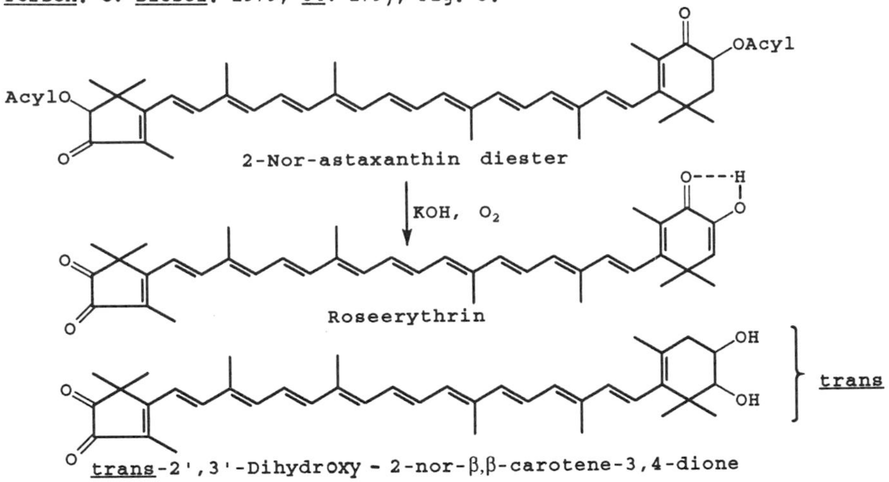

2-Nor-astaxanthin diester

KOH, O_2

Roseerythrin

<u>trans</u>-2',3'-Dihydroxy - 2-nor-β,β-carotene-3,4-dione   <u>trans</u>

Fig. 5

De <u>novo</u> biosynthesis of the second type ii) is restricted to
dinoflagellates.   The allenic peridinin (H.H. Strain <u>et</u> <u>al</u>., <u>Acta Chem.
Scand</u>., 1976, <u>B30</u>: 109; H. Kjøsen <u>et</u> <u>al</u>., <u>ibid</u> 1976, <u>B30</u>: 157; J.E.
Johansen and S. Liaaen-Jensen, <u>Phytochem</u>., 1980, <u>19</u>: 441; J. Krane <u>et</u>
<u>al</u>., <u>Magn. Res. Chem</u>., 1992, <u>in</u> press) and the acetylenic pyrrhoxanthin
(J.E. Johansen <u>et</u> <u>al</u>., <u>Phytochem</u>., 1974, <u>13</u>: 2261; G. Englert <u>et</u> <u>al</u>.,
<u>Magn. Res. Chem</u>., 1993, <u>in</u> preparation) have been extensively studied
during the last two decades and represented the first carotenoid lactones.
Peridinin has been converted <u>in</u> <u>vitro</u> to pyrrhoxanthin, both protected as
3'-acetates (T. Aakermann and S. Liaaen-Jensen, <u>Phytochem</u>., 1992, <u>31</u>:
1779), Fig. 6.

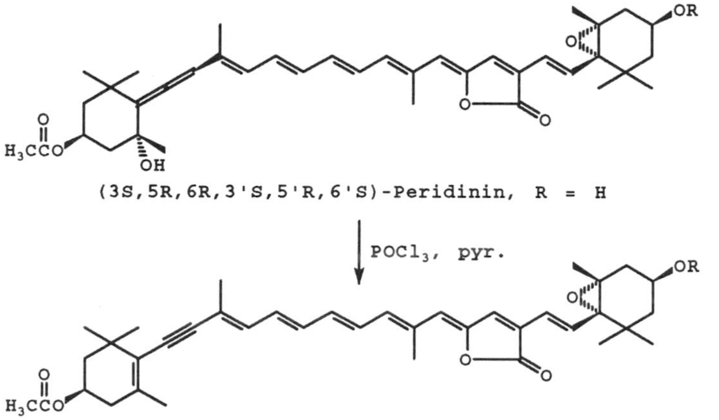

(3S,5R,6R,3'S,5'R,6'S)-Peridinin, R = H

↓ POCl₃, pyr.

(3S,3'S,5'R,6'S)-Pyrrhoxanthin, R = H

Fig. 6

Peridinin (Section 5) and pyrrhoxanthin (Ito et al., J. Chem. Soc. Perkin Trans.I, 1990: 197) have both been prepared by total synthesis, pyrrhoxanthin, so far, in the optically inactive form.

The recently reported C_{40}-skeletal type gelliodenone (Y. Tanaka and T. Inoue, Nippon Suisan Gakkaishi, 1987, 53: 1271) is formally a 17'-nor-carotenoid and represents a likely aldol condensation product of gelliodesxanthin (Fig. 7) and acetone.

(c) Apo-carotenoids (< C_{40}). According to IUPAC apo-carotenoids are carotenoids where the carbon skeleton has been shortened by the formal removal of fragments from one or both ends.

To the fifty or so apo-carotenoids listed in Key to Carotenoids (Birk-häuser, 1987) belong some of the classical carotenoids such as bixin and azafrin. The chirality of azafrin is settled (W. Eschenmoser and C.H. Eugster, Helv. Chim. Acta 1975, 58: 1722), Fig. 7. Some apo-carotenoids were discussed according to functional groups in the previous review (J.B. Davis, First Supplement Vol. IIB, pp. 191-357), and some of the methyl ketones listed are now thought likely to be isolation artifacts (see Section 9). Acetylenic, allenic and epoxidic representatives are considered separately below. This leaves a few rather well-stablished apo-carotenoids to be mentioned here, such as the methoxylated Thiothece-460 (A.G. Andrewes and S. Liaaen-Jensen, Acta Chem. Scand., 1972, 26: 2194) and an azafrin-related carboxylic acid (W. Eschenmoser et al., Helv. Chim. Acta, 1982, 65: 353), Fig. 7.

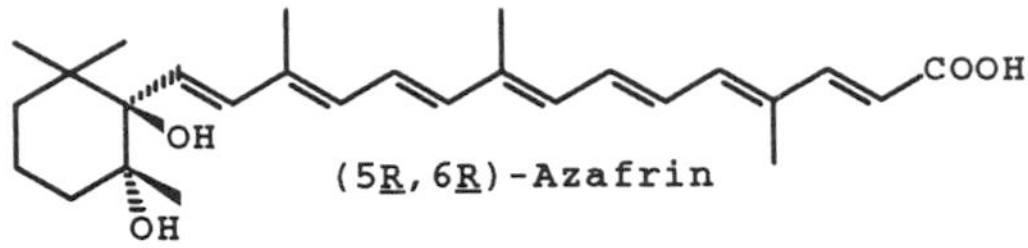

72

(5R,6R)-Azafrin

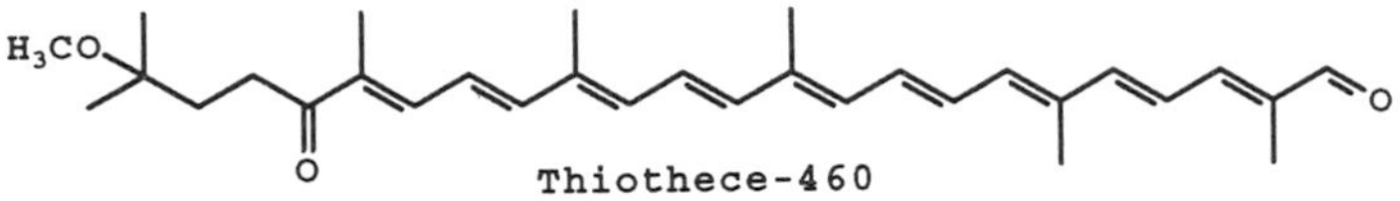

Thiothece-460

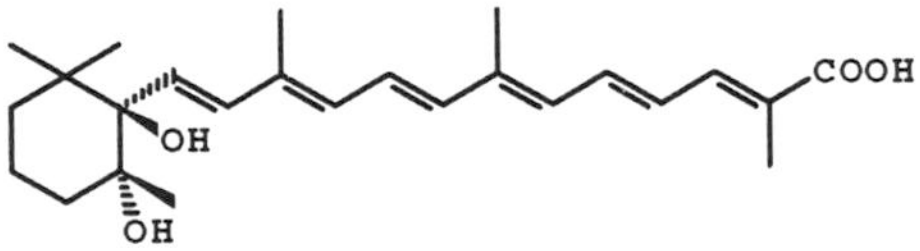

(5R,6R)-5,6-Dihydroxy-5,6-dihydro-
12'-apo-β-caroten-12'-oic acid

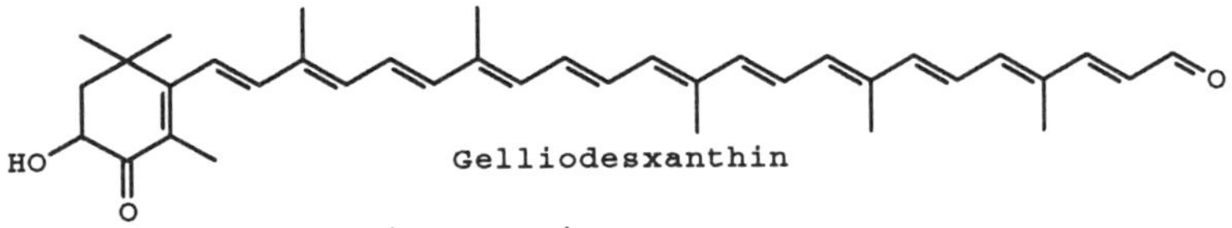

Gelliodesxanthin

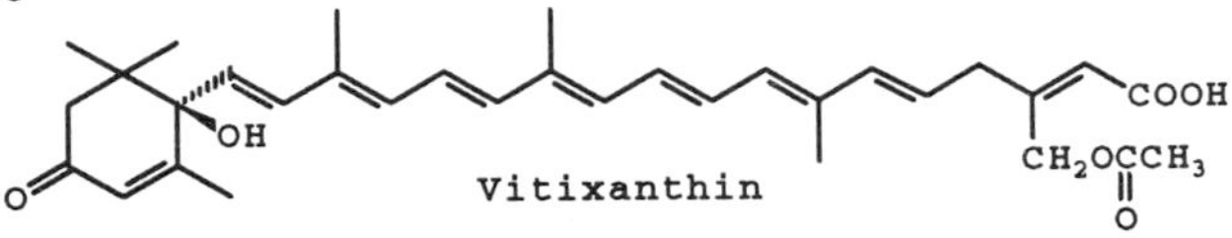

Vitixanthin

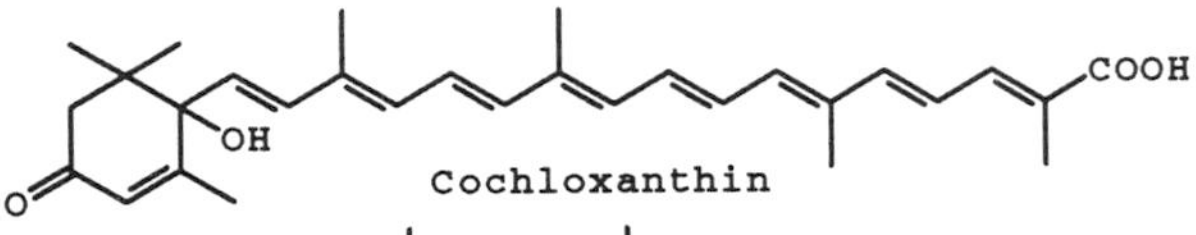

Cochloxanthin

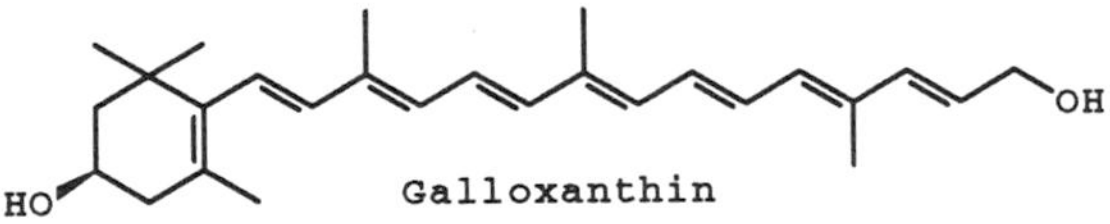

Galloxanthin

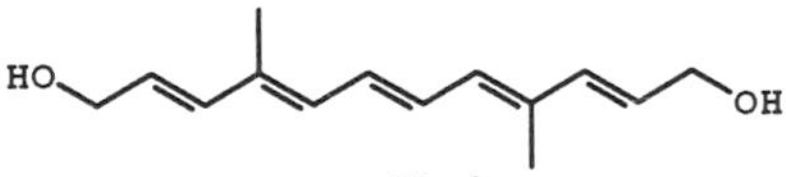

Rosafluin

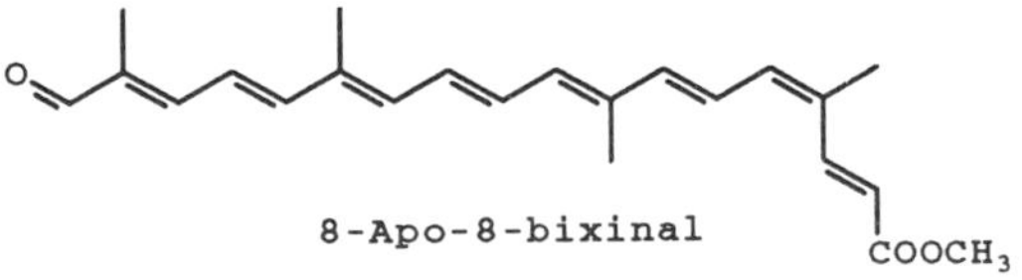

8-Apo-8-bixinal

Fig. 7

More recently some ten new apo-carotenoids have been reported. These include (see Fig. 7) the apo-astaxanthin gelliodesxanthin (Y. Tanaka and T. Inoue, Nippon Suisan Gakkaishi, 1987, 53: 1271), the remarkable viti xanthin and its 4,5-dihydro derivative (H. Achenbach et al., Tetrahedron Lett., 1989, 30: 3059) and the structurally related cochloxanthin and its 4,5-dihydro derivative (B. Diallo et. al., Phytochem., 1987, 26: 1491), se veral apo-carotenols such as galloxanthin (E. Märki-Fischer and C.H. Eugster, Helv. Chim. Acta, 1987 70: 1988) and the interesting natural microcarotenoid, the diapocarotenediol rosafluin (E. Marki-Fischer and C.H. Eugster Helv. Chim. Acta, 1988, 71: 1491) and finally an apocarotenoid derived from bixin (J.O. Iondiko and G. Pattenden, Phytochem., 1989, 28: 3159).

(d) C_{30}-Carotenoids (diapo). A new series of interesting C_{30}-carotenoids, classified as diapo-carotenoids, but which biosynthetically appears

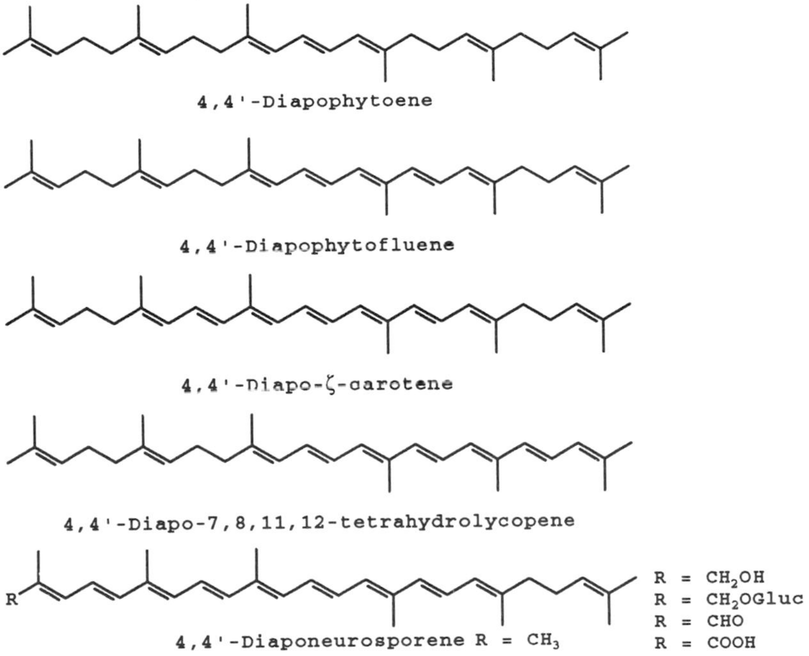

4,4'-Diapophytoene

4,4'-Diapophytofluene

4,4'-Diapo-ζ-carotene

4,4'-Diapo-7,8,11,12-tetrahydrolycopene

4,4'-Diaponeurosporene R = CH₃

R = CH₂OH
R = CH₂OGluc
R = CHO
R = COOH

Fig. 8

to be formed in some bacteria by an independent pathway via dehydrosqualene, have been reported. Thus the entire series dehydrosqualene, 4,4'-
diapophytofluene, 4,4'-diaponeurosporene, 4,4'-diapo-7,8,11,12-tetrahydro-
lycopene have been encountered, as well as hydroxylated and glycosylated
representatives, aldehydes and carboxylic acids, see Fig. 8, (R.F. Taylor
and B.H. Davies Biochem. Soc. Trans., 1973, 1: 1091; Biochem. J., 1974,
139: 751; ibid., 1974, 139:761, ibid., 1976, 153: 233; Can J. Biochem.,
1982, 60: 675; ibid., 1983, 61: 892; B.H. Davies and R.F. Taylor Can
J. Biochem., 1982, 60: 684).

(e) Sulfates. The ionized sulfate function represents a novel functional
group in the carotenoid context. First detected in marine sponges with
the bastaxanthins in the late seventies (S. Hertzberg et al., Acta Chem.

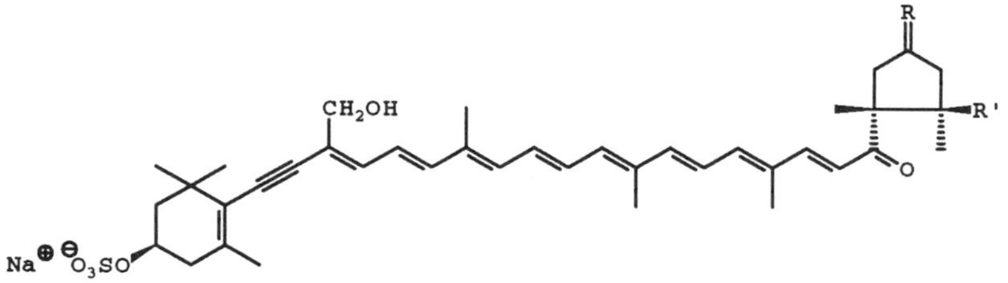

Bastaxanthin b	R = O	R' = CHO
Bastaxanthin c	R = O	R' = CH$_2$OH
Bastaxanthin d	R = H,OH	R' = CH$_2$OH
Bastaxanthin e	R = O	R' = COOH
Bastaxanthin f	R = H,OH	R' = COOH

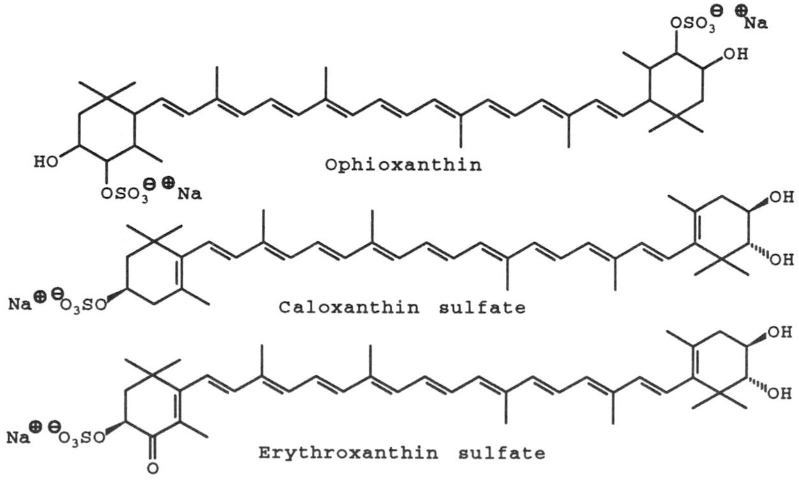

Ophioxanthin

Caloxanthin sulfate

Erythroxanthin sulfate

Fig. 9

Scand., 1983, B37: 267; Biochem. Syst. Ecol., 1983, 11: 267; ibid., 1989, 17: 51), other examples have followed. Ophioxanthin (M.V. D'Auria et al., Tetrahedron Lett., 1985, 26: 1871), caloxanthin sulfate and erythro-xanthin sulfate (S. Takaichi et al., Phytochem., 1991, 30: 3411) have extended the source to other marine animals and bacteria, see Fig. 9.

The partial syntheses and solvolytic reactions of several model carotenoid sulfates have been studied (S. Hertzberg and S. Liaaen-Jensen, Acta Chem. Scand., 1985, B39: 629; S. Hertzberg et al., ibid., 1985, B39: 725). The sulfate function has a great impact on the polarity and water solubility of carotenoid sulfates (S. Liaaen-Jensen et al., 1st Conference Chem. Biotechn. Biol. Active Nat. Prod., Varna, 1980, Vol. 2, pp. 150-164).

(f) Glycosides and glycosyl esters. Carotenoids are bound to sugar as glycosides of prim., sec. or tert. carotenols or as glycosyl esters of carotenoid carboxylic acids. Progress during the last two decades is largely due to the availability of high field ^{1}H NMR, allowing unequivocal assignment of the carbohydrate moiety and the stereochemistry of the glycosidic linkage.

Thus it appears that not rhamnose, but the C-2 epimer chivonose, occasionally in mixture with fucose, is a common carbohydrate in myxo-xanthophyll and oscillaxanthin from blue-green algae (Cyanobacteria) (R. Riesen et al., Abstr. 7th Int. IUPAC Carotenoid Symp., 1984, München P.40; P. Foss et al., Phytochem., 1986, 25: 1127; T. Aakermann et al., Biochem. Syst. Ecol., 1992, in press). Moreover, an O-methyl methyl-pentoside, glycosidically bound to myxol and oscillol in some blue-green algae, has been identified as 3-O-methyl-fucose (G.W. Francis et al., Phytochem., 1970, 9: 629; Foss et al., Phytochem., 1986, 25: 1127).

Rhamnosides and glucosides seem rather common in bacteria, e.g. zeaxanthin dirhamnoside (G. Nybraaten and S. Liaaen-Jensen, Acta Chem. Scand., 1974, B28: 1219), Fig. 10, or tert. rhamnosides (H. Kleinig et al., Arch. Microbiol., 1971, 78: 224), and the first lactoside (Fig. 1) has been demonstrated.

The absolute stereochemistry of myxoxanthophyll (Fig. 10) has been settled by a CD-correlation (H. Rønneberg et al., Phytochem., 1980, 19: 2167). Partial syntheses of carotenoid glycosides are still hampered by the low yield of the Koenigs-Knorr reaction. Enzymatic hydrolysis of carotenoid glycosides requires further elaboration, since acidic hydrolysis generally destroys the aglycone.

Turning now to glycosyl esters, particular carbohydrate constituents have been encountered in esterified apocarotenoic acids, such as gentiobiose and neapolitanose (H. Pfander and F. Wittwer, Helv. Chim. Acta, 1975, 58: 2233; M. Rychener et al., ibid., 1984, 67: 386) besides se-

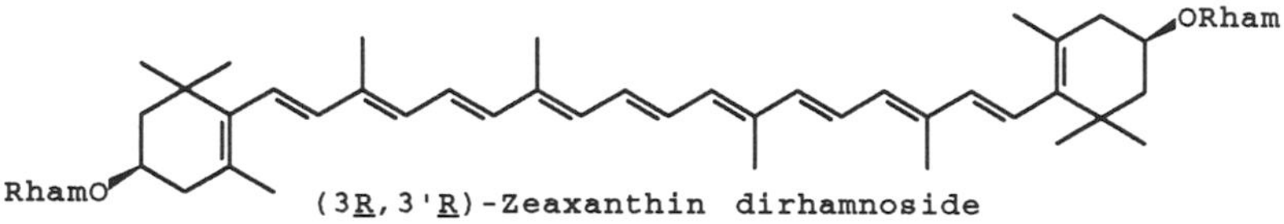

(3R,3'R)-Zeaxanthin dirhamnoside

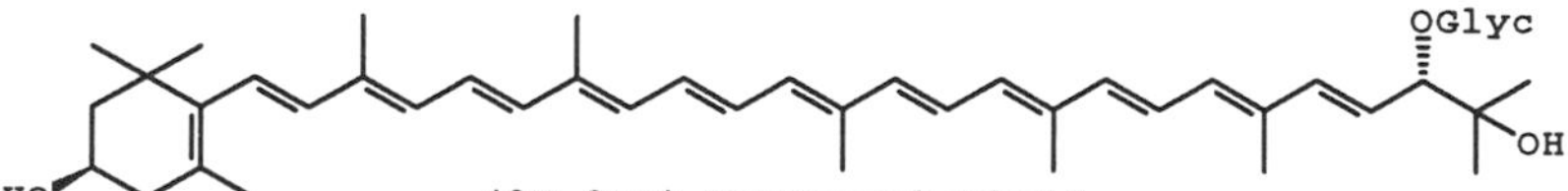

(3R,2'S)-Myxoxanthophyll

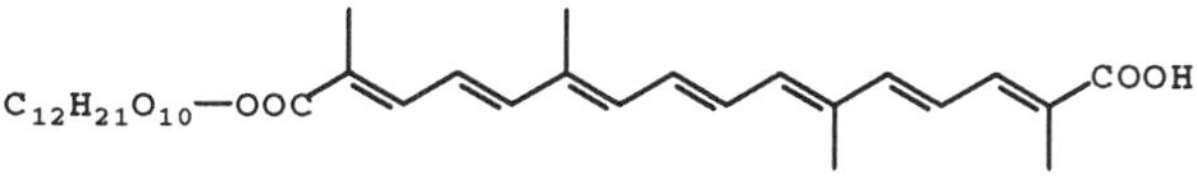

Crocetin monogentiobiosyl ester

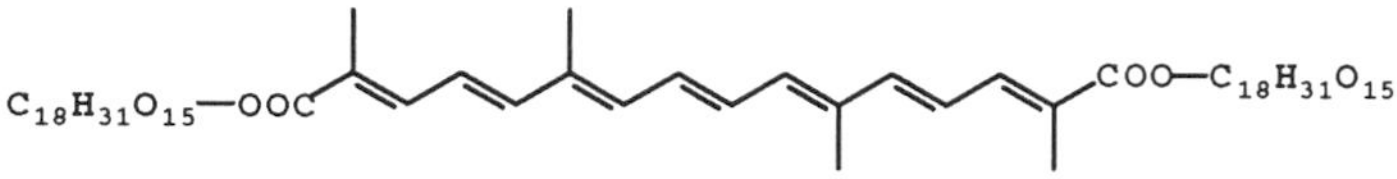

Dineapolitanosyl 8,8'-diapocarotene-8,8'-dioate

Fig. 10

veral glucosyl esters (R.F. Taylor and B.H. Davies, Biochem. J., 1974, 139: 761; Can J. Biochem. Cell Biol., 1983, 61: 892; H. Pfander and F. Wittwer, Helv. Chim. Acta, 1975, 58: 1608, ibid., 1979, 62: 1944), for examples see Fig. 10.

A good method for the partial synthesis of carotenoid glycosyl esters without protection of hydroxy groups in the carbohydrate moiety is available (H. Pfander and F. Wittwer, Helv. Chim. Acta, 1979, 62: 1944). As esters may carotenoid glycosyl esters be hydrolyzed under strong alkaline conditions.

(g) Epoxides and oxabicycloheptanes. 5,6-Epoxidized β-rings is a common structural element in carotenoids of algae and higher plants, known since Karrer's pioneering work. The corresponding 5,8-furanoxides are common isolation artifacts (see Section 9). Some thirty different 5,6-epoxides have been characterized. More recent 5,6-epoxides include latoxanthin

(Märki-Fischer et al., Helv. Chim. Acta, 1984, 67: 461), and less characterized 2-hydroxy-5,6-epoxides (G. Nybraaten and S. Liaaen-Jensen, Acta Chem. Scand., 1974, B28: 483; ibid., 1974, B28: 485); Fig. 11. In the former type a relative trans configuration is general for the 3-OH/epoxide functions, whereas in the latter type a cis configuration for the 2-OH/epoxide groups is reported.

4,5-Epoxidized ε-rings have recently been encountered in carotenoids, namely 7',8'-dihydroprasinoxanthin-4',5'-epoxide (P. Foss et al., Phytochem., 1986, 25: 119), Fig. 11.

Moreover, some five 1,2-epoxides of carotenoids with an aliphatic end group have been described in recent years (A. Ben-Aziz et al., Phytochem., 1973, 12: 2759; G. Britton and T.W. Goodwin, ibid., 1975, 14: 2530; D. Berset and H. Pfander, Helv. Chim. Acta, 1984, 67: 964), as well as the aliphatic lycopene 5,6-epoxide (G. Britton and T.W. Goodwin, Phytochem., 1975, 14: 2530). Chiral lycopene 1,2, 1',2'-diepoxides have been synthesized (H. Meier et al., Helv. Chim. Acta, 1986, 69: 106). Alternative syntheses of (S)-1,2-epoxy-lycopene and (S)-1',2'-epoxy-β,Ψ-carotene have been reported (M. Kamber et al., Helv. Chim. Acta, 1984, 67: 968; H. Pfander, Pure Appl. Chem., 1991, 63: 23).

Novel types are 3,6-epoxides of oxabicycloheptane first encountered in eutreptiellanone (A. Fiksdahl et al., Phytochem., 1984, 23: 649), were followed by β-cryptoeutreptiellanone and α-cryptoeutreptiellanone (T. Bjørnland et al., Phytochem., 1986, 25: 201). Subsequently a related 3,6-epoxidic end group was reported for capsantin-3,6-epoxide (K.F.B. Parkers et al., Tetrahedron Lett., 1986, 27: 2535), and for cucurbitaxanthins A and B (T. Matsuno et al., Phytochem., 1986, 25: 2827), as well as for cycloviolascin (J. Deli et al., Helv. Chim. Acta 1991, 74: 819, see Fig. 11. A 3-hydroxy-5,6-epoxy β-ring is a likely precursor of the latter oxabicyclo [2.2.1] heptane end group. As tetrahydrofuran derivatives the 3,6-epoxides represent stable structural elements, e.g towards acids or complex metal hydride reduction. Similar 2,5-epoxides have not been encountered in naturally occurring carotenoids, but were obtained by the treatment of β,β-carotene-2,2'-diol with BF₃. This is accompanied by a retro shift of the polyene system (K. Aareskjold et al., Tetrahedron Lett., 1981, 22: 4541).

Selected examples illustrating the structural diversity of recent carotenoid epoxides are given in Fig. 11.

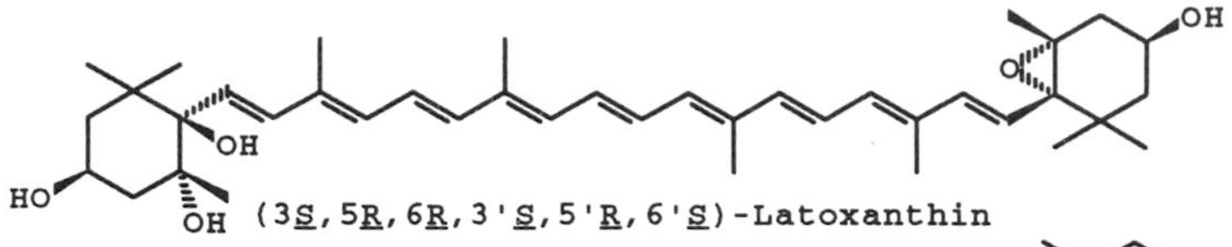

(3*S*,5*R*,6*R*,3'*S*,5'*R*,6'*S*)-Latoxanthin

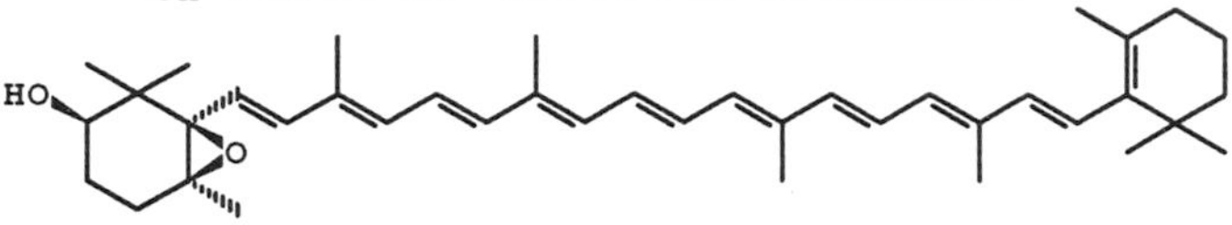

(2*R*,5*S*,6*R*)-5,6-Epoxy-5,6-dihydro-β,β-caroten-2-ol

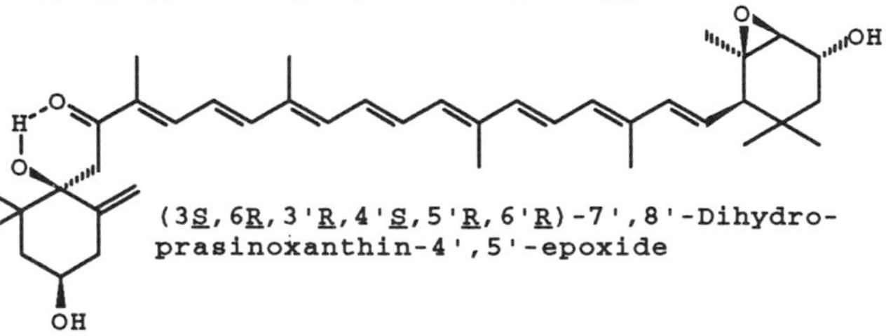

(3*S*,6*R*,3'*R*,4'*S*,5'*R*,6'*R*)-7',8'-Dihydro-
prasinoxanthin-4',5'-epoxide

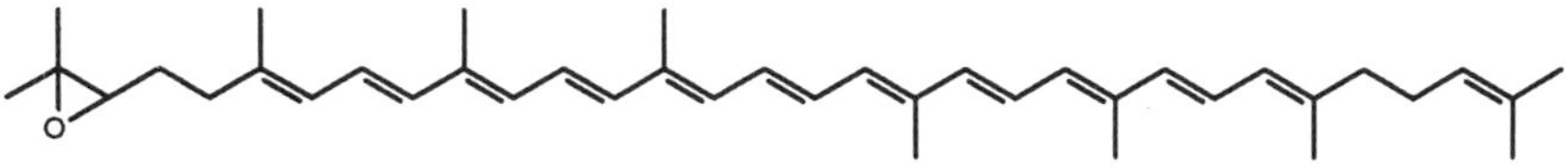

Lycopene 1,2-epoxide

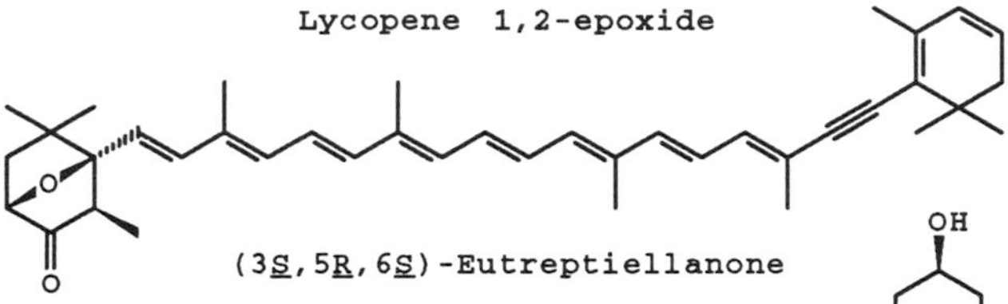

(3*S*,5*R*,6*S*)-Eutreptiellanone

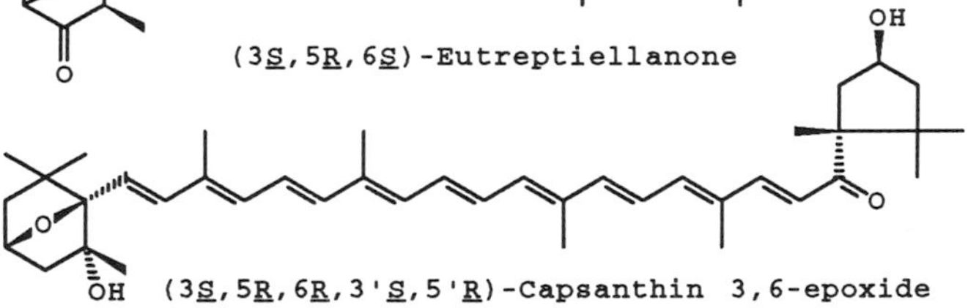

(3*S*,5*R*,6*R*,3'*S*,5'*R*)-Capsanthin 3,6-epoxide

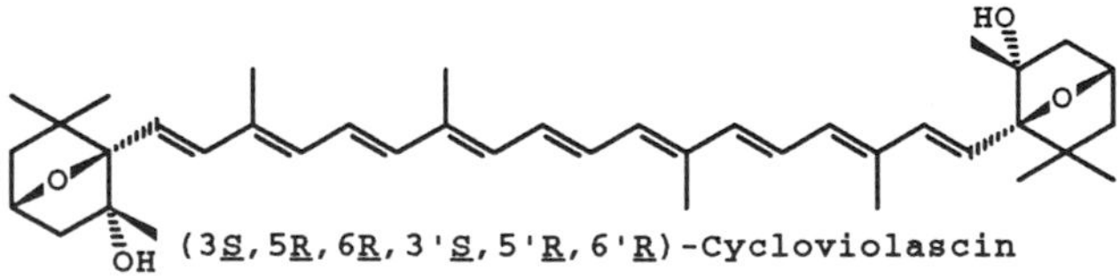

(3*S*,5*R*,6*R*,3'*S*,5'*R*,6'*R*)-Cycloviolascin

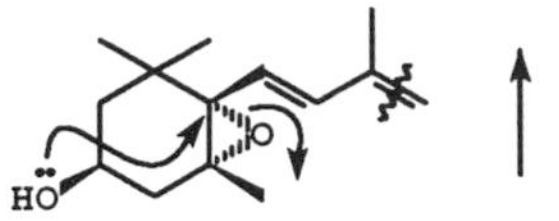

Fig. 11

The identification of the C-8 epimeric furanoid rearrangement products obtained from common 5,6-epoxides upon treatment with weak acid is now possible using high field ^{1}H NMR. Thus (8'R)- and (8'S)-luteoxanthin and (8R,8'R)-auroxanthin, as well as the (8R,8'S)- and (8S,8'S)-diastereomers have been thoroughly characterized (Märki-Fischer et al., Helv. Chim. Acta, 1984, 67: 2143), see Fig. 12.

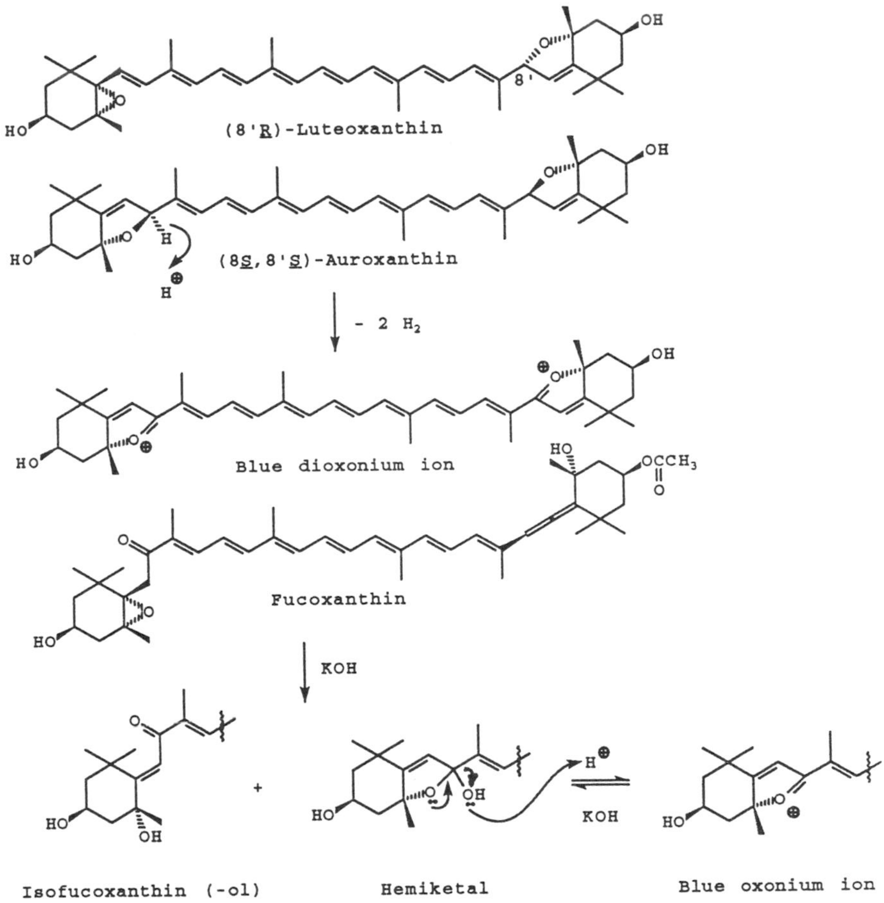

(8'R)-Luteoxanthin

(8S,8'S)-Auroxanthin

Blue dioxonium ion

Fucoxanthin

Isofucoxanthin (-ol)

Hemiketal

Blue oxonium ion

Fig. 12

Recently insight has been gained into the classical colour reaction of carotenoid 5,6-epoxides, providing blue colours, now identified as being due to oxonmium ions (J.A. Haugan and S. Liaaen-Jensen, Abstr, 9th Int, IUPAC Carotenoid symp., Kyoto, 1990, p . 113).   Moreover, the alkali lability of the keto-epoxide fucoxanthin has now been rationalized. Isofucoxanthin type and hemiketal products have been characterized (J.A. Haugan et al., Acta Chem, Scand., 1991, in press).   The latter are extremely acid sensitive, providing blue oxonium ions, Fig. 12.

The total synthesis of carotenoid 5,6-epoxides including optically active peridinin (M. Ito et al., J, Chem, Soc, Perkin Trans 1, 1990: 197) and neoxanthin (A. Baumeler and C.H. Eugster, Helv, Chim, Acta, 1992, 75; 773) have been successfully performed.

(h) In-chain-methyl oxidized carotenoids.   Whereas structural modifications most frequently occur in the two end groups, certain carotenoids undergo oxidation of in-chain-methyl groups, resulting in the introduction of prim. hydroxy, aldehyde or lactone functions.

19(19')-Carotenols have been known for some time (loroxanthin = lutein-19-ol and vaucheriaxanthin = neoxanthin-19'-ol). The absolute stereochemistry of loroxanthin has been confirmed as (3$\underline{R}$,3'$\underline{R}$,6'$\underline{R}$) by a synthetic approach (E. Märki-Fischer et al., Helv, Chim, Acta, 1988, 66: 1175), Fig. 13. The bastaxanthins (Fig. 9) are new 19-ols, and 19'-hexanoyloxyfucoxanthin (N. Arpin et al., Phytochem., 1976, 15: 529) and 19'-butanoyloxyfucoxanthin (T. Bjørnland et al., Phytochem., 1989, 28: 3347) represent new naturally occurring fucoxanthin derivatives, Fig. 13. The first 19-al, micromonal (E.S. Egeland and S. Liaaen-Jensen, Abstr, 7th Int, Symp Marine Nat, Prod., Capri, 1992) with noteworthy 13'-$\underline{Z}$ configuration has recently been encountered. Ternstroemiaxanthin (K. Kikuchi and M. Yamaguchi, Bull, Chem, Soc, Jap., 1974, 47: 885) is a new, terminal 18'-al, Fig. 13. Moreover, 20(20')-als, referred to as cross-conjugated carotenals have been further characterized and the preferred cis-configuration of the α,β-double bond proved by detailed ^{1}H NMR analysis (G. Englert et al., Magn, Res, Chem., 1988, 26: 55), see also P457 (Fig. 1). Other E/$\underline{Z}$ isomers were also characterized by the same workers for the synthetic molecule renierapurpurin-20'-al, Fig. 13.

Lactone functions are encountered in peridinin and pyrrhoxanthin (Fig. 6) and their derivatives that have been isolated from dinoflagellates. A more recent example from the Prasinophyceae is uriolide (P. Foss et al., Phytochem., 1986, 25: 119), Fig. 13, in which the butenolide ring occupies the same position. The lactones can be regarded as being formed through formal oxidation of the 19'-CH_3 group.

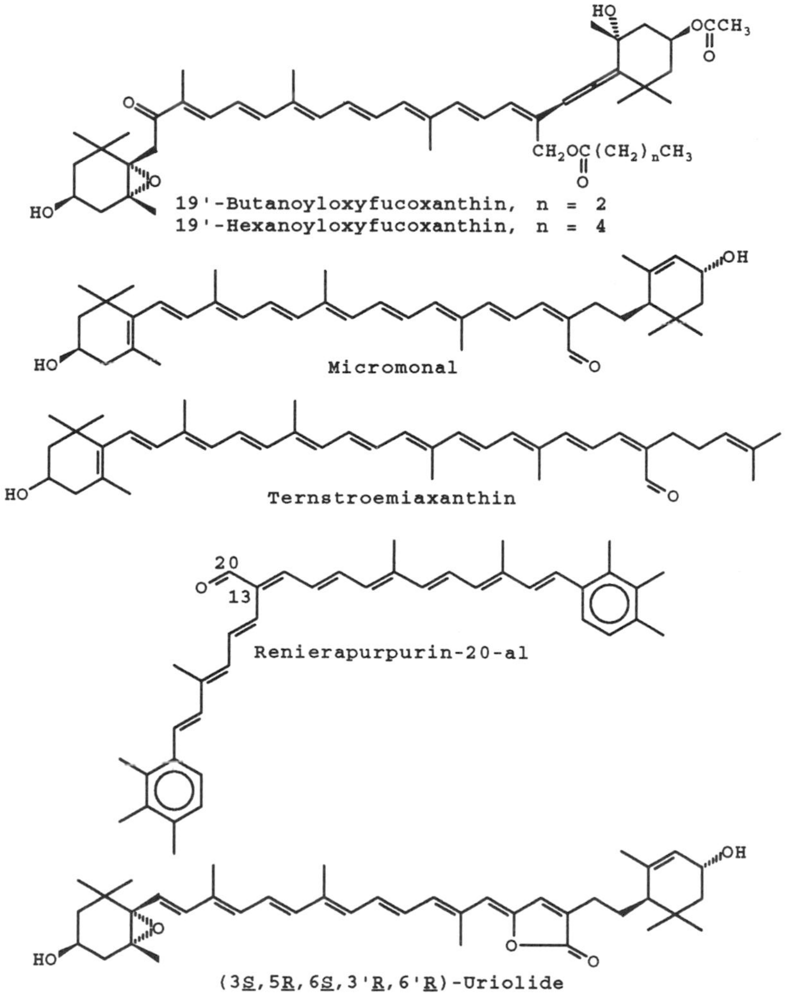
HO
OCCH3
O
CH2OC(CH2)nCH3
O
HO
19'-Butanoyloxyfucoxanthin, n = 2
19'-Hexanoyloxyfucoxanthin, n = 4
OH
HO
Micromonal
O
HO
Ternstroemiaxanthin
O
20
O
13
Renierapurpurin-20-al
OH
HO
O
O
(3S,5R,6S,3'R,6'R)-Uriolide

Fig. 13

(i) <u>Allenes</u>. Allenic carotenoids were treated separately by J.B. Davis (this series, First Supplement Vol. IIB, 1973, pp. 288-300). At that time nine compounds were recognised. Since then nostoxanthin and caloxanthin have been assigned non-allenic structures (R. Buchecker <u>et</u> <u>al</u>., <u>Phytochem</u>., 1976, <u>15</u>: 1015) and the structure of vaucheriaxanthin (H. Nitsche, <u>Z. Naturforsch</u>., 1973, <u>C28</u>: 641) has been revised, Fig. 14. The deacetylated fucoxanthin (Fig. 2) and peridinin (Fig. 3) derivatives fucoxanthinol and peridininol are encountered as minor carotenoids in algae, (H. Pfander, <u>Key to Carotenoids</u>, Birkhäuser, 1987), whereas dinoanthin is shown to be the 3-acetate of neoxanthin, Fig. 3, (J.E. Johansen <u>et</u> <u>al</u>., <u>Phytochem</u>., 1980, <u>19</u>: 441). The 19'-butenoyloxy- and 19'-hexanoyloxy derivatives of fucoxanthin were already mentioned above, Fig. 13, and 19-hexanoyloxyparacentrone (N. Arpin <u>et</u> <u>al</u>., <u>Phytochem</u>., 1976, <u>15</u>: 529), Fig. 14, has been isolated. Modern studies on peridinin (Fig. 6) are cited above. Attention should be paid to amarouciaxanthin A (T. Matsuno <u>et</u> <u>al</u>., <u>J. Nat. Prod.</u>, 1985, <u>48</u>: 606), Fig. 14, 7',8'-di-hydroneoxanthin-20'-al 3'-lactoside (S. Liaaen-Jensen, <u>New J. Chem</u>., 1990, <u>14</u>: 747; G. Englert, <u>Pure & Appl. Chem</u>., 1991, <u>63</u>: 59), Fig. 1 and the allenic/acetylenic gyroxanthin ester (T. Bjørnland <u>et</u> <u>al</u>., <u>Abstr.</u> <u>8th Int. Carotenoid Symp</u>., Boston 1987, P2), Fig. 14.

A comprehensive review on allenic and acetylenic carotenoids was recently given by the present author (in T.I. Atta-ur-Rahman ed. <u>Natural</u> <u>Product Chemistry</u>, Elsevier, 1990, pp. 133-169), covering structural and chemical aspects including total syntheses, distribution, algal chemosystematics and metabolism. It should be pointed out that allenic carotenoids appear to be the most abundant allenes in Nature and that fucoxanthin (Fig. 2) and peridinin (Fig. 3) are the carotenoids bio-synthesized in largest quantity. <u>De</u> <u>novo</u> biosynthesis of allenic carotenoids is restricted to certain algae and higher plants. Procaryotes do not have the ability to synthesize allenic carotenoids. The metabolic conversion of allenic to acetylenic carotenoids (see. Fig. 6) by various marine animals is noteworthy. In addition to peridinin and neoxanthin, Fig. 3, the diallenic mimulaxanthin has recently been synthesized enan-tiomercally pure (A. Baumeler and C.H. Eugster, <u>Helv. Chim. Acta</u>, 1991, <u>74</u>: 469), Fig. 14.

(j) <u>Acetylenes</u>. Acetylenic carotenoids were also treated separately in the previous review in this series (J.B. Davis, First Supplement, Vol.

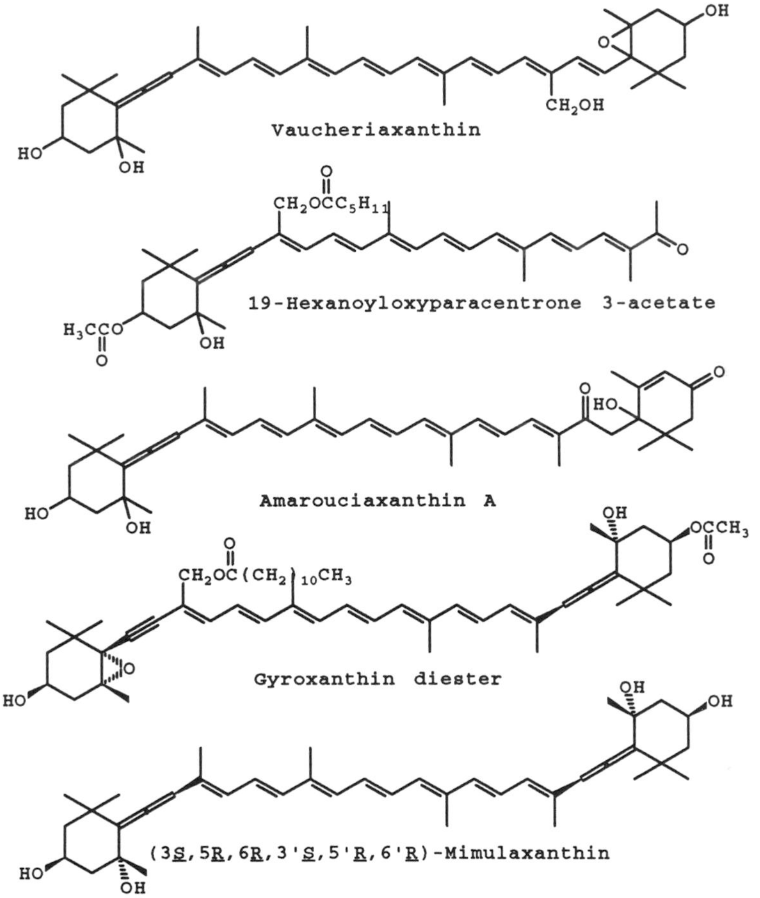

Fig. 14.

IIB, 1973, pp. 300-309), covering some ten acetylenic carotenoids. The number of known, naturally occurring acetylenic carotenoids has now increased significantly to around fifty (H. Pfander, Key to Carotenoids, Birkhäuser, 1987; Appendix Carotenoids, Vol. 1A. Isolation and analysis, eds. G. Britton, S. Liaaen-Jensen and H. Pfander, Birkhäuser, 1993). The reader is referred to the recent review by the present author (in T.I. Atta-ur-Rahman ed. Natural Product Chemistry, Elsevier 1990, pp. 147-169), covering various aspects of acetylenic carotenoids. De novo biosynthesis of acetylenic carotenoids appears to be restricted to certain classes of eucaryotic algae, but have not been detected within the Chlorophyceae and

Prasinophyceae green algae. They are encountered along the marine food chain, where their presence may be due to resorption of microalgal acetylenic carotenoids or metabolic modification of allenic carotenoids. In natural carotenoids the triple bond is always located in 7,8(7',8')-position, and Z-isomerization of the neighbouring Δ9(9') double bond is facile.

Selected new acetylenic carotenoids in addition to the acetylenic carotenoids mentioned previously (pyrrhoxanthin Fig. 6, the bastaxanthins Fig. 9, eutreptiellanone Fig. 11, gyroxanthin ester Fig. 14 and a diacetylenic carotene Fig. 17) are given in Fig. 15 A and B.

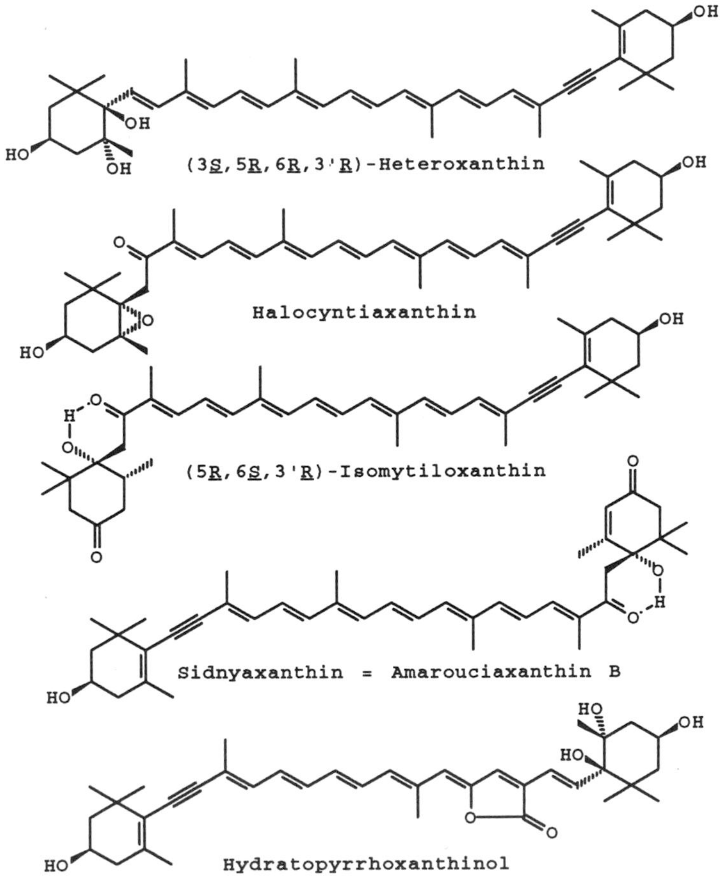

(3S,5R,6R,3'R)-Heteroxanthin

Halocyntiaxanthin

(5R,6S,3'R)-Isomytiloxanthin

Sidnyaxanthin = Amarouciaxanthin B

Hydratopyrrhoxanthinol

Fig. 15A

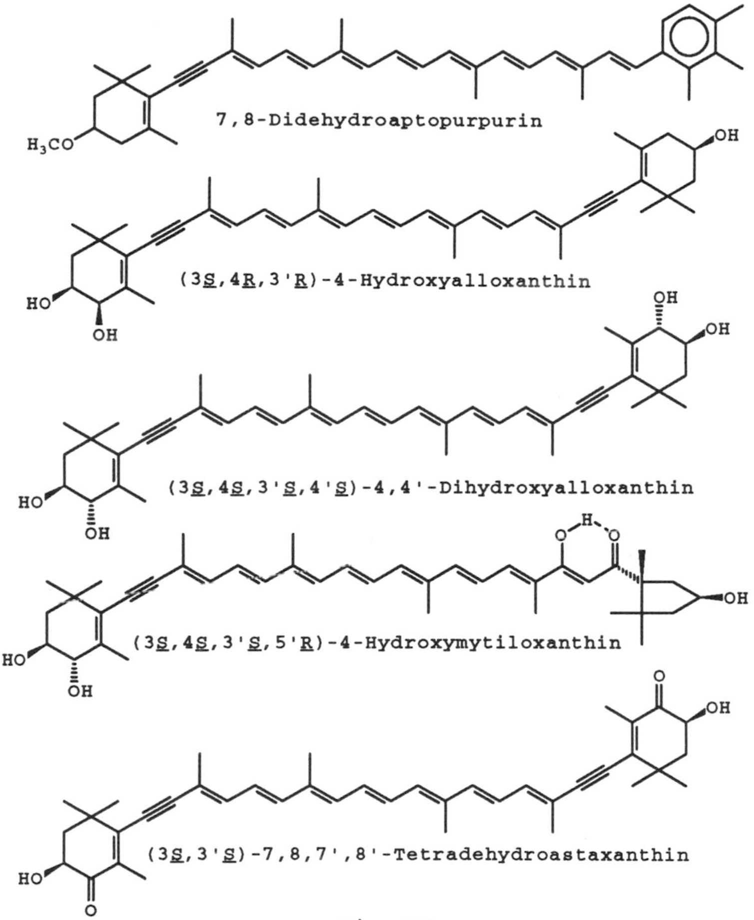

7,8-Didehydroaptopurpurin

(3S,4R,3'R)-4-Hydroxyalloxanthin

(3S,4S,3'S,4'S)-4,4'-Dihydroxyalloxanthin

(3S,4S,3'S,5'R)-4-Hydroxymytiloxanthin

(3S,3'S)-7,8,7',8'-Tetradehydroastaxanthin

Fig. 15B

Included are the revised structure of heteroxanthin (R. Buchecker et al., Phytochem., 1977, 16: 729; Helv. Chim. Acta, 1984, 67: 2043), halocynthiaxanthin (T. Matsuno et al., Chem. Pharm. Bull., 1984, 32: 4309) which is the acetylenic analogue of fucoxanthinol, and a likely metabolic precursor of isomytiloxanthin (A. Khare et al., J. Chem. Soc. Perkin Trans, I , 1988: 1 389), sidnyaxanthin = amarouciaxanthin B (G. Belaud and M. Guyot, Tetrahedron Lett., 1984, 25: 3087; T. Matsuno et al., J.Nat.Prod., 1985, 42: 606) and hydratopyrrhoxanthinol (S. Hertzberg et al., _Acta Chem. Scand._, 1988, B42: 495). To the acetylenic carotenoids reported during the last five years must be added the methoxylated compound 7,8-didehydroaptopurpurin (Y. Tanaka and T. Inoue, _Nippon Suisan Gakkaishi_, 1988, _54_: 1 55, a diacetylenic triol (T. Maoka and T. Matsuno, _Nippon Suisan Gakkaishi_, 1988, _54_: 1443) and tetrol (T. Maoka et al., _Comp. Biochem. Physiol._, 1989, _9 3B_: 829) and a mytiloxanthin-related enolized β-diketone (T. Maoka et al., _loc. cit._).

The total synthesis of pyrrhoxanthin has already been mentioned (M. Ito et al., _J.Chem. Soc. Perkin Trans._, _1990_: 197), Fig. 6, and attention is focused upon the first synthesis of all-_trans_ 7,8,7',8'-tetradehydroastaxanthin (G. Bernhard et al., _Helv. Chem. Act a_, 1980, _63_: 1473), Fig. 15, and that of a _cis_-isomer of mytiloxanthin, _cf._, Fig. 15, (A.K. Chopra et al., _J. Chem. Sos. Perkin Trans. I_, _1988_: 1383) as selected examples of modern total syntheses of acetylenic carotenoids.

(k) 7,8-Dihydrocarotenoids. In line with the step-wise dehydrogenation of phytoene to lycopene during the biosynthesis of coloured carotenoids, the existence of 7,8(7',8')-dihydrocarotenes including phytofluene, ζ-carotene, neurosporene, α—zeacarotene etc. (H. Pfander, Key to Carotenoids, Birkhäuser, 1987), have long been known. Highly oxygenated C_{40}-carotenoids with short chromophores containing 7,8(7',8')-dihydro features are now being detected in microalgae, suggesting a branching of the biosynthetic pathway, prior to lycopene formation (E.S. Egeland and S. Liaaen-Jensen, _Abstr. 7th Int. Symp. Marie Natural Prod._, Capri 1992). Relevant examples are P457 (Fig. 1), 7',8'-dihydroprasinoxanthin-4',5'-epoxide (Fig. 11), micromonal and uriolide (Fig. 13), and other new representatives under current investigation, _e.g_ anhydromicromonol (E.S. Egeland and S. Liaaen-Jensen, _loc. cit._, Fig. 16).

To the 7,8(7',8')-dihydro category also belong parasiloxanthin and 7',8'-dihydroparasiloxanthin (T. Matsuno and S. Nagaka, _Nippon Suisan Gakkaishi_, 1980, _46_: 1191), Fig. 16. In fish the 7,8-dihydro feature may be of metabolic origin.

(l) Other carotenes and xanthophylls. Among the carotenes should be mentioned 1,2-dihydrophytoene and 1,2-dihydrophytofluene (H.C. Malhotra et al., _Internat. Z. Vitaminforsch._, 1970, _40_: 315; G. Britton et al., _Phytochem._, 1977, _16_: 1561), besides other carotenoids of the 1,2-dihydro series discussed in the 1973 review (J.B. Davis, this series, First Supplement Vol. IIB, pp. 191-357), see Fig. 17. A new acetylenic carotene

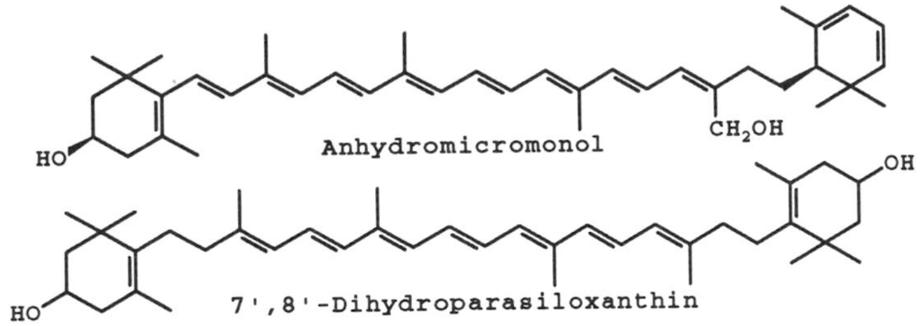

Fig. 16

from microalgae has been characterized (A. Fiksdahl and S. Liaaen-Jensen, Phytochem., 1988, 27: 1447). The asolute stereochemistry of ϵ,ϵ-carotene from avian retina has been settled as 6S,6'S (K. Schiedt et al., Pure & Appl. Chem., 1991, 63: 89). Also de novo biosynthesis in a marine chrysophyte provides (6S,6'S)-ϵ,ϵ-carotene (T. Bjørnland et al., Phytochem., 1989, 28: 3347) according to chiroptical properties.

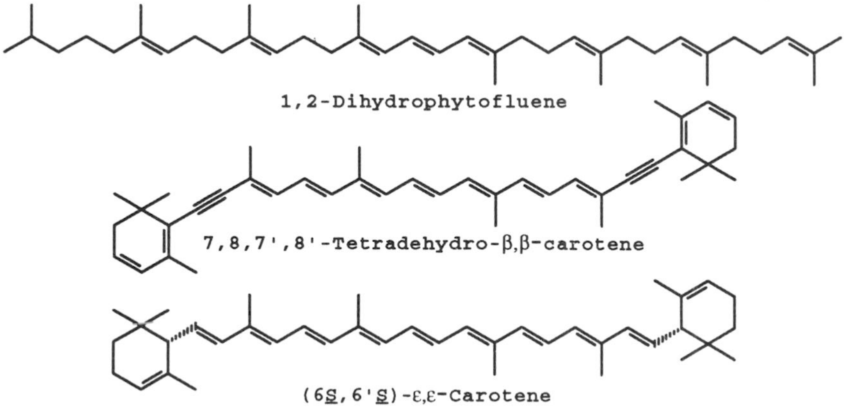

Fig. 17

Among the numerous new xanthophylls a representative series of examples will be selected, aiming at illustrating the types of structural variety encountered.

Since the classical period 3-hydroxylated β-rings or ε-rings have been recognised as a common structural element in xanthophylls. New diastereomers are being detected, e.g. lutein G (T. Matsuno et al., Comp. Biochem. Physiol., 1986, 85B: 77), see Fig. 16 and also under Optical isomerism below.

Examples of 3,4-anhydro structures such as anhydrodiatoxanthin (A. Fiksdahl et al., Phytochem., 1984, 23: 649), Fig. 18, and eutreptiellanone (A. Fiksdahl et al., loc. cit.,) are now known, Fig. 11, and also 2,3-anhydro structures including anhydromicromonal (E.S. Egeland and S. Liaaen-Jensen, Abstr. 7th Int. Symp. Marine Nat. Prod., Capri, 1992), Fig. 18.

With improved methodology several 2-hydroxylated β-rings have been encountered in carotenoids, such as β,β-carotene-2,2-diol (H. Kjøsen et al., Acta Chem. Scand., 1972, 26: 3053)´and caloxanthin (R. Buchecker et al., Phytochem., 1976, 15: 1015). 2-Hydroxyechinenone (P. Foss et al., Acta Chem. Scand., 1986, B40: 157), as a β-hydroxyketone, is readily dehydrated by treatment with base. A 4-keto-β-end group is encountered in a 2,3- diol and two tetrols (P. Beyer et al., Helv. Chim. Acta, 1979, 62: 2551). These examples are given in Fig. 18.

As to 2-hydroxylated aliphatic end groups (2'R)-chirality has been determined for plectaniaxanthin and 2'-hydroxyplectaniaxanthin by CD-correlations (H. Rønneberg et al., Phytochem., 1985, 24: 309), whereas the (2'R)-chirality of aleuriaxanthin was confirmed by a total synthesis (W. Eschenmoser et al., Helv. Chim. Acta, 1982, 66: 82).

The 4S(4'S) chirality of natural isocryptoxanthin and isozeaxanthin (K. Schiedt, in Carotenoids Chemistry and Biology, eds. N.I. Krinsky, M.M. Mathews-Roth and R.T. Taylor, Plenum, 1989, p. 247) was solved via a synthetic approach (A. Haag and C.H. Eugster, Helv. Chim. Acta, 1982, 65: 1795), Fig. 19. Other 4-hydroxylated β-rings are encountered in carotenoid triols and tetrols. Such triols comprise pectenol, of established configuration (T. Matsuno et al., Nippon Suisan Gakkaishi, 1981, 47: 143, 377, 385, 501), Fig. 19, idoxanthins (K. Schiedt et al., Helv. Chim. Acta, 1988, 71: 881) lilixanthin (E. Märki-Fischer and C.H. Eugster, Helv. Chim. Acta, 1985, 68: 1708) and a β,ε-carotene-triol, (M. Tsushima et al., Comp. Biochem. Physiol. C. Comp. Biochem., 1989, 93B: 665.

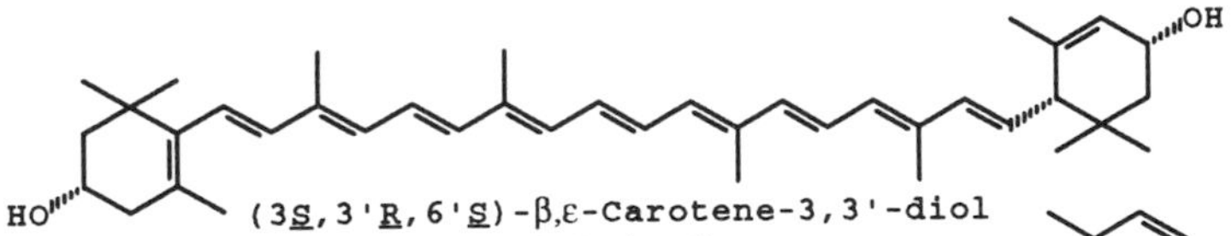

(3S,3'R,6'S)-β,ε-Carotene-3,3'-diol
Lutein G

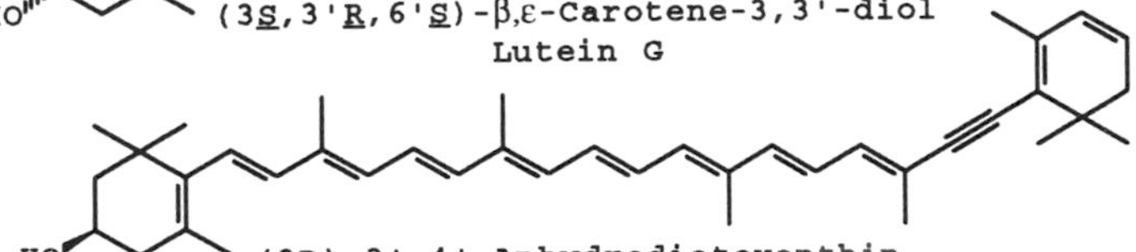

(3R)-3',4'-Anhydrodiatoxanthin

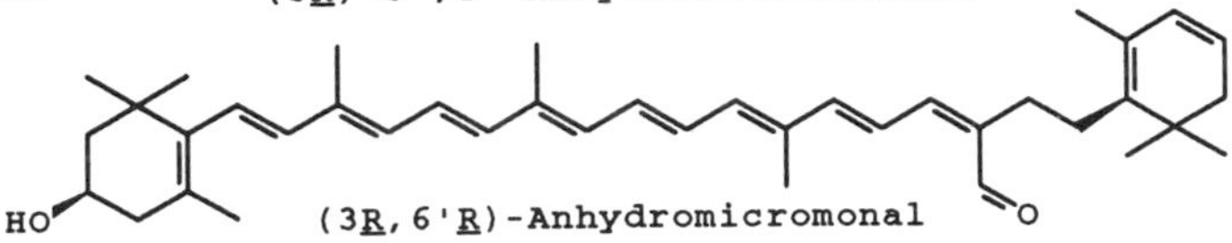

(3R,6'R)-Anhydromicromonal

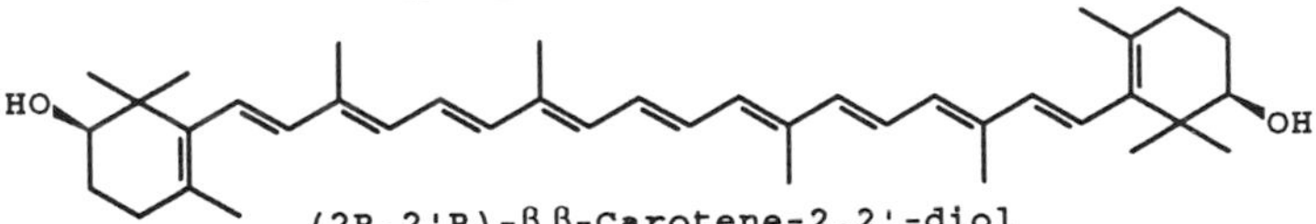

(2R,2'R)-β,β-Carotene-2,2'-diol

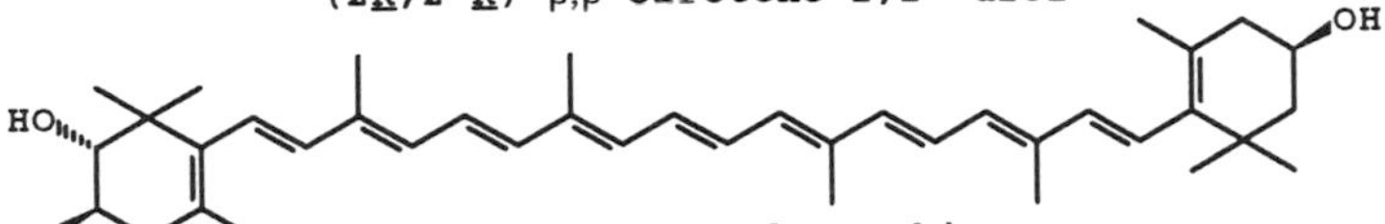

(2R,3R,3'R)-Caloxanthin

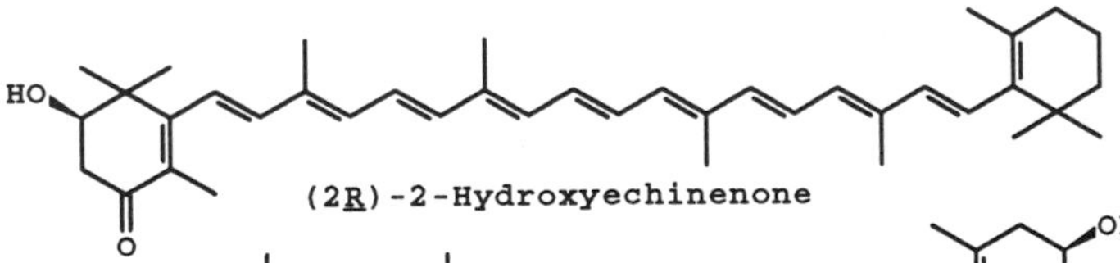

(2R)-2-Hydroxyechinenone

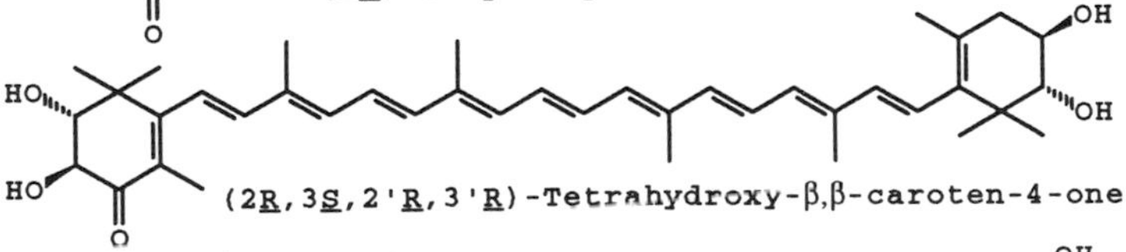

(2R,3S,2'R,3'R)-Tetrahydroxy-β,β-caroten-4-one

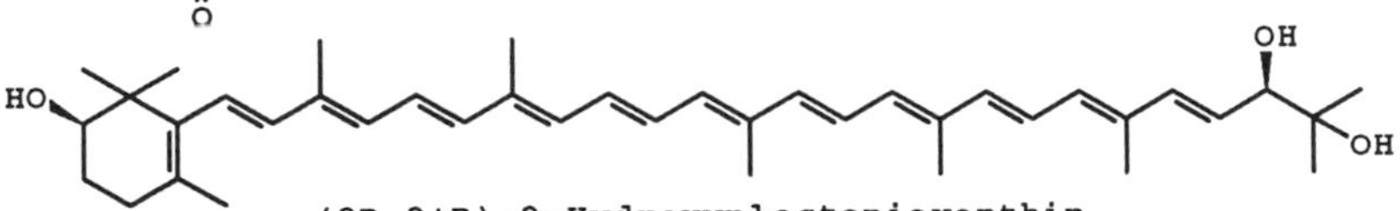

(2R,2'R)-2-Hydroxyplectaniaxanthin

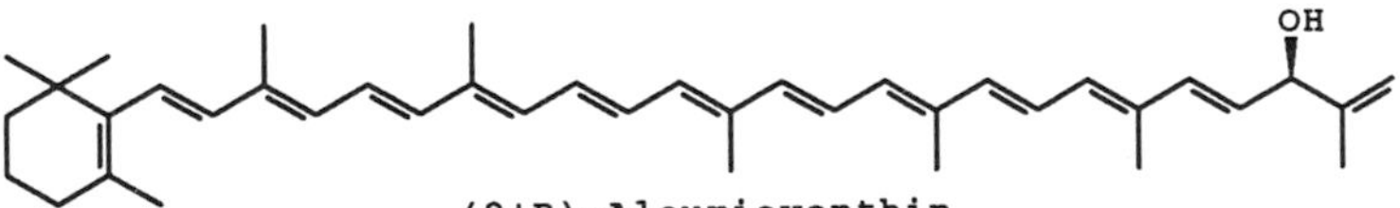

(2'R)-Aleuriaxanthin

Fig. 18

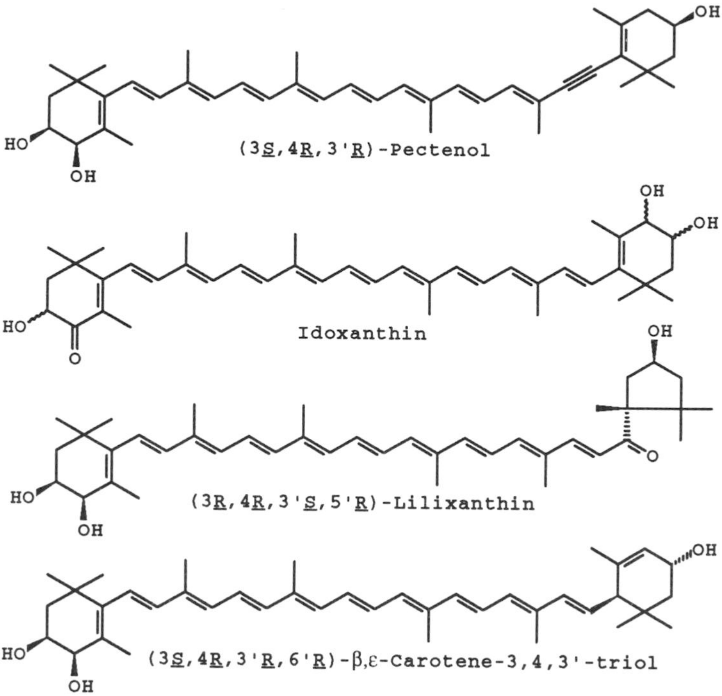

Fig. 19

Within tetrols should be mentioned crustaxanthin of variable configuration. Other recent tetrols of established configuration with triol end group are karpoxanthin and 6-epi-karpoxanthin (E. Märki-Fischer and C.H. Eugster, Helv. Chim. Acta, 1984, 67: 2143). Finally attention is directed to an allenic pentol (R. Buchecker et al., Helv. Chim. Acta, 1984, 67: 2043) and the hexol mactraxanthin (R. Buchecker et al., loc. cit.), see Fig. 20.

Dicyclic xanthophylls with keto groups in 2- and 3-positions are known, supplementing 4(4') and 3(3') keto-carotenoids which have been known for many years. The new compounds are represented by more than ten dicyclic 2(2')-keto carotenoids. Previously only carotenoids with 2(2')-keto functions in aliphatic end groups had been recognised. Selected examples, see Fig. 21 are insect carotenoids such as the conjugated

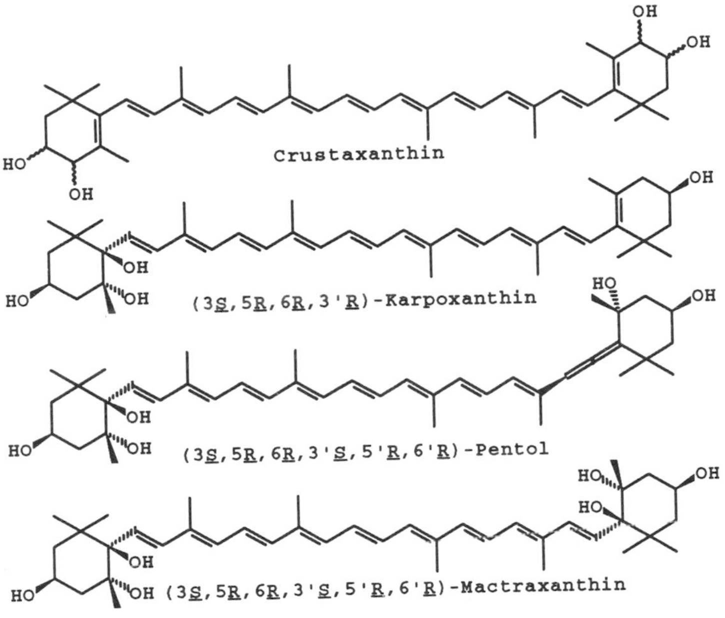

Fig. 20

ω,ω'-dione (H. Kayser, Comp. Biochem. Physiol., 1982, 72B: 427) and less dehydrogenated carotenoids such as the dione (H. Kayser, loc. cit.) and the corresponding optically inactive monool (H. Kayser et al., Insect Biochem., 1984, 14: 51). The most recent example in this series is the conjugated 2-one (T. Matsuno et al., Comp. Biochem. Physiol., 1990, 95B: 583).

Among dicyclic 3(3')-keto carotenoids several new carotenoids have been characterized, for selected examples, see Fig. 22. The configuration has been assigned of philosamiaxanthin (R. Buchecker and C.H. Eugster, Helv. Chim. Acta, 1979, 62: 2871). (6S,3'R,6'R)-3'-Hydroxy-ε,ε-caroten-3-one (T. Matsuno et al., Comp. Biochem. Physiol., 1986, 84B: 477). Its 3'- and 6'-epimer are also known (T. Matsuno et al., Comp. Biochem. Physiol, B : Comp. Biochem., 1985, 80B: 779). Both enantiomers and the meso form of ε,ε-carotene-3,3'-dione has been reported (Y. Ikuno et al., Nippon Suisan Gakkaishi, 1985, 51: 2033; J. Chromatogr., 1985. 51: 2033; Matsuno et al., loc. cit.). The stereochemistry of the α-ketol papilio-

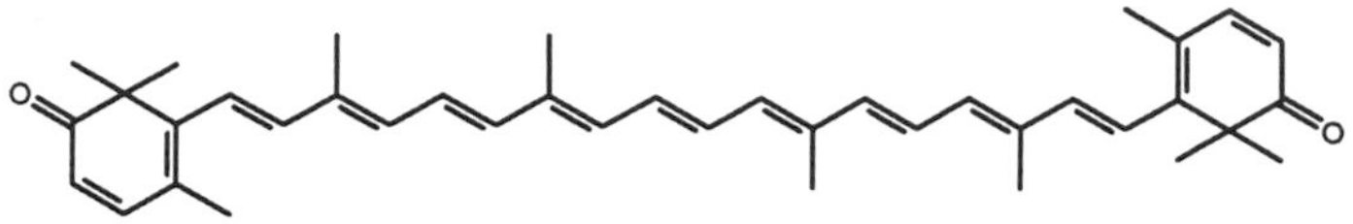

3,4,3',4'-Tetradehydro-β,β-carotene-2,2'-dione

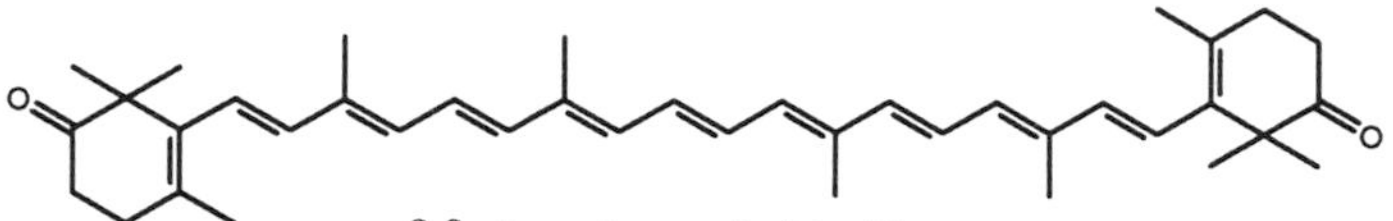

β,β-Carotene-2,2'-dione

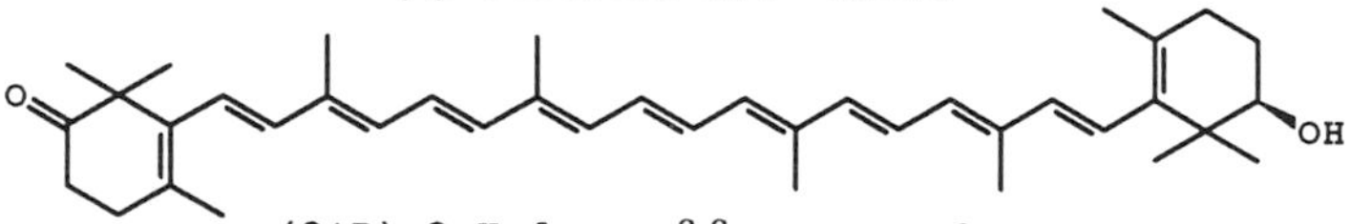

(2'R)-2-Hydroxy-β,β-caroten-2-one

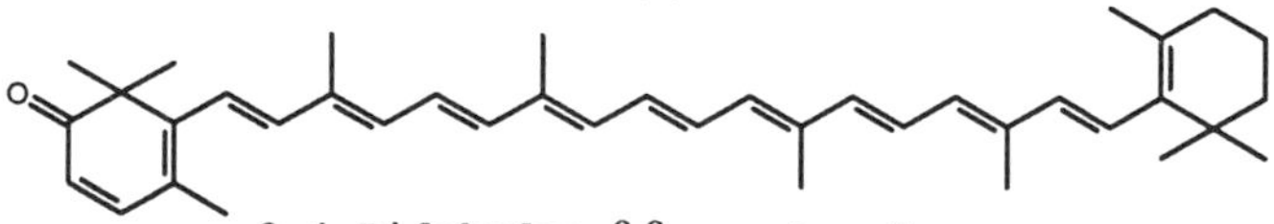

3,4-Didehydro-β,β-caroten-2-one

Fig. 21

erythrinone (K. Harashima et al., Agric. Biol. Chem., 1976, 40: 711) and
the retro-carotenoid loniceraxanthin (A.K. Rahman and K. Egger, Z.
Naturforsch., 1973, C28: 434) has not yet been assigned. Finally capso-
rubone (A. Rüttimann et al., Helv. Chim. Acta, 1983, 66: 1939) represents
the cyclopentanone carotenoids.

Some aryl-carotenoids yet remain to be mentioned, Fig. 23. The
previously known phenolic carotenoid 3-hydroxyisorenieratene and 3,3'-
dihydroxyrenieratene have been further examined (G. Nybraaten and S.
Liaaen-Jensen, Acta Chem. Scand., 1974, B28: 584; W. Kohl et al.,
Phytochem., 1983, 22: 207). Whereas tert. methoxy groups are still a
feature exclusive to carotenoids from photosynthetic bacteria,
methoxylated carotenoids such as aaptopurpurin (Y. Tanaka and Y. Ito,
Nippon Suisan Gakkaishi, 1985, 51: 1743) are encountered in marine
sponges. The aryl carotenoid trikentriorhodin, which is an enolized
β-diketone, has been prepared by total synthesis (A.K. Chopra et al., J.
Chem. Soc. Chem. Commun., 1977: 467).

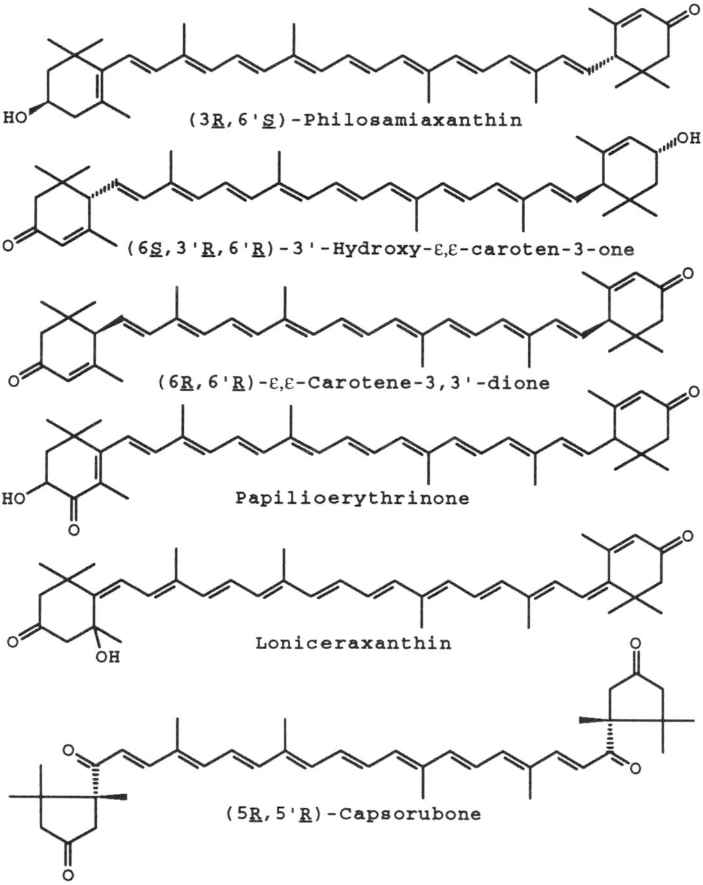

(3R,6'S)-Philosamiaxanthin

(6S,3'R,6'R)-3'-Hydroxy-ε,ε-caroten-3-one

(6R,6'R)-ε,ε-Carotene-3,3'-dione

Papilioerythrinone

Loniceraxanthin

(5R,5'R)-Capsorubone

Fig. 22

(m) Carotenoproteins. Continuous progress is being made in this field. The reader is referred to review articles by P. Zagalsky (Pure Appl. Chem., 1976, 47: 103-120; Oceanis, 1983, 9: 73-90; J.B.C. Findlay et al., in Carotenoids Chemistry and Biology, eds. N.I. Krinsky, M.M. Mathews-Roth and R.T. Taylor, Plenum, 1989, 75-105). The complete characterization of the protein part, including amino acid sequencing, is described. Similarly recombination experiments between the colourless apoprotein and appropriate ketocarotenoids, and recent progress in the

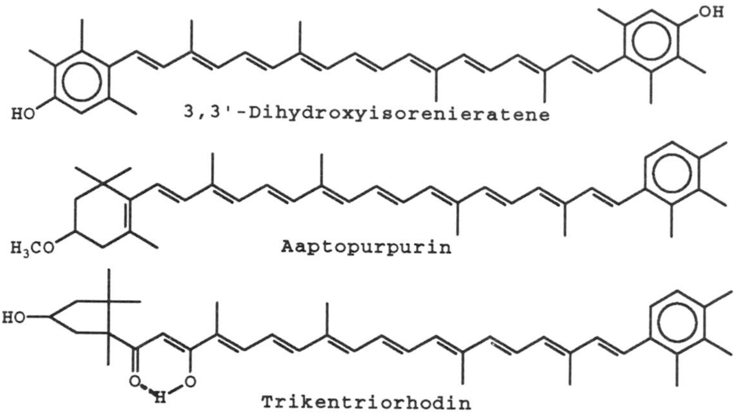

3,3'-Dihydroxyisorenieratene

Aaptopurpurin

Trikentriorhodin

Fig. 23

spectroscopic characterization of coloured carotenoproteins by Resonance Raman and NMR methods are detailed (P. Zagalsky, Pure & Appl. Chem., 1993, to be published).

7. Optical isomerism including allene isomerism

Until the early seventies it was assumed that a given carotenoid constitution, possessing one or more chiral centers, occurred in Nature with one preferred configuration. Whereas this simple picture still appears to be true for the large majority of cases involving de novo biosynthesis of carotenoids, in the last two decades there have been ample illustrations of i) chiral carotenoids occurring with different chirality in different organisms and ii) chiral carotenoids occurring as optically impure mixtures of different optical isomers in the same organism. The last phenomenon is now generally associated with organisms in which the metabolism of optically pure dietary carotenoids takes place, or in organisms further along the food chain resorbing such optical impure mixtures.

As examples of the former category the natural occurrence of pure enantiomers of (3S,3'S)-astaxanthin in microalgae may be noted (B. Renstrøm and S. Liaaen-Jensen, Phytochem., 1981, 20: 2561) and of (3R,3'R astaxanthin in yeast (A.G. Andrewes et al., Phytochem., 1976, 15: 1003), Fig. 24.

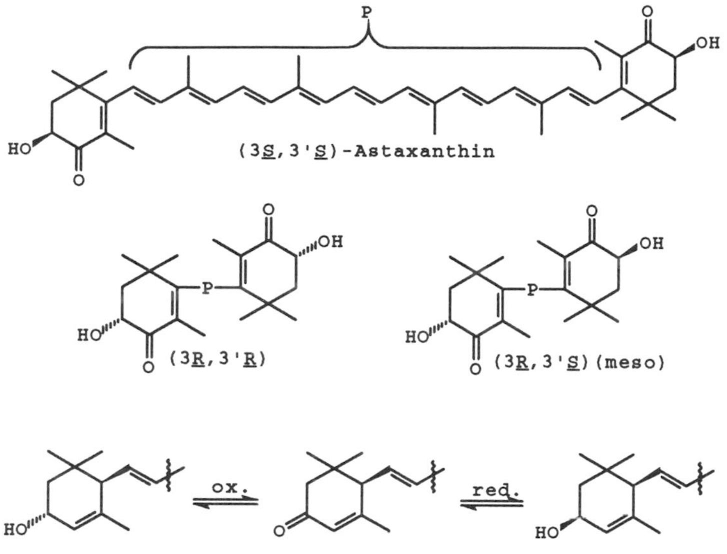

Fig. 24

Particularly within marine animals α-ketols such as astaxanthin may occur as mixtures of both enantiomers and the meso form (H. Rønneberg et al., Helv. Chim. Acta, 1980, 63: 711), Fig. 24. Similarly for carotenoids of marine animal origin with 3-hydroxy- and 3-keto-ε-end groups there is wide stereochemical variety: the allylic hydroxy groups may undergo oxidation and then reduction, and other modifications resulting in metabolites with 6R- or 6S-configurations. Some examples were already mentioned under xanthophylls, Fig. 22. Also for insect carotenoids with 2-keto/hydroxy groups optical impure carotenols are encountered (H. Kayser et al., Insect Biochem. 1984, 14: 51).

Yet another type of optical isomerism, associated with allene groups has received much interest. According to a biogenetic hypothesis (S. Isoe et al., Tetrahedron Letters, 1971: 1089) the allenic (6'S)-isomer of fucoxanthin, Fig. 25, was postulated to be an unstable biosynthetic precursor of the common (6'R)-allenic isomer (K. Bernhard et al., Tetrahedron Letters, 1974: 3899). Indeed the allenic (6'S)-isomer was identified as a minor carotenoid in brown algae and also as a product isolated from the iodine catalyzed stereomutation mixture of fucoxanthin (K. Bernhard et al., loc. cit.).

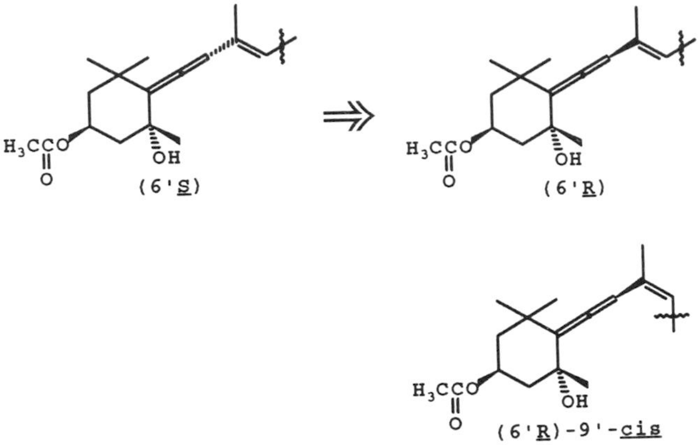

Fig. 25

However, this compound was subsequently shown to be the geometrical (6'R)-9'-cis isomer (T. Bjørnland et al., Tetrahedron Letters, 1989: 2577; G. Englert et al., Magn. Res. Chem., 1990, 28: 519). Moreover, it was demonstrated that no allenic (6'S)-isomers would be present in amounts exceeding 1% of the iodine catalyzed stereomutation mixture of fucoxanthin (J.A. Haugan et al., Acta Chem. Scand., 1992, 46: 389). However, very recently M. Ito et al., (Tetrahedron Lett., 1992, 33:2991),have reported the isolation of the allenic (6S)-isomer of peridinin, cfr. Fig. 3 and Fig. 25, as ca. 9% of the iodine catalyzed isomerization mixture. This is verified by detailed NMR studies (G. Englert et al., to be published).

8. Geometrical isomerism

A recent trend in carotenoid research is increased interest in geometrical isomerism. Whereas it was previously assumed that the all-trans isomer represented the thermodynamically most stable, natural isomer there are now several examples of the natural occurrence of i) preferred cis (Z)-isomers and ii) mixtures of geometrical isomers. As already pointed out in Section 3 the developments of high field ^{1}H NMR spectroscopy and HPLC instruments have been imperative for the recognition of these phenomena.

Within category i) important topics are the dominance and stability of 13'-_cis_ configuration in cross-conjugated 20'-als (G. Englert _et_ _al._, Magn, Res, Chem., 1988, 26: 55), the occurrence of sterically hindered 7-_cis_ aryl carotenoids (renieracistene = 7-_cis_-renieratene (Y. Tanaka _et_ al., Nippon Suisan Gakkaishi, 1982, 48: 1651), (S. Hertzberg _et_ al., Bull, Soc, Chim, Belg., 1986, 95: 801) and of 9'-_cis_-neoxanthin, the occurence Fig. 3, (T. Bjørnland _et_ al., Abstr, 8th Internat, Carotenoid Symp., Boston, 1987, P29) in particular sources.

Category ii) is exemplified by the natural occurrence of several geometrical isomers of bacterioruberin, Fig. 3, in halophilic bacteria (M. Rønnekleiv and S. Liaaen-Jensen, Acta Chem, Scand., B38, 1992, 871 and the presence of _cis_-isomers in human plasma (K. Khachik _et_ al., J. Chromatogr. In press).

9. Artifacts

Increased understanding of the chemistry of carotenoids has contributed to the recognition of various types of artifacts encountered during isolation procedures. In a comprehensive treatment (S. Liaaen-Jensen in N.I. Krinsky, M. Matthews-Roth and R.F. Taylor eds. Carotenoid Chemistry & Biochemistry, Plenum, 1989, pp. 149-165) the following topics are considered i) pre-extraction artifacts, comprising enzyme catalyzed reactions, plant acid catalyzed reactions etc, and ii) work-up artifacts. The most common work-up artifacts are due to stereoisomerization in solution to give other geometrical isomers, epoxide-furanoid rearrangements and alkali catalyzed artifact formation. The last type comprise the hydrolysis of esters, aldol condensations, particularly of carotenals during saponification in the presence of acetone, retroaldol cleavage and miscellaneous other reactions. Further work-up artifacts are ascribed to air oxidation, reactions on active surfaces, and thermal effects.

Artifact candidates recognised in recent years are gelliodesxanthin, Fig. 7, (Y. Tanaka and T. Inoue, Nippon Suisan Gakkaishi, 198 , 53: 1271) and a furanoid derivative of peridinin, Fig. 3, (A. Padilla _et_ al., Abstr, 7th Int, Symp, Marine Nat, Prod,, Capri, 1992, P.47).

10. Interdisciplinary fields

As already mentioned a special trend in carotenoid chemical research has been an increased interdisiplinary approach.

(a) Biosynthesis. For a recent update see G. Britton (Pure and Applied Chem., 1991, 63: 101). Comments were already made to the fundamental breakthrough in the field of carotenoid biosynthesis by genetic engineering in Section 1. As in other fields of natural product chemistry an interpretation of biosynthetic evidence is closely tied to structure determination.

(b) Chemosystematics. Chemosystematics implies the application of secondary metabolites for classification purposes as an additional tool, besides morphological, physiological and other biological criteria. The use of carotenoids has been dealt with by T.W. Goodwin (The Comparative Biochemistry of the Carotenoids, Vol. I Plants and Vol. II Animals, Chapman & Hall, 1980 and 1984) and S. Liaaen-Jensen (Pure and Applied Chem., 1978, 51: 661). Particularly useful results are obtained for microalgae (T.W. Goodwin, loc. cit., S. Liaaen-Jensen in Marine Natural Products Chemistry, eds. D.J. Faulkner and W.H. Fenical, Plenum, 1977, 239-59; in Marine Natural Products, ed. P. Scheuer, Academic Press, 1978, 2-73; T. Bjørnland and S. Liaaen-Jensen in The Chromophyte Algae : Problems and Perspectives, eds. J.C. Green, B.S.C. Leadbeater and W.I. Diver, Clarendon Press, Oxford, 1989, pp. 37-61). Structural evidence including absolute stereochemistry is a requirement for chemosystematic evaluations. Biosynthetic considerations are also essential for deciding which structural elements are of major importance.

Chemosystematics based on carotenoids has proved useful for the classification of carotenogenic organisms carrying out de novo biosynthesis. However, for organisms where metabolic modifications of existing carotenoids occur the picture is frequently complicated, cf. marine sponges (S. Liaaen-Jensen et al., Biochem. Syst. Ecol., 1982, 10: 167).

(c) Metabolism. Several animals have the ability to modify the structures of carotenoids obtained from their diet. Carotenoid metabolism will not be fully covered here, cf. Section 1. However, it is pointed out that striking progress within this field has been made based largely on the availability of isotopically labelled carotenoids and synthetic reference carotenoids, besides improved analytical methods. Detailed progress reports, particularly covering carotenoid metabolism in fish and chicken are available by Schiedt (K. Schiedt et al., Pure and Appl. Chem., 1985, 57: 685-692; K. Schiedt in Carotenoids Chemistry and Biology, eds. N.I. Krinsky, M.M: Mathews-Roth, R.T. Taylor, Plenum, 1989, pp. 247-268; K. Schiedt et al., Pure and Applied Chem., 1991, 63: 89).

A selected example of combined structural and metabolic studies illustrates the state of the art: the discovery of a novel, naturally occurring isoastaxanthin, (6S,6'S)-4,4'-dihydroxy-ε,ε-carotene-3,3'-dione, offered an explanation for the in vivo racemization of astaxanthin, proved to occur in the prawn Panaeus japonicus after administration of optically active [³H]-labelled (3S,3'S)-astaxanthin, Fig. 26.

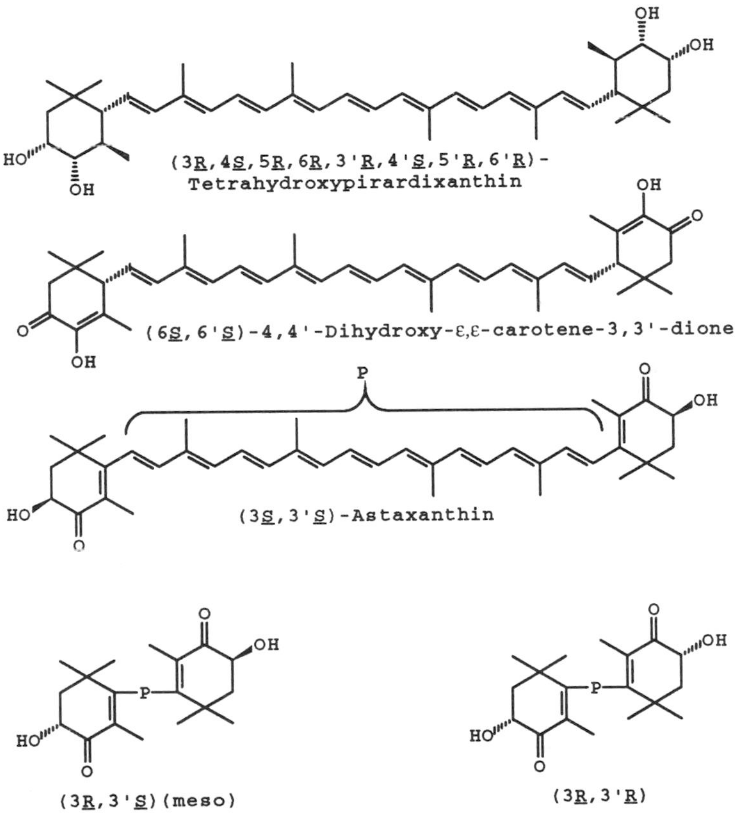

(3R,4S,5R,6R,3'R,4'S,5'R,6'R)-
Tetrahydroxypirardixanthin

(6S,6'S)-4,4'-Dihydroxy-ε,ε-carotene-3,3'-dione

(3S,3'S)-Astaxanthin

(3R,3'S)(meso)

(3R,3'R)

Fig. 26

Other relevant metabolites were structurally identified using (3R,4S,5R,-6R,3'R,4'S,5'R,6'R)-tetrahydroxypirardixanthin, Fig. 26, prepared by a total synthesis (U. Hengartner, Abstr. 9th Int. Symp. Carotenoids, Kyoto, 1990, p.22) as a spectroscopic model, Fig. 26.

The metabolism of carotenoids to apocarotenoids, Vitamin A and to smaller metabolites such as abscisic acid is being further investigated. The biodegradation of carotenoids to aromatics has been reviewed (C.R. Enzell, Pure and Applied Chem., 1985, 57: 693).

(d) Functions.   Functional aspects were reviewed by N.I. Krinsky in 1971 (In O. Isler, Carotenoids, Birkhäuser, Basel, pp. 669-716) and is being updated in 1993 (Pure and Applied Chem., to be published). Again, studies on the diverse functions of carotenoids is based on interdisiplinary approach.

(e) Photochemistry and Photoprotection.   Carotenoids play an important role in photosynthesis and photoprotection. (R.J. Cogdell, Pure and Appl. Chem., 1985, 57: 723; E.L. Schrott ibid 729). Recent results on the func tion of carotenoids in photosynthesis have been discussed by T.A. Moore et al., (In N.I. Krinsky, M.M. Mathews-Roth and R.F. Taylor eds., Carotenoids Chemistry and Biology, Plenum, 1989: 223); H.A. Frank et al., Pure and Applied Chem., 1991, 63: 109).   Interestingly 15Z-isomers play an essential role in the photosynthetic reaction center (Y. Koyama in Carotenoids Chemistry and Biology loc. cit., 1989: 207); R.   Gebhard et al., Pure and Appl. Chem., 1991, 63: 115). The syntheses of isotopically labelled ca rotenoids are important for current studies, which also involve advanced spectroscopy.

(f) Vision.   The visual process was reviewed by J.B. Davis in this series (First Supplement Vol. IIB, 1973, pp. 228-241).   Much interest has since been centered around synthetic and structural studies of visual pigments including point charge models for binding retinals to the protein (R.S.H. Liu in Carotenoid Chemistry and Biochemistry, eds. G. Britton and T.W. Goodwin, Pergamon, 1982, 253; K. Nakanishi, Pure and Applied Chem., 1991, 63: 161). Bacteriorhodopsin in the halophilic bacterium (Halobacterium halobium functions as a photosynthetic energy source (K. Nakanishi, loc. cit.,).

(g) Aquaculture.   Extensive development of fish farming of salmonids has created a large market for astaxanthin as a required feed ingredient (O.J. Torrissen et al., in CRS Marine Sciences, 1989, pp. 209-225), K. Bernhard in Carotenoids Chemistry and Biochemistry, eds. N.I. Krinsky, M.M. Mathews-Roth and R.T. Taylor, Plenum, 1989, 337). Industrial synthesis of astaxanthin, feeding experiments including resorption and metabolic studies are different interdisiplinary aspects (K. Bernhard loc. cit., K. Schiedt et al., Pure and Appl. Chem., 1985, 57: 685; K. Schiedt in Carotenoid Chemistry and Biochemistry, loc. cit., p. 247).

(h) Medical applications. The medical application of carotenoids other than as a dietary source of Vitamin A is currently receiving increasing interest. β,β-Carotene is successfully administered to patients with certain dermatological diseases to alleviate the photosensitivity associated with these conditions. (M.M. Mathews-Roth in Carotenoid Chemistry and Biochemistry, eds. G. Britton and T.W. Goodwin, Pergamon, 1982, 297). The application of carotenoids as cancer-preventive agents is being actively investigated (M.M. Mathews-Roth, Pure and Appl. Chem., 1985, 57: 717; ibid 1991, 63: 147; J.S. Bertram in Vitamins and Minerals in the Prevention and Treatment of Cancer, ed. M.M. Jacobs, pp. 38-50, CRS Press, London 1991).

*Second Supplements to the 2nd Edition of Rodd's Chemistry
of Carbon Compounds, Vol. II B(Partial), C, D and E,* edited by M. Sainsbury
© 1994 Elsevier Science B.V. All rights reserved.

Chapter 8a

THE CYCLOHEPTANES AND CYCLOOCTANES

D.F.EWING

1. Introduction

The search for synthetic methods for the formation of seven and eight membered carbocyclic systems has grown remarkably in the last 15 years. This is due largely to the increase in the number of known natural products (terpenes) which contain a seven or eight membered ring as a structural unit. This has stimulated interest in these new terpenoids especially since many of these compounds have been found to exhibit significant biochemical effects. It is likely that there is enormous therapeutic potential in this natural reservoir of structures and hence there is great interest in obtaining synthetic material for extensive investigation. Thus there is an obvious need for efficient, stereospecific procedures to construct these compounds, which often have a very complex molecular architecture. Unlike the situation for five and six membered ring systems where the diversity of synthetic procedures is legion, any approach to the formation of larger rings has, until recently, been limited by the variety of available methods and the intrinsically unfavourable thermodynamics associated with the formation of C_7 and C_8 rings (the entropic factor and the existence of transannular interactions).

Due to restrictions on space the following material is primarily concerned with the chemistry of monocyclic compounds with seven and eight membered rings, although some of the more general work on bi- and even polycyclic systems is included. Complete reaction sequences for the total or partial synthesis of natural products are not discussed. Even with this restricted coverage it has been necessary to concentrate mostly on recent work. Cyclooctatetraene (8-annulene) and other fully conjugated systems are also excluded for the most part.

Some useful general discussion of cycloheptane and cyclooctane can be found in a book in the functional group series (*The Chemistry of the Alkanes and Cycloalkanes,* Eds. S.Patai and Z.Rappoport, Wiley, 1992). Physical properties are covered extensively, and there is some discussion of synthetic methods, reactions including electrophilic and organometallic chemistry, radical reactions, electrochemistry and biochemical and toxicological effects, but the main emphasis is on alkane chemistry.

2. Preparation of cycloheptanes and cyclooctanes

The following material is arranged by synthetic method. No attempt is made to discuss the two ring sizes separately since many procedures are applicable to both.

The synthesis of terpenoid natural products containing the cyclooctane ring has been reviewed (N.A.Petasis and M.A.Patane, *Tetrahedron*, 1992, **48**, 5757), as have ring closure methods generally, in natural product chemistry (Q.C.Meng amd M.Hess, *Top. Curr. Chem.* 1992, **161**, 107).

(a) Ring Expansion Methods

(i) Cope Rearrangement

The [3,3] sigmatropic rearrangement of *cis*-divinylcyclopropane and *cis*-divinylcyclobutane to 1,4-cycloheptadiene and 1,5-cyclooctadiene respectively, has long been established as a convenient synthetic route to these ring systems (for a general survey of the Cope rearrangement see R.K.Hill in *Comprehensive Organic Synthesis,* Eds. B.M.Trost and I.Fleming, Pergamon, 1991, Vol 5, Chap. 7.1; S.J.Rhoads and N.R.Raulins, *Org. React.,* 1975, **22**, 1). The inherent strain in C_3 and C_4 rings ensures that the thermodynamic position of the rearrangement is favourable to the expansion to C_7 and C_8 rings and many applications are reported (for a survey of cyclopropane rearrangements see E.Piers, in *Comprehensive Organic Synthesis,* Eds. B.M.Trost and I.Fleming,

Pergamon, 1991, Vol 5, Chap. 8.2; H.-U.Reissig in *The Chemistry of the Cyclopropyl Group,* Ed. Z.Rappoport, Wiley, 1987).

The Cope rearrangement is stereospecific (due to the boat shape of the transition state) and both the *trans* isomer (1) and the *cis* isomer (2) of a divinylcyclopropane give the corresponding *cis,cis*-1,4-cycloheptadiene (3) (Scheme 1), although more forcing conditions are required for thermal cyclisation of the *trans* isomer. Thus *cis*-divinylcyclopropane has a half-life of only 90 s at 35 °C with respect to rearrangement (J.M.Brown, B.T.Golding and J.J.Stofko, *J. Chem. Soc., Perkin Trans. 2,* 1978, 436) whereas the *trans* isomer is stable at room temperature. On heating it rearranges to the *cis* isomer via a radical intermediate (M.R.Schneider and A.Rau, *J. Am. Chem. Soc.,* 1979, **101,** 4426).

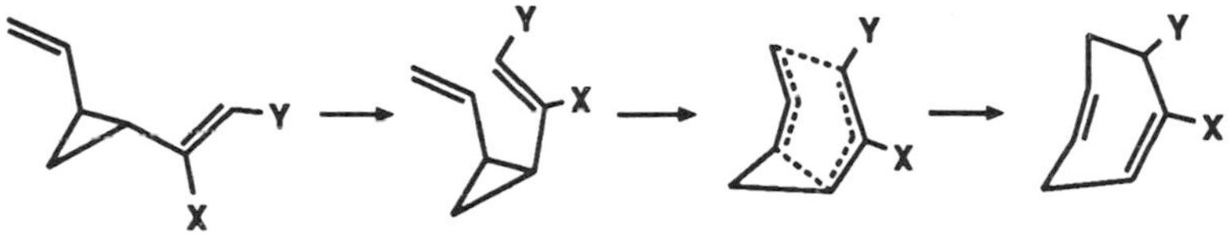

Scheme 1.

The natural product, dictyopterene D (3, X = H, Y = *cis*-1-butenyl) is obtained quantitatively from the appropriate *trans*-cyclopropane by heating at 120 °C for 18 hours (W.D.Abraham and T.Cohen, *J. Am. Chem. Soc.,* 1991, **113,** 2313). In contrast an analogous *cis*-cyclopropane was converted to optically pure (−)-dictyopterene C (3, X = H, Y = *n*-butyl) in 5 hours at 75 °C (T.Schotten, W.Boland and L.Jaenicke, *Tetrahedron Lett.,* 1986, **27,** 2349). A *trans*-cyclopropane was rearranged thermally to the cycloheptadiene derivative (3, X = CO$_2$Et, Y = H or Ph) a precursor to the tropone derivative, nezukone (E.Wenkert, R.S.Greenberg and H.-S.Kim, *Helv. Chim. Acta,* 1987, **70,** 2159).

Several types of bicyclic compound containing the cycloheptadiene ring are also accessible by this method (J.D.Marino and T.Kameko, *J. Org. Chem.,* 1974, **39,** 3175; E.Peirs *et al., Can. J. Chem.,* 1983, **61,** 1239).

The usefulness of the Cope rearrangement has been extended by the development of new synthetic methods for the formation of the cyclopropane ring. One notable method is the cyclopropanation of alkenes by rhodium(II)-catalysed decomposition of vinyldiazomethanes $RC(N_2)CH=CH_2$ (H.M.L.Davies, T.J.Clark and H.D.Smith, *J. Org. Chem.*, 1991, **56**, 3817 and references therein). Using $Rh_2(OAc)_4$ as catalyst, a vinylcarbenoid is generated and the cyclopropanation reaction and Cope rearrangement can occur in tandem to provide a convenient stereoselective route to cycloheptadienes, bicyclo[3.2.1]octadienes and other fused ring systems containing a C_7 ring (W.R.Cantrell and H.M.L.Davies, *J. Org. Chem.*, 1991, **56**, 723; H.M.L.Davies, M.J.McAfee and C.E.M.Oldenburg, *J. Org. Chem.*, 1989, **54**, 930).

A similar reaction has been found to occur between a metal carbene and a suitable enyne (Scheme 2). Initial attack of the molybdenum carbene at the terminal yne group is followed by cyclopropanation to give a bicyclic compound containing a vinylcyclopropane moiety. If a second *cis* vinyl group is present then [3,3] rearrangement affords a hexahydroazulene (D.F.Harvey and K.P.Lund, *J. Am. Chem. Soc.*, 1991, **113**, 5066). A clever extension of this reaction is the generation, *in situ*, of an intramolecular molybdenum carbene, which can then be taken through to a tricyclic system structured round a cycloheptadiene ring (D.F.Harvey and M.F.Brown, *J. Org. Chem.*, 1992, **57**, 5559).

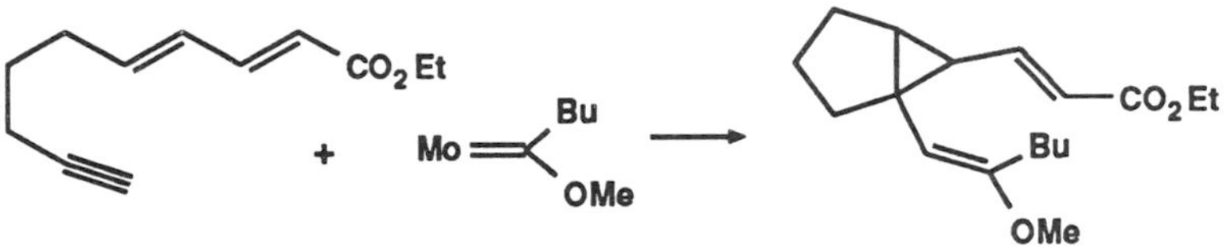

Scheme 2.

An example of an alternative approach to divinylcyclopropanes is shown in Scheme 3. Bisalkenylation of a protected cyclopropenone (4, R,R = $-CH_2CMe_2CH_2-$) is achieved using a copper alkyl, such as (*t*-BuCH=CHCH$_2$)$_2$CuLi. The initial carbocupration product (5) reacts with a suitable alkenyl electrophile (*t*-BuCH=CHCH$_2$) in presence of Pd(PPh$_3$)$_4$ at −70 °C and the *cis*-dialkenylcyclopropanone derivative (6) rearranges directly to a protected 6,7-dibutyl-1,4-cyclopheptadien-3-one (7) (E.Nakamura, M.Isaka and S.Matsuzawa, *J. Am. Chem. Soc.*, 1988, **110**, 1297).

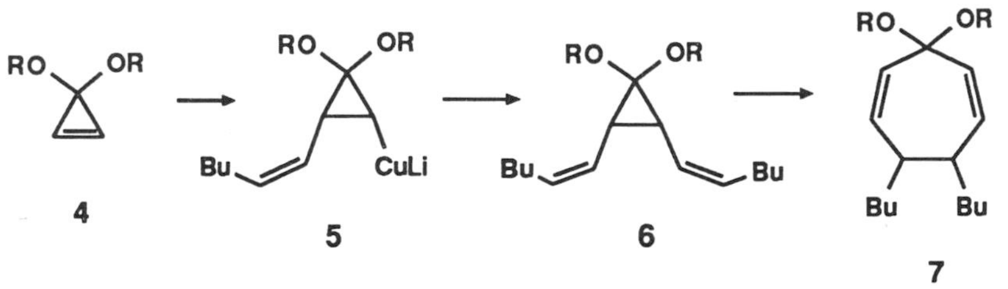

Scheme 3.

The thermal Cope rearrangement of a *cis*-divinylcyclobutane gives *cis*-1,5-cyclooctadiene in high yield but the corresponding *trans* isomer may give other products including that from a [1,3]-sigmatropic rearrangement. The generation of divinylcyclobutanones by [2 + 2] annulation of a 1,3-diene with a vinylketene has been described (R.L.Danheiser, S.K.Gee and H.Sard, *J. Am. Chem. Soc.*, 1982, **104**, 7670). A Cope rearrangement occurs *in situ* to afford cyclooctadienones in variable yield, an overall [4 + 4] cyclisation. A similar methodology for fused cyclooctadienes (Scheme 4) has been explored by P.A.Wender and C.R.D.Correia (*J. Am. Chem. Soc.*, 1987, **109**, 2523). The stereochemistry of the ring junction is *cis* but an analogous reaction promoted by catalysis with Ni(COD)$_2$ is much less stereospecific (P.A.Wender and N.C.Ihle, *J. Am. Chem. Soc.*, 1986, **108**, 4678). This reaction is particularly useful as an entry to the complex ring systems of many natural products.

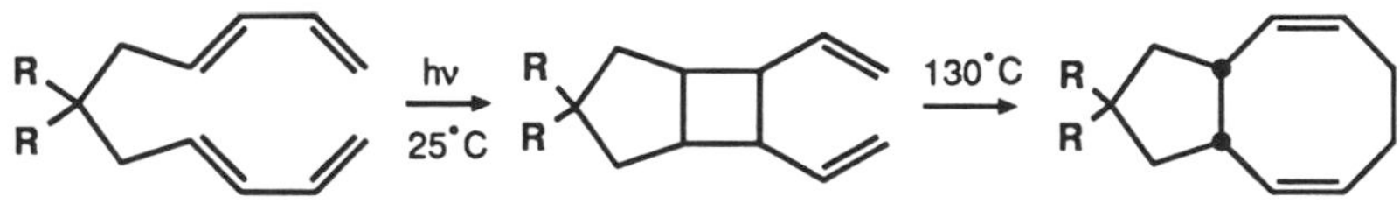

Scheme 4.

The intrinsic thermodynamic preference for the ring expanded product relative to the small ring precursor means that the normal rate enhancement of the thermal oxy-Cope and anionic oxy-Cope variants of this rearrangement are less advantageous and these variants of the basic sigmatropic rearrangement have been less exploited for the formation of C_7 and C_8 rings than is the case for larger rings (see this Volume, Chapter 8b). The thermal procedure is illustrated by the formation of a 4-cycloheptenone from the *O*-silyl enol of a *cis, trans* mixture of (vinylcyclopropyl)ketones (Scheme 5) (P.A.Wender and M.P.Filosa, *J. Org. Chem.*, 1976, **41**, 3490).

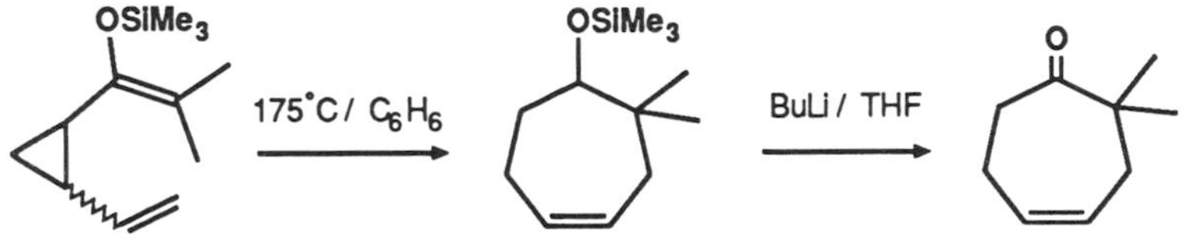

Scheme 5.

A range of substituted 4-cycloheptenones has been obtained similarly in high yield (>85%), the vinylcyclopropyl ketones being accessed by direct acylation (E.Peirs, M.S.Burstmeister and H.-U.Reissig, *Can. J. Chem.*, 1986, **64**, 180). A more recent example is the cyclopropanation of a silyloxy triene catalysed by $Rh_2(OAc)_4$, the product rearranging immediately to a chlorocycloheptadienone (A.de Meijere *et al.*, *Synthesis*, 1991, 547).

Rather more attention has been directed to the sigmatropic ring enlargement of dialkenylcyclobutanols to 4-cyclooctenone by an anionic oxy-Cope rearrangement. Initial observations (R.C.Gadwood and R.M.Lett, *J. Org. Chem.*, 1982, **47**, 2268; T.A.Lyle *et al.*, *Helv. Chim. Acta*, 1984, **67**, 774) suggested that both *cis* and *trans* isomers could participate in an anionic oxy-Cope rearrangement with potassium hydride but later work using a range of compounds, indicated that only the *cis*-disubstituted cyclobutane (9) gave the [3,3] rearrangement product under thermal or anionic conditions (J.P.Barnier, J.Ollivier and J.Salaun,

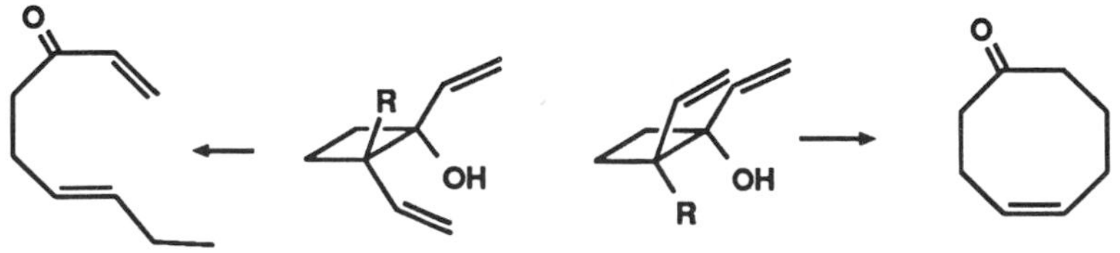

8   9

Tetrahedron Lett., 1989, **30**, 2525). The *trans* isomer (8) would only undergo retro-ene fragmentation to an acyclic dienone. For recent work on the synthesis of divinylcyclobutanes see J.J.Bronson and R.L.Danheiser in *Comprehensive Organic Synthesis*, Eds. B.M.Trost and I.Fleming, Pergamon, 1991, Vol 5, Chap. 8.3.

An anion accelerated oxy-Cope rearrangement is usually preferable to thermal rearrangement for cyclobutanes particularly when alternative, competing sigmatropic pathways are available and this approach has been applied to the formation of fused cyclooctene systems (G.Majetich and K.Hull, *Tetrahedron Lett.*, 1988, **29**, 2773; B.B.Snider and R.B.Real, *J. Org. Chem.*, 1988, **53**, 4508).

(ii) Claisen rearrangement

The [3,3]-sigmatropic rearrangement of vinyl allyl ethers such as (10, R = CO_2Et) is a simple two-atom ring expansion providing a route to the cyclooctenone system but this reaction has had surprisingly little application (A.Johns and J.A.Murphy, *Tetrahedron Lett.*, 1988, **29**, 837). An elegant example of this method is the synthesis of (11) (*en route* to the bicyclic natural product precapnelladiene) by placing the usually exocyclic vinyl group of the Claisen precursor within a cyclopentane ring (N.A.Petasis and M.A.Patane, *Tetrahedron Lett.*, 1990, **31**, 6799). Paquette has described a version of this two-carbon expansion methodology in his work on the development of route to terpenoids such as (+)-ceroplastol (C.M.G.Philippo, N.H.Vo and L.A.Paquette, *J. Am. Chem. Soc.*, 1991, **113**, 2762; L.A.Paquette, T.-Z.Wang and N.H.Vo, *J. Am. Chem. Soc.*, 1993, **115**, 1676).

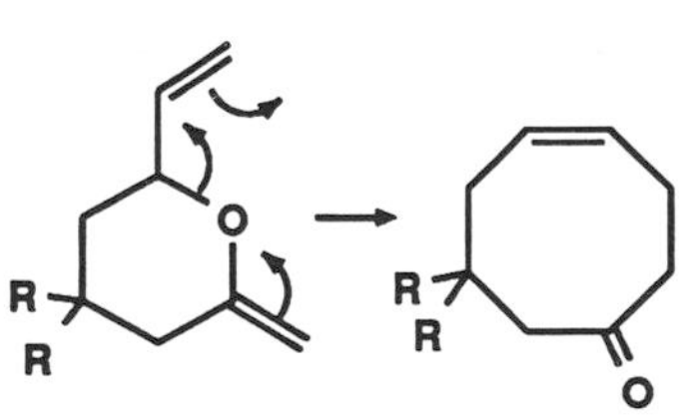

10 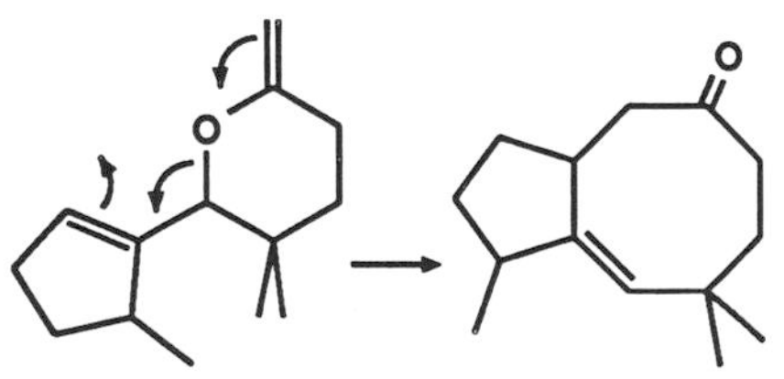   11

The Claisen rearrangement has been more commonly applied to the formation of the C_7 and C_8 rings by contraction of a larger lactone, usually using the Ireland modification (R.E.Ireland, R.H.Mueller and A.K.Willard, *J. Am. Chem. Soc.*, 1976, **98**, 2868), where the lactone is converted to its *O*-silyl ester enolate. The rearrangement occurs cleanly in high yield (80 - 90%) (R.L.Funk, M.M.Abelman and J.D.Munger, *Tetrahedron*, 1986, **42**, 2831) even in the case of complex systems (M.J.Bigeley, A.G.Cameron and D.W.Knight, *J. Chem. Soc., Perkin Trans. 1*, 1986, 1933).

(iii) Intramolecular reaction with a nucleophilic side chain

Additive alkylation of a cyclic ketone with dibromomethyllithium gives a 1-alkyl derivative in high yield (Scheme 6). This product, a β-dibromoalcohol, reacts with two moles of butyllithium to form a β-oxido carbenoid which undergoes a one-carbon insertion reaction to give the corresponding ring expanded ketone (H.Taguchi, H.Yamamoto and H. Nozaki, *J. Am. Chem. Soc.*, 1974, **96**, 6510; *Bull. Chem. Soc. Jpn.*, 1977, **50**, 1592). The extent to which the intermediate develops full carbene character is not well established but simple C_7 and C_8 ketones are obtained in 70-80% yield by this route. The regiospecificity of this rearrangement is dependent on reaction conditions, and the observed distribution of products in more complex cases is in keeping with the

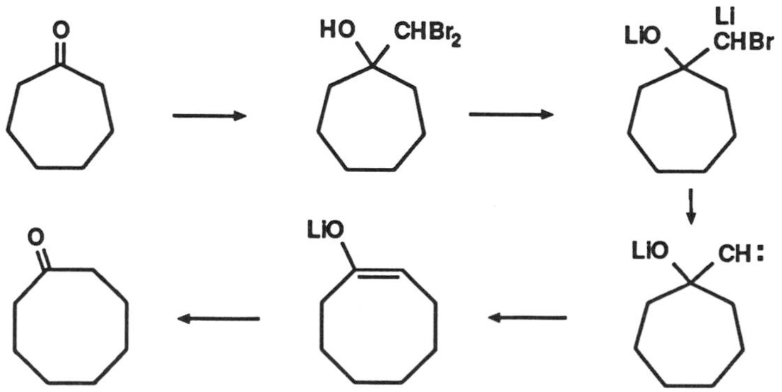

Scheme 6.

development of electrophilic character in the carbenoid species (H.D.Ward, D.S.Teager and R.K.Murray, *J. Org. Chem.*, 1992, **57**, 1926).

Similar reactions are possible with a variety of alkylation reagents to give in each case, the corresponding α-substituted cyclic alcohol, which undergoes a homologation rearrangement, via a carbenoid intermediate, to yield an α-substituted cyclic ketone. The mechanistic details vary depending on the nature of the heterosubstituents on the alkyl group and the reagent used to effect the rearrangement. Thus the use of $LiCH(SPh)_2$ affords 2-phenylsulphenylcycloheptanone (55%) and the analogous cyclooctanone (54%) (W.D.Abraham, M.Bhupathy and T.Cohen, *Tetrahedron Lett.*, 1987, **28**, 2203) and $LiCH(NO_2)SPh$ gives similar compounds with high regiospecificity, the leaving group being NO_2 (S.Kim and J.H.Park, *Chem. Lett.*, 1988, 1323). With $LiCH(SPh)(SO_2Ph)$ the rearrangement step requires catalysis by Et_2AlCl since the leaving group is SO_2Ph but the yield of the 2-phenylsulphenylcycloheptanone is 70% (B.M.Trost and G.M.Mikhail, *J. Am. Chem. Soc.*, 1987, **109**, 4124). The corresponding 2-phenylsulphinyl cyclic ketones (47%) are accessible with LiCHCl(SOPh) (T.Satoh *et al.*, *Tetrahedron Lett.*, 1992, **33**, 7181) and the alkylating reagent $LiCH_2SePh$ leads to homologation without the introduction of a hetero atom substituent (S.Uemura, K.Ohe and N.Sugita, *J. Chem. Soc., Chem. Commun.*, 1988, 111).

The reaction of cyclohexanones with $LiC(Me_2)SeMe$ has been studied in great detail by A.Krief and coworkers. The mechanism is not well established but generally the most substituted carbon migrates with retention of configuration. However both the regiochemistry and stereochemistry of the cycloheptanone product are sensitive to steric and other effects including the nature of the deselenation reagent (dichlorocarbene or silver tetrafluoroborate) (A.Krief *et al.*, *Tetrahedron Lett.*, 1989, **30**, 575 and references therein). This reaction has not been widely applied to C_7 and C_8 systems. One carbon expansion with the introduction of an α-alkyl group can be accomplished with the reagent LiCRClS(O)Ar, since desulphinylation occurs readily. Thus several alkylcycloheptanones are obtained (65-70%) (T.Satoh *et al.*, *Tetrahedron Lett.*, 1992, **33**, 7543).

Alkylation of 2-nitrocyclohexanone via Michael addition of a vinyl ketone gives an α-substituted cyclic ketone (12) which undergoes an intramolecular aldol reaction in presence of *t*-BuOK. Cleavage of the bridging bond in the bicyclic intermediate occurs directly to produce the substituted cyclooctanone (13) (Y.Nakashita and M.Hesse, *Helv. Chim. Acta*, 1983, **66**, 845). The analogous three carbon expansion of a cyclopentanone also gives a cyclooctanone (Z.-F.Xie, H.Suemune and K.Sakai, *J. Chem. Soc., Chem. Commun.*, 1988, 612). Although this is a general method for ring expansion it appears not to have been applied to the formation of C_7 rings.

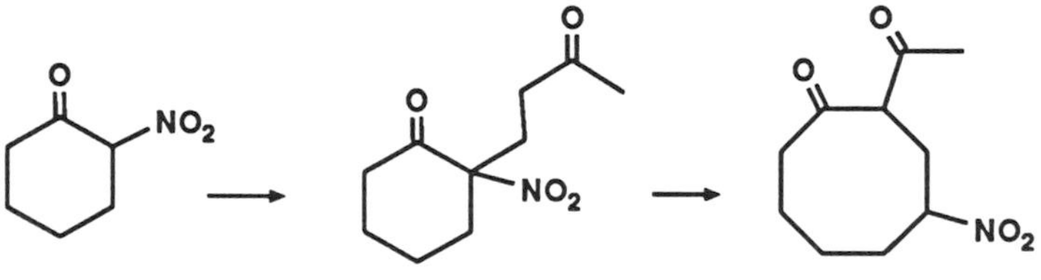

12   **13**

A similar type of cyclisation is observed for the compound with a silyl side chain (14). Fluoride ion promoted cleavage of the $SiMe_3$ group allows nucleophilic cyclisation to a bicyclic alcohol. Treatment with base results in ring opening, removal of the phenylsulphonyl group and double bond isomerism to give the 2,4-cyclooctadienone (15) (B.M.Trost and J.E.Vincent, *J. Am. Chem. Soc.*, 1980, **102**, 5680).

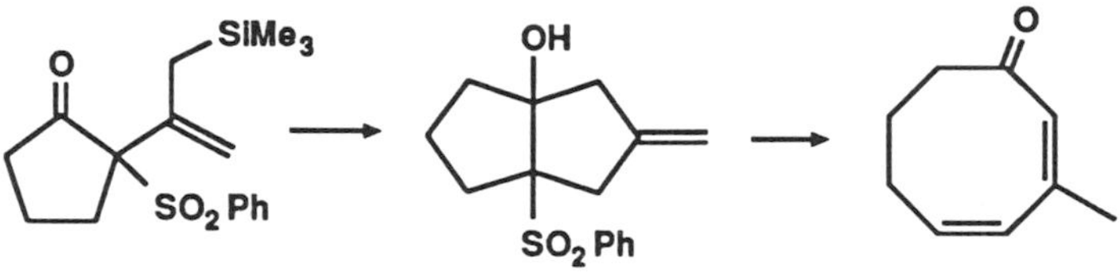

14   **15**

Reaction of cyclohexanone (as an enamine derivative) with $Ph_2P(O)CH=CH_2$ results in alkylation at the α-position. A two-carbon ring expansion can be achieved in the absence of an activating group by first converting to a cyclic lactone (Baeyer-Villiger oxidation). The diphenylphosphinoyl group is removed with KOH to afford 4-hydroxy-cyclooctanone in about 20% overall yield from cyclohexanone (P.Wallace and S.Warren, *J. Chem. Soc., Perkin Trans. 1*, 1992, 3169).

(iv) Intramolecular reaction with a radical side chain

One carbon ring expansion by intramolecular attack of a radical at carbonyl carbon has been found to be extremely useful for large rings and is also successful for seven and eight membered rings (Scheme 7). Thus a bromo- or iodomethyl group is introduced by alkylation at the ester carbonyl of a cyclic β-ketoester. Standard radical generation with Bu$_3$SnH gives the corresponding C$_7$ and C$_8$ ketoesters in about 70% yield (P.Dowd and C.-C.Choi, *Tetrahedron*, 1992, **48**, 4773 and references therein). Similar two and three carbon expansions of an appropriate cyclopentanone ester have also been reported (A.L.J.Beckwith, D.M.O'Shea and S.W.Westwood, *J. Am. Chem. Soc.*, 1988, **110**, 2565). This methodology has also been effective for the synthesis of a bicyclic system containing a cyclooctanone ring (G.Mehta, N.Krishnamurth and S.R.Karra, *J. Am. Chem. Soc.*, 1991, **113**, 5765).

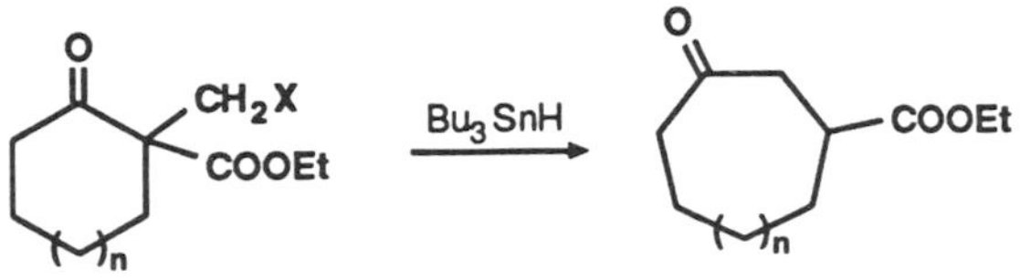

Scheme 7

A more indirect approach to ring expansion by 3 or 4 carbons is shown in Scheme 8. This radical reaction produces several products as might be expected but the ring expanded product predominates, 80% for n = 5 and 43% for n = 8 (P.Dowd and W.Zhang, *J. Am. Chem. Soc.*, 1991, **113**, 9875). This procedure is also effective for the ring expansion of spirobutanones (W.Zhang and P.Dowd, *Tetrahedron Lett.*, 1992, **33**, 3285).

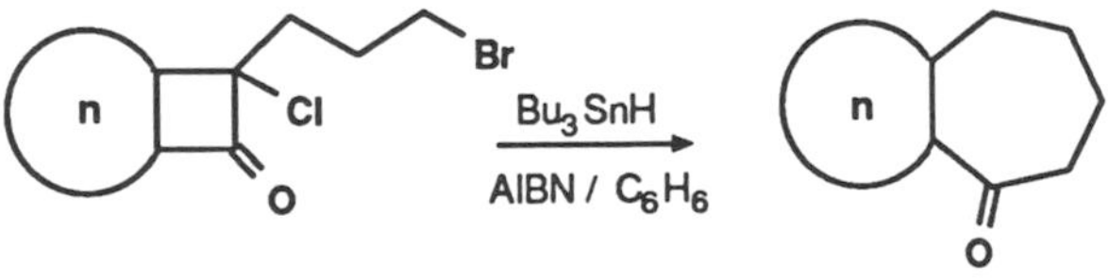

Scheme 8.

(v) Intramolecular reaction with an electrophilic side chain

Kuwajima and coworkers have investigated methods for one carbon ring expansion involving the generation of a cationic side chain stabilised by the presence of a β-silicon atom. Thus the cyclohexane carbaldehyde (16, X = CH_2, Y = Me, R = H) is converted quantitatively to the corresponding cycloheptanone (17) by treatment with $MeAlCl_2$ (K.Tanino, T.Katoh and I.Kuwajima, *Tetrahedron Lett.*, 1989, **30**, 1815). The silyloxy analog (16, X = O, Y = Me_2CH, R = Me) is converted by $FeCl_3$ to the corresponding cyclic ketone (89%) with very high regiospecificity (T.Matsuda, K.Tanino and I.Kuwajima, *Tetrahedron Lett.*, 1989, **30**, 4267).

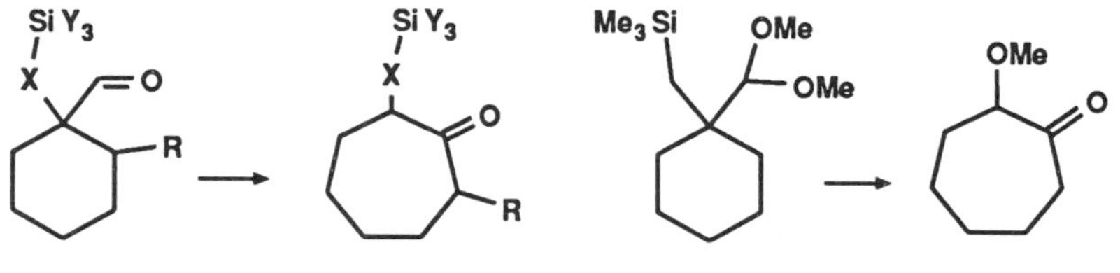

This electrophilic ring expansion reaction can take place with concomitant functionalisation of the carbonyl group. With $Me_3SiOSO_2CF_3$ (TMSOTf) as catalyst and Me_3SiOMe as alkylating reagent, a cationic methoxy derivative of the aldehydo group is formed, but ring expansion of this species only occurs for saturated rings with eight or more carbons. However, if the methoxy acetal (18) is preformed, methoxycycloheptanone (19) is obtained using zinc bromide as catalyst.

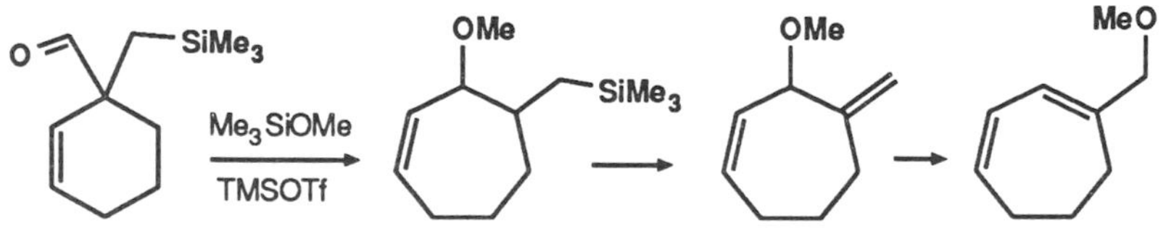

The analogous cyclohexene (20) derivative is more reactive and expands readily to a methoxycycloheptene (21), which eliminates trimethylsilane and undergoes a [1,3] rearrangement to afford (22) (T.Katoh, K.Tanino and I.Kuwajima, *Tetrahedron Lett.*, 1988, **29**, 1819). The thiomethyl analogue is formed similarly (83%) with Me₃SiSMe and TMSOTf as catalyst (K.Tanino, K.Sato and I.Kuwajima, *Tetrahedron Lett.*, 1989, **30**, 6551). The alkylation of a cyclohexanone with LiCH(OMe)SPh (Scheme 9) is only effective as a one carbon expansion reaction, for non-enolisable cyclohexanones. It probably involves intermediate cationic cyclopropanation catalysed by mercurinium ion with anchimeric assistance from the butylthiomethylene group and is not likely to be of wide ranging synthetic value. However a usefully functionalised cycloheptanone is produced in 58% yield (A.Guerrero, A.Parrilla and F.Camps, *Tetrahedron Lett.*, 1991, **32**, 1051).

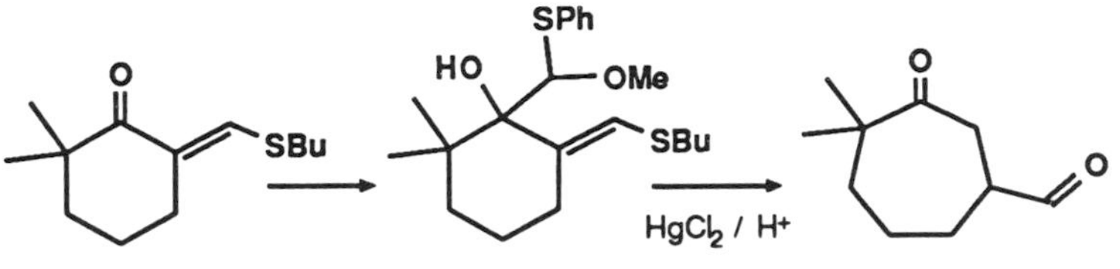

Scheme 9.

The vinylated cycloalkyl silylether (23) undergoes a facile one carbon ring expansion in presence of mercurinium ion to give the corresponding 2-methylene ketone in 65% yield (73% for C_8 product) (S.Kim and K.H.Uh, *Tetrahedron Lett.*, 1992, **33**, 4325). The mercurinium ion mediation probably generates a cationic centre and this reaction has obvious potential.

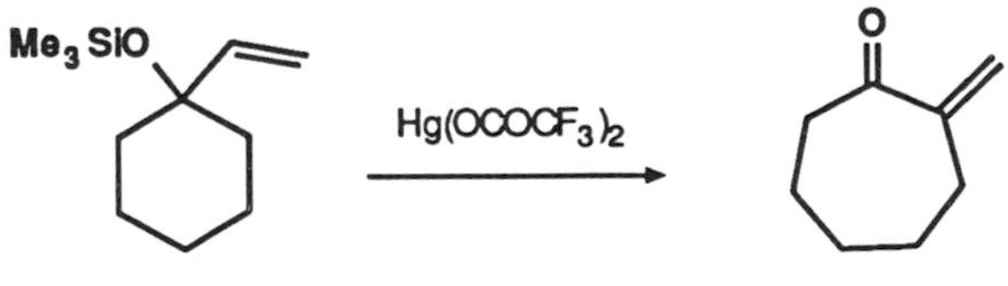

23

(vi) Other reactions

The application of standard procedures for cyclopropanation to cyclic alkenes followed by cleavage of the bridging bond is a useful one-carbon expansion route to C_7 and C_8 rings. This methodology is particularly attractive for the homologation of a cyclic ketone through conversion to a silyl enol ether as shown in Scheme 10 for the synthesis of eucarvone (L.Blanco *et al.*, *Tetrahedron Lett.*, 1981, **22**, 645 and references therein; E. Wenkert *et al.*, *J. Org. Chem.*, 1990, **55**, 1185).

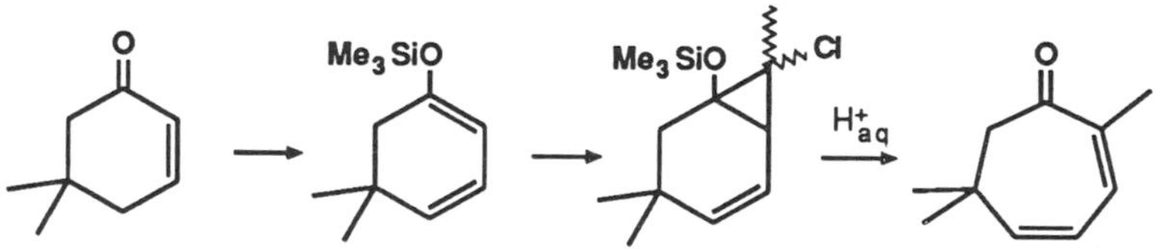

Scheme 10.

Acid or base catalysed ring-opening of the bicycloalkane results in elimination of HCl to form the α,β-unsaturated ketone directly, but the use of $FeCl_3$ in DMF allows the isolation of the intermediate chloroketone. (Y.Ito *et al.*, *Org. Synth.*, 1979, **59**, 113; M.Asaoka, K.Takenouchi and H.Takei, *Chem. Lett.*, 1988, 921). Treatment of the intermediates with sodium acetate gives 2-cycloheptenone (80%) or 2-cyclooctenone (92%).

Cyclopropanation of the disilylether (24) followed by oxidative cleavage with $FeCl_3$ affords cyclohepta-1,3-dione (Y.Ito, S.Fujii and T.Saegusa, *J. Org. Chem.*, 1976 **41**, 2073). Cycloocta-1,3-dione was obtained similarly (M.C.Pirrung and N.G.Webster, *J. Org. Chem.*, 1987, **52**, 3603). An interesting variant is the use of the -$OCH_2CH_2OSiMe_3$ group as the ether substituent. After cyclopropanation of the cycloalkenyl ether with CH_2Br_2 to give (25), the $SiMe_3$ group was removed and a photochemically induced ring expansion led directly to the ketal (26) (60%) (P.G.Gassman and J.Burns, *J. Org. Chem.*, 1988, **53**, 5576). The C_8 analog was obtained similarly (67%). Simple enol ethers where rearrangement to a ketal is not possible give much lower yield of ring expanded product.

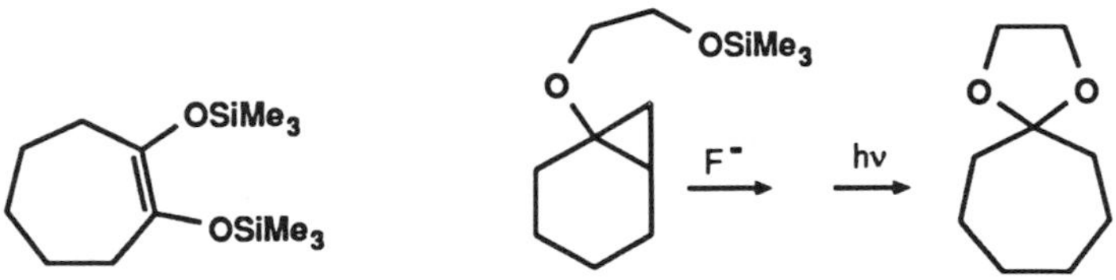

24　　　　**25**　　　　**26**

The protected bicyclo[4.2.0]octane hydroxyaldehyde (27) can undergo cationic rearrangement to a corresponding bicyclo[3.2.1]octane derivative. This species is unstable with respect to a retroaldol reaction and mild oxidation affords the cycloheptane dicarboxylic acid (28). This is a convenient route to a functionalised C_7 ring from a readily accessible cyclohexane derivative (B.C.Ranu, D.C.Sarker and N.K.Basu, *Tetrahedron*, 1989, **45**, 3107).

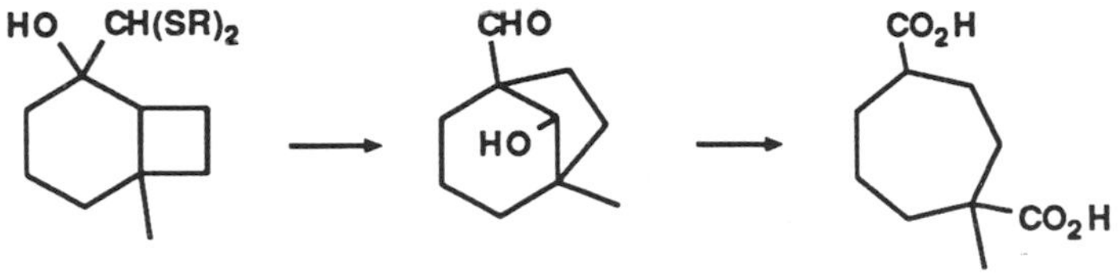

27　　　　　　　　**28**

Macrocyclic lactones (macrolides) are of great importance as antibiotics and the chemistry of lactonisation has been intensively investigated (T.Mukaiyama, *Angew. Chem., Int. Ed. Engl.*, 1979, **18**, 707; K.C.Nicolaou, *Tetrahedron*, 1977, **33**,683; S.Masamune, G.S.Bates and W.J.Corcoran, *Angew. Chem., Int. Ed. Engl.*, 1977, **16**, 585). Hence access to a keto lactone of type (29) is assured and the contraction of this system to a cyclooctene derivative is an attractive route to C_8 rings (M.R.Karim and P.Sampson, *Tetrahedron Lett.*, 1988, **29**, 6897). The transannular aldol reaction is completely specific, that is the ketone enolate reacts with the ester carbonyl, with no evidence for the alternative ester enolate attack on ketone carbonyl. This specificity of pathway is probably the result of stereochemical control arising in the conformational rigidity of the macrocyclic ring.

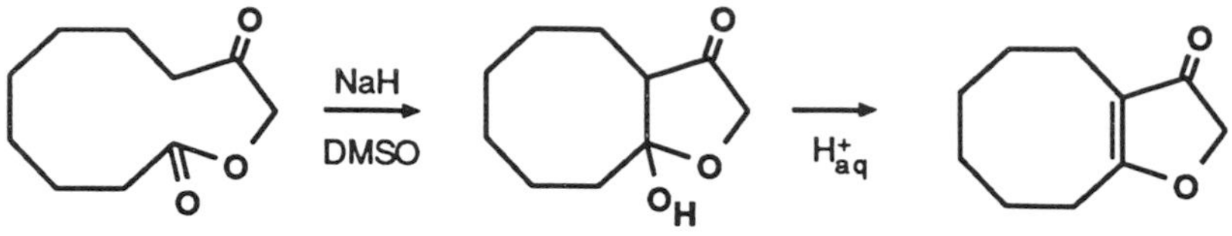

29

(b) Cyclisation Reactions

(i) Intramolecular radical reactions

Intramolecular pinacol coupling of dicarbonyl compounds has been reviewed (J.E.McMurry, *Chem. Rev.*, 1989, **89**, 1513). This reaction proceeds by reductive coupling of two carbonyl groups in presence of $TiCl_3/Zn$-Cu to give a cyclic vicinal diol which is usually deoxygenated directly by the organometallic reagent to a cyclic alkene. Several 1,2-dialkylcyclooctenes are obtained this way. The intermediate 1,2-cyclooctandiol has also been isolated (82%) (J.E.McMurry and J.G.Rico, *Tetrahedron Lett.*, 1989, **30**, 1169). The usual titanium reagents are ineffective for the coupling of oxygenated substrates but potassium graphite (C_8K) reacts with $TiCl_3$ to give a reagent which (in 16 mole excess) promotes the cyclisation of (30) to form a seven membered ring in high yield (D.L.Clive *et al.*, *J. Org. Chem.*, 1991, **56**, 6447).

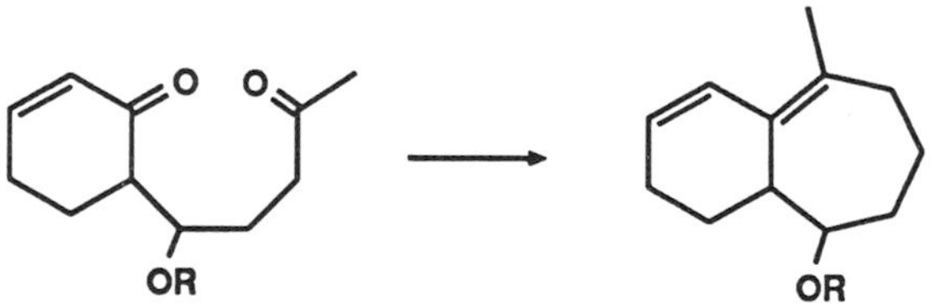

30

A more complex radical cyclisation leading to C_7 and C_8 rings has been discovered by C.M.Thompson and S.Docter, (*Tetrahedron Lett.*, 1988, **29**, 5213). This involves the generation of a radical at the α-site of a β-ketoester (31) followed by attack of the radical at an ω double bond. The yield is low (<20%) due to further oxidation at the reactive α-site, but yields are improved if this site is blocked with Me (30 - 40 %) or Cl (*ca* 50%) (31, X = Me, Cl). Since the chloro group is easily removed by reduction (Zn in acetic acid) an economical route to a cycloheptenone ester is available.

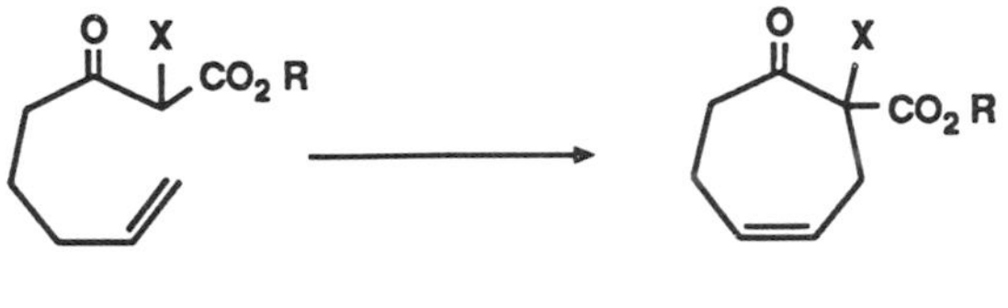

31

(ii) Intramolecular electrophilic cyclisation

The ene cyclisation of ω-unsaturated aldehydes finds application for the synthesis of seven-membered rings only if there is a suitable structural constraint which places the reacting ene and carbonyl groups in juxtaposition. Thus it is only useful for the formation of bicyclic compounds such as the hydroazulenols (Scheme 11) where the C_7 ring can be annulated on to a C_5 ring (J.A.Marshall, N.H.Andersen and P.C.Johnson, *J. Org. Chem.*, 1970, **35**, 186; B.B.Snider and Y.Ke, *Tetrahedron Lett.*, 1989, **30**, 2465; G.Mehta, N.Krishnamurthy and S.R.Karra, *J. Am. Chem. Soc.*, 1991, **113**, 5765) and other constrained systems (Y.Horiguchi, T.Furukawa and I.Kuwajima, *J. Am. Chem. Soc.*, 1989, **111**, 8277).

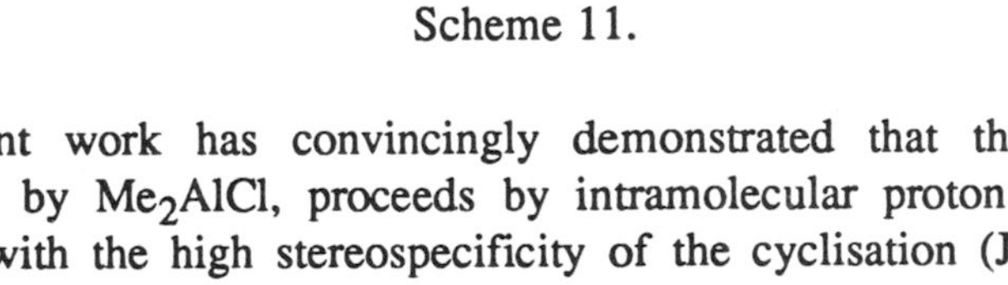

Scheme 11.

Recent work has convincingly demonstrated that this reaction, promoted by Me_2AlCl, proceeds by intramolecular proton transfer in keeping with the high stereospecificity of the cyclisation (J.A.Marshall and M.W.Andersen, *J. Org. Chem.*, 1992, **57**, 5851). An ene reaction has been reported (N.Kato *et al.*, *Chem, Lett.*, 1989, 91) between two vinyl groups resulting in the formation of a C_8 ring as the central ring in a tricyclic compound. Here the constraint imposed by the presence of two C_5 rings as part of the eight-carbon chain, is crucial to successful closure of the chain.

It is interesting that the aldol reaction between a protected aldehyde and the silylether of an enol can be used to achieve a C_8 ring closure, seemingly without need for the usual geometric constraints. Thus compound (32) gives the corresponding cyclooctanone (33) in 67% yield, in a Lewis acid catalysed reaction (G.S.Cockerill, P.Kociencki and R.Treadgold, *J. Chem. Soc., Perkin Trans. 1*, 1985, 2101). Although other Lewis acids were investigated, including $TiCl_4$ for which a template effect has been invoked, $BF_3.Et_2O$ was the most effective. For a compound with geminal dimethyl substitution at another site in the ring the yield is low, suggesting that conformational effects may be important. Further work is require to confirm the efficacy of this reaction but as a cyclisation of a notionally saturated chain this is a reaction of some importance.

The cyclisation of ω-unsaturated allylic silanes can be a complex reaction leading to several bicyclic products. However catalysis by Lewis acids ensures complete specificity and closure to a C_7 ring is the sole pathway (G.Majetich *et al., Tetrahedron Lett.,* 1985, **23**, 2747; G.Majetich, J.Defauw and C.Ringold, *J. Org. Chem.,* 1988, **53**, 50). This is a versatile reaction which has been applied to the formation of many bicyclic conpounds containing a seven or eight membered ring (G.Majetich *et al., J. Org. Chem.,* 1993, **58**, 1030 and references therein). Nucleophilic cyclisation of allyl silanes (promoted by fluoride ion) is less selective and usually gives rise to two or more products.

34

A combination of stabilisation by silicon and favourable conformation effects is probably responsible for the electrophilic cyclisation of the ene epoxide (34) (D.Wang and T.H.Chan, *J. Chem. Soc., Chem. Commun.*, 1984, 1273).

The cyanohydrin of a cyclic ketone will react with diazomethane derivatives, then eliminate N_2 to give a cationic species which rearranges to the ring expanded ketone (Tiffeneau-Demjanov and related reactions). This reaction now has limited application in the synthesis of C_7 and C_8 rings but 3-isopropylcycloheptanone has been obtained in 73% yield (M.Miyashita, S.Hara and A.Yoshikoshi, *J. Org. Chem.*, 1987, **52**, 2602). The trimethylsilyl ether of the cyanohydrin can be used with advantage in the conversion of a benzocycloheptanone to the corresponding benzocyclooctanone (R.W.Theis and E.P.Seitz, *J. Org. Chem.*, 1978, **43**, 1050). For applications to the generation of C_7 and C_8 rings in bi- or polycyclic compounds see P.M.Wovkulich in *Comprehensive Organic Synthesis*, Eds. B.M.Trost and I.Fleming, Pergamon, 1991, Vol 1, Chap. 3.3).

(iii) Intramolecular nucleophilic reactions

Base catalysed intramolecular aldol reaction is especially useful for the annulation of a C_7 ring on to a C_5 ring and hence is applied successfully to the synthesis of pseudoguaianes , Scheme 12 (A.G.Schultz and L.A.Motyka, *J. Am. Chem. Soc.*, 1982, **104**, 5800; J.H.Hutchinson, T.Money and S.E.Piper, *Can. J. Chem.*, 1986, **64**, 1404; P.E.Grieco *et al.*, *J. Am. Chem. Soc.*, 1982, **104**, 4233). Some derivatives of the bicyclo[3.2.1]octane system have been obtained by a similar type of reaction. Cleavage of the one carbon bridge gave a cycloheptane derivative. Thus a cyclopentadione was converted to a cycloheptadione (H.Schick *et al.*, *Leibigs Ann. Chem.*, 1992, 419). An intramolecular aldol condensation, catalysed by chromium chloride has found application to the formation of a C_8 ring in a tricyclic compound. (N.Kato, S.Tanaka and H.Takeshita, *Bull. Chem. Soc. Jpn.*, 1988, **61**, 3231).

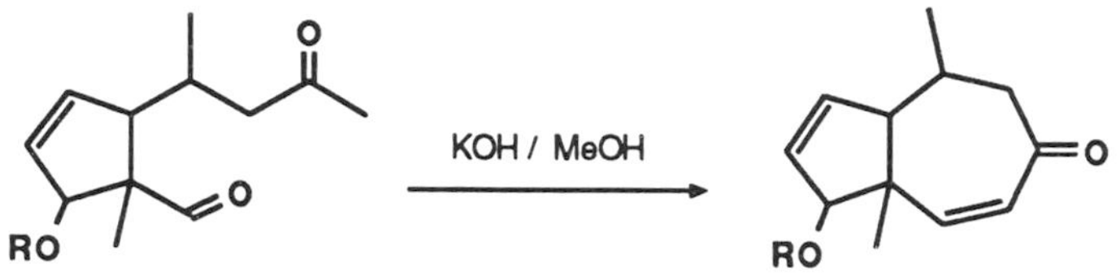

Scheme 12.

Another type of nucleophilic ring closure is shown in Scheme 13. Fluoride ion induced cleavage of the SiMe$_3$ group leads to nucleophilic attack by the β-carbon of the butenyl sidechain at the β-site of the dienone moiety resulting in formation of a divinylcyclobutane ring. Anionic oxy-Cope cyclisation, probably promoted by fluoride ion, then gives the fused cyclooctene in moderate yield (60%) (G.Majetich *et al.*, *J. Am. Chem. Soc.*, 1991, **56**, 3958 and references therein). The geometric constraints imposed by the presence of the C_6 ring are probably crucial to the first cyclisation step.

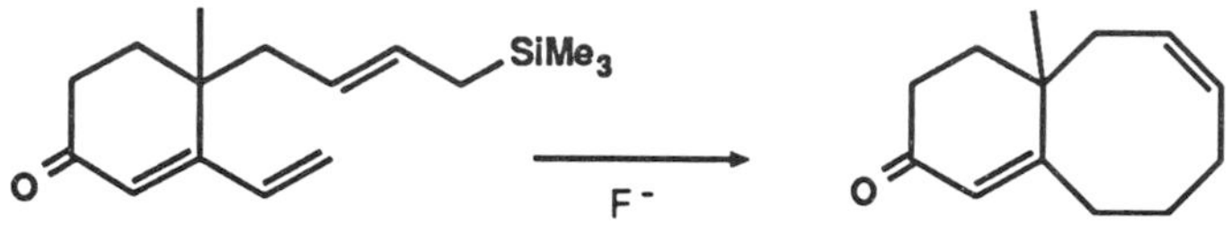

Scheme 13.

Samarium iodide has been found to be an efficient new catalyst for an intramolecular Reformatsky reaction leading to a C_8 ring (Scheme 14). The coupling of an ω-aldehydo-α-bromo ester gives the cyclooctane in 68% yield. Although this reaction is also effective for larger rings the formation of a samarium chelate is especially important for the closure of the eight-membered ring (J.Inanaga *et al.*, *Tetrahedron Lett.*, 1991, **32**, 6371).

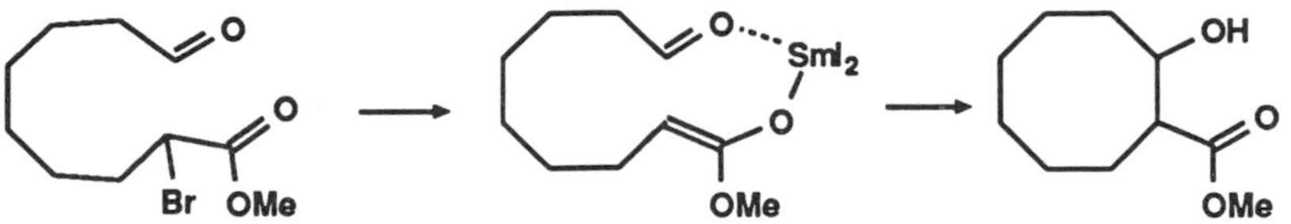

Scheme 14.

(iv) Other types of intramolecular cyclisation

An unusual approach to the formation of bicyclic systems containing a C_8 ring is shown in Scheme 15. The starting point is a pair of pyridone molecules joined by a three or four carbon chain. The intermolecular photodimerisation of pyridones is a well know reaction, the major product being the head-to-tail adduct, a reaction which leads to stereospecific generation of four new sp^3 centres in a cyclooctadiene ring. In the present case the novel use of tethered pyridones results in an intramolecular [4 + 4] photocycloaddition, an entry to the bicyclic carbon skeletons of either the 8-5 or 8-6 type, compounds with obvious capacity for elaboration to a variety of natural product classes (S.M.Sieburth and J.-L.Chen, *J. Am. Chem. Soc.*, 1991, **113**, 8163). The full potential of this methodology has not yet been established but the presence of a substituent in the bridging chain does influence the stereochemistry of the addition reaction (S.M.Sieburth and P.V.Joshi, *J. Org. Chem.*, 1993, **58**, 1661).

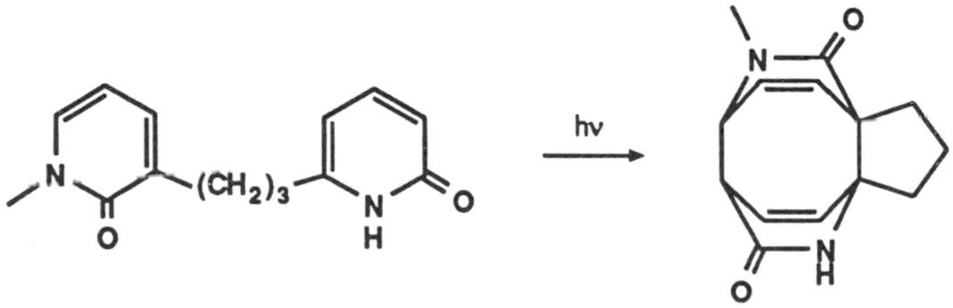

Scheme 15.

Although nickel catalysed dimerisation of butadiene to cyclo-octadiene has been known for a long time, this reaction has been of little general value, due to low yields and a multiplicity of products. However if the double bonds are suitably disposed in one molecule then the intra-

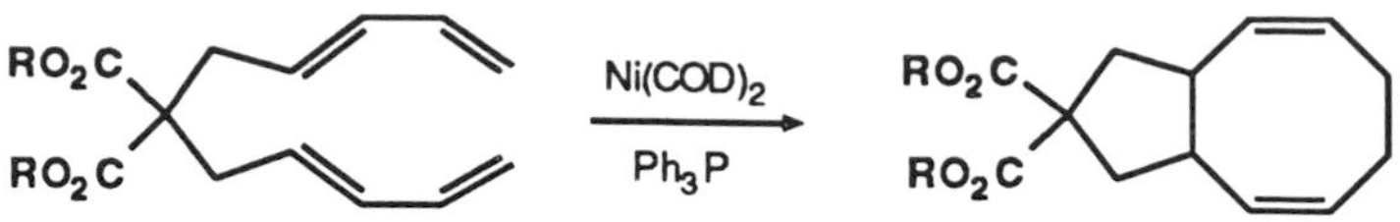

Scheme 16

molecular [4 + 4] cyclisation shown in Scheme 16 becomes the principal reaction, giving the cyclooctadiene product in 70% yield (P.A.Wender and N.C.Ihle, *J. Am. Chem. Soc.*, 1986, **108**, 4678). The catalyst can determine the product type and the stereochemistry at the ring junction but generally dienes with a three atom bridge give *cis*-fused compounds, a four atom bridge leading to *trans*-fused compounds. In the case of polyenes with a substituent in the allylic position the cycloaddition exhibits high stereoselectivity, probably due to face-selective coordination by the catalyst. This methodology has been applied successfully to the synthesis of (+)-asteriscanolide, a sesquiterpene lactone (P.A.Wender and N.C.Ihle and C.R.D.Correia, *J. Am. Chem. Soc.*, 1988, **110**, 5904). A further extension of this procedure involves moving the bridge to the β-carbon of one of the diene moieties. The [4 + 4] cyclisation of such a tetraene with Ni(COD)₂/tri-*o*-biphenyl phosphite in toluene at 85 °C, forms a cyclooctadiene ring as part of a bicyclo[5.3.1]undecane system (P.A.Wender and M.J.Tebbe, *Synthesis*, 1991, 1089).

Some recent work has illustrated the use of intramolecular 1,3-dipolar cycloaddition (Scheme 17). The highly functionalised oxime is oxidised to a nitrile oxide which rearranges to an oxazoline and thereby creates a seven membered ring (O.Duclos, A.Duréault and J.C.Depezay, *Tetrahedron Lett.*, 1992, **33**, 1059).

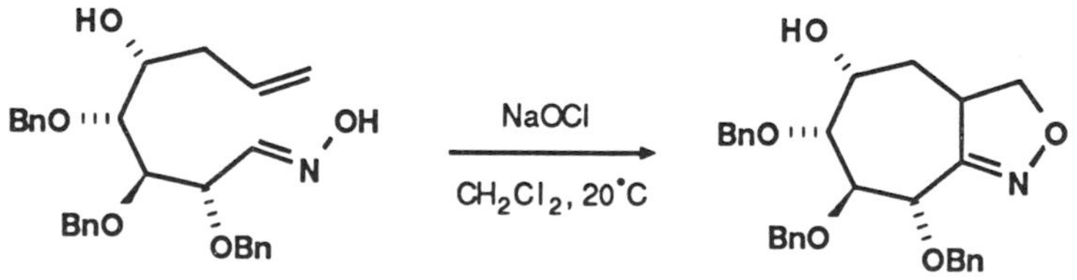

Scheme 17

(v) Intermolecular reactions

A double nucleophilic substitution which may have interesting potential is shown in Scheme 18. This reaction of a diketone with a suitable 1,3-dianionic substrate generates the C_7 or C_8 ring with an oxygen bridge in excellent yield (80 - 90%) (G.A.Molander and D.C.Schubert, *J. Am. Chem. Soc.*, 1987, **109**, 6877; see also S.D.Lee and T.H.Chan, *Tetrahedron*, 1984, **40**, 3611). The regiochemisty of this cyclisation has been explored using the bis(trimethylsilyl) enolate of methyl acetoacetate as the dianion synthon, with $Me_3SiOSO_2CF_3$ as catalyst (G.A.Molander and K.O.Cameron, *J. Org. Chem.*, 1991, **56**, 2617; *J. Org. Chem.*, 1993, **58**, 830). Initial attack always occurs at the most hindered carbonyl site of the diketone, presumably due to specific activation by the catalyst. Using a variety of dielectrophiles and dinucleophiles these workers have shown that diastereoselectivity is excellent in this type of annulation and hence controlled generation of multiple stereogenic centres is possible.

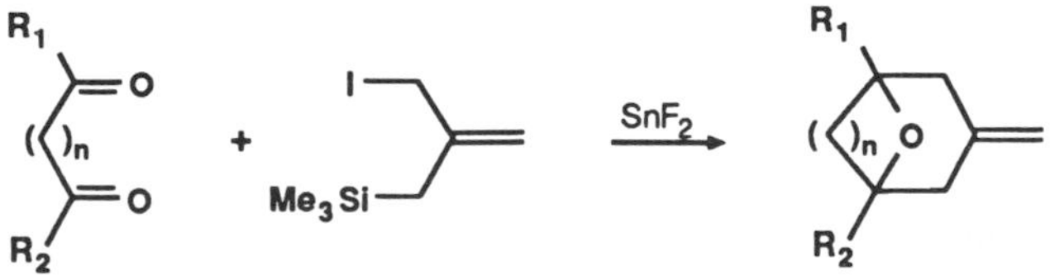

Scheme 18

The allyl bis(stannane), $(Me_3SnCH_2)_2C=CH_2$, functions as a dianion equivalent and reacts with diacid chlorides, usually under catalysis by Pd(0), to afford access to substituted cycloheptadiones (A.Degl'Innocenti *et al.*, *Synthesis*, 1991, 267).

An intriguing method for the construction of a C_7 ring has been discovered recently by J.W.Herndon and coworkers (*J. Am. Chem. Soc.*, 1991, **113**, 7808). Reaction of a cyclopropylcarbene complex of tungsten (35) with an alkyne results in cycloaddition with incorporation of a carbonyl group, a [4 + 2 + 1] cyclisation. The initial cycloheptadienone isomerises when heated, but this isomer (36) is obtained directly in 55% yield if the reaction is carried out in refluxing toluene, in presence of

1,2-di(Ph$_2$P)C$_6$H$_4$. Regioselectivity is observed in the addition step and in the opening of the cyclopropane ring.

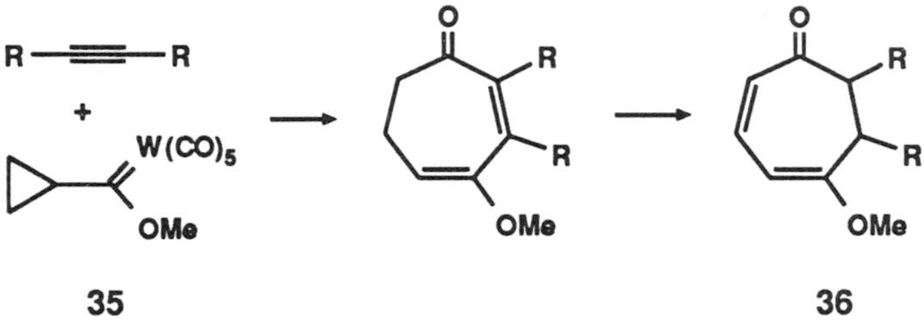

35   **36**

Addition of an allyl cation to any simple 1,3-diene can lead to a seven membered ring by [4 + 3] concerted or stepwise mechanisms, Scheme 19. The stability of the cation is important and allylic cations with an oxygen substituent on the central carbon have been used commonly, especially since these are now readily available due to the discovery of new synthetic methods (A.Hosomi and Y.Tominaga, *Comprehensive Organic Synthesis*, B.M.Trost and I.Fleming, Eds, Pergamon, 1991, Vol 5, Chap. 5.1).

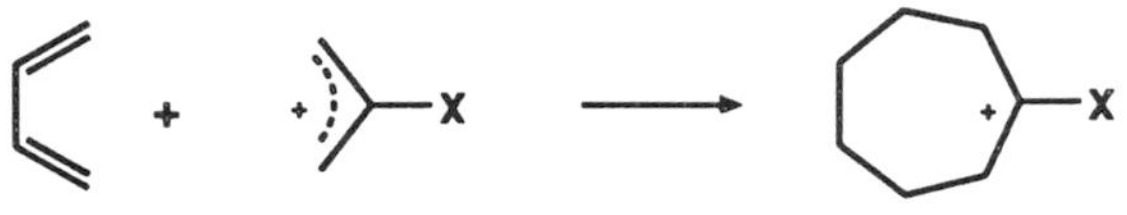

Scheme 19.

The nucleophilicity of the allylic substituent is important and the mechanistic implications of the nature of this substituent have been reviewed in detail (H.M.R.Hoffmann, *Angew. Chem., Int. Ed. Engl.,* 1984, **23**, 1). An early application of interest was the cycloaddition of a 1,3-diene to an α,α'-dibromoketone to give a 4-cycloheptenone in fair yield (T.Takana *et al., J. Am. Chem. Soc.,* 1974, 100, 1765 and references therein). One interesting recent application is the formation of the monoterpene karaharaenone (37) and its isomer (38) in 71% overall yield (H.Sakurai, A.Shirahata and A.Hosomi, *Angew. Chem., Int. Ed. Engl.,* 1979, **18**, 163).

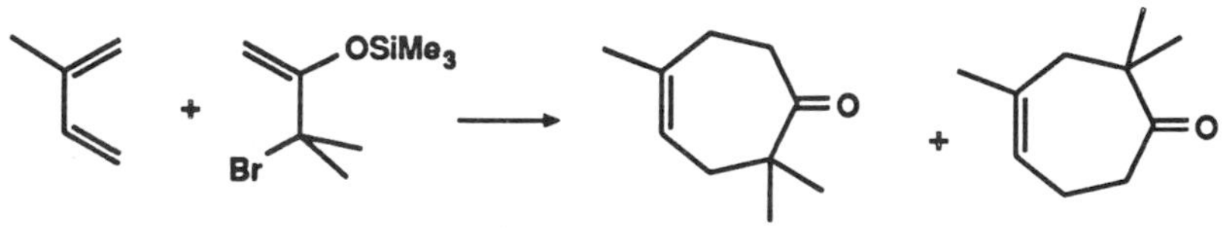

37   38

Most cycloadditions of this type have involved cyclic dienes leading to bi- and polycyclic systems (J.Mann, *Tetrahedron*, 1984, **47**, 4611; R.Noyori and Y.Hayakawa, *Org. React.*, 1983, **29**, 163; M.Harmata, V.R.Fletcher and R.J.Claassen, *J. Am. Chem. Soc.*, 1991, **113**, 9861). The use of furan as the 4π addend leads to oxa-bridged cycloheptane derivatives and recent improvements to this cycloaddition have been reported (J.Mann and L.C. de A.Barbosa, *J. Chem. Soc., Perkin Trans. 1,* 1992, 787).

An interesting ene cyclisation (Scheme 20) has been investigated as a [4 + 3] route to annulated cycloheptenes. The dialkylidenecyclopentane precursors (X = electron acceptor group) are obtained by cyclisation of an enyne under catalysis by Pd(OAc)$_2$. This catalyst is converted *in situ* to Pd(0) by appropriate reduction and in this form it promotes the second cyclisation step. Both steps proceed in about 70 - 80% yield to give the octahydroazulene derivatives, although a small amount of a five membered ring product can be formed in the second step (B.M.Trost and D.T.MacPherson, *J. Am. Chem. Soc.*, 1987, **109**, 3483).

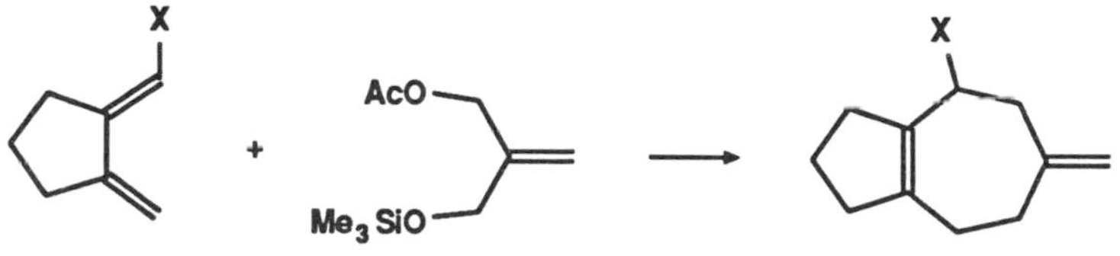

Scheme 20.

(c) Functional group modification

Since seven and eight membered rings are around the limit of ring size which will allow the formation of stable compounds containing acetylenic and other types of unsaturation, there is a continuing interest in methods for the introduction of multiple bonds into C_7 and C_8 rings.

Dehydrohalogenation of *trans*-3-methoxy-1-bromocyclooctene with potassium *t*-butoxide can give the corresponding cyclooctyne in moderate yield if exposure to the base is restricted to 5 - 10 seconds (C.B.Reese and A.Shaw, *J. Chem. Soc., Perkin Trans. 1*, 1976, 596). This short reaction time prevents base catalysed dimerisation of the product. An alternative method of generating the triple bond is the thermal decomposition of a selenadiazole, obtained from the cyclic ketone (Scheme 21).

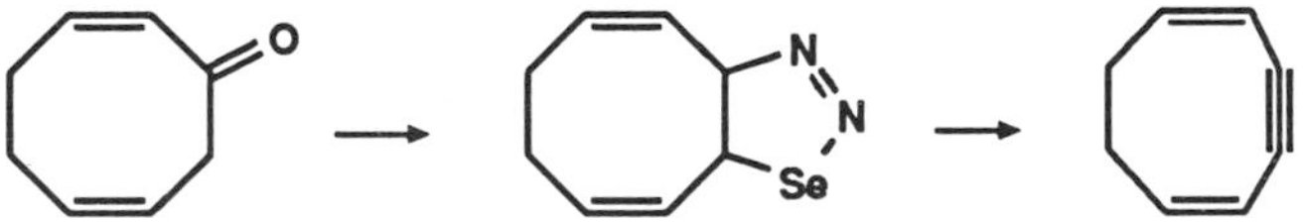

Scheme 21.

This reaction was developed for acylic compounds but is very effective for strained cyclic systems. Thus H.Petersen and H.Meier (*Chem. Ber.*, 1980, **113**, 2383) obtained 3-, 4- and 5-octenynes either by thermolysis or by treatment of the selenadiazole with BuLi at −78 °C. Substituted octynes have been obtained similarly (H.Meier and H.Petersen, *Synthesis*, 1978, 596) as have the more highly strained compounds 1,3-cyclooctadien-6-yne and 1,5-cyclooctadien-3-yne (T.Ecter and H.Meier, *Chem. Ber.*, 1985, **118**, 182). Cycloocta-1,5-diyne is formed by dimerisation of butatriene (E.Kloster-Jensen and J.Wirz, *Helv. Chim. Acta*, 1975, **58**, 162).

1,2-Cyclooctadiene (37) is highly strained and can only be isolated as a dimer. R.P.Johnson has reviewed work on this and other related cumulenes (*Chem. Rev.*, 1989, **89**, 1111). Dimerisation is prevented by a bulky group at the 1-position and the 1-*t*-butyl derivative has been prepared from the corresponding cycloheptene by cyclopropanation with CH_2Br_2 then dehydrohalogenation with MeLi. The allene moiety has a 22° bend in this strained molecule. (J.D.Price,and R.P.Johnson, *Tetrahedron Lett.*, 1986, **27**, 4679). However it is sufficiently stable to be purified by GC and some photochemical studies are reported (J.D.Price,and R.P.Johnson, *J. Org. Chem.*, 1991, **56**, 6372). 1-Methyl-1,2-cyclooctadiene is too unstable to be isolated and has been characterised by its dimerisation products (J.Pietruszka *et al.*, *Chem. Ber.*, 1993, **126**, 159).

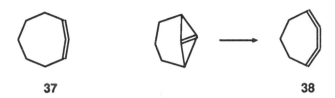

37   38

The very highly strained compound 1,2,3-cycloheptatriene (38) has been obtained by rearrangement of an even more strained tricyclic compound (H.G.Koch *et al.*, *Chem. Ber.*, 1983, **116**, 2285). It can only be detected by formation of an adduct with various 1,3-dienes.

A convenient method for the formation of a cycloheptadiene carboxylate from cycloheptanone has been reported using standard procedures (J.L.Pawlak and G.A.Berchtold, *J. Org. Chem.*, 1988, **53**, 4063). Electroreduction of 1,3,5-cycloheptatriene gives an anionic species which in presence of an alkyl halide gives the corresponding 6-alkyl-1,3-cycloheptadiene. The regiospecificity is reduced by a 1-methoxy substituent and two isomeric products are obtained (T.Shono *et al.*, *Tetrahedron Lett.*, 1991, **32**, 1051). A series of papers has dealt with the preparation of polyfluorinated cycloheptanes (P.L.Coe, A.W.Mott and J.C.Tatlow, *J. Fluorine Chem.*, 1985, **30**, 297 and references therein).

3. Properties and reactions

(a) Conformation and structure

The conformation of seven and eight membered carbocyclic rings has been reviewed thoroughly by T.A.Crabb and A.V.Patel, (in *Second Supplement to the 2nd Edition of Rodd's Chemistry of Carbon Compounds*, Ed. M.Sainsbury, Elsevier, Amsterdam, 1992, Volume IIA/B, Ch 1) and discussion here is limited to more recent work.

A comprehensive conformational analysis of saturated carbocyclic rings by molecular mechanics methods includes cycloheptane and cyclooctane (I.Kolossváry and W.C.Guida, *J. Am. Chem. Soc.*, 1993, **115**, 2107). This study evaluates the energy of all saddle points as well as minima in an exploration of the potential energy hypersurface of both molecules. In this way a full understanding of all conformational interconversion pathways can be developed for these two cyclic systems.

The strain resulting from the incorporation of one acetylenic bond and variable numbers of double bonds in an eight membered ring has been examined by molecular orbital methods (H.Meier, H.Petersen and H.Kolshorn, *Chem. Ber.*, 1980, **113**, 2398; H.Kolshorn and H.Meier, *Chem. Ber.*, 1985, **118**, 176). The calculated heats of formation and strain energies for octenynes show that introduction of a double bond reduced the conformational mobility and there is about 100 kJ mol^{-1} increase in strain energy between cyclooctyne and the fully unsaturated cyclooctatrienyne. For a series of octynes with varying degrees of additional ring strain (simple double bonds or annulated benzene rings) the stretching frequency $v(C \equiv C)$ and the ^{13}C chemical shift of the acetylenic carbon are linearly related to the bond angle at the triple bond. UV spectra and reactivity have also been discussed (H.Meier, H.Petersen and H.Kolshorn, *Chem. Ber.*, 1980, **113**, 2398).

Cyclooctα-1,5-diyne is planar but each *sp* centre has a 21° bend (E.Kloster-Jensen and J.Wirz, *Helv. Chim. Acta*, 1975, **58**, 162). The effect of angle strain has been evaluated by photoelectron spectroscopy (G.Bieri *et al.*, *Helv. Chim. Acta*, 1974, **57**,1265). For a general survey of angle strain in cycloalkynes see A.Krebs and J.Wilke, *Top. Curr. Chem.*, 1983, **109**, 189, and for a review of the chemistry of strained cycloalkenynes see H.Meier *et al.*, *Tetrahedron*, 1986, **42**, 1711).

Some useful ^{13}C NMR data have been accumulated for cyclooctane derivatives by K.G.Penman, W.Kitching and A.P.Wells, *J. Chem. Soc., Perkin Trans. 1*, 1991, 721 and G.Read and J.Shaw, *J. Chem. Soc., Perkin Trans. 1*, 1988, 2287.

(b) Reactions

(i) General chemistry

Much of the chemistry of cycloheptane and cyclooctane derivatives involves unexceptional functional group modification and such material is not discussed here. Recently functionalisation of cyclooctanone has been developed using simple bromination and oxidation reactions to afford dialkoxycyclooctatetraenes (L.A.Paquette, T.J.Watson and D.Friedrich, *J.*

Org. Chem., 1993, **58**, 776). Stereochemical control of functionalisation of the cycloheptane ring is often difficult and this aspect has been studied by A.J.Pearson and H.S.Bansal (*Tetrahedron Lett.*, 1986, **27**, 287). The isomerisation of cycloheptatriene to norcaradiene has long been of interest, at least from the theoretical point of view. This skeletal rearrangement can be studied directly since a convenient route to a homo-*o*-benzoquinone has been reported. In this case rearrangement gives an α-tropolone (M.G.Banwell and M.P.Curtis, *J. Chem. Soc., Chem. Commun.*, 1991, 1343). Studies of the nucleophilic substitution of tropone continue, including the reaction with tropylium ion to form a bicycloheptane derivative (H.Miyano and M.Nitta, *J. Chem. Soc., Perkin Trans. 1*, 1990, 839, 999).

(ii) Cycloadditions

The pericyclic reactions of nπ systems (n = 2, 4 or 6) are of key importance to the elaboration of monocyclic and polycyclic compounds and the appropriate unsaturated cycloheptanes have been widely studied in this respect. Of particular interest are [4 + 4] and [6 + 4] cycloadditions and tropone has been investigated extensively as a triene or diene addend. This area has been reviewed recently (J.H.Rigby, *Comprehensive Organic Synthesis*, B.M.Trost and I.Fleming, Eds, Pergamon, 1991, Vol 5) and only significant synthetic developments are mentioned here.

Cycloaddition reactions of cycloheptatriene suffer from lack of periselectivity since this substrate can act as a 2π, 4π or 6π addend and a mutiplicity of products is usually obtained. However the potential synthetic utility of [6 + n] cyclisation is so high that several groups workers have studied this reaction in detail in recent years.

Cycloheptatriene will dimerise in presence of a titanium complex to give two pentacyclic compounds. However with the Ziegler catalyst, $TiCl_4$-Et_2AlCl, dimerisation is minimal and [6 + 2] cycloaddition occurs with simple dienes or alkynes, to give several polycyclic species but mainly derivatives of bicyclo[4.2.1]nonadiene, Scheme 22 (K.Mach *et al.*, *Tetrahedron*, 1984, **40**, 3295). According to symmetry rules this cycloaddition is a forbidden concerted reaction but coordination of the triene to the catalyst probably alters the molecular symmetry. The photochemically promoted addition of cycloheptatriene to napthoquinone has been studied by A.Mori and H.Takeshita, *Chem. Lett.*, 1975, 599.

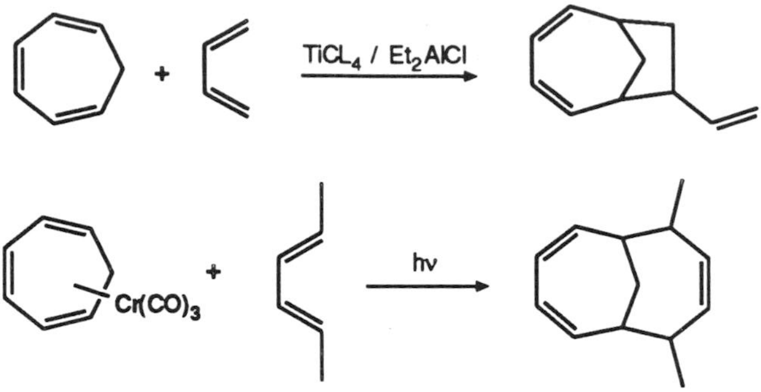

Scheme 22.

Following from some early work of Kreiter and his coworkers (*Adv. Organomet. Chem.*, 1986, **26**, 297) a recent study by J.H.Rigby *et al.* (*J. Am. Chem. Soc.*, 1993, **115**, 1382) has shown that the photocycloaddition of dienes to cycloheptatriene is efficiently mediated by chromium(0). Thus the $Cr(CO)_3$ complex of cycloheptatriene undergoes [6 + 4] addition to give derivatives of the bicyclo[4.4.1]undecatriene system (Scheme 22). The yield is in the range 65 - 95% and shows little dependence on the nature of substituents in the diene addend. Likewise the regioselectivity is poor, indicating negligible effect of substituents in the triene addend. Tropone gives similar adducts but the yield is generally lower. In all cases the corresponding molybenum complexes were less effective and no reaction occurred with tungsten complexes.

Thermally promoted cycloaddition of tropone to a wide variety of dienes has provided much insight into mechanistic aspects of this class of reaction. The normal product is the result of [6 + 4] cycloaddition with *endo* stereochemistry, similar to a Diels Alder reaction, but with a reaction temperature above 100 °C, the *endo* [4 + 2] adduct is formed (M.E.Garst, V.A.Roberts and C.Prussin, *Tetrahedron*, 1983, **39**, 581). In contrast to the transition metal mediated reaction discussed above the thermal cycloaddition has high sensitivity to substituents in both addends and the peri-, regio- and stereoselectivity can be affected ((M.E.Garst *et al., J. Am. Chem. Soc.,* 1984, **106**, 3882).

The photocyclisation of pentenyltropones is a key step in the synthesis of (±)-dactylol and this reaction has received extensive exploration as a route to the bicyclo[6.3.0]undecane system (K.S.Feldman, M.-J. Wu and D.P.Rotella, *J. Am. Chem. Soc.*, 1990, **112**, 8490 and references therein). The stereochemistry of the reaction of iodocarbene addition to cyclooctadiene and cyclooctatetraene has been examined in detail (E.V.Dehmlow *et al.*, *Chem. Ber.*, 1993, **126**, 499).

(iii) Metal Complexes

A very large number of complexes have been prepared, containing C_7 or C_8 rings as ligands of the ene, diene, triene or tetraene type. In the main these cyclic ligands have very similar properties to acyclic olefin ligands. The organometallic chemistry of simple olefin complexes is well understood and mainly centres round the variability of the hapticity of the bonding in polyene complexes. A very useful review of this area had been published (G.Deganello, *Transition Metal Complexes of Cyclic Polyolefins*, Academic Press, New York, 1979) and numerous references to synthesis and reactions of complexes involving a C_7 or C_8 polyene are found in *Comprehensive Organometallic Chemistry*, G.Wilkinson, F.G.A.Stone and A.W.Abel, Eds, Pergamon, Oxford, 1982.

Cyclooctyne is not very stable and trimerises in presence of many transition metals (for a review see M.A.Bennett and H.P.Schwemlein, *Angew. Chem., Int. Ed. Engl.*, 1989, **28**, 1296). If the metal carries carbonyl ligands these may be co-oligomerised to form ketonic compounds. Some stable complexes can be formed with Mo and W, and complexes of cycloheptyne are formed with platinum. Treatment of a mixture of isomers of bromocycloheptatriene with t-BuOK in presence of $Pt(PPh_3)_3$ gives the platinum complex of the cyclic allene (39). Use of a stronger base (LDA) gives the analogous dienyne complex (40) (Z.Lu, K.A.Abboud and W.M.Jones, *Organometallics*, 1993, **12**, 1471). Hydride

39   **40**   **41**

abstaction from (40) leads to the complex of a delocalised cation (41), this ion being the tropylium analogue of benzyne (Z.Lu, K.A.Abboud and W.M.Jones, *J. Am. Chem. Soc.*, 1992, **114**, 10991).

Iron complexes of 1,2-cycloheptadiene have a nonplanar conformation for the allene fragment with the metal coordinated to one double bond (42, $Fe^* = FeCp(CO)_2$) and as a result the compound shows normal fluxional behaviour. In contrast the analogous iron complex of 1,2,4,6-cycloheptatetraene (43) adopts a planar conformation due to the reduction in strain and introduction of cyclic resonance (F.J.Manganiello *et al.*, *Organometallics*, 1985, **4**, 1069; S.M.Oon and W.M.Jones, *Organometallics*, 1988, **7**, 2172).

42   **43**

The complexation of a metal to cycloheptene or 1,3-cycloheptadiene imposes a stereochemical constraint which can be of great value in functionalisation of the ring. There are several early reports of substitution reactions of unsaturated C_7 and C_8 rings when ligated to various metals but the most thorough investigation has been carried out by Pearson and his coworkers. This group has shown that hydride abstraction from the iron complex of 1,3-cycloheptadiene (44, $Fe^* = Fe(CO)_3$ or $Fe(CO)_2(PR_3)$) gives a cationic complex (45) which reacts stereospecifically with a range of carbon nucleophiles, the substituent being introduced *trans* to the metal. The product is a dienyl complex and the overall reaction is effectively a nucleophilic substitution. Before oxidative demetallation is carried out the monosubstituted cycloheptadiene complex can be taken through a second cycle of the reaction sequence to afford a disubstituted cycloheptadiene complex (46), with complete stereo- and regiospecificity (A.J.Pearson, S.L.Kole and J.Yoon, *Organometallics*, 1986, **5**, 2075). This methodology has been extended to substitution reactions of methyl 1,3-cycloheptadiene carboxylate (A.J.Pearson and M.P.Burello, *Organometallics*, 1992, **11**, 448) and to trimethylsilyl and trimethylstannyl derivatives of 1,3-

cycloheptadiene (A.J.Pearson abd M.S.Holden, *J. Organomet. Chem.*, 1990, **383**, 307).

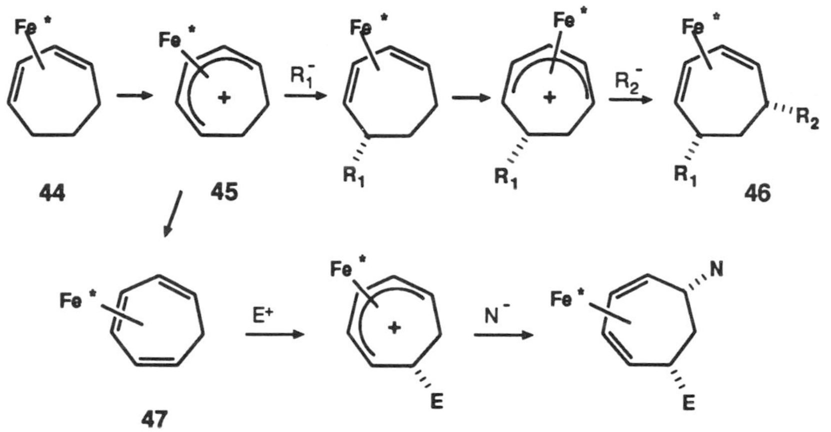

Scheme 23.

Extension of this methodology (iron complex as a template) to electrophilic substitution chemistry is easily achieved (Scheme 23). Proton abstraction from the cation complex (45) gives a neutral cycloheptatriene complex (47) which is an excellent substrate for Friedel/Craft and related reactions. These two types of substitution (nucleophilic and electrophilic) can be combined in tandem fashion to give access to a wide range of stereospecifically di- and trisubstituted cycloheptane systems (A.J.Pearson and K.Srinivasan, *Synlett,* 1992, 983). Further examples of stereocontrolled substitution, using an iron complex as template, are reported for cycloheptadienone and cycloheptatrienone (tropone) as substrates (A.J.Pearson and K.Chang, *J. Org. Chem.*, 1993, **58** 1228; A.J.Pearson and K.Srinivasan, *J. Org. Chem.*, 1992, **57**, 3965). This chemistry has not yet been fully developed but it is clear that it can ultimately lead to acyclic compounds with several contiguous chiral centres, each with an unambiguous configuration, and may offer access to novel alditols and similar compounds, Scheme 24.

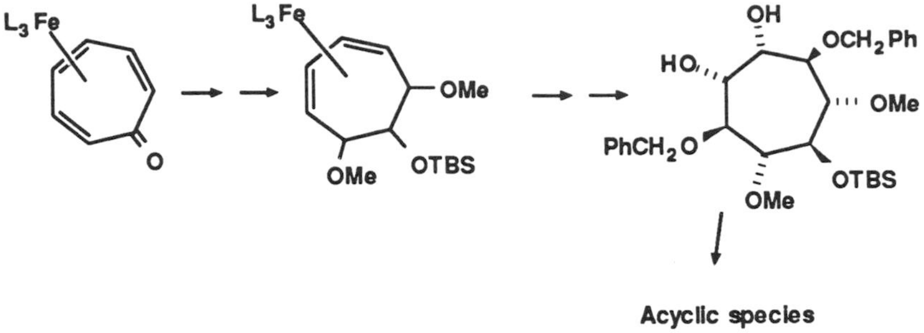

Acyclic species

Scheme 24.

The complex η^3-cycloheptenyl-Mo(CO)$_2$Cp readily undergoes hydride abstaction to afford an η^4-diene complex, and nucleophilic addition can be effected with a stereospecificity similar to that observed for the analogous iron complex. A second cycle of these reactions gives more complex results since the regiocontrol is less effective and a mixture of dialkylated products is formed in most cases (A.J.Pearson and N.I.Khan, *J. Org. Chem.*, 1985, **50**, 5276). These reactions are less useful than might be assumed since demetallation requires rather more complex chemistry than is the case for the corresponding iron complexes. Some nucleophilic addition reactions of similar manganese complexes are reported by E.D.Honig and D.A.Sweigart (*J. Organomet. Chem.*, 1986, **308**, 229; *Organometallics*, 1985, **4**,871)

The reaction of cycloheptatriene with the complex Fe(CO)$_3$(methyl acrylate) followed by photochemical rearrangement gives compound (48) (R.Goddard, F.-W.Grevels and R.Schrader, *Angew. Chem., Int. Ed. Engl.*, 1985, **24**, 353), a limited but valuable method for the functionalisation of the C$_7$ ring.

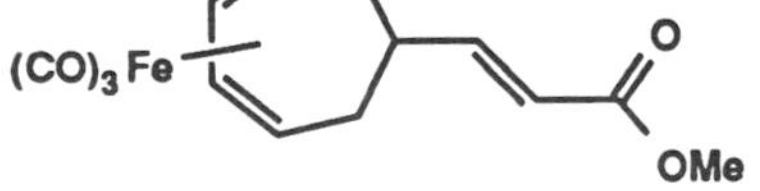

48

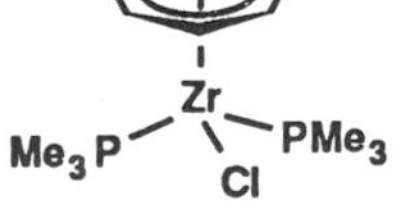

49

Complexes of cycloheptatriene are of particular interest since hydride abstraction generates the cycloheptatrienyl cation (tropylium ion), a 6π-electron ligand. This ligand bonds in a symmetrical η^7 mode with a cone angle of 154° (*cf* 110° for the cyclopentadienyl ligand) and many tropylium complexes of the half-sandwich type (49) are known particularly, for the early transition metals. From available X-ray data for molybdenum complexes, the metal is 1.61Å from the ring centroid compared to 2.01Å in analogous cyclopentadienyl complexes (A.G.Orpen *et al.*, *J. Chem. Soc., Dalton Trans.*, 1989, S1)

A sandwich complex of the type $(\eta^6\text{-}C_7H_8)M(\eta^6\text{-}C_7H_8)$ can be prepared directly by reacting $ZrCl_4$ with cycloheptatriene. This complex is stable for M = Zr but the isomeric form $(\eta^7\text{-}C_7H_7)M(\eta^5\text{-}C_7H_9)$ in which the tropylium ion is generated by intramolecular hydride transfer, is the preferred structure for M = Ti, Zr, Hf and Mo (J.C.Green, M.L.H.Green and N.M.Walker, *J. Chem. Soc., Dalton Trans.*, 1991, 173 and references to early work), and for M = Nb (H.O.van Oven, C.J.Groenenboom and H.J.De Liefde Meijer, *J. Organomet. Chem.*, 1975, **81**, 379). This (trienyl)M(dienyl) type of complex is an excellent precursor to tropylium half sandwich compounds of molybdenum (M.L.H.Green, D.K.P.Ng and R.C.Tovey, *J. Chem. Soc., Chem. Commun.*, 1992, 918; M.L.H.Green and D.K.P.Ng, *J. Chem. Soc., Dalton Trans.*, 1993, 11). Other recent work includes tropylium half sandwich complexes of zirconium and hafnium (G.M.Diamond *et al.*, *J. Chem. Soc., Dalton Trans.*, 1992, 417), niobium (M.L.H.Green *et al.*, *Polyhedron*, 1991, **10**, 389; M.L.H.Green *et al.*, *J. Chem. Soc., Dalton Trans.*, 1992, 1591) and tungsten (M.L.H.Green *et al.*, *J. Chem. Soc., Dalton Trans.*, 1988, 2851).

The effect of substituents in a tropylium ligand is likely to be significant but virtually no work has yet been done with such derivatives of cycloheptatriene apart from the recent study of complexes with the $\eta^7\text{-}C_7H_3Me_4$ ligand (M.L.H.Green and D.K.P.Ng, *J. Chem. Soc., Dalton Trans.*, 1993, 17).

The exploration of the nucleophilic and electrophilic substitution reactions of iron complexes of 1,3-cyclooctadiene and 1,3,5-cyclo-octatriene has occurred sporadically in recent years but a comprehensive study is underway (A.J.Pearson and K.Srinivasan, *Tetrahedron Lett.*, 1992, **33**, 7295 and references therein to early work). It is evident that this is a versatile route to a range of cyclooctyl derivatives with well-defined stereochemistry at several contiguous centres. Functionalisation of 1,5-cyclooctadiene *via* a molybdenum allyl complex has been studied recently and this is also a useful entry to cyclooctyl derivatives with multiple stereospecific substitution (A.J.Pearson *et al.*, *J. Org. Chem.*, 1992, **57**, 2910).

Deprotonation of complexes of the type $(\eta^4\text{-}C_7H_8)FeL_3$ (Scheme 25) occurs much more readily than is the case for the free ligand, cycloheptatriene. This corresponds to an effective acidity enhancement of about 17 pK units, the stabilisation being achieved through η^3-bonding to the metal atom (R.C.Kerber, *J. Organomet. Chem.*, 1983, **254**, 131). Electophilic substitution of the anionic complex usually takes place at the ring, providing another route to substituted cycloheptatrienes (see J.G.A.Reuvers and J.Takats, *Organometallics*, 1990, **9**, 578 and references therein).

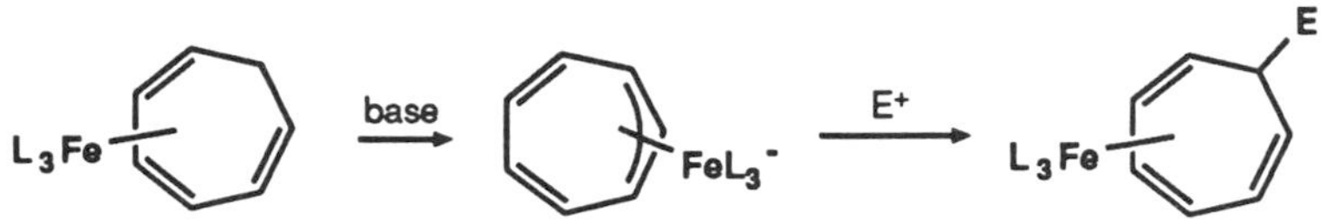

Scheme 25.

4. Biochemistry

Many hundreds of naturally occurring terpenoid compounds which contain a seven or eight membered ring have been isolated and characterised. These materials are products of plants, fungi or marine organisms and vary extensively in the structure of the polycycle (usually two or three fused rings) and the degree of oxygenation, (hydroxyl, keto, ester or lactone groups are found, involving up to six or more oxygen atoms).

Of the many terpenoid systems containing C_7 and C_8 rings, only some representative types are indicated here. For a larger list of structures containing a cyclooctane ring see N.A.Petasis and M.A.Patane, *Tetrahedron*, 1992, **48**, 5757; J.S.Glasby, *Encyclopaedia of Terpenoids*, Wiley-Interscience, New York, 1982. The diversity of terpenoid systems is amazing and not only are new substituent variations of known systems being discovered but quite new systems are reported frequently, even in the small subclass made up of those containing a seven or eight membered ring.

The three systems, carotane (50), guaiane (51) and pseudoguaiane (52) constitute a large class of 5-7 sesquiterpenes, occurring widely in plants with a very wide range of biochemical effects. New members of this class are reported frequently in the phytochemical literature.

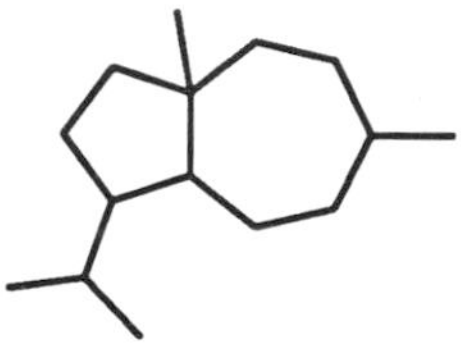

50 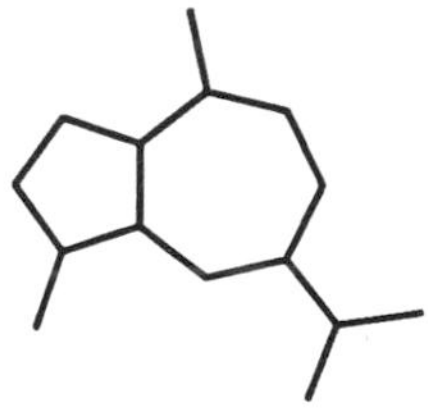   51 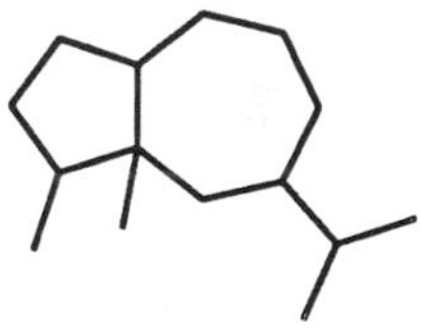   52

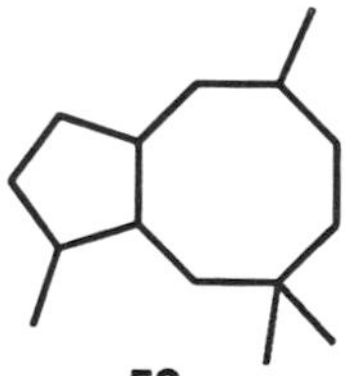

53 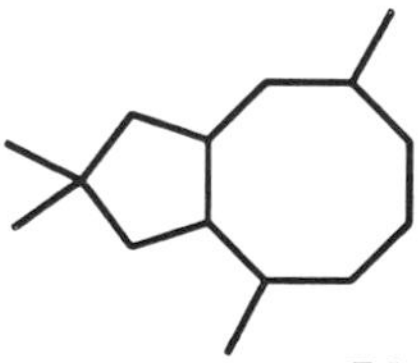   54 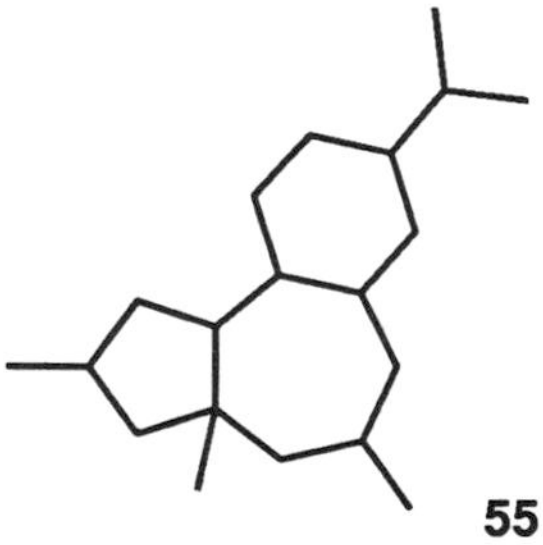   55

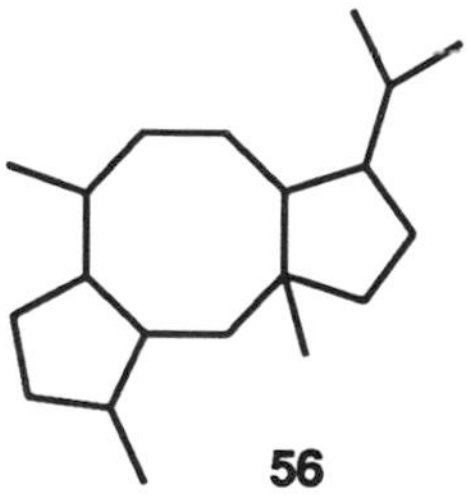

56 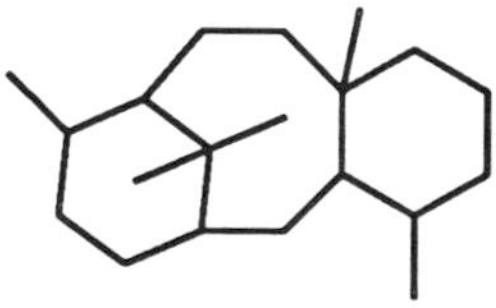   57

Precapnellane (53) is an unusual 5-8 ring combination. A diene of this class has been isolated from soft coral (E.Ayanoglu et al., Tetrahedron, 1979, 35, 1035) and other repesentatives of this family include dactylol and poitediol (G.Mehta and A.N.Murthy, J. Org. Chem.,

1987, **52**, 2875). Astericane (54) is a recently discovered plant sesquiterpene of the 5-8 type.

Daphnane (55) and the closely related tigliane system includes natural phorbol esters, (*Naturally Occurring Phorbol esters, F.J.Evans, Ed.,* CRC Press, Florida, 1984) and have been investigated intensively for their role in plant cellular biochemistry (A.T.Evans *et al., J. Pharm. Pharmacol.,* 1992, **44**, 361). Members of this class are found in Chinese medicinal plants (W.Adolf and E.Hecker, *Z. Naturforsch., Teil B,* 1993, **48**, 364).

Fusicoccane (56) and other related 5-8-5 diterpene systems such as ophiobolane are known as ceroplastins, and this class has a very diverse range of biochemical interactions, including phytotoxic effects (L.A.Paquette, T.-Z.Wang and N.H.Vo, *J. Am. Chem. Soc.,* 1993, **115**, 1676; K.D.Barrow *et al., J. Chem. Soc., Perkin Trans. 1,* 1975, 877).

Taxane (57) is perhaps the most widely known of all the terpenoid systems included here since it includes taxol and related compounds obtained from the yew family (R.W.Miller, *J. Nat. Prod.,* 1980, **43**, 425). Taxol has potent anticancer effects and is an especially effective agent for the control of ovarian cancer (S.Blechert and A.Kleine-Klausing, *Angew. Chem., Int. Ed. Engl.,* 1991, **30**, 412; K.Sakan, *et al., J. Org. Chem.,* 1991, **56**, 2311; G.Samaranayake *et al., J. Org. Chem.,* 1991, **56**, 5114; J.Riondel *et al., Cancer Chemother. Pharmacol.,* 1986, **17**, 137).

Other new classes with the 6-8-6 combination of rings include compounds such as vinigrol, an antihypertensive agent of fungal origin (I.Uchida *et al., J. Org. Chem.,* 1987, **52**, 5292) and variecolin, which inhibits angiotensin II receptor binding (O.D.Hensens, *et al., J. Org. Chem.,* 1991, **56**, 3399).

Second Supplements to the 2nd Edition of Rodd's Chemistry of Carbon Compounds, Vol. II B(Partial), C, D and E, edited by M. Sainsbury

Chapter 8b

LARGE ALICYCLIC RING SYSTEMS

D.F.EWING

1. Introduction

The development of new synthetic methodology for the formation of medium sized carbocyclic rings has enjoyed a resurgence in the last 10 to 15 years. Renewed interest in compounds containing rings in the C_{10} to C_{20} range stems from the discovery of interesting biological activity in many natural products, combined with increasing commercial need for synthetic substitutes for natural perfume constituents.

As part of the intense activity which has developed, the chemistry of medium ring systems has expanded in new directions and many older synthetic strategies have been replaced. Generally, direct cyclisation of a long molecule suffers from an unfavourable entropic factor which can only be offset partially by the introduction of structural features such as a *cis* double bond. Fortunately the range of methods described below includes many new reactions or reagents which give access to any ring size in the C_9 to C_{20} range, with essentially any distribution of useful functional groups such as keto or alkene.

Although cyclisation chemistry is often very similar for macrocyclic lactams and lactones to that for carbocyclic systems (that is when the ring closure step does not involve the heteroatom) these heterocycles have been excluded from this survey. For an entry to this area see D.L.Boger and R.J.Mathvinic, *J. Am. Chem. Soc.,* 1990, **112**, 4008 and references therein. The presence of a fused ring often has little effect on cyclisation chemistry but bicyclic and related fused ring systems have been excluded in general. Annulenes, cyclic polyacetylenes and related fully conjugated systems are not included since such compounds will be dealt with elsewhere.

A useful book in the functional group series has appeared very recently (*The Chemistry of the Alkanes and Cycloalkanes,* Eds. S.Patai and Z.Rappoport, Wiley, 1992). This book gives very extensive coverage of all kinds of physical properties, synthetic methods, electrophilic chemistry, organometallic chemistry, radical reactions, electrochemistry and biochemical and toxicological effects, with the main emphasis on alkane chemistry.

2. Preparation of macrocyclic hydrocarbons

The following material is organised in terms of synthetic strategy rather than ring size since the majority of synthetic procedures are versatile and will accommodate a variety of ring sizes. Some schemes are themselves cyclic giving rise to the possibility of repeated increments in ring size.

Reviews of interest in this area include (a) ring closure methods in natural product chemistry (Q.C.Meng amd M.Hess, *Top. Curr. Chem.* 1992, **161**, 107), (b) general synthetic methods (P.Kocovsky, F.Turecek, J.Jajicek, *Synthesis of Natural Products : Problems of Stereochemistry,* CRC, Florida, 1986, Vol 1), (c) synthetic methods for muscone (B.D.Mookherjee and R.A.Wilson, in *Fragrance Chemistry,* Ed. E.T.Theimer, Academic Press, New York, 1982; G.Ohloff, *Helv. Chim. Acta,* 1992, **75**, 2041), (d) synthetic methods for cembranes (M.A.Tius, *Chem. Rev.,* 1988, **88**, 719).

(a) Ring Expansion Methods

(i) Cope Rearrangement

Although the use of [3,3] sigmatropic transformations in acyclic systems has been widely studied (R.K.Hill in *Comprehensive Organic Synthesis,* Eds. B.M.Trost and I.Fleming, Pergamon, 1991, Vol 1, Chap. 7.1) it is only recently that this class of reaction has been applied extensively to ring expansion. An exhaustive survey of the Cope and Claisen rearrangements (S.J.Rhoads and N.R.Raulins, *Org. React. (N.Y.),*

1974, **22**, 1) includes very few references to the synthetic strategy shown in Scheme 1, but the survey of the oxy-Cope rearrangement by L.A.Paquette (*Angew. Chem., Int. Ed. Engl.,* 1990, **29**, 609) illustrates its application to many systems.

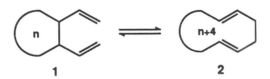

Scheme 1.

The attractiveness of the Cope rearrangement is the expansion of a ring by four carbon atoms in a single step but the usefulness of this reaction is reduced by the fact that the equilibrium position is delicately balanced and for simple systems lies far to the left in Scheme 1. Thus divinylcyclohexane (1, n = 6) is not a suitable starting point for the formation of 1,5-cyclohexadecadiene (2, n = 6). The reverse of this rearrangement (ring contraction) has been used with advantage to reduce the ring size from 10 to 6, as in the conversion of the germacrane system to the elemane sytem (K.Takeda and I.Horibe, *J. Chem. Soc., Perkin Trans. 1*, 1975, 870; K.Takeda, *Tetrahedron*, 1975, **30**, 1525). In contrast, the ring strain in a cyclobutane (1, n = 4) drives the Cope rearrangement to the right and *cis,trans,cis*-tetravinylcyclobutane undergoes a double Cope rearrangement on heating at 163 °C. Sigmatropic rearrangement of both pairs of *cis* vinyl groups gives a cyclododecatetraene (63%), a single step eight carbon (two by four) expansion. Some rearrangement of an intermediate single Cope product also occurs (K.Gubernator and R.Gleiter, *Angew. Chem., Int. Ed. Engl.,* 1982, **21**, 686).

Annelation of a furanoid ring appears to stabilise the expanded ring relative to the divinylcyclohexane and a standard Cope rearrangement, Scheme 2, is the key step in the first total synthesis of (+)-costunolide (P.A.Grieco and M.Nishizawa, *J. Org. Chem.* 1977, **42**, 1717).

When the initial ring size is larger than six the Cope rearrangement favours the ring expanded product and this reaction becomes synthetically

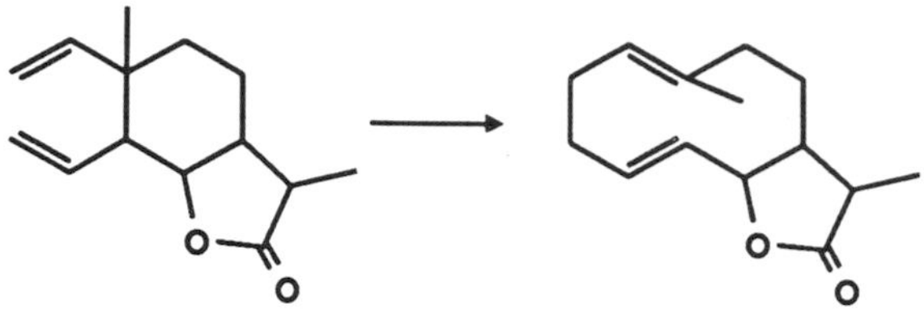

144

Scheme 2.

useful for macrocyclic compounds. For example an unsaturated β-keto-ester such as (3, n = 9, 12 or 15) is converted by standard reactions to a divinylcycloalkane derivative (4) which easily gives the corresponding (n+4) cycloalkane (5) by a Cope rearrangement (J.Bruhn, H.Heimgartner and H.Schmid, *Helv. Chim. Acta.*, 1979, **62**, 2630). It is notable that the product has the same key structural features as the starting compound (3) which could permit the whole sequence to be repeated.

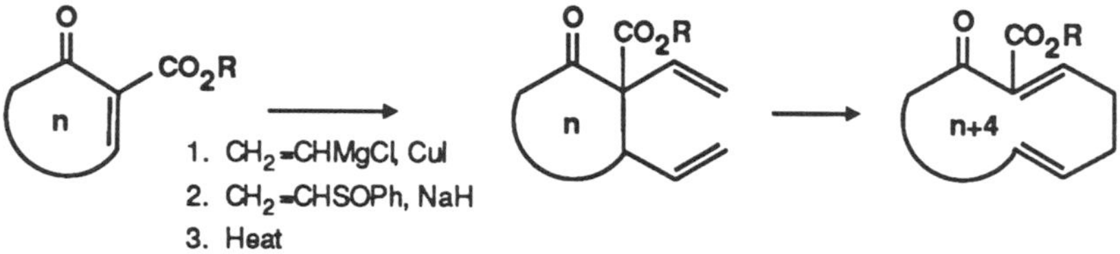

1. CH₂=CHMgCl, CuI
2. CH₂=CHSOPh, NaH
3. Heat

3   **4**   **5**

Another strategy to overcome the thermodynamic difficulty outlined above for the Cope rearrangement is to ensure the removal of the reaction product from the Cope reaction surface directly it is formed. An elegant illustration of this approach is the tandem Cope-Claisen rearrangement shown in Scheme 3 for a cyclodecadiene system (F.E.Ziegler and J.J.Piwinski, *J. Am. Chem. Soc.*, 1980, **102**, 880, 6576). The nature of the substituent group X in the vinyl ether group can effect the relative $\Delta G^{\ddagger}$ values for the two rearrangements (S.Raucher *et al.*, *J. Am. Chem. Soc.*, 1981, **103**, 1853).

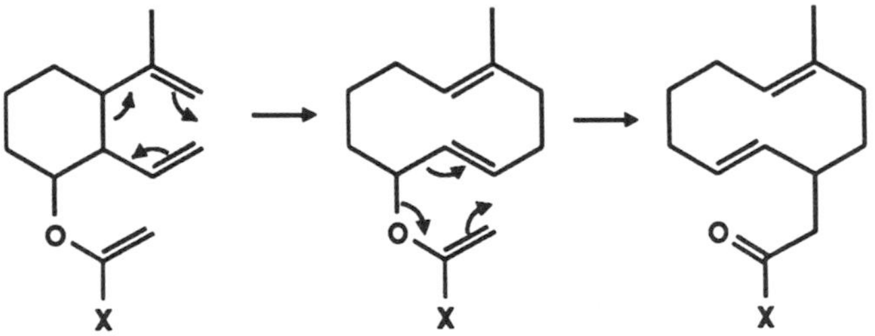

Scheme 3.

A well established variant of the Cope rearrangement requires the presence of a hydroxy or silyloxy group at the site of bond cleavage. This so-called oxy-Cope reaction (Scheme 4) involves a thermal [3,3]-sigmatropic rearrangement in which the ring expanded product (7) is favoured, and the forward reaction may be further enhanced by the tautomerisation of (7, R=H) to a stable ketone (8). Starting from a silyl allyl ether the required divinylcyclohexyl derivative (6, R=SiMe₃) can be obtained (65%) in four steps. This is converted by heating to the ring expanded compound (7, R=SiMe₃) (90%) with complete stereospecifity. It is notable that (7) can be subjected to another cycle of the same sequence of reactions and the corresponding cycloeicosane derivative was obtained by this methodology (D.L.J.Clive *et al., J. Chem. Soc., Chem. Commun.*, 1982, 828). An analogous thermal rearrangement of divinylcyclopentanol gives 5-cyclononenone (T.Kato *et al., Bull. Chem. Soc. Jpn.*, 1980, **53**, 2958). The oxy-Cope rearrangement can be accelerated by the presence of mercuric trifluoroacetate (N.Bluthe, M.Malacria and J.Gore, *Tetrahedron*, 1984, **40**, 3277).

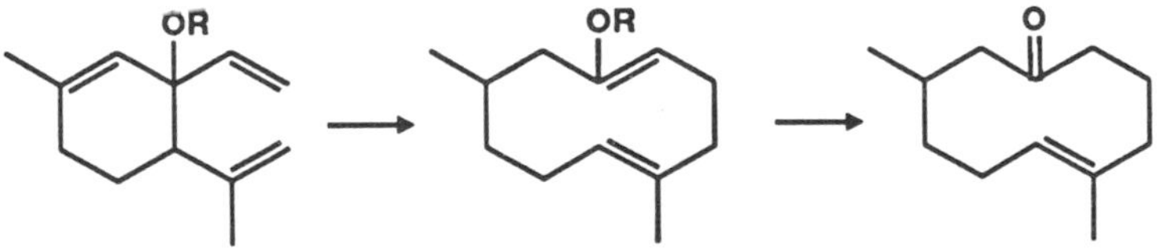

Scheme 4.

A limitation of the thermal oxy-Cope reaction is the high temperature required and the discovery by D.A.Evans and A.M.Golob (*J. Am. Chem. Soc.*, 1975, **97**, 4765) that the rate of this rearrangement is enhanced enormously if the starting alcohol is first converted to its anion (6, R=K$^+$) has provided a versatile, stereoselective entry to the 5-cyclodecenone system (W.C.Still, *J. Am. Chem. Soc.*, 1977, **99**, 4186; L.Jisheng, T.Gallardo and J.B.White, *J. Org. Chem.*, 1990, **55**, 5426). This reaction, using potassium hydride with 18-crown-6 in THF between 30-60 °C, has found application in the synthesis of several important terpenoids, including periplanone-B, the sex pheromone of the American cockroach (W.C.Still, *J. Am. Chem. Soc.*, 1979, **101**, 2493). The steric requirements of this [3,3]-sigmatropic rearrangement have been studied in detail (L.A.Paquette and Y.-J.Shi, *J. Am. Chem. Soc.*, 1990, **112**, 8478), and an interesting adaptation involves the rearrangement of a bicyclic divinyl species (9) as a route to the dienone (10) (S.L.Schreiber and C.Santini, *Tetrahedron Lett.*, 1981, **22**, 4651).

For an example of the rearrangement of a 1,2-divinylcycloheptanol see T.Kato *et al.*, *Bull. Chem. Soc. Jpn.*, 1980, **53**, 2958. An oxy-Cope rearrangement will also occur with acetylenic carbinols, and a 3,5-cyclodecadienone was obtained in this way (K.Thangaraj, P.C.Srinivasan and S.Swaminathan, *Synthesis*, 1984, 1010).

9   **10**

P.A.Wender and his coworkers (*Tetrahedron*, 1981, **37**, 3967) have investigated an attractive application of the anion-assisted oxy-Cope rearrangement involving an eight carbon expansion of 1,2-dibutadienyl-cyclohexanol (11), giving access to the 3,5,7-cyclotetradecatrienone system (12). This reaction is thought to be a [5,5]-sigmatropic rearrangement rather than two [3,3] rearrangements and is a key step in the synthesis of (−)-(3-Z)-cembrene A ((P.A.Wender and D.A.Holt, *J. Am. Chem. Soc.*, 1985, **107**, 7771). In a similar reaction a C$_7$ to C$_{15}$

conversion was also successful (P.A.Wender, P.A.Holt and S.M.Sieburth, *J. Am. Chem. Soc.*, 1983, **105**, 3348).

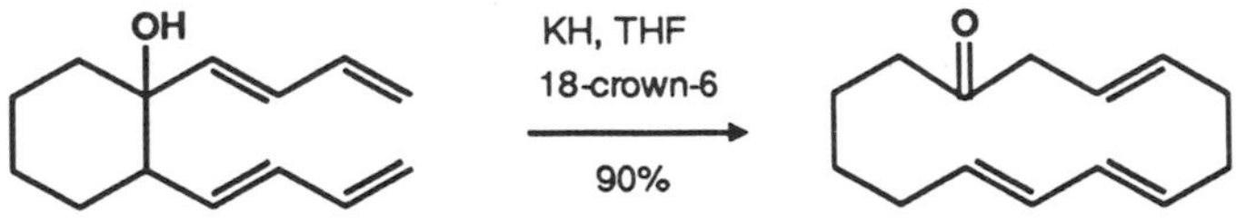

11   **12**

The oxy-Cope rearrangement (thermal or anionic) can also be envisaged for a compound with one endocyclic and one exocyclic vinyl group (14). A [3,3]-sigmatropic shift (14 → 15) does not increase the ring size but the competing [1,3] shift (14 → 13) gives the corresponding (n+2) system. For silyl ether compounds of this type with C_9 and C_{10} rings the [1,3] shift is predominant (70%) under thermal conditions but for the cyclotridecadiene analogue (14, n = 13) the two processes occur to a similar extent (R.W.Thies and R.E.Bolesta, *J. Org. Chem.*, 1976, **41**, 1233 and references therein). Geometry constraints in the transition state are probably responsible for this dependence of selectivity on ring size.

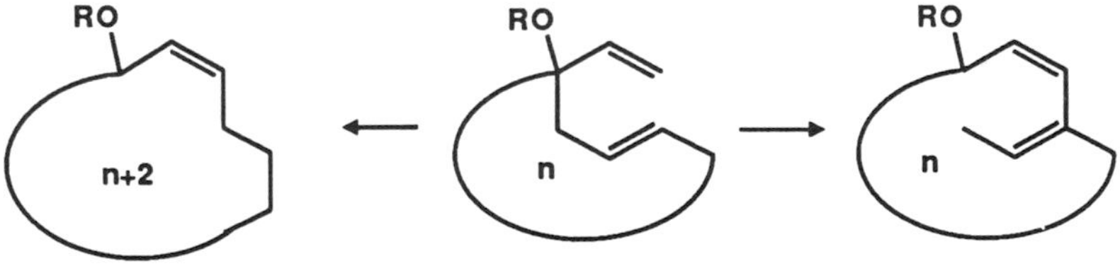

13   **14**   **15**

Under anionic conditions only the product from a [1,3] shift can be isolated (R.W.Thies and E.P.Satz, *J. Org. Chem.*, 1978, **43**,1050). Substituents at the terminal site of the exocyclic vinyl group in (14) can enhance the selectivity for [1,3] rearrangement under thermal conditions, but the anionic reaction is not affected significantly (R.W.Thies and K.P.Daruwala, *J. Org. Chem.*, 1987, **52**, 3798).

148

(ii) Claisen rearrangement

As an alternative to the Cope rearrangement, the Claisen rearrangement has been applied as a means of changing ring size in a variety of ways. The crucial structural element for this type of rearrangement, a vinyl allyl ether or equivalent, can be built on to the cyclic precursor in a variety ways and hence the product carbocycle can be smaller, larger or the same size as the initial cyclic ether. This last type of structural change is shown in Scheme 5. The most efficacious Claisen rearrangement involves the use of the Ireland modification (R.E.Ireland, R.H.Mueller and A.K.Willard, *J. Am. Chem. Soc.,* 1976, **98**, 2868), in which the vinyl ether derives from *O*-silylation of an ester or lactone (ester enolate).

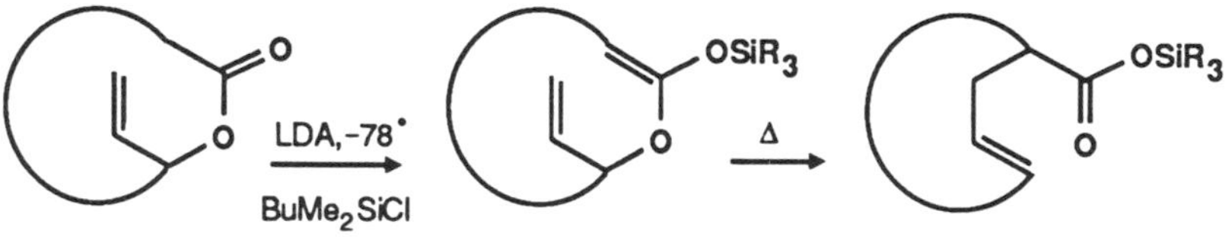

Scheme 5.

Since a macrocyclic lactone is the most usual precursor (Scheme 5), the application of the Ireland-Claisen method has benefited from the synthetic activity associated with the macrolide antibiotics. Thus many methods are available for the formation of cyclic lactones (T.Mukaiyama, *Angew. Chem., Int. Ed. Engl.,* 1979, **18**, 707; K.C.Nicolaou, *Tetrahedron,* 1977, **33**, 683; S.Masamune, G.S.Bates and W.J.Corcoran, *Angew. Chem., Int. Ed. Engl.,* 1977, **16**, 585). Using the Claisen rearrangement methodology R.K.Brunner and H.-J.Borschberg (*Helv. Chim. Acta,* 1983, **66**, 2608) have obtained (±) muscone in 24% overall yield, starting from an acyclic hydroxy acid. Similar work illustrates the formation of esters containing the cyclononane and cycloundecane rings (by a four atom reduction in ring size) and the cycloundecene and cyclotridecene rings (by replacement of C-O in the ring with C-C) (A.G.Cameron and D.W.Knight, *J. Chem. Soc., Perkin Trans. 1,* 1986, 161; see also R.L.Funk, M.M.Abelman and J.D.Munger, *Tetrahedron,* 1986, **42**, 2831).

(iii) Intramolecular reaction with a nucleophilic side chain.

Ring expansion of cyclic ketones by alkylation of the carbonyl group (via intramolecular nucleophilic attack) has been a widely used procedure and very many variants are known, particularly for the synthesis of medium ring ketones ($C_9 - C_{12}$). The strategy involved is shown in Scheme 6.

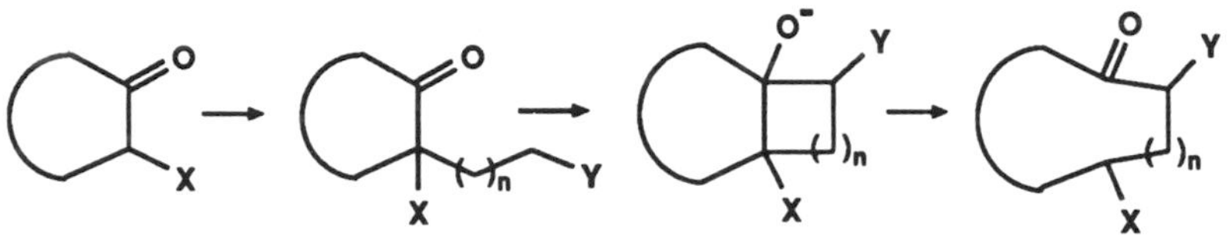

Scheme 6

Activation of the ketone is desirable and this is achieved by the inclusion of an acceptor substituent X at the 2-position. This auxiliary group has commonly been one of the set COR, NO_2, SO_2Ph or CN, and the chosen group determines, in some measure, the facility with which the sidechain can be introduced, the extent of activation at C-1, the promotion of C-1, C-2 bond cleavage in the intermediate bicyclic compound and the protocol required for subsequent removal or chemical modification of the auxiliary group. Many of these considerations involve general synthetic chemistry and are not dicussed here. The early work has been surveyed by H.Stach and M.Hesse (*Tetrahedron*, 1988, **44**, 1573).

Although an activating substituent X is commonly used it is not essential as is shown by the high yield cyclisation (94%) of a cyclododecanone with an iodopropyl sidechain using BuLi in THF or SmI_2 in THF/HMPA. The intermediate bicyclic compound is formed and the bridging bond cleaved with HgO/I_2 in benzene followed by photolysis to afford an iodocyclopentadecanone in 96% yield (H.Suginome and S.Yamada, *Tetrahedron Lett.*, 1987, **28**, 3963).

The sidechain is usually designed to result in a ring expansion by three or four carbon atoms and the nucleophilic site is activated (incipient carbanion) in the normal way by a terminal group Y such as CO_2Me or NO_2, which is retained in the ring-expanded compound. For a recent example of cyclisation with a keto group in the sidechain see Z.F.Xie, H.Suemune and K.Sakai (*J. Chem. Soc., Chem. Commun.*, 1988, 612). With an aldehyde as the sidechain activating group conversion to the corresponding enamine (pentylamine in ethanol) gives a highly reactive species and cyclisation occurs at ambient temperature, giving a 90% yield of the corresponding ring expanded enamine (S.Bienz, and M.Hesse, *Helv. Chim. Acta*, 1988, **71**, 1704; F.Hadj-Abo and H.Hess, *Helv. Chim. Acta*, 1992, **75**, 1834).

Another type of activation is observed with the alkenyl sidechain $CH_2C(Z)=CH_2$ where Z is CH_2SiMe_3. Fluoride ion induced cleavage of the $SiMe_3$ group leads to nucleophilic attack by the terminal vinyl carbon resulting in a three carbon expansion. This reaction has been applied to the formation of the C_{15} ring in muscone (B.M.Trost and J.E.Vincent, *J. Am. Chem. Soc.*, 1980, **102**, 5680).

A similar strategy to that shown in Scheme 6, involves the use of a cyclic lactone rather than a ketone. This has the advantage that the intramolecular displacement requires no activation. Thus Baeyer-Villiger oxidation of the bromopropylcyclododecanone (16) gives a lactone which can be treated with lithium in THF to give the hydroxycyclopentadecanone (17) in 78% yield. Replacement of Br by the SO_2Ph group in (16, n = 12), cyclisation is accomplished with any strong base (C.Fehr, *Helv. Chim. Acta*, 1983, **66**, 2512).

16   17

(iv) Intramolecular reaction with radical sidechain

Radical alkylation at the carbonyl carbon of a β-keto ester has proved to be an excellent method for the homologation of large rings (P.Dowd and C.-C.Choi, *Tetrahedron*, 1992, **48**, 4773 and references therein). Synthesis of the required iodomethyl or bromomethyl keto esters is readily accomplished (Scheme 7) and reduction with Bu_3SnH under radical conditions gives the corresponding one carbon expanded keto ester (30 — 75% for C_{13}, C_{15} and C_{16} products). Attempted extension of this reaction to a cyclododecanone ester with a three carbon sidechain resulted in ring contraction to a cycloundecanone. However in the absence of the ester group the radical reaction took its usual course and the corresponding C_{15} ketone (muscone) was obtained albeit in low yield (15%). *R*-(−)-muscone was also prepared by this expeditious route.

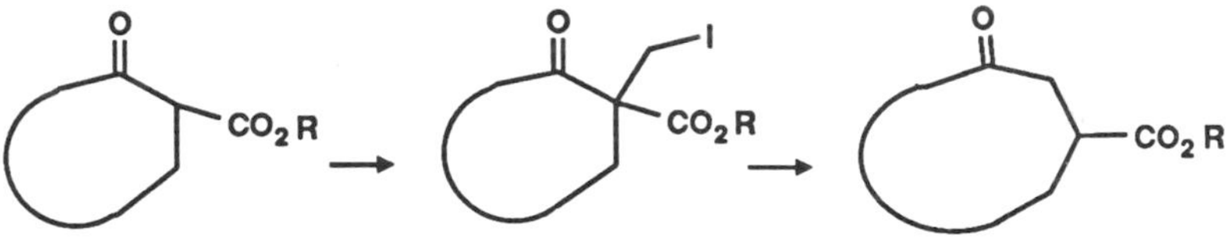

Scheme 7.

A entropically less favourable radical ring expansion by four carbon atoms is reported for cyclohexanone derivatives of type (18). Initial cyclisation of the butyl side chain to the carbonyl carbon gives bicyclo[4.4.0]decane system and the cleavage of the bridging bond is assisted by the homolytic loss of the Bu_3Sn group to afford the cyclodecanone in 70% yield. (J.E.Baldwin, R.M.Adlington and J.Robertson, *Tetrahedron*, 1989, **45**, 909; 1991, **47**, 6795).

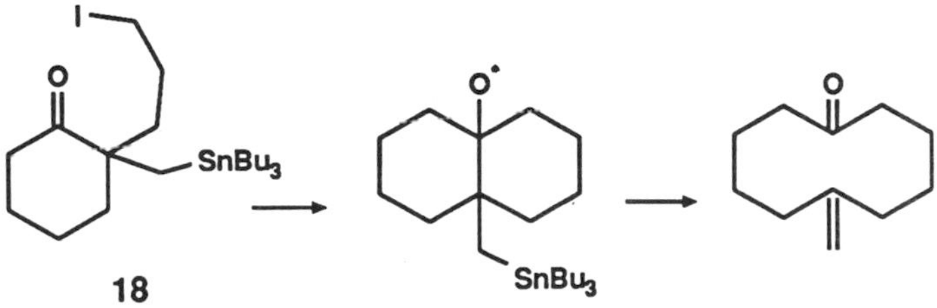

(v) Other rearrangements

A cationic centre can be stabilised by a β-silicon atom and this observation has been developed into a simple one-carbon ring expansion reaction. Thus a cyclic carbaldehyde of type (20) in presence of Al(OPh)Cl$_2$ is converted in high yield (>85%) to the ring expanded ketone (19) (K.Tanino, T.Katoh and I.Kuwajima, *Tetrahedron Lett.*, 1989, **30**, 1815). This reaction can be modified to produce initial *O*-alkylation of the carbonyl group with Me$_3$SiOMe. With Me$_3$SiOSO$_2$CF$_3$ as catalyst ring expansion occurs with concomitant elimination of Me$_3$SiH to produce the α-methylenecycloalkyl methyl ether (21) (T.Katoh, K.Tanino and I.Kuwajima, *Tetrahedron Lett.*, 1988, **29**, 1819). If the aldehyde acetal is preformed then the conversion to (21) is effected by ZnBr$_2$ to give the C$_{12}$ system in quantitative yield. If the reactant is Me$_3$SiSMe, with Me$_3$SiOSO$_2$CF$_3$ as catalyst the thiomethyl analogue of (21) is obtained. With Me$_3$SiOSO$_2$CF$_3$ as reactant and catalyst an OSO$_2$CF$_3$ derivative is produced. (K.Tanino, K.Sato and I.Kuwajima, *Tetrahedron Lett.*, 1989, **30**, 6551).

Addition of dibromocarbene to a cycloalkene gives the corresponding annulated cyclopropane (23), a bicyclo[n.1.0]alkane. This readily undergoes solvolytic disrotatory opening of the bridge in presence of AgF or AgOTs to produce a substituted cycloalkene expanded by one carbon (22). The stereochemistry of the product is controlled in part by the nucleophilicity of the anion X (fluoride or tosylate) and nine and ten membered rings were obtained successfully (H.J.J.Loozen, W.M.M.Robben and H.M.Buck, *Recl. Trav. Chim. Pays-Bas,* 1976, **95**, 248; H.J.J.Loozen *et al.*, *J. Org. Chem.*, 1976, **41**, 384).

Treatment of (23) with methyllithium results in double elimination of HBr with concomitant cleavage of the bridging bond to give the cyclic allene (24). This method has been applied to the formation of 2,3-cyclo-alkadienols (24, R = OH, n = 9, 11, 13) from which the corresponding ketones were obtained by oxidation with MnO$_2$ (J.-L.Luche, J.-C.Damiano and P.Crabbé, *J. Chem. Res. (S),* 1977, 32). The isomeric

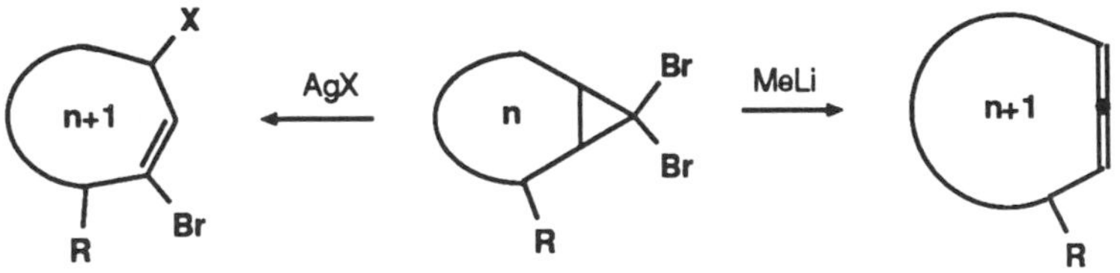

Scheme 8.

3,4- and 4,5-cyclononadienones were obtained in a similar way (G.H.Perez and P.Weyerstahl, *Synthesis*, 1985, 174) as was (*E*)-1,2,6-cyclononatriene (A.C.Connell and G.H.Whitham, *J. Chem. Soc., Perkin Trans. 1*, 1983, 995). This method is also a useful entry to various isomeric 1,5-cyclotridecadienes and 1,5,9-cyclotridecatrienes (K.Rissanen and G.Haufe, *Z. Chem.*, 1988, **28**, 365; G.Haufe and H.Trauer, *J. Chem. Res. (M)*, 1990, 1581; *(S)*, 1990, 210). Repetition of the sequence (addition of :CBr$_2$ then dehydrobromination with MeLi) can give the corresponding 1,2,3-triene. The cyclononatriene of this type is probably the smallest ring which can include the butatriene moiety (R.O.Angus and R.D.Johnson, *J. Org. Chem.*, 1984, **49**, 2880).

A bicyclo[n.1.0]alkane can be derived from a cyclic ketone by cyclopropanation of the silyl enol ether. Oxidative ring opening with FeCl$_3$ gives the (n + 1) α-chloroketone which readily eliminates HCl (sodium acetate). Thus 2-cyclotridecenone is obtained from cyclododecanone in 81% yield (Y.Ito, S.Fujii and T. Saegusa, *J. Org. Chem.*, 1976, **41**, 2073). This methodology has also been applied to the formation of the 10-π electron aromatic cyclononatetraene anion and even the analogous dianion. These nucleophiles react with electrophilic reagents to give substituted cyclononatrienes (T.S.Cantrell and A.C.Allen, *Tetrahedron Lett.*, 1985, **26**, 5351; G.Sabbioni and M.Neuenschwander, *Helv. Chim. Acta*, 1985, **68**, 623).

The Beckmann rearrangement of an *O*-mesyl oxime (25) to the expected amide can be intercepted at the intermediate cation stage by the presence of a suitable allylic sidechain. Cyclisation occurs quantitatively to give the N-bridged macrocycle (26) which is readily hydrogenated to (27). Removal of the nitrogen bridge requires several steps but the C$_{15}$ ketone muscone has been obtained *via* a route involving this three carbon

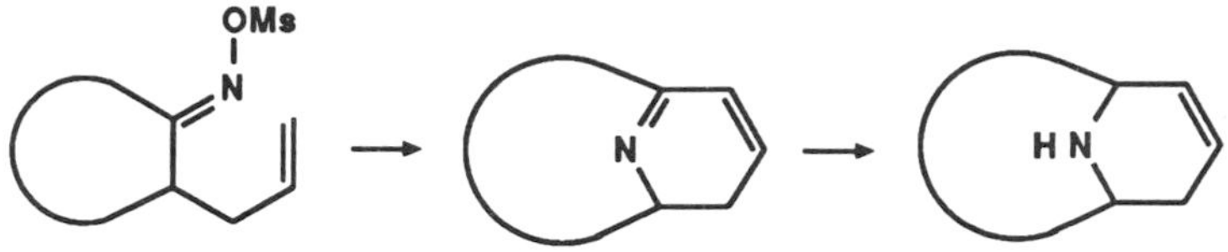

154

OMs

25   26   27

expansion procedure (S.Sakane, K.Marouka and H.Yamamoto, *Tetrahedron Lett.*, 1983, **24**, 943).

Direct intramolecular alkylation at a keto group by an adjacent vinyl group without activation of either site is desirable but difficult reaction. Several Lewis acid reagents have been investigated by V.Rautenstrauch, R.L.Snowden and S.M.Linder (*Helv. Chim. Acta*, 1990, **73**, 896) who found that Me_2AlCl was an effective catalyst for the cyclisation of a 1,4-enone, to form a bicyclic enol. Cleavage of the bridge gave the ring enlarged enone, but the overall yield for two steps was only 20%.

Oxidative cleavage of a suitable α,β-unsaturated tosylhydrazone can generate a cyclic ynone by fragmentation at the carbon-carbon double bond and this is an attractive method for expansion of cyclododecanone to cyclopentadecanone. However competition from simple reversion to the original enone is a problem. Using excess N-bromosuccinimide in butanol at −15 °C gives a 71% yield of cleavage products with a 19:1 preference for the vinylogous fragmentation route (C.Fehr, G.Ohloff and G.Büchi, *Helv. Chim. Acta*, 1979, **62**, 2655). Electrochemical oxidation is much less selective (L.L.Limacher *et al.*, *Helv. Chim. Acta*, 1989, **72**, 1383).

Scheme 8.

M.Karpf and A.S.Dreiding (*Helv. Chim. Acta*, 1977, **69**, 3045) have shown that gas phase thermolysis (at 550 °C) of a 1-alkynyl-2-methyl-1,2-epoxycycloalkane results in rearrangement to a ring expanded alkynone, such as cyclopentadec-4-ynone (Scheme 8). This reaction is unlikely to find wide application, due to the need for an enyne as precursor and the high temperature required for cyclisation.

Scheme 9 shows a very unusual route to cyclic tetraketones. It involves photochemical conversion of an oxanorbornadiene derivative to an oxaquadricyclane which rearranges to an oxepine. Oxidation of the latter with RuO_4 gives the tetraketone in 40-60% yield. There are several variants on the final step and a variety of products are possible (W.Tochtermann *et al.*, *Chem. Ber.*, 1991, **124**, 915; 1992, **125**, 923 and references therein). The oxanorbornadiene derivative is usually obtained from cycloundecanone and the chemistry is overall a very adaptible ring expansion. Another variant of this methodology transforms cycloundeca-none to [10]paracyclophane (W.Tochtermann and M.Haase, *Chem. Ber.*, 1984, **117**, 2293).

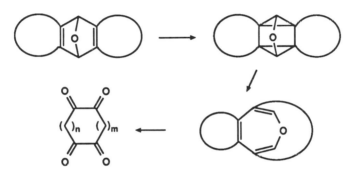

Scheme 9

(b) Ring contraction methods

(i) Wittig rearrangement

The sigmatropic rearrangement of an α-metallated allylic ether has been developed as a convenient route to carbon macrocycles from larger macrolides (Scheme 10). Successful transannular [2,3] Wittig cyclisation relies on selectivity in the formation of the carbanion, and selectivity for a [2,3]-shift. Thus the diallylic cyclic ether (28) can be converted (85%) to a C_{10} ring compound (29), with high stereoselectivity. This is an entry to the germacrane system and has been used successfully for the synthesis of

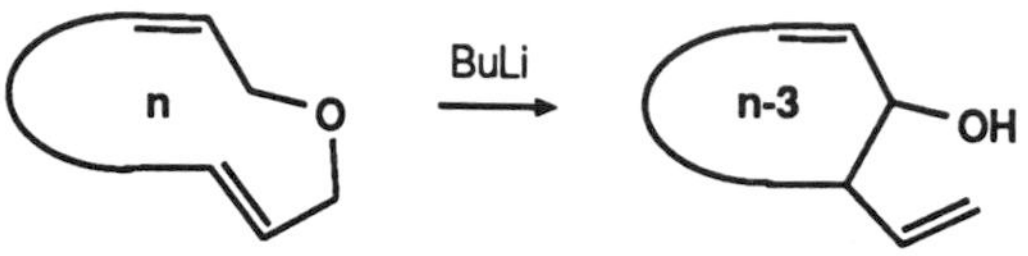

Scheme 10.

costunolide and haageanolide, and the observed stereoselectivity has been rationalised with the aid of molecular mechanics calculations (T.Takahashi *et al.*, *Tetrahedron*, 1987, **43**, 5499 and references therein). A [2,3] Wittig contraction of a cyclic propargylic allylic ether is a key step in the formation of a highly strained cyclononadiyne, *en route* to analogues of esperimicin (T.Doi and T.Takahashi, *J. Org. Chem.*, 1991, **56**, 3465) and a similar reaction has afforded a cyclotetradecyne derivative with excellent stereoselectivity (J.A.Marshall, T.M.Jenson and B.S.DeHoff, *J. Org. Chem.*, 1986, **51**, 4316; J.A.Marshall and D.J.Nelson, *Tetrahedron Lett.*, 1988, **29**, 741). When the analogous conversion of a 13-membered cyclic ether to a C_{10} carbocycle was carried out in the presence of a chiral auxiliary base in solution the enantioselectivity was further enhanced (J.A.Marshall and J.Lebreton, *Tetrahedron Lett.*, 1987, **28**, 3323).

The importance of the stereoselectivity of the Wittig contraction is emphasised by its application to the synthesis of cembratriene diols (tumour inhibitory constituents of tobacco) (J.A.Marshall, E.D.Robinson and J.Lebreton, *J. Org. Chem.*, 1990, **55**, 227) and the synthesis of aristolactone and its epimer (J.A.Marshall *et al.*, *J. Org. Chem.*, 1987, **52**, 3883). A recent attempt to apply the [2,3] Wittig rearrangement to the bicyclic system of the enediyne antibiotics has shown that the corresponding [1,2] shift can compete (T.Skrydstrup *et al.*, *Tetrahedron Lett.*, 1992, **33**, 4563).

(ii) Other methods

Recently the very old (the Reformatsky reaction) has combined with the very new (samarium iodide reagent) to furnish an intramolecular cyclisation reaction which generates cyclic α-hydroxy esters directly,

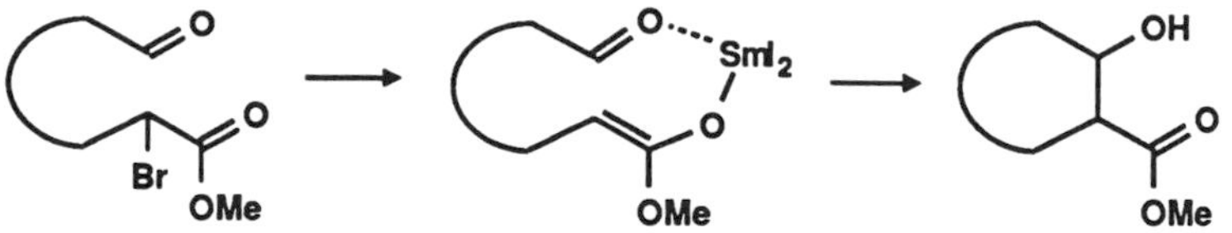

Scheme 11.

from an ω-oxo-α-bromo ester (Scheme 11). This starting compound is conveniently made from a macrocyclic ketone (C_{12}, C_{15} or C_{16}) so the overall process involves ring contraction by one carbon (J.Inanaga et al., *Tetrahedron Lett.*, 1991, **32**, 6371). The cyclisation probably proceeds through a samarium enolate and a chelated transition state, accommodated by the large ionic radius and high oxophilicity of this metal.

Another old reaction, the generation of a double bond by elimination from an α-chlorosulphone (Ramberg-Backlund reaction) has been applied to the synthesis of the series of C_{10} to C_{16} conjugated cyclic enediynes in 20-80% yield (K.C.Nicolaou et al., *J. Am. Chem. Soc.*, 1988, **110**, 4806).

(c) Cyclisation Reactions

(i) Intramolecular nucleophilic substitution

An intensive study has been made of cyclisation by S_N2 attack of malonate anion on allylic (30) or propargylic halides, at medium dilution in THF/DMF using K_2CO_3 or Cs_2CO_3 as the base (P.Deslongchamps, S.Lamothe and H.-S.Lin, *Can. J. Chem.*, 1984, **62**, 2395; D.Brillon and P.Deslongchamps, *Can. J. Chem.*, 1987, **65**, 43, 56, 1298). The relative rates of monomeric cyclisation, dimeric cyclisation and polymerisation were examined for the effect of variation in the number and type of unsaturated bonds in the chain. Over seventy compounds were obtained as a series with $C_{10} - C_{13}$ rings containing none, one or two double or triple bonds, and the corresponding dimers with twice as many unsaturated bonds. In most cases the ring has four CO_2Et substituents. In the smaller rings the yield of monomer is very low for the saturated rings but this rises to about 80% for the diacetylenic compound where the geometry is favourable in the acylic species and intraannular interactions are

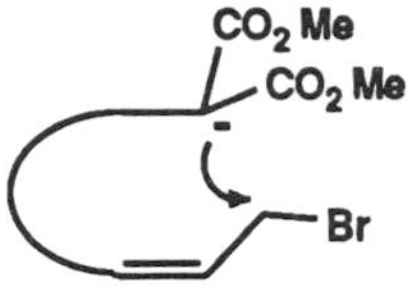

30 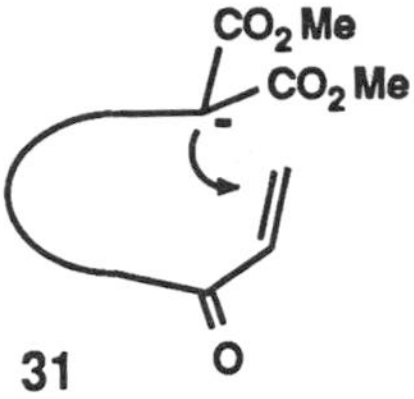   31

minimised in the cyclic product. Larger rings show less discrimination, the monomer predominating in all cases.

The influence of ring size has been determined from rate studies of ring closure of ω-bromo malonates, to give saturated compounds $(CH_2)_nC(CO_2Et)_2$ (M.A.Cadadei, C.Galli and L.Mandolini, *J. Am. Chem. Soc.*, 1984, **106**, 1051). Although cyclisation is extremely fast when a five membered ring is generated (favourable transition state), the rate decreases markedly with ring size to reach a minimum at ten membered rings. For rings larger than C_{15} the rate of cyclisation increases by a factor of about 10^3.

A variant on this reaction is the cyclisation of malonate anion on to an enone or ynone (31) by a Michael reaction (P.Deslongchamps and B.L.Roy, *Can. J. Chem.*, 1986, **64**, 2068).

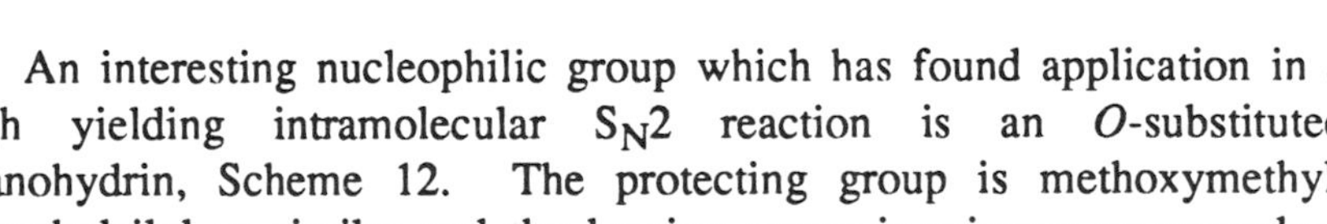

Scheme 12.

An interesting nucleophilic group which has found application in a high yielding intramolecular S_N2 reaction is an *O*-substituted cyanohydrin, Scheme 12. The protecting group is methoxymethyl, trimethylsilyl or similar and the leaving group is primary or secondary chloride or tosyl. The product is a cyclic cyanohydrin which is easily deprotected and converted to the corresponding ketone. This initial masking of the keto function prevents enone rearrangement and this reaction is particularly useful when it is essential to minimise rearrangement of double bonds in the ring, as in the formation of (*E,E*)-

2,6-cyclodecadienone (T.Takahashi, H.Nemoto and J.Tsuji, *Tetrahedron Lett.*, 1983, **24**, 2005, 3489), (*E,E,E*)-2,6,9-cycloundecatrienone (T.Takahashi, K.Kitamura and J.Tsuji, *Tetrahedron Lett.*, 1983, **24**, 4695), a C_{14} trienone (T.Takahasi *et al.*, *Tetrahedron Lett.*, 1983, **24**, 3485) and a C_{14} tetraenone (H.Takayanagi, Y.Kitano and Y.Morinaka, *Tetrahedron Lett.*, 1990, **31**, 3317).

The stabilisation of an allylic anion by a phenylthio or phenyl-sulphonyl group provides a good nucleophile for cyclisation under high dilution conditions. A variety of leaving groups have been employed such as epoxide (63% cyclisation to a C_{14} ring, M.Kodoma, Y.Matsuki and S.Ito, *Tetrahedron Lett.*,1975, 3065), allylic bromide (53% for a C_{14} triene, J.A.Marshall and D.G.Cleary, *J. Org. Chem.*, 1986, **51**, 858), tosyl (23% for a C_{14} triene, W.G.Dauben, R.K.Saugier and I.Fleischhauer, *J. Org. Chem.*, 1985, **50**, 3767). In some cases the regiochemistry of this allylic alkylation is determined by the geometry (*E* or *Z*) of a second double bond in the ring. The yield (15-60%) can also depend on ring size, C_9 to C_{11} (S.Usui *et al.*, *Chem. Lett.*, 1992, 527).

Intramolecular Wittig cyclisation of phosphonate esters has been described by several workers. The selection of an appropriate base and the control of the stereochemistry of the double bond which is formed are both difficult. Mixtures of *E* and *Z* isomers are obtained in applications to the synthesis of the C_{14} ring in methyl ceriferate, a sesterpene from scale insects (M.Kodama *et al.*, *J. Org. Chem.*, 1988, **53**, 1437), and the synthesis of asperdiol (M.A.Tius and A.H.Fauq, *J. Am. Chem. Soc.*, 1986, **108**, 1035, 6389). High *Z*-selectivity has been reported by J.A.Marshall and B.S.DeHoff (*Tetrahedron*, 1987, **43**, 4849). A general review (B.E.Maryanoff and A.B.Reitz, *Chem. Rev.*, 1989, **89**, 863) gives details of the Horner-Wadsworth-Emmons modification of the Wittig olefination reaction.

The synthesis of the polyacetylene compounds of type (32), a representative of the [n]pericyclynes where n is the number of –CMe_2–C≡C– units making up the ring, relies on repetitive coupling of a copper acetylide with propargyl bromide. The smallest system accessible so far has n = 5 (L.T.Scott *et al.*, *J. Am. Chem. Soc.*, 1985, **107**, 6546). This methodology has recently been extended to the synthesis of analogous compounds with the methyl groups replaced by a spirocyclopropyl group (A.de Meijere *et al.*, *J. Am. Chem. Soc.*, 1991, **113**, 3935).

160

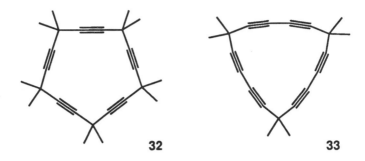

| 32 | 33 |

Cyclic polydiacetylenes such as (33), with a repeating –CMe$_2$–C≡C–C≡C– unit, (and analogues with 4, 5 and 7 such units) are readily obtained by a similar strategy employing a copper diacetylide and the corresponding bromide. Each diacetylene unit is significantly bent in the smallest system, 5,5,10,10,15,15-hexamethyl-1,3,6,8,11,13-cyclopentadecahexyne (33) (L.T.Scott, M.J.Cooney and D.Johnels, *J. Am. Chem. Soc.,* 1990, **112**, 4054). Generation of the acetylide nucleophile with KN(SiMe$_3$)$_2$ allows direct cyclisation by attack at an ω-aldehydo group, a method applied to the formation of a C$_{10}$ enyne (S.J.Danishefsky, N.B.Mantlo and D.S.Yamashita, *J. Am. Chem. Soc.,* 1988, **110**, 6890).

(ii) Intramolecular electrophilic addition

Addition to an aldehyde by an α-alkoxyallylstannane under thermal or Lewis acid conditions (Scheme 11) is ineffective for cyclisation. However the presence of a triple bond conjugated with the carbonyl group leads to a facile addition at −78 °C in presence of BF$_3$.Et$_2$O under high dilution conditions to afford the C$_{14}$ ring of the cembranoid system, with good stereochemical control (*Z/E* ratio is 9:1) (J.A.Marshall, S.L.Crooks

Scheme 13.

and B.S.DeHoff, *J. Org. Chem.*, 1988, **53**, 1616). An attempt to form a C_{10} ring by this method was unsuccessful (presumably due to transannular steric effects), coupling taking place at the α-allylic site to form a C_{12} ring. Complexation of the acetylenic bond with $Co_2(CO)_6$ failed to relieve the steric constraints sufficiently and only the C_{12} ring was obtained (J.A.Marshall and W.Y.Gung, *Tetrahedron Lett.*, 1989, **30**, 309).

Although the success of the above reaction is due in part to the activating effect of the OR and $SnBu_3$ groups, neither is actually essential. In the presence of the particular Lewis acid $EtAlCl_2$, in dilute solution at -78 °C, a simple alkene group $-CH=CMe_2$ adds to an intramolecular propargylic aldehyde to form a C_{14} cyclic alcohol in high yield (80%) but with no stereoselectivity (J.A.Marshall and M.W.Anderson, *J. Org. Chem.*, 1992, **57**, 2766). A similar reaction is reported for the protected aldehyde shown in Scheme 14. It is probable that the vinylcyclopropane fragment is an efficient trap for the cation generated by removal of a methoxy group and with CF_3SO_3H as catalyst, the C_{11} product is formed in 84% yield (P.G.Gassman and R.J.Riehle, *Tetrahedron Lett.*, 1989, **30**, 3275). In the presence of of $BF_3.OEt_2$, this cyclisation gives the methoxycycloundecadiene (34) in an unprecedented 92% yield, as a 4:1 mixture of diastereoisomers.

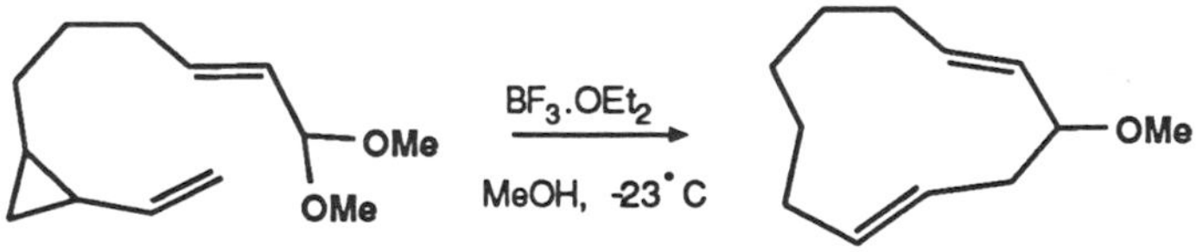

Scheme 14.

(iii) Intramolecular radical reactions

Until recently macrocyclisation by radical methods was unknown. However N.A.Porter *et al.* (*J. Am. Chem. Soc.*, 1988, **110**, 3554 and references therein) have shown that under suitable conditions the Bu_3SnH mediated ring closure of ω-iodoalkenones (35, X = H) occurs in reasonable yield with high regioselectivity to afford cycloalkanones (36) directly. The reaction typically requires a low concentration of the substrate in benzene with a slight excess of Bu_3SnH and AIBN as initiator and refluxing for several hours. These conditions gave, for example, cyclopentadecanone in 65% yield. Various other ring sizes were also accessible and the C_{14} cembranoid system has been obtained in 52% yield by cyclisation of an allylic radical (N.J.G.Cox, S.D.Mills and G.Pattenden, *J. Chem. Soc., Perkin Trans.1*, 1992, 1313). Similar radical cyclisation has been reported for an ω-phenylselenoalkene (M.P.Astley and G.Pattenden, *Synlett.*, 1991, 335). If the alkene is further activated by substitution with an ester or amide function (35, X=COOEt or $CONH_2$) the cyclisation has high yield (80%), high regioselectivity (ester end of the double bond) and high stereoelectivity (N.A.Porter *et al., J. Am. Chem. Soc.*, 1989, **111**, 8309).

35   **36**

Cyclisation by radical attack on an unactivated alkene requires a strongly 'electrophilic' radical and this can be provided by an iodo-malononitrile group. The $RC^{\bullet}(CN)_2$ radical can be generated thermally but in presence of an excess of Bu_3SnH one cyano group is also removed from the cyclised product, thus providing a route to a cyanocycloalkane (D.P.Curran and C.M.Seong, *J. Am. Chem. Soc.*, 1990, **112**, 9401).

Effective pinacol coupling of dialdehydes, involving the reaction of ketyl radicals generated by the action of low valence titanium, probably Ti(0), is known as the McMurry reaction. The preferred reagent is the ether solvate $TiCl_3(DME)$, in presence of a Zn–Cu couple, and initial reductive coupling of the aldehyde groups gives a diol which is usually deoxygenated directly by the reagent to give an alkene. This is an extremely versatile reaction and many ring systems are accessible by this route including all the simple cyclic ketones (C_9 to C_{20}) (J.E.McMurry, *Chem. Rev.*, 1989, **89**, 1513). The scope of this procedure for intra-molecular reaction has been explored in detail, including the tolerance for other functional groups. The isolation of the intermediate cyclic diols for the ring systems in the range C_8 to C_{14}, has shown a dependence of the stereochemistry of the vicinal diol on ring size, the *cis*-form being predominant for small rings and the *trans*-form for large rings. Applications include the synthesis of many macrocyclic terpenes including sarcophytol B (J.E.McMurry and J.G.Rico, *Tetrahedron Lett.*, 1989, **30**, 1169), (±)-isolobophytolide and (±)-crassin (J.E.McMurry and R.G.Dushin, *J. Am. Chem. Soc.*, 1990, **112**, 6942; W.G.Dauben, T.-Z.Wang and R.W.Stephens, *Tetrahedron Lett.*, 1990, **31**, 2393) and flexibilene (J.E.McMurry *et al.*, *Tetrahedron Lett.*, 1982, **23**, 1777).

Analogous cyclisation of a ketoester leads to formation of a cyclic ketone and alkyl derivatives of the C_{10} to C_{13} cycloalkanones are obtained in 50-65% yield using $TiCl_3/LiAlH_3/Et_3N$ as the promoter (J.E.McMurry and D.D.Miller, *J. Am. Chem. Soc.*, 1983, **105**, 1660). Pinacol coupling of $PhCO(CH_2)_{10}COPh$ to afford a diphenylcyclododecene (89%) has been achieved with the reagent $TiCl_3/C_8K$ (1:3) (A.Fürstner *et al.*, *J. Chem. Soc., Perkin Trans. 1*, 1988, 1729).

Recently the reagent $[V_2Cl_3(THF)_6][Zn_2Cl_6]$ has been used for intramolecular carbonyl coupling of a propargylic aldehyde, dec-2,8-diyn-1,10-dial. This is a milder reducing reagent which can thereby tolerate more sensitive functionality in the substrate (A.G.Myers and P.S.Dragovich, *J. Am. Chem. Soc.*, 1992, **114**, 5859).

(iv) Intermolecular nucleophilic substitution

Ring closure by a double intermolecular nucleophilic substitution reaction is particularly attractive for the synthesis of cyclic diynes since the terminal yne groups are readily metalated (Scheme 15). Reaction with α,ω-dihalogenoalkanes has afforded cyclic diynes with C_{19} to C_{44} rings (G.Schill, E.Logemann and H.Fritz, *Chem. Ber.*, 1976, **107**, 497). A series of cyclic compounds with two double bonds and two triple bonds separated by a single methylene group (so-called skipped enynes, important models for the study of homoconjugation) have been obtained by optimisation of this method (R.Gleiter, R.Merger and B.Nuber, *J. Am. Chem. Soc.*, 1992, 114, 8921). These workers find that the presence of a copper salt is deleterious. The double bonds are either endo- or exo-cyclic for ten compounds based on cyclodeca-1,6-diyne, cycloundeca-1,7-diyne and cyclododeca-1,7-diyne. The yield was around 5% in most cases. In some cases successive intermolecular condensations occurred followed by

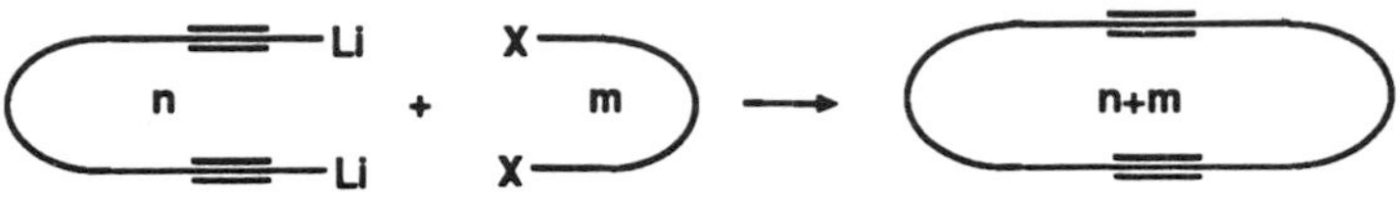

Scheme 15.

an intramolecular step to afford access to the dimeric systems, cyclo-eicosatetrayne and cyclodocosatetrayne.

A double condensation has also been achieved between a bis aldehyde and a bis phosphonate, a novel application of the Horner-Emmons-Wittig reaction. Two routes to the keto phosphonate (37) are described (G.Büchi and H.Wüest, *Helv. Chim. Acta*, 1979, **62**, 2661) and condensation of this compound with 1,12-dodecanedial gives a C_{15} (*E,E*)-dienone (50%). Small amounts of the *E,Z* isomer and the corresponding dimeric C_{30} tetraendione are also formed. This method was also applied to the formation of civetone (38, n = 6) the final step requiring selective hydrogenation with Pd/BaSO$_4$ in pyridine.

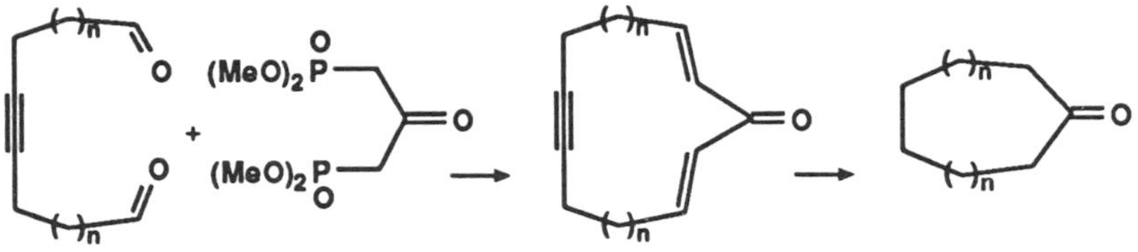

37   38

T.Takahashi and his coworkers (*Tetrahedron Lett.*, 1992, **33**, 7561) have extended their protected cyanohydrin methodology (see Scheme 12) to the formation of the C_{15} ring system by a double S_N2 reaction, *ie* intermolecular coupling of a dicyanohydrin with a dichloride or ditosylate. The yield of cyclic diketone can be enhanced by the incorporation of one or more double bonds in either open chain fragment so as to enhance the likelihood of the desired intermolecular reaction. The optimum position of the double bond was identified by examination of the low energy conformations (by molecular mechanics) of target cyclic systems. Thus the protected dicyanohydrin from (*E,E*)-2,6-decadiene-1,9-dial with the bistosylate of pentane-1,5-diol, in presence of LiN(SiMe$_3$)$_2$ gave the C_{15} dienedione (64%).

(v) Allylic cyclisation catalysed by palladium(0)

Activation of an allylic group (to nucleophilic attack) by coordination to a Pd(0) complex such as $Pd(Ph)_4$, is a well established process. When applied to intramolecular reactions the regiochemistry of the substitution (ie. which end of the allylic group reacts) is crucial since this dictates the size of the ring which is formed and subtle control of the regioselectivity and stereoselectivity of this reaction can be achieved by variation in the structure of the palladium complex and the nature of the solvent. Early work has been briefly reviewed (B.M.Trost, *Angew. Chem., Int. Ed. Engl.,* 1989, **28**, 1173). A notable example from the early work is the use of the palladium(0) complex of an allylic acetate to promote the alkylation of a β-ketoester to form the C_{11} ring system of humulene (Scheme 16) (Y.Kitagawa *et al., J. Am. Chem. Soc.,* 1977, **99**, 3864).

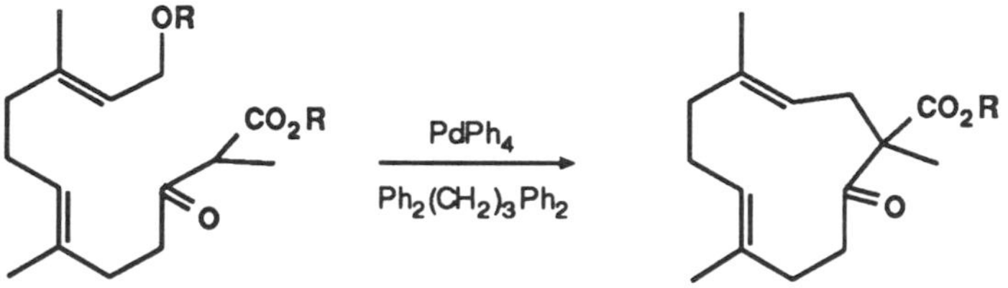

Scheme 16.

The high dilution requirement for intramolecular cyclisation is a serious shortcoming and a clever way of avoiding this problem is the use of a palladium catalyst bound to a polymer support (B.M.Trost and R.W.Warner, *J. Am. Chem. Soc.,* 1982, **104**, 6112). The catalytic sites are at low dilution but the reacting species (39) is in the concentration range 0.1 - 0.5 M. This pseudodilution technique is effective for allylic alkylation by the reaction shown in Scheme 17, to give C_{10} and C_{15} cycloalkenes (40, n = 4 or 9) in excellent (70-80%) yield. For smaller rings the substrate concentration is less critical and homogeneous catalysis is adequate for the formation of the cyclononene analogue. The nucleophilicity of the alkylation site can have a drastic effect on the regiochemistry of small ring formation, that is the ratio of C_7 to C_9 rings. (B.M.Trost, *Tetrahedron Lett.*, 1992, **33**, 717).

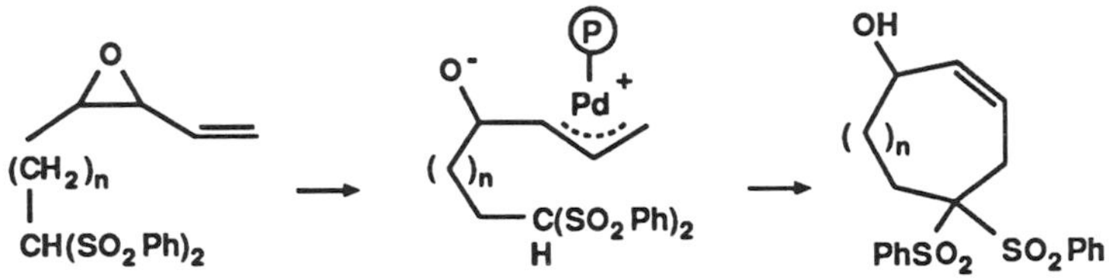

Scheme 17.

Palladium catalysed reaction of a vinyl bromide with an acetylenic tin derivative has been used as a method of closing a C_{10} diyne ring (M.Hirama *et al.*, *J. Am. Chem. Soc.*, 1989, **111**, 4120). Successful reaction required catalysis by $Pd(PPh_3)_4$ for 86 h at 50 °C, and may have undeveloped potential for macrocyclisation.

(vi) Cyclisation catalysed by chromium(II)

Allylic alkylation of an aldehyde is efficiently catalysed by $CrCl_2$ (Scheme 18), with high stereospecificity. This selectivity is probably due to the formation of a cyclic transition state or intermediate involving the solvated catalyst (C.T.Buse and C.H.Heathcock, *Tetrahedron Lett.*, 1978, 1685). This reaction has been applied to the formation of the strained C_9

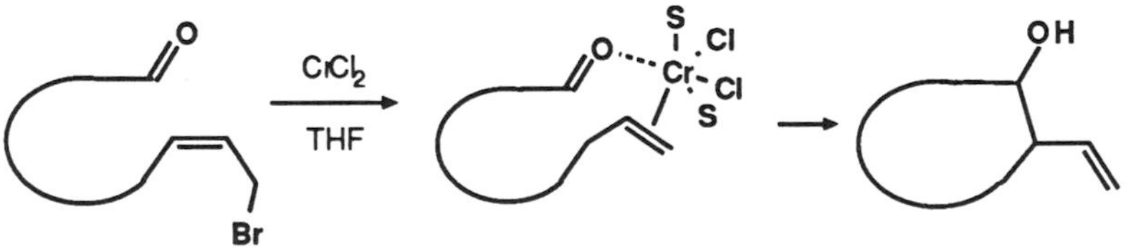

Scheme 18.

diyne ring in the neocarzinostatin chromophore (P.A.Wender, J.A.McKinney and C.Mukai, *J. Am. Chem. Soc.*, 1990, **112**, 5369) and the synthesis of the C_{10} system of costunolide (H.Shibuya *et al.*, *Chem. Lett.*,

1986, 85) and is crucial in the generation of the required stereochemistry of the C_{14} diene ring in the cembranoid antitumour agent, asperdiol (W.C.Still and D.Mobilio, *J. Org. Chem.*, 1983, **48**, 4785).

(vii) Cyclisation promoted by cobalt complexation

Intramolecular cyclisation of acetylenic compounds can be promoted by complexation of the triple bond with $Co_2(CO)_6$. The η^2 complex has a bent bond (ca 143°) which effectively reduces the entropic factor. Several reactions have been promoted in this way, including an intramolecular aldol reaction (P.Magnus, H.Annoura and J.Darling, *J. Org. Chem.*, 1990, **55**, 1709; P.Magnus and T.Pitterna, *J. Chem. Soc., Chem. Commun.*, 1991, 541; M.E.Maier and T.Brandstetter, *Tetrahedron Lett.*, 1991, **32**, 3679), and Lewis acid catalysed carbocation attack on a double bond (P.Magnus *et al.*, *J. Am. Chem. Soc.*, 1992, **114**, 2544).

(d) Functional group modification

H.Meier (*Synthesis*, 1972, 235) has reviewed early work on methods for the formation of cycloalkynes, including dehalogenation, dehydro-halogenation, and elimination of nitrogen from bistosylhydrazones (oxidatively or photolytically). The synthesis of 1,5,9-cyclododecatriyne from corresponding triene by bromination, mild dehydrobromination then forcing dehydrobromination (NaOMe in MeOH, ten fold excess, 72 hour reflux) has been reported (A.J.Barkovich and K.P.C.Vollhardt, *J. Am. Chem. Soc.*, 1976, **98**, 2667). The rapid dehydrohalogenation of a *trans*-1-bromocycloalkene (5 - 10 second exposure to potassium *t*-butoxide) is effective for the formation of 1,2-cyclononadiene and the C_{10} analogue without causing dimerisation of these very reactive compounds (C.B.Reese and A.Shaw, *J. Chem. Soc., Perkin Trans.1*, 1976, 890; see also R.P.Johnson, *Chem. Rev.*, 1989, **89**, 1111).

Gleiter and his group have applied the method shown in Scheme 19 to the formation of cyclic diynes from readily available cyclic ketones. Thermolysis of a selenadiazole as a means of generating a carbon-carbon triple bond was first described for conjugated arylalkynes (I.Lalezari, A.Shafiee and M.Yalpani, *Angew. Chem., Int. Ed. Engl.*, 1970, **9**, 404) but the method works well for the generation of isolated triple bonds such as

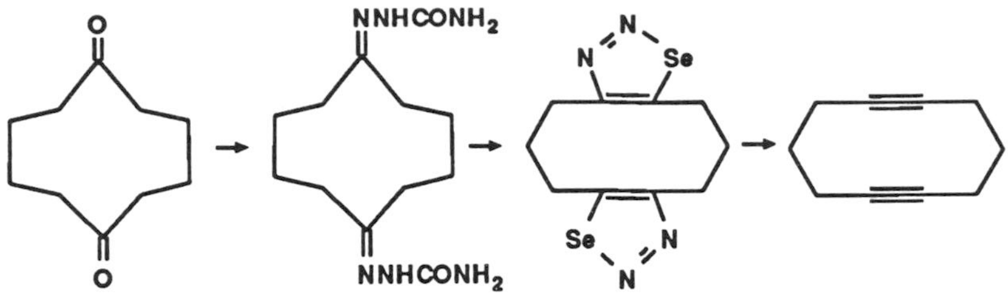

Scheme 19

those in 1,6-decadiyne (R.Gleiter *et al.*, *Chem. Ber.*, 1988, **121**, 735), and similar 1,5- and 1,6-diynes based on C_9 to C_{12} rings (R.Gleiter *et al.*, *J. Am. Chem. Soc.*, 1991, **113**, 9258 and references therein). It has also been applied to the synthesis of the highly strained compound (*E*)-1-cyclodecen-3-yne (which has a half-life of 2.2 hours) (H.-J.Bissinger, H.Detert and H.Meier, *Leibigs Ann. Chem.*, 1988, 221), and similar compounds H.Meier *et al.*, *Tetrahedron*, 1986, **42**, 1711).

Stereochemically controlled conjugate addition to the double bond of (*E*)-2-cyclopentadecenone offers a novel route to enantiomerically pure muscone. Thus methylation with a chiral dimethylcuprate reagent has provide both enantiomers with optical purity up to 100%, the chiral ligands (amino alcohols) being derived from camphor (K.Tanaka *et al.*, *J. Chem. Soc., Perkin Trans. 1* , 1991, 1445).

Barton and coworkers have exhaustively investigated the oxidation of cycloalkanes to the corresponding ketone by a developing series of catalytic systems based on Fe(V). The bulk of this work is not of direct synthetic application on the laboratory scale although viable industrial catalysts will undoubtedly evolve. For an entry to these studies see E.About-Jandet *et al.*, *Tetrahedron Lett.*, 1990, **31**, 1657. An improved synthesis is reported for cyclodecane-1,3-dione and the C_{12} analogue by standard chemistry (E.M.Beccelli, L.Majori and A.Marchesini, *J. Org. Chem.*, 1981, **46**, 222).

3. Properties and reactions

Much of the chemistry of larger cyclic compounds is not significantly different from that of acyclic analogues and the following section is restricted to those aspects of macrocyclic chemistry where the existence of a ring structure is important.

(a) Conformation and Structure

The conformation of carbocyclic compounds has been reviewed thoroughly by T.A.Crabb and A.V.Patel, (in *Second Supplement to the 2nd Edition of Rodd's Chemistry of Carbon Compounds,* Ed. M.Sainsbury, Elsevier, Amsterdam, 1992, Volume IIA/B, Ch 1) and only brief discussion of more recent work is included here.

Searching for stable conformations of highly flexible rings has been attempted by several computational procedures and for cycloheptadecane it has been established that 262 conformers exist with energies within 3 kcal mol^{-1} of the global minimum, with smaller numbers of conformers for C_9-C_{12} systems (M.Saunders, *J. Comput. Chem.,* 1991, **12**, 645; M.Saunders *et al., J. Am. Chem. Soc.,* 1990, **112**, 1429). New developments to the search algorithms have been evaluated using cycloheptadecane (H.Goto and E.Osawa, *Tetrahedron Lett.,* 1992, **33**, 1343). The importance of locating saddle points as well as minima has been stressed recently in a comprehensive conformational analysis study of C_n rings (n = 4 to 12) (I.Kolossváry and W.C.Guida, *J. Am. Chem. Soc.,* 1993, **115**, 2107). An exploration of the potential energy hypersurface can then develop into a full understanding of all conformational interconversion pathways.

41   **42**

The lowest energy conformation of *cis,cis*-1,6-cyclodecadiene is the armchair form (41, X = H$_2$) on the basis of molecular mechanics calculations and this is confirmed by the first X-ray analysis of this system, on the 4,9-dimethylene derivative (41, X = CH$_2$) (R.G.Gleiter, H.Irngartinger and R.Merger, *J. Org. Chem.*, 1993, **58**, 456). In contrast, detailed NMR investigation of the diester (42, X = CO$_2$Me) shows that this derivative has the hammock (or crown) conformation (D.P.G.Hamon and G.Y.Krippner, *J. Chem. Soc., Chem. Commun.*, 1992, 1507). Evidently the two forms are close in energy and the conformational balance is sensitive to subtle substituent effects.

Cyclic compounds containing two or more triple bonds are of interest for several reasons as indicated in a recent review (R.Gleiter, *Angew. Chem., Int. Ed. Engl.*, 1992, **31**, 27). Homoconjugation between double and triple bonds has been investigated for several C$_{10}$, C$_{11}$ and C$_{12}$ cyclic systems (R.Gleiter, R.Merger and H.Irngartinger, *J. Am. Chem. Soc.*, 1992, **114**, 8927). Compounds (43) and (44) both have a chair form (X-ray analysis) and the conformation has been established for ten other analogues by molecular mechanics calculations. These compounds have

43   **44**

sufficient conformational rigidity to allow a significant interaction between the π-systems. Generally the triple bonds are nearly parallel to each other and the separation between the end carbon atoms in nearest neighbour multiple bond is 2.4-2.5 Å. Many of the bands in the photoelectron spectra of these compounds have been assigned and the observed high energy shifts (relative to simple cycloalkadiynes) confirm the homoconjugative interaction between the multiple bonds (see also R.Gleiter, *Acc. Chem. Res.*, 1992, **31**, 27). The all *cis* isomer of 1,4,7,10-cyclododecatetraene adopts a crown conformation in the solid state. This conformation allows a maximum homoconjugative interaction as indicated by the photoelectron spectrum (A.Krause *et al.*, *Angew. Chem., Int. Ed. Engl.*, 1989, **28**, 1379).

The pericyclynes, such as (32) and (33) are members of an interesting class of polyunsaturated compounds which could show the participation of homoconjugation. The application of molecular orbital and molecular mechanics methods have suggested that the class of compounds with single acetylenic bonds between the saturated centres, type (32), are not planar when the number of repeating units is greater than four, but the various possible conformations differ in energy only to a negligible extent (K.N.Houk *et al.*, *J. Am. Chem. Soc.*, 1985, **107**, 6556). Both through-bond and through-space interactions are likely between the π-systems and for decamethyl[5]pericyclyne the experimental heat of hydrogenation is about 35 kJ mol^{-1} less than expected by comparison with similar acyclic compounds. Such thermodynamic stabilisation suggests that this system is homoaromatic (L.T.Scott *et al.*, *J. Am. Chem. Soc.*, 1988, **110**, 7244). The photoelectron spectra of the series with two acetylenic groups bridging the saturated centres indicate that homoconjugation may be a general feature of these compounds but is particularly important for the smallest ring size (15 atoms, three diacetylene units) (L.T.Scott, M.J.Cooney and D.Johnels, *J. Am. Chem. Soc.*, 1990, **112**, 4045).

Some aspects of strain in cycloalkenynes have been discussed (H.Meier *et al.*, *Tetrahedron*, 1986, **42**, 1711). The smallest ring size for which there can be a strain-free conformation is ten, for rings containing one triple bond, or one triple and one unconjugated double bond. For the case of a conjugated enyne this minimum size is thirteen atoms and for a conjugated dienyne, it is seventeen atoms. The photochemistry of strained 1,2-cycloalkadienes has been studied in detail (T.J.Steirman, W.C.Shakespeare and R.P.Johnson, *J. Org. Chem.*, 1990, **55**, 1043 and references therein). Bending of the 1,2,3-butatriene fragment when incorporated into a ring has been assessed by MO calculations. The smallest stable cyclic system is likely to be the C_9 compound, 1,2,3-cyclononatriene (R.O.Angus and R.P.Johnstone, *J. Org. Chem.*, 1984, **49**, 2880). A complex of this triene has been isolated (R.O.Angus *et al.*, *Organometallics*, 1987, **6**, 1909).

The conformation of cyclododecanone has been reinvestigated by NMR using deuterated isotopomers (T.N.Rawdah, *Tetrahedron*, 1991, **47**, 8579). Earlier crystal structure studies established a [3333] conformation for this compound in the solid state and the same conformation is adopted in solution. Cyclohexacosa-1,14-dione exists as two polymorphs. One crystallises in plates and the crystal structure indicates a [3 10 3 10] conformation (using the nomenclature of J.Dale, *Top. Stereochem.*, 1976, **9**, 199) whereas the other form (needles) has complementary zig-zag chains (T.J.Lewis *et al.*, *J. Am. Chem. Soc.*, 1991, **113**, 8180). These two forms also have quite different photochemical behaviour, giving respectively *cis* and *trans* cyclobutanols by transannular cyclisation. This photochemistry is characteristic of compounds with a close C=O···H contact, and is also observed for the analogous C_{16} ketone (T.J.Lewis *et al.*, *J. Am. Chem. Soc.*, 1990, **112**, 3679).

Measurements of ^{13}C NMR data are reported for cycloalkanes C_6 to C_{44} (H.Fritz *et al.*, *Chem. Ber.*, 1989, **109**, 1258). The chemical shift has a minimum value (24 ppm) for cyclododecane. A combined IR/Raman analysis of the C_{14}, C_{16} and C_{22} cycloalkanes indicates great conformational complexity (V.L.Shannon *et al.*, *J. Am. Chem. Soc.*, 1989, **111**, 1947). Variable temperature ^{13}C MAS NMR spectroscopy has been applied to the study of cyclohexatriacontane. Below the melting point several different phases are detected (Q.Chen *et al.*, *J. Mol. Struct*, 1992, **265**, 153).

(b) Transannular reactions

Cyclic alkenes have the potential for ring closure to bicyclic (or polycyclic) molecules and this kind of rearrangement has long been of theoretical interest. It also offers a route to fragments of importance in natural product chemistry and is relevant to the study of biosynthetic pathways for polycyclic sesquiterpenes. For a recent review of the cyclisation reactions in cyclodecane and cycloundecane systems see D.C.Harroven and G.Pattenden in *Comprehensive Organic Synthesis*, Eds. B.M.Trost and I.Fleming, Pergamon, 1991, Vol 3, Ch 1.10). Photochemically induced reactions of simple cycloalkenes (C_9 to C_{12}) have been studied in great detail (G.Haufe, M.W.Tubergen and P.J.Kropp, *J. Org. Chem.*, 1991, **56**, 4292 and references therein) as have the thermal cyclisations of trienes with similar ring size (W.G.Dauben, D.M.Michno and E.G.Olsen, *J. Org. Chem.*, 1981, **46**, 687). The product distribution and stereochemistry is largely controlled by conformation.

A detailed study of the oxy-anion assisted rearrangement of cyclononadienols has verified that [3,3]-sigmatropic carbon shifts are usually preferred over [1,5]-sigmatropic hydrogen shifts but substituents can exert subtle effects and this preference may be reversed or other reactions may predominate in some cases (L.A.Paquette, G.D.Crouse and A.K.Sharma, *J. Am. Chem. Soc.*, 1982, **104**, 4411).

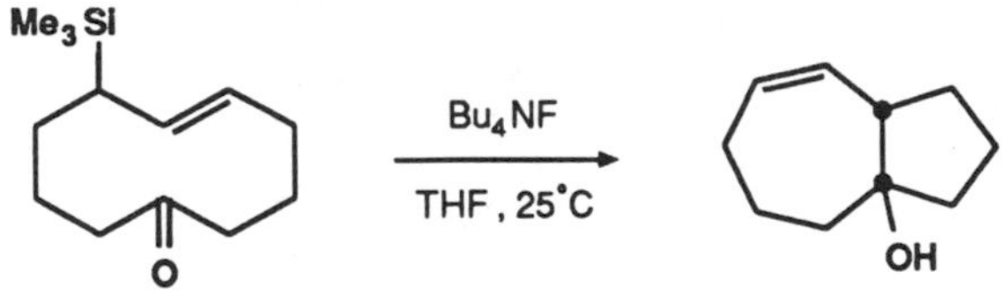

Scheme 20.

The rearrangement of 5-cyclodecenone affords a *cis*-fused bicyclo[5.3.0]decen-1-ol (hydroazulenol), Scheme 20, especially in the presence of a trimethylsilyl group at the allylic site adjacent to the incipient bridgehead carbon (T.Jisheng, T.Gallardo and J.B.White, *J. Org. Chem.*, 1990, **55**, 5426). The rearrangement of the allyl silane is promoted by fluoride ion. A similarly placed tributylstannyl substituent allows greater versatility in reaction conditions and the stereochemistry at the bridge can be controlled so as to give either *cis* or *trans* ring fusion (W.Fan and J.B.White, *Tetrahedron Lett.*, 1993, **34**, 957). It is well known that 5-cyclodecynone rearranges in presence of acid to bicyclo[4.4.0]dec-1-en-2-one and the mechanism has now been examined in detail (C.E.Harding and G.R.Stanford, *J. Org. Chem.*, 1989, **54**, 3054). The keto oxygen atom is transferred across the ring by intermediate formation of an oxete. Likewise the mechanistic details of the rearrangement of the cyclic bisallene, 1,2,6,7-cyclodecatetraene (45), have been established from a very detailed study of radical trapping behaviour (W.R.Roth, T.Schaffers and M.Heiber, *Chem. Ber.*, 1992, **125**, 739). This is probably a non-concerted Cope reaction.

45

Electroreduction of (*Z,E*)-4,8-cyclododecadienone produces a ketyl radical which rearranges to bicyclic alcohols. The regiochemistry and stereochemistry of the ring closure have been determined (F.Lombardo, R.A.Newmark and E.Kariv-Millar, *J. Org. Chem.*, 1991, **56**, 2422).

174

Due to the discovery of the highly potent anticancer antibiotic properties of compounds containing a 9- or 10-membered enediyne ring (see section 4 below), there has been an upsurge in interest in the cycloaromatisation reaction of simple model compounds, the Bergman reaction, Scheme 21 (R.G.Bergman, *Acc. Chem. Res.*, 1973, **6**, 25). The parent cyclodecenediyne has a barrier of 23.6 kcal mol^{-1} to cyclisation and hence can be isolated but the corresponding C_9 system is too unstable and can only be obtained in the aromatised form. For extensive discussion of mechanism and molecular mechanics calculations see J.P.Snyder, *J. Am. Chem. Soc.*, 1990, **112**, 5367; 1989, **111**, 7630; K.C.Nicolaou *et al.*, *J. Am. Chem. Soc.*, 1988, **110**, 4866, 7247 and references therein.

Scheme 21.

Triene cyclisation by intramolecular Diels-Alder is complicated by stereochemical considerations since the two new bonds can create four chiral centres (Scheme 22). Both synthetic and theoretical studies are reported for the C_{14} triene, including the preparation and transannular cyclisation of all eight geometric isomers of the macrocyclic triene. For the two isomers with the most favourable geometry the Diels-Alder reaction occurs spontaneously, in the other cases Cs_2CO_3 in THF/DMF at 80°C was effective (T.Takahashi *et al.*, *J. Am. Chem. Soc.*, 1988, **110**, 2674; T.Takahashi, Y.Sakamoto and T.Doi, *Tetrahedron Lett.*, 1992, **33**,

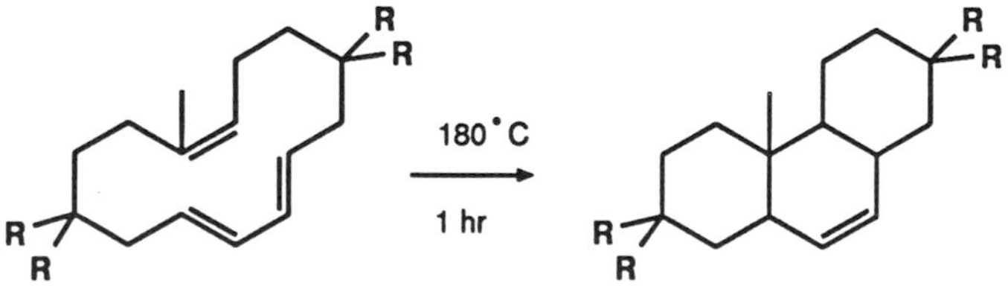

Scheme 22.

3519; S.Lamothe, A.Ndibwami and P.Deslongchamps, *Tetrahedron Lett.,* 1988, **29**, 1639, 1641) and for the analogous C_{13} triene (K.Baettig *et al., Tetrahedron Lett.,* 1987, **28**, 5249, 5253).

The isomerisation of 1,5,9-cyclododecatriyne to hexaradialene (Scheme 23) has been known for some time (A.J.Barkovich, E.S.Strauss and K.P.C.Vollhardt, *J. Am. Chem. Soc.,* 1977, **99**, 8321), but the mechanism has not been established. With the aid of ^{13}C-labelled material the involvement of a tricyclobutabenzene intermediate structure can be ruled out. This rearrangement probably proceeds by successive sigmatropic shifts (W.V.Dower and K.P.C.Vollhardt, *J. Am. Chem. Soc.,* 1982, **104**, 6878).

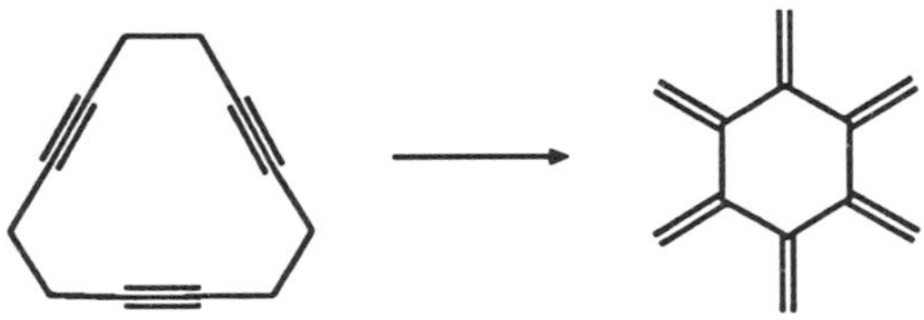

Scheme 23.

G.Haufe has shown that hydroxybromination of cycloalkenes with N-bromosuccinimide leads to transannular *O*-heterocyclisation (*J. Chem. Res. S,* 1987, 100 and references therein).

(c) Complex formation

Although macrocyclic olefins could be expected to form metal complexes with great facility their use as ligands has been surprisingly limited. Compounds reported prior to 1980 include platinum complexes of C_9 dienes and iron complexes of C_9-tetraenes (see to *Comprehensive Organometallic Chemistry,* Eds. G.Wilkinson, F.G.A.Stone and E.W.Abel, Pergamon Press, 1982).

Cyclododecatetraene forms an η^1,η^3-bridging complex with a Ru_3 cluster and with a Ru_3-Au_3 double cluster (M.I.Bruce, O.B.Shawkataly and B.K.Nicholson, *J. Organomet. Chem.,* 1984, **275**, 223). Similar isomeric complexes with a Ru_4 cluster are also reported (M.I.Bruce *et al., New J. Chem.* 1988, **12**, 595). Iron tricarbonyl Complexes of 1,3,5,7-

176

cyclononatetraene, (olefin)Fe(CO)$_3$, are η^4-bonded and two isomeric forms have been isolated (E.Meier, R.Roulet and A.Scrivanti, *Inorg. Chim. Acta,* 1980, **45**, L267).

The formation of a cyclobutadiene ring by complexation of two acetylenic groups to cobalt is a well established reaction. Gleiter and his coworkers have explored this chemistry with cyclic diynes and found that the ring size can influence the nature of the product. All diynes in the range C$_{10}$ to C$_{13}$ react with a cyclopentadienyl-cobalt carbonyl complex to form mononuclear complexes in which the two yne groups react on the Co atom in the usual way to form a tricyclic ligand (46) containing a cyclobutadienyl ring (R.Gleiter, *et al., Tetrahedron Lett.,* 1992, **33**, 1733 and references therein). However, 1,6-cyclodecadiyne reacts so as to form the cyclobutadienyl ring intermolecularly, giving rise to a superphane structure capped with two CpCo moieties (47).

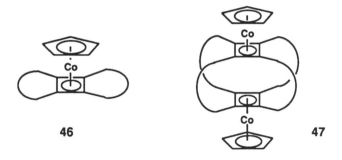

46   **47**

Cyclic diynes also form complexes with AgSO$_3$CF$_3$. The C$_{10}$, C$_{12}$ and C$_{14}$ diyne complexes and the analogous copper complexes are polymeric by X-ray analysis (R.Gleiter *et al., Chem. Ber.,* 1990, **123**, 1461).

β-Diketones complex to many metals by chelation of the enolate anion, and cyclotridecane-1,3-dione likewise forms many metal complexes. In contrast cyclodecane-1,3-dione will complex only with Fe(III) and Cu(III). Evidently ring strain inhibits formation of the enol (C.L.Modenbach *et al., Inorg. Chem.,* 1983, **22**, 4083).

4. Biochemistry

Neocarzinostatin is a protein/chromophore combination which has antitumour antibiotic activity (I.H.Goldberg, *Acc. Chem. Res.*, 1991, **24**, 191). The active entity is the chromophore which has, at its core, a highly strained cyclononenediyne ring (annelated to a C_5 ring, forming a bicyclo[7.3.0]decane) (48). This substance was first isolated in 1965 but its structure and biochemistry have been elucidated only in the last 15 years. The antitumour activity of neocarzinostatin arises from the possibility for aromatisation of the enediyne moiety with concomitant formation of a biradical species. This biradical can become involved in DNA chemistry and ultimately causes DNA scission. Recently a new compound has been reported, kedarcidin, with the same core cyclononenediyne ring and a similar potent antitumour effect (J.E.Leet, *J. Am. Chem. Soc.*, 1992, **114**, 7946.

Another class of compounds, esperamicin, calicheamicin and dynemicin (each itself a subclass with several variants) has a structure based on a bicyclo[7.3.1]undecane ring system, which contains a cyclodecenediyne ring (49). Although these compounds are completely unrelated to neocarzinostatin they have the same potent antitumour effect. This is probably due to scission of DNA by interaction with the biradical from rearrangement of the enediyne moiety. For reviews and leading references to the chemistry of these antitumor antibacterial compounds see M.D.Lee, G.A.Ellstad and D.B.Borders, *Acc. Chem. Res.*, 1991, **24**, 235; M.D.Lee *et al.*, *J. Am. Chem. Soc.*, 1992, **114**, 985; P.Magnus *et al.*, *J. Am. Chem. Soc.*, 1992, **114**, 2544; K.C.Nicolaou and W.-D.Dai, *Angew. Chem., Int. Ed. Engl.*, 1991, **30**, 1387; K.C.Nicolaou *et al.*, *J. Am. Chem. Soc.*, 1992, **114**, 8890.

48

49

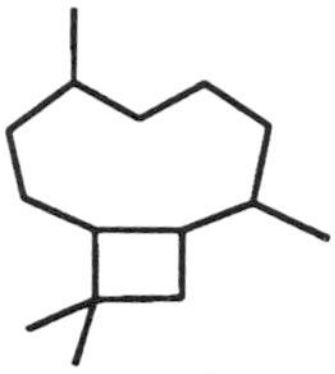

50 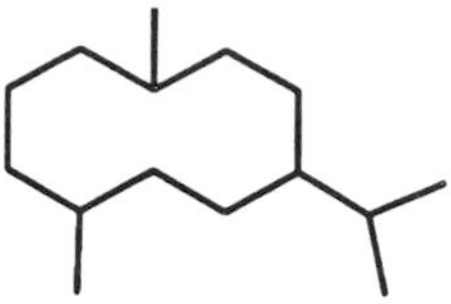   51 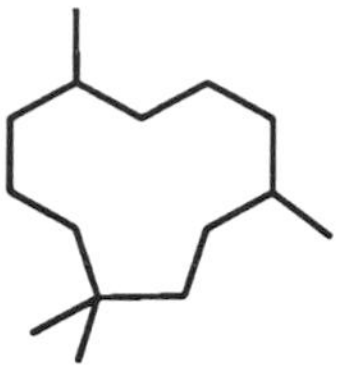   52

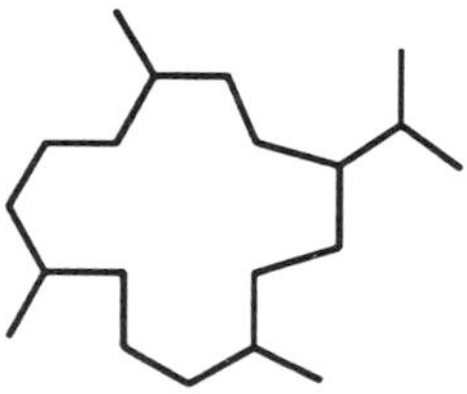

53 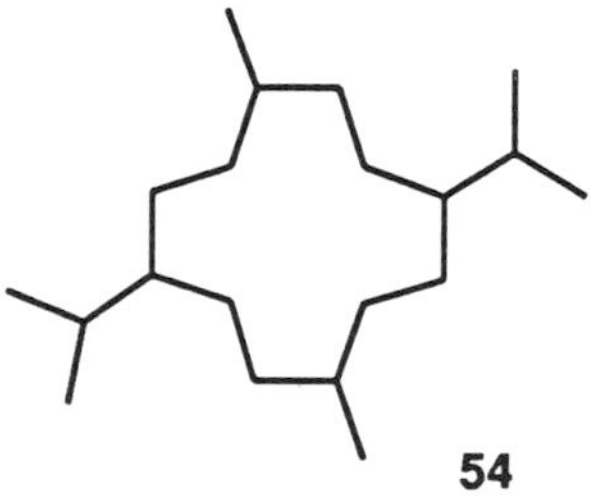   54

There are several different classes of terpenoid which contain a macrocyclic structure, and a very large number compounds in these classes have been isolated. Only representative systems are indicated here. Caryophyllane (50), germacrane(51), and humulane (52) are ubiquitous sesquiterpene systems, occurring widely in many species of plant. These systems contain C_9, C_{10} and C_{11}, rings respectively. Many thousands of derivatives have been isolated, with a very diverse range of biochemical effects. The chemistry and biochemistry of these natural products are covered in depth in the phytochemical literature and will not be discussed further (N.H.Fischer, E.T.Olivier and H.D.Fischer, *Prog. Chem. Org. Nat. Prod.*, 1979, **38**, 71; P.Kocovsky, F.Turecek, J.Jajicek, *Synthesis of Natural Products : Problems of Stereochemistry*, CRC, Florida, 1986, Vol 1).

The cembranes are a class of diterpenes based on a cyclotetradecane ring (53). This core structure can be functionalised variously with ene, hydroxy and epoxy groups and may contain annulated furan rings. Cembranoids have been isolated from pine resins and related plant sources but are found predominantly in marine organisms notably the soft corals. They exhibit a wide range of biological activity, but are of particular interest for their potent cytotoxic and anti-inflammatory effects. For recent work see A.J.Weinheimer, C.W.J.Chang and J.A.Matson, *Fortschr.*

Chem. Org. Naturst. 1979, **36**, 285 (naturally occuring cembranes); I.Wahlberg *et al.*, *J. Org. Chem.*, 1985, **50**, 4527 (tobacco cembranes); J.A.Marshall, M.J.Coghlan and M.Watanabe, *J. Org. Chem.*, 1984, **49**, 747 (synthetic procedures); M.A.Tius, *Chem. Rev.*, 1988, **88**, 719 (chemical review); L.A.Paquette and P.C.Astles, *J. Org. Chem.*, 1993, **58**, 165 (total synthesis).

The pseudopterane system (54) is closely related to the cembrane system. It is based on a cyclododecane ring, but this is a minor structural difference, corresponding notionally to an allylic shift. The pseudopteranoids are otherwise functionalised in a very similar way to the natural cembranoids and these compounds occur naturally in the same marine species (M.M.Bandurraga *et al.*, *J. Am. Chem. Soc.*, 1982, **104**, 6463; W.F.Tinto *et al.*, *Tetrahedron*, 1991, **47**, 8679; L.A.Paquette, A.M.Doherty and C.M.Rayner, *J. Am. Chem. Soc.*, 1992, **114**, 3910, 3926).

Second Supplements to the 2nd Edition of Rodd's Chemistry of Carbon Compounds, Vol. II B(Partial), C, D and E, edited by M. Sainsbury

181

Chapter 9

POLYCARBOCYCLIC COMPOUNDS WITH SEPARATE RING SYSTEMS, AND SPIRO COMPOUNDS

MARTIN WILLS

1. Introduction

The classification and nomenclature of compounds reviewed in this chapter follows that used in the second edition.

2. Compounds with rings joined directly or through a carbon chain

(a) Polycyclopropanes and polycyclopropenes

(i) Two directly connected three membered rings:
Treatment of butadiene with excess N_2CHCO_2Et in the presence of $CuSO_4$ or CuO gives (1) directly as a mixture of isomers (I. E. Dolgii, E. A. Shapiro and O. M. Nefedov, *Izv. Akad. Nauk SSSR, Ser. Khim.,* 1979, 1282-6). The analogous reaction of acetylene (2) gives (3), which has been employed as a dienophile (E. A. Shapiro, G. V. Lun'kova, I. E. Doglii and O. M. Nefedov, ibid., 1981, 1316).

The reaction of dichlorocarbene with dienes remains a popular method for the synthesis of dicyclopropanes and numerous examples of its application have been reported (G. V. Kryshtal, V. F. Kucherov and L. A. Yanovskaya, ibid., 1978, 2803, 2806; A. Kh. Khusid, Kryshtal, V. F. Kucherov et al., ibid., 1977, 2135; N. I. Yakushkina, L. F. Germanova, V. D. Klebanova, L. I. Leonova and I. G. Bolesov, ibid., 1976, 2141; Y. M. Slobodin, I. Z. Egenburg and A. S. Khachaturov, ibid., 1974, 21; R. R. Kostikov, V. S. Aksenov and I. A. D'yakonov, *Zh. Org. Khim.,* 1974, 2099; F. Kaspar, T. Beier, *Z. Chem.,* 1976, **16**, 435; L. A. Khachatryan, R. A. Kazaryan, G. S. Grigoryan et al., *Arm. Khim. Zh.,* 1989, **42**, 762).

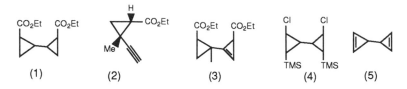

(1) (2) (3) (4) (5)

The reaction of chlorocarbene with 1,4-bis(trimethylsilyl)butadiene resulted in the formation of (4) from which bicyclopro-2-enyl (5) was generated by treatment with tetrabutylammonium fluoride (W. E. Billups and M. M. Haley, *Angew. Chem.*, 1989, **101**, 1735).

Two improved methods for the synthesis of bicyclopropylidene (6) rely on the cyclopropanation of an unsaturated nitrile (7) (A. H. Schmidt, U. Schirmer and J. M. Conia, *Chem. Ber.*, 1976, **109**, 2588) and a silyl enol ether (8) (W. Weber and A. DeMeijere, *Synth. Commun*, 1986, **16**, 837) respectively, followed by elimination to form the double bond.

(6) (7) X=CN (9) (10) (11)
     (8) X=OTMS

The reaction of a carbene containing a cyclopropyl group with an alkene provides an alternative entry to dicyclopropyl systems (V. V. Ragulin and E. N. Tsvetkov, *Izv. Akad. Nauk. SSSR, Ser. Khim.*, 1988, 2652), however in some cases exact control of reaction conditions is essential.

Photolysis of the diazomethane (9) in isobutylene at 25°C yields only a trace of the adduct (10) along with 1-phenylcyclobutene (63%) and phenylacetylene (16%), however the same reaction at -128°C gave (10) in 42% yield and less than 14.4% of the other products (R. A. Moss and W. P. Wetter, *Tetrahedron Lett.*, 1981, **22**, 997).

The use of cyclopropyl chromium pentacarbonyl carbenes (11) in reactions with electron poor alkenes has also been reported to provide an efficient access to the target systems (J. W. Herndon and S. U. Tumer, *J. Org. Chem.*, 1991, **56**, 286). Carbene (12), generated from the the diazo precursor, gave bicyclopropyl compound (13) in 64-68.5% yield and with less than 3% of the spiro side product (14) (A. R. Kraska, L. I. Cherney and H. Schechter, *Tetrahedron Lett.*, 1982, **23**, 2163). Reactions of a *gem*-dinucleophile with 1,2 dibromides has also been used for the synthesis of cyclopropanes as illustrated by the synthesis of cyclopropane amino carboxylate (15) from (16) and $CNCH_2CO_2Et$ (M. C. Pirrung and G. M. McGeehan, *Angew. Chem., Int. Edn. Engl.*, 1985, **24**, 1044).

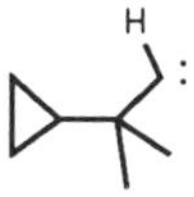

(12)    (13)    (14) 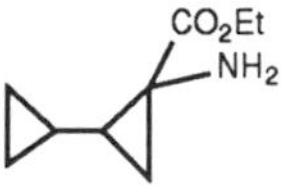   (15) 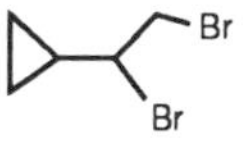   (16)

Direct coupling of two cyclopropyl rings may be achieved by the direct anodic oxidation of carbethoxycyclopropanecarboxylic acids (radical coupling) (K. Kimura, S. Horie, I. Minato et al., *Chem. Lett.,* 1973, 1209) or by the reaction of cyclopropane rings with organometallic reagents. The latter approach is illustrated by the reaction of methylmagnesium iodide with 1-methylcyclopropene followed by treatment with acetaldehyde to yield 30% of bicyclopropyl alcohol (17), in addition to the expected monocyclopropyl alcohol products (scheme 1) (O. A. Nesmeyanova and T. Y. Rudashevskaya, *Izv. Akad. Nauk., SSSR, Ser. Khim.,* 1978, 1562;O. A. Nesmeyanova, T. Y. Rudashevskaya and V. I. Grinberg, *ibid.,* 1977, 2590).

In an unusual dehydrohalogenation reaction the bromide (18, R=H or an n-alkyl chain) was converted in 65-80% yield into dimer (19) upon treatment with potassium hydroxide in DMSO containing a crown ether (A. I. D'yachenko, S. Agre, A. M. Taber et al., *ibid.,* 1984, 955). One example of a Favorskii reaction of cyclobutanones has been reported as effective for the synthesis of dicyclopropyl systems (N. A. Donskaya, A. G. Bessmertnykh and Y. S. Shabarov, *Zh. Org. Khim.,* 1989, **25**, 332).

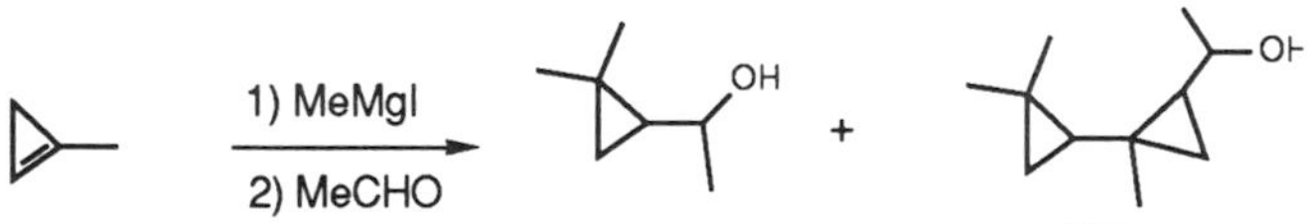

Scheme 1

(18) 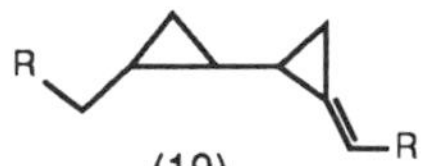   (19)    (20) 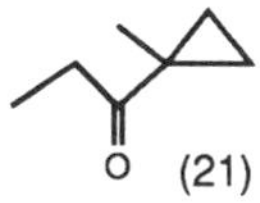   (21)

Reactions of dicyclopropyl systems generally depend on the energetic preference for opening of the strained three membered ring. Treatment of (20) with sodium hydroxide/methanol results in ring opening to the ketone (21) in 95% yield (C. Girard and J. M. Conia, *Tetrahedron Lett.,* 1974, 3333). Similarly the copper (I) catalysed reaction of (22) generates a carbene which reacts with cyclopentadiene and derivatives to give (23). This compound undergoes thermal ring expansion to (24) (Y. V.

Tomilov, V. G. Borkakov, N. M. Tsvetkova, et al., *Izv. Akad. Nauk SSSR, Ser. Khim.*, 1983, 336).

Attempted dihydroxylation (osmium tetroxide) of 2,2,2',2'-tetramethylbis(cyclopropylidene) (25) results in formation of a mixture of 3,3,5,5-tetramethylcyclohexane-1,2-dione and 2-*t*-butyl-3,3-dimethyl cyclopropane-1-carboxaldehyde whereas in contrast the same reaction on substrates with a larger ring (four to six membered) at one end of the double bond gave the expected diols (A. Tubul, M. Bertrand and C. Ghiglione, *Tetrahedron Lett.*, 1979, 2381).

The reaction of cyclopropylcyclopropene (26) with methyllithium at low temperature (-70°C) gives the ring-opened allenic product (27) (N. I. Yakushkina and I. G. Bolesov, *Zh. Org. Khim.*, 1978, **14**, 1788). Reaction of bicyclopropenyl compound (28) with bromine results in ring opening to a mixture of cyclobutene (29) (15%) and triene (30) (60%) (R. Weiss and H. P. Kempcke, *Tetrahedron Lett.*, 1974, 155). An unusual double intramolecular cyclisation was observed when dicyclopropane (31) was treated with *n*-butyllithium or LDA; the structure of the major product (32) was confirmed by X-ray crystallography (K. Kratzat, F. W. Nader and T Schwarz, *Angew. Chem., Int. Edn. Engl.*, 1981, **20**, 611).

(22)　　　(23)　　　(24)　　　(25)

(26)　　　(27)　　　(28)　　　(29)

Bicyclopropylidene (6) undergoes cyclisation reactions in at least six different modes according to a recent comprehensive study (A. De Meijere, I. Erden, W. Weber et al., *J. Org. Chem.*, 1988, **53**, 152). Another unprecedented reaction is the addition of (6) to tetrachloropropene to give the tricyclopropane (33) (W. Weber, U. Behrens and A. De Meijere, *Chem. Ber.*, 1981, **114**, 1196). The reaction of bicyclopropenyl derivative (34) with a tetrazene (35) is reported to yield two semibullvalene derivatives (36), which were characterised by X-ray crystallography (N. S. Zefirov, T. S. Kuznetsova and S. I. Kozhushkov, *Zh. Org. Chem.*, 1983, **19**, 1599).

(30)　　　　(31)　　　　(32)　　　　(33)

(34)　　　(35)　　　(36)　　　(37)　　　(38)
　　　　　　　　　　R=CN, CO$_2$Et

(ii) Two cyclopropane rings connected by a carbon chain

The reaction of cyclopropanecarboxaldehyde with Ph$_3$P=CH(CH$_2$)$_2$Br gave (37) in 55-60% yield. Treatment with bromine, followed by dehydrobromination by pyridine gave dicyclopropylidenemethane (38) in 58% yield (R. Kopp and M. Hanack, *Angew. Chem., Int. Edn. Engl.,* 1975, **14**, 4389). The isomeric allene has also been prepared (T. Satoh, Y. Kawase and K. Yamakawa, *Bull. Chem. Soc. Jpn.,* 1991, **64**, 1129).

An efficient synthesis of ketone (39) was achieved by the reaction between 1,2,3-tri-*t*-butylcyclopropenium tetrafluoroborate and Me$_3$SiCHN$_2$, followed by photolysis of the intermediate azine (G. Maier, I. Bauer, D. Born et al., *Angew. Chem.,* 1986, **98**, 1132). Selective functionalisations of 1,1-dicyclopropylalkenes may often be carried out without ring opening as illustrated by the conversion of (40) to bromohydrin (41) using aqueous bromine (N. A. Donskaya, E. V. Shulishov and Y. S. Shabarov, *Zh. Org. Chem.,* 1981, **17**, 2102) and the photooxygenation of (42) to a 75:25 mixture of allyl alcohols (43) and (44), after reductive workup (G. Rousseau, P. Le Perchec and J. M. Conia, *Tetrahedron Lett.,* 1977, 45). In contrast ring opening in the related methylenecyclopropanes (45) promoted by *n*-butyllithium leads to the formation of spiro[2,4]heptanes (46) (T. V. Akhachinskaya, N. A. Donskaya, Y. K. Grishia et al., *Zh. Org. Chem.,* 1982, **18**, 458; idem., ibid., 1987, **23**, 332). [2+2] Cycloaddition reactions of the same systems with ketenes have been reported (N. A. Donskaya, A. G. Bessmertnykh, O. E. Grushina et al., ibid., 1987, **23**, 2369).

The preparation of bis(2,3-diphenylcyclopropylium)methane bis(tetrafluoroborate) (47) via a process of hydride abstraction and then protonation of the neutral precursor (48) has been reported (K. Komatsu, K. Masumoto and K. Okamoto, *J. Chem. Soc., Chem. Commun.,* 1977, 232).

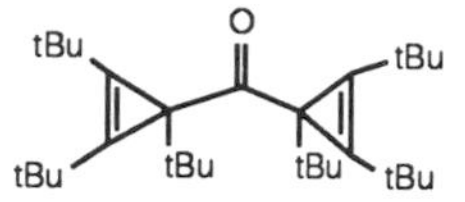

(39)

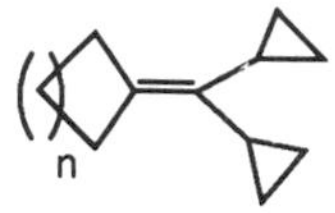

(40) n=1
(42) n=2

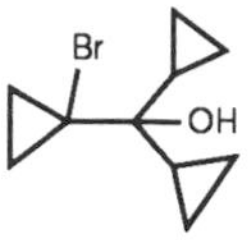

(41)

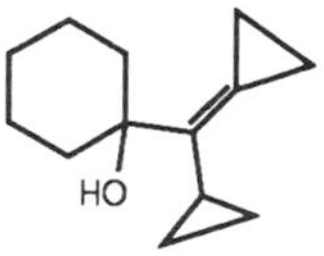

(43)

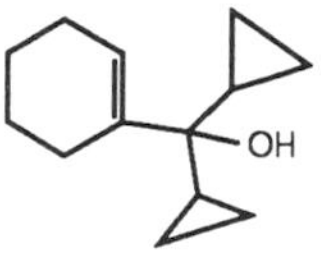

(44)

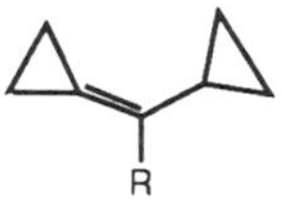

(45)
R=Me,c-C$_3$H$_5$

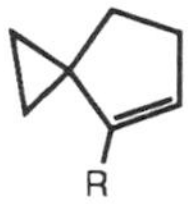

(46)

(47)

(48)

(49)

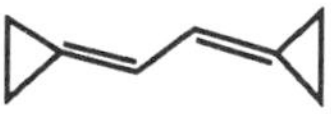

(50)

The synthesis and properties of cyclopropyl rings linked by an acetylene group (49) have been reported (O. M. Nefedov, I. E. Dolgii, I. B. Shvedova et al., *Izv. Akad. Nauk. SSSR, Ser. Khim.*, 1978, 39; A. L. Ivanov and I. N. Domnin, *Zh. Org. Khim.*, 1988, **24**, 2547; A. V. Tarakanova, S. V. Baranova, T. Pehk et al., ibid., 1987, **23**, 515) as has the isomeric di(cyclopropylidene)ethane (50) system ((L. A. Paquette, G. J. Wells and G. Wickham, *J. Org. Chem.*, 1984, **49**, 3618).

(iii) Three or more three-membered rings

Although the self condensation reaction of ethyl cyclopropanecarboxylate is reported to lead to the formation the trimer (51) (H. W. Pinnick, Y-H. Chang , S. C. Foster et al., *J. Org. Chem.*, 1980, **45**, 4505), these compounds are commonly prepared via a carbene insertion reaction onto an alkene, as illustrated by the conversion of (52) to (53) (M. E. Alonso and M. Matilde, *Tetrahedron Lett.*, 1979, 2763) and other related reactions (M. I. Komendantov, V. N. Pronyaev and R. R. Bekmukhametov, *Zh. Org. Chem.*, 1979, **15**, 328; O. M. Nefedov, I. E. Dolgii, E. V. Bulusheva et al., *Izv. Akad. Nauk. SSSR, Ser. Khim.*, 1979, 1535; idem, ibid., 1976, 1901; R. R. Kostikov and A. P. Molchanov, *Zh. Org. Chem.*, 1978, **14**, 879).

A synthesis of the tricyclopropylcyclopropenyl cation (54) via addition of cyclopropylchlorocarbene to dicyclopropylacetylene has been independantly reported by two groups (K. Komatsu, I. Tomioka and K.

Okamoto, *Tetrahedron Lett.*, 1980, **21**, 947; R. A. Moss and R. C. Munjal, *Tetrahedron Lett.*, 1980, **21**, 1221).

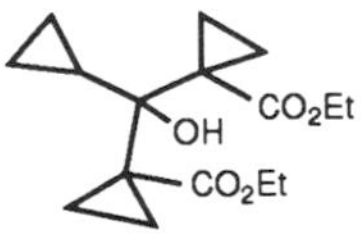

(51)    (52) 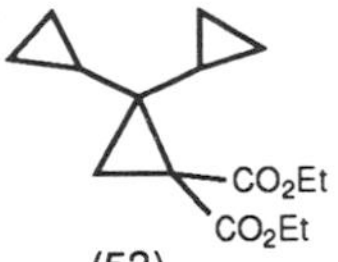   (53) 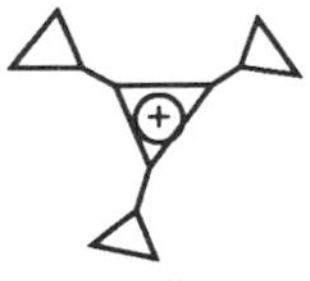   (54)

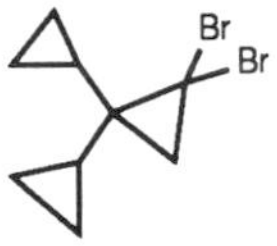

(55) 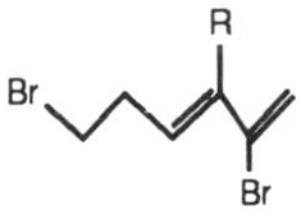   (56) 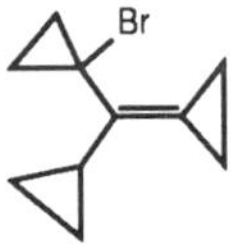   (57)    (58)

Ring opening reactions of these materials dominate their chemistry; hence heating of *gem* dibromide (55) results in rearrangement to a mixture of *cis* and *trans*- (56) in 70% yield (R. R. Kostikov and A. P. Molchanov, *Zh Org. Chem.*, 1978, **14**, 1108), whilst solvolysis of (57) leads to the ring expanded product (58) (N. A. Adamchyan and Y. S. Shabarov, *ibid.*, 1983, **19**, 1769). A double ring expansion was observed in the conversion of (59) to (60) using bromine/triphenylphosphine (64%) (N. S. Zefirov, S. I. Kozhushkov and T. S. Kuznetsova, *ibid.*, 1988, **24**, 447).

Photolysis of diazenes results in nitrogen extrusion to give hydrocarbons, as illustrated by the conversion of (61) to (62) (J. W. Timberlake and Y. M. Jun, *J. Org. Chem.*, 1979, **44**, 4729; M. Bruch and J. W. Timberlake, *J. Org. Chem.*, 1986, **51**, 2969). A similar process of nitrogen and sulphur extrusion (using tributylphosphine) permitted the conversion of (63) into the alkene (64) (K. Reinholdt and J. P. Aune, *J. Mol. Catal.*, 1983, **22**, 73).

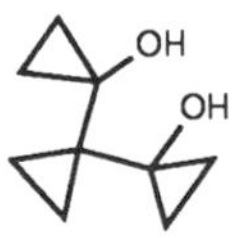

(59)    (60) 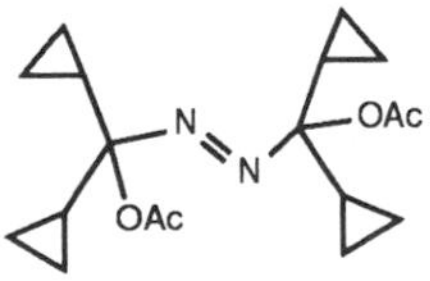   (61) 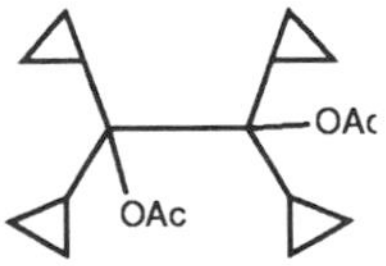   (62)

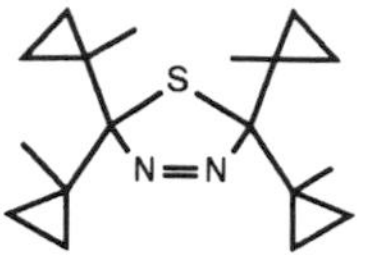

(63) 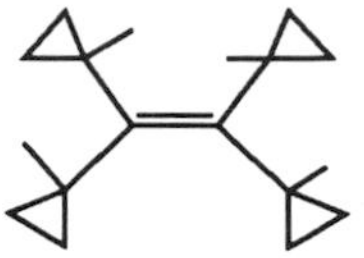   (64)

(b) Cyclopropylcyclobutanes and cyclopropylbutenes

Cyclopropanation of the allene (1) using diazomethane in the presence of palladium(II) gives cyclopropylidenecyclobutane (2) in 70% yield (N. S. Zefirov, K. A. Likin and A. Y. Timofeeva, *Zh. Org. Khim.,* 1987, **23**, 2545). A [2+2] cycloaddition of dichlorokene with an olefin was employed for the synthesis of (3) (A. G. Bessmertnykh, Y. N. Bubnov, T. I. Voevodsdkaya et al., *Zh. Org. Khim.,* 1990, **26**, 2348) in which the chlorine atoms could be substituted with methoxy groups by reaction with sodium methoxide. These reactions proceed without opening of any of the rings (N. A. Donskaya, A. G. Bessmertnykh, V. A. Drobysh et al., ibid., 1987, **23**, 751).

The solvolysis of cyclopropylcyclobutenyl triflates (4) (prepared from the corresponding ketones) in fluorinated alcohols takes place through an S_N1 mechanism involving a intermediate non classical 1-cyclobutenyl cation (G. Auchter and M. Hanack, *Chem Ber.,* 1982, **115**, 3402). A [3+4] cycloaddition was observed in the conversion (thermal) of dispiro alkanes (5) to the corresponding seven-membered ring products (6) (S. Sarel, A. Felzenstein and J. Yovell, *J. Chem. Soc., Chem. Comun.,* 1974, 753).

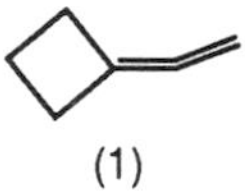

(1) 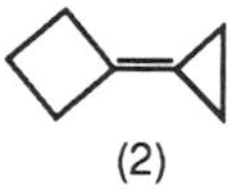   (2) 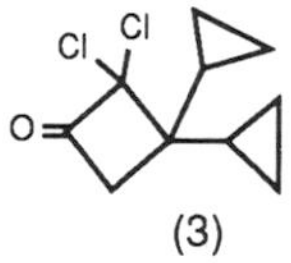   (3) 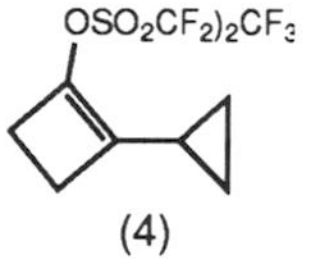   (4)

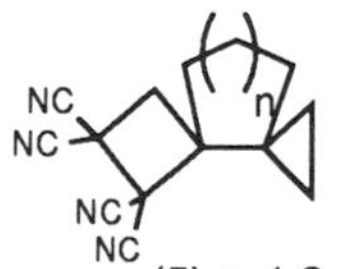

(5) n=1,2 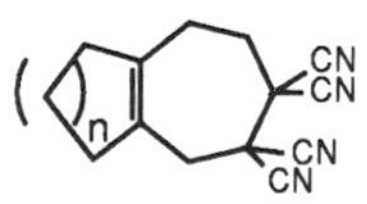   (6) n=1,2

(c) Polycyclobutanes

The synthesis, epoxidation and cyclopropanation reactions of bicyclobutylidene (1) have been reported in detail ((L. K. Bee, J. Beeby, J. W. Everett et al., *J. Org. Chem.*, 1975, **40**, 2212; P. A. Krupcho and E. G. E. Johngen, *J. Org. Chem.*, 1974, **39**, 1650), as has an electron diffraction study of the same molecule (V. S. Mastryukov, N. A. Tarasenko, L. V. Vilkov, et al, *Zh. Strukt. Chem.*, 1981, **22**, 57). A number of catalysts for the rearrangement of (1) to bicyclo[3,3,0]octene (2) have been examined of which Al_2O_3 gave the best yield (70%) (E. S. Finkel'shtein, B. S. Strel'chik, V. M. Volovin et al., *Dokl. Akad. Nauk. SSSR,* 1975, **220**, 131).

Directly attached four-membered rings have been prepared by the reductive coupling of cyclobutanones using a Mg-Hg couple (H. Rupp, W. Schwarz and H. Musso, *Chem. Ber.,* 1983, **116**, 2554). The elusive strained diene dicyclobutylideneethane (3) has been successfully prepared by the reaction between cyclobutylideneacetaldehyde and cyclobutylidene triphenylphosphorane (G. Wickham, G. J. Wells, L. A. Paquette et al., *J. Org. Chem.*, 1985, **50**, 3485).

A remarkable co-metathesis reaction in the presence of Re_2O_7/Al_2O_3 catalyst was effective for the conversion of cyclopentene and dicyclobutylidene to the diene (4) (V. M. Vodovin, S. V. Kotov, E. V. Portnykh et al., *Neftekhimiya,* 1988, **28**, 61). A number of cationic reactions of methylenecyclobutane have been reported, one of which is dimerisation catalysed by boron trifluoride etherate in 30% yield to (5) (E. S. Finkel'shtein, A. I Mikata, E. M. Sire et al., *Dokl. Akad. Nauk. SSSR,* 1976, **228**, 1123).

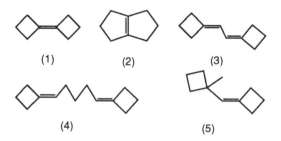

(1)   (2)   (3)

(4)   (5)

(d) Polycyclopentanes and polycyclopentenes

Cyclopentane dimers are most commonly prepared by the direct coupling of two monocyclic precursor radicals (A. A. Stepanov, I. N. Rozhkov and I. L. Knunyants, *Izv. Akad. Nauk. SSSR, Ser. Khim.,* 1981, 701; A. A. Kamyshova, E. T. Chukovskaya and R. K. Freidlina, *ibid.,*1985, 1795; M. Kasai, M. Funamizu, M. Oda et al., *J. Chem.*

Soc., Perkin Trans. I, 1977, 1660) as illustrated by the conversion of cyclopentadiene into diamine (1) using hydroxyamine hydrochloride in the presence of zinc metal and titanium tetrachloride (T. A. Gadzhiev, A. M. Mustafaev, U. K. Agaev et al., *Zh. Prikl. Chem., (Leningrad),* 1976, **49**, 933). The same overall coupling may be achieved in the combination of electrophilic and nucleophilic reaction partners in an aldol condensation between cyclopentanones (A. S. Koyasimov, M. M. Movsumade and Z. A. Safarov, *Dolk. Akad. Nauk. Az. SSSR,* 1979, **35**, 31; A. H. Schmidt and M. Ross, *Chem.-Ztg.,* 1979, **103**, 183) or a nitrogen, or sulphur/selenium, extrusion reaction from an appropriate precursor (E. R. Cullen, F. S. Guziec Jr. and C. J. Murphy, *J. Org. Chem.,* 1982, **47**, 3563; A. Krebs and W. Rueger, *Tetrahedron Lett.,* 1979, 1305; A. Krebs, W. Rueger and W. U. Nickel, *Tetrahedron Lett.,* 1981, **22**, 4937).

Each of the above reactions generates a product bearing an unsaturated linkage between the rings which may be employed for further functionalisation reactions. Bicyclopentane functions may be also be created by a number of cyclisation reactions as illustrated by the conversion of the dienolate derived from (2) into (3) (K. Antczak, J. F. Kingston, S. J. Alward et al., *Can. J. Chem.,* 1984, **62**, 829).

Cyclisation reactions were also employed for the synthesis of (4) from (Z)-8-*t*-butyl-6,11-dimethyl-5,10-dodecadienal (5) with tin tetrachloride catalysis (P. Missiaen, P. J. DeClercq, L. Van Meervelt and G. S. D. King, *Bull. Chem. Soc. Belg.,* 1988, **97**, 993) and in the remarkable conversion of 1,3-dichloro-2,4-dimethyl-2,4-pentadiene to (6), which is believed to proceed through the intermediacy of cation (7) (Y. Gaoni, *Tetrahedron Lett.,* 1977, 371).

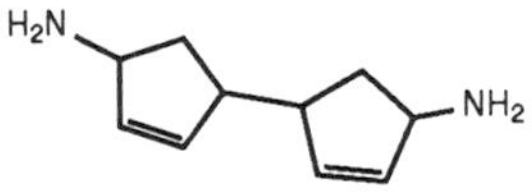

(1) 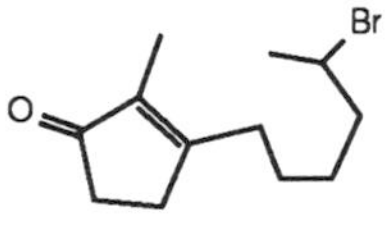   (2) 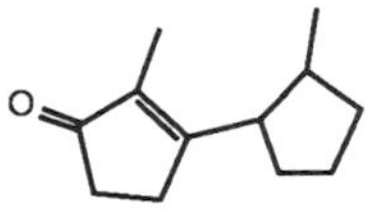   (3)

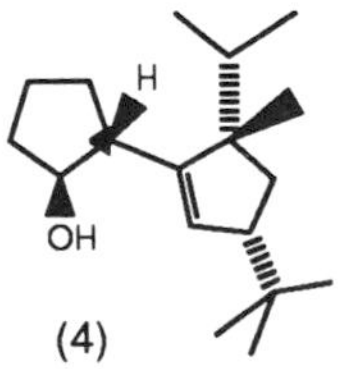

(4) 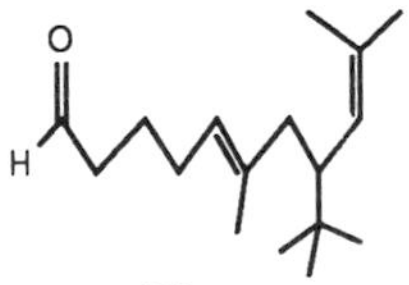   (5) 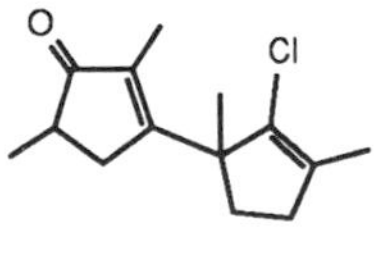   (6) 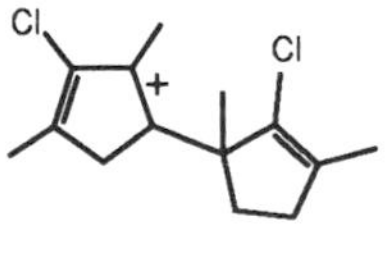   (7)

Ozonolysis of the enone produced by self-condensation of cyclopentanone leads to an efficient synthesis of 1,2-cyclopentanedione

(J. Wrobel and J. M. Cook, *Synth. Commun.*, 1980, **10**, 333). The stereochemistry of the diepoxidation reaction of 1,1'-dicyclopentene (8) (F. Plenat, F. Pietrasanta, M. R. Darvich et al., *Bull. Chem. Soc. Fr.*, 1975, 2227) and multiple cycloaddition reactions of 9,10-dihydrofulvalene (9) with acetylenes (D. McNeil, B. R. Vogt, J. J. Sudol et al., *J. Am. Chem. Soc.*, 1974, **96**, 4673) have been described. Pentafulvalenes (10) contain a full system of unsaturated carbon atoms and may be prepared by the oxidative coupling reaction of monomeric sodium cyclopentadienide precursors (W. Rutsch, A. Escher and M. Neuenschwander, *Chemia*, 1983, **37**, 160; idem, *Helv. Chim. Acta*, 1986, **69**, 1644; R. Brand, H. P. Krimmer, H. J. Lindner, V. Sturm and K. Hafner, *Tetrahedron Lett.*, 1982, 5131) and are known to undergo cycloaddition reactions with a number of substrates as well as each other.

Perchlorinated derivatives (G. I. Fray, G. M. Hearn and J. C. Petts, *Tetrahedron Lett.*, 1984, **25**, 2923; G. I. Fray, C. B. Jones and J. C. Petts, *J. Chem. Soc., Res. Synop.*, 1990, 260; J. Niessing and D. Fenske, *Monatsh. Chem.*, 1982, **113**, 1225) also exhibit a rich variety of reactivity as both dienes and dienophiles. The synthesis of a series of higher order cross-conjugated pentafulvalenes has also been reported (N. Schweikert, T. Netscher, G. L. McMullen et al., *Chem. Ber.*, 1984, **117**, 2006).

(8)   (9)   (10)   (11)

A molecule containing the hirsutene carbon skeleton (11) may be assembled through photocyclisation and *in situ* methanolysis of the initial [2+2] cycloaddition product of di(cyclopent-1-enyl)methane (12) (J. S. H. Kueh, M. Mellor and G. Pattenden, *J. Chem. Soc., Chem. Commun.*, 1978, 5; idem, *J. Chem. Soc., Perkin Trans. 1*, 1981, 1052). Certain hydrocarbon molecules, for example (13), derived from hydrogenation of (14), in this class possess distinctive odours and have application to the perfume industry which in this case led to the granting of several patents (T. Kawanobe, N. Noboru and K. Kojo, *JP. 80 35,041*, 1980; K. Tokoro, T. Kobayashi, S. Muraki et al., *JP. 79,125,645*, 1979,; E. Sundt and R. Chappaz, *Swiss App. 79/7,422*, 1979; B. J. Willis, J. M. Yurekko Jr., *US App. 77, 037* 1979).

Dilithiobis(butylcyclopentadienyl)ethane (15) has been used for the synthesis of a symmetric dicyclopentacyclooctene via reaction with glyoxal sulphate (K. Hafner, G. F. Thiele and C. Mink, *Angew. Chem.*, 1988, **100**, 1213). A year long experiment involving the use of (16) as a hydrogen absorbing material has been reported (E. S. W. Kong, D. H.

Doughty, J. E. Clarkson et al., *Natl. SAMPE Symp. Exhib. [Proc.],* 1985, 660).

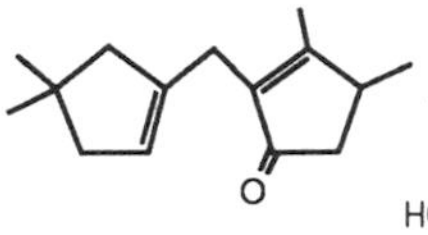

(12) 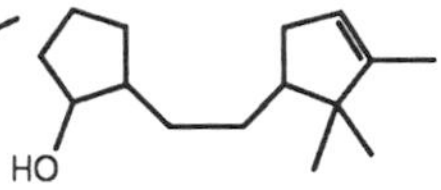   (13) 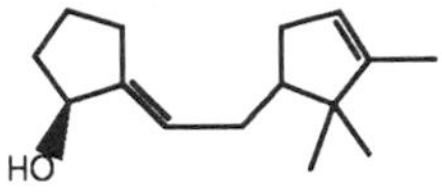   (+-)-(14)

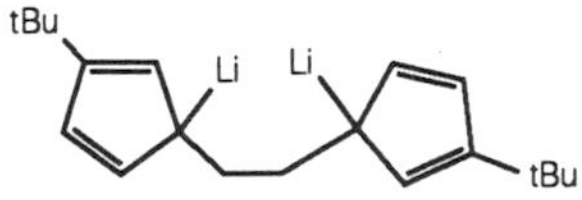

(15) 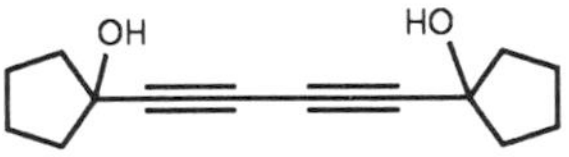   (16)

(e) Cyclopropylcyclopentanes and cyclopropylcyclopentenes

The reaction of the cyclopentadienyl anion with an appropriate cyclopropenyl electrophile has been employed for the synthesis of calicenes of general structure (1). For example, the reaction of sodium cyclopentadienide with 1,2-bis(*t*-butylthio-3,3-dichlorocyclopropene) (2) at -60°C gives (3) in 70-80% yield. A small amount of dimer was also formed (Z. Yoshida, *JP. 79, 115,357,* 1979; Z. Yoshida, *JP. 75, 126,643,* 1975). Similarly the addition of lithiated cyclopropane (4) to cyclopentanone (and a number of other ketones) gave (5) (E. W. Thomas, *Tetrahedron Lett.,* 1983, **24,** 1467) whilst the related lithiated species (6), generated from the dibromide, added in a 1,2 fashion to cyclopentenone to give alcohol (7) (G. Sabbione, A. Weber, R. Galli et al., *Chemia,* 1981, **35,** 95).

The reaction of a cyclopentanone with the metallated hemiacetal (8) gave adduct (9) in good yield, which was subsequently converted into fused bicyclic ketones via cyclopentane ring expansion (J. T. Carey and P. Helquist, *Tetrahedron Lett.,* 1988, **29,** 1243).

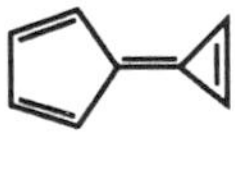

(1) 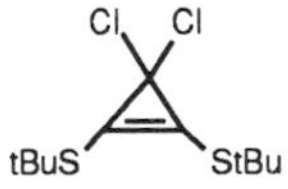   (2) 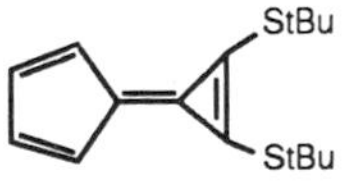   (3)    (4) 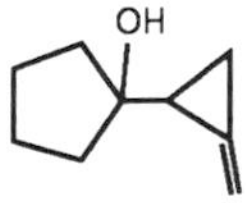   (5)

A titanium tetrachloride catalysed intramolecular cyclisation reaction was employed for the synthesis of (10) from the linear precursor (11) (Y. Hatanaka and I. Kuwajima, *Tetrahedron Lett.*, 1986, **27**, 719), whilst a remarkable intramolecular cyclopalladation reaction, followed by coupling to iodobenzene was effective for the conversion of (12) into the cyclopropylcyclopentane (13) (G. Fourney, G. Balme, J. J. Barieux et al., *Tetrahedron*, 1988, **44**, 5821).

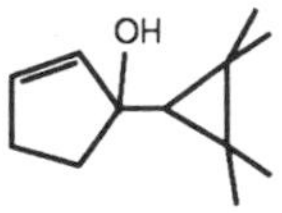

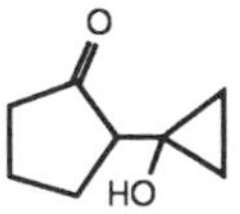

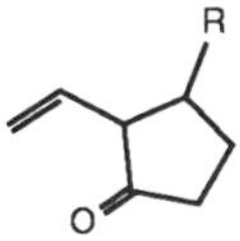

The synthesis of the first acetylenic fulvene; cyclopropylethynylfulvene (14), has been reported (I. E. Dolgii, K. N. Shavrin and S. V. Pavlycheva, *Izv. Akad. Nauk. SSSR, Ser. Khim.*, 1985, 2145). Cycloaddition reactions of cyclopropylethylenes (15) with tetracyanoethylene generally give [2+2] cycloadducts although in the case of (15, R=cyclopropyl or phenyl) the products are those of addition to the five membered ring i.e. (16) (S. Nishida, *Angew. Chem., Int. Edn. Engl.*, 1972, **11**, 328).

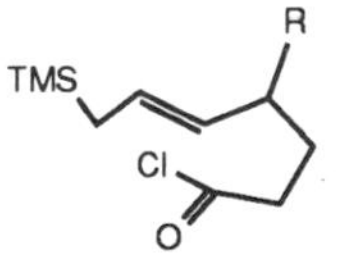

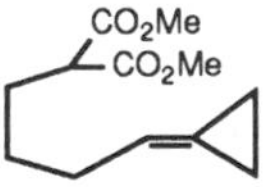

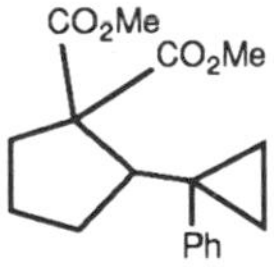

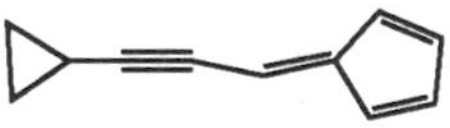

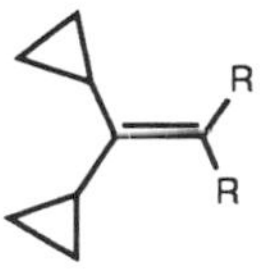

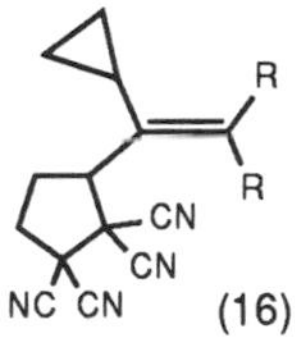

Reactions of compounds in this class generally involve ring opening processes of the three membered ring, which can lead to ring expansion as in the pyrolytic conversion of (17) to (18) (R. D. Miller, *J. Chem. Soc., Chem. Commun.*, 1976, 277; A. J. Barker, J. S. H. Kueh, G. Pattenden et al., *Tetrahedron Lett.*, 1979, 1881).

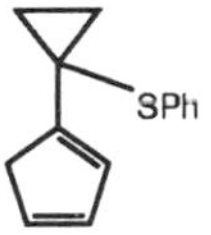

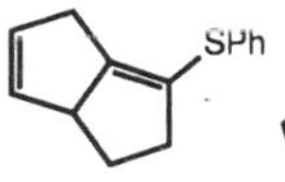

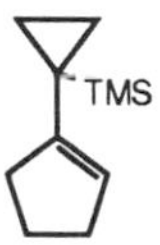

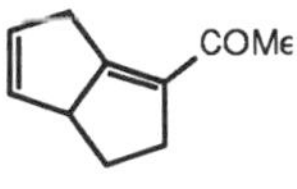

The inclusion of the trimethylsilyl group on the reagent (19) permits the direct synthesis of enone (20) in a pyrolysis reaction, followed by treatment with acetyl chloride and aluminium trichloride (L. A. Paquette,

G. J. Wells, K. A. Horn and T.-H. Yan, *Tetrahedron Lett.*, 1982, **23**, 263). In some cases however a ring opening to a diene can take place (G. Pattenden and D. Whybrow, *J. Chem. Soc., Perkin Trans. I.*, 1981, 3147). Larger ring expansions are possible when an appropriate double bond is located on one of the rings, for example in the rearrangement of (21) to the hydroazulene (22) (J. P. Marino and L. J. Browne, *Tetrahedron Lett.*, 1976, 3245).

In an example of an alternative mode of ring opening reaction of (23) with bromine gives the spiro compound (24) (B. M. Trost and M. K. T. Mao, *J. Am. Chem. Soc.*, 1983, **105**, 6753). Prostaglandins and prostanoic acids are important classes of products with numerous physiological properties which contain a five membered carbocyclic core. Numerous derivatives have been prepared including several in which a three membered ring is attached either directly (K. Sakai, K. Inoue, Y. Tajima et al., *JP. 73 62,123*, 1973; K. Inoue, S. Amemiya, Y. Tajima et al., *Chem. Lett.*, 1978, 1121; H. Vorbrueggen, U. Mende, B. Raduechel et al., *Ger. Offen. 2,460,711,* 1976; R. S. Aries, *Fr. Demande 2, 4000,905,* 1979) or at a further position (E. W. Walker, *Ger. Offen., 2,534,990,* 1976; K. Sakai, K. Inoue and T. Yusa, *JP. 76 13,757,* 1976; M. Baumgarth, D. Orth and H. E. Radunz, *Ger. Offen., 2,737,027,* 1979; I. Ohno, T. Nishioka, H. Takeda and N. Itaya, *JP. 77 07,427,* 1977; S. Kori, I. Oyama and T. Tanouchi, *JP. 76, 101,959,* 1976).

(21)　　　　　(22)　　　　　(23)　　　　　(24)

(f) Cyclopropylcyclohexanes

The formation of the three membered ring portion of this class of compound has been achieved by a selective cyclopropanation of (1) using diiodomethane and tri(isobutyl)aluminium to give (2) (K. Maruato, S. Sakane and H. Yamamoto, *Org. Synth.,* 1989, **67**, 176) and through the reaction of $Me_2S(O)CH_2$- with the terminal double bond in (3) to give (4) (J. C. Chalchat, R. Garry, A. Michet et al., *C. R. Hebd. Seances Acad. Sci. Ser C,* 1978, **286**, 329). A similar product (5) was obtained in the reaction of (6) with an appropriate electron poor double bond (J. P. Marino, T. Takushi, *Tetrahedron,* 1973, 3971).

An alkylidiene carbene generated from the dichloride (7) has been employed for the synthesis of cyclopropylcyclohexylidene (L. Xu, F. Tao and S. Wu, *Huaxue Xuebao,* 1984, **42**, 347; L. Xu, F. Tao, S. Wu et al., *Kexue Tongbao,* 1983, **28**, 447). Formation of the six membered ring via a Diels-Alder reaction (F. Kataoka, S. Nishida, T. Tsuji et al., *J. Am. Chem. Soc.,* 1981, **103**, 6878) or an intramolecular aldol condensation reaction (P. A. Grieco and Y. Ohfune, *J. Org. Chem.,* 1978, **43**, 2720) of a substrate containing an intact cyclopropyl group has also been reported.

(1) (2) (3)

(4) R=H, R'=H
(5) R=CO₂Et, R'=H
(8) R=H, R'=TMS

(6) (7) (9)

The cyclopropyl group in (4) remains unopened in its reaction with cuprates, addition occuring exclusively to the 4- position (T. Ibuka, E. Tabushi and K. Yasuda, *Chem. Pharm. Bull.,* 1983, **31**, 128). The silyl group in (8) prevents enolisation to this position hence allowing enolisation reactions to take place with high regioselectivity (L. A. Paquette, C. Blankenship and G. J. Wells, *J. Am. Chem. Soc.,* 1984, **106**, 6442).

Cyclopropylhexylidene itself is reported to undergo ene reaction with singlet oxygen at -75°C to give the alcohol (9) whilst at -50°C decomposition is observed (C. J. M. Van den Heuvel, H. Steinberg and T. J. De Boer, *Recl. Trav. Chim. Pays-Bas,* 1985, **104**, 145). Ring expansion reactions of derivatives of general structure (10) have been extensively studied, several examples of applications to the synthesis of 5-5 fused rings (B. M. Trost and D. E. Keeley, *J. Am. Chem. Soc.,* 1976, **98**, 248; J. C. Chalchat, R. P. Garry, B. Lacroix et al., *C. R. Acad. Sci., Ser. 2,* 1984, **298**, 395) and 5-7 fused rings (J. P. Marino and T. Kaneko, *Tetrahedron Lett.,* 1973, 3975; E. Piers, H. E. Morton, I. Nagakura et al., *Can. J. Chem.,* 1983, **61**, 1226) having been reported.

(g) Cyclobutylcyclohexanes

Ultrasound promoted dichloroketene-olefin cycloadditions have been employed for the synthesis of this class of hydrocarbon (G. Mehta, R. Prakash and H. Surya, *Synth. Commun.*, 1985, **15**, 991), as has the ring expansion reactions of cyclopropylcyclohexyl methyl systems. An illustration is the conversion of (1) to (2) (T. Cohen and L. Brockunier, *Tetrahedron*, 1989, **45**, 2917; J. A. Kaydos, J. H. Byaers and T. A. Spencer, *J. Org. Chem.*, 1989, **54**, 4698). The conversion of (2) into decalins by a further ring expansion has also been described (J. R. Matz and T. Cohen, *Tetrahedron Lett.*, 1981, **22**, 2459).

(1)   (2)

(h) Cyclopentylcyclohexanes

Aldol condensations between ketones often result in the formation of complex mixtures of products, however the combination of cyclohexanone and cyclopentanone in the presence of CaC_2 has been shown to give 65% 2-cyclohexylidenecyclopentanone (1), along with 33% of the corresponding dicyclopentylidene product (R. D. Sands, *Org. Prep. Proced. Int.*, 1974, **6**, 153).

The condensation between a cyclopentadienyl anion and cyclohexanone provides an efficient method for the synthesis of fulvenes (H. P. Figeys, M. Destrebecq and G. Van Lommen, *Tetrahedron Lett.*, 1980, **21**, 2369). In some cases the same reaction has been shown to be more efficient using imine derivatives of cyclohexanones (F. G. Fick, K. Hartke and R. Matusch, *Chem. Ber.*, 1975, **108**, 2593). The reaction of cuprates, for example (2), with cyclopentenones gives the expected products of 1,4 addition (3) (H. C. Arndt, W. G. Biddlecom, H. C. Kluender et al., *Prostaglandins*, 1975, **9**, 521; I. Tomoskozi, L. Gruber, G. Kovacs, V. Simonidesz, S. Virag and M. Szentivanyi, *Ger. Offen.*, 2,642,558, 1977; idem, *BP.*, 1,515,789, 1978). A remarkable titanium tetrachloride catalysed cyclisation of vinylsilane (4) to (5) provides a means for the rapid synthesis of the trichothecane carbon framework (E. Nakamura, K. Fukazaki and I. Kuwajima, *J. Chem. Soc., Chem. Commun.*, 1983, 499).

The reaction of β-pinene with cyclopentanone in the presence of Cu(OAc)$_2$ gave 44% (6) as the major product via a radical addition reaction (J. Y. Lallemand, *Tetrahedron Lett.*, 1975, 1217). The condensation of (7) with cyclohexanone yields the non-conjugated isomer (8) as the major product. This compound is reduced to the corresponding alcohol, which has applications as a scent (T. Yoshida, B. D. Mookherjee, V. Kamath et al., *U.S. 4,173,585,* 1979, idem, *U. S. 4,278,569,* 1981; E. J. Brunke and E. Klein, *Ger. Offen., 2,935,683,* 1979). The use of catalytic palladium methodology in conjunction with the enamine of cyclohexanone promoted the cyclisation of the tetraene (7) to (8) in high yield (J. M. Takacs and J. Zhu, *Tetrahedron Lett.,* 1990, **31**, 1117).

(1) (2) (3) (4)

(5) (6) (7) E=CO$_2$Et (8) E=CO$_2$Et

(i) Polycyclohexanes and polycyclohexenes

Self condensation of cyclohexanones provides the simplest method for the synthesis of compounds in this class. In this reaction the conjugated product is often not the major one, and a mixture may be formed (A. Kalm, J. Kaminski and W. Kolodziejski, *Acta Chim. Hung.,* 1988, **125**, 141; A. M. Kim, A. F. Markov, V. I. Mamatyuk et al., *Izv. Vyssh. Zaved., Khim. Khim. Tekhol,* 1986, **29**, 37).

Decarboxylative dehydration of the adduct formed between the dianion of 4,4-(ethylenedioxy)cyclohexanecarboxylic acid and a cyclohexanone yields (1), which after reduction and ketal hydrolysis, forms the basis for another iteration of cyclohexane ring addition (E. G. Nidy, D. R. Graber, P. A. Spinelli et al., *Synthesis,* 1990, 1053).

1,3-Dicarbonyl compounds may also be used as suitable substrates for condensation reactions in the synthesis of cyclohexylidene systems (V. A. Nikanorov, S. V. Sergeev, V. I. Rozenberg et al., *Izv. Akad. Nauk. SSSR, Ser. Khim.* 1988, 925). Diphenoquinone (2) was formed in 95% yield in the reaction of (3) with potassium *t*-butoxide (P. Bartolmei and P.

Boldt, *Angew. Chem.*, 1975, **87**, 39). The addition of cyclohexanone enolates to cyclohexyl electrophiles has also been examined for the synthesis of polycyclohexyls (A. A. Akhrem, F. A. Lakhvich, A. N. Pyrko, *Zh. Org. Khim.*, 1983, **19**, 2322). The use of palladium catalysed additions of allylic acetates represents a particularly useful method (J. C. Fiaud and J. L. Malleron, *J. Chem. Soc., Chem. Commun.*, 1981, 1159).

The direct reductive coupling of cyclic ketones to give alkenes using titanium reagents (the McMurry coupling) has now become an established methodology in synthesis (J. E. McMurry and M. P. Fleming, *J. Org. Chem.*, 1976, **41**, 896; J. E. McMurry, T. Lectka and J. G. Rico, *J. Org. Chem.*, 1989, **54**, 3748; B. P. Mundy, D. R. Bruss, Y. Kim et al., *Tetrahedron Lett.*, 1985, **26**, 3927; J. Janssen and W. Luettke, *Chem. Ber.*, 1982, **115**, 1234).

Diketone (4) (S. E. Denmark, C. J. Cramer and J. A. Sternberg, *Tetrahedron Lett.*, 1986, **27**, 3693) has been prepared by the reductive (electrochemical) coupling of either 2-bromo or 2-chloro-cyclohexanone (I. Carelli, A. Curulli, A. Inesi et al., *J. Chem. Res., Synop.*, 1990, 74), the reaction of lead tetraacetate or $PhI(OAc)_2$ with the triethylstannyl enol ether of cyclohexanone (A. N. Kaskin, M. L. Tul'chinskii, N. A. Bumagin et al., *Zh. Org. Chem.*, 1982, **18**, 1588) and the direct dehydrodimerisation of cyclohexanone using nickel peroxide (E. G. E. Hawkins and R. Large, *J. Chem. Soc., Perkin Trans 1*, 1974, 280). Ferric chloride has been employed for a similar transformation (R. H. Frazier and R. L. Harlow, *J. Org. Chem.*, 1980, **45**, 5408).

An unusual dilithiated enolate equivalent (5) is reported to react with cyclohexanone to yield, after hydrolysis, the enone (6) in high yield ((**129**) C. J. Kowalski and K. W. Fields, *J. Am. Chem. Soc.*, 1982, **104**, 1777). The Claisen rearrangement of the ester (7) yielded (8) (S. E. Denmark and M. A. Harmata, *Tetrahedron Lett.*, 1984, **25**, 1543). As for the formation of related ring systems, extrusion processes have been successfully employed for the synthesis of dicycloalkylidenes (H. Redlof, F. Voeggtle, H. Puff et al., *J. Chem. Res., Synop.*, 1984, 314). One report has appeared describing the trapping of a cyclohexyne generated *in situ* by cyclohexanone to give a bicyclic system (B. Fixari, J. J. Brunet and P. Caubere, *Tetrahedron,* 1976, **32**, 927).

O O OLi O O O

(4)   (5)   (6)   (7)

CO₂Me O O O O O O

(8)   (9)   (10)   tBu⟷tBu (11)

Several studies have been carried out on the selective oxidation of unsaturated cyclohexylcyclohexane compounds to form epoxides: hence the diepoxidation of 1,1'-bicyclohexenyl has been shown to give a mixture from which a meso product (9) could be separated from a (racemic) diastereoisomer (10). Subsequent reductive ring opening gave the corresponding secondary alcohols (F. Plenat, F. Pietrasanta, M. R. Darvich et al., *Bull. Soc. Chim. Fr.,*1975, 361). Several transformations of the epoxidation product (11) from the appropriate enone have been reported (E. G. E. Hawkins and R. Large, *J. Chem. Soc., Perkin Trans. 1,* 1973, 2169; M. M. Movsumzade, Z. A. Safarova and A. S. Kyazimov, *Dokl. Akad. Nauk. Az. SSR,* 1978, **34**, 54).

The stereochemistry of the addition of singlet oxygen to *syn* and *anti* isomers (11) have been studied but no strong pattern of stereoselectivities has been observed (H. S. Dang and A. G. Davies, *J. Chem. Soc., Perkin Trans. 1,* 1991, 721). The stereochemistry of hydroboration of unsaturated bicyclohexanes has also been studied (T. W. Bell, J. R. Vargas and G. A. Crispino, *J. Org. Chem.,* 1989, **54**, 1978; T. W. Bell, *Tetrahedron Lett.,* 1980, **21**, 3443).

In an unusual reaction alcohols (12) or (13) (or a mixture of both), upon treatment with formic acid and sulphuric acid gave [1,1'-bi(cyclohexyl)-1-carboxylic acid (14) (H. Kagawa, and Y. Kimizuka, *Ger. Offen.,* 3,213,255, 1983; *JP. 82,122,051,* 1982; *JP. 58, 118,546,* 1983). An intramolecular version of the McMurry coupling was effective for the conversion of (15) into (16) (J. E. McMurry, G. J. Haley, J. R. Matz et al., *J. Am. Chem. Soc.,* 1986, **108**, 2932). An intramolecular photo-Fries rearrangement of amide (17) resulted in formation of the condensed spiro structures (18) (C. Bochu, A. Couture, P. Grandclaudon et al., *J. Chem. Soc., Chem. Commun.,* 1986, 839).

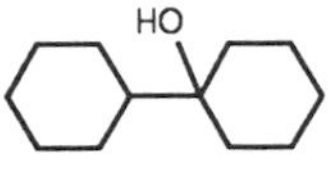

(12) 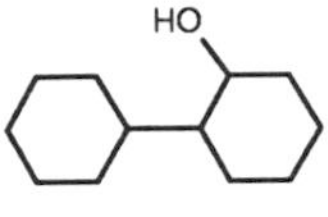　　　　(13) 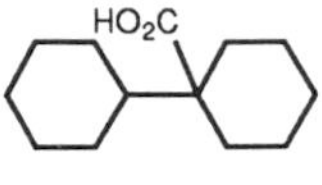　　　　(14) 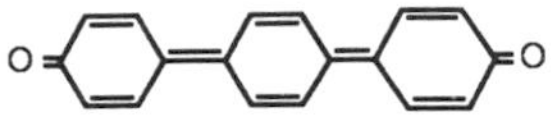　　　　(15)

(16) 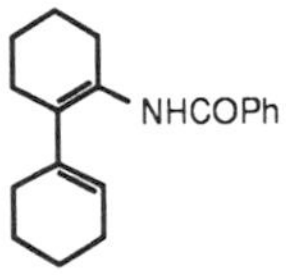　　　　(17) 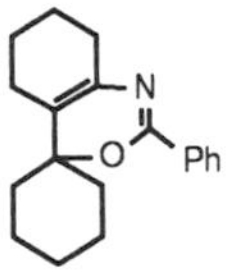　　　　(18) 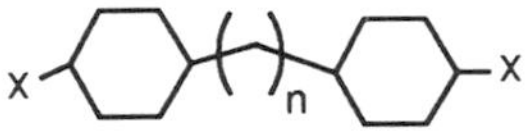　　　　(19) X=NH₂
(20) X=OH

Compounds containing cyclohexyl groups linked by carbon chains are most simply prepared by reduction of the aromatic precursor. This has proved to be a most fruitful approach for the synthesis of diamines (19) or diols (20) (O. Weissel, *Ger. Offen., 2,502,893, 1976*; H. Kamamota and F. Hayano, *JP. 75 39,660, 1975,* idem, *JP 75 11,375, 1975;* Y. Koneko, M. Shimada and H. Yasuda, *JP. 73 34,735, 1973;* Y. Shiraki, *JP. 03,157,342, 1991;* K. Mizuno, T. Tanitsu and M. Kawamura, *JP. 62,281,833, 1987;* K. Mizuno, T. Tanitsu and M. Kawamura, *JP. 62,281,832, 1987;* G. F. Allen, *Eur. Appl. 66,210 to 212, 1982*). The synthesis and X-ray crystal structure of 1,1,3,3-bis(1',1',5',5'-tetramethylpentamethylene)allene (21) has been reported (H. Irngartinger, E. Kurda, H. Rodewald et al., *Chem. Ber., 1982, 115, 967*).

Oxidative coupling of cyclic enones is effective for the synthesis of six-membered rings linked by an alicyclic two carbon chain (H. Hart and R. Willer, *Tetrahedron Lett., 1977, 2307*). The related dienes (21) have been prepared by the oxidative dimerisation of a vinyl cuprate (R. B. Banks and H. M. Walborsky, *J. Am. Chem. Soc., 1976, 98, 3732*) and via a Julia olefination process (P. J. Kocienski, B. Lythgoe and S. Ruston, *J. Chem. Soc., Perkin Trans. 1, 1978, 829*).

Thermal rearrangement of (23) furnishes the acetylenic product (24) (R. Roemer, J. Harnish, A. Roeder et al., *Chem Ber., 1984, 117, 925*). A study of the cyclodimerisation reaction of hexapentaene (25) has been reported (M. Iyoda, Y. Masahiko, M. Oda et al., *Angew. Chem., 1990, 102, 1077*).

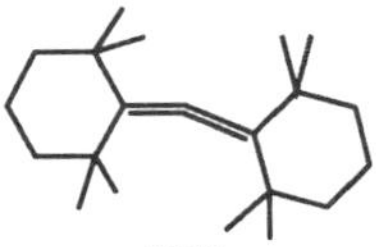

(21) 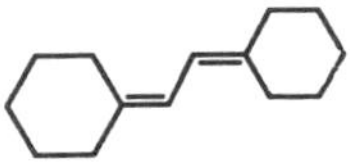　　　　(22) 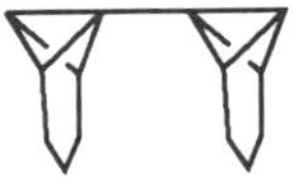　　　　(23) 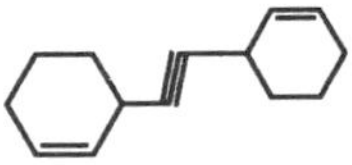　　　　(24)

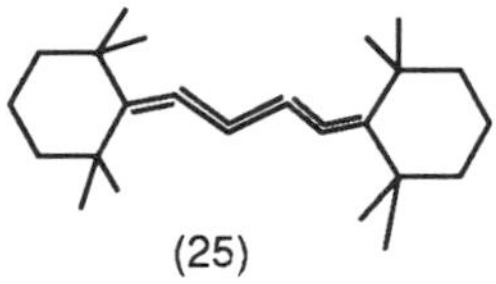

(25)

(j) Cyclopentylcycloheptanes and cyclopentylcycloheptenes

The tin tetrachloride catalysed reaction of 1,2-bis(ethylthio)cycloheptane with the trimethylsilyl enol ether of cyclopentanone furnishes the sulphide (1) which may be converted into the enone via the sulphoxide followed by elimination of sulphenic acid (R. D. Bach and R. C. Klix, *J. Org. Chem.*, 1985, **50**, 5438). A conjugate addition to an enone was employed for the synthesis of (2) from a cyclopentenone (W. D. Woessner, W. G. Biddlecom, H. C. Arndt et al., *U.S. 4,029,670,* 1977).

Fulvene (3) was prepared by the condensation reaction between a Grignard reagent and the appropriate tropylium salts (P. Boenzli, M. Neuenschwander and P. Engel, *Helv. Chim. Acta,* 1990, **73**, 1685). The fully unsaturated compound (4) was prepared by a similar condensation process (N. Shimazaki, T. Toda and T. Mukai, *Koen Yoshishu - Hibenzenkei Hokozoku Kagaku Toronkai 12th,* 1979, 113) although (5) was obtained via ring opening of quadricyclanone (6) (O. Schweikert, T. Netscher, L. Knothe et al., *Chem Ber.,* 1984, **117**, 2027).

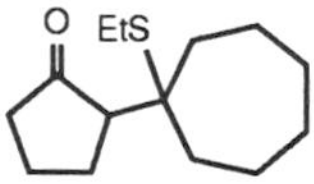

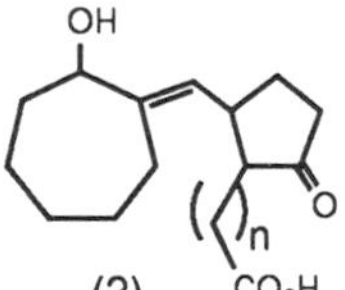

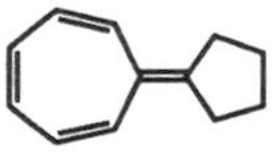

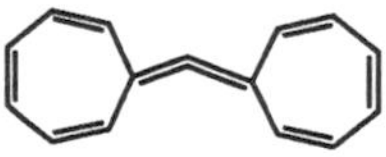

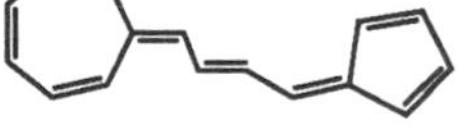

(k) Cyclohexylcycloheptadienes

The novel diphenoquinone (1) contains a seven membered quinonoid structure and has redox properties (M. Iyoda, K. Sato and M. Oda,

Tetrahedron Lett., 1987, **28**, 625). Similar properties are exhibited by the tropoquinone (2), which was prepared from a tropone monomer (K. Mori, Y. Goto and H. Takeshita, *Bull. Chem. Soc. Jpn.*, 1987, **60**, 2497).

(1)   (2)

(l) Polycycloheptenes

The McMurry coupling reaction previously described has been successfully applied to the dimerisation of cycloheptanone (J. E. McMurry, *Ger. Offen.*, *2,641,075, 1977*).

Coupling of perchlorocycloheptatriene can be achieved by treatment with butyllithium in hexane, presumably via chlorine/lithium exchange, followed by nucleophilic attack upon a second molecule (H. M. Hugel, E. Horn and M. R. Snow, *Aust. J. Chem.*, 1985, **38**, 383). A nickel complex is the catalyst of choice for the ring opening dimerisation of cyclopropylbenzene to give (1). Initially the product complexes with the catalyst and acid treatment is requried to release it (R. Mynott, R. Neidlein, H. Schwager et al., *Angew. Chem., Int. Edn. Engl.*, 1986, **25**, 367). The synthesis of the conjugated compound 8-(heptafulven-8-yl)-*p*-tropoquinone methide (2) from a lithiated monotropone has been reported (K. Takahashi, N. Namekata, K. Takase et al., *J. Chem. Soc., Chem. Commun.*, 1987, 935). The preparation of higher vinylogous heptafulvalenes has also been recorded (O. Schweikert, T. Netscher, L. Knothe et al., *Chem. Ber.*, 1984, **117**, 2045).

(m) Miscellaneous bicyclic compounds

Cyclobutylcyclopentanes: Fluoride initiated (CsF) reaction of a mixture of perfluorocyclo-butene and -propene gave (1). Reagents containing different combinations of ring sizes afforded a variety of related products of varying ring-size combinations and double bond positions. The position of the double bond in the product depended on conformational interactions and varied for different ring sizes (R. Chambers, G. Taylor and R. L.Powell, *J. Chem. Soc., Perkin Trans 1,* 1980, 429). Compound (2) is shown to have a quintet ground state, by EPR spectroscopy, which is unusual as the molecule does not contain a delocalised π-system (J. A. Novak, R. Jain and D. A. Dougherty, *J. Am. Chem. Soc.*, 1989, **111**, 7618). The thermal rearrangement of conjugated cyclopropylidenecyclobutanone (3) gives (4) in 50% yield (M. Bertrand, G. Gil, A. Junino et al., *Tetrahedron Lett.*, 1977, 1779).

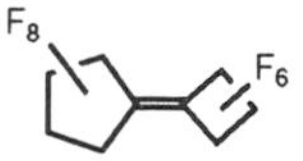

(1) 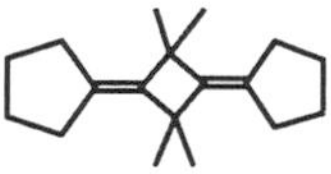   (2) 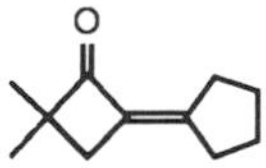   (3) 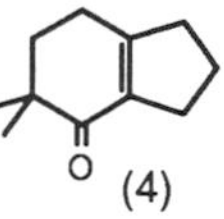   (4) 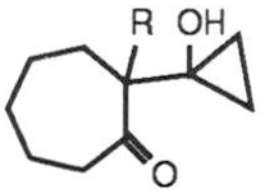   (5), R=Et, Pr, allyl, Bn

Cycloheptylcyclopropanes: Ketone (5), a substrate for base promoted ring opening rearrangement reactions, has been prepared from cycloheptanone derivative (6) and metallated cyclopropanone hemiketal (7) (V. Reydellet and P. Helquist, *Tetrahedron Lett.*, 1989, **30**, 6837).

A spiro-fused cyclobutanone (8) was the product of the reaction between trimethylsilyl enol ether (9) and PhCH(OMe)$_2$ catalysed by TMSOTf (B. M. Trost and A. Brandi, *J. Am. Chem. Soc.*, 1984, **106**, 5041). Cycloheptadienequinone (10) contains an electron donating diaminocyclopropenylidene moiety and has been shown to exist in the delocalised dipolar form illustrated (K. Takahashi, N. Namekata, K. Takase et al., *Tetrahedron Lett.*, 1987, **28**, 5683).

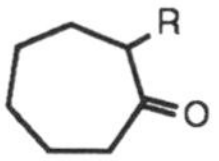

(6)    (7) 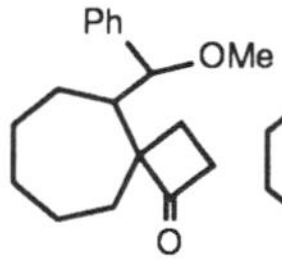   (8) 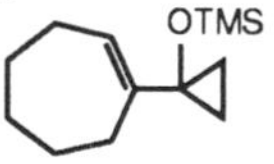   (9) 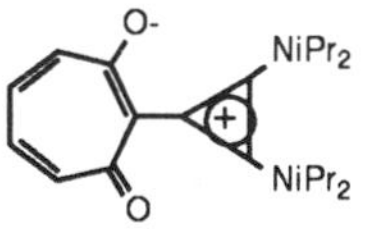   (10)

Cyclooctylcyclopropyl compounds: The nickel catalysed dimerisation of cyclopropyl substituted dienes with unsubstituted dienes has been studied. The products are (11) and (12). (U. M. Dzhemilev, R. I. Khusnutdinov and V. A. Dokichev, *Izv. Akad. Nauk. SSSR, Ser. Khim.*, 1979, 2634). The related diene (13) underwent palladium catalysed intramolecular transannular ring closure to give propellanes (14) (S. Yamago and E. Nakamura, *J. Chem. Soc., Chem. Commun.*, 1988, 1112; idem, *Tetrahedron*, 1989, **45**, 3081).

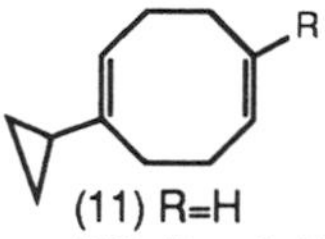

(11) R=H
(12) R=c-C$_3$H$_5$ (13) R=H, CO$_2$Et   (14)

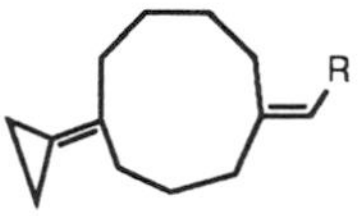

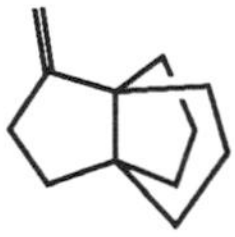

Bicyclononyl compounds: The synthesis and reations of bicyclonona-1,3,5,7-tetraene and its fulvalene derivative have been reported (A, Escher, M. Neuenschwander and P. Engel, *Helv. Chim. Acta,* 1987, **70**, 1623; T. S. Cantrell and A. C. Allen, *Tetrahedron Lett.,* 1985, **26**, 5351; K. Hakner, S. Braun, T. Nakazawa et al., *Tetrahedron Lett.,* 1975, 3507).

Cyclononylcyclopropanes: The synthesis and reactions of a number of members of this class of compound have been reported (A. Escher, M. Neuenschwander, *Angew. Chem., Int. Edn. Engl.,* 1984, **23**, 973; idem., *Helv. Chim. Acta,* 1987, **70**, 49).

3. Spiro compounds; spiranes

(a) The spiro[2,2]pentane group

Spiro[2,2]pentanes (1) may be most conveniently prepared by the reaction of carbenes with methylene cyclopropanones (2) (E. Dunkelblum, and B. Singer, *Synthesis,* 1975, 323; R. R. Kostikov, A. P. Molchanov and M. I. Komendantov, *Zh. Org. Khim.,* 1979, **15**, 1437) or directly from allenes (T. Greibrokk, *Acta. Chem. Scand.,* 1973, **27**, 3207. Dihalo carbenes react in high yield and are most conveniently generated, however they may be easily converted into hydrocarbons via reduction with zinc in a protic solvent (A. A. Formanovskii and I. G. Bolesov, *Zh. Org. Khim.,* 1982, **18**, 2299). Although dichloro and dibromocarbenes are most commonly used, difluorocarbene (W. R. Dolbier, S. F. Sellers and B. E. Smart, *Tetrahedron Lett.,* 1981, **22**, 2953) and chloromethylcarbene (K. A. Lukin, A. Y. Masunova, S. L. Kozhushkov et al., *Zh. Org. Chem.,* 1991, **27**, 488) have also been used for cyclopropanations.

The generation of carbenes by the thermally promoted loss of nitrogen from diazoacetates has been successfully applied to the synthesis of spiropentanes, as illustrated by the conversion of (3) to (4) (M. I. Komendantov, T. K. Klindukova, G. N. Suvorova et al., *Zh. Org. Khim.,* 1979, **15**, 2076; R. H. Davis and R. J. G. Searle, *Ger.,* 2,712,333, 1977).

Spiropentane (1) is the major product (90%) from the reaction of (5) with zinc in alcoholic potassium hydroxide (A. I. D'yachenko, E. L. Protasova and O. M. Nefedov, *Izv. Akad. Nauk. SSSR, Ser. Khim.,* 1979, 1166). Unsaturated derivatives of this class of compound may contain double bonds in a ring, as in the case of spiropentene (5), which undergoes reactions with dienes in Diels-Alder reactions (R. Bloch and J. M. Denis, *Angew. Chem.,* 1980, **92**, 969), or on a substituent, as in the case of vinyl spiropentane (6). The last compound may be prepared by elimination from an appropriate precursor without opening of the three membered rings (N. S. Zefirov, S. Kozhushkov, T. S. Ketmetsova, et al., *Zh. Org. Khim.*, 1988, **24**, 673).

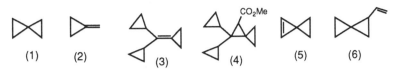

(1)   (2)   (3)   (4)   (5)   (6)

Reactions of dispiropentanes generally involve ring opening as illustrated by the treatment of dimethylspiropentane (7) with potassium *t*-butoxide to give acetylene (8) in 30% yield (A. A. Formanovskii and I. G. Bolesov, *Zh. Org. Khim.*, 1978, **14**, 884). The propensity of the spiropentane system to act as a "tensile spring" is emphasised by a X-ray structural study of compounds (9) and (10) which shows the bending of the spiro moeity is greatest in the aromatic derivative (R. Boese, D. Blaeser, K. Gomann et al., *J. Am. Chem. Soc.*, 1989, **111**, 1501).

The synthesis and reactivity of polyspiropropanes, e.g. (11) (A. Tubul and M. Betrand, *Tetrahedron Lett.*, 1984, **25**, 2219; K. A. Lukin, A. A. Andrievskii, A. Y. Masunova et al., *Dokl. Akad. Nauk. SSSR*, 1991, **316**, 379; L. Fitjer and J. M. Conia, *Angew. Chem., Int. Edn. Engl.*, 1973, **12**, 3), (12) (I. Erden, *Synth. Commun.*, 1986, **16**, 117; K. A. Lukin, S. I. Kozhushkov, A. A. Andrievskii et al., *J. Org. Chem.*, 1991, **56**, 6176; N. S. Zefirov, K. A. Lukin, S. I. Kozhushkov et al., *Zh. Org. Khim.*, 1989, **25**, 312) and higher homologues (N. S. Zefirov, S. I. Kozhushkov, T. S. Kuznetsova, et al., *J. Am. Chem. Soc.*, 1990, **112**, 7702) have been reported.

A process of iterative cyclopropanation of an alkene with MeClC: and dehydrochlorination was employed for the synthesis of a number of derivatives based on (12) and also for the synthesis of a bismethylene derivative of polyspirocyclopropanes (13), where n=1-4 (K. A. Lukin, A. Y. Masunova, B. I. Ugrak and N. S. Zefiorov, *Tetrahedron,* 1991, **47**, 5769). The synthesis and iodination reactions of the 1,2-dinitrospirane (14) have been reported (P. A. Wade, P. A. Kondracki and P. J. Carroll, *J. Am. Chem. Soc.*, 1991, **113**, 8807). Treatment of dibromoalkene (15) with phenyllithium results in cyclisation to (16) in 55% yield (L. Fitjer, *Angew. Chem.*, 1976, **88**, 803).

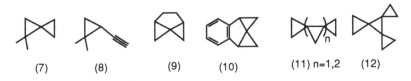

(7)   (8)   (9)   (10)   (11) n=1,2   (12)

(13) n=1,2   (14)   (15)   (16)

(b) The spiro[2,3]hexane group

Cyclopropanation of cyclobutylidene derivatives is a convenient method for the synthesis of certain spiro[2,3]hexanes: In the conversion of (1) to (2), using a copper catalysed formation of a carbene from a diazo compound, the reaction takes place exclusively on the double bond attached to the cyclobutane group (R. H. Davis and R. J. G. Searle, *U.S. 3,961,070,* 1976). Polycyclopropyl rings may be attached to cyclobutylene rings, as illustrated by (3), and its lower homologues via carbene reactions (L. Trabert, H. Hopf and D. Schomberg, *Chem. Ber.,* 1981, **114**, 2405; J. M. Denis, P. LePerchec and J. M. Conia, *Tetrahedron,* 1977, **33**, 399). Addition of dibromocarbene to diene (4), followed by tributyltin hydride reduction and dehydrobromination results in the formation of trispirooctadiene (5). This product has been the subject of rearrangement studies (F. C. Peelen, I. J. Landheer, W. H. DeWolf et al., *Recl. Trav. Chim. Pays-Bas,* 1986, **105**, 326).

Bicyclobutylidene reacts with cyclopropylidene to give trispiro compound (6) in 31% yield (L. K. Bee, J. W. Everett and P. J. Garratt, *Tetrahedron,* 1977, **33**, 2143). At least one example of the inverse of this process i.e. the addition of cyclobutylidene to an unsaturated material (in this case butyne) has been reported (U. H. Brinker and J. Weber, *Tetrahedron Lett.,* 1986, **27**, 5371). A [2+2] cycloaddition may be employed to construct the four membered ring in this class of compound (A. DeMeijere, H. Wenck, F. Seyed-Mahdavi et al., *Tetrahedron,* 1986, **42**, 1291) as illustrated by the head to head dimerisation of (7) to give dispirooctanes (8) (X= Br, Cl, OEt) (A. T. Bottini and L. J. Cabral, *Tetrahedron,* 1978, **34**, 3187).

The same mode of dimerisation is observed for diene (9), which gives the *trans*-adduct (10), a cyclooctadiene precursor, upon dimerisation (F. Kienzle and J. Stadlwieser, *Tetrahedron Lett.,* 1991, **32**, 551). In contrast dimerisation of carbonylcyclopropane at 500°C gives dispiro-[2,1,2,1]-octane-4,8-dione (11) (G. J. Baxter, R. F. C. Brown, F. W. Eastwood et al., *Tetrahedron Lett.,* 1975, 4283). A valuable nickel catalysed variant of this reaction has been reported (P. Binger, A. Bringmann and P. Wedemann, *Chem. Ber.,* 1983, **116**, 2920), however it should be noted that this reaction often gives side products due to opening of cyclopropane rings and rearrangements (W. Weber, I. Erden and A. DeMeijere, *Angew. Chem.,* 1980, **92**, 387; D. J. Pasto and D.

Wampfler, *Tetrahedron Lett.*, 1974, 1933; S. Nishida, *Angew. Chem. Int. Edn. Engl.*, 1972, **11**, 328). Photooxygenation of bicyclopropylidene gave 1-oxo-spiro[2,3]hexane via the corresponding epoxide (I. Erden, A. DeMeijere, G. Rousseau et al., *Tetrahedron Lett.*, 1980, **21**, 2501).

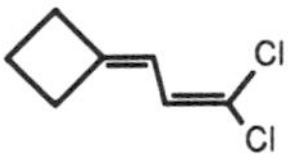

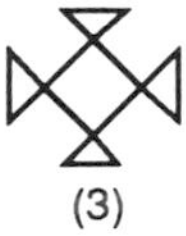

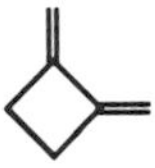

Reductive coupling of the corresponding ketone with titanium trichloride and lithium aluminium hydride gave (12), in which no cyclopropane rings have been opened, as the major product in 60% yield (H. Wenck, A. DeMeijere, F. Gerson et al., *Angew. Chem.*, 1986, **98**, 343). Carbene (13) can be generated and does not dimerise. Instead it undergoes ring opening enlargement to give the reactive alkene (14) (A. DeMeijere, H. Wenck and J. Kopf, *Tetrahedron*, 1988, **44**, 2427).

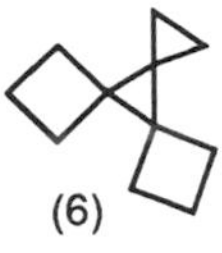

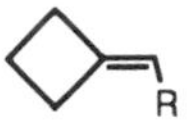

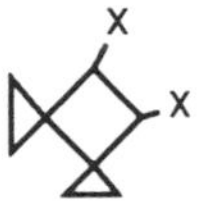

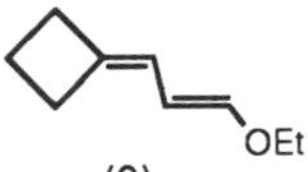

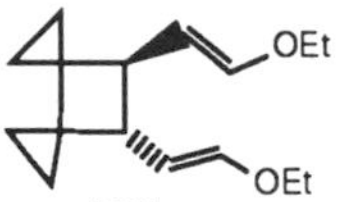

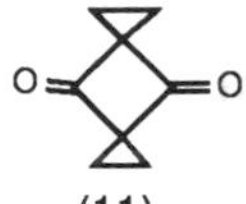

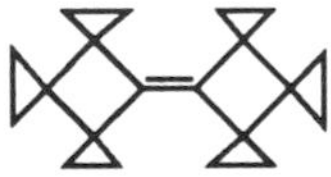

(c) The spiro[2,4]heptane group

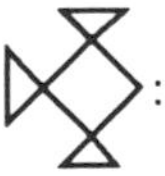

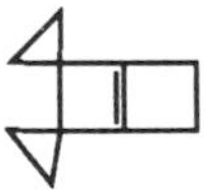

The spiroalkylation reaction of 1,3-cyclopentanediones is an attractive method for the synthesis of spiro[2.4]heptanes, however this reaction is often restricted to benzo fused derivatives (N. S. Zefirov, T. S. Kuznetsova and S. I. Kozhushkov, *Zh. Org. Khim.*, 1983, **19**, 1599). The related cyclopropanation of cyclopentadienyl anions using dibromoethane derivatives has proved to be a more versatile method (A. I. D'yachenko, L. G. Menchikov and O. M. Nefedov, *Izv. Akad. Nauk. SSSR, Ser. Khim.*, 1985, 709; idem, ibid., 1984, 1664; idem, ibid., 1985, 949; F. Naef and R. Decorzant, *Helv. Chim. Acta.*, **61**, 2524). As with other ring systems carbene addition may be used to construct the three membered ring (L. Jenneskens, W. W. DeWolf and F. Bickelhaupt, *Angew. Chem.*, 1985, **97**, 568; A. Amaro and K. Grohmann, *J. Am.*

208

Chem. Soc., 1975, **97**, 3830; A. A. Arbale, R. H. Naik and G. H. Kulkarni, *Indian J. Chem. Sect. B.*, 1990, **29B**, 568; G. A. Tolstikov, F. Z. Gakin, V. N. Iskandarova et al., *Izv. Akad. Nauk. SSSR, Ser. Khim.*, 1990, 1373), as may selenium (N. N. Magdeisieva, T. A. Sergeeva and N. V. Averina, *Zh. Org. Khim.*, 1986, **22**, 2114) and sulphur (W. W. Brand, *Ger. 2,724,734,* 1977) ylids.

The Khand-Pauson reaction of methylenecyclopropane with an acetylene has proved to be an effective method for the synthesis of spirocyclopropylcyclopentenones (1) and (2) (V. A. Smit, S. L. Kireev, O. M. Nefedov et al., *Tetrahedron Lett.*, 1989, **30**, 4021). Nickel is the metal of choice for the catalysis of the dimerisation of methylenecyclopropane (3) to (4) (P. Binger, *Synthesis,* 1973, 427) whilst palladium has been used to mediate the [2+2] cycloaddition of (3) to alkenes giving methylenecyclopropanes (P. Binger and U. Schuchardt, *Angew. Chem.*, 1977, **89**, 254).

A novel method for the synthesis of of spiro-methylenecyclopropanes (5) via the cyclisation of allyl epoxides (6) has been described (T. Satoh, Y. Kawase and K. Yamakawa, *Tetrahedron Lett.*, 1990, **31**, 3609). The larger ring in this reaction may be up to seven membered.

(1)   (2)   (3)   (4)   (5)   (6)

Ring opening photoreactions of methylene spiranes such as (7) with iron pentacarbonyl give mainly bicyclic enones as products, although the related six/three membered homologue yields an iron tricarbonyl complex (S. Sarel, A. Felzenstein, R. Victor et al., *J. Chem. Soc., Chem. Commun.,* 1974, 1025). Cyclopropenyl compound (8) gives a Dewar benzene (9) and indane upon treatment with silver perchlorate ((I. Landheer, W. H. DeWolf and F. Bickelhaupt, *Tetrahedron Lett.*, 1975, 349).

The mechanisms of the thermal and photochemical rearrangement processes of cyclopropyl ketones have been investigated (E. Lee-Ruff and P. G. Khazanie, *Can. J. Chem.*, 1978, **56**, 803). Iron pentacarbonyl was employed for the ring opening reaction of divinylcyclopropane (10), followed by carbonyl incorporation to give cyclooctadienone (11) (S. Sarel and M. Langbeheim, *J. Chem. Soc., Chem. Commun.,* 1977, 827). Compound (10) could also be employed for ring expansions to give nine membered carbocyclic ring systems (S. Sarel and M. Langbeheim, *J. Chem. Soc., Chem. Commun.,* 1977, 593).

(7)   (8)   (9)   (10)   (11)

(d) The spiro[2,5]octane group

Spiro[2,5]octanes may be prepared by carbene cyclopropanation of methylene cyclohexane rings (R. A. Moss and C. B. Mallon, *J. Org. Chem.* 1975, **40**, 1368; R. C. Ronald, S. M. Ruder and T. S. Lillie, *Tetrahedron Lett.*, 1987, **28**, 131; E. Wenkert and J. Rego de Sousa, *Synth. Commun.*, 1977, **7**, 457; E. Wenkert, T. E. Goodwin and B. C. Ranu, *J. Org. Chem.*, 1977, **42**, 2137). Formation of a spiro[2,5]octane from a cyclic ketone has been achieved by the use of $ClCH_2CH_2S^+Me_2$ in the presence of *t*-butoxide anion (S. M. Ruder and R. C. Ronald, *Tetrahedron Lett.*, 1984, **25**, 5501). The key step in the synthesis of the insecticidal compounds (1) was the cyclisation of enolates derived from (2) (G. H. Kulkarni and A. A. Arbale, *Synth. Commun.*, 1988, **18**, 2147).

Phenol rings are the target of a carbene in the intramolecular conversion of the diazo compound (3) to spiro- compound (4) in which the non-aromatic derived ring may be up to five membered in size (D. J. Beames and L. N. Mander, *Aust. J. Chem.*, 1974, **27**, 1257). An unusual cyclopropation reaction was achieved in the conversion of nitroepoxide (5) to (6) upon treatment with methyl sodiocyanoacetate (Y. Yokomori and R. Tamura, *J. Chem. Soc., Chem. Commun.*, 1991, 159). A [4+2] cycloaddition reaction between an acetylene and the trimethylsilyl enol ether of an appropriate cyclopropylidene ketone (T. Thiemann, S. Kohlstruk, G. Schwaer and A. DeMeijere, *Tetrahedron Lett.*, 1991, **32**, 3483). The synthesis, spectroscopic analysis and photooxidation studies of dispiro[2,0,2,4]deca-7,9-diene (7) and its derivatives have been reported (A. DeMeijere, D. Kaufmann and I. Erden, *Tetrahedron*, 1986, **42**, 6487; A. DeMeijere, *Chem. Ber.*, 1974, **107**, 1684; idem, ibid., 1974, **107**, 1702; L. Fitjer, H. J. Scheurmann, U.Klages et al., *Chem Ber.*, 1986, **119**, 1144).

(1)   (2)   (3)   (4)

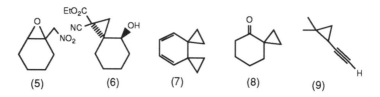

(5)   (6)   (7)   (8)   (9)

Synthetic transformations of this class of spiro compound do not necessarily depend on cyclopropane ring opening (R. L. Danheiser and A. C. Savoca, *J. Org. Chem.*, 1985, **50**, 2401; J. M. Denis, R. Niamayoua, M. Vata et al., *Tetrahedron Lett.*, 1980, **21**, 515). For example, oxidation of the parent spiro[2,5]octane with ozone is reported to give ketone (8) and the corresponding symmetric 1,3-diketone (E. Proksch and A. DeMeijere, *Angew. Chem.*, 1976, **88**, 802).

Acid promoted cyclopropyl ring opening of these compounds has been studied in detail (L. H. Schwartz and V. P. Gully, *J. Org. Chem.*, 1974, **39**, 219; E. Egert and G. M. Sheldrick, *Tetrahedron Lett.*, 1986, **27**, 3603), as have a number of reactions requiring the use of a base, for example potassium *t*-butoxide is reported to give acetylene (9) upon reaction with (10) (A. A. Formanovskii and I. G. Bolesov, *Zh. Org. Khim.*, 1978, **14**, 884).

Cyclopropyl rings adjacent to ketones are known to be reactive towards nucleophiles and this represents another mode of reactivity of spiro[2,5]octanes (P. H. Hefferty, P. Mahler and P. Yates, *J. Chem. Soc., Chem. Commun.*, 1979, 1084; K. Ohkata, T. Sakai, Y. Kubo et al., *J. Chem. Soc., Chem. Commun.*, 1974, 581). Irradiation of azobis[4-spiro[2,5]octane] provides an ideal method for the generation of radicals adjacent to cyclopropyl rings (M. Suzuki, S. Murahashi, A. Sonoda et al., *Chem. Lett.*, 1974, 267). The carbene generated by treatment of dibromide (11) with methyllithium shows a ninefold preference for insertion into a secondary (to give (12) rather than (13)) rather than into a primary carbon-hydrogen bond (R. B. Reinharz and G. J. Fonken, *Tetrahedron Lett.*, 1974, 441).

(10)   (11)   (12)   (13)

(e) The spiro[2,6]nonane group

A number of methods for the synthesis of this class of spirane have been reported, which most commonly require the synthesis of a cyclopropane ring on an intact seven-membered ring (R. A. LaBar, *Org.*

Photochem. Synth., 1976, **2**, 90; M. Exkert-Maksic, S. Zollner, A. DeMeijere et al., *Chem. Ber.,* 1991, **124**, 1591).

(f) The spiro[3,3]heptane group

Compounds in this class may be conveniently prepared by the [2+2] cycloaddition reaction of dichlorocarbene with methylene cyclobutene (I. Erden and E. M. Sorenson, *Tetrahedron Lett.,* 1983, **24**, 2731). Enzymic ester hydrolysis has been employed for the resolution of racemic 2,6-diacetoxyspiro[3,3]heptanes with high efficiency (K. Naemura and A. Furutani, *J. Chem. Soc., Perkin Trans 1,* 1990, 3215). The synthesis and reactions of spiroheptanes containing three appended cyclobutane rings have been described (T. G. Archibald, L. C. Garver, K. Baum et al., *J. Org. Chem.,* 1989, **54**, 2869; C. M. Sharts, M. E. McKee, R. F. Steed, D. F. Shellhamer, A. C. Greeley, R. C. Green and L. G. Sprague, *J. Fluorine Chem.,* 1979, **14**, 351). An example is provided by the conversion of the tetrachloride (1) to the bicyclo[1,1,0] derivative (2) upon treatment with sodium or lithium metal (V. N. Borodin and M. I. Komendantov, *Zh. Org. Khim.,* 1980, **16**, 2010).

Cyclobutane ring opening takes place in the cascade rearrangement of dispiro[3,0,4,2]undecene (3) to the [3,3,3]propellane (4) (L. Fitjer, M. Majewski and A. Kanachia, *Tetrahedron Lett.,* 1988, **29**, 1263; L. Fitjer, A. Kanschik and M. Majewski, *Tetrahedron Lett.,* 1985, **26**, 5277). Spiro[3,3]heptane-2-carboxylic acid has been prepared and used in the synthesis of prostaglandin derivatives (H. C. Arndt, W. G. Biddlecom and W. Gerard, *Ger. Offen.,* *2,705,590,* 1977).

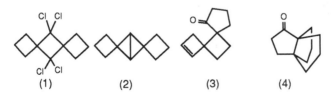

(1)   (2)   (3)   (4)

(g) The spiro[3,4]octane group

A number of disparate strategies have been developed for the synthesis of members of this class of compound. These include the reaction of a cyclopropyl sulphur ylid to a cyclopentanone, followed by ring expanding rearrangement (B. M. Trost and L. H. Latimer, *J. Org. Chem.,* 1978, **43**, 1031), the [2+2] cycloaddition reaction of methylene cyclopropanones (H. Frey, G. Behmann and G. Kaupp, *Chem. Ber.,* 1987, **120**, 387), an intramolecular cyclopropenone cyclisation (also applicable to larger spiro systems) (M. Pohmakotr, S. Popuang and S. Chancharunee, *Tetrahedron Lett.,* 1989, **30**, 1715), a photochemical

Norrish type-II reaction (H. R. Sonawane, B. S. Nanjundiah, S. I. Rajput et al., *Tetrahedron Lett.*, 1986, **27**, 6125) and a photochemical ring contraction reaction of a pyran (D. Armesto, W. H. Horspool, N. Martin et al., *J. Org. Chem.*, 1989, **54**, 3069).

The synthesis and characterisation of unsaturated derivatives (1) and (2) have been reported (R. D. Miller and M. Schneider, *Tetrahedron Lett.*, 1975, 1557; A. DeMeijere and U. Lueder, *Chem. Ber.*, 1977, **110**, 2561; A. DeMeijere and U. Lueder, *Angew. Chem.*, 1973, **85**, 908). An investigation of the effect of spiro ring size on the photochemical ring expansion of cyclobutane-1,3-dione derivatives (3) revealed that the product acetal was obtained in a low yield when the spiro ring was five- or seven-membered (K. Kimura, M. Takamura, S. Koshibe et al., *Bull. Chem. Soc. Jpn.*, 1976, **49**, 741).

(1)   (2)   (3) n = 0 to 2

(h) The spiro[3,5]nonane group

The four membered ring in this class of structure may be prepared using a [2+2] cycloaddition between an alkylidene cyclohexane and an alkene (B. B. Snider and E. A. Deutsch, *J. Org. Chem.*, 1983, **48**, 1822) or a ketene (W. T. Bailey and P. L. Ting, *Tetrahedron Lett.*, 1974, 2619; M. I. Komendantov, Y. B. Koptelov, R. R. Kostikov et al., *Zh. Org. Khim.*, 1987, **23**, 986; T. Strelow, J. Voss and G. Adiwidaja, *J. Chem. Res., Synop.*, 1989, 136).

Trispiro[2,1,5,0]-4-oxadecane (1) is cleaved by sodium phenylselenide in ethanol to give hydroxy selenides which, after oxidation, undergo rearrangement to a 13:87 mixture of (2) and (3), a selectivity opposite to that observed for analogous acid catalysed rearrangements (B. M. Trost and P. H. Scudder, *J. Am. Chem. Soc.*, 1977, **99**, 7601; R. D. Miller and D. R. McKean, *Tetrahedron Lett.*, 1980, **21**, 2639).

A spiro[3,5]nonane is a key intermediate in a new cyclohexenone annulation thus reaction of 1-lithio-1-methoxycyclopropane (4) with α-butylthiomethylenecyclohexanone gives the product (5), which is subsequently converted into (6) (J. H. Byers and T. A. Spencer, *Tetrahedron Lett.*, 1985, **26**, 713). Intramolecular cyclisation of 1-tosyloxymethyldicyclohexyl ketone was effected by treatment with methyllithium, resulting in the formation of spiro compounds (7) (H. Marschall and W. B. Muehlenkamp, *Chem. Ber.*, 1976, **109**, 2785).

Synthetic studies on this class of compound include an investigation of nucleophilic (thiophenol/HCl/zinc dibromide) ring opening of spiroalkanones (R. D. Miller and D. R. McKean, *Izv. Akad. Nauk. SSSR, Ser. Khim.*, 1979, 1419) and the preparation of dimethylene-1,3-dione (which is trapped *in situ* by dienes) by retro Diels-Alder reaction of (J. L. Ripoll and M. C. Lasne, *Tetrahedron Lett.*, 1978, 5201).

(1) (2) (3) R=Me (4) (5)
           (8) R=H

(6) (7)

(i) The spiro[4,4]nonane group

1-Spirononanone may be simply prepared by the reaction between cyclopentanone and 1,4-dibromobutane using potassium *t*-butoxide as the base with the aid of ultrasound (T. Fujita, S. Watanabe, M. Sakamoto et al., *Chem. Ind.*, 1986, 427). Palladium catalysis has proved effective for the generation of enolates by decarboxylation of allyl β–ketocarboxylates (e.g. (1)) which subsequently cyclise in an intramolecular fashion onto ketones or enones to give spiro products (2) (J. Nokami, T. Mandai, H. Watanabe et al., *J. Am. Chem. Soc.*, 1989, **111**, 4126; J. Nokami, H. Watanabe, T. Mandai et al., *Tetrahedron Lett.*, 1989, **30**, 4829). Similar transformations may be achieved by the use of radical cyclisations onto double bonds (L. Set, D. R. Cheshire and D. L. J. Clive, *J. Chem. Soc., Chem. Commun.*, 1985, 1205; B. B. Snider and B. O. Buckmann, *Tetrahedron*, 1989, **45**, 6969), or an ene reaction (H. Strickler and K. P. Dastur, *Swiss 621,760*, 1981). Spiro[4,4]nonane-1,6-dione (3) may be prepared by the sulphonic acid catalysed cyclisation of 4-(2-oxocyclopentyl)butyric acid (W. Carruthers and A. Orridge, *J. Chem. Soc., Perkin Trans. 1*, 1977, 2411), whilst the related dieneone (4) has been prepared in optically enriched form from the diene (5) (M. Sumiyoshi, H. Kuritani and K. Shingi, *J. Chem. Soc., Chem. Commun.*, 1977, 812). Other unsaturated derivatives of spiro[4,4]nonanes include (6) and (7) (M. F. Semmelheck, J. S. Foos and S. Katz, *J. Am. Chem. Soc.*, 1973, **95**, 7325).

It is possible to construct the spiro skeleton in one step from an acyclic precursor as illustrated by the vapour phase pyrolysis of (8) to give (9) (mixture of isomers) (F. Leyendecker, F. Drouin and J. M. Conia, *Tetrahedron Lett.*, 1974, 2931) and the di-isobutylaluminiumhydride promoted conversion of (10) to (11) (P. W. Chum and S. E. Wilson, *Tetrahedron Lett.*, 1976, 1257).

(1)   (2)   (3)   (4)   (5)   (6)

(7)   (8)   (9)   (10)   (11)

(j) The spiro[4,5]decane group

Spiro-alkylation of cyclohexan-1,3-diones with 1,4-dibromobutane (DMSO, excess potassium hydroxide) has been employed for the efficient synthesis of some members of spiro[4,5]decane-6,10-diones (N. S. Zefirov, T. S. Kuznetsova and S. I. Kozhushkov, *Zh. Org. Khim.*, 1983, **19**, 1599). The double addition of an enamine to a dibenylidene acetone derivative achieves the same result (L. Anandan and G. K. Trivedi, *Indian J. Chem., Sect. B*, 1978, **16B**, 428), however more commonly, a formylcyclohexanone, or a synthetic equivalent, (1) is reacted with an methyl vinyl ketone to give an intermediate which is cyclised using a Claisen condensation (Scheme 1) to (2). Many variations on this process have been reported (N. R. Natale and R. O. Hutchins, *Org. Prep. Proced. Int.*, 1977, **9**, 103; A. DeGroot and B. J. M. Jansen, *Tetrahedron Lett.*, 1976, 2709; P. E. Eaton and P. G. Jobe, *Synthesis*, 1983, 796).

An alternative approach uses a Dieckmann cyclisation to complete the six membered ring as in the conversion of (3) to (4) (M. A. Metwally, E. Afsah and M. M. Khalifa, *J. Prackt. Chem.*, 1987, **329**, 732). A sequence of radical additions to alkenes are involved in the conversion of 3-bromo-2-methyl-cyclohexenone to (5). This is achieved by a reaction with two equivalents of methyl acrylate in the presence of tributyltin hydride (E. Lee and D. S. Lee, *Tetrahedron Lett.*, 1990, **31**, 4341).

Construction of the five membered ring is often the preferred approach to spiro[4,5]decanes (S. F. Martin, T-S. Chou and C. W. Payne, *J. Org. Chem.*, 1977, **42**, 2520) and may be achieved by an intramolecular 1,4-

addition to an enone as illustrated by the tin tetrachloride catalysed conversion of tributyltin compound (6) to spirodecane (7) (T. L. MacDonald and S. Mahalingam, *J. Am. Chem. Soc.,* 1980, **102**, 2113).

The selective addition of Me_2CuLi to an enone is an excellent method for the generation of an enolate, which subsequently participates in an intramolecular reaction with a ketone to form a spiro system (F. Naef, R. Decorzant and W. Thommen, *Helv. Chim. Acta.,* 1975, **58**, 1808). A highly impressive application of a copper reagent however, is illustrated by the addition reaction of the dicuprate (8) to 3-chlorocyclohexenones (9) to give spiranes (10) directly (P. A. Wender and D. L. Eck, *Tetrahedron Lett.,* 1977, 1245; P. A. Wender and A. W. White, *J. Am. Chem. Soc.,* 1988, **110**, 2218). Several other transition metal catalysed reactions have found application in this area, for example the cobalt promoted carbonylation/cyclisation of 1,1-divinyl cyclohexane to give ketone (11) ((**263**) P. Eilbracht, M. Acker and I. Haedrich, *Chem. Ber.,* 1988, **121**, 519) and the palladium catalysed intramolecular cyclisation of allylic acetate (12) to spirane (13) (S. A. Godleski and R. S. Valpey, *J. Org. Chem.,* 1982, **47**, 381).

Palladium was also the catalyst of choice for the cyclisation of iodide (14) to (15) (E. Negishi, Y. Yantao and B. O'Conner, *Tetrahedron Lett.,* 1988, **29**, 2915). Treatment of methylenecyclohexane with diethyl cyclopropane-1,1-dicarboxylate in the presence of ethylaluminium dichloride catalysis furnishes the corresponding spirane. This process involves initial opening of the cyclopropane ring by the Lewis acid (R. B. Beal, M. A. Dombroski and B. B. Snider, *J. Org. Chem.,* 1986, **51**, 4391).

(1)   (2)   (3) R=CH_2CH_2COMe (4)

(5)   (6)   (7)   (8)   (9) R=H,Me

OAc

(10) R=H,Me   (11)   (12)   (13)

(14)   (15)   (16)   (17)

In common with other bicyclic systems a number of synthetic approaches require the use of a Claisen rearrangement to set up the required quaternary carbon centre for spirane targets. An example of this is the conversion of acetylene (16) (itself derived from cyclohexanone) to the allenic aldehyde (17), a key step for the synthesis of (18) via an ene reaction (M. L. Roumestant, B. Cavallin and M. Bertrand, *Bull. Soc. Chim. Fr.* 1983, 309; F. E. Zeigler and J. J. Mencel, *Tetrahedron Lett.*, 1984, **25**, 123; S. D. Burke, C. W. Murtiashaw and J. O. Saunders, *J. Org. Chem.*, 1981, **46**, 2400; A. S. Kende and R. C. Newbold, *Tetrahedron Lett.*, 1989, **30**, 4329).

An intramolecular variation of the Sakurai reaction was used for the conversion of cyclohexenone (19) to the vinylic spirane (20) as a mixture of diastereoisomers (D. Schinzer, *Angew. Chem.*, 1984, **96**, 292; G. Stork, Y. Kobayashi, T. Suzuki and K. Zhoa, *J. Am. Chem. Soc.*, 1990, **112**, 1661). Acid catalysed intramolecular cyclisations have been employed for the syntheses of other targets in this series (H. Wolf, M. Kolleck and W. Rascher, *Chem. Ber.*, 1976, **109**, 2805; H. Wolf, W. Juerss and K. Claussen, *Chem. Ber.*, 1974, **107**, 2887; P. Naegeli, *Tetrahedron Lett.*, 1978, 2127).

The intramolecular reaction of an electrophilic group (C. Iwata, T. Tanaka, T. Fusaka et al., *Chem. Pharm. Bull.*, 1984, **32**, 447; A. S. Kende and K. Koch, *Tetrahedron Lett.*, 1986, **27**, 6051), or a carbene (C. Iwata, K. Miyashita, T. Imao et al., *Chem. Pharm. Bull.*, 1985, **33**, 853; C. Iwata, M. Yamada, Y. Shinoo et al., *J. Chem. Soc., Chem. Commun.*, 1977, 888; C. Iwata, T. Fusaka, K. Tomita and M. Yamada, *J. Chem. Soc., Chem. Commun.*, 1991, 463), located at the γ-position of alkylated phenol affords spirodienones. In turn these may be converted into a range of spiranes. Selective ring cleavage in more complex polycyclic compounds has been developed as a new method for spirane synthesis (J. F. Ruppert and J. D. White, *J. Am. Chem. Soc.*, 1981, **103**, 1808; J. F. Ruppert and J. D. White, *J. Chem. Soc., Chem. Commun.*, 1976, 976; D. D. K. Duc, J. Ecoto, M. Fetizon H. Colin and

J. C. Diez-Masa, *J. Chem. Soc., Chem. Commun.*, 1981, 953). A number of other miscellaneous method have been developed which cannot be fully described in this review, but are worthy of citation (E. Nakamura and I. Kuwajima, *Org. Synth.*, 1987, **65**, 17; J. Ficini, G. Revial and J. P. Genet, *Tetrahedron Lett.*, 1981, **22**, 629; J. E. Nystrom, T. D. McCanna, P. Helquist et al., *Tetrahedron Lett.*, 1985, 5393; V. Nair and T. S. Jannke, *Tetrahedron Lett.*, 1984, **25**, 3547). One example of a spiro[4,5]decane, (21) has been resolved into enantiomers (J. Brugidou, H. Chrisol and R. Sales, *C. R. Akad. Sci., Ser. C.*, 1974, **278**, 725).

(18)   (19)   (20)   (21)

(k) The spiro[5,5]undecane group

Many methods for the synthesis of this class of compounds mirror those in section 2(j). Thus the reaction of formylcyclohexane or a synthetic equivalent with methyl vinyl ketone and base results in formation of the spiro-appended cyclohexanone via the intermediate aldehyde (V. T. Ravi Kumar, S. Swaminathan and K. Rajagopalan, *Indian J. Chem. Sect. B*, 1985, **24B**, 1168; S. F. Martin, *J. Org. Chem.*, 1976, **41**, 3337). The use of proline to promote the reaction is reported to give enantiomerically enriched products (N. Ramamurthi and S. Swaminathan, *Indian J. Chem. Sect. B*, 1990, **29B**, 401). An equivalent reagent to methyl vinyl ketone, [3-(phenylthio)-1,3-butadienyl]triphenylphosphonium chloride, has been developed for use in this class of spirocyclisation reaction (R. J. Pariza and P. L. Fuchs, *J. Org. Chem.*, 1983, **48**, 2304). The related reaction which involves the use of two equivalents of methyl vinyl ketone to dimedone enolate, followed by intramolecular aldol condensation between the two ketone chains has also been shown to be an efficient method (Y. Houbrechts, P. Laszlo and P. Pennetreau, *Tetrahedron Lett.*, 1986, **27**, 705).

A spiro[5,5]undecane product, (1) is formed in the reaction between pyrrolidine, acetone and sodium iodide, presumably via aldol-type condensations of intermediate enamine adducts (M. Miocque, M. Duchon d'Engenieres and O. Lafont, *Tetrahedron*, 1976, 2133). The base promoted reaction of cyclohexanone with dibenzylidene acetone efficiently furnishes the desired spirane (D. Loganathan, T. Varghese and G. K. Trivedi, *Org. Prep. Proced. Int.*, 1984, **16**, 115; D. N. Rele, H. H. Mathur, G. K. Trivedi et al., *J. Chem. Res., Synop.* 1990, 140). After hydrolytic work-up, intramolecular Michael-type alkylation of imine (2) gave the spiro adduct (3) with a high asymmetric induction (J.

D'Angelo, C. Ferroud, C. Riche and A. Chiaroni, *Tetrahedron Lett.*, 1989, **30**, 6511). Trimethylstannyl aldehyde (4) undergoes a carbocyclisation reaction to give (5) upon activation with a Lewis acid (T. L. MacDonald, C. M. Delahunty, K. Mead et al., *Tetrahedron Lett.*, 1989, **30**, 1473).

A photoreductive cyclisation of acetylenic aldehyde (6) gives spirane (7) in 68% yield (J. Cossy, J. P. Pete and C. Portella, *Tetrahedron Lett.*, 1989, **30**, 7361). Diels-Alder reactions have been employed for the synthesis of spiro[5,5]undecanes (P. M. McCurry and R. K. Singh, *J. Chem. Soc., Chem. Commun.*, 1976, 59), as have intramolecular cyclisations of carbenes onto phenols (K. K. Bhattacharya and P. K. Sen, *Bull. Chem. Soc. Jpn.*, 1979, **52**, 2173; J. M. Roper, *U. S. 4,562,293*, 1985; J. M. Roper, *U.S. 4,480,133*, 1984). The photochemical intramolecular reactions between ketonic and olefinic groups of functionalised spiranes in this class have been examined (J. Zizuashvili, S. Abramson, U. Shmueli et al., *J. Chem. Soc., Chem. Commun.*, 1982, 1375) as have a number of rearrangement reactions involving appended vinylic (C. S. S. Rao, G. Kumar, K. Rajagopalan et al., *Tetrahedron*, 1982, **38**, 2195) and ethynylic (Y. Houbrechts, P. Laszlo and P. Pennetreau, *Tetrahedron Lett.*, 1986, **27**, 705) groups.

(1)   (2)   (3)   (4)

(5)   (6)   (7)

(l) Miscellaneous groups

The spiro[5,6]dodecane group: Details of a synthetic approach to compounds of this type from 1,2-divinylcyclopropane have been reported (E. Piers, M. S. Burmeister and H. U. Reissig, *Can. J. Chem.*, 1986, **64**, 180).

The spiro(5,7]tridecane group: The reaction between methyl vinyl ketone and the enamine from cyclooctanecarboxaldehyde and piperidine gives a 44-49% yield of spiro[5,7]trideca-1,4-dien-3-one (V. V. Kane and M. Jones Jr., *Org. Synth.*, 1983, **61**, 129).

The spiro[4,6]undecane group: Synthetic approaches to spiro[4,6]undecanes via β-hydroxy selenides (L. A. Paquette, J. R. Peterson and R. J. Ross, *J. Org. Chem.*, 1985, **50**, 5200), via transannular opening of smaller rings (H. Duerr, M. Kausch and H. Kober, *Angew. Chem.*, 1974, **86**, 739) and pinacol rearrangement (B. P. Mundy and R. Srinivasa, *Tetrahedron Lett.*, 1979, 2671; B. P. Mundy, R. Srinivasa, R. D. Otzenberger et al., *Tetrahedron Lett.*, 1979, 2673) have been described.

Spiro[5,7] and [5,9] compounds: Syntheses of these compounds have been described (V. Dave and J. S. Whitehurst, *Tetrahedron,* 1974, **30**, 745).

(m) Polyspiranes

By definition polyspiranes contain several carbocyclic rings attached to a central carbon ring. Syntheses of these compounds generally follow conventional methods such as carbene reactions into alkenes. Examples of thus molecular type include members of the trioxotrispirane (1) (H. M. R. Hoffman, A. Walenta, U. Eggert et al., *Angew. Chem., Int. Edn. Engl.,* 1985, **24**, 607; H. Hoffmann, R. Martin, U. Eggert et al., *J. Org. Chem.,* 1989, **54**, 6096; E. Proksch and A. DeMeijere, *Tetrahedron Lett.,* 1976, 4851) and hexaspiranes (2) (L. Fitjer, *Angew. Chem.,* 1976, **88**, 804; L. Fitjer, U. Klages, W. Kuehn et al., *Tetrahedron,* 1984, **40**, 4337; W. Wehle, N. Schormann and L. Fitjer, *Chem. Ber.,* 1988, **122**, 2171; L. Fitjer, M. Giersig, D. Wehlem et al., *Tetrahedron,* 1988, **44**, 393; L. Fitjer, U. Klages, D. Wehle et al., *Tetrahedron,* 1988, **44**, 405; L. Fitjer, K. Justus, P. Puder et al., *Angew. Chem., Int. Edn. Engl.,* 1991, **30**, 436). Several examples of reactions have been published, most of which examine the remarkable cascade ring openings which can take place, for example the acid catalysed conversion of (3) into (4) (L. Fitjer and D. Wehle, *Angew. Chem., Int. Edn. Engl.,* 1979, **18**, 868; L. Fitjer, D. Wehle, M. Noltemeyer et al., *Chem. Ber.,* 1984, **117**, 203; L. Fitjer, M. Giersig, W. Clegg et al., *Tetrahedron lett.,* 1983, **24**, 5351; L. Fitjer and D. Wehle, *Angew. Chem.,* 1987, **99**, 135; L. Fitjer and U. Quabeck, *Angew. Chem.,* 1987, **99**, 1054; L. Fitjer and U. Quabeck, *Angew. Chem., Int. Edn. Engl.,* 1989, **28**, 55; L. Fitjer, *Chem. Ber.,* 1982, **115**, 1047).

220

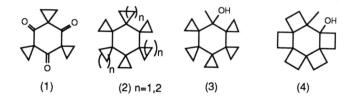

(1)   (2) n=1,2   (3)   (4)

Second Supplements to the 2nd Edition of Rodd's Chemistry
of Carbon Compounds, Vol. II B(Partial), C, D and E, edited by M. Sainsbury   221
© 1994 Elsevier Science B.V. All rights reserved.

Chapter 10

POLYCYCLIC COMPOUNDS. FUSED OR CONDENSED CYCLIC SYSTEMS

MALCOLM SAINSBURY

1. Bicyclo[n.1.0]alkanes

(a) General physical and spectroscopic properties

(i) Molecular structure

Ab initio SCF calculations, with force relaxation using a 4-31 G basis, have been used to achieve complete geometry optimisation of bicyclo[n.1.0]alkanes, where n = 1-3. The results provide information about the local symmetry of the methylene groups in these molecules (P.N.Skancke, *Theochem.*, 1982, **3**, 255).

The heats of vaporisation of the *cis* isomers of bicyclo[n.1.0]alkanes (n = 2-6) and some of their 1-methyl derivatives have been calculated from Benson structural contribution factors. The calculated values agree well with those obtained experimentally (R.M.Varushenchenko and A.I.Druzhinina, *Russ. J. Phys. Chem.*, 1975, **49**, 124).

Molecular mechanics (MM2) methods have been used to compare the stabilities of *cis* and *trans* isomers of bicyclo[n.1.0]alkanes. It is concluded that the *trans* forms of bicyclo[4.1.0]heptane and bicyclo[3.1.0]hexane are 40-126 kJ mol^{-1} less stable than the *cis* isomers, but the second member of the series, bicyclo[2.1.0]pentane, has no minimum corresponding to a *trans* isomer. Conversely, for bicyclo[1.1.0]butane a *trans* form is possible (V.A.Svyyatin, A.I.Ioffe, and O.M.Nefedov, *Izv. Akad. Nauk. SSSR., Ser. Khim.*, 1988, 78). MM2 methods have also been employed to determine the barriers to planarity and/or pseudorotation for small ring systems, including bicyclo[3.1.0]hexanes (R.L.Rosas, C.Cooper, and J.Laane, *J. Phys. Chem.*, 1990, **94**, 1830). The hydrocarbon parent and the derivatives 2,4-dithia-, 2,4-dioxa-, 6-oxa-, and 6-thia-bicyclo[3.1.0]hexanes have also been subjected to *ab initio* calculations designed to ascertain relative total energies and geometries for the various boat and chair forms. The results agree with experimental observations and indicate that the boat form of bicyclo[3.1.0]hexane itself, and those of its mono-heteroatom analogues, are preferred. However, a chair form is favoured for 2,4-dioxabicyclo[3.1.0]hexane.

Chair   Boat

In the case of 2,4-dithiabicyclo[3.1.0]hexane both forms are estimated to have equal energy. However, for bicyclo[3.2.0]hept-6-ene calculations predict that the boat form is much more stable than the chair. This result is in conflict with data from experimental work (R.Okazaki, J.Niwa, and S.Kato, *Bull. Chem. Soc. Japan,* 1988, **61**, 1619). Conformational analyses of many diverse types of related bicycloalkanes have also been reported (see N.L.Allinger, *J. Amer. Chem. Soc.,* 1977, **99**, 8127; C.Jaime and E.Osawa, *Tetrahedron,* 1983, **39**, 2769).

(ii) <u>Spectroscopy and spectrometry</u>
The ^{13}C N.M.R. spectra of bicyclo[n.1.0]alkanes have been recorded in order to correlate ^{13}C-^{1}H coupling constants and ring strain (M.Christl, *Chem. Ber.*, 1975, **108**, 2781; see also P.Brun *et al.*, *Compt. rendu. Ser. C,* 1979, **288**, 201), and the ^{13}C N.M.R. spectra of 70 known *cis* and *trans* fused bicyclo[4.n.0]alkanes have been analysed in order to provide a data base for predicting the geometry of the ring fusion in compounds where this is unknown (P.Metzer, E.Casadevall, and M.J.Pouet, *Org. Magn. Reson.,* 1982, **19**, 229).
As might be expected, ring size and the orientation of substituents bonded to the framework carbon atoms determine the fragmentation behaviour of the molecular ions of bicyclo[n.1.0]alkanes (R.Herzschuh and K.Epsch, *J. Prakt. Chem.,* 1980, **322**, 28).   However, some basic 'rules' have been deduced which are helpful in the interpretation of the mass spectra of

these types of molecules (J.R.Dias and C.Djerassi, *Org. Mass Spectrom.*, 1973, **7**, 753). A compilation of the mass spectrometric fragmentation patterns of a number of bicyclo[n.1.0]alkanes and their halogeno derivatives is also available (E.S.Agavelyan, A.K.Mal'tsev, and O.M.Nefedov, *Ivz. Akad. Nauk. SSSR., Ser. Khim.,* 1977, 91).

(b) *General chemical reactions*

(i) Hydrogenolysis
The regiochemistry of the hydrogenolysis of bicycloalkanes (1, X = H) in the presence of platinum, or palladium on carbon, as catalysts is determined by the ring size. For some hydrocarbons the central bond is cleaved, whereas for others a peripheral C-C bond of the cyclopropyl unit is broken. Thus for (1, X = H), where n = 5 or more, the products are methylcycloalkanes of the same ring size, for example, (1, X = H; n = 5) yields methylcycloheptane, and (1, X = H; n = 6) affords methylcyclooctane. Hydrogenolysis of (1, X = H; n = 4), however, gives methylcyclohexane, as well as 2-7% cycloheptane, and (1, X = H; n = 3) produces methylcyclopentane and 5-20% cyclohexane. In the case of bicyclo[2.1.0]pentane, however, only cyclopentane is produced, while bicyclo[1.1.0]butane, yields butane (98%) containing 1-2% of cyclobutane (K.L.Stahl, W.Hertzsch, and H.Musso, *Annalen,* 1985, 1474).

The hydrogenations of dihalobicycloalkanes (1, X = F, Cl, Br; n = 1, 2, 4) have been similarly studied. When (1, X = F; n = 1) is hydrogenated, over either a palladium or a nickel catalyst, ring opening occurs affording (2, X = H, F; n = 1). However, (1, X = F; n = 2, or 4) only gives (2, X = H,F; n = 2, or 4) when palladium is the catalyst. The chloro analogue (1, X = Cl; n = 1) yields (2, X = Cl; n = 1) with palladium as catalyst, as well as some monochlorinated starting material. Indeed, reductive monodechlorination is noted for all the other dichlorobicycloalkanes when the hydrogenation experiment is carried out with a nickel catalyst in the presence of ethylamines. In addition, methylcycloalkanes (3) are commonly produced through both hydrogenolysis of the C-Cl bonds and reductive cleavage of a peripheral bond of the cyclopropane unit. It is noted that ethylamines (EtNH$_2$, Et$_2$NH, or Et$_3$N) retard the rates of these reactions to a degree corresponding to the binding affinity of the amine to the catalyst.

Reductive monodebromination also occurs when the bicycloalkanes (1, X = Br; n = 2, or 4) are hydrogenated over either palladium, or nickel, in the presence of an amine (K.Isogai and J.Sakai, *Nippon Kagaku Kaishi,* 1986, 650).

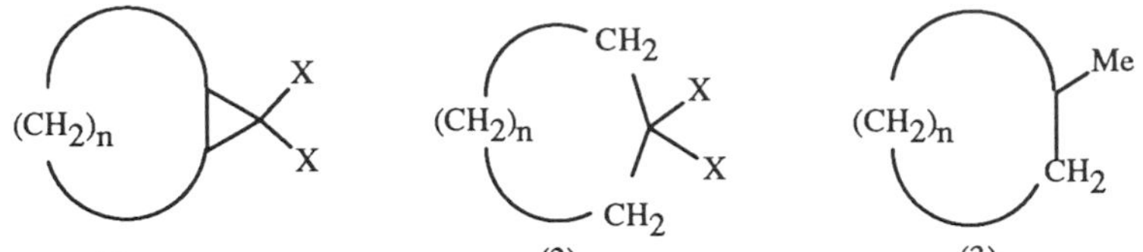

(1)   (2)   (3)

(ii) Reduction by triethylsilane

Reduction of bicyclo[n.1.0]alkanes (n = 3, or 4) or their 1-methyl derivatives with triethylsilane - trifluoroacetic acid also proceeds with ring opening of the cyclopropane ring. With two mol. equivalents of trifluoroacetic acid bicyclo[3.1.0]hexane gives methylcyclopentane (16%), and cyclohexane (1-2%), plus the trifluoroacetates of hexanol (10%), and of *cis*- and *trans*- methyl-2-cyclopentanols (35%). With 4 mol. equivalents of the acid the same hydrocarbons are formed in 28% and 2% yields, respectively, together with 70% of the mixed trifluoroacetates. Bicyclo[4.1.0]heptane with two mol. equivalents of trifluoroacetic acid gives methylcyclohexane (36%), cycloheptane (1.5%), and 62% of the trifluoroacetates of heptanol and 2-methylhexanols. With 4 mol. equivalents of trifluoroacetic acid the yields are 79% and 21%, respectively (G.A.Khotimskaya *et al.*, *Izv. Akad. Nauk. SSSR., Ser. Khim.*, 1972, 1989).

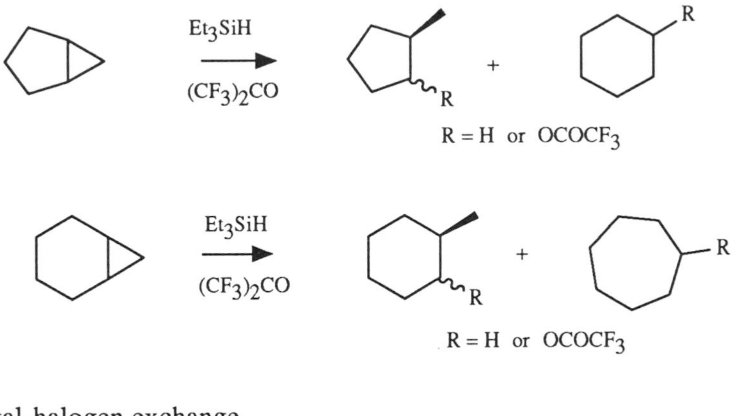

R = H or OCOCF$_3$

R = H or OCOCF$_3$

(iii) Metal-halogen exchange

When dibromobicyclo[n.1.0]alkanes (1; X = Br, n = 4, or 6) are treated with butyllithium the cyclopropylidene-bromolithium carbenoids (4), or (5) respectively, are formed. These can be detected as stable species by ^{13}C N.M.R. spectroscopy (D.Seebach *et al., Angew. Chem. Int. Ed. Engl.,* 1979, **18**, 784).

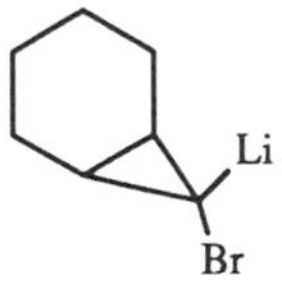

(4) 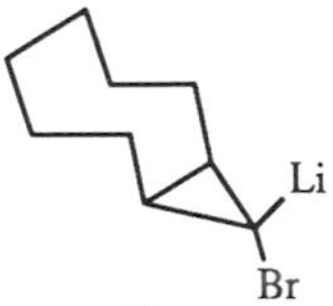   (5)

(iv) Reactions with olefin-metathesis catalysts

The behaviour of bicyclo[n.1.0]alkanes with tungsten hexachloride/ Lewis acids (WCl_6/Ph_4Sn and $WCl_6/EtAlCl_2$) has been studied. Although bicyclo[3.1.0]hexane and bicyclo[4.1.0]heptane are inert to WCl_6-Ph_4Sn, higher members of the series do react, but cycloalkenes containing one less carbon atom - the products of reterocarbene addition reactions - are not formed. All the carbon atoms of the parent hydrocarbon are retained, and bicyclo[6.1.0]nonane, for example, gives methylenecyclooctane and a mixture of isomeric methylcyclooctenes (A.Uchida and K.Hata., *J. Chem. Soc. Dalton Trans.,* 1981, 1111).

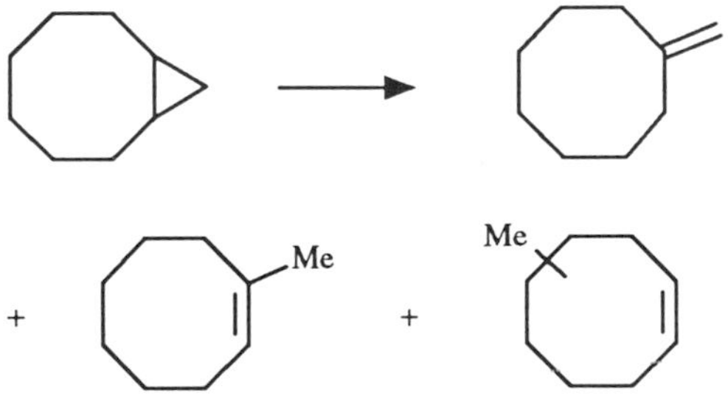

(v) Reaction with helium tritide ions

Radio gas chromatography of the products of the reactions of bicyclo[n.1.0]alkanes with HeT^+ (from the β-decay of molecular tritium) reveals that some tritiated hydrocarbons retaining the original bicyclic structure are produced. Bicycloalkylium ions, formed by triton transfer from HeT^+ to the substrates, are intermediates, but because of the exothermic nature of the reactions many of the ions formed undergo fragmentations and isomerisations involving C-C bond cleavages; mainly in the cyclopropyl unit (F.Cacarce, A.Guarino, and M.Speranza, *J. Chem. Soc. Perkin Trans*.2, 1973, 66).

(vi) Acetolysis

The rate at which bicyclo[n.1.0]alkanes undergo acetolysis is not directly related to relief of ring strain, but there does appear to be a correlation between susceptibility towards acetolysis and ionization potential. Thus the degree of acid mediated polarisation of the C-C bonds relates well with the reactivities observed, and is a rate controlling factor (K.B.Wiberg *et al.*, *J. Amer. Chem. Soc.*, 1985, **107**, 1003). In line with this bicyclo[2.1.0]pentane undergoes acetolysis 88 times more rapidly than bicyclo[3.1.0]hexane. This is much slower than predicted if relief of ring strain was the influential factor (K.B.Wiberg, S.R.Kass, and K.C.Bishop III, *ibid.*, 1985, **107**, 996). The uncatalysed acetolysis of bicyclo[2.1.0]pentane gives cyclopentyl acetate (47%) and cyclopentene (53%). However, in the presence of 4-toluenesulphonic acid in acetic acid the reaction is accelerated and cyclopentyl acetate (30%), cyclopentyl tosylate (24%, and cyclopentene (6%) are produced. If DOAc replaces acetic acid the reaction is retarded by a factor of 2.55, suggesting that proton transfer is at least partly rate determining. Evidence is also presented to show that cyclopentyl acetate is formed largely through inversion at both C-1 and C-4, whereas cyclopentyl tosylate arises through inversion at the site of protonation, migration of the endo C-5 proton, and capture of the tosylate anion. *Cis*-bicyclo-[5.1.0]octane undergoes acetolysis of one of the peripheral cyclopropyl C-C bonds, but the *trans* isomer reacts through cleavage of the central bond.

(vii) Oxymercuration

Bicyclo[2.1.0]pentane undergoes exclusive *anti*-Markownikoff oxymercuration with cleavage of the highly reactive central bond, however, in general this type of reaction becomes less important as ring strain reduces, i.e. as the size of the non-cyclopropane unit is expanded. Thus oxymercuration-acetylation-demercuration of bicyclo[3.1.0]hexane gives both cyclohexyl acetate and some 2-methylcyclopentenyl acetate.

The first product is that expected from the cleavage of the central bond, whereas the second arises through scission of a peripheral C-C bond.

The presence of a *cis-* or a *trans-*3-hydroxyl group promotes the cleavage of the central bond and oxymercuration-acetylation-demercuration of *cis-*3-hydroxybicyclo[3.1.0]hexane gives mainly *cis-*1,3-diacetoxyhexane. No rate enhancement is noted if an acetoxy group replaces the hydroxyl group, and the promotional effect of a neighbouring hydroxyl group does not extend to the next highest homologue. Indeed, bicyclo[4.1.0]heptane and its derivatives, undergo oxymercuration-acetylation-deoxymercuration with exclusive fission of a peripheral cyclopropane C-C bond. 3-Hydroxybicyclo[4.1.0]heptane affords 4-methyl-*trans-*1,3-diacetoxycyclohexane and 2-methyl-*trans-*1,4-diacetoxycyclohexane in the ratio 2:3 (99%). In each of these structures the relative orientations of the methyl substituent is assumed, and is considered to follow from the stereochemistry of the starting material (R.G.Salomon and R.D.Gleim, *J. Org. Chem.*, 1976, **41**, 1529).

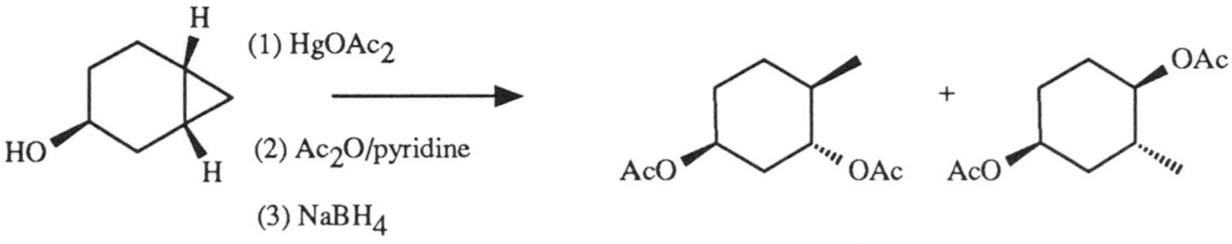

By way of contrast, bicyclo[4.1.0]heptane gives 3-(3-methoxycycloheptanyl)propionitrile when it is reacted with mercuric acetate and the product is reduced with sodium borohydride (B.Giese and W.Zwick, *Tetrahedron Letters,* 1980, **21**, 3569). A radical chain mechanism may be involved in the first of the two reactions.

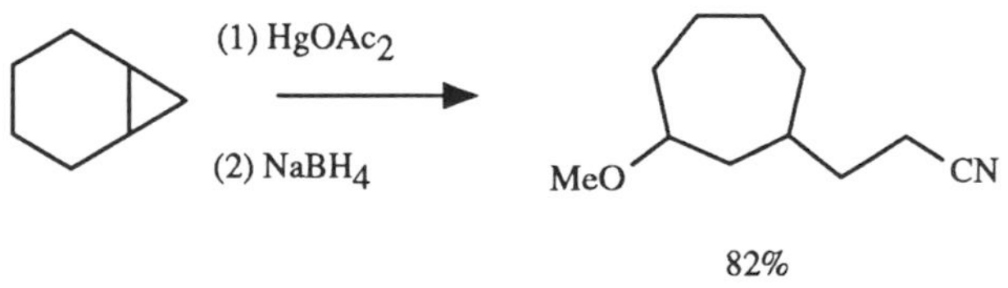

82%

(viii) Acylation
Acetylations of bicyclo[3.1.0]hexane and bicyclo[5.1.0]octane with

pivaloyl tetrafluoroborate in methyl nitrate at -30°C afford enones and their hydration products (Yu.V.Tomilov, V.A.Smit, and O.M.Nefedov, *Izv. Akad. Nauk. SSR. Ser. Khim.*, 1975, 2614).

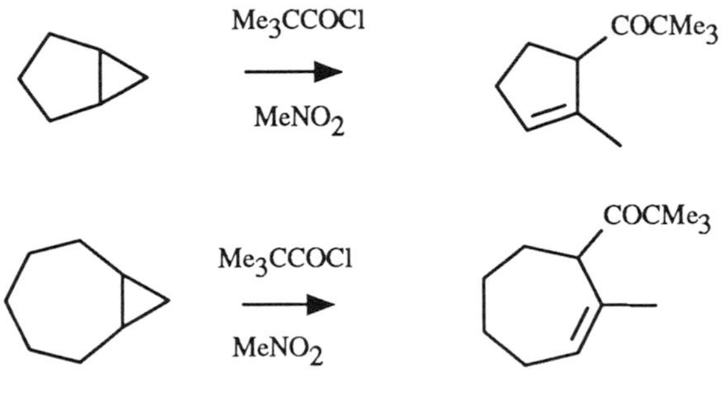

(ix) Radical and photochemical reactions

Bicyclo[n,1.0]alk-2-yl radicals are the primary products when the parent alkanes (n = 0-5) undergo hydrogen abstraction by reactions with ^{t}butoxyl radicals. For the hydrocarbon radicals, where n = 2-5, rearrangement then occurs through β-scission of the outer cyclopropane bonds. This yields cycloalkenylmethyl radicals, but for the lower members of the series (n = 0 or 1) the radicals rearrange through fission of the bridging bonds affording ring expanded cycloalkenyl radicals.

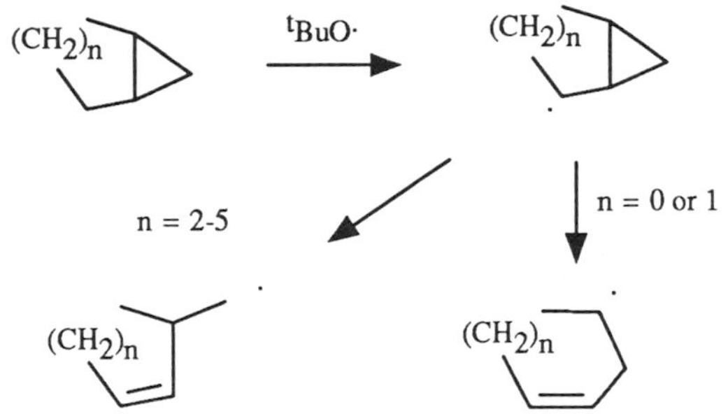

Photobromination of bicyclo[n.1.0]alkanes has also been studied: for those substrates where n = 3 or 4, bimolecular homolytic substitution (S_H2) by bromine atoms occurs at one of the cyclopropane carbon atoms, however, there is an increase in the ease of hydrogen abstraction with decreasing ring size [(C.Roberts and J.C.Walton, *J. Chem. Soc. Perkin Trans. 2*, 1983, 879; *J. Chem. Soc. Chem. Commun.*, 1984, 1109); see also section 4b(ii)].

Photolyses of bicyclo[n.1.0]alkanes, through irradiation at 185nm in pentane solution, are characterised by two isomerisation processes. In the case of the alkanes, where n = 3, 4, or 5, two bonds of the cyclopropane ring are cleaved to give α,ω-dienes. In addition, the central bond is also fragmented to produce cyclic dienes. This is observed in all cases, except when n = 4 (R.Srinivasan and J.A.Ors, *J. Amer. Chem. Soc.*, 1978, **100**, 7089).

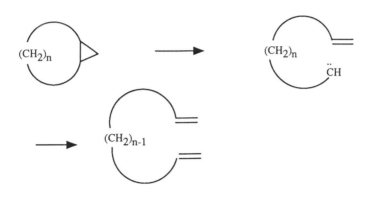

2. Bicyclo[n.1.0]alk-2-enes

(a) *Synthesis*

Bicyclo[n.1.0]alk-2-enes (n = 7 and 8) can be obtained from the corresponding 2-ketones through formation of the tosylhydrazones, which are then treated with butyl lithium. Hydrogenation of the alkene (1, n =8), over platinum on carbon, gives *cis*- (7%), and *trans*-bicyclo[8.1.0]undecane (2)(68%), and methylcyclodecane (3)(25%). Surprisingly, hydrogenation of bicyclo[n.1.0]alk-2-enes (n =3-7) affords only bicyclo[n.1.0]alkanes. No cleavage of the cyclopropane ring is observed (T.L.Levashova, O.V.Semeikin, and E.S.Balenkova, *Vestn. Mosk. Univ. Ser. 2, Khim.*, 1981, **22**, 587).

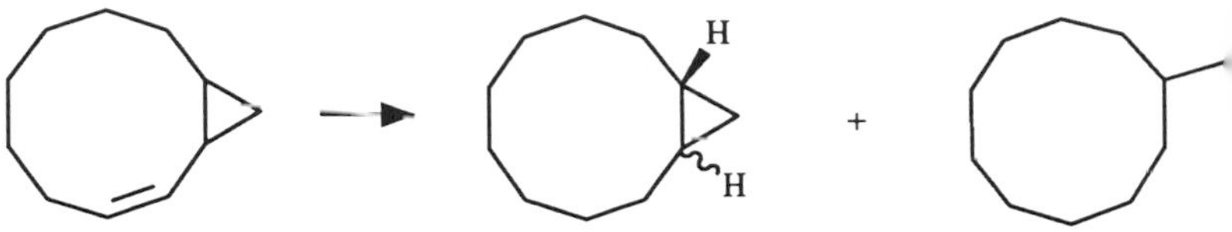

(1)   (2)   (3)

3. The bicyclo[1.1.0]butane group

(a) Synthesis

When treated with zinc dust suspended in methanol chloromethyl-cyclopropanes (1), formed from the addition of dibromocarbene to an appropriate alkene, undergo monodebromination. The monobromo products if reacted further with magnesium in tetrahydrofuran, or with sodium in dioxane, afford bicyclobutanes (2) (A.I.D'yachenko *et al.*, *Izv. Akad. Nauk. SSR. Ser. Khim.* 1985, 2043).

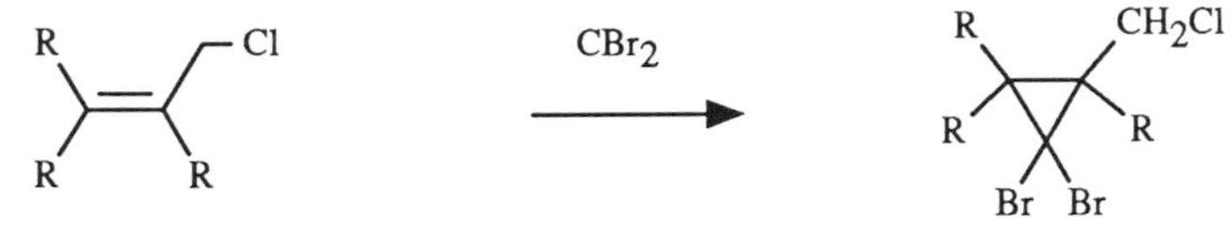

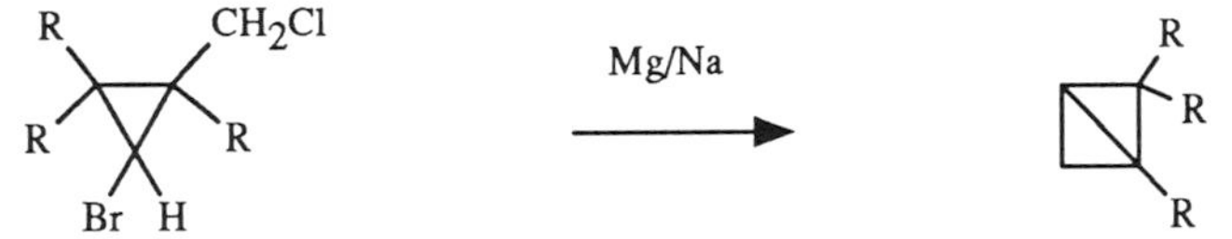

The synthesis is applicable to bicyclo[1.1.0]butane itself, here the alkene used is 3-chloroprop-1-ene (N.M.Abramova, and S.V.Zotova, *ibid.*, 1979, 697). Similarly, the reaction of chloroacetyl bromide with ethoxyethene affords the adduct (3), which on reduction with lithium aluminium hydride gives 1-ethoxy-3-hydroxycyclobutane (4). When this product is reacted with thionyl chloride and then with magnesium it yields bicyclo[1.1.0]butane (*idem, ibid.,* 1981, 439).

(3)   (4)

(b) Reactions

(i) Ionisation potentials and anodic oxidation.
The effect of substituents upon the reactivity of the cyclo[1.1.0]butane carbon framework towards electrophiles has been calculated. It is concluded that most common substituents, other than amino, diminish reactivity, although the central C-C bond is weakened by the presence of NO or NO_2 functions at the bridgehead positions (P.Politzer, G.P.Kirschenheuter, and J.Alster, *J. Amer. Chem. Soc.,* 1987, **109**, 1033). Similarly, the ionisation potentials for all the possible mono- methyl, amino, hydroxy, fluoro and cyano derivatives of bicyclo[1.1.0]butane have been calculated. The methyl group lowers the ionisation potential, especially when it is bonded to the bridgehead position. The amino group exerts the same effect, but the situation is complicated by the fact that initial ionisation of one of the lone pair electrons may occur.

Substitution at the bridgehead by a hydroxyl group similarly leads to a reduction in the ionisation potential, and electron loss usually takes place from the central bond. A fluorine atom at the bridgehead also causes a decrease in the ionisation potential, but if it is present as an *endo* or *exo* substituent then there is an increase. A cyano substituent also causes an increase in the ionisation potential, although the effect is least when the group occupies a bridgehead site (S.C.Richtmeier, P.G.Gassman, and D.A.Dixon, *J. Org. Chem.,* 1985, **50**, 311).

In practice electron transfer from 1,2,2-trimethylbicyclo[1.1.0]butane (5) to a platinum anode leads to the radical cation (6), as expected if the HOMO is mainly associated with the C_1-C_3 bond. In an electrolyte consisting of 50:1 methanol:pyridine and tetraethylammonium perchlorate four oxidation products (9), (10), (11), and (12) are formed.

These all stem from the radical cation (6) which reacts with methanol and deprotonates to give the cyclobutyl radical (7). Further electron transfer affords the 'classical' cyclobutyl cation (8), which decomposes *via* four different reaction pathways. Thus deprotonation leads to the methylenecyclobutane (9), whereas quenching by methanol yields the *trans*-1,3-dimethoxycyclobutane (10). The remaining acyclic products (11) and (12) are assumed to arise through two alternative fragmentations (a) and (b), respectively, of the cyclobutyl cation. In either case the initial product is a homoallyl cation which then reacts with methanol (P.G.Gassman and G.T.Carroll, *J. Org. Chem.,* 1984, **49**, 2074).

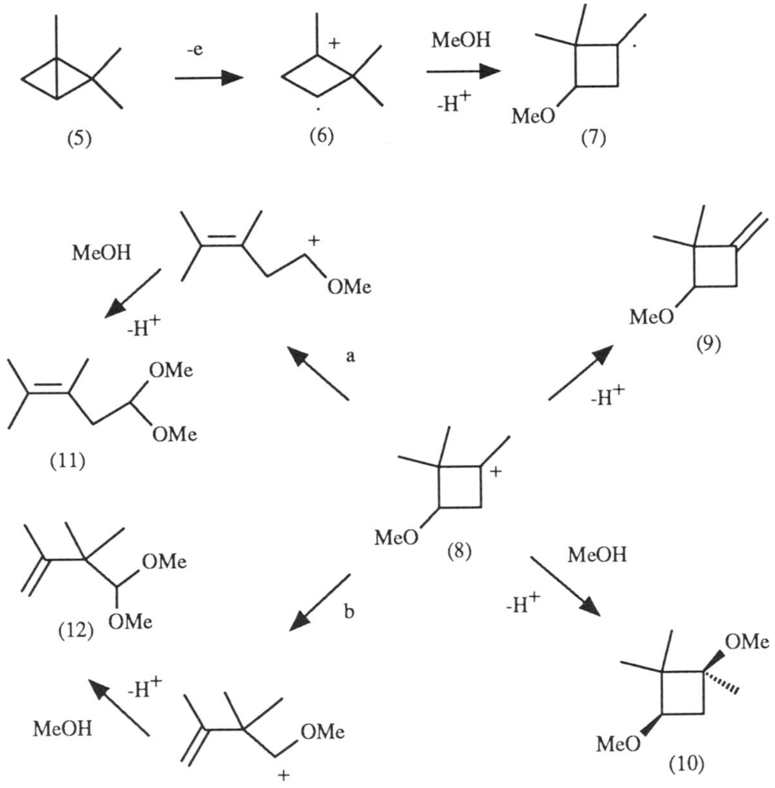

(ii) Reactions with electrophiles and organometallic reagents

When bicyclo[1.1.0]butane is reacted with mercuric chloride in dichloromethane, at 40°C, 1-acetoxy-3-(acetoxymercuri)cyclobutane (13) is formed, this gives 3-acetoxy-1-chloromercuricyclobutane (14) by further reaction with potassium chloride. Product (14) can be converted into a mixture of the bicyclobutane (15) and 3-acetoxycyclobutene (16) by treatment with lithium chloride and palladium dichloride (N.M.Abramova, S.V.Zotova, and O.A.Nesmeyanova, *Izv. Akad. Nauk. SSR. Ser. Khim.,* 1984, 2813).

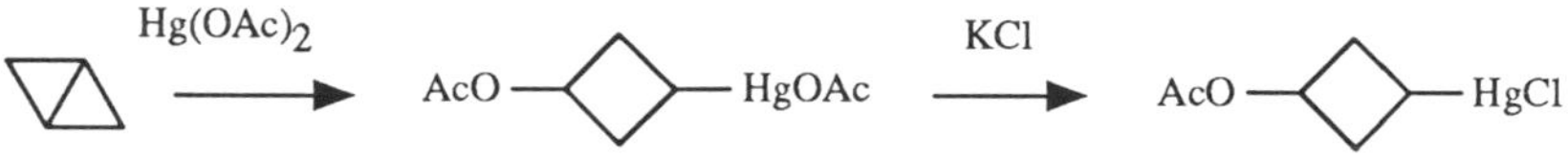

(13)   (14)

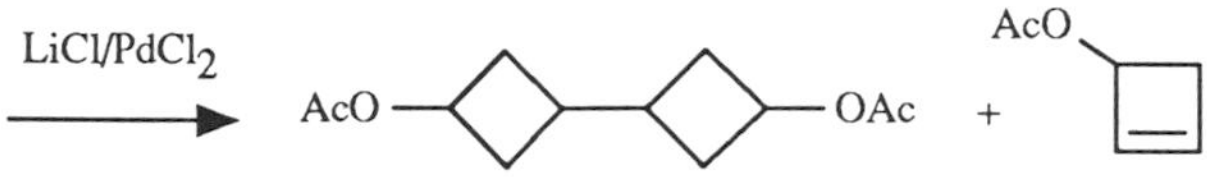

(15)   (16)

Treatment of 1-methylbicyclo[1.1.0]butane with norbornadiene rhodium chloride dimer [Rh(nor)Cl₂] in chloroform gives a mixture of the alkenes (17), and (18). Repetition of the reaction in the presence of methyl acrylate gives the same products accompanied by the *cis-* and *trans*-cyclopropanes (19). The formation of the cyclopropanes suggests that a rhodium-carbene complex may participate in the reaction (P.G.Gassman and R.R.Reitz, *J. Organometal. Chem.,* 1973, **52**, C51-C54).

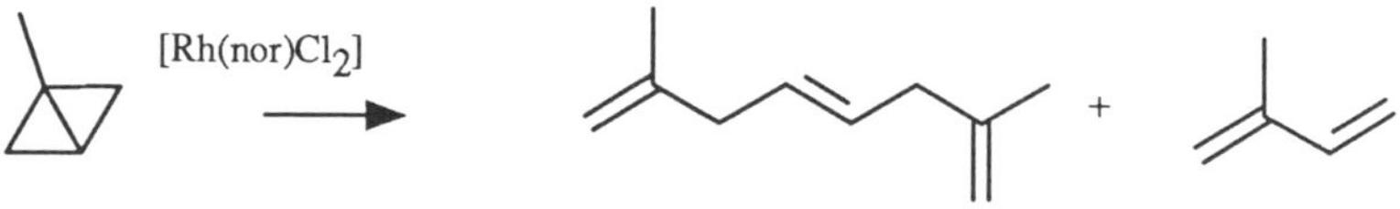

(17)   (18)

Organoboranes cleave bicyclo[1.1.0]butane to give 4-borylbutenes (B.A.Kazansky *et al., Tetrahedron Letters,* 1974, 567).

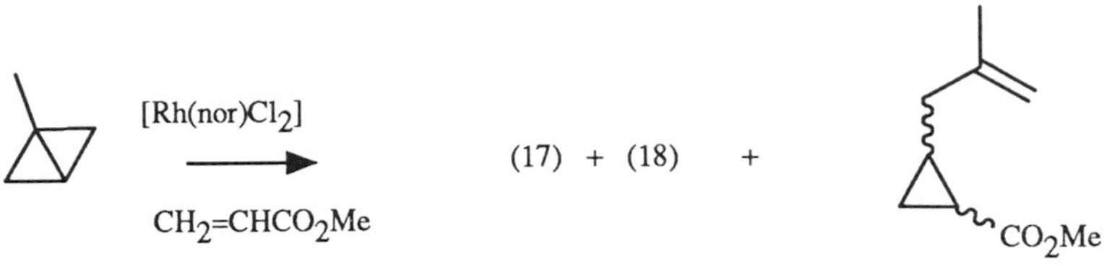

(17) + (18) +

(19)

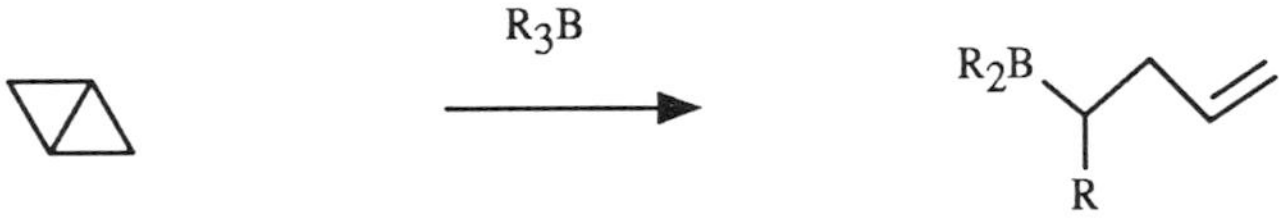

(iii) <u>Reactions with dienophiles</u>

The arylonitrile (20) adds to 1-cyano-1-methylbicyclo[1.1.0]butane to give the butylthiodicyanomethyltricyclohexane (21) (A.De Meijere *et al.*, *Tetrahedron*, 1986, **42**, 1291).

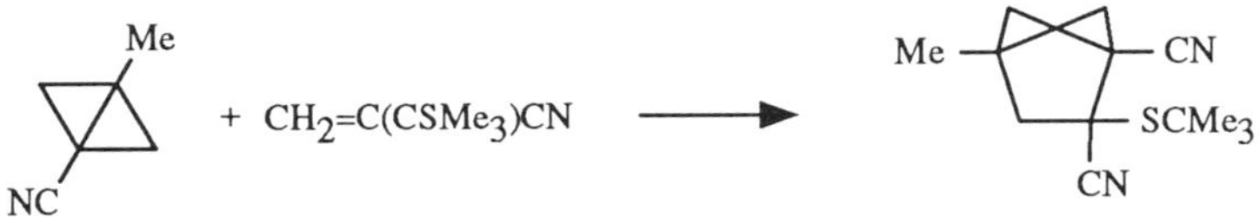

(20)   (21)

The very strained molecule [1.1.1]propellane (22) reacts with the dienophile tetracyanoethene to give the 1:1 adduct (23, R=CN). However, with di(methoxycarbonyl)ethyne both the 1:1 adduct (23, R=CO_2Me) and the 2:1 adducts (24, R=CO_2Me) and (25, R=CO_2Me) are formed. Interestingly, bicyclo[1.1.0]butane is even more reactive and it reacts with di(methoxycarbonyl)ethyne to give 1:1:1 adducts (K.B.Wiberg and S.T.Sherman, *Tetrahedron Letters*, 1987, **28**, 151).

(22) 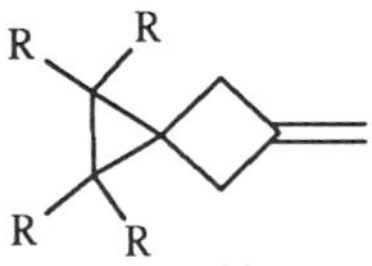   (23)

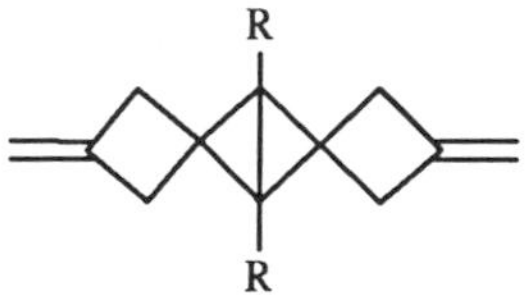

(24) 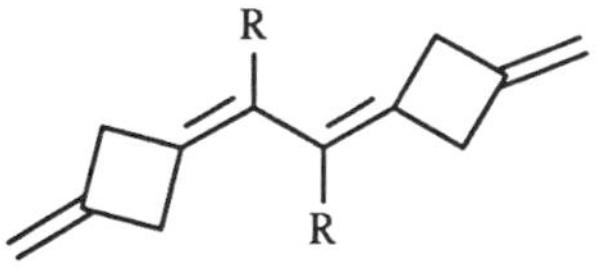   (25)

(iv) Radical reactions and photochemistry

In radical reactions, however, the reverse is true. Thus Wiberg *et al.* (*idem*, 1986, **27**, 1553) have also shown that [1.1.1] propellane is more reactive than bicyclo[1.1.0]butane, and much more so than bicyclo[2.1.0]pentane. From this it is argued that the enhanced reactivity is not determined by ring strain, nor the energy of the HOMO.

Relative rate constants have been determined in competitive addition reactions of the phenylsulphide radical to the C_1-C_3 bond of the bicyclo[1.1.0]butane derivatives (26), where R = H, Me, or CN, and R^1, R^2, R^3 = H or Me. Electron acceptors at C_1 slow the addition, whereas electron donating groups enhance the reaction rate (A.Schlosser *et al.*, *Chem. Ber.*, 1980, **113**, 1053).

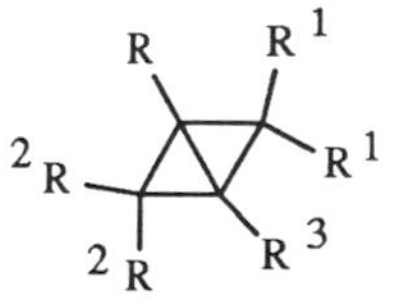

(26)

When bicyclo[1.1.0]butane is photolysed through irradiation at 185nm cyclobutene and 1,3-butadiene are produced in the ratio 10:1 (W.Adam, T.Oppenländer, and G.Zang, *J. Amer. Chem. Soc.*, 1985, **107**, 3921). 1,3-Butadiene is generated by two mechanisms: the major one, which occurs for two thirds of the time, involves a carbene pathway in which the C_1-C_3 and the C_2-C_3 bonds are cleaved, and a nonspecific H-shift takes place. The less important route, which takes place for the remainder of the time, requires a direct rearrangement.

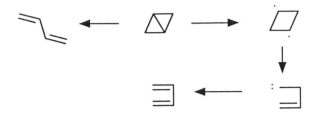

Cyclobutene formation may occur through a H-shift in a concerted reaction, or alternatively may involve a 1,3-diradical which originates from the carbene implicated in the formation of 1,3-butadiene by the main route (A.F.Becknell, J.A.Berson, and R.Srinvasan, *J. Amer. Chem. Soc.,* 1985, **107,** 1076)

Bicyclo[1.1.0]butane reacts with the oxygen radical anion in the gas phase to give the radical anion of bicyclo[1.1.0]but-1(3)-ene (P.K.Chou and S.R.Kass, *J. Amer. Chem. Soc.,* 1991, **113,** 697).

A radical chain reaction is involved in the photo-initiated addition of acetone across the C_1-C_3 bond of bicyclo[1.1.0]butane. α-Keto radicals are the chain initiating species, as well as being responsible for chain propagation. When a methyl group is bonded to one of the bridgehead positions only *anti*-Markownikoff addition is observed, thus 1-methylbicyclo[1.1.0]butane affords the isomeric cyclobutylpropan-2-ones (27)and (28), but in only 21% yield. Productivity is enhanced when methyl groups are absent from the bridgehead sites and, for example, 2,2,4,4-tetramethylbicyclo[1.1.0]butane gives the tetramethyl ketone (29) in 68%.

Some other alkanones react with bicyclobutanes in the same way, however, there is a major competitive process involving a Norrish type 1 cleavage of the ketone to produce an acyl radical. For example, in a photoaddition reaction with 3-pentanone 2,2,4,4-tetramethylbicyclo[1.1.0]-butane gives the expected product (30) in 43% yield plus the lower ketone (31) in 12% (P.G.Gassman and G.T.Carroll, *J. Org. Chem.,* 1984, **49,** 2074).

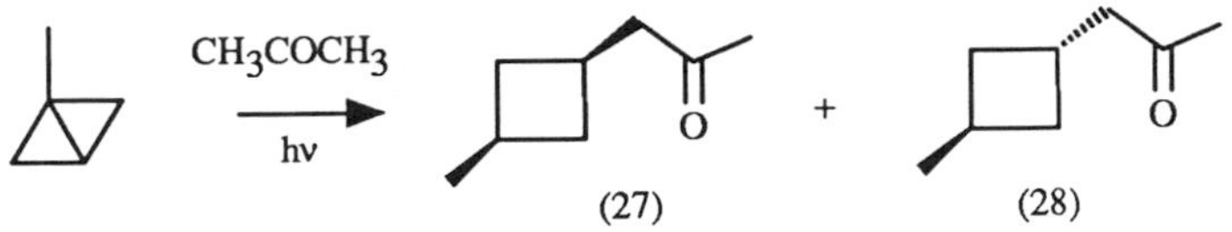

(27)   (28)

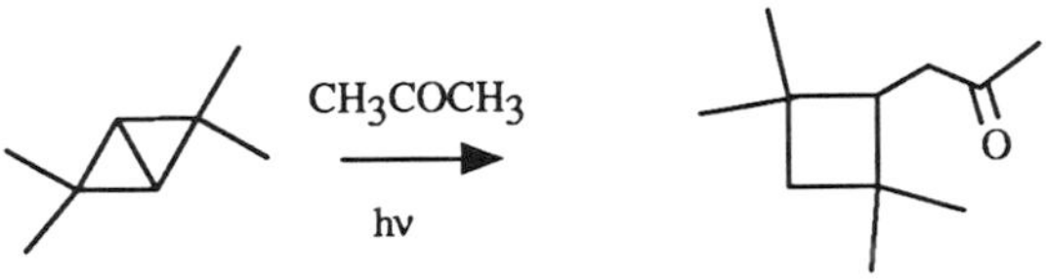

(29)

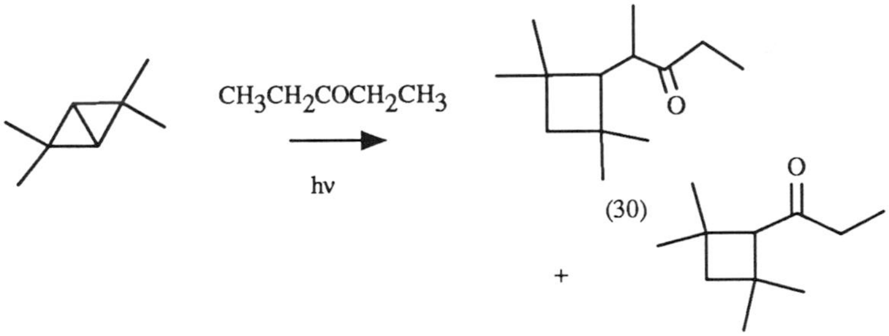

(30)

(31)

(v) Reactions with thiols and alcohols
Thiols add to bicyclo[1.1.0]cyclobutanes to give cyclobutyl sulphides.

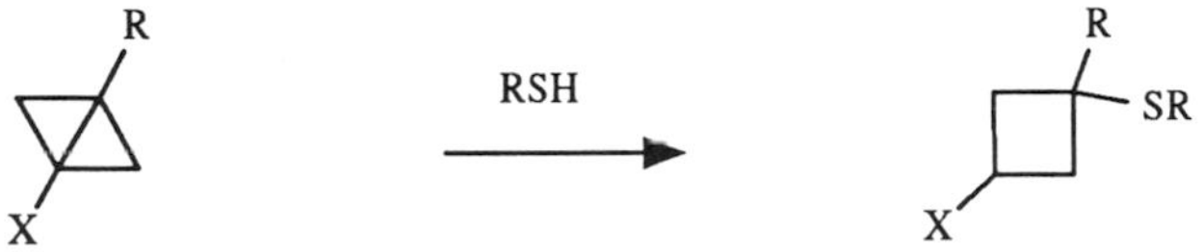

A radical chain reaction is involved which opens the central bond giving an intermediate thiyl radical with inversion at the site of initial attack. The stability of the thiyl radical seems to be less important than steric factors in determining the regioselectivity of the reactions of thiols with 1,3-disubstituted bicyclobutanes.

When 1-substituted tricyclo[4.1.0.0^{2,7}]heptanes (32) are reacted in this way, with for example benzenethiol, the same stereoregulation applies, and the reagent approaches the less hindered bridgehead position to generate two products (33) and (34) (G.Szeimies *et al.*, *Chem. Ber.*, 1978, **11**, 1922).

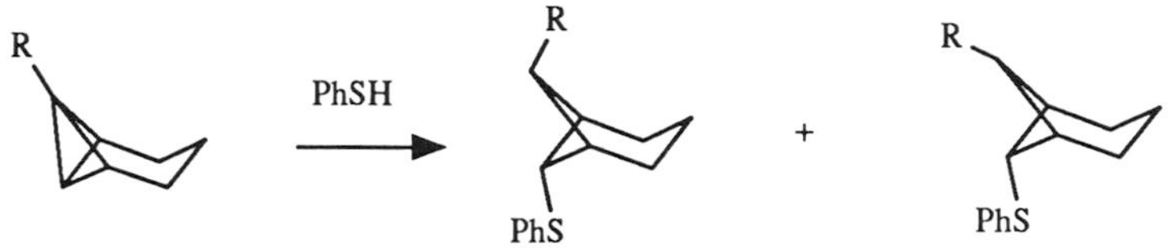

(32)   (33)   (34)

Methanol reacts with 1-methoxycarbonylbicyclo[1.1.0]butane to give 1-, 2-, and 3-methoxy-1-methoxycarbonylcyclobutane, as well as 1- and 2-(methoxymethyl)cyclopropane-1-carboxylates (V.V.Razin and M.V.Eremenko, *Zh. Org. Khim.*, 1978, 14, 1475).

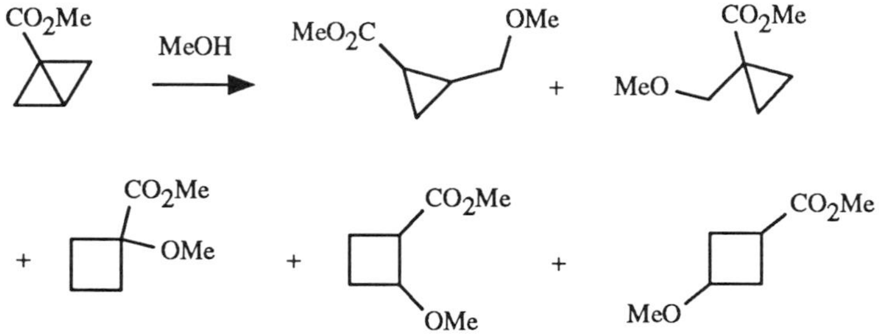

4. Bicyclo[2.1.0]pentane

(a) *General properties*

Bicyclo[2.1.0]pentane is also a highly strained hydrocarbon with a calculated strain energy of 238 kJ.mol^{-1}. The central bond contains the greatest strain energy and when it is cleaved *ca.* 209 kJ. mol^{-1} of energy is released. Despite this the molecule is thermally stable and only decomposes when heated at temperatures over 300°C; then it isomerises into cyclopentene. The mechanism of this reaction involves a hydrogen, rather than a carbon atom, migration, a fact revealed by a deuterium labelling experiment (J.E.Baldwin and G.D.Andrews, *J. Org. Chem.*, 1973, **38**, 1063).

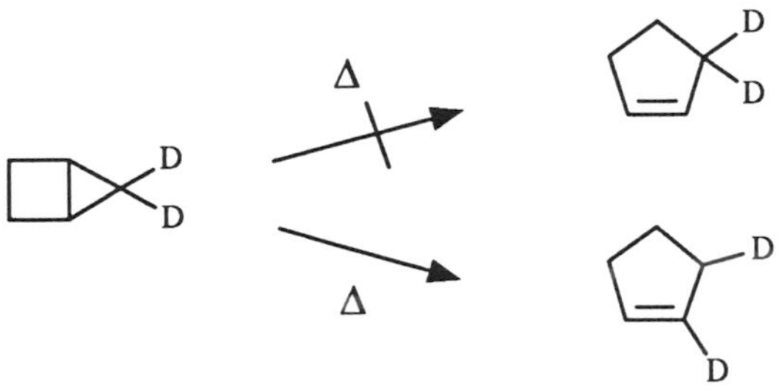

(b) *Reactions*

(i) Reactions with dienophiles

The central bond of bicyclo[2.1.0]pentane is, however, kinetically labile and readily reacts with electrophiles to give monocyclic products. Thus with *trans*-1,2-dicyanoethene at 160°C in the course of two days a variety of products are formed (P.G.Gassman, *Acc. Chem. Res.*, 1971, **4**, 128)

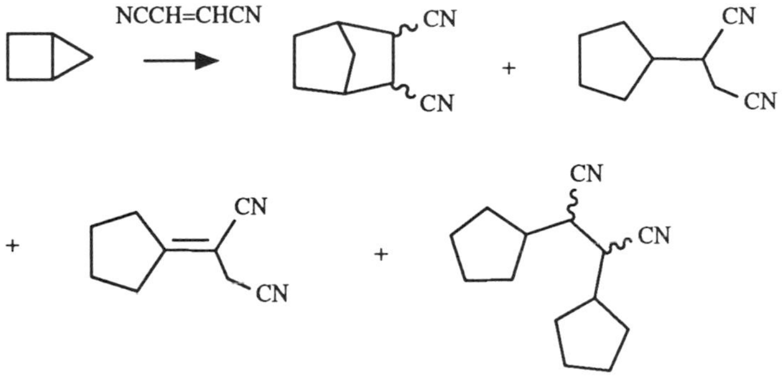

At 70°C, and in the presence of a catalytic amount of bis(acrylonitrile)nickel(0), bicyclo[2.1.0]pentane reacts with acrylonitrile to give the nitrile adducts (1), (2), and (3).

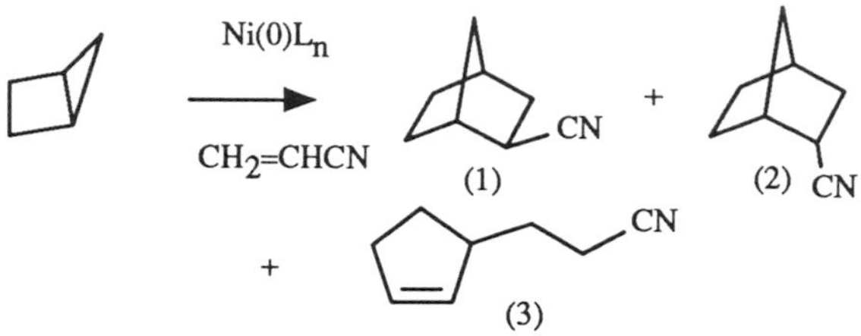

Methyl acrylate acts in an analogous manner, and it is proposed that the nickel(0) complex interacts with the central bond, which has considerable π-character, to give a second complex (4). Oxidative addition of the strained σ-bond to the nickel(0) atom (d^{10}-d^8 conversion) forms a metallobicyclic intermediate, and insertion of the coordinated alkane (Z=Z $= CH_2=CHCN$, or $CH_2=CHCO_2Me$) into the Ni-C σ-bond produces a new nickel organo species (5). This then undergoes a reductive elimination of nickel(0) (d^8-d^{10}conversion) affording the formal [2+2] adduct (6). Concurrently, an intramolecular β-metal-hydride elimination, followed by reductive elimination of the nickel(0) complex, leads to the cyclopentene (7) (H.Takaya *et al., J. Org. Chem.,* 1981, **46**, 2846).

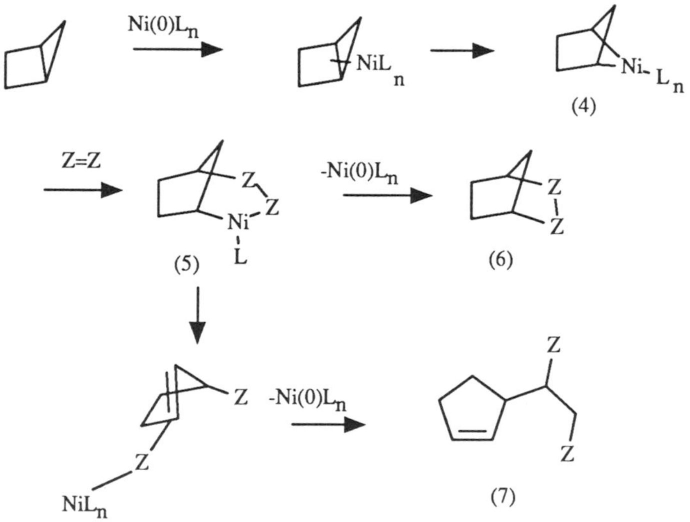

(ii) <u>Radical reactions</u>

Radical bromination of bicyclo[2.1.0]pentane with molecular bromine occurs rapidly at 0°C in carbon tetrachloride solution, without the necessity to initiate the reaction by exposure to light. The products are bromocyclopentane, *trans*-1,2-dibromopentanes, and *cis*- and *trans*-1,3-dibromocyclopentanes. It should be noted that a complementary electrophilic bromination reaction, if conducted in chloroform solution, yields *trans*-1,2-dibromocyclopentane as the main product. No 1,3-dibromopentanes are detected. Therefore it seems probable that the formation of 10-15% *trans*-1,2-dibromocyclopentane in the first reaction arises through an electrophilic pathway, operating concurrently with the main radical bromination. However, the principal reaction products are 1,3-dibromopentanes which are generated by a bimolecular radical substitution (S_H2) reaction, which occurs at $C_{(1)}$, followed by reaction with a bromine atom. Bromocyclopentane is derived from the same intermediate radical which then abstracts a hydrogen atom from the starting material (C.Jamieson, J.C.Walton, and K.U.Ingold, *J. Chem. Soc. Perkin Trans.* 2, 1980, 1366).

Photolysis of bromotrichloromethane gives the trichloromethyl radical and Jamieson, Walton, and Ingold (*loc. cit.*) have shown that this species reacts with bicyclo[2.1.0]pentane to afford a mixture of products including cyclobutene, *cis*- and *trans*-1,3-dibromocyclopentanes, *trans*-1-bromo-2-trichloromethylcyclopentane, 3-chlorocyclopentene, 3-bromocyclopentene, 4-chlorocyclopentene, 4-bromocyclopentene, *cis*-1,2- and *trans*-1,2-dibromocyclopentanes, *cis*- and *trans*-1-bromo-3-chlorocyclopentanes, *trans*-1-bromo-2-(trichloromethyl)cyclopentane, and cyclopentene.

The presence of *cis*- and *trans*-1,3-dibromocyclopentanes in this mixture indicates that bromine atoms take part in the same type of reaction as previously described in the radical bromination of the hydrocarbon. *Cis*- and *trans*-1-bromo-3-chlorocyclopentanes are also important products in the reaction, and these may form through 3-bromocyclopentyl radicals abstracting a chlorine atom from trichlorobromomethane. However, a more likely route is *via* a S_H2 abstraction reaction between bicyclopentane and a chlorine atom, followed by a bromine abstraction. 3-Bromocyclopentene and 3-chlorocyclopentene may be produced, in part, from the corresponding 3-halogenocyclopentyl radicals, and, in part, through reactions of bicyclopentane with hydrogen halides. Cyclopentene and 4-bromocyclopentene originate by first hydrogen atom abstraction from the staring material, to produce a cyclopenten-4-yl radical, and then either abstraction of a hydrogen atom from another molecule of the hydrocarbon, or a removal of a bromine atom from trichlorobromomethane respectively. Finally *trans*-1-bromo-2-trichloromethylcyclopentane is formed by a photochemical addition of trichloromethane to cyclopentene.

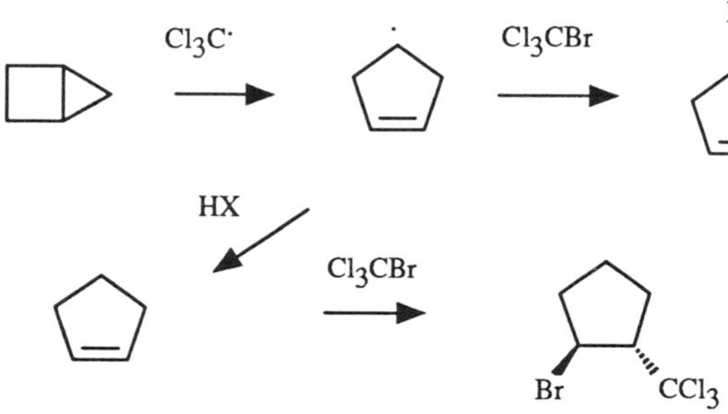

When irradiated at 185nm bicyclo[2.1.0]pentane isomerises mainly to 1,4-pentadiene and cyclopentene. A secondary product is methylene-cyclobutane. From the results of deuterium labelling experiments the 1,3-cyclopentadiyl radical is considered to be a common intermediate leading to cyclopentene and to 1,4-pentadiene, however, the last compound is also formed as the result of a direct [2+2] cycloreversion reaction (W.Adam and T.Oppenländer, *J. Amer. Chem. Soc.*, 1985, **107**, 3924).

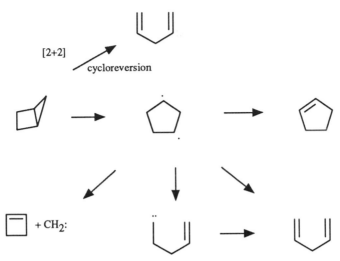

Oppenländer and Zang (*J. Amer. Chem. Soc.*, 1985, **107**, 3921) further present evidence to show that the inefficiency of the photochemical stereomutation of bicyclo[2.1.0]pentane is due to rearrangement of the intermediate 1,3-pentadiyl radical into cyclopentene, and also its fragmentation into 1,4-pentadiene.

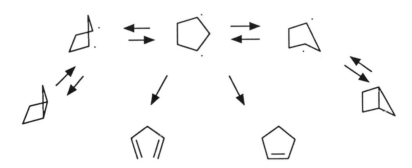

(iii) <u>Ozonolysis</u>

When bicyclo[2.1.0]pentane is reacted with ozone, adsorbed on silica, a primary ozonide, or 1,2,3-trioxane, is formed. This rearranges to a secondary ozonide which undergoes fragmentation into succindialdehyde and formaldehyde. A competing reaction is the production of cyclopentylacetaldehyde, which is assumed to form through the intermediacy of a zwitterion, produced directly from the primary ozonide (T.Preuss, E.Proksch, and A.De Meijere, *Tetrahedron Letters,* 1978, 833).

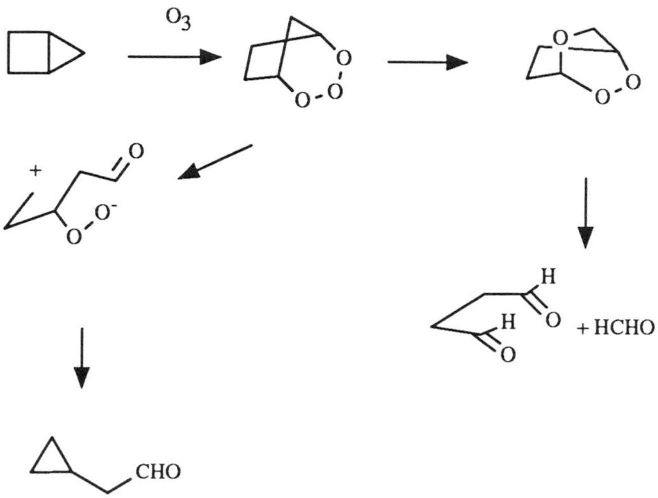

(iv) Rearrangements

Dehydronoriceane (8) also has the bicyclo[2.1.0]pentane unit as an integral part of its structure. This compound rearranges when treated with 3.6 mol.% of silver fluoroborate in chloroform to give 2,4-etheno-noradamantane (9). Bicyclo[2.1.0]pentane itself does not react under these conditions and it is suggested that the greater reactivity of dehydronoricene is due to the formation of an intermediate which is stabilised by hyperconjugation (T.Katsushima *et al., J. Chem. Soc. Chem. Commun.,* 1976, 39).

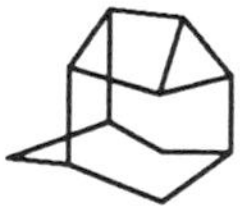

(8)

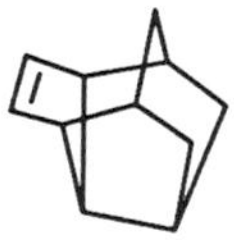

(9)

(v) Reactions with carbenes

Whereas bicyclo[2.1.0]pentane reacts with either phenylcarbene or di(methoxycarbonyl)carbene by a simple insertion reaction giving (10, R=PhCH$_2$) and [10,R=(MeO$_2$C)$_2$CH] respectively, difluorocarbene affords 1,1-difluoro-1,5-hexadiene (11) and 2,2-difluorobicyclo[3.1.0]hexane (12) (G.H.Shuie *et al., Tetrahedron Letters,* 1985, **26**, 5399).

(10) 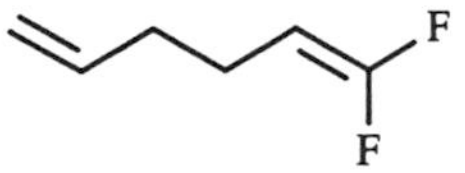   (11) 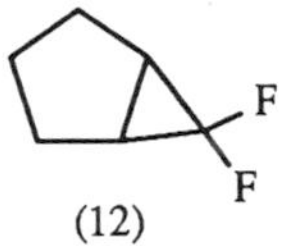   (12)

(vi) <u>Reaction with chlorosulphonylisocyanate</u>

A cyclopentano-β-lactam (14) is formed when bicyclo[2.1.0]pentane is reacted with chlorosulphonylisocyanate (CSI), followed by reductive dechlorosulphonylation. A possible mechanism for the first reaction requires that a zwitterion (13) is produced, which undergoes a 1,2-proton shift and then cyclises. A similar reaction with 1,3-dimethylbicyclo-[1.1.0]butane gives the lactam (15) (W.E.Voltz *et al., Chem. Ind.,* 1974, 771).

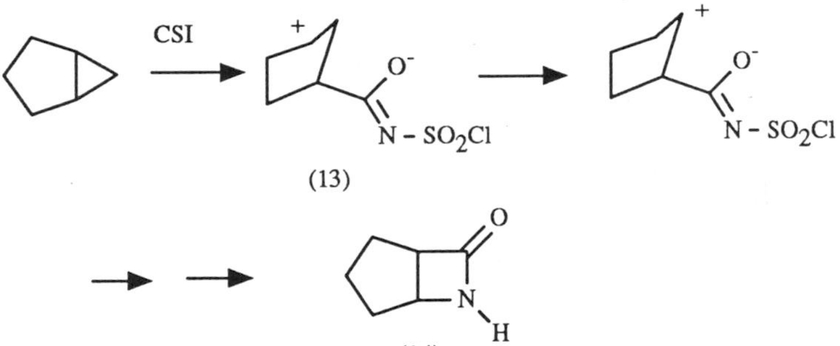

(13)

(14)

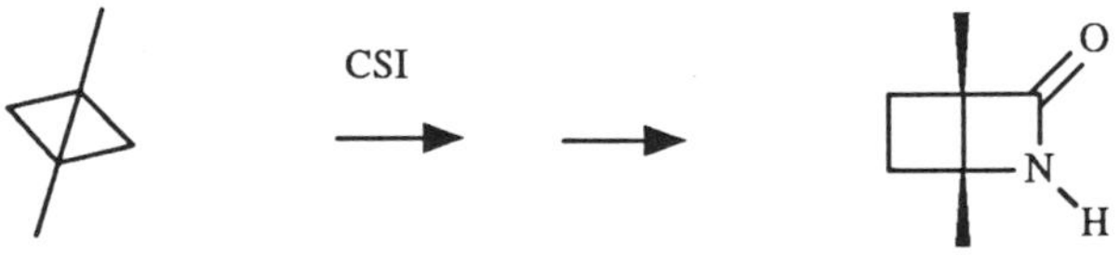

(15)

5. Bicyclo[3.1.0]hexanes

(a) Synthesis

Bicyclo[3.1.0]hexane and methylenecyclopentane are formed from the intramolecular cyclisation of 6-chloro-5-hexen-1-yllithium. It is suggested that the reactions proceed *via* a common carbenoid intermediate (W.R.Dolbier Jr. and C.Yaxiong, *J.Org.Chem.* 1992, **57**, 1947)

(b) *Structure and spectra*

The microwave spectrum of bicyclo[3.1.0)hexane in the range 26.5-40.0GMz shows A = 5542.955 ∓0.074, B = 4236.818 ∓0.007, and C = 3127.040 ∓0.007. The results were deduced from a least squares fit to give 22 lines including both a- and c- type low-J R-branch transitions. The data indicate that the molecule adopts a boat conformation and allow the determination of dipole moment components [μa] = 0.093 ∓0.001D and [μc] = 0.168 ∓0.002D in turn yielding a total dipole moment μ = 0.192 ∓0.003D (R.L.Cook and T.B.Malloy Jr., *J. Amer. Chem. Soc.*, 1974, **96**, 1703). Electron diffraction and additional microwave spectra data support this conclusion and show the boat form to exhibit flap angles of 70.6° (1.1) and 25.2° (2.8), for the cyclopropane and cyclopentane units respectively (V.S.Mastryukov *et al.*, *J. Amer. Chem. Soc.*, 1977, **99**, 6855).

Bicyclo[3.1.0]hexane, 3-oxabicyclo[3.1.0]hexane, and 3,6-dioxabicyclo-[3.1.0]hexane exhibit a series of Q branches in the frequency range 250-125 cm^{-1} of their far infra red spectra. These correspond to single quantum transitions of a one dimensional ring-puckering vibration governed by a potential function of the form $V(cm^{-1}) = A(Z4 + BZ2 + CZ3)$, where Z is a reduced ring-puckering coordinate (R.C.Malloy and B.Thomas Jr., *J. Mol. Spectrosc.*, 1973, **46**, 358). In the mass spectrum the bicyclo[3.1.0]hexane molecular ion isomerises to the molecular ion of methylcyclopentene, prior to the loss of a methyl radical (P.Wolkogg and J.L.Holmes, *Can. J. Chem.*, 1979, **57**, 348).

(c) *Chemical reactions*

(i) Protonation and related reactions
Cis-bicyclo[3.1.0]hex-3-ol when reacted with $FSO_3H-SbF_5-SO_2$ at -78°C gives the 1-methylcyclopentenyl cation (1) (G.A.Olah, G.Liang, and K.S.Mo, *J. Amer. Chem. Soc.*, 1972, **94**, 3544).

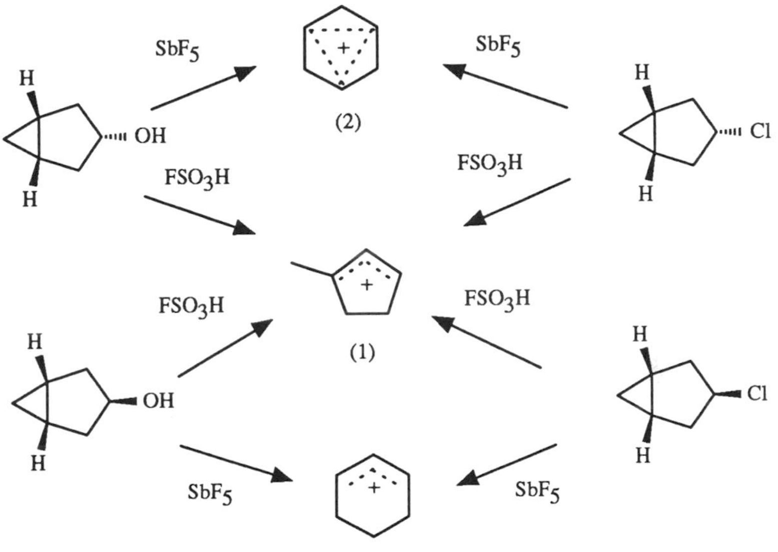

(2)

(1)

(3)

Subsequently, S.Masamune *et al.* (*Can. J. Chem.*, 1974, **82**, 3544) showed that the trishomocyclopropenium ion (2) is formed from *cis*-3-chlorobicyclo[3.1.0]hexane when it is treated with SbF_5-SO_2ClF at this temperature. This reagent combination also converts the alcohol into the trishomocyclopropenium ion. Similar reactions between either the *trans* alcohol, or the *trans* chloride with fluorosulphonic acid afford the 1-methylcyclopentenyl ion (1). When the reagent is antimony pentafluoride, however, the cyclohexenium ion (3) is formed in preference to the trishomocyclopropenium ion (J.C.Rees and D.Whittaker, *J. Chem. Soc. Perkin Trans. 2*, 1981, 948).

Deuterium labelling experiments demonstrate that the formation of the 1-methylcyclopentenyl cation from the *cis* alcohol, through reaction with fluorosulphonic acid, involves initial cleavage of the cyclopropane ring, followed by protonation and elimination of a mol. of water. On the other hand, the first reaction between the *trans* alcohol and antimony pentafluoride is protonation at the hydroxy group, then loss of water, and finally rearrangement to the cyclohexenium ion.

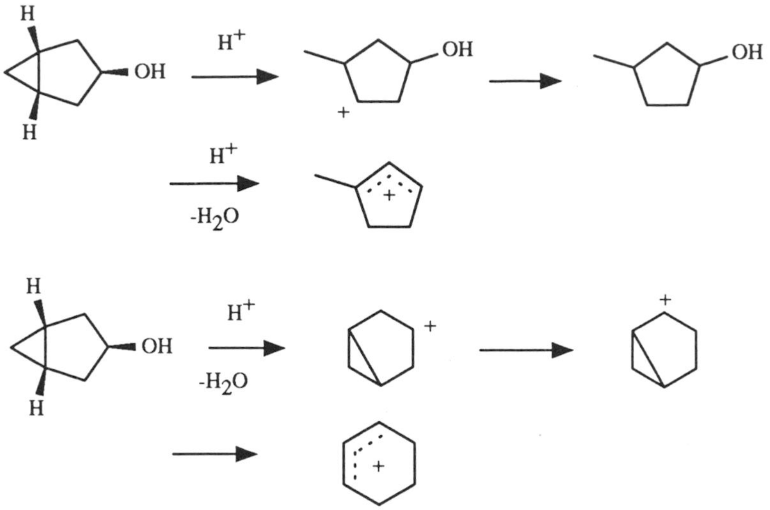

6. Bicyclo[4.1.0]heptanes

(a) *Synthesis*

The parent compound, bicyclo[4.1.0]heptane, is also known as norcarane. It is normally obtained through the cyclopropanation of cyclohexene. Thus titanium tetrachloride catalyses a reaction which uses dibromomethane and zinc dust in diethyl ether to generate carbene. In this case the yield of norcarane from cyclohexene is 58%, but if more than 2 mol.% of the catalyst is used the reaction becomes uncontrollable, in addition the zinc residue left after the reaction represents a fire hazard (E.C.Friedrich, S.E.Lunetta, and E.J.Lewis, *J. Org. Chem.*, 1989, **54**, 2388). An alternative method requires the sonication of a mixture of dibromomethane, zinc-copper(I) chloride and diethyl ether. Now the yield of norcarane is 60% (E.C.Friedrich, J.M.Domek, and R.V.Pong, *J. Org. Chem.*, 1985, **50**, 4640).

Diphenylsulphonium methylide, generated *in situ* by the reaction of diphenyl sulphide and methyl tetrafluoroborate with sodium hydride in tetrahydrofuran in the presence of copper diacetylacetonate, also reacts with cyclohexene to give norcarane. This is a general reaction which provides a laboratory model for the biosynthesis of cyclopropanoalkanes, since in Nature these compounds are considered to form through a transition metal-induced methylene transfer from the ylide of *S*-adenosylmethionine to unactivated alkenes (T.Cohen *et al., J. Amer. Chem. Soc.,* 1974, **96**, 5627).

Another useful propanation reaction involves the reactions of cycloalkenes with an organic *gem*-dihalide and copper powder in a hydrocarbon solvent. The reactions appear to involve organocopper intermediates, rather than carbenes, since isomeric insertion products arising from reactions at the C-H bonds are not detected. Stereospecificity is also noted and, for example, 3-methoxycyclohexene when reacted with diiodomethane gives *cis*- 2-methoxybicyclo[4.1.0]heptane. In this case it seems that coordination of the reacting species to the oxygen atom initiates the reaction and dictates the observed *syn* addition (N.Kawabata, I.Kamemura, and M.Naka, *J. Amer. Chem. Soc.,* 1979, **101**, 2139.

MeO   CH$_2$I$_2$ MeO   H H

Cu

H

72%

7,7-Dichloronorcarane is easily synthesised from cyclohexene by the reaction of chloroform with lithium 3-ethylpentoxide in hexane (R.H.Prager and H.C.Brown., *Synthesis,* 1974, 736).

Electrocatalytic reduction of 7,7-dichloronorcarane in dimethylformamide, containing lithium chloride and magnesium chloride as supporting electrolytes, at either an aluminium or a graphite cathode, affords norcarane in 76% yield (V.A.Afanas'ev *et al., Izv. Akad. Nauk. SSSR., Ser. Khim.,* 1982, 2637).

(b) *Structure*

Calculations predict that the preferred conformation for norcarane is a half-chair, but equilibration with other forms cannot be excluded (R.Todeschini, D.Pitea, and G.Flavini, *J. Mol. Struct.,* 1981, **71**, 279).

(c) *Reactions*

(i) Isomerisation
The kinetics of the gas phase thermal isomerisation of bicyclo[4.1.0]hept-

ane at 708-769°K and at pressures between 1 and 17 torr, have been documented. The principal reaction occurs by a first-order, homogeneous, nonradical mechanism giving 1-methylcyclohex-1-ene, methylenecyclohexane, and cycloheptene as products (M.C.Flowers and D.E.Penny, *Int. J. Chem. Kinet.*, 1973, **5**, 469).

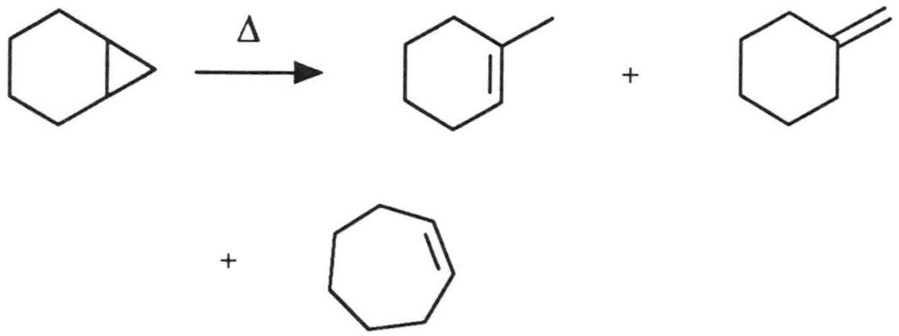

(ii) Hydroxybromination

Hydroxybromination of norcarane with *N*-bromosuccinimide, occurs in a Markownikoff manner to produce (1) (52%), and (2) (7%). The cyclohexene (3) is also formed, together with its reaction products with excess *N*-bromosuccinimide (G.Haufte, *J. Prakt. Chem.*, 1982, **32**, 896).

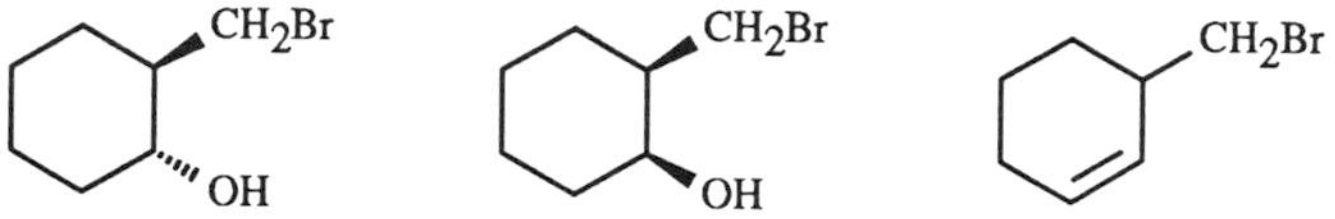

(iii) Reactions with iodonium cations

Reaction of norcarane with either iodine azide, or iodine monochloride, gives a variety of reaction products depending upon the conditions employed. For example, with iodine azide in acetonitrile at room temperature the products are the isomers (4) (24%) and (5) (6%), and 1-azido-3-iodocycloheptane (6) (5%) (R.C.Cambie *et al., J. Chem. Soc. Perkin Trans.*1, 1982, 961).

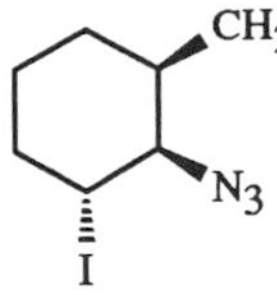

(4) 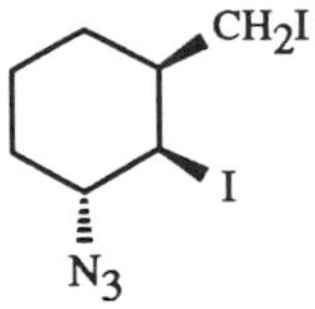 (5) 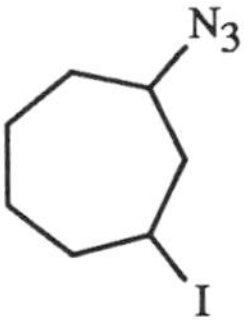 (6)

(iv) Acylation

The acylation of norcarane by pivaloyl tetrafluoroborate at -50°C in dichloromethane-dichloroethane, followed by the addition of water, leads to the cyclohexene (7) and the cyclohexanol (8). At -25°C an additional product (9), is produced. If sodium borohydride, or tributyltin hydride, is added instead of water the fused tetrahydropyran (10) is formed (Y.Tomilov, V.A.Smit, and O.M.Nefedov, *Ivz. Akad. Nauk. SSSR. Ser. Khim.*, 1974, 1439; 1976, 2512).

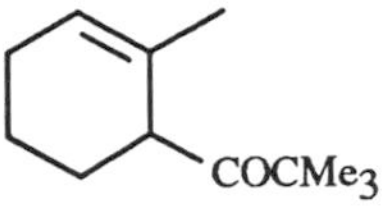

(7) 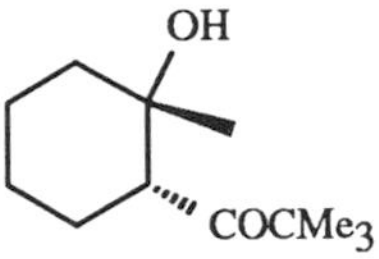 (8) 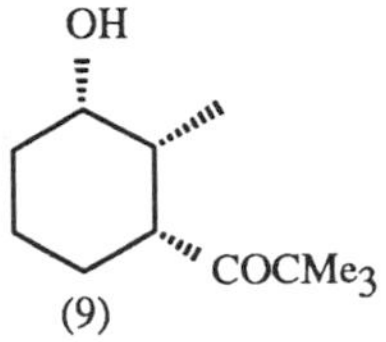 (9)

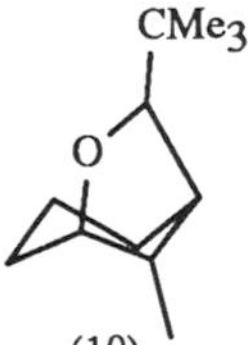

(10)

(v) Hydrogenation

Hydrogenations of 7,7-difluoro-, 7-chloro-7-fluoro-, and 7-bromo-7-fluorobicyclo[4.1.0]heptanes over palladium on carbon, either at atmospheric pressure and room temperature or at 50kg/cm^2 and 80°C, has been carried out in order to investigate the effect of the halogen atoms upon the bond fissions within the molecules. In all cases the C_1-C_6 bond was cleaved selectively to give ring enlarged products. The *endo*-fluoro isomers give only cycloheptane, whereas the *exo*-fluoro isomers yield both cycloheptane and fluorocycloheptane. When amines are added the ring-breaking reactions are retarded, but in the hydrogenolysis of 7-bromo-7-fluorobicyclo[4.1.0]heptane the rate at which the C-Br bond is cleaved is accelerated. The hydrogenolyses of 7-fluoro-, 7-chloro- and 7-bromo-bicyclo[4.1.0]heptanes have also been studied. The *endo* isomers of the 7-fluoro- and 7-chloro compounds undergo reductive scission at the C_1-C_6 bond, whereas the *exo*-analogues resist hydrogenolysis. *Endo*-7-bromo-bicyclo[4.1.0]heptane is hydrogenolysed with fragmentation of the central bond, but this is also accompanied by cleavage of the C-Br bond. In the case of the *exo*-bromo isomer C-Br cleavage and rupture of the C_1-C_7 bond are noted (J.Sakai *et al., Nippon Kagaku Kaishi,* 1991, 748).

(vi)Reactions with lithium or magnesium

7,7-Dibromobicyclo[4.1.0]heptane reacts with lithium, or magnesium, in the presence of cyclohexene, vinyl ethyl ether, or dihydropyran, to give the adducts (11), (12), or (13) respectively. The reactions are initiated by ultrasound (L.Xu, F.Tao, and T.Tu, *Ziran Zazhi,* 1986, **9**, 315).

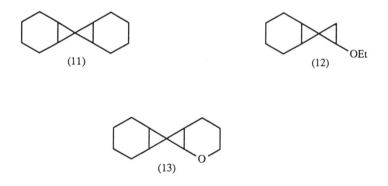

(11)

(12)

(13)

(vii)Dimerisation

1-Trimethylsilyl-7-chlorobicyclo[4.1.0]heptane undergoes gas-phase elimination of trimethylsilyl chloride when it is passed over glass helices coated with tetrabutylammonium fluoride. The product is bicyclo[4.1.0]-hept-1(7)-ene. This compound dimerises at -90°C *via* an ene reaction and then dimerises again to afford two compounds (14) and (15) (W.E.Billups *et al., J. Amer. Chem. Soc.,* 1991, **113**, 7980).

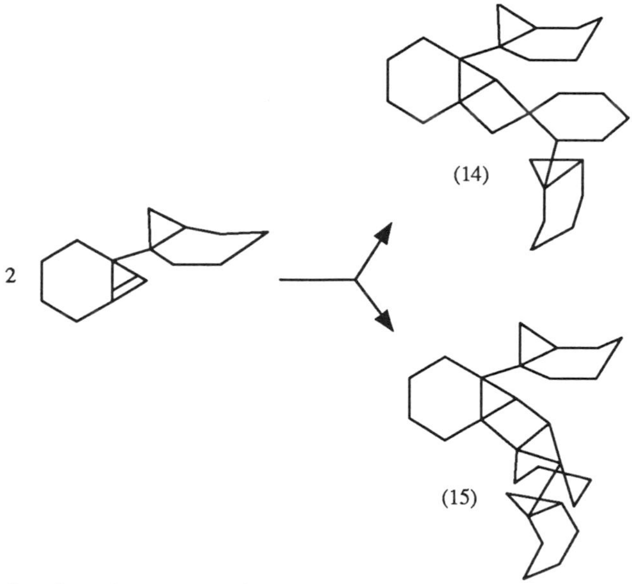

(14)

2

(15)

(viii)Reductive ring-expansion

The butynylbicycloheptanone (16) undergoes a radical induced ring opening reaction, mediated by samarium diiodide in the presence of 1,3-dimethyl-3,4,5,6-tetrahydropyrimidin-2(1H)-one, to give the spiro ketone (17) in 57% yield (R.A.Batey and W.Motherwell, *Tetrahedron Letters,* 1991, **32**, 6211).

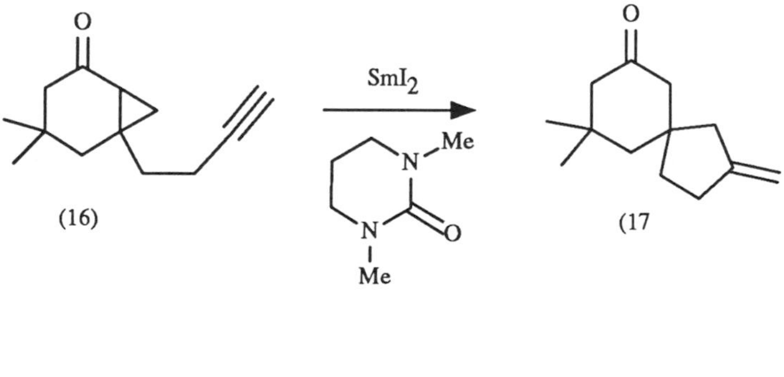

(16)   (17)

7. Bicyclo[5.1.0]octanes and bicyclo[6.1.0]nonanes

(a) *Structure and physical properties*

The heats of combustion and formation of bicyclo[5.1.0]octane have been measured (V.A.Luk'yanova *et al., Zh. Fiz. Khim.,* 1991, **65**, 824). Calculations suggest that the most stable conformation of *cis*-bicyclo[5.1.0]octane is the chair-chair form, it is estimated that the heat of isomerisation into the *trans*-isomer is 3.8 Kcal. mol.$^{-1}$ (R.Todeschini and G.Flavini, *J. Mol. Struct.,* 1980, **64**, 47). The electronic polarisation, dipole moment, and molar Kerr constant of bicyclo[5.1.0]octane have also been determined (D.Pitea, G.Moro, and G.Favini, *J. Chem. Soc. Perkin Trans.2,* 1987, 313).

Cis- and *trans*-bicyclo[6.1.0]nonanes have almost the same heats of formation, but the *cis-* and *trans*-isomers of cyclooctene differ in energy by 37.5kJmol^{-1}. The bicyclononanes are less reactive than the isomers of bicyclo[5.1.0]octane towards electrophiles. Although *cis*-bicyclo[5.1.0]octane undergoes acetolysis at one of the peripheral cyclopropane bonds, the *trans*-isomer reacts through scission of the central bond (K.B.Wiberg, *Angew. Chem. Int. Ed. Engl.,* 1972, **11**, 332; Wiberg *et al., J. Amer. Chem. Soc.,* 1984, **106**, 1740).

Calculations give the surprising result that in the gas phase the *trans*-isomer of bicyclo[6.1.0]nonane should be more stable than the *cis*-form, with a strain energy of 55.8 kJmol^{-1}, only 28.7kJmol^{-1} greater than cyclopropane itself and considerably less than the combined strain energies of cyclopropane and cyclooctane (R.Corbally *et al., J. Chem. Soc. Chem. Commun.,* 1978, 778).

(b) *Synthesis*

Bicyclo[6.1.0]nonane is obtained in 88% yield when cyclooctene is reacted with either dibromomethane or diiodomethane, using zinc dust and copper(I) chloride in diethyl ether. The yield is reduced to 73% if titanium tetrachloride is used as the 'promoter' (E.C.Friedrich and E.J.Lewis, *J. Org. Chem.,* 1990, **55**, 2491). 1,5-Cyclooctadiene when reacted with diazomethane at -10°C - 0°C gives 35% of the monoadduct (1), 27% of the *cis* diadduct (2) and 25% of the *trans* diadduct (3) (O.M.Nefedov *et al., Ivz. Akad. Nauk. SSSR, Ser. Khim.,* 1984, 119).

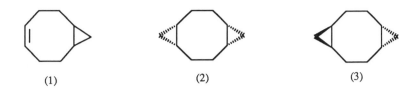

(1)　　　　　　(2)　　　　　　(3)

If cyclooctatetraene is the substrate the reaction is neither regio- nor stereo-selective, giving a mixture of all five possible products and their stereomers (Y.Tomilov *et al.*, *ibid.*, 1986, 77).

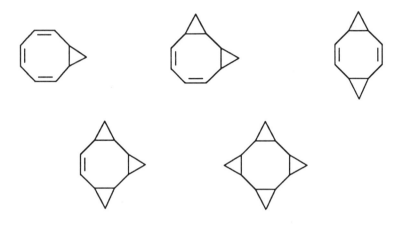

(c) *Reactions*

1,2-Cyclononadiene (3) on direct solution phase irradiation yields bicyclo[6.1.0]-1-ene (4), the tricycle (5), and cyclononyne as the primary photochemical products in the ratio 94:3:3. Benzene-sensitised vapour phase irradiation of the same starting material yields 89% of the tricycle and three bicyclo[4.3.0]nonenes (T.J.Stierman and R.P.Johnson, *J. Amer. Chem. Soc.*, 1985, **107**, 3971).

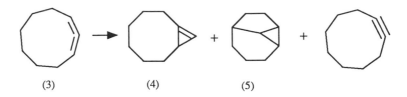

(3)   (4)   (5)

Photolyses of *cis*- and *trans*-bicyclo[6.1.0]nonanes at 185nm give 1,8-nonadiene as the major product and a mixture of *cis*- and *trans*-cyclononenes as the minor component. Both substrates photolyse in the same quantum yield, but give different ratios of the cyclononenes. It is suggested that both *cis*- and *trans*-isomers decompose *via* common intermediates, and that small differences in the energy of a biradical may explain the varying amounts of cyclononenes produced (R.Srinivasan, J.A.Ors, and T.Baum., *J. Org. Chem.*, 1981, **46**, 1950).

When heated at 70°C in benzene containing 4-toluenesulphonic acid *cis*-bicyclo[6.1.0]nonane undergoes isomerisation into alkylcycloalkenes. Cyclononene is not formed, but in a similar reaction *cis*-bicyclo[6.1.0]-nonan-2-one gives both 3- and 4-cyclononenones (E.S.Balenkova and M.A.Gorokhova, *Veatn. Mosk. Univ. Khim.*, 1975, **16**, 735).

8. Bicyclo[7.1.0]decene

Bicyclo[7.1.0]dec-1(10)-ene is a photochemical product obtained through the irradiation of 1,2-cyclodecadiene in pentane solution at wavelength 220nm. It is accompanied by cyclodecyne. However, if the irradiation experiment is carried out in benzene as the solvent and the wavelength is switched to 254nm then the tricyclic isomers (1), (2), (3), and (4) are produced instead (J.D.Price and R.P.Johnson, *J. Amer. Chem. Soc.*, 1985, **107**, 2187).

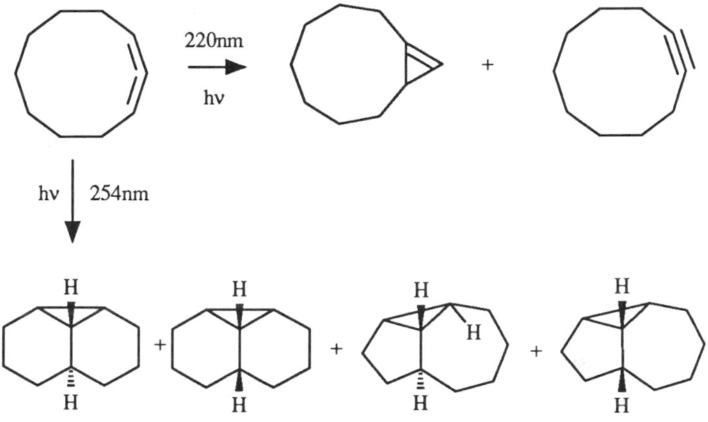

9. Bicyclo[10.1.0]tridecane

Bicyclo[10.1.0]tridecane is present in a mixture of compounds produced when 1,2-cyclotridecanediol is subjected to hydroboronation-oxidation (I.Mehrotra, M.M.Bhagwat, and D.Devaprabhakara, *Proc. Indian Acad. Sci. Ser. A,* 1977, **86A**, 15).

10. Bicyclo[m.n.0]alkanes

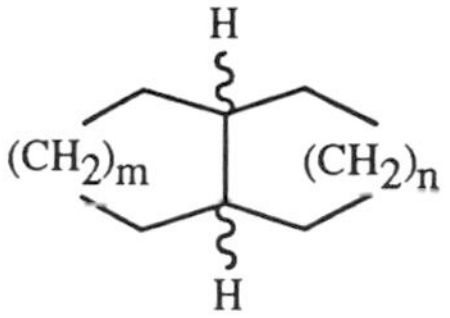

(a) *Spectrometry spectroscopy and general chemistry*

The mass spectra of bicyclo[m.n.0]alkanes where m = 1-3, and n = 1-4 have been recorded and the results correlated with ring strain energies (R.Herzschuh and G.Mann, *J. Prakt. Chem.,* 1980, **322**, 37; 450).

Data taken from the ^{13}C N.M.R. spectra of both *cis* and *trans* fused bicyclo[4.n.0]alkanes show that the six-membered ring adopts a basic chair conformation. However, as predicted, this is distorted to a degree dependant upon the size of the other ring (P.Metzer *et al., Org. Magn. Reson.,* 1982, 19, 144).

Bicyclo[m.n.0]-1-alkenes (3; where m, n = 3, 4, or 5) can be synthesised *via* intramolecular Wittig reactions. The starting materials α,ω-(2-oxo-cycloalkyl)alkyl bromides (2) are prepared by treating the anions of the appropriate ethyl 2-oxocycloalkanyl carboxylates (2) with an excess of a α,ω-dibromoalkane, followed by hydrolysis and decarboxylation of the products. Reaction of the ω-(2-oxocycloalkyl)alkyl bromides with triphenylphosphine and sodium hydride in dimethylsulphoxide gives the bicyclo[m.n.0]alk-$^{1,2}\Delta$-enes (K.B.Becker, *Helv. Chim. Acta,* 1977, **60**, 68).

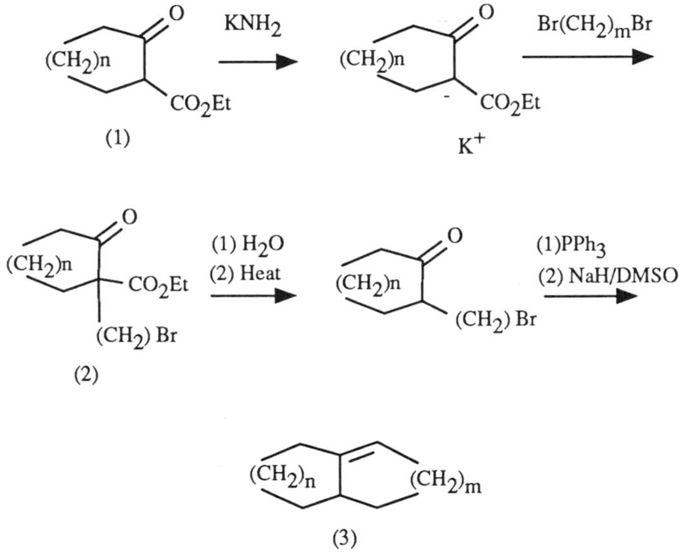

11. Bicyclo[2.2.0]hexanes and bicyclo[3.2.0]heptanes

(a) *Structure*

Electron diffraction techniques have shown that the molecular structure of bicyclo[3.2.0]heptane exhibits C-C bond lengths in the 5-membered ring shorter than those in the 4-membered unit. The central bridging bond is the longest in the molecule. The cyclobutane system is planar, but both *endo*- and *exo*-conformers exist in the ratio 9:1. For Δ^6-bicyclo[3.2.0]heptene calculations indicate that the molecule exists almost exclusively as the *endo*-conformer (R.Glen *et al.*, *Acta Chem. Scand. Ser. A*, 1983, **A37**, 853).

An X-ray crystallographic determination shows that in the case of the ketone (1) the cyclopentane ring is in an approximate *endo*-conformation, and that both the hydroxyl group and the bromine atoms occupy pseudo-axial positions (A.Brown *et al.*, *J. Chem. Soc. Chem. Commun.*, 1979, 1178).

E.S.R. spectroscopy reveals that radiolytic oxidation of bicyclo[2.2.0]-hexane in a haloethane matrix generates the cyclohexane-1,4-diyl radical. The same radical is formed from 1,5-hexadiene. If *exo-cis*-2,3-dideuterio-[2.2.0]hexane is oxidised in this way a product with a six line E.S.R. spectrum is obtained, contrasting with the seven line signal given by the cyclohexane-1,4-diyl radical. This shows that one deuterium atom must occupy an axial position (F.Williams *et al.*, *J. Amer. Chem. Soc.*, 1989, **111**, 4133).

(b) *Reactions*

Bicycloalkanes containing cyclobutane rings, like those containing cyclopropane units, undergo a series of unusual reactions. In particular unexpected selectivities are shown which lead on to novel rearrangements. These effects are caused by the rigid molecular framework and by the large strain energy inherent in such compounds.

For example, radical abstraction of hydrogen from bicyclo[2.2.0]hexane occurs at both bridgehead and bridge positions, however, while the bicyclo[2.2.0]hexan-1-yl radical can be observed by E.S.R. spectroscopy, the bicyclo[2.2.0]hex-2-yl radical rapidly rearranges into the 3-cyclohexen-yl radical. In the case of bicyclo[3.2.0]heptane, radical abstraction of hydrogen only occurs at the methylene groups of the C_5 ring giving bicyclo[3.2.0]heptan-2-yl radicals, these rearrange to 2-(2-cyclopent-enyl)ethyl radicals. Normally cyclobutylmethyl radicals rearrange by β-scission, and it is argued that this occurs because the SOMO overlaps the orbitals of the β,γ-bond. Thus rearrangement of the bicyclo[2.2.0]hexan-2-yl radical would be expected to form the cyclobutenylethyl radical. The fact that this stereo-electronically favoured rearrangement is not followed suggests that the β-scission of the central bond is much more exothermic than the alternative cleavage of the C_1-C_6 bond (J.C.Walton *J. Chem. Soc. Perkin Trans.2*, 1988, 1371).

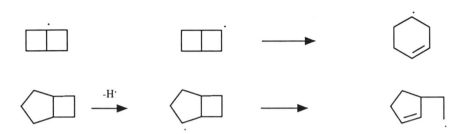

2,6-Diphenylhepta-1,6-diene and 1,5-diphenylbicyclo[3.2.0]heptane are thermally interconvertible through a continuous diradical or a di-radical-as-transition-state mechanism, closely related to a Cope rearrangement. Thus the activation parameters relate to a transition state which approximates to the 1,5-diphenylcyclohepta-1,5-diyl diradical (W.R.Roth *et al.*, *J. Amer. Chem. Soc.,* 1990, **112**, 1722).

Rhodium complexes catalyse the rearrangement of bicyclo[2.2.0]hexane into cyclohexene, and it appears likely that the reaction proceeds through a rate determining oxidative addition of Rh (I), across the strained C_1-C_4 bond, followed by a (β-C → Rh) hydrogen shift and a reductive elimination reaction of the resulting metallocyclic intermediate (M.Sohn, J.Blum, and J.Halpern, *J. Amer. Chem. Soc.,* 1979, **101**, 2694).

12. Bicyclo[2.2.0]hexenes and bicyclo[3.2.0]heptenes

Bicyclo[2.2.0]hex-1(4)-ene (1) has been prepared by dehalogenation of 1-bromo-4-chlorobicyclo[2.2.0]hexane. When this product is treated with bromine, or with a peracid, spiro[2.3]hexanes are formed, whereas cycloaddition reactions afford [m.2.2]propellanes (K.B.Wiberg *et al., Tetrahedron,* 1986, **42**, 1895).

The vibrational spectrum of bicyclo[2.2.0]hex-1(4)-ene has been determined and *ab initio* calculations to determine strain energy of the molecule have been performed by Hrovat and Borden (*J. Amer. Chem. Soc.,* 1988, **110**, 4710).

Tricyclo[4.1.0.0^{2,7}]heptane on heating between 110-230°C rearranges into bicyclo[3.2.0]hept-6-ene. The reaction involves an asynchronous initial ring opening to a cycloheptatriene, which then undergoes rapid conrotatory cyclisation to the heptene. Other derivatives of the starting material such as the 1- and 2-phenyl analogues rearrange in a similar manner (M.Christl, R.Stang and H.Jelmek-Fink, *Chem. Ber.*, 1992, **125**, 485).

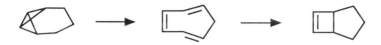

13. Bicyclo[4.2.0]octanes and bicyclo[5.2.0]nonanes

When bicyclo[4.2.0]octane is photolysed through irradiation at 185nm it gives 1,7-octadiene, cyclohexene, and ethane as products, possibly *via* the participation of an intermediate radical cation (W.Adam and T.Oppenländer, *Angew. Chem. Int. Ed. Engl.*, 1984, **23**, 641).

Bicyclo[4.2.1]nonanes can be synthesised through the acid promoted rearrangement of 5,6-disubstituted bicyclo[4.2.0]octanones. Thus the starting materials (1, $R^1 = R^2 = R^3 = Me$; $R^4 = H$), (1, $R^1 = R^2 = Me$; $R^3 = R^4 = H$), and [1, $R^1 = R^2 = (CH)_{2n}$; $R^3 = R^4 = H$) give the corresponding bicyclo-nonanones (3) in high yield, although for (1, $R^1 = R^2 = R^3 = Me$; $R^4 = H$) some of the bicyclo[3.3.0]octanone (4) is formed as a minor product. The mechanism of the reaction seems to be the formation of a zwitterion (2) which undergoes internal migration of R^3 (path a) and cyclisation to the bicyclononanone (3). Bicyclooctanones (4) are produced from the same intermediate through an alternate process (path b), wherein a proton from C-5 is transferred to the cationic site prior to cyclisation. The first reaction is stereochemically sensitive. Thus when the compounds (5) and (7) were each reacted with aluminium trichloride the first gave only a bicyclononane (6). On the other hand, its isomer yielded a bicyclooctanone (8). In the first compound the *endo* hydrogen atom of the cyclopentano unit is aligned almost antiperiplanar to the central cyclobutane bond facilitating a 1,2-shift, but in the second the hydrogen atom is *exo* orientated and is not favourably orientated for such a migration. In this case the hydrogen atom at C-5 is transferred preferentially (K.Kakiuchi *et al.*, *J. Org. Chem.*, 1991, **56**, 6742).

262

(7)   (8)

14. Bicyclo[3.3.0]octane (hexahydropentalene)

The ^{13}C N.M.R. spectra of 52 bicyclo[3.3.0]octanes have been analysed in order to provide an accurate prediction of chemical shifts for poly-substituted bicyclo[3.3.0]octanes. Further, this data in combination with statistical calculations can be used to predict regio- and stereochemical assignments for other compounds of this type (J.K.Whitesell and R.S.Randall, *J. Org. Chem.,* 1977, **42**, 3878).

Cis-bicyclo[3.3.0]octane is formed by hydrogenation of the dihydropent-alenes (7), (8), and (9), which are the thermolysis products of tetracyclo[3.3.0.0^{2,4}.0^{3,6}]oct-7-ene (1). The hydrogenolysis reaction is carried out in a flow system using nitrogen as the carrier gas at 2 Torr pressure and at 200°C, and, although the mechanism for the production of the dihydropentalenes is uncertain, there is evidence to suggest that they are the result of a reverse Diels-Alder reaction which affords a cyclopropylcyclopentadiene (2). Very fast 1,5 hydrogen shifts then equilibrate this compound with its isomers (3) and (4). Further bond shifts generate (5) and (6) and thence the isomers (7), (8), and (9) (J.Stapersma, I.D.C.Rood, and G.W.Klump, *Tetrahedron*, 1982, **38**, 2201).

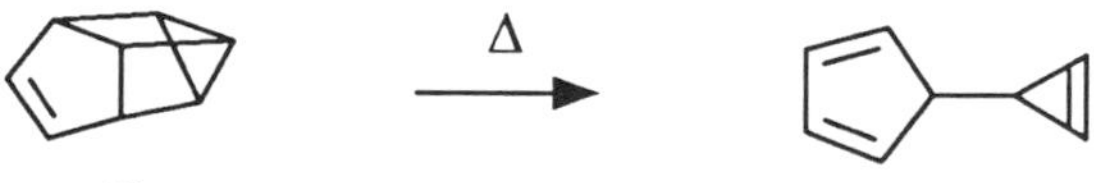

(1)   (2)

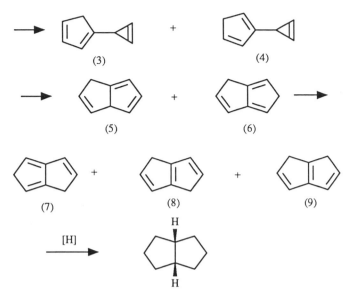

(3)   +   (4)

(5)   +   (6)

(7)   +   (8)   +   (9)

[H]

1-Acetyl-*cis*-bicyclo[3.3.0]octane (10, R = Ac) has been converted into its oxime (10, R = CMe:NOH). A Beckmann rearrangement of the latter compound gave *N*-(*cis*-1-bicyclo[3.3.0]octyl)acetamide (10, R = NHAc), as well as a minor amount of *cis*-pentalane (10, R = H). It is suggested that the Beckmann rearrangement proceeds through a two step fragmentation-recombination mechanism in strong acid (PCl$_5$, H$_2$SO$_4$ etc.). A bridgehead tertiary carbocation (11) is produced first, together with acetonitrile. These species recombine in a Ritter type reaction to give the product (R.R.Rao *et al.*, *J. Indian Chem. Soc.*, 1983, **22B**, 1122).

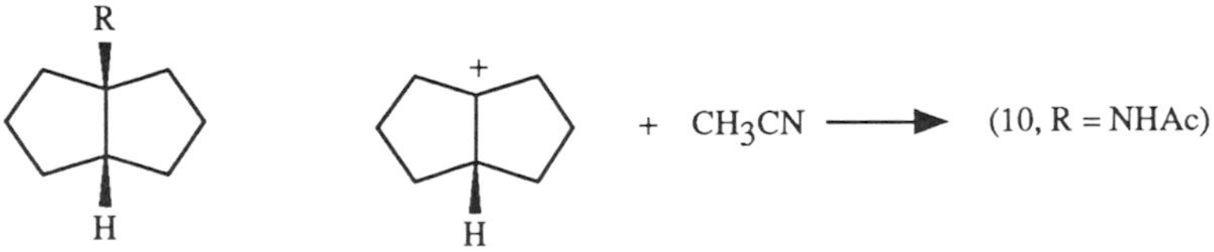

+   CH$_3$CN   ⟶   (10, R = NHAc)

(10)   (11)

Cis-bicyclo[3.3.0]octane-1-formyl (or acetyl) compounds (13), can be obtained through the reactions of 1-chloro-9-hydroxy (or alkoxy) -bicyclo-[3.3.1]nonanes (12), with potassium hydroxide in ethanol (A.Gambacorta *et al., Tetrahedron*, 1991, **47**, 9097).

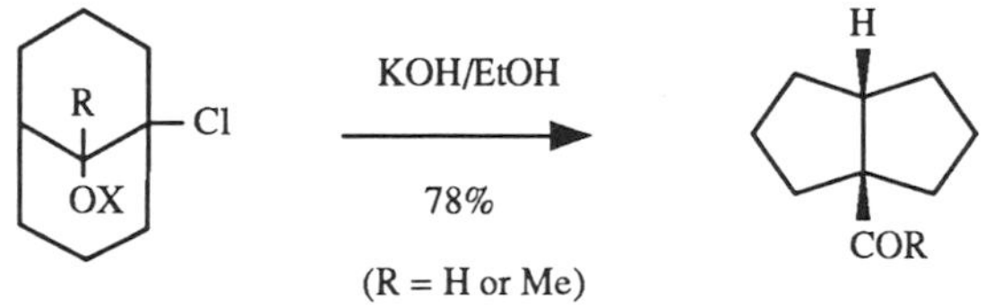

(12) → (13)

KOH/EtOH

78%

(R = H or Me)

Ethyl 2-butoxycyclopropane carboxylate in the presence of tin (IV) chloride reacts as a 1,3-zwitterion (14) and undergoes cycloaddition with *O*-trimethylsilyl-1-hydroxycyclopentene to yield the adduct (15). This is a general reaction which can be used to synthesise other substituted bicyclo[n.3.0]alkanes (where n = 4 or 5), from the appropriate cyclic enolsilyl ethers (M.Komatsu *et al., Synlett.*, 1991, 771).

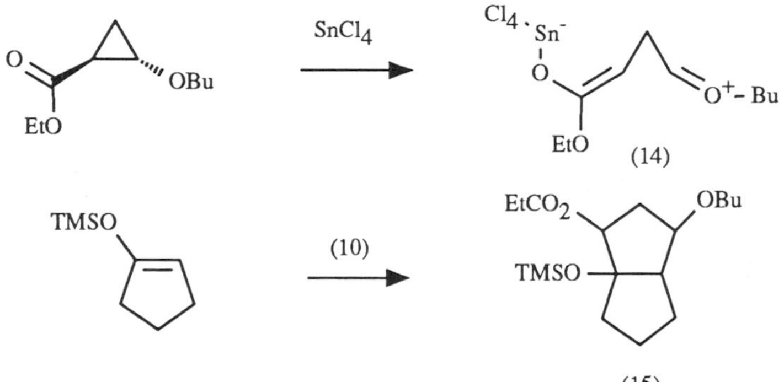

Bicyclo[3.3.0]octenones are available from hept-1-en-6-ynes through the Pauson-Khand procedure, which uses dicobalt octacarbonyl as reagent (L.Daalman *et al., J. Chem Res.(M)*, 1984, 3131).

Dilithium diphenylcyanocuprate reacts with bicyclo[3.3.0]octadienedione (16, R = H, or Me) through conjugate addition on one side of the molecule only to affords the corresponding adducts (17). However, if boron trifluoride diethyletherate is present then the diphenylated products (18) are produced. In the case of the related bicycloctanediones (19), phenylmetal reagents afford a mixture of *exo, exo-; exo, endo-; and endo, endo*-isomers (20). The recommended reagent for this reaction is phenylcerium dichloride, which is superior to both phenyllithium and phenylmagnesium bromide (H.Quast *et al., Annalen*, 1992, 495).

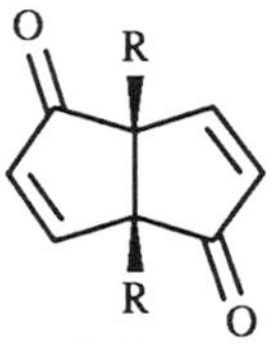

(16) 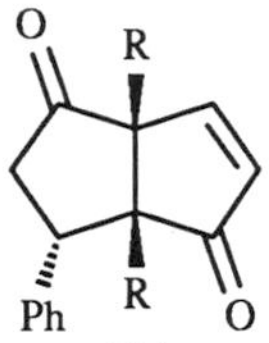   (17) 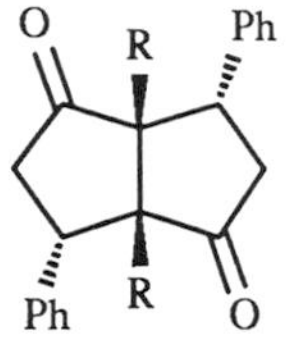   (18)

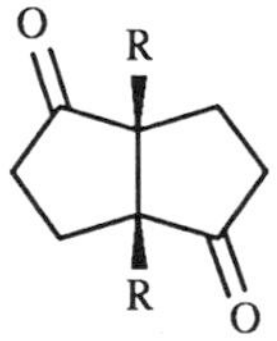

(19) 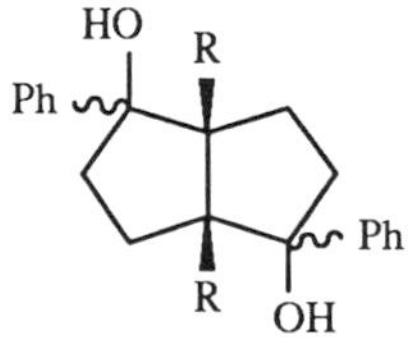   (20)

The spirocyclopropane (21), synthesised in three steps from hexahydro-pentalenone, undergoes thermally induced isomerisation at 500°C to the tricycle (22). Oxidative ring opening through reaction with ruthenium(III) chloride and sodium iodate then affords bicyclo[6.3.0]undecane-2,6-dione (23).

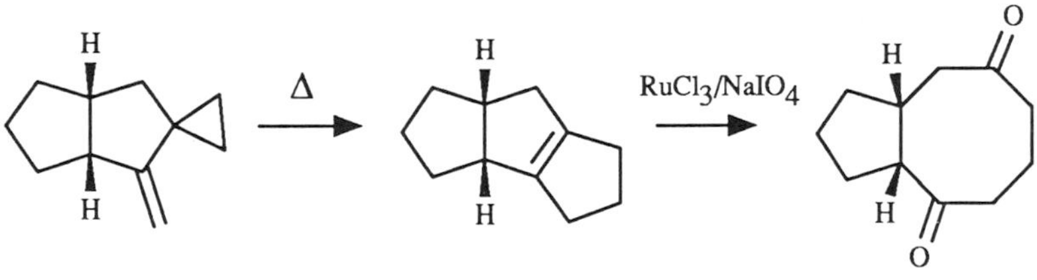

(21)   (22)   (23)

The reagents for the last step are also useful in cleaving the central double bonds of other fused bicyclocycloalkenes, hence it is possible to convert bicyclo[10.3.0]pentadecanene into cyclopentadecan-1,5-dione (H.R.Son-awane, B.S.Nanjundiah, and G.M.Nazeruddin, *Tetrahedron Letters,* 1992, **33**, 1645).

15. Bicyclo[4.3.0]nonane (octahydro-1H-indene), bicyclo[5.3.0]decane (decahydroazulene), and bicyclo[6.3.0]undecane

The energy separation between the appearance potentials of the m/z 96 peaks in the mass spectra of *cis*- and *trans*-bicyclo[4.3.0]nonanes is the same as the difference between the heats of formation of these isomers. A similar correlation is also noted for *cis*- and *trans*-bicyclo[4.4.0]decanes. The fragmentation patterns in the mass spectra of the bicyclononanes isomers are characteristic, and may be used as indicators of the type of ring-fusion geometry present in the parent molecules (A.I.Mikaya, and V.G.Zaikin, *Izv. Akad. Nauk. SSSR, Ser. Khim.*, 1980, 1286).

Cis-bicyclo[4.3.0]nonane is formed by the irradiation of cyclononene in pentane solution. *Cis*-bicyclo[6.3.0]undecane can similarly be synthesised from cycloundecene (G.Haufe, M.Tubergen, and P.J.Kropp, *J. Org. Chem.*, 1991, **56**, 4292).

Octahydroazulenediones and heptahydroazulenones can be prepared by the thermal [2+2] cycloaddition of dichloroketene to 3-trimethylsilyloxybi-cyclo[3.3.0]oct-2-ene. The dichloro adduct (1) may then be reductively dechlorinated by reaction with tributyltin hydride affording the cyclobutanone (2). The octahydroazulenedione (3) can be synthesised from the cyclobutanone (2) by treatment with tetrabutylammonium fluoride, while the heptahydroazulenone (4) is available when the butanone is reacted with methylmagnesium bromide, prior to reaction with methanesulphonyl chloride and triethylamine (C.S.Pak, S.K.Kim, and H.K.Lee, *Tetrahedron Letters*, 1991, **32**, 6011).

Iodination of 3,4-dimethyl-*cis*-bicyclo[4.3.0]nona-3,7-diene with iodine in carbon tetrachloride affords the iodobrexanes (5) (V.A.Andreev *et al.*, *Zhur. Org. Chem.*, 1991, **27**, 1450). Other iodinating agents such as iodine/pyridine, or iodine monochloride give products (6) and (7) respectively, arising through selective additions to the Δ^7-bond:

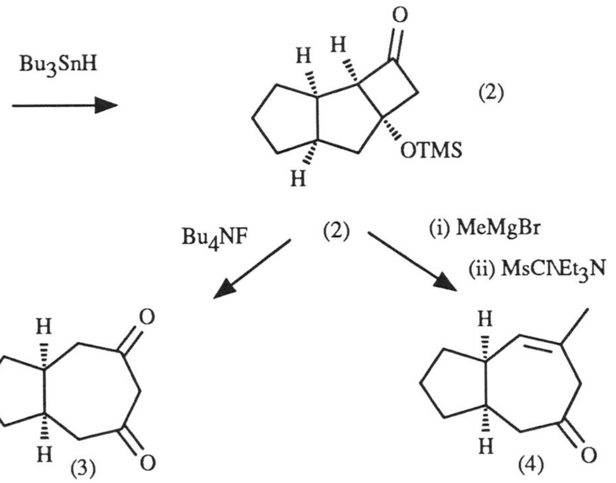

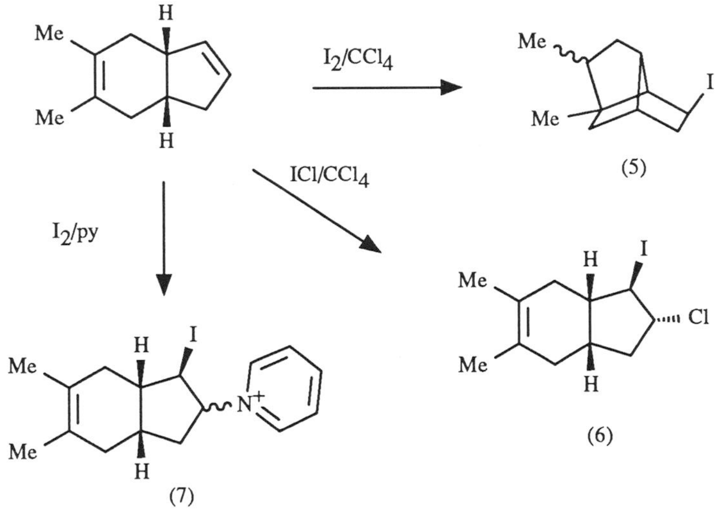

Similary bicyclo[5.3.0]nona-3,7-diene (8) reacts with iodine to give a mixture of iodides which include *ca.* 40% *endo,exo*-4,5-diiodobrexane (9). Reduction of the product mixture with lithium aluminium hydride in dry tetrahydrofuran affords 50-60% brexane, 25-35% *cis*-bicyclo[4.3.0]nonane, and smaller amounts of *cis*-bicyclo[4.3.0]non-3-ene, and *cis*-bicyclo[4.3.0]-non-7-ene (E.V.Lukovskaya *et al., Zh. Org. Khim.*, 1988, **24**, 1457).

(8) I_2 (9)

16. Bicyclo[4.4.0]decanes (decalins)

Bicyclo[4.4.0]decane, much better known as decalin, exists in both *cis*- and *trans*-fused forms. Its chemistry is normally reviewed alongside that of the parent arene, naphthalene. Only a short summary of the literature is presented here.

(a) Structure and stereochemistry

Much experimental and theoretical effort has been devoted to analyses of the conformational and general stereochemistry of the decalins (see, for example, J.M.A.Baas, *et al.*, *J. Amer. Chem. Soc.*, 1981, **101**, 5014; J.Curtis, D.M.Grant, and R.J.Pugmire, *ibid.*, 1989, **110**, 7716). Conformational preferences have also been assigned for many *cis*-decalins through considerations of ^{13}C N.M.R. data (L.M.Browne, R.E.Klinck, and J.B.Stothers, *Org. Magn. Reson.*, 1979. **12**, 561; *Can. J. Chem.*, 1979, **57**, 803). ^{2}H N.M.R. data have also been accumulated which indicate the extent of gauche interactions between vicinal C-C and C-D bonds (J.Curtis, D.M.Grant, and R.J.Pugmire, *J. Amer. Chem. Soc.*, 1989, **110**, 7711).

It is concluded that in the gas phase both the *cis*- and *trans*-decalins contain cyclohexane rings which adopt distorted chair conformations. For the *cis* form the ring is flattened along the ring junction, but more twisted elsewhere. For *trans*-decalin the reverse situation appertains (L.Van den Enden and H.J.Geise, *J. Mol. Struct.*, 1981, **74**, 309).

The order of the stability of the molecular ions of *cis*- and *trans*-bicyclo-[4.4.0]decanes are matched by the thermodynamic stability of the hydrocarbons themselves (Y.Demisov, I.M.Sokolova, and A.A.Petrov, *Neftekhimiya I*, 1977, 491).

The nature of the ring fusion influences the reactivity of the decalins and in ozonisation, for example, the *cis*-isomer is more reactive than the *trans*-form (N.N.Binov *et al.*, *Kinet. Katal.*, 1979, **20**, 1429).

(b) Reactions

The isomeric bridgehead decalin radicals have been obtained through the reactions of the parent hydrocarbons with di-^{t}butyl radicals. E.S.R. spectroscopy indicates that these radicals are distinct from one another and hence non-planar, although the tetrahedral configuration at the radical centres is not ideal and a somewhat flattened geometry is assumed (R.V.Lloyd and R.V.Williams, *J.Phys. Chem.*, 1985, **89**, 5379).

When *trans*-decalin is oxidised under 'Gif' conditions (*i.e.* triplet oxygen, in pyridine containing trifluoroacetic acid and 2,2′-dipyridyl and in the presence of $Fe_3O(OAc)_6$ pyridine$_{3.5}$), but in a divided electrochemical cell, rather than in contact with zinc, several products result. These include the diastereomeric alcohols (1), (2), and (3), plus the ketones (4) and (5) (G.Balavoine *et al.*, *Tetrahedron Letters*, 1986, **27**, 2849).

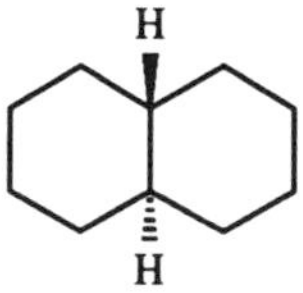

(1) 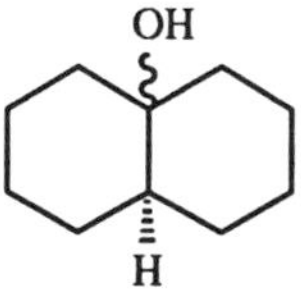   (2) 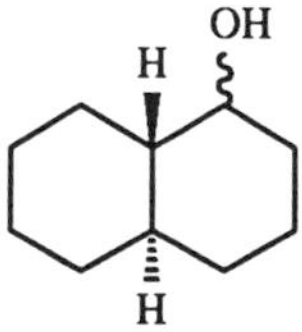   (3)

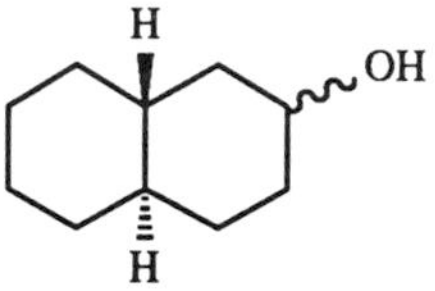

(4) 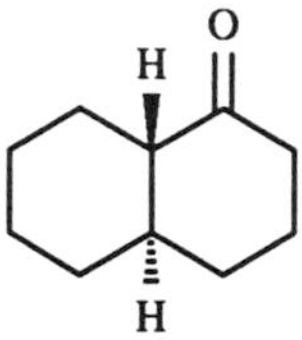   (5) 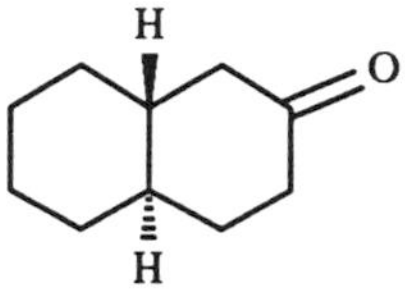   (6)

When cyclodecanone tosylhydrazone is treated with base in an aprotic solvent the carbene (7) is produced, which in turn affords mainly *cis*-bicyclo[5.3.0]decane (8) and a mixture of the *Z*- and *E*-cycloalkenes (9) and (10), respectively. However, if the cycloalkenes are irradiated, the only product is *trans*-decalin (11). Presumably a triplet carbene is involved in this second reaction, but it is uncertain why the *trans* form of the bicyclo[5.3.0]decane is not observed, especially if it is argued that a stepwise biradical intermediate participates and cyclises under thermodynamic rather than kinetic control (P.K.Kropp *et al.*, *Can. J. Chem.*, 1985, **63**, 1845).

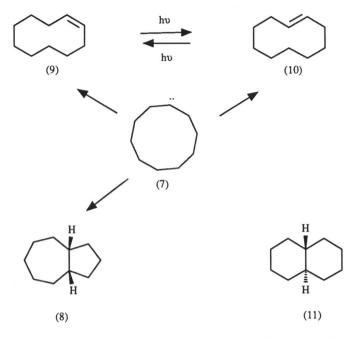

(9)

hυ

hυ

(10)

(7)

H

H

(8)

H

H

(11)

Heating *cis*-decalin over gallium (III) chloride at 115°C for 10 h causes extensive decomposition and at least 42 products can be identified. These include 1,5-dimethylbicyclo[3.2.1]octane (the main product), *exo*-1,3-di-methylbicyclo[3.3.0]octane, *endo*-1,4-dimethylbicyclo[3.2.1]octane, and *exo*-1,6-dimethylbicyclo[3.3.0]octane (V.E.Kovyazin, Ya M.Slobodin, and M.B.Temyanko, *Neftekhimiya*, 1980, **20**, 64). Similar products are formed when *cis*-decalin is heated over aluminosilicates at 250°C (M.B.Temyanko, *Ts. Vses. Neft. Nauchno-Issled. Geolograzed. Inst.*, 1974, **353**, 128).

Addendum

The above chapter was written in September 1992. Between this date and the end of the year many relevant papers were published. Some of the more important include the following:

(a) Bicyclo[3.3.0]octane group

Zirconocene induced cyclisations of 1,6- or 1,7-dienes provide new routes to bicycles, thus treatment of the enyne (1) with dibutylzirconocene, (generated from Cp_2ZrCl_2 and 2 equivalents of butyllithium), followed by exposure of the product zirconocycle to carbon monoxide affords the bicyclo[3.3.0]oct-2-en-3-one (2) (G.Agnel, Z.Owczarczyk, and E.-I.Negishi, *Tetrahedron Letters*, 1992, **33**, 1543).

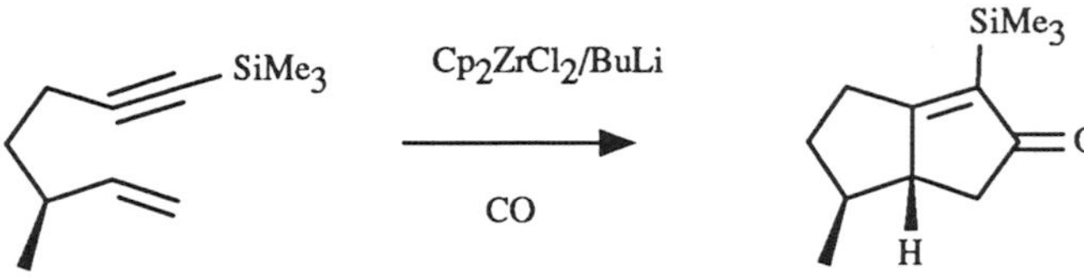

Similarly, treatment of the 1,6-heptadiene (3) with dibutylzirconocene gives the zirconacyclopentane (4). This is thermally stable, but reacts with 'butylisonitrile to give the iminoacyl complex (5). This last compound rearranges into the tricycle (6) which reacts with alkynes to give the 3-amino-3-vinylbicycles (7). In this last reaction when unsymmetrical alkynes are used the larger of the two substituents on the acetylene is bonded closest to the zirconium atom in the intermediate and so resides on the carbon atom next to the spiro centre in the final product (J.M.Davis, R.J.Whitby, and A.J.-Chamiec, *Tetrahedron Letters*, 1992, **33**, 5655).

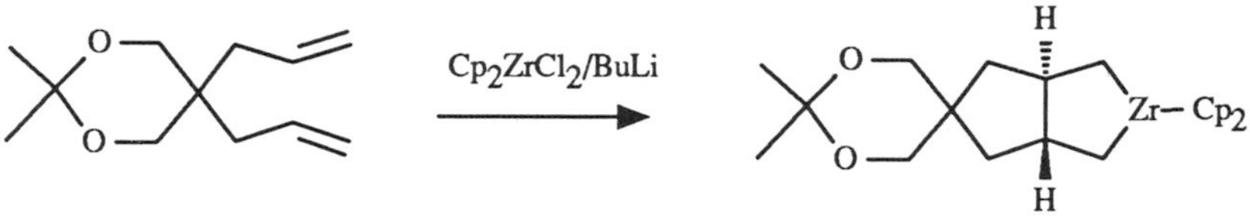

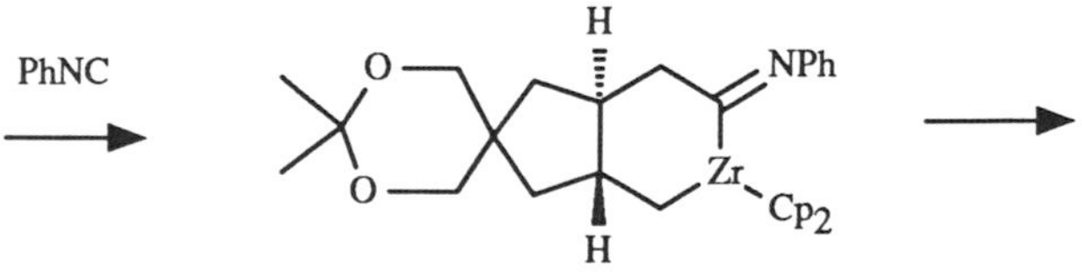

(6)   R = Pr, Ph, Pr and H, Me₃Si and H   (7)

Cyclopentenones are conveniently synthesised by the Pauson-Khand cycloaddition reaction (see N.E.Shore, *Org. Reactions*, 1991, **40**, 1). For intramolecular variants coordinating ligands in the homo- and bishomo-propargylic positions of 1,6-enynes enhance the rate of cyclisation. Thus for [(8) L = EtOCH₂ or EtSCH₂] the products are the corresponding bicyclo[3.3.0]oct-1-en-3-ones (9) (M.E.Krafft, I.L.Scott, and R.H.Romero, *Tetrahedron Letters,* 1992, **33**, 3829).

(8)   (9)

Radical mediated cyclisations of halogeno dienes and enynes have become of major importance in organic synthesis and have been used to generate numerous natural products bearing bicyclic and polycyclic systems (it should be noted that other 'easily cleaved' groups can be used in place of halogen atoms in such reactions). The cyclisations proceed by homolysis of the carbon-halogen bond using a radical initiator such as AIBN (2,2′-azobisisobutyronitrile), and they are frequently terminated by hydrogen abstraction from tributyltin hydride. Most often there is a preference for an *endo*-, rather than an *exo*-, mode of cyclisation which occurs *via* a chair-like transition state. These points are illustrated by the cyclisation of the iodo compound (10) to the natural product hirsutene (11). For a recent review, which includes a mechanistic interpretation of this and many other related AIBN/tributyltin reactions, see C.P.Jasperse, D.P.Curran, and T.L.Fevig (*Chem. Reviews*, 1991, **91**, 1237).

A recent example of this type is the synthesis of 1-methoxycarbonylbicyclo[3.3.0]heptane (13) through treatment of the mesitylenesulphonyl hydrazone (12) with AIBN and tributyltin (S.K.Kim and J.R.Cho, *Synlett.*, 1992, 629).

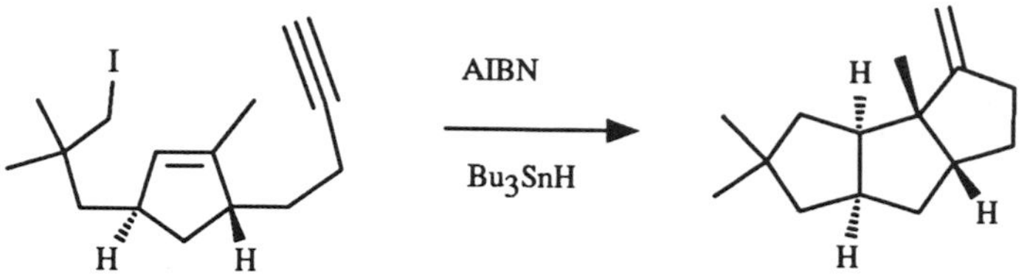
I
AIBN
Bu3SnH
H
H
H
H
H
H

(10) (11)

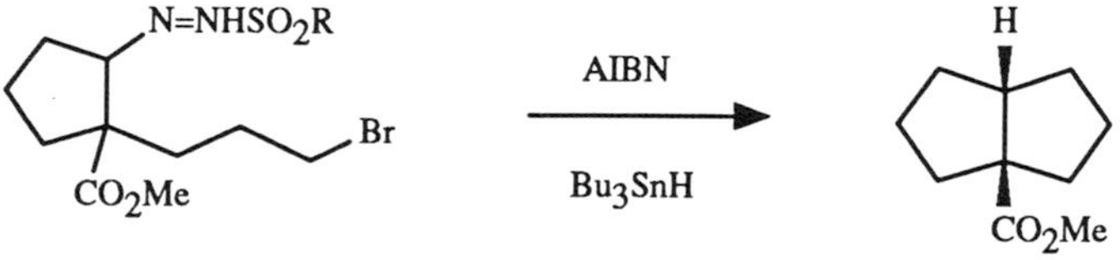
N=NHSO2R
Br
CO2Me
AIBN
Bu3SnH
H
CO2Me

(12) (13)

R = mesitylenyl

Other substrates can also cyclised through initial homolysis of a 'weak bond': thus epoxy ketones (14) undergo cyclisation to 6-hydroxybicyclo[4.3.0]nonan-2-ones (15) (S.Kim and J.S.Koh, *J. Chem. Soc. Chem. Commun.,* 1992, 1377).

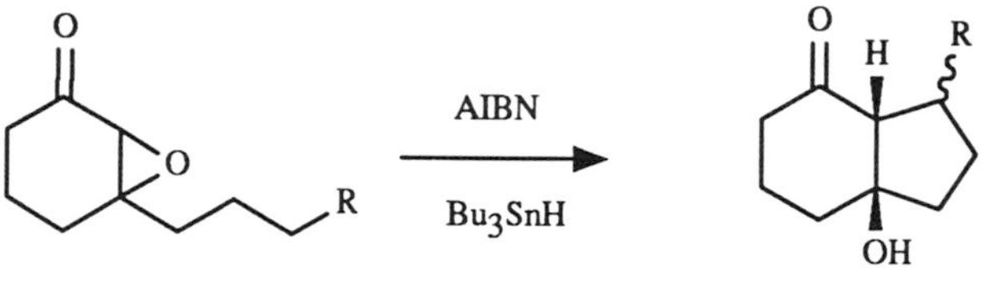
O
O
R
AIBN
Bu3SnH
O
H
R
OH

(14) (15)

Another well used approach utilises palladium catalysed cyclisations of dienes and related compounds. A recent illustration is the oxidative conversion of *cis*-1,2-divinylcyclohexane (16) into 7-oxycarbonyl-9-methylidenebicyclo[4.3.0]nonanes (17). Here the reagents are palladium(II) acetate, manganese(IV) oxide, 1,4-benzoquinone, and a chiral acid (such as the enantiomers of various lactic or mandelic acids). Good diastereoselectivity is noted especially if the reactions are conducted in the presence of molecular sieves (L.Tottie *et al., J. Org. Chem.,* 1992, **57**, 6579).

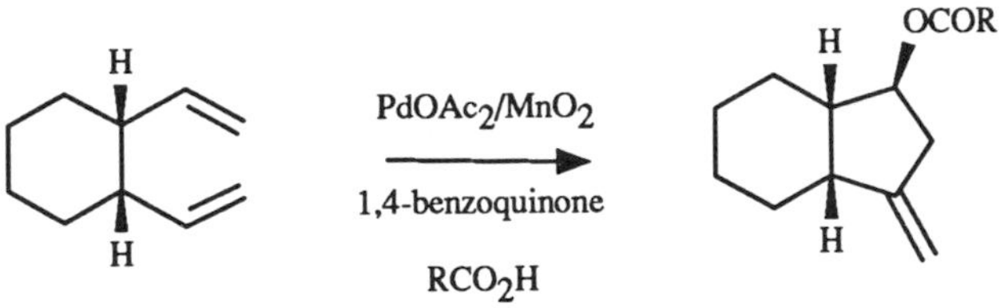

(16)   (17)

Meso-7-substituted bicyclo[3.3.0]octan-3-ones (18) can be converted into chiral bicyclo[3.2.1]octenes by reactions with chiral 1,2-cycloalkanediols, such as *S,S*-cyclohexane-1,2-diol or *R,R*-cycloheptane-1,2-diol, in the presence of boron trifluoride. The Lewis acid mediated ring expansions proceed with high diastereoselectivity, and the esters (19) may then be reductively cleaved by reactions with lithium aluminium hydride to give the appropriate chiral alcohols (20) (T.Yamamoto, H.Suemune, and K.Sakai, *J. Chem. Soc. Chem. Commun.*, 1992, 1482).

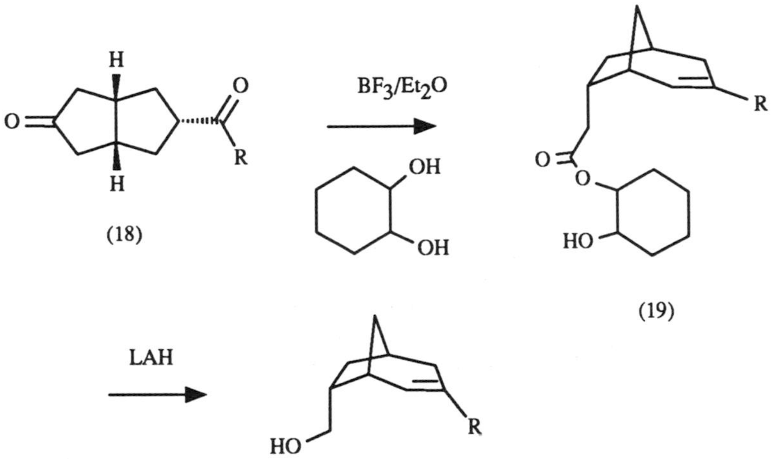

(18)   (19)

(20)

(b) Bicyclo[4.3.0]nonanes, bicyclo[4.4.0]decanes, bicyclo[3.1.0]hexanes, and bicyclo[4.1.0]heptanes

1-Acylbicyclo[4.2.0]oct-3-enes (1) when reacted with ethylaluminium dichloride ring expand to yield 6-alkylbicyclo[4.3.0]non-3-en-7-ones (2) (T.Fujiwara *et al., Tetrahedron Letters,* 1992, **33**, 2583).

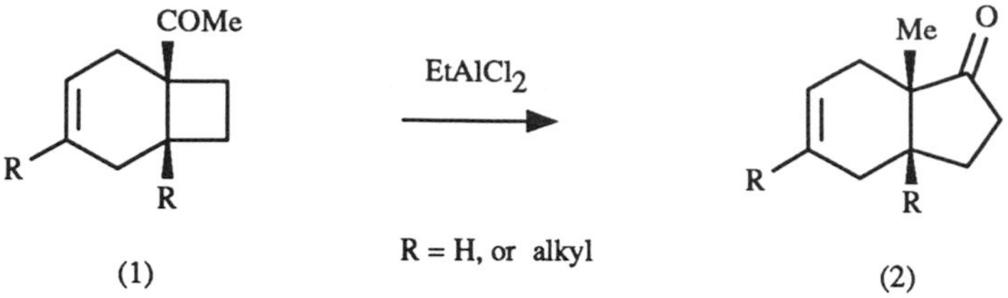

(1)   EtAlCl$_2$ →   (2)

R = H, or alkyl

Trans-fused decalins (4) are synthesised through the intramolecular cycloaddition of α,β-unsaturated Fischer carbene complexes (3) (G.Müller and G.Jas, *Tetrahedron Letters,* 1992, **33**, 4417).

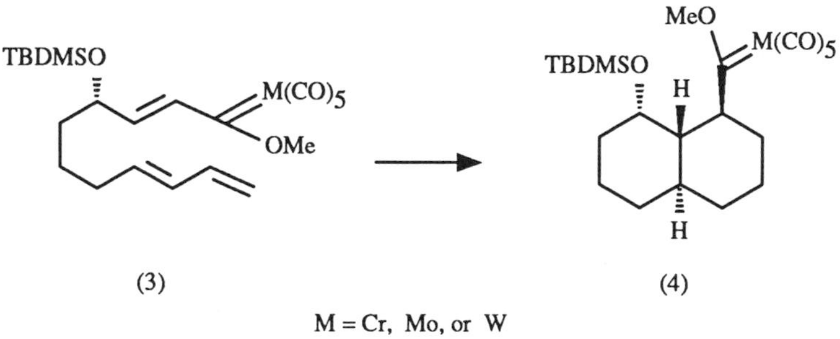

(3)   →   (4)

M = Cr, Mo, or W

Related chemistry is involved in the reactions of enynes (5, n = 1, or 2) with pentacarbonyl(butylidenemethoxycarbene) molybdenum(0) in boiling benzene. This affords the corresponding bicyclo[3.1.0]hexanes (6, n = 1), or bicyclo[4.1.0]heptanes (6, n = 2) (D.F.Harvey, K.P.Lund, and D.A.Neil, *J. Amer. Chem. Soc.,* 1992, **114**, 8424).

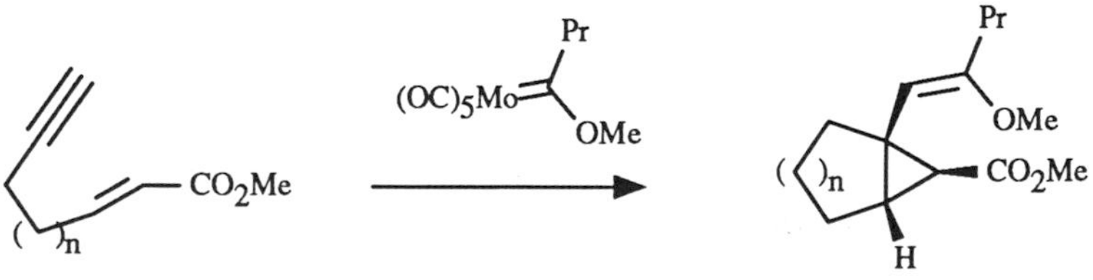

(5)   →   (6)

Second Supplements to the 2nd Edition of Rodd's Chemistry of Carbon Compounds, Vol. II B(Partial), C, D and E, edited by M. Sainsbury
© 1994 Elsevier Science B.V. All rights reserved.

Chapter 11

Polycarbocyclic Bridged Ring Compounds

A. P. Marchand

This chapter reviews significant advances in the synthesis and chemistry of polycarbocyclic bridged ring compounds that have accrued since the appearance of Rodd, Volume IIC, Supplement to the 2nd Edition (1974). References to Rodd's C.C.C. relate to the relevant page in that volume.

1. Introduction

Recent years have witnessed an explosive growth in the number of publications that deal with the synthesis and chemistry of polycarbocyclic bridged ring compounds. Members of this class of compounds frequently possess rigid structures whose carbocyclic frameworks contain varying degrees of steric strain. In recent years, highly regio- and stereoselective methods have been developed for synthesizing structurally complex bridged ring systems of known molecular geometry. Compounds of this type constitute a valuable class of synthetic intermediates that have been utilized extensively in, e.g., polyquinane synthesis (G. Mehta et al., Tetrahedron, 1981, 37, 4543) and natural product synthesis [Natural product synthesis through pericyclic reactions by G. Desimoni et al., ACS Monograph 180 (American Chemical Society: Washington, D. C., 1983); G. Mehta et al., J. Chem. Soc., Chem. Commun., 1982, 540; 1983, 824; 1984, 1058; J. Org. Chem., 1987, 52, 2875; 1990, 55, 3568; Tetrahedron 1987, 28, 1467]. These compounds also have been used as templates for the construction of a new class of molecular clefts that are of interest as potential host molecules for use in host-guest complexation studies and potential ion transport agents [(a) R. P. Thummel, Tetrahedron, 1991, 47, 6851; SYNLETT 1992, 1; (b) J.-L. Lim, S. Chirayil, and R. P. Thummel, J. Org. Chem. 1991, 56, 1492; (c) A. P. Marchand et al., Tetrahedron, 1990, 46, 5077].

Ring-opening olefin metathesis polymerization (ROMP) occurs when α,β,α'-trisubstituted bis(η^5-cyclopentadienyl)titanacyclobutanes derived from strained polycyclic bridged alkenes, like norbornene, are cleaved thermally in the presence of excess strained alkene (R. H. Grubbs et al., J. Am. Chem. Soc., 1986, 108, 733; Macromolecules, 1989, 22, 3205). In addition, tungsten, molybdenum, and tantalum complexes have been applied similarly to promote ROMP of norbornene (R. R. Schrock et al., J. Am. Chem. Soc., 1988, 110, 1423; Macromolecules, 1987, 20, 448, 1169, 2640),

As a consequence of their rigid carbocyclic frameworks, polycyclic bridged alkenes have proved valuable as substrates for mechanistic studies. Of particular interest in this connection is the use of appropriately substituted polycyclic bridged systems as substrates for the study of diastereofacial selectivity in such reactions as, e.g., (i) electophilic additions to C=C double bonds (G. Mehta and F. A. Khan, J. Chem. Soc., Chem. Commun., 1991, 18), (ii) nucleophilic additions to ketone carbonyl groups [(a) A. S. Cieplak, J. Am. Chem. Soc., 1981, 103, 4540; (b) W. J. le Noble et al., J. Am. Chem. Soc., 1986, 108, 1598; J. Org. Chem., 1989, 54, 3836; (c) Y.-D. Wu, J. A. Tucker, and K. N. Houk, J. Am. Chem. Soc., 1991, 113, 5018; (d) S. S. Wong and M. N. Paddon-Row, J. Chem. Soc., Chem. Commun., 1991, 327; (e) G. Mehta and F. A. Khan, J. Am. Chem. Soc., 1990, 112, 6140], (iii) pericyclic processes such as Diels-

Alder cycloadditions and sigmatropic shifts (Cope rearrangement) [(a) K. N. Houk et al., J. Am. Chem. Soc., 1980, 102, 6482; Tetrahedron Lett., 1990, 31, 7285, 7289; (b) J. M. Coxon et al., J. Org. Chem., 1987, 52, 4726; 1991, 56, 2542; (c) T. Tsuji, M. Ohkita, and S. Nishida, J. Org. Chem., 1991, 56, 997; (d) W. J. le Noble et al., J. Org. Chem., 1989, 54, 998; 1990, 55, 3597; 1991, 56, 5932; J. Am. Chem. Soc., 1988, 110, 7882; (e) H. Prinzbach et al., Tetrahedron Lett., 1991, 32, 5935, 5939].

Norbornane-derived spacers of variable length provide a family of rigid, σ-bonded carbocyclic frameworks of known molecular geometry (M. N. Paddon-Row et al., Tetrahedron, 1986, 42, 1779; R. N. Warrener et al., J. Chem. Soc., Chem. Commun., 1983, 1340; Tetrahedron Lett., 1991, 32, 1885, 1889). Long-range electron transfer between π-systems attached to these spacers has been studied by a variety of spectroscopic methods, including photoelectron spectroscopy (H.-D. Martin and B. Mayer, Angew. Chem., Int. Ed. Engl., 1983, 22, 283; M. N. Paddon-Row et al., J. Am. Chem. Soc., 1981, 103, 5575; Tetrahedron Lett., 1983, 24, 5415; J. Chem. Soc., Chem. Commun., 1984, 564), electron transmission spectroscopy (M. N. Paddon-Row et al., J. Am. Chem. Soc., 1987, 109, 6957), and electron paramagnetic resonance (EPR/ENDOR) spectroscopy (M. N. Paddon-Row et al., Helv. Chim. Acta, 1990, 73, 1586).

The steric strain contained within polycarbocyclic bridged ring systems (E. M. Engler, J. D. Andose, and P. von R. Schleyer, J. Am. Chem. Soc., 1973, 95, 8005) renders this class of hydrocarbons (G. W. Burdette, H. T. Lander, and J. R. McCoy, J. Energy, 1979, 2, 289) and their corresponding polynitro derivatives (A. P. Marchand, Tetrahedron, 1988, 44, 2377; P. R. Dave et al., J. Org. Chem., 1990, 55, 4459) of interest as a potential new class of energetic materials.

4-Amino derivatives of D₃-trishomocubane have been found to possess anticataleptic and anticholinergic activity

(D. W. Oliver et al., J. Med. Chem., 1991, 34, 851). In addition, this class of compounds has been shown to possess *in vivo* antiviral activity (D. W. Oliver et al., Arzneim.-Forsch., 1991, 41, 549).

In addition, intramolecular [2 + 2] photocyclizations of polycyclic dienes to form cage compounds has been studied as a means to store solar energy. Once solar energy has been "locked" into the cage molecules, it later can be released in the form of heat via catalytic {2 + 2] cycloreversion, thereby regenerating the original diene system. Several diene-cage interconversions have been studied in this regard, e.g., (i) norbornadiene - quadricyclane [C. Kutal, Adv. Chem. Ser., 1978, 168, 158; G. Jones, II, S. Chiang, and P. T. Xuan, J. Photochem., 1979, 10, 1; K.-i. Hirao et al., J. Chem. Soc., Chem. Commun., 1984, 300; K. Maruyama, H. Tamiaki, and S. Kawabata, J. Org. Chem., 1983, 50, 4742; K. Maruyama, K. Terada, and Y. Yamamoto, J. Org. Chem., 1981, 46, 5294; R. B. King, R. M. Hanes, and S. Ikai, Adv. Chem. Ser. 1979, 173, 344; Bren', V. A et al, Russ. Chem. Rev. 1991, 60, 451]; (ii) *cis,cisoid,cis*-tricyclo[6.3.0.0^{2,6}]-undeca-4,9-diene-3,11-dione - pentacyclo[5.4.0.0^{2,6}.0^{3,10}.-0^{5,9}]undecane-8,11-dione [Y. Yamashita and T. Mukai, Chem. Lett., 1984, 1741; G. Mehta, D. S. Reddy, and A. V. Reddy, Tetrahedron Lett., 1984, 25, 2275; T. Hamada et al., J. Chem. Soc., Chem. Commun., 1980, 696; T. Mukai and Y. Yamashita, Tetrahedron Lett., 1978, 357].

The foregoing examples illustrate (i) the wide variety of synthetic and mechanistic problems to which polycarbocyclic bridged ring systems have been applied, (ii) the use that has been made of these systems to develop new polymeric materials via the ROMP process, new energetic materials, and new classes of pharmacologically active amines, and, finally, (iii) their utility as repositories of solar energy. In addition to the references cited above, the following reviews deal with various aspects of the synthesis and chemistry of polycarbocyclic bridged ring compounds:

"Strained Organic Molecules" (A. Greenberg and J. F. Liebman, Academic Press: New York, 1978). "Synthesis of Non-natural products: challenge and reward" (P. E. Eaton, Editor, Tetrahedron Symposia-in-Print No. 26, Tetrahedron, 1986, 42, 1549-1915). Polyquinane chemistry; synthesis and reactions by L. A. Paquette and A. M. Doherty (Springer-Verlag, Heidelberg, 1987). Strained organic compounds by J. Michl and J. A. Gladysz, Editors (Chem. Rev., 1989, 89, 973-1270). The chemistry of pentacyclo-[5.4.0.0^{2,6}.0^{3,10}.0^{5,9}]undecane (PCUD) and related systems by A. P. Marchand (SYNLETT, 1991, 73). Recent studies of carbocations by M. Saunders and H. A. Jiminez-Vazquez (Chem. Rev., 1991, 91, 375).

2 General Methods of Synthesis

(a) From simple monocyclic and nonbridged polycyclic compounds

(1) By inter- and intramolecular Diels-Alder reactions (E. Ciganek, Org. React., 1984, 32, 1-374; G. Brieger and J. N. Bennett, Chem. Rev., 1980, 80, 63), A key step in a synthesis of racemic α- and β-cedrenes proceeds via intramolecular Diels-Alder reaction (E. G. Breitholle and A. G. Fallis, J. Org. Chem., 1978, 43, 1964).

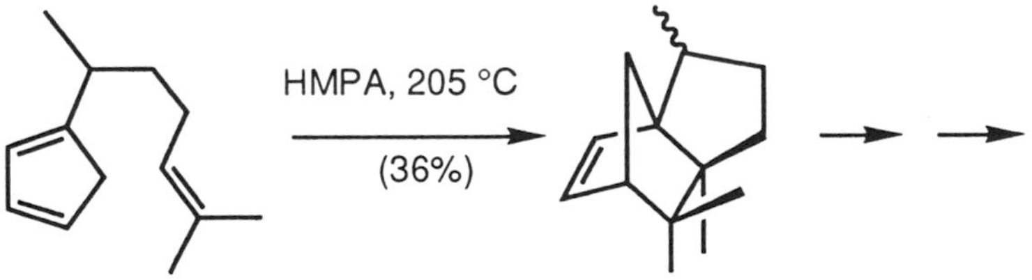

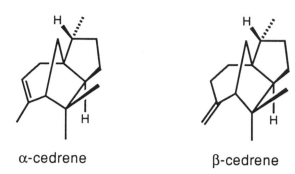

α-cedrene   β-cedrene

Intermolecular [4 + 2] cycloadditions of cyclopentadiene to monosubstituted alkenes frequently afford mixtures of *exo-* and *endo*-5-substituted norborn-2-enes. The rates of these reactions as well as the product stereochemistry (i.e., *exo:endo* ratio) can be affected dramatically by a variety of added catalysts and solvent systems, i.e., when thermal Diels-Alder reactions are performed:

(i) in the presence of any of a variety of added Lewis acid and/or organometallic catalysts, e.g., γ-alumina (G. W. Kabalka et al., Tetrahedron Lett., 1990, 31, 5433), LiClO$_4$ (P. A. Grieco, et al., J. Am. Chem. Soc., 1990, 112, 4595; M. A. Forman and W. P. Dailey, J. Am. Chem. Soc., 1991, 113, 2761), metalloporphyrins (D. W. Bartley and T. J. Kodadek, Tetrahedron Lett., 1990, 31, 6303);

(ii) in water (R. Breslow and D. Rideout, J. Am. Chem. Soc., 1980, 102, 7816) or in nonaqueous polar solvent (R. Breslow and T. Guo, J. Am. Chem. Soc., 1988, 110, 5613; D. Liotta et al., Tetrahedron Lett., 1988, 29, 3745);

(iii) in micellar media (V. K. Singh et al., Synth. Commun., 1988, 18, 567);

(iv) inside crystalline cyclodextrins, via ternary complex formation (cyclodextrin + diene + dienophile) in the solid state (D. L. Wenick et al., J. Chem. Soc., Chem. Commun., 1990, 956).

Reactions of pentamethylcyclopentadiene with ketenes, when performed in the presence of ammoniumyl salts, occur rapidly at 0 °C to afford primarily the products of [4 + 2] rather than [2 + 2] cycloaddition (M. Schmittel and H. von Seggern, Angew. Chem., Int. Ed. Engl., 1991, 30, 999).

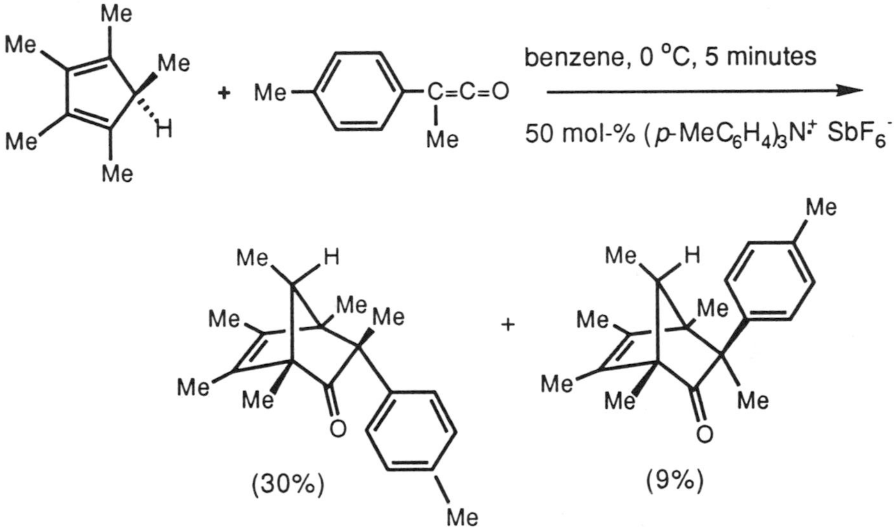

"Domino Diels-Alder" reactions occur when 9,10-dihydrofulvalene is heated with dimethyl acetylene-dicarboxylate (DMAD, L. A. Paquette et al., J. Am. Chem. Soc., 1978, 100, 5845; Org. Synth., 1989, 68, 198). Although several products result from this Diels-Alder reaction, the "internal diester" can be isolated readily by selectively hydrolyzing the mixture of product esters. Subsequent acidic hydrolysis of the "internal ester" affords the corresponding "internal diacid", which has been further elaborated into a variety of complex polycarbocyclic compounds, including dodecahedrane and a number of substituted dodecahedranes (L. A. Paquette et al., J. Am. Chem. Soc., 1982, 104, 4503; 1986, 108, 2343, 1716; J. Org. Chem., 1983, 48, 3282; 1987, 52, 1265).

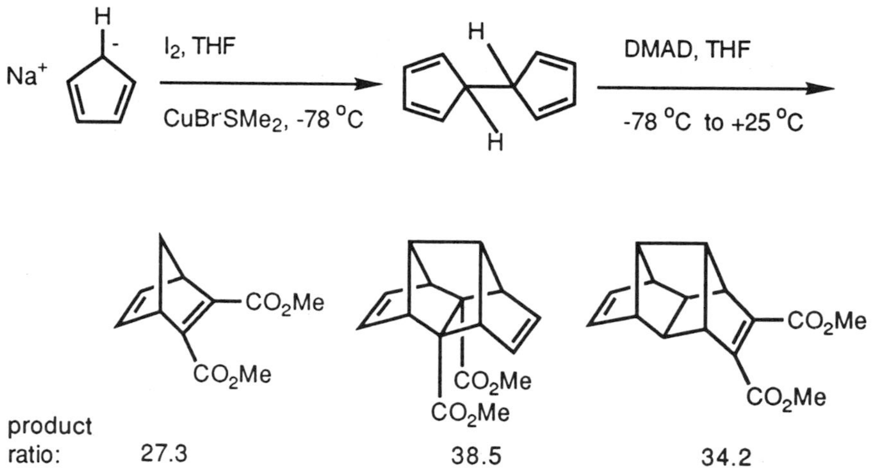

Domino Diels-Alder reactions also have been studied with a "syn-ortho,ortho'-dibenzene" as the diene system by using a variety of dienophiles (H. Prinzbach et al., Tetrahedron Lett., 1989, 30, 3133, 3137). In the case of reactions with acetylenic dienophiles, two products are possible, the external (domino) adduct, or the internal ("pincer") adduct. The results shown below suggest that the relative amount of pincer product formed in these reactions is influenced markedly by the steric bulk of substituents on the dienophile.

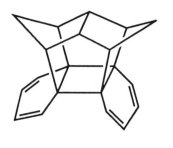

dienophile

$$\xrightarrow{\hspace{3cm}}$$

XCH=CHY or
XC≡CY

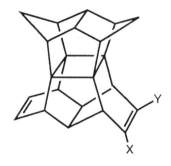

+

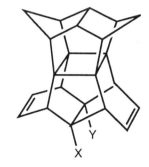

"domino" adduct (with
alkene or alkyne dienophile

"pincer" adduct (with
alkyne dienophile only)

Dienophile	Conditions	Ratio (domino/pincer)	Yield (%)
maleic anhydride	80 °C, 12 h	100:0	100
p-benzoquinone	100 °C, 24 h	100:0	79
dimethyl acetylene-dicarboxylate	100 °C, 80 h	97:3	89
dicyanoacetylene	100 °C, 4 h	26:74	70

(2) By a Wurtz-type coupling reaction that occurs when α,ω-di-Grignard reagents are treated with Ag(I) salts (G. M. Whitesides and F. D. Gutowsky, J. Org. Chem., 1976, 41, 2882).

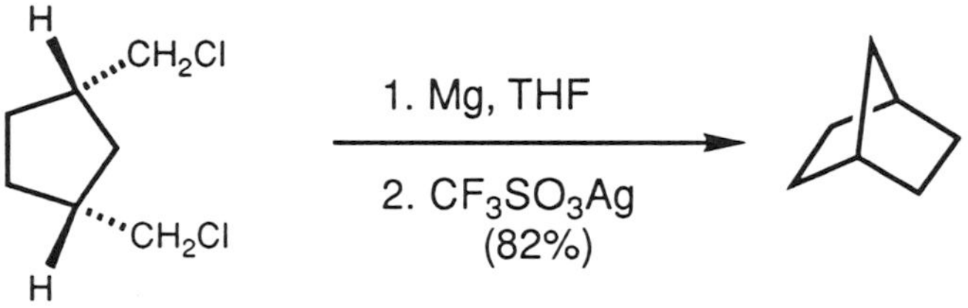

(3) By intramolecular radical cyclization [(a) M. Ramaiah, Tetrahedron, 1987, 43, 3541; (b) D. P. Curran, Synthesis, 1988, 417, 489]. Bicyclo[3.2.1]octan-3-ones have been synthesized from (S)-(+)-carvone via an intramolecular radical cyclization strategy (A. Srikrishna and P. Hemamalini, J. Org Chem, 1990, 55, 4883).

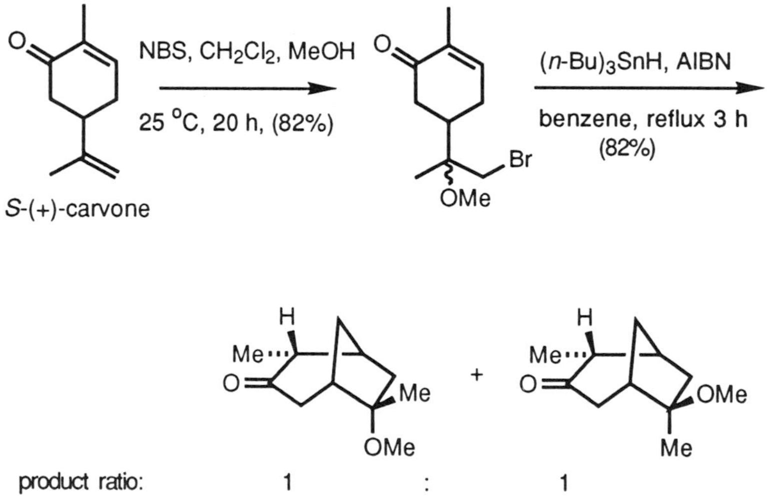

(4) By intramolecular [2 + 2] alkene photocylization. Irradiation of acetone solutions of 4,10-dichloro- and of 4,-10-dibromo-*cis,cisoid,cis*-tricyclo[6.3.0.0^{2,6}]undecane- 4,9 -diene-3,11-dione proceeds via enone-enone [2 + 2] photocyclization (in 85% and 94% yield, respectively). In each case, the product was isolated in admixture with its corresponding diketone hydrate (A. P. Marchand and G. S. Annapurna, Synth. Commun. 1989, 19, 3477).

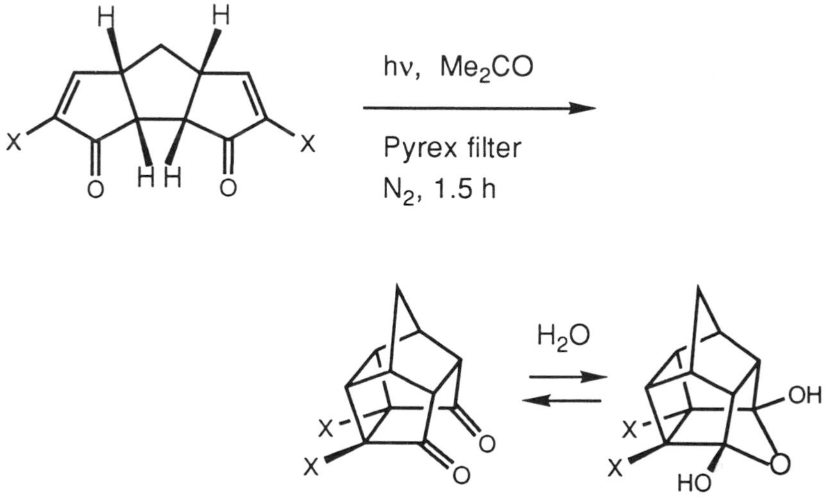

hv, Me$_2$CO

Pyrex filter

N$_2$, 1.5 h

H$_2$O

(X = Br: 94%)

(X = Cl: 85%)

Enone-alkene cyclization has been utilized as a key step in an efficient syntheses of longifolene (W. Oppolzer and T. Godel, J. Am. Chem. Soc., 1978, 100, 2583; Helv. Chim. Acta, 1984, 67, 1154).

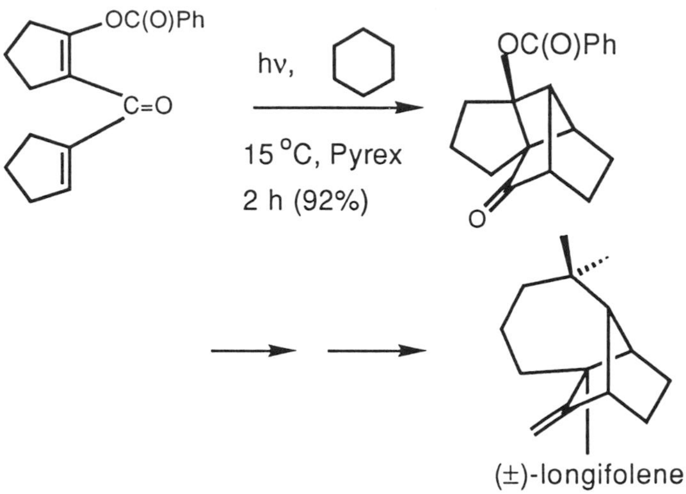

(±)-longifolene

(5) By arene-olefin 1,3-photocycloadditions. This reaction, an intermolecular variant of which is depicted below, can be used to produce a tricyclic system which can contain as many as six stereocenters (T. Wagner-Jauregg, Synthesis, 1980, 165; H. Morrison, Acc. Chem. Res., 1979, 12, 383).

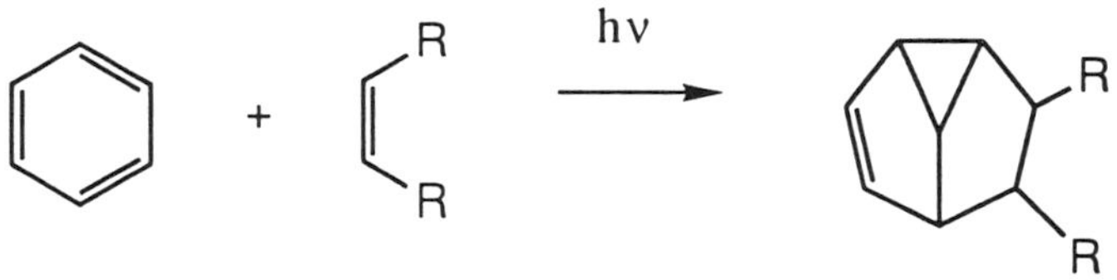

A particularly elegant application of an *intra*molecular variant of this reaction is exemplified by Wender's four step total synthesis of (±)-α-cedrene (P. A. Wender and J. J. Howbert, J. Am. Chem. Soc., 1981, 103, 688).

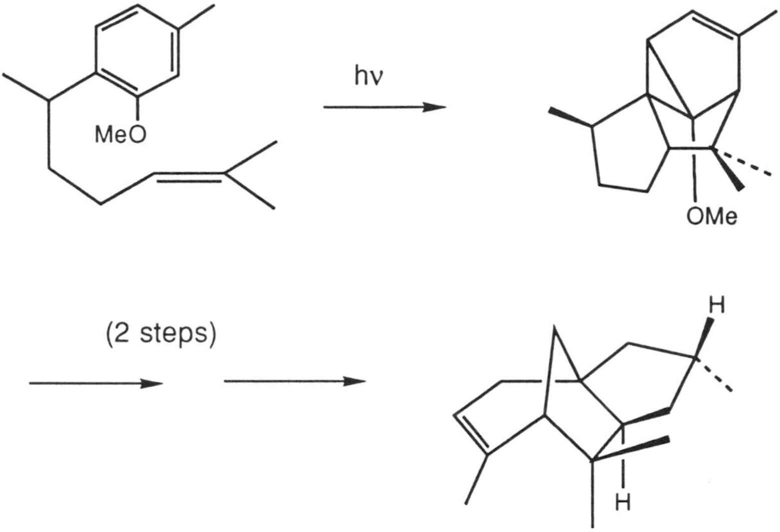

α-cedrene

(6) By intramolecular diyl trapping reactions (R. D. Little et al., J. Org. Chem., 1990, 55, 2742; SYNLETT, 1992, 107). Little and coworkers have shown that intramolecular trapping of triplet diyls afford bridged products, whereas linearly fused products arise via intramolecular trapping of either the triplet or the corresponding singlet diyl. The presence of an electron withdrawing substituent attached to the diylophile affords linearly fused products (via selective stabilization and trapping of the singlet diyl). However, when this diylophile subsituent is a bulky alkyl group (as is the case in the example shown below), the corresponding bridged product predominates.

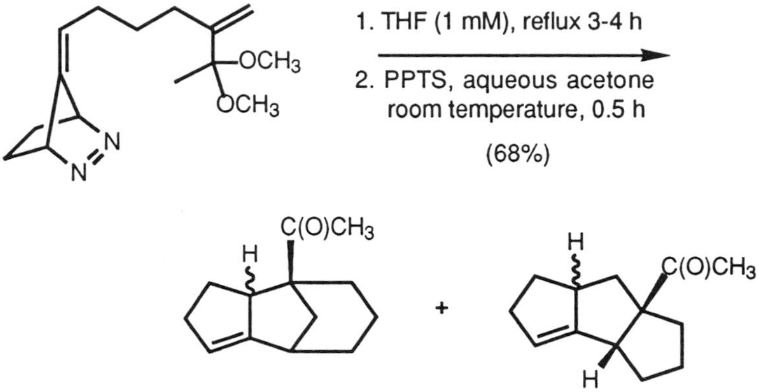
OCH3
OCH3
N
N
1. THF (1 mM), reflux 3-4 h
2. PPTS, aqueous acetone
room temperature, 0.5 h
(68%)
C(O)CH3
H
+
H
C(O)CH3
H

product ratio: 15.7 : 1

(7) By application of the Weiss reaction (J. M. Cook et al., Tetrahedron, 1991, 42, 3665), This approach has been utilized to synthesize [3.3.3]propellane (J. M. Cook et al., J. Org. Chem., 1983, 48, 139.

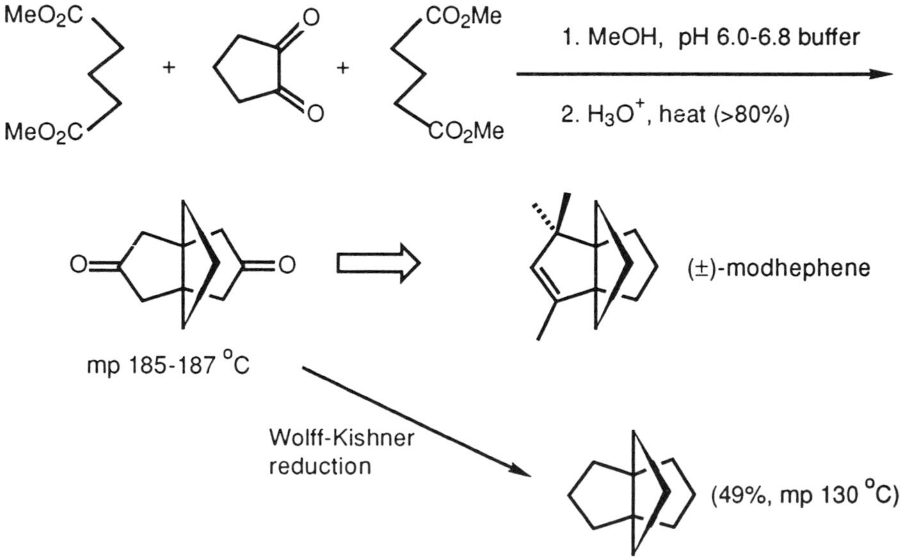
MeO2C
MeO2C
+
O
O
+
CO2Me
CO2Me
1. MeOH, pH 6.0-6.8 buffer
2. H3O+, heat (>80%)
O
O
(±)-modhephene
mp 185-187 °C
Wolff-Kishner
reduction
(49%, mp 130 °C)

(b) Elaboration of bridged bicyclic and tricyclic compounds into higher polycyclics

(1) By intramolecular [2 + 2 + 2] ("homo-Diels-Alder") cycloadditions. Norbornadiene undergoes homo-Diels-Alder cycloaddition to alkynes when the reactants are heated in the presence of a Co/Al catalyst (J. E. Lyons et al., Ann. N. Y. Acad. Sci., 1980, 333, 273). The corresponding reactions, when performed with electron deficient alkynes, occur thermally in the absence of catalyst (T. Sasaki et al., J. Org. Chem., 1972, 37, 2317).

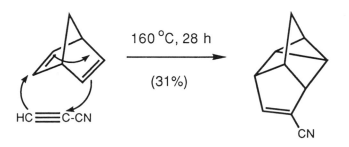

bp 122-124 °C (27 torr)

(2) By Ramberg-Bäcklund rearrangement (L. A. Paquette, Org. React., 1977, 25, 1-71). Cyclic α-halosulfones undergo base promoted ring contraction in the manner illustrated below (L. A. Paquette et al., J. Am. Chem. Soc., 1971, 93, 4516):

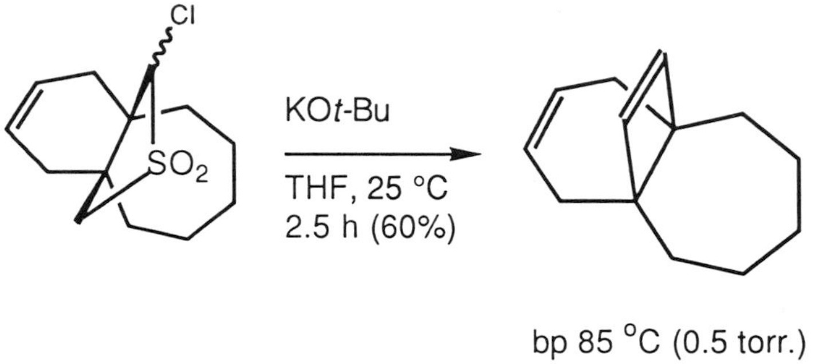
Cl
SO2
KOt-Bu
THF, 25 °C
2.5 h (60%)
bp 85 °C (0.5 torr.)

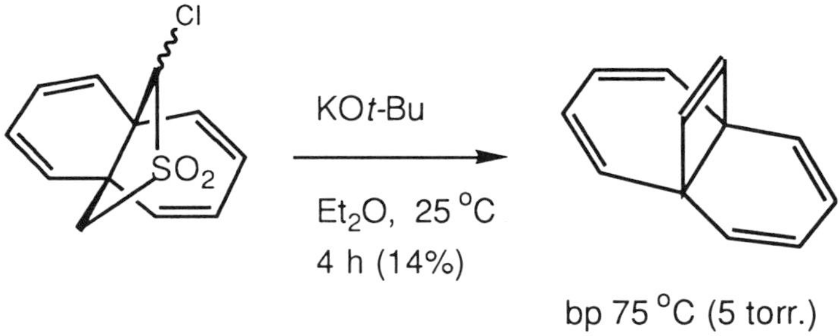
Cl
SO2
KOt-Bu
Et2O, 25 °C
4 h (14%)
bp 75 °C (5 torr.)

(3) By intramolecular reactions of diazocarbonyl compounds (S. D. Burke and P. A. Grieco, Org. React., 1979, 26, 361-475). By refluxing a benzene solution of the diazoketone derived from bicyclo[3.2.2]nona-2,6,8-triene-4-carboxylic acid with Cu bronze, a carbenoid is produced which undergoes subsequent intramolecular alkene cycloaddition to a neighboring C=C double bond in the substrate, thereby affording the corresponding enone in 43% yield (E. Vedejs and R. A. Shepherd, J. Org. Chem., 1976, 41, 742).

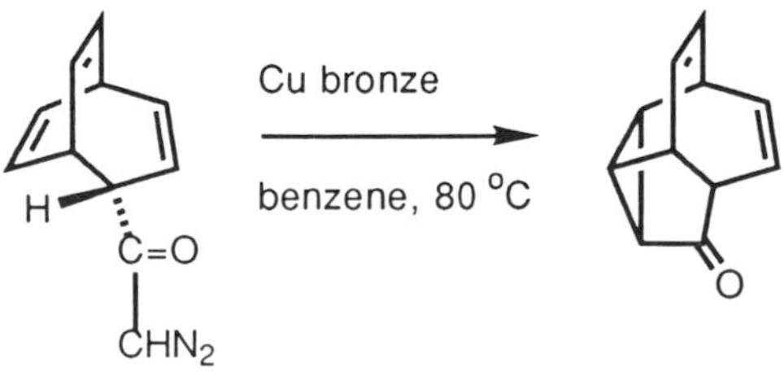
H
C=O
CHN2
Cu bronze
benzene, 80 °C
O

(4) By intramolecular [2 + 2] enone-alkene photocyclo-addition (M. R. Crimmins, Chem. Rev., 1988, <u>88</u>, 1453). Substituted cyclopentadienes undergo intermolecular Diels-Alder cycloaddition to substituted *p*-benzoquinones to afford the corresponding substituted *endo*-tricyclo-[6.2.1.0^{2,7}]undeca-4,9-diene-3,6-diones. Subsequent irradiation of the resulting adducts affords the corresponding substituted pentacyclo[5.4.0.0^{2,6}.0^{3,10}.0^{5,9}]undecane-8,11-diones via intramolecular [2 + 2] enone-alkene photocyclo-addition (G. Mehta et al., Tetrahedron, 1981, <u>37</u>, 4543).

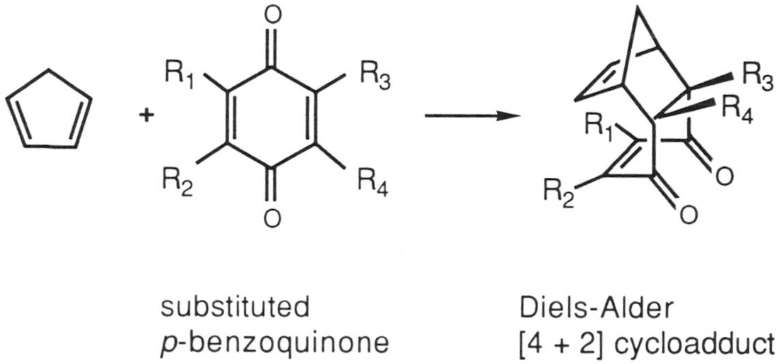

	substituted *p*-benzoquinone			Diels-Alder [4 + 2] cycloadduct	
<u>R$_1$</u>	<u>R$_2$</u>	<u>R$_3$</u>	<u>R$_4$</u>	<u>Yield (%)</u>	<u>mp (°C)</u>
H	H	H	H		76
Me	Me	H	H		46
Cl	Cl	H	H		109-110
Me	H	H	Me		64-65
Cl	H	H	Cl		113-114
Br	H	H	Br	90	130-131
Cl	OMe	Cl	OMe	97	94-95
Cl	OMe	OMe	Cl	95	79-80
Cl	Cl	Cl	Cl	96	145-146
Br	Br	Br	Br	90	164-165

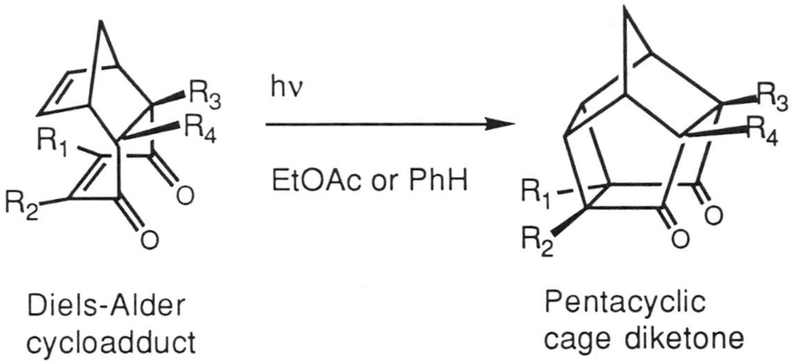

Diels-Alder
cycloadduct

Pentacyclic
cage diketone

R_1	R_2	R_3	R_4	Yield (%)	mp (°C)
H	H	H	H	92	233-234
Me	Me	H	H	98	108
Cl	Cl	H	H	95	203-205
Me	H	H	Me	90	72-73
Cl	H	H	Cl	92	179-180
Br	H	H	Br	94	185-186
Cl	OMe	Cl	OMe	90	(oil)
Cl	OMe	OMe	Cl	92	147-148
Cl	Cl	Cl	Cl	97	215 (dec.)
Br	Br	Br	Br	95	238-240

(5) By transition metal promoted coupling and dimerization reactions. 7-Substituted norbornadienes react thermally with stoichiometric quantities of transition metal carbonyls to afford a variety of novel bridged polycyclic compounds. Thus, cyclodimerization occurs when substituted norbornadienes are heated with $Mo(CO)_6$ (T. J. Chow et al., J. Am. Chem. Soc., 1987, 109, 797) or with $Fe(CO)_5$ (A. P. Marchand et al., J. Org. Chem., 1984, 49, 1660; 1985, 50, 396; 1986, 51, 4096). In addition to a heptacyclic dimer, reactions of substituted norbornenes and norbornadienes with $Fe(CO)_5$ also produce a variety of polycyclic ketones which result via Fe(0)-promoted

dimerization with concomitant insertion of carbon monoxide (E. Weissberger and P. Laszlo, Acc. Chem. Res., 1976, 9, 209; J. Mantzaris and E. Weissberger, J. Am. Chem. Soc., 1974, 96, 1873, 1880).

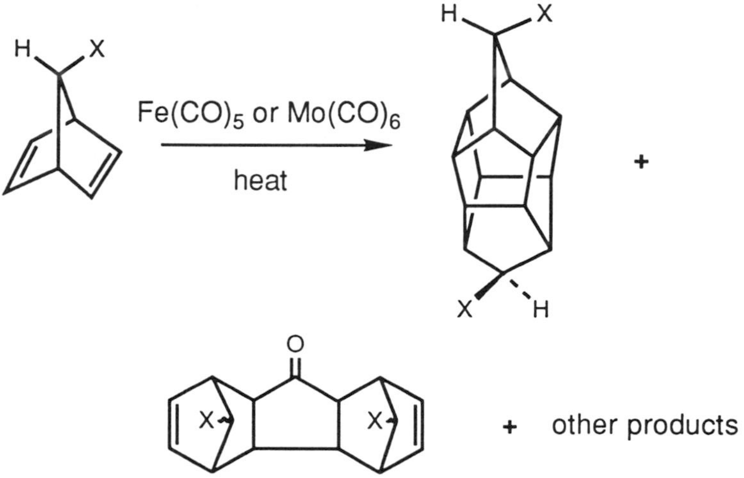

The Pauson-Khand reaction (N. E. Schore, Chem. Rev., 1988, 88, 1081; Org. React., 1991, 40, 1-90) has been used to synthesize a variety of bridged polycarbocyclic ring systems. Thus, norbornene reacts with Co-complexed acetylene to afford *exo*-tricyclo[5.2.1.0^{2,6}]dec-4-en-3-one (74%), mp 32 °C (P. L. Pauson et al., J. Chem. Soc., Perkin Trans. 1, 1973, 977).

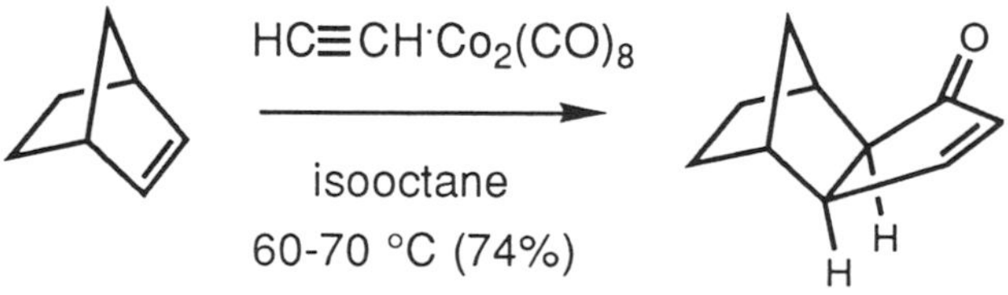

(6) By Lewis acid promoted thermodynamic cationic rearrangement of polycyclic hydrocarbons. This procedure has been used with notable success for synthesizing adamantane and higher "adamantalogues" such as diamantane ("congressane": M. A. McKervey et al., Tetrahedron Lett., 1971, 1671; J. Chem. Soc., Perkin Trans. 1, 1972, 2691). In such cases, AlBr$_3$-promoted rearrangement of polycyclic hydrocarbons frequently leads to the formation of the thermodynamically most stable isomer, i.e., the "stabilomer" (M. A. McKervey, Chem. Soc. Rev., 1974, 3, 479; P. von R. Schleyer et al., Progr. Phys. Org. Chem., 1981, 13, 63-117). This approach has been used, e.g., to synthesize D$_3$-trishomocubane, mp 150-152 °C, the pentacyclic C$_{11}$ stabilomer (P. von R. Schleyer et al., J. Chem. Soc., Chem. Commun., 1974, 976; J. Org. Chem., 1977, 42, 3852).

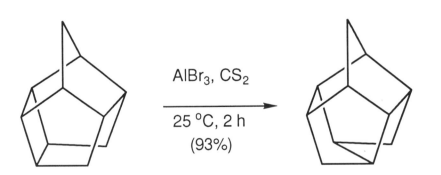

(7) By tandem vicinal difunctionalization (M. J. Chapdelaine and M. Hulce, Org. React., 1990, 38, 225-653). The kinetic enolate of 2-cyclohexanone adds in Michael fashion to methyl acrylate. The resulting conjugate enolate then undergoes a second (intramolecular) Michael addition to afford a substituted bicyclo[2.2.2]octan-2-one (R. A. Lee, Tetrahedron Lett., 1973, 3333).

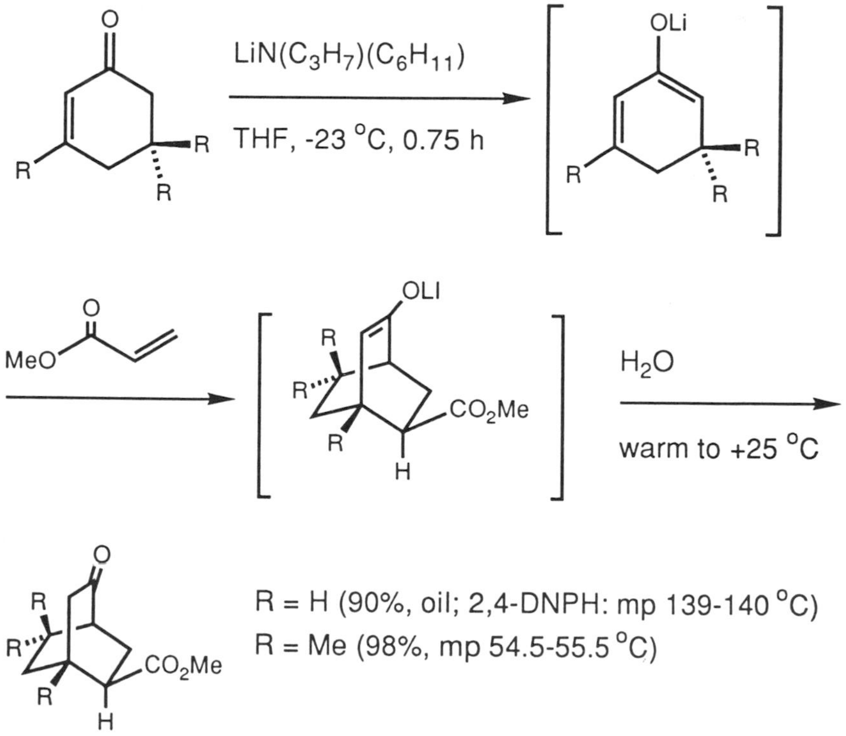

R = H (90%, oil; 2,4-DNPH: mp 139-140 °C)
R = Me (98%, mp 54.5-55.5 °C)

(8) By intramolecular [3 + 3] sigmatropic shifts (e.g., Claisen and Cope rearrangements; S. J. Rhoads and N. R. Raulins, Org. React., 1975, 22, 1-252). When heated at 200 °C in a sealed tube for 24 h, (7-cycloheptatrienyl) allyl ether undergoes a series of [1,5] sigmatropic hydrogen migrations followed by a Claisen rearrangement and intramolecular Diels-Alder reaction, thereby affording a 1:1 mixture of isomeric tricyclic ketones (C. A. Cupas et al., J. Am. Chem. Soc., 1970, 92, 3237).

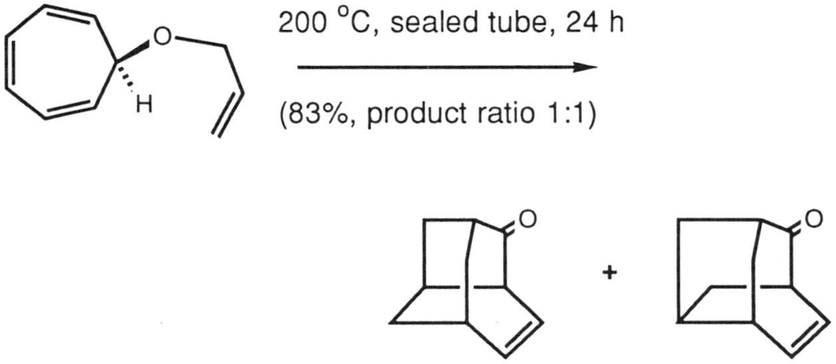

3. Strained polycarbocyclic alkenes

Rigid polycarbocyclic compounds provide σ-bonded frameworks into which highly strained C=C double bonds can be incorporated. Two important classes of strained alkenes have been studied extensively in such situations. The first of these contains a bridgehead C=C double bond and thus violates Bredt's rule (Rodd's C. C. C., Suppl. to Vol. II C, p. 25). Such π-bonds are often severely twisted, thereby resulting in substantial deviation from planarity of the sp^2 hybridized centers which comprise the double bond. As a further consequence of this C=C distortion, significant levels of steric strain are introduced into the molecule, the HOMO-LUMO energy difference is decreased, and the resulting bridgehead C=C frequently possesses "significant dira-dicaloid character" (W. F. Maier and P. von R. Schleyer, J. Am. Chem. Soc., 1981, 103, 1891). Several alkenes of this type either have been synthesized (i.e., isolated and char-acterized) or have been implicated as highly reactive inter-mediates via the results of intra- or intermolecular trap-ping experiments (G. Köbrich, Angew. Chem., Int. Ed. Engl., 1973, 12, 464; R. Keese, Ibid., 1975, 14, 528; G. L. Buchanan, Chem. Soc. Rev., 1974, 3, 41; K. J. Shea, Tetra-hedron, 1980, 36, 1683).

A second class of strained alkenes includes situations in which the C=C double bond is incorporated into a polycarbocyclic framework in such a manner that devi-ation from planarity of the sp^2 hybridized centers arises via pyramidalization (rather than twisting) effects. Two familiar examples in this regard are norbornene and norbornadiene, both of which are pyramidalized slightly toward the endo direction (K. N. Houk et al., J. Am. Chem. Soc., 1980, 102, 6582; 1981, 103, 2435).

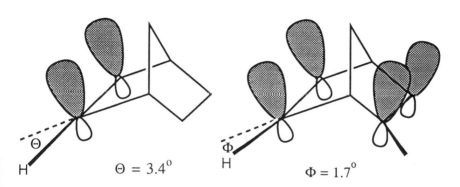

$$\Theta = 3.4°$$

$$\Phi = 1.7°$$

More recently, a number of complex polycarbocyclic systems which contain "pyramidalized alkenes" have been studied (Aspekte von Kraftfeldrechnungen, by O. Ermer, Wolfgang Baur Verlag: Munich, 1981; Stereochemistry and reactivity of systems containing π-electrons, by W. H. Watson, Editor, Verlag Chemie International: Deerfield Beach, FL, 1983; W. T. Borden, Chem. Rev., 1989, 89, 1095). Some of the methods which have been used to synthesize anti-Bredt systems and pyramidalized alkenes are presented below.

(a) Syntheses of anti-Bredt systems

(1) By eliminations on polycarbocyclic β-halolithium compounds. Reactions of n-BuLi with 1-iodo-2-halonorbornenes results in the formation of norborn-1-ene, which can be trapped in situ with furan (R. Keese and E. P. Krebs, Angew. Chem., Int. Ed. Engl., 1971, 10, 262; 1972, 11, 518).

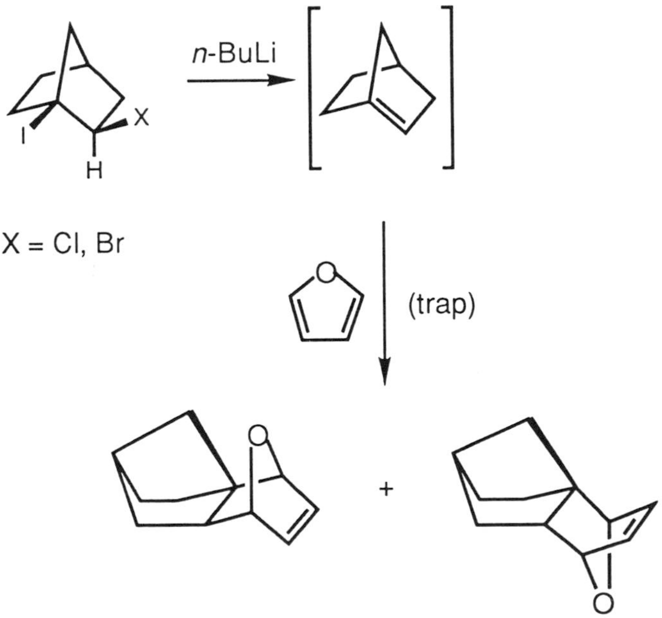

X = Cl, Br

(trap)

(2) By eliminations on polycarbocyclic β-halosilanes. A substituted dibenzobicyclo[2.2.2]oct-1-ene has been generated via this route. This strained alkene rearranges spontaneously to the corresponding dibenzobicyclo[3.2.1]octan-2-ylidenecarbene, which can be trapped subsequently by dimethylsulfoxide or by benzonitrile (T. H. Chan and D. Massuda, J. Am. Chem. Soc., 1977, 99, 936).

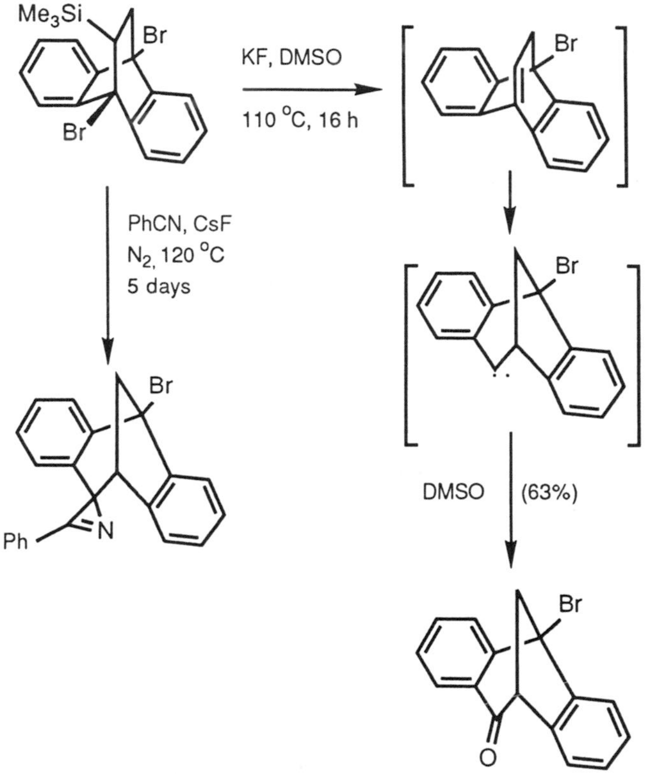

(3) By organolithium promoted elimination of HX from 1-halopolycycloalkanes. Reaction of 1-iodo- or 1-bromoad-

amantane with alkyllithium reagents proceeds via elimination of HX to form adamantene, which subsequently dimerizes spontaneously to afford a mixture of $C_{20}H_{28}$ dimers (R.
L. Cargill and A. B. Sears, Tetrahedron Lett., 1972, 3255).

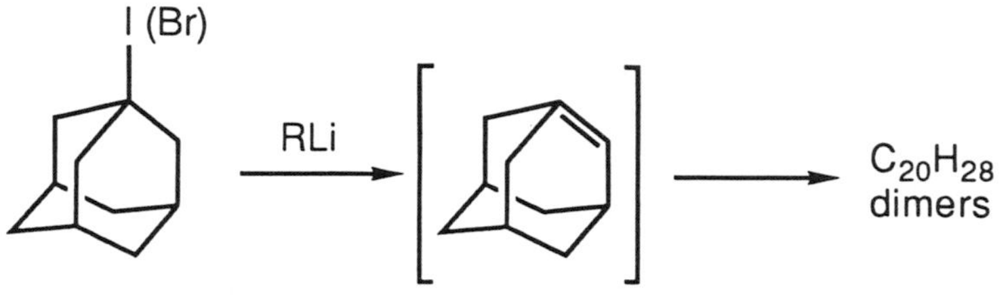

(4) By thermal cationic cyclopropyl-allyl rearrangement. 9,9-Dichlorotricyclo[4.2.1.0^{1,6}]non-3-ene spontaneously rearranges at room temperature to afford 6,9-
dichlorobicyclo[4.2.1]octa-1(9),3-diene, which spontaneously dimerizes (P. Warner et al., J. Am. Chem. Soc., 1972,
94, 7607).

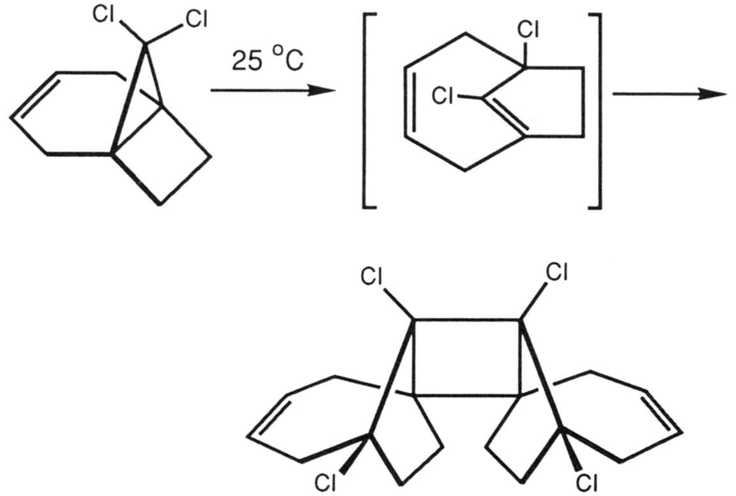

(50-60%, mp 228 °C, dec.)

(5) By thermolysis of appropriately constructed bicyclic dienes which contain a 1,5-hexadiene moiety. Compounds of this type can undergo formal [3,3] sigmatropic rearrangement to afford bridgehead dienes.

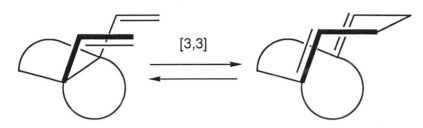

This approach has been utilized, e.g., to convert 1,6-divinylbicyclo[4.1.0]heptane into the corresponding bicyclo-[4.4.1]undecadiene (K. J. Shea et al, Tetrahedron Lett., 1983, 24, 4173).

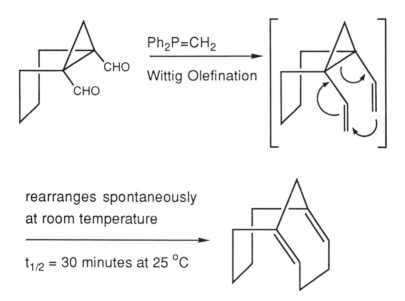

304

(b) Syntheses of pyramidalized alkenes

(1) By thermal decarboxylation of polycarbocyclic β-lactones. Tricyclo[3.3.2.0^{3,7}]dec-3(7)-ene has been synthesized via pyrolysis of the corresponding β-lactone in refluxing tetraglyme (W. T. Borden et al., Tetrahedron, 1986, 42, 1581). In addition, thermal extrusion of CO_2 from this same β-lactone occurs in a flow system at temperatures above 450 °C. The resulting alkene was trapped by matrix isolation techniques (J. Michl et al., J. Am. Chem. Soc., 1986, 108, 3544).

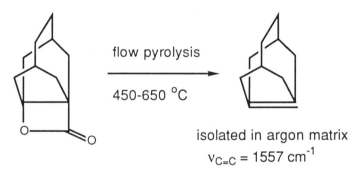

flow pyrolysis

450-650 °C

isolated in argon matrix

$v_{C=C} = 1557 \text{ cm}^{-1}$

(2) By cycloelimination performed on suitably functionalized polycarbocyclic *vic*-diols. Borden and coworkers have utilized this approach to synthesize 9,10-benzotricyclo[3.3.2.0^{3,7}]dec-3(7)-ene, which was trapped *in situ* by diphenylisobenzofuran (W. T. Borden et al., J. Am. Chem. Soc., 1977, 99, 1664).

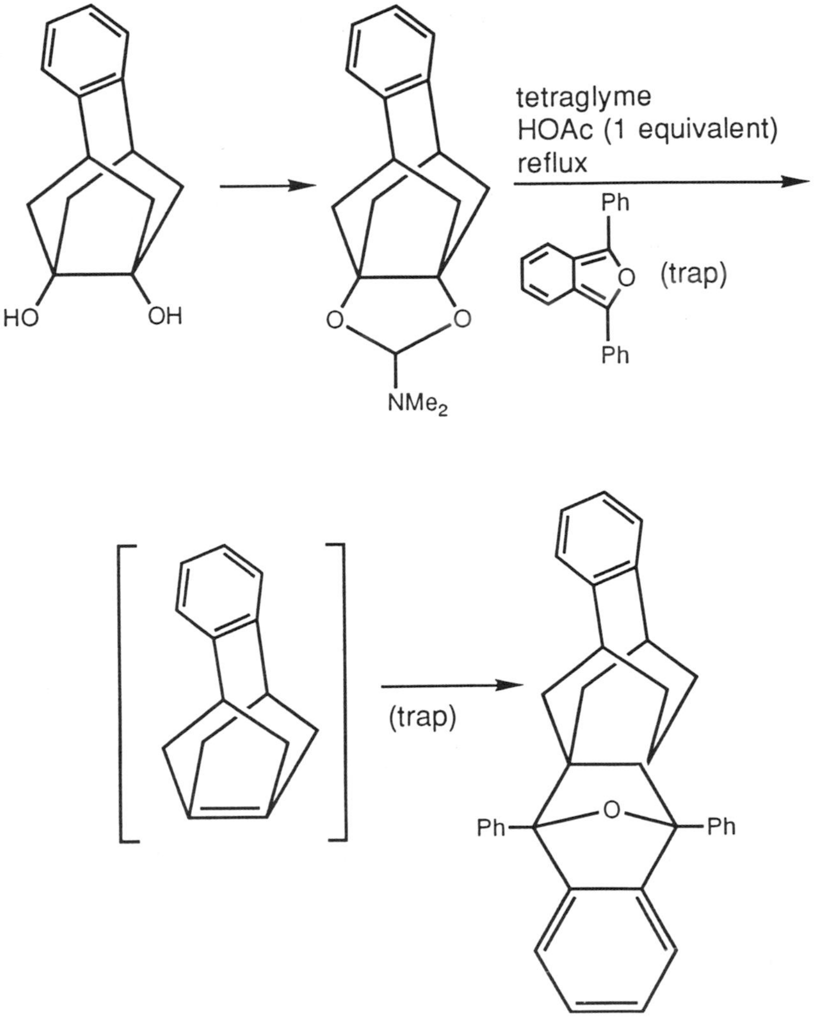

(3) By organolithium promoted reductive dehalogenation of vicinal dihalopolycyclic compounds. Reaction of 4-bromo-5-iodohomocubane with *n*-BuLi results in dehalogenation, thereby affording homocub-4(5)-ene. This highly strained intermediate reacts further with *n*-BuLi to afford 4-*n*-butyl-5-lithiohomocubane which subsequently under-

goes Li-I exchange to give 4-*n*-butyl-5-iodohomocubane
(D. A. Hrovat and W. T. Borden, J. Am. Chem. Soc., 1988,
110 (7229).

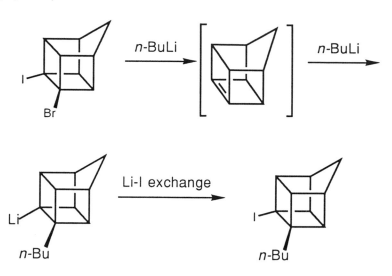

(4) By base promoted dehalogenation of halopolycyclic
compounds. Treatment of 1-chlorotricyclo[4.1.0.0^{2,7}]hep-
tane with excess *n*-BuLi results in dehydrohalogenation,
thereby affording tricyclo[4.1.0.0^{2,7}]hept-1(7)ene, a severely bent bicyclo[1.1.0]but-1(3)-ene. In the absence of a
trapping agent, this highly strained, pyramidalized alkene
reacts further with *n*-BuLi, ultimately affording 1-*n*-
butyltricyclo[4.1.0.0^{2,7}]heptane. The intermediate alkene
can be trapped by addition of D$_2$O, thereby producing 1-*n*-
butyl-7-deuteriotricyclo[4.1.0.0^{2,7}]heptane (G. Szeimies, J.
Harnisch, and O. Baumgärtel, J. Am. Chem. Soc., 1977, 99,
5183).

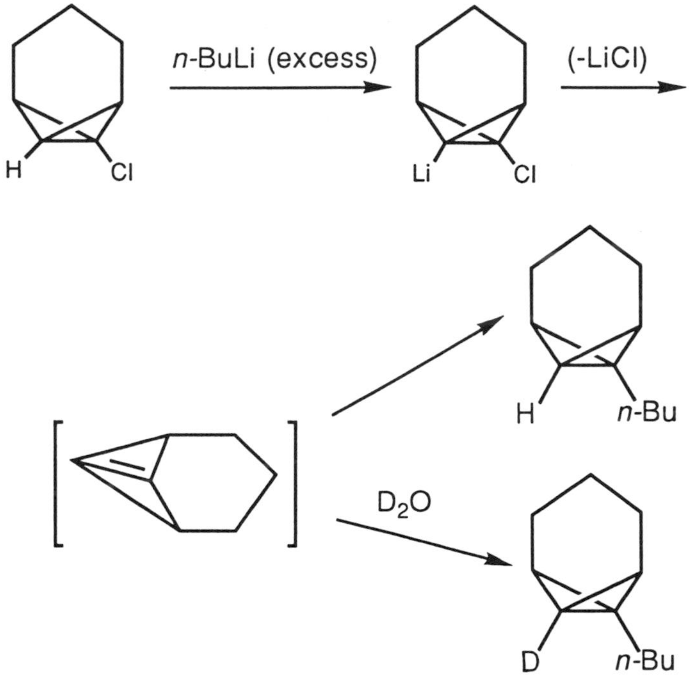

Greene and coworkers have employed base promoted dehalogenation of 9,10-dibromodianthracene (the photo-dimer of 9-bromoanthracene) to synthesize 9,9',10,10'-tetradehydrodianthracene, mp 388 °C (sealed, evacuated capillary; F. D. Greene et al., J. Am. Chem. Soc., 1974, 96, 4342). Some years later, Haddon performed theoretical calculations on several pyramidalized alkene systems. On the basis of his results, he declared 9,9',10,10'-tetradehydrodianthracene to be "the most highly pyramidalized alkene known" at that time (R. C. Haddon, J. Am. Chem. Soc., 1990, 112, 3385).

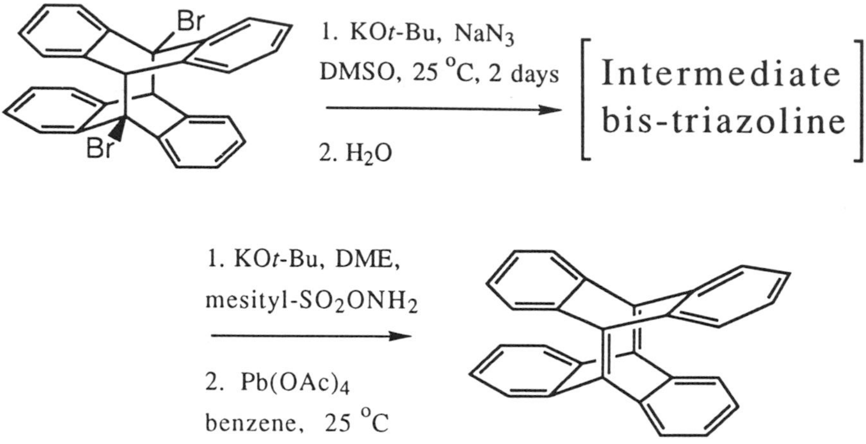

1. KO*t*-Bu, NaN$_3$
   DMSO, 25 °C, 2 days

2. H$_2$O

$$\left[\begin{array}{c}\text{Intermediate}\\\text{bis-triazoline}\end{array}\right]$$

1. KO*t*-Bu, DME,
   mesityl-SO$_2$ONH$_2$

2. Pb(OAc)$_4$
   benzene, 25 °C

This unusual diene is a member of a class of compounds known as "superphanes". The simplest member of this series, tricyclo[4.2.2.2^{2,5}]dodeca-1,5-diene (i.e., the "debenzo" analog of 9,9',10,10'-tetradehydrodianthracene), has been prepared via *in situ* dimerization of bicyclo-[2.2.0]hex-1(4)-ene followed by spontaneous [2 + 2] cycloreversion (K. B. Wiberg et al., J. Am. Chem. Soc., 1984, 106, 2194, 2200).

(5) By fluoride ion promoted dehalosilylation of vicinal halotrimethylsilyl substituted polycyclic compounds. Tricyclo[4.1.0.0^{2,7}]hept-1(7)-ene has also been prepared by Szeimies and coworkers, who heated 1-bromo-7-(trimethylsilyl)tricyclo[4.1.0.0^{2,7}]heptane with KF in DMSO at 55 ºC for 24 h. The intermediate pyramidalized alkene was trapped *in situ* by 9-methoxyanthracene (G. Szeimies et al., Angew. Chem., Int. Ed. Engl, 1981, 20, 877; Chem. Ber., 1983, 116, 2285).

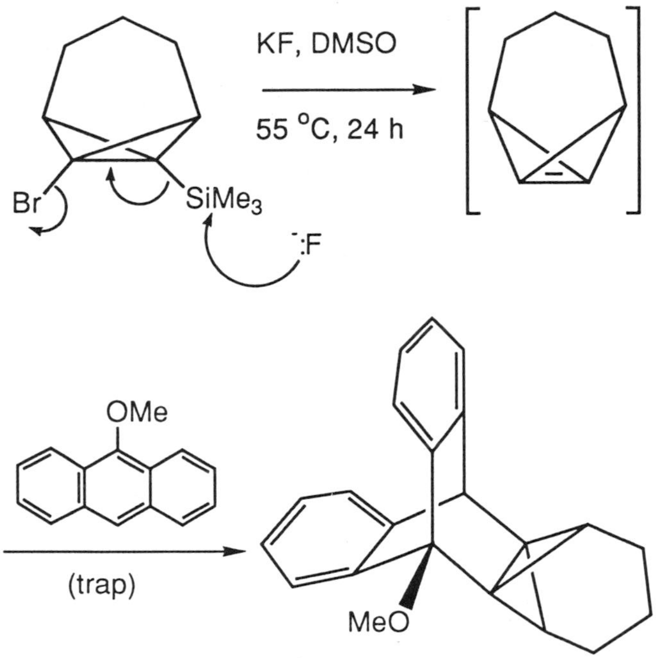

(6) By titanium promoted intramolecular reductive coupling of polycyclic diketones (J.-M. Pons and M. Santelli, Tetrahedron, 1988, 44, 4295; J. E. McMurry, Chem. Rev., 1989, 89, 1513; C. Betschart and D. Seebach, Chimia, 1989, 43, 39; D. Lenoir, Synthesis, 1989, 883). McMurry and coworkers have utilized this approach to synthesize tetracyclo[8.2.2.2^{2,5}.2^{6,9}]octadeca-1,5,9-triene (J. E. McMurry et al., 1984, 106, 5018; 1986, 108, 2932).

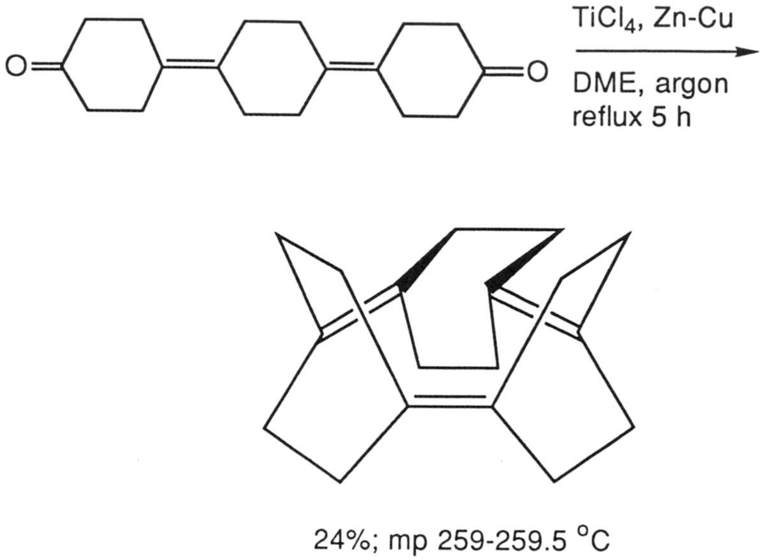

310

TiCl$_4$, Zn-Cu
$\xrightarrow{\hspace{2cm}}$
DME, argon
reflux 5 h

24%; mp 259-259.5 °C

(7) By Diels-Alder reactions of substituted "isodicyclo-pentadienes". Paquette and coworkers have utilized this approach to synthesize "*syn*-sesquinorbornenes", a new class of highly pyramidalized alkenes. The example shown below is the first synthesis of a "shelf-stable syn-sesqui-norbornatriene". X-ray crystallographic analysis reveals that the central C=C double bond in this molecule possesses a "pyramidalization angle" of 32.4° (L. A. Paquette, C.-C. Shen, and J. A. Krause, J. Am. Chem. Soc., 1989, 111, 2351).

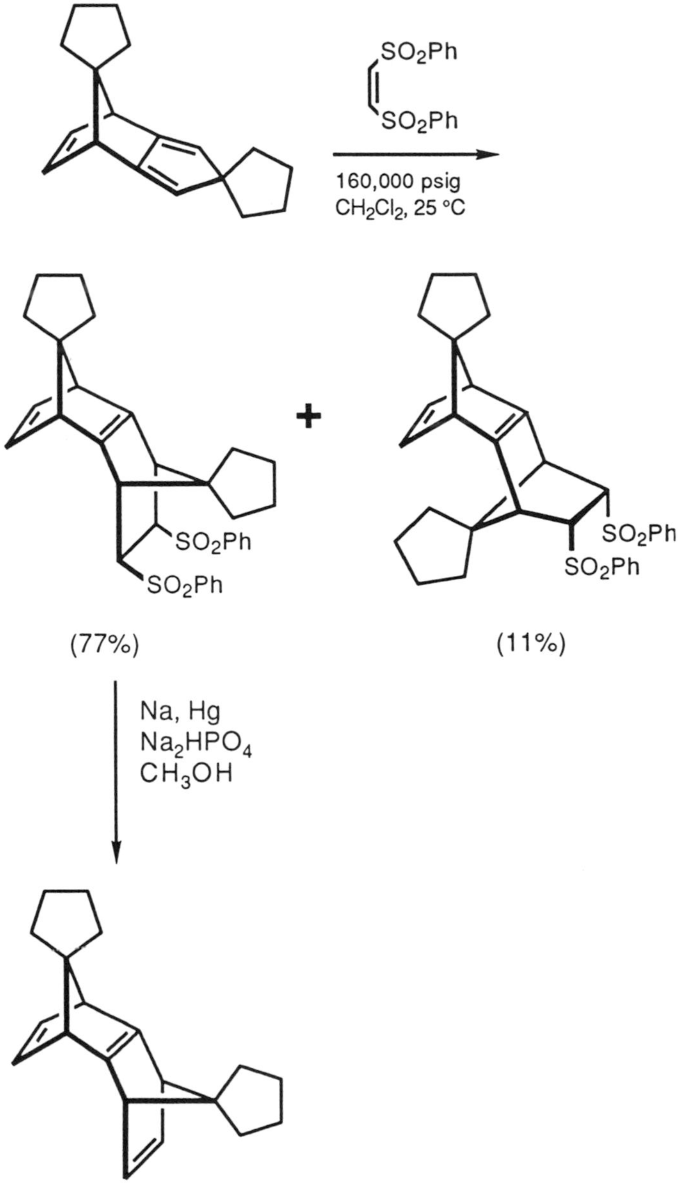

(77%)

+

(11%)

Na, Hg
Na$_2$HPO$_4$
CH$_3$OH

(87%, mp 86 °C)

4. Special Topics

(a) Propellanes

(i) Introduction

The name "propellane" was coined by Ginsberg with reference to tricyclic compounds which possess one zero bridge and three nonzero bridges [(a) Propellanes-structure and reactions by D. Ginsburg, (Verlag Chemie: Weinheim, 1975). (b) Propellanes-structure and reactions, sequel I by D. Ginsburg (Department of Chemistry, Technion: Haifa, 1980). (c) Propellanes-structure and reactions, sequel II by D. Ginsburg (Department of Chemistry, Technion, Haifa, 1985). (d) [m.n.1]Propellanes by D. Ginsburg (in Z. Rappoport, Editor, The chemistry of the cyclopropyl group, Wiley: Chichester, 1987, Chapter 20, pp. 1193-1221)].

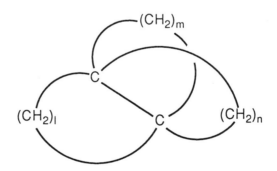

There is considerable theoretical interest in compounds of this type due to the fact that the two bridgehead carbon atoms in small-ring propellanes possess inverted geometries. The synthesis and chemistry of small-ring propellanes has been reviewed recently (K. B. Wiberg, Acc. Chem. Res., 1984, 17, 379; Chem. Rev., 1989, 89, 975).

The smallest member of this series of compounds, i.e., [1.1.1]propellane, was synthesized by Wiberg and Walker in 1982 (K. B. Wiberg and F. H. Walker, J. Am. Chem. Soc., 1982, 104, 5239) and subsequently by Szeimies and coworkers, who at that time also prepared bridged [1.1.1]propellanes (i.e., 2,4-bridged tricyclo[1.1.1.0^{1,3}]pentanes; G. Szeimies et al., J. Am. Chem. Soc., 1985, 107, 6410). The question of the nature of the central C(1)-C(3) bond in [1.1.1]propellane has been the subject of numerous experimental [(a) N. S. Zefirov et al., J. Org. Chem., USSR, 1990, 26, 2002; (b) K. B. Wiberg and S. T. Waddell, J. Am. Chem. Soc., 1990, 112, 2194] and theoretical investigations [(a) M. D. Newton and J. M. Schulman, J. Am. Chem. Soc., 1972, 94, 773. (b) W.-D. Stohrer and R. Hoffmann, J. Am. Chem. Soc., 1972, 94, 779. (c) J. E. Jackson and L. C. Allen, J. Am. Chem. Soc., 1984, 106, 591. (d) R. F. Messmer and P. A. Schultz, J. Am. Chem. Soc., 1986, 108, 7407. (e) K. B. Wiberg et al., J. Am. Chem. Soc., 1987, 109, 985, 1001. (f) D. Feller and E. R. Davidson, J. Am. Chem. Soc., 1987, 109, 4133; (g) P. Seiler et al., Helv. Chim. Acta, 1988, 71, 2100.] The molecular structure of [1.1.1]propellane has been determined by X-ray crystallographic methods at 138 K (P. Seiler, Helv. Chim. Acta, 1990, 73, 1574).

(ii) Synthesis of propellanes

(1) By inter- and intramolecular carbene additions to the C=C double bond in bicyclic alkenes. [m.n.1]Propellanes have been prepared via addition of carbenes (or carbenoids), :CXY, to the C=C double bond in bicyclic alkenes in the manner shown below.

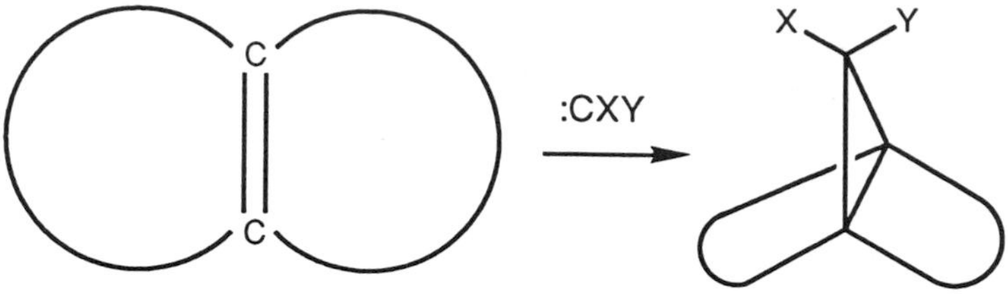

314

Thus, :CCl$_2$ adds to the more highly strained tetrasubstituted C=C double bond in bicyclo[4.2.0]hexa-1(6),3-diene to afford 9,9-dichlorotricyclo[4.2.1.0^{1,6}]non-3-ene (P. Warner and R. La Rose, Tetrahedron Lett., 1971, 2141).

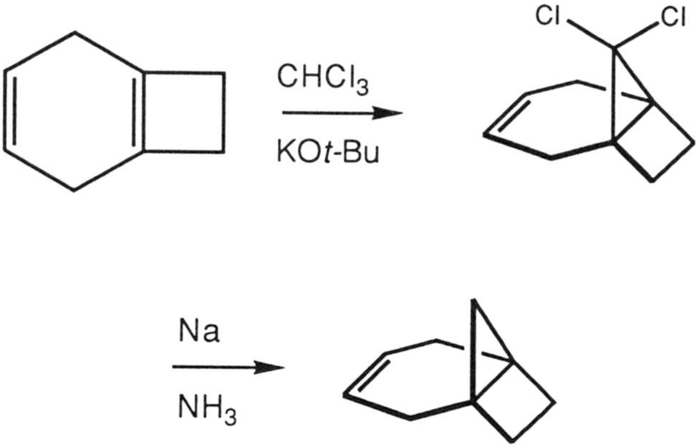

Similarly, 2,4-methano-2,4-dehydroadamantane, a substituted [3.1.1]propellane, has been synthesized by intramolecular addition of 4-methylene-2-adamantylidene (K. Mlinaric-Majerski and Z. Majerski, J. Am. Chem. Soc., 1980, 102, 1418).

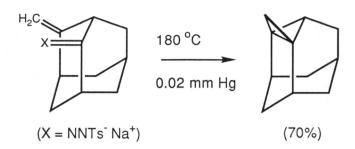

H_2C

X⚌

180 °C

0.02 mm Hg

(X = NNTs⁻ Na⁺)   (70%)

This same approach was used subsequently to synthesize 2,6-methano-2,6-dehydronorbornane, an exceptionally highly strained [3.1.1]propellane (V. Vincovic and Z. Majerski, J. Am. Chem. Soc., 1982, 104, 4027).

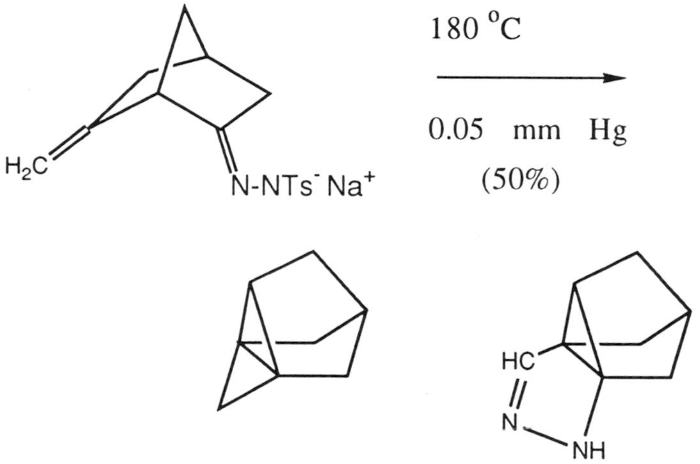

180 °C

0.05 mm Hg

(50%)

H_2C

N-NTs⁻ Na⁺

HC

N

NH

product ratio: 2   :   1

A novel class of structurally complex propellanes has been synthesized via addition of :CCl₂ to the C=C double bonds of "bis-secododecahedradiene" (H. Prinzbach et al., Angew. Chem., Int. Ed. Engl., 1987, 26, 455).

316

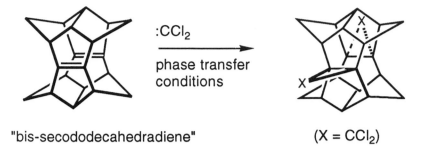

:CCl₂

phase transfer
conditions

(X = CCl₂)

"bis-secododecahedradiene"

(2) By organolithium promoted reductive dehalogenation of appropriately constructed 1,4-dihalobicyclo[2.m.n]-alkanes. The general reaction is shown below:

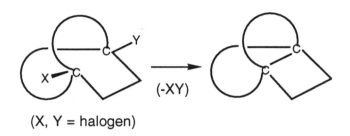

(X, Y = halogen)

Thus, reductive dehalogenation of 1,5-diiodobicyclo[3.2.1]-octane with *t*-BuLi gives the corresponding propellane, tricyclo[3.2.1.0^{1,5}]octane, in 59% yield (K. B. Wiberg, W. E. Pratt, and W. F. Bailey, J. Am. Chem. Soc., 1977, <u>99</u>, 2297).

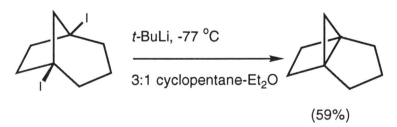

t-BuLi, -77 °C

3:1 cyclopentane-Et₂O

(59%)

(3) By intramolecular transannular carbene or carbenoid insertion into a carbon-hydrogen bond. Paquette and coworkers successfully employed this approach to effect cyclopropanation of dodecahedrane, thereby affording a complex polycarbocyclic [3.3.1]propellane (L. A. Paquette, T. Kobayashi, and J. C. Gallucci, J. Am. Chem. Soc., 1988, 110, 1305).

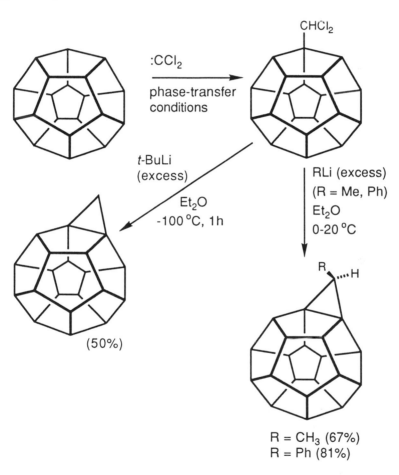

(50%)

R = CH$_3$ (67%)
R = Ph (81%)

(4) By dimerization of highly strained bicyclic alkenes. Reductive elimination of halogen occurs when 1,5-dibromobicyclo[3.1.0]hexane is reacted with *t*-BuLi in Et$_2$O at -78 °C, thereby affording a dimer (which, in this instance, is a bis-propellane) along with 1-*t*-butyl-bicyclo[3.1.0]-hexane (K. Wiberg and G. Bonneville, Tetrahedron Lett., 1982, 23, 5385).

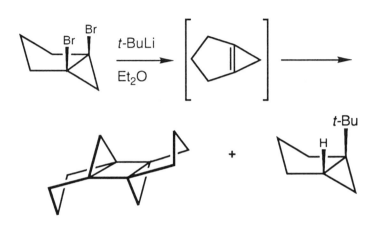

(5) By Diels-Alder cycloadditions of dienes to highly strained bicyclic alkenes. Bicyclo[2.2.0]hex-1(4)-ene is a highly reactive dienophile which can be trapped by various dienes to afford [m.2.2]propellanes (K. B. Wiberg et al., J. Am. Chem. Soc., 1971, 93, 246; Tetrahedron, 1986, 42, 1895). (The ability of bicyclo[2.2.0]hex-1(4)-ene to undergo *in situ* dimerization to afford the corresponding "superphane" has been noted earlier; see Section 3.b.4).

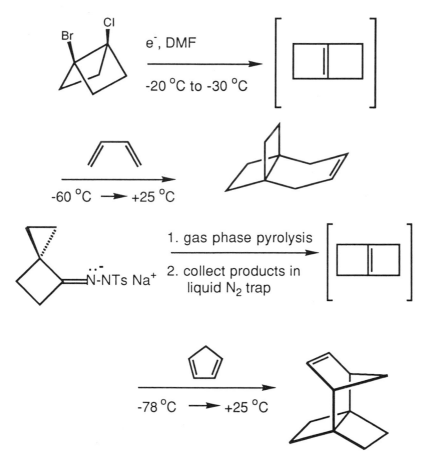

More recently, Eaton and Maggini have generated 1,2-dehydrocubane ("cubene") via reaction of 1,2-diiodocubane with *t*-BuLi. This unusual, highly pyramidalized alkene was trapped *in situ* by using a polycyclic diene, thereby generating a polycyclic [4.2.2]propellane (P. E. Eaton and M. Maggini, J. Am. Chem. Soc., 1988, 110, 7230).

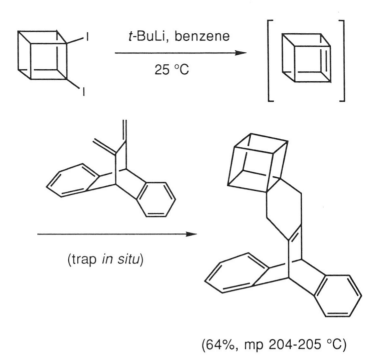

(64%, mp 204-205 °C)

(6) By addition of diazoalkanes to the C=C double bond of polycarbocyclic alkenes followed by photodenitrogenation of the resulting cycloadduct. Cyclopropanation of the central C=C double bond in a substituted *syn*-sesquinorbornadiene followed by photodenitrogenation of the resulting cycloadduct afforded the corresponding propellane in good overall yield (L. A. Paquette, M. A. Kesselmayer, and R. D. Rogers, J. Am. Chem. Soc., 1990, 112, 284).

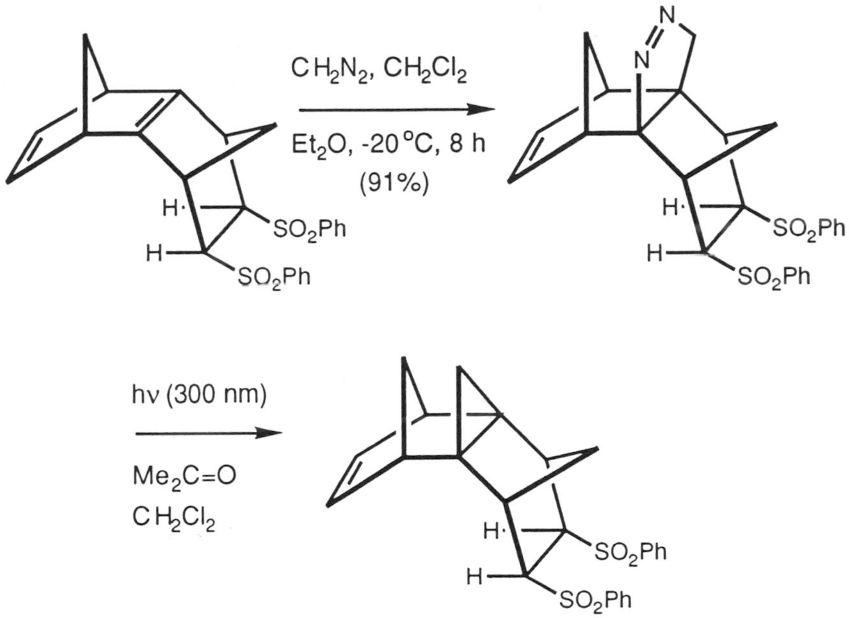

(7) By carbocationic rearrangements. When treated with HSO_3F at -78 °C, benzyl carbinols undergo carbocationic skeletal rearrangement to afford bridged polycarbocyclic compounds (J. M. Coxon et al., J. Org. Chem. 1989, 54, 2542).

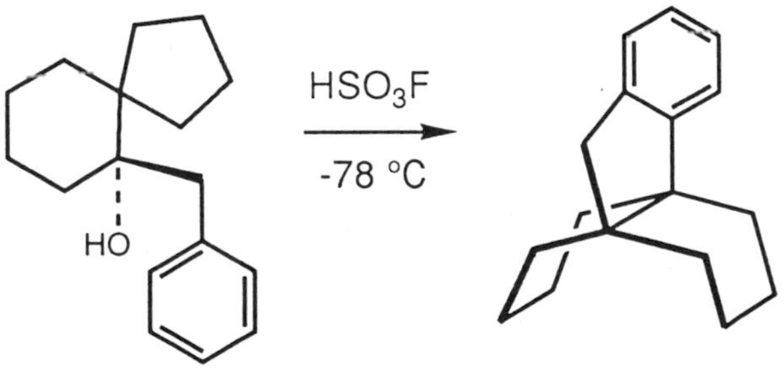

(80%, mp 41-42.5 °C)

322

Treatment of functionalized dispiro[3.0.3.3]undecanes with TsOH-benzene results in cascade rearrangements which lead to the corresponding [3.3.3]propellane (L. Fitjer, A. Kanschik, and M. Majewski, Tetrahedron Lett., 1985, 26, 4277).

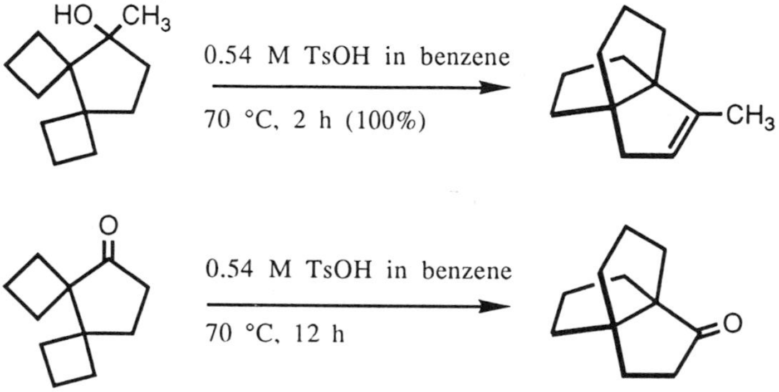

Extension of this methodology to the corresponding, substituted tetraspiroheptadecanes provides access to bridged pentacyclic compounds (L. Fitjer and U. Quabeck, Angew. Chem., Int. Ed. Engl., 1987, 26, 1023; 1989, 28, 94).

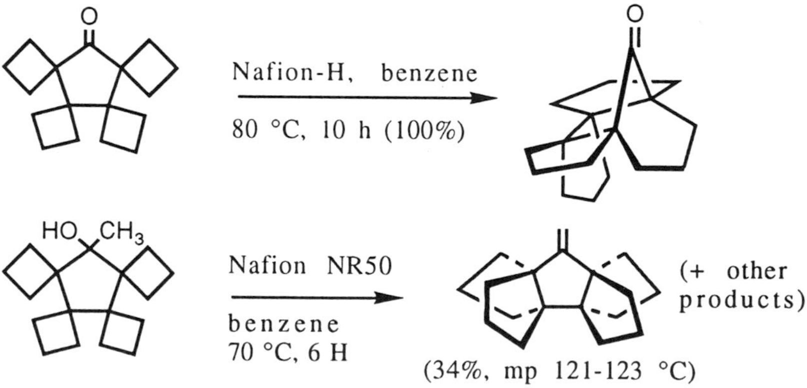

(b) the tricyclo[3.3.1.1³·⁷]decane (adamantane) group

(i) Mechanism and structure

Thermodynamically driven cationic rearrangements of polycarbocyclic hydrocarbons to adamantanoid systems (Rodd's C. C. C., 2nd Edn., Suppl. to Vol. II C, pp. 47-48) have been subjected to computer-assisted graph theoretical analysis (P. von R. Schleyer et al., J. Am. Chem. Soc., 1975, <u>97</u>, 743; 1977, <u>99</u>, 5361). The first X-ray crystal structure of a salt of a stable carbocation was reported by Laube for 3,5,7-trimethyl-1-adamantyl undecafluorodiantimonate (T. Laube, Angew. Chem., Int. Ed. Engl., 1986, <u>25</u>, 349). A four-center, two-electron dication which possesses tetrahedral topology is produced when 5,7-difluoro-1,3-dehydroadamantane is dissolved in SbF_5-SO_2ClF at -80 °C (P. von R. Schleyer et al., Angew. Chem., Int. Ed. Engl., 1987, <u>26</u>, 761). The results of MINDO/3 calculations suggest that this "nonclassical" four-center dication is 47 kcal/mol more stable than the "classical" 1,3-adamantanediyl dication and 1,3-dehydroadamantane. Schleyer and coworkers suggested that the unusually high stability of this dication is due to the fact that it possesses "three-dimensional aromaticity".

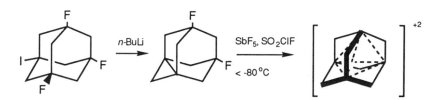

Substituted adamantanes have been utilized extensively as substrates for the study of facial selectivity in a number of reactions which take place at an unsaturated center, C=X, which is situated at the 2-position. Examples in this regard include: (i) nucleophilic attack upon the carbonyl group in 5-substituted 2-adamantanones (W. J. le Noble et al., J. Am. Chem. Soc., 1986, 108, 1598; 1987, 109, 7239; J. Org Chem., 1988, 53, 5155; 1989, 54, 3836); (ii) electrophilic attack upon a C=C group in 5-substituted-2-methyleneadamantanes (S. Srivastava and W. J. le Noble, J. Am. Chem. Soc., 1987, 109, 5874); (iii) capture of radicals generated via Hunsdiecker reactions of E- and Z-2-(5-phenyl)adamantanecarboxylic acids (V. R. Bodepudi and W. J. le Noble, J. Org. Chem., 1991, 56, 2001); (iv) [3.3]sigmatropic rearrangements of allyl vinyl sulfoxides (A. Mukherjee, E. M. Schulman, and W. J. le Noble, J. Org. Chem., 1992, 57, 3120) and in the oxy-Cope rearrangement (M.-h. Lin and W. J. le Noble, J. Org. Chem., 1989, 54, 997; 1990, 55, 3597). The conformational rigidity of the adamantane ring system, together with the fact that the two faces of the C=X group in these 2,5-disubstituted adamantanes are equivalent sterically render them particularly well suited as substrates for such mechanistic studies.

(ii) Syntheses of adamantanes and related compounds

(1) By Lewis acid promoted rearrangment of polycyclic hydrocarbon precursors to adamantanoid systems. Higher adamantalogues of adamantane have been synthesized by using this approach. Thus, diamantane ("congressane", Rodd's C. C. C., 2nd Edn., Suppl. to Vol. II C, p. 51) has been prepared in 60-62% yield via rearrangement of "tetrahydro-Binor-S" by using $AlBr_3$-CS_2 (or $AlBr_3$-

cyclohexane) sludge (T. M. Gund, W. Thielecke, and P. von
R. Schleyer, Org. Synth., 1973, 53, 30).

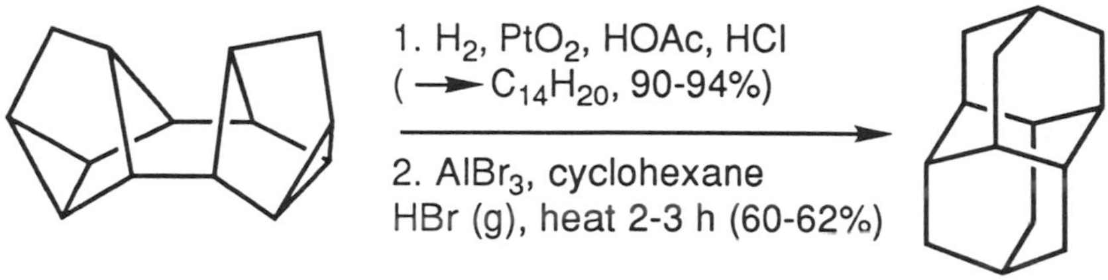

1. H_2, PtO_2, HOAc, HCl
($\rightarrow C_{14}H_{20}$, 90-94%)

2. $AlBr_3$, cyclohexane
HBr (g), heat 2-3 h (60-62%)

"Binor-S"

Diamantane
("Congressane")

Gas phase thermal rearrangement of a mixture of
polycyclic $C_{22}H_{28}$ dienes over Pt-impregnated silica gel at
350 °C affords *anti*-tetramantane, mp 173.5-174.0 °C in
10% yield (M. A. McKervey et al., J. Chem. Soc., Chem.
Commun., 1976, 893; J. Am. Chem. Soc., 1978, 100, 906).

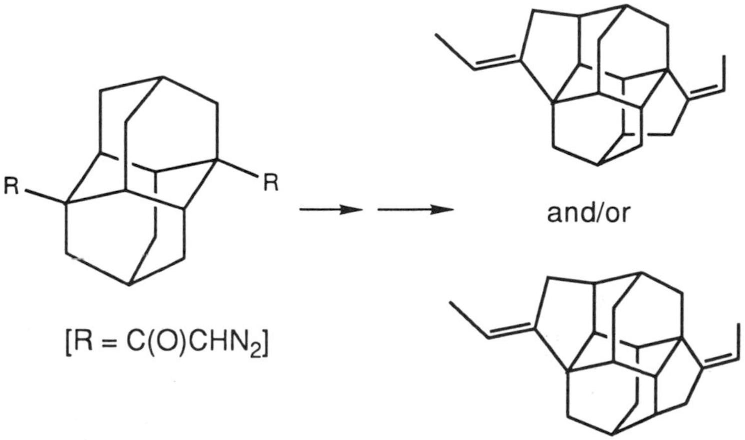

and/or

[R = C(O)CHN$_2$]

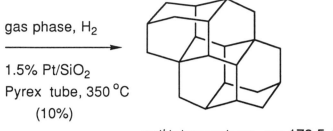

gas phase, H_2

⟶

1.5% Pt/SiO_2
Pyrex tube, 350 °C
(10%)

anti-tetramantane, mp 173.5-174.0 °C

(2) By reductive isomerization of strained unsaturated polycyclic compounds. Olah and coworkers have used $NaBH_4-CF_3SO_3H$ at -30 °C to +25 °C to promote reductive isomerization of the mixture of $C_{18}H_{22}$ heptacyclooctadecenes which results from "C_4-elaboration of Binor-S" (M. A. McKervey et al., J. Org. Chem., 1980, 45, 4954). Application of this single step procedure afforded triamantane in 92% yield (G. A. Olah et al., J. Org. Chem., 1989, 54, 1450).

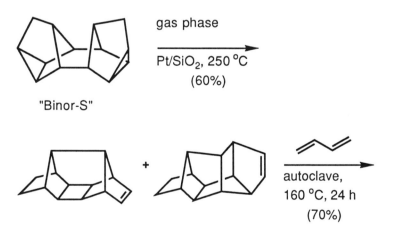

"Binor-S"

gas phase

⟶

Pt/SiO_2, 250 °C
(60%)

+

autoclave,
160 °C, 24 h
(70%)

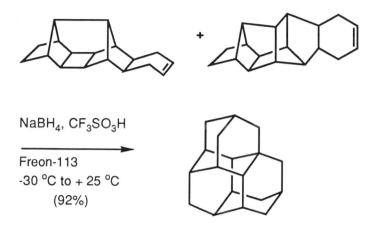

triamantane

(3) By a two-step fragmentation transannular cyclization reaction sequence. When heated with zinc, a dimethylformamide (DMF) solution of 1,4-dibromoadamantane suffers loss of bromine with concomitant fragmentation, thereby affording 3-methylenebicyclo-[3.3.1]non-6-ene (57%, mp 55-57 °C). Reaction of this bicyclic diene with I_2-CCl_4 results in addition with concomitant intramolecular cyclization, thereby affording *trans*-1,4-diiodoadamantane (25%, mp 122-123 °C; A. G. Yurchenko et al., J. Org. Chem. USSR, 1974, 10, 1139).

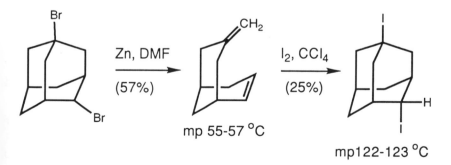

This same approach has been used to synthesize more highly functionalized adamantanes (A. G. Yurchenko et al., J. Org. Chem. USSR, 1986, 22, 407).

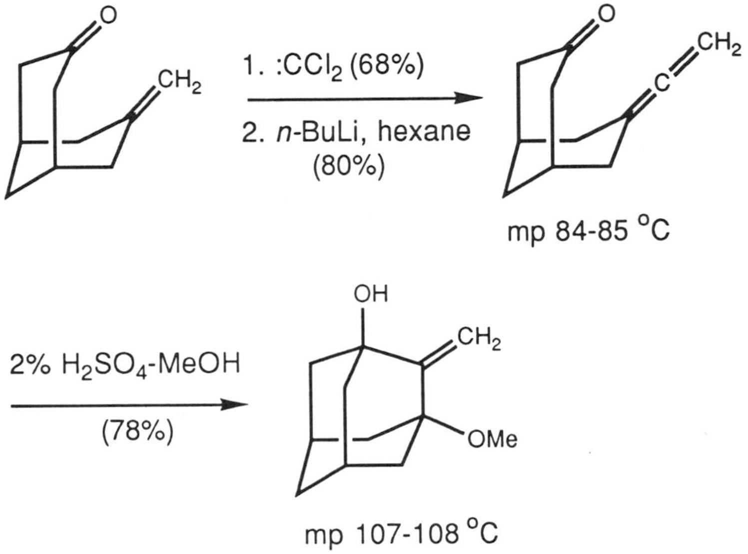

1. :CCl$_2$ (68%)

2. n-BuLi, hexane (80%)

mp 84-85 °C

2% H$_2$SO$_4$-MeOH

(78%)

mp 107-108 °C

The following reviews deal with the synthesis and chemistry of adamantanoid systems:

Recent developments in the chemistry of adamantane and related polycyclic hydrocarbons by R. C. Bingham and P. von R. Schleyer (Top. Curr. Chem., 1971, 18, 1-102). Adamantane rearrangements by M. A. McKervey (Chem. Soc. Rev., 1974, 3, 479). Adamantane: the chemistry of diamondoid molecules by R. C. Fort, Jr. (Marcel Dekker: New York, 1976). Structures and stereochemistry of diamondoid polycycloalkanes by S. Hala (Chem. Listy, 1977, 71, 8). Synthetic approaches to large diamondoid hydrocarbons by M. A. McKervey, Tetrahedron, 1980, 36, 971. My thirty years in hydrocarbon cages: from adamantane to dodecahedrane by P. von R. Schleyer (in G. A. Olah, Editor,

Cage Hydrocarbons, Wiley-Interscience, New York, 1990, pp. 1-38). Catalytic routes to adamantane and its homologues by M. A. McKervey and J. J. Rooney (in G. A. Olah, Editor, Cage Hydrocarbons, Wiley-Interscience, New York, 1990, pp. 39-64). The superacid route to 1-adamantyl cation by T. S. Sorensen and S. M. Whitworth (in G. A. Olah, Editor, Cage Hydrocarbons, Wiley-Interscience, New York, 1990, pp. 65-102). Fragmentation and transannular cyclization routes to cage hydrocarbons by A. G. Yurchenko (in G. A. Olah, Editor, Cage Hydocarbons, Wiley-Interscience, New York, 1990, pp. 155-188). A new approach to the adamantane rearrangements by C. Ganter (in E. Osawa and Y. Yonemitsu, Editors, Chemistry of Three-Dimensional Polycyclic Molecules, VCH, New York, 1992, In press).

5. Acknowledgment

This review was written in part during the tenure (Summer, 1992) of an Erskine Fellowship in the Department of Chemistry, University of Canterbury, Christchurch, New Zealand. I am grateful to the University of Canterbury for the award of this fellowship and to my host, Professor James M. Coxon, and his colleagues in the Department of Chemistry for their having kindly extended warm hospitality and the use of their excellent facilities to me during this period. In particular, I thank Professor Coxon and Dr. Peter J. Steel for having constructively criticized the manuscript of this review.. Our work on the synthesis and chemistry of novel polycarbocyclic compounds has received generous financial support from the Air Force Office of Scientific Research, the Office of Naval Research, and the Robert A. Welch Foundation (Grant B-963). Finally, I thank my wife, Dr. Nancy Wu Marchand, for her love and continuing support.

Second Supplements to the 2nd Edition of Rodd's Chemistry of Carbon Compounds, Vol. II B(Partial), C, D and E, edited by M. Sainsbury
© 1994 Elsevier Science B.V. All rights reserved.

Chapter 12

BICARBOCYCLIC NATURAL PRODUCTS

R. LIVINGSTONE

1. Introduction

Recent advances in the chemistry of bicarbocyclic natural products are concerned in the main with the preparation and properties of some of their derivatives. The order of compounds described and the nomenclature adopted, follow the pattern utilized in the second edition, but where convenient for further reference names are used from the original publication.

2. The thujane group

(a) Hydrocarbons

(i) Thuj-3-ene (α-thujene)
Selenium dioxide oxidation of thuj-3-ene (1) gives 4-formyl-1-
-isopropylbicyclo[3.1.0]hex-3-ene and chromium trioxide affords umbellulone, also obtained by the autoxidation of (1) in the presence of copper or cobalt abietate [A.N.C. Catalan and J.A. Retamar, An. Acad. Bras. Cienc., 1972, 44 (Suppl.), 360; Essenze Deriv. Agrum., 1974, 44, 35]. Photochemical hydroperoxidation-isomerization of thuj-3-ene gives *trans*-4-hydroperoxythuj-
-2-ene (Lion Corp., Japan P. 81 75,472, 1981).

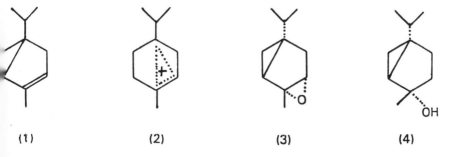

(1)   (2)   (3)   (4)

332

Acid catalyzed hydration of thuj-3-ene and thuj-4(10)-ene (5) proceeds through the same intermediate (2) to yield a mixture of *a*-terpinene, γ--terpinene, terpinolene, and terpinen-4-ol (M.A. Cooper *et al.*, J. Chem. Soc., Perkin II, 1973, 665). For results of the based catalyzed isomerisation see A. Ferro and Y.R. Naves (Helv., 1974, 57, 1152).

Photochemical epoxidation of thuj-3-ene in the presence of *a*-diketones, with the introduction of oxygen furnishes *trans*-3,4-epoxythujane (3), which on LiAlH$_4$ reduction yields *trans*-thuj-4(10)-hydrate (*trans*-thujan-4-ol) (4) (Lion Corp., Japan P. 80 51,081, 1980; S. Shinpo *et al.*, *ibid.*, 80 51,030, 1980), also obtained by the catalytic reduction, using Raney nickel of the *trans*-4--hydroperoxide obtained by the photochemical oxidation of thuj-3-ene (T.Shimpo *et al.*, *ibid.*, 80 28,965, 1980).

(ii) Thuj-4(10)-ene (sabinene)

Treatment of thuj-4(10)-ene (5) with 3-chloroperbenzoic acid gives 4,10--epoxythujane (spiro[5-isopropylbicyclo[3.1.0]hexan-2,2'-oxirane]) (6) (M. Higo *et al.*, *ibid.*, 79 66,663, 1979). Epoxides (6 and 7) are obtained by the photochemical epoxidation of thuj-4(10)-ene in the presence of *a*-diketones (Lion Corp., *ibid.*, 80 51,082, 1980). Thuj-4(10)-ene along with hedycaryol have been shown by chromatography and mass spectroscopy to be the main components of the seed oil from *Thujopsis dolabrata* (S. Hasegawa and Y. Hirose, Phytochem., 1981, 20, 508). Thuj-4(10)-ene on hydrozirconation followed by oxidation with Me$_3$COOH gives a mixture of thujan-10-ol and isothujan-10-ol (M.S. Miftakhov, N.N. Sidorov, and G.A. Tolstikov, Izv. Akad. Nauk SSSR, Ser. Khim., 1979, 2748).

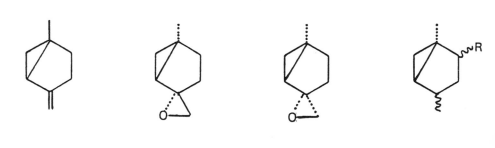

(5) (6) (7)

(8) R = OH
(9) R = PhCH = CHCO$_2$

(b) Alcohols

(i) Thujan-3-ol (thujyl alcohol)

The four isomeric thujan-4-ols (8) have been prepared from (-)-isothujone, isolated from *Thuja occidertalis* [C.A.N. Catalan, Arch. Bioquim., Quim. Farm., 1977 (Pub. 1978), 20, 25], cinnamates (V. Hach and H.G. Higson, Can. Pat. 914,214, 1972; U.S. Pat. 3,708,521, 1973). The conformation and reactivity of (-)-neoisothujan-3-ol, (-)-isothujan-3-ol, (+)-thujan-3-ol, and (+)-neothujan-3-ol have been studied by lanthanide shift reagent-induced ^{1}H NMR, rate of chromium trioxide oxidation, and rate of acetic acid-pyridine acetylation; shifts induced by tris(2,2,6,6-tetramethyl-3,5--heptanedionato)europium indicated a boat-like conformation (V. Hach, J. Org. Chem., 1977, 42, 1616). (-)-Neoisothujan-3-ol, (+)-thujan-3-ol, (+)--trans-thuj-4(10)-en-3-ol (sabinol), and (-)-cis-thuj-4(10)-en-3-ol ^{1}H NMR spectral data (T. Norin, S. Stromberg, and M. Weber, Acta Chem. Scand., 1973, 27, 1579). Heterolysis of the C-O bond of the thujan-3-ols or their derivatives can give rise to either an undelocalized cation or a trishomocyclopropyl cation, depending on the alignment of the leaving group. In either case, a 1,2-hydride shift converts the ion into a bishomocyclopropyl cation (2) (C.M. Holden and D. Whittaker, J. Chem. Soc., Perkin II, 1976, 1345). Reaction of the isomeric thujols with FSO_3H-SO_2 at -78°C has been shown by deuterium studies to involve at least two routes, giving the cyclopentenium ion (10), which subsequently rearranges to ion (11) (J.C. Rees and D. Whittaker, *ibid.*, 1981, 953). For the hydroalumination of bicyclic monoterpenes, including formation of isomeric thujanols see V.P. Yur'ev *et al.*, (Zh. Obshch. Khim., 1974, 44, 2084).

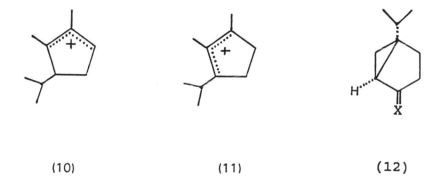

(10)   (11)   (12)

(ii) thujan-10-ol

Hydroalumination of thuj-4(10)-ene followed by oxidation yields isomeric thujan-10-ol (12) (X = equatorial-CH_2OH, axial-H; axial-CH_2OH, equatorial-H) (A.V. Kuchin *et al.*, *ibid.*, 1979, <u>49</u>, 1567).

(c) Ketones

(i) Thujan-3-one (thujone)

Various names have been used for the diastereoisomers of thujan-3-one, *viz.*, (-)-*a*-thujone and (+)-ß-thujone, more generally called (-)-thujone (13) and (+)-isothujone (14). The latter two names are used in this section. A mixture of (-)-thujone and (+)-isothujone on sequential treatment with HCHO/HCl, Jones oxidation, and decarboxylation is epimerized to give (+)- -isothujone (90% from mixture) (C.H. Brieskorn and W. Schwack, Tetrahedron Letters, 1980, <u>21</u>, 255; Chung Kyu Ryu and Sun Hye Park, Nonchong-Han'guk Saenghwal Kwahak Yonguwon, 1988, <u>42</u>, 129). For isomerization under basic and acidic conditions see V. Hach, R.W. Lockhart, and D.M Cartlidge (U.S. Pat. 3,709,943, 1973) and C.A.N. Catalan and J.A. Retamar [Rev. Fac. Ing. Quim. (Univ. Nac. Litoral), 1978 (Pub. 1979), <u>43</u>, 53], respectively.

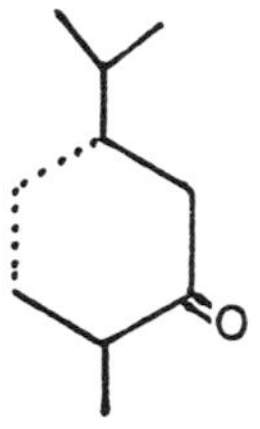

(13) 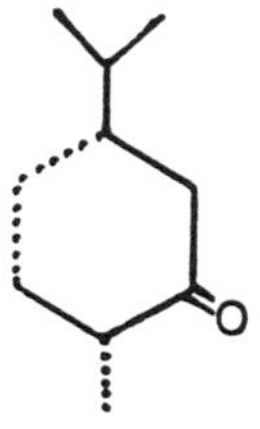   (14)

The stereochemistry of the reduction of some bicyclo[3.1.0]hexanone analogues of thujone has been discussed (M. Walkowicz and B. Laczynska--Przepiorka, Pol. J. Chem., 1985, <u>59</u>, 801). Addition of butyl- and phenyl--lithium to (-)-thujone and (+)-isothujone produces optically active tertiary alcohols (N.S. Nudelman, Z. Gatto, and L. Bohé, J. Org. Chem., 1984, <u>49</u>, 1540). Preparation of 4-methylthujone (C.H. Brieskorn, Ber., 1981, <u>114</u>, 1993).

The configuration of the oximes and lactams involved in the Beckmann rearrangement of a number of oximes, including that of thujone have been determined from NMR and IR spectral data (A. Zabza, Pr. Nauk Inst. Chem. Org. Fiz. Politech. Wroclaw., 1971, 3, 1). Condensation of the thujones with biogenic amines and amino acids [C.H. Brieskorn and W. Schwack, Arch. Pharm. (Weinheim, Ger.), 1983, 316, 552]; synthesis of pyrethroid analogues *via* chiral cyclopropanation of various isoprenoid units derived from thujone (A. Becalski *et al.*, Canad. J. Chem., 1977, 66, 3108), and insect juvenile hormone analogues and pyrethrin analogues from derivatives of *α*- and ß-thujaketonic acids (J.P. Kutney *et al.*, Bioorg. Chem., 1978, 7, 289; Canad. J. Chem., 1981, 59, 3162); stereochemistry and mechanism of the formation of (±)-*cis*-5,6-dibromo-3-(1-bromo-1-methylethyl)-6-methylcyclo-hex-2-enone (tribromothujone) (15), m.p. 122°C (decomp.) from (+)-thujone and absolute configuration of a number of halogenated compounds derived from (+)-thujone (W. Cocker, P.V.R. Shannon, and M. Dowsett, J. Chem. Soc., Perkin I, 1988, 1527) have been reported.

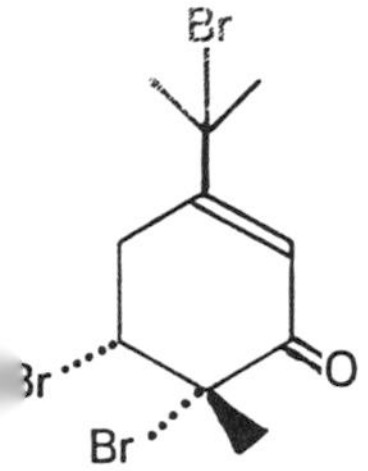

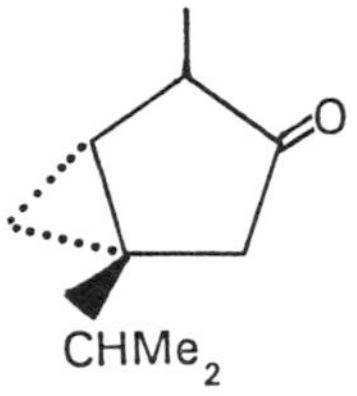

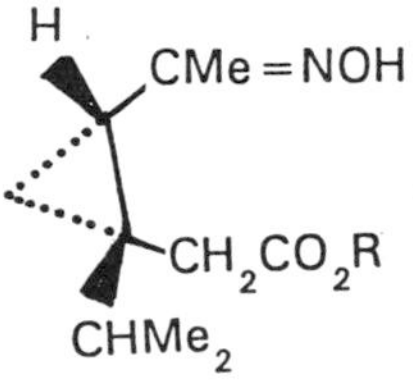

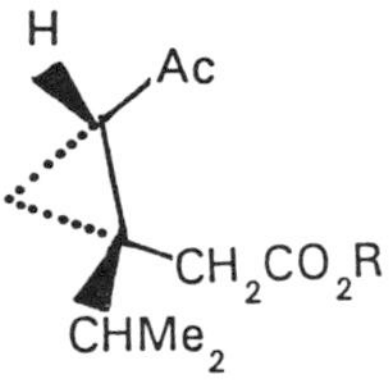

R = Me, Et

(15)   (16)   (17)   (18)

Base-catalyzed cleavage of thujone (16) by ethyl nitrite in sodium methoxide or ethoxide gives ketones (18) *via* intermediate oximes (17) [C.H. Brieskorn and W. Schwack, Arch. Pharm. (Weinheim, Ger.), 1982, 315, 207].

Mass (I. Nykanen and L. Nykanen, Finn. Chem. Letters, 1983, 31) and ^{13}C and ^{1}H NMR (M.I. Burgar, D. Kikelj, and D. Karba, Vestn. Slov. Kem. Drus., 1981, 28, 97) spectral data of (+)- and (-)-thujone have been reported.

(ii) Thuj-3-en-2-one (umbellone)
Umbellone (22) has been obtained by reacting cyclopropane-1,2-
-dicarboxylic anhydride (19) with Me_2CuLi, MeMgBr, or Me_2Cd to give
lactones (20 and 21), followed by treatment of lactone (21) with
$(EtO)_2P(O)CH_2^-$ (S. Benyache *et al.*, Riv. Ital. Essenze, Profumi, Piante Off.,
Aromat., Syndets, Saponi, Cosmet., Aerosols, 1978, 60, 118).

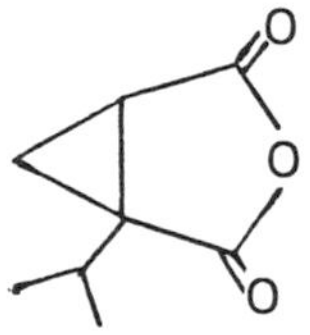

(19) 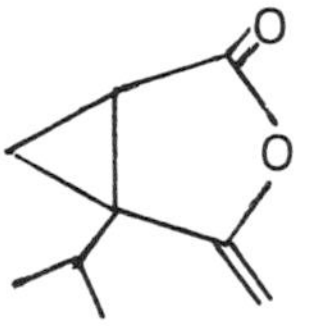   (20) 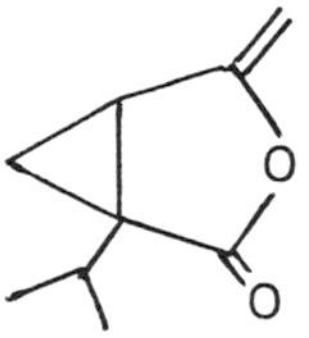   (21) 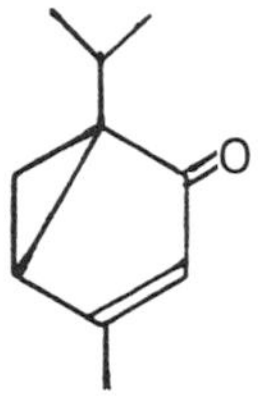   (22)

The stereochemistry of the reduction of umbellone and isodihydro-
umbellone by metal alkoxides and $LiAlH_4$ or $NaBH_4$ has been discussed (C.M.
Holden *et al.*, J. Chem. Soc., Perkin II, 1976, 1342).

(iii) Sabina ketone (5-isopropylbicyclo[3.1.0]hexan-2-one)
Treatment of sabina ketone (23) with MeLi gives mostly *cis*-sabinene
hydrate (*cis*-thuj-4(10)-hydrate, *cis*-thujan-4-ol) (24) (Y. Gaoni, Tetrahedron,
1972, 28, 5525). The synthesis of sabina ketone and 5-alkylbicyclo[3.1.0]-
hexan-2-ones (D.P.G. Hamon and N.J. Shirley, Austral. J. Chem., 1987, 40,
1321) have been described, as has the low-resolution microwave (rotational)
spectra of thujone, isothujone, umbellone, dihydroumbellone, sabinene, and
sabina ketone (Z. Kisiel and A.C. Legon, J. Amer. Chem. Soc., 1978, 100,
8166).
A revised structure (25) with (1*R*,5*S*,6*S*) absolute stereochemistry has
been proposed for artemeseole isolated from *Artemesia tridentata* (T.A. Noble
and W.W. Epstein, Tetrahedron Letters, 1977, 3931).

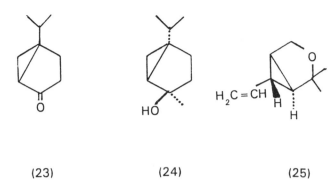

(23) (24) (25)

3. The carane group

(a) Hydrocarbons

(i) Carane

Thermal isomerization of (-)-*cis*-carane (1) at 400-450°C gives a mixture of products, containing (+)-*trans*-carane (2) and various menthenes (I.I. Bardyshev, E.F. Buinova, and B.G. Udarov, Zh. Org. Khim., 1973, 9 1670) and *cis*- and *trans*-carane in the presence of Zeocar 2 at 80° afford a mixture of menthenes (G.V. Kalechits and M.F. Rusak, Zh. Obshch. Khim., 1986, 56, 2132), also obtained on heating the caranes with 4-MeC₆H₄SO₃H in a sealed tube at 80° (I.I. Bardyshev and G.V. Deshits, Vestsi Akad. Navuk BSSR, Ser. Khim. Navuk, 1975, 89).

(1) (2)

The autoxidation of *cis*-carane gives a mixture of hydroperoxycaranes (I.I. Bardyshev *et al.*, Doklady Akad. Nauk BSSR, 1974, 18, 913) and the hydrogenolysis of *cis*- and *trans*-caranes over Pt/C yields mixtures of menthanes with cymenes being formed at temperatures >180°C (I.I. Bardyshev, G.V. Deshits, and B.G. Udarov, Vestsi Akad. Navuk BSSR, Ser. Khim. Navuk, 1976, 67). The hydrochlorination of *cis*-carane (I.I. Bardyshev and E.F. Buinova, *ibid.*, 1973, 94), hydrobromination (E.F. Buinova *et al.*, Zh. Org. Khim., 1985, 21, 2015) and hydroxymercuration of *cis*- and *trans*- -caranes (I.I. Bardyshev *et al.*, *ibid.*, 1975 11, 1424; G.V. Kalechits *et al.*, *ibid.*, 1989, 25, 1402) have been discussed. Enthalpies of combustion and formation of *cis*- and *trans*-caranes (M.P. Kozina *et al.*, Vestsi Akad. Navuk BSSR, Ser. Khim. Navuk, 1976, 14).

(ii) Car-3-ene

The essential oil from Siberian larch needles contains car-3-ene (45.1%) and pin-2-ene (19.9%) and from the bark (58.1-66.1%) and (6.3-13.2%), respectively [R.I. Deryuzhkin *et al.*, Biol. Nauki (Moscow), 1975, 18, 83]. Car-3-ene isomerizes over activated MgO and CaO to car-2-ene (K. Shimazu, H. Hattori, and K. Tanabe, J. Catal., 1977, 48, 302) and the basic-catalyzed isomerization of car-3-ene and -2-ene affords a 55:45 mixture of the two (A. Ferro and Y.R. Naves, Helv., 1974, 57, 1152). Catalytic isomerization of car-3-ene in the vapour phase [V.V. Bazyl'chik *et al.*, Zh. Prikl. Khim. (Leningrad), 1975, 48, 582] and in the presence of TiO_2-H_2SO_4 to give a number of products (M. Dul and M. Bukala, Chem. Stosow., 1973, 17, 19), and performance studies of chromia, chromia-alumina (V. Krishnasamy and L.M. Yeddanapalli, Canad. J. Chem., 1976, 54, 3458; V. Krishnasamy, Austral. J. Chem., 1980, 33, 1313) and platinum-alumina catalyzed (V. Krishnasamy, P. Mathur, and K. Balasubramanian, Chem. Eng. Comm., 1983, 23, 115) have been discussed.

The oxidation of car-3-ene with $Pb(OAc)_4$ in benzene (B.A. Arbuzov, V.V. Ratner, and Z.G. Isaeva, Izv. Akad. Nauk SSSR, Ser. Khim., 1973, 45), red lead in acetic acid and acetic anhydride (Y. Matubara *et al.*, Yukagaku, 1974, 23, 420), or manganese triacetate (K. Witkiewicz and Z. Chabudzinski, Rocz. Chem., 1977, 51, 825) gives a mixture of products and some novel products are obtained with pyridinium fluorochromate in acidic medium (R. Varadarajan and R.K. Dhar, Indian J. Chem., 1986, 25B, 971). Oxidation with thallium (III) nitrate affords the rearranged product (3) (A.V. Pol, V.G. Naik, and H.R. Sonawane, *ibid.*, 1980, 19B, 603) and *m*- and *p*-cymenols as the main products with thallium (III) acetate (K. Pandita *et al.*, *ibid.*, 1984, 23B, 763), other products have been reported from the latter oxidation (V.V. Ratner *et al.*, Izv. Akad. Nauk SSSR, Ser. Khim., 1983, 1136); SeO_2 in pyridine yields mainly (+)-*p*-menth-1,5-dien-8-ol (D.A. Baines and W. Cocker, J. Chem. Soc., Perkin I, 1975, 2232), and $Na_2Cr_2O_7$-AcOH gives cymenol (4) as the major product (P.P. Pai *et al.*, Curr. Sci., 1979, 48, 155). For the autoxidation of car-3-ene see M. Nomura *et al.*, (Chem. Abs., 1981, <u>95</u>, 220152c; Yukagaku, 1983, <u>32</u>, 113), S. Bhaduri, V. Khanwalkar, and D. Mukesh (J. Mol. Catal., 1988, <u>48</u>, 3) and D.A. Baines and W. Cocker (*loc. cit.*) and ozonolysis in the syntheses of odouriferous substances J. Kulesza and J. Kula (Chem. Abs., 1976, <u>84</u>, 105788a; <u>85</u>, 33199e; Riechst., Aromen, Koerperpflegem., 1975, <u>25</u>, 317; 1976, <u>26</u>, 278). The hydrogenation of car-3-ene over Pt and Pt/C at different temperatures gives mixtures of products (E.N. Manukov, V.A. Chuiko, and O.G. Vyglazov, Khim. Prir. Soedin., 1982, 259; E.N. Manukov and V.A. Chuiko, Khim. Drev., 1983, 48).

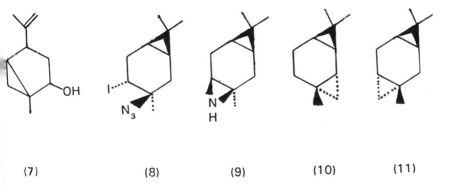

| (3) | (4) | (5) | (6) |

The addition of AcBr to car-3-ene in CCl_4 gives a mixture of acetates (5 and 6). Treatment of (6) with KOH-EtOH affords caran-4-one, car-3(10)-en-4--ol, and alcohol (7) (Z.G. Isaeva, E.Kh. Kazakova, and R.R. D'yakonova, Chem. Abs., 1973, <u>78</u>, 30007k).

| (7) | (8) | (9) | (10) | (11) |

Reaction of car-3-ene with IN_3 in DMF gives derivative (8), which with $LiAlH_4$ yields the aziridine (9) (B. Bochwic and G. Kuswik, Rocz. Chem., 1974, 48, 793), car-2- and -3-ene and pin-2-ene with CH_2I_2 in presence of Zn(Cu) yield derivatives (10 and 11) and the corresponding pinane derivative (I.I. Bardyshev and V.I. Lysenkov, Vestsi Akad. Navuk BSSR. Ser. Khim. Navuk, 1975, 114), car-3-ene with N_2CHCO_2Et affords an epimeric mixture of tricyclic esters related to (11) [D.A. Baines et al., Proc. R. Ir. Acad., 1977, 77B, 323; P.A. Krasutskii et al., Ukr. Khim. Zh. (Russ. Ed.), 1989, 55, 842], and car-3-ene with $ClSO_2NCO$ gives the adduct (12), which can be converted to ß-lactam (13) and azetidine (14) (T. Sasaki, S. Eguchi, and H. Yamada, J. Org. Chem., 1973, 38, 679). Addition reactions between car-3-ene and R_2S_2 (R = Me, Et) (L.E. Nikitina and V.V. Plemenkov, Khim. Prir. Soedin., 1990, 624) and RSCl (R = Ph, Me) (K. Takabe et al., Shizuoka Daigaku Kogakubu Kenkyu Hokoku, 1984, 35, 25), and its hydrochlorination (J. Arct and M. Bukala, Chem. Stosow, 1972, 16, 267), hydration (Y. Matsubara, K. Tanaka, and M. Kenbou, Nippon Kagaku Kaishi, 1974, 1590; M. Nomura and Y. Fujihara, Kinki Daigaku Kogakubu Kenkyu Hokoku, 1985, 19, 1), and hydroboration (I. Uzarewicz and A. Uzarewicz, Rocz. Chem., 1975, 49, 1113; 1976, 50, 1315) have been discussed.

(12) X = O, R = SO_2Cl
(13) X = O, R = H
(14) X = H_2, R = H

(15)   (16)

Acetylation of car-3-ene gives acetyl derivatives (15 and 16) [P. Richter and M. Muehlstaedt, Mezhdunar. Kongr. Efirnym Maslam, [Mater], 4th 1968 (Pub. 1971), 1, 276. Ed. P.V. Naumenko, "Pishchevaya Promyshlennost": Moscow, USSR). The Friedel-Crafts reaction with Ac_2O in presence of $ZnCl_2$ at 80-100° gives 4-acetyl-7,7-dimethyl-3-methylenebicyclo[4.1.0]heptane (17) (L.N. Misra and M.C. Nigam, Chem. Ind., 1980, 294) in addition to derivatives (15 and 16).

(17) (18) R = H (20) (21)
      (19) R = Ac

Related products are obtained by the reaction with $(RCO)_2O/ZnCl_2$ (R = Et, Pr) (*idem*, Indian Perfum., 1981, 25, 29). Prins reaction of (+)-car-3-ene in AcOH-paraformaldehyde yields the carenemethanol (18), its acetate (19), and other products (N.G. Bhat, P.P. Pai, and G.H. Kulkarni, Indian J. Chem., 1980, 19B, 316; O.G. Vyglazov *et al*., Khim. Drev., 1990, 78). For the reaction with isoamyl nitrite - HNO_3 - AcOH at -20° see J. Harmatha *et al.* (Acta Chem. Scand., 1982, B36, 459), properties of car-3-ene bisnitrosochloride (20), C.P. Mathew and J. Verghese (Indian J. Chem., 1977, 15B, 1081), and the isomerization of the nitrosochloride (+)-car-3-ene to a *a*-chlorooxime and subsequent dehydrochlorination to the *aß*-unsaturated oxime (21), A.V. Tkachev *et al.* (Zh. Org. Khim., 1990, 26, 1939).

6,6-Dimethylbicyclo[3.1.0]hexane derivatives (M. Walkowicz, S. Lochynski, and J. Gora, Perfum. Flavor., 1981, 6, 21; 24), 4-(1--acyloxyalkyl)-3,7,7-trimethylbicyclo[4.1.0]hept-3-ene and related compounds (L.N. Misra and M.C. Nigam, Indian J. Chem., 1983, 22B, 453), 3-methoxy--3,7,7-trimethylbicyclo[4.1.0]heptane and related compounds (L.N. Misra, M.C. Nigam, and M.S. Siddiqui, Indian Perfum., 1983, 27, 14), 3,4--dihydroxy-7,7-dimethyl-3-phenylbicyclo[4.1.0]heptane (G.S. Joshi and G.H. Kulkarni, Chem. Ind., 1988, 370), 5-methylene-1,8,8-trimethylcycloocta--1,3,6-triene (O.G. Vyglazov, E.N. Manukov, and V.A. Chuiko, Zh. Org. Khim., 1989, 25, 1571), and (1R,2R,3R,5R)-(-)-2-acetyl-3-hydroxy-6,6--dimethylbicyclo[3.1.0]hexane (M. Kozlowska and W. Sobotka, Bull. Pol. Acad. Sci., Chem., 1986, 34, 333) have all been synthesized from car-3-ene.

(+)-Car-2- and -3-ene have been converted stereospecifically into optically active seven-membered ring systems (P. Eilbracht and I. Winkels, Ber., 1991, 124, 191) and it has been shown that (+)-car-3-ene occurring in the oleoresin from *Pinus roxburghii* is optically pure (L.N. Misra, R. Soman, and S. Dev, Tetrahedron, 1988, 44, 6941). Regioselective routes to nucleophilic optically active car-2- and -3-ene systems have been reported

(L.A. Paquette, R.J. Ross, and Y.J. Shi, J. Org. Chem., 1990, 55, 1589).

Car-3-ene has been converted into a mixture of monocyclic hydrocarbons (V.V. Bazyl'chik, Chem. Abs., 1977, 86, 155799s), pine oil (V. Krishnasamy, Chem. Age India, 1976, 27, 710), (+)-carvone (M.H. Shastri et al., Tetrahedron, 1985, 41, 3083), and by halogenation-dehydrohalogenation into p-cymene (L.N. Misra, M.S. Siddiqui, and M.C. Nigam, Riv. Ital. EPPOS, 1980, 62, 359). ^1H NMR and conformation (R.J. Abraham, M.A. Cooper, and D. Whittaker, Org. Magn. Reson., 1973, 5, 515) and ^{13}C NMR (F. Fringuelli et al., Gazz., 1975, 105, 1215) of car-3-ene; synthesis of 4--methylcar-3-ene and some of its derivatives (L.K. Novikova et al., Izv. Akad. Nauk SSSR, Ser. Khim., 1978, 605) have been discussed.

(iii) Car-2-ene

The preparation of car-2-, -3-, and -3(10)-ene (W. Cocker, N.W.A. Geraghy, and D.H. Grayson, J. Chem. Soc., Perkin I, 1978, 1370), the total synthesis of (+)-car-2-ene (S. Yamada, N. Takamura, and T. Mizoguchi, Chem., Pharm. Bull., 1975, 23, 2539), and an improved preparation of (+) and (-)-car-2-ene (P. Brunetti, F. Fringuelli, and A. Taticchi, Gazz., 1977, 107, 433) have been reported. Car-2-ene has also been obtained from car-3-ene (K. Tanabe, K. Shimazu, and H. Hattori, Chem. Letters, 1975, 507) and (+)--car-2-ene from R-(+)-limonene (J.D. Fourneron, L.M. Harwood, and M. Julia, Tetrahedron, 1982, 38, 693).

Oxidation of car-2-ene with aqueous $KMnO_4$ at 0° gives caran-2α,3α-diol (22) (B.A. Arbuzov et al., Izv. Akad. Nauk SSSR, Ser. Khim., 1974, 2762) and Tl (OAc)$_3$ furnishes a mixture of products (V.V. Ratner et al., ibid., 1983, 1824). The carbonylation of (+)-car-2-ene induced by iron pentacarbonyl yields (-)-(1R)-3,8,8-trimethylbicyclo[4.1.1]oct-3-en-7-one (~ 50%) and (+)--(1R,7S)-3,8,8-trimethylbicyclo[4.1.1]oct-3-en-7-ol (~ 20%) along with other products (C. Santelli-Rouvier, M. Santelli, and J.P. Zahra, Tetrahedron Letters, 1985, 26, 1213). Prins reaction of car-2-ene with CH_2O in AcOH yields car-3-enes (23; R = H, Ac) along with three cyclic ethers (E.N. Manukov et al., Khim. Prir. Soedin, 1989, 193) and the hydroxymercuration--demercuration of car-2-ene by Hg(OAc)$_2$ gives a mixture of products containing trans-caran-3-ol (B.A. Arbuzov et al., Izv. Akad. Nauk SSSR, Ser. Khim., 1979, 1049). The stereochemistry and mechanism of the reaction between car-2-ene and $PdCl_2(MeCN)_2$ in a nonnucleophilic solvent (D. Wilhelm et al., Organometallics, 1985, 4, 1296), and the thermal rearrangement of car-2-ene borate esters (E.N. Manukov, O.G. Vyglazov, and V.A. Chuiko, Vestsi Akad. Navuk BSSR, Ser. Khim. Navuk, 1989, 58) have been investigated. Acetolysis of trans-4-(tosyloxymethyl)car-2-ene (24) affords the bicycloheptene derivative (25) (86%) (E.N. Manukov et al., Zh. Org. Khim., 1988, 24, 1449). For a molecular orbital study of (+)-car-2-ene see H. Tylli (Theochem., 1988, 50, 11).

R = H,Ac

(22)   (23)   (24)   (25)

(iv) Car-4-ene and cara-3(10),4-diene

For the thermal isomerization of 3-acetylcar-4-ene in liquid phase see J. Kulesza, J. Podlejski, and J. Kula (Rocz. Chem., 1973, <u>47</u>, 657). The dehydration of 4-hydroxymethylcar-2-ene over KOH at 250° gives 4--methylcara-3(10),4-diene, 1,1,4,5-tetramethylcyclohepta-2,4,6-triene, and other products (E.N. Manukov *et al.*, Zh. Org. Khim., 1985, <u>21</u>, 2089). Dehydrochlorination of 4-chlorocar-3(10)-ene by 3M KOH at 80° yields cara--3(10),4-diene (8.5%) and other products (E.N. Manukov, V.A. Chuiko, and O.G. Vyglazov, Zh. Org. Khim., 1983, <u>19</u>, 662). The isomerization of 4--methylcara-3(10),4-diene under various conditions has been studied (*idem*, Vestsi Akad. Navuk BSSR, Ser. Khim. Navuk, 1988, 40).

(b) Alcohols and epoxides

The cyclization of dehydrolinalyl acetate in the presence of $ZnCl_2$ yields a mixture of mono- and bicyclic derivatives, including 2-acetoxycar-2-ene (H. Strickler, J.B. Davis, and G. Ohloff, Helv., 1976, <u>59</u>, 1328). The reductive amination of *cis*-caran-*trans*-2-ol affords a mixture of products (G.V. Kalechits *et al.*, Zh. Obshch. Khim., 1983, <u>53</u>, 203).

(-)-Car-4-en-3α-ol, (-)-car-4-en-3ß-ol (R.B. Mitra *et al.*, Synth. Comm., 1984, <u>14</u>, 101), 3ß-acetyl-car-4-en-3α-ol (I.P. Povodyreva, R.R. Shagidullin, and T.N. Ivaseva, Izv. Akad. Nauk SSSR, Ser. Khim., 1972, 2618), *trans*-car--2-en-4-ol, (-)-*cis*-carane-*trans*-3-ol, carane-3ß,4α-diol (I. Uzarewicz, E. Zientek, and A. Uzarewicz, Rocz. Chem., 1976, <u>50</u>, 1515), car-3(10)-en-4--ols (M. Lajunen and J. Kujala, Acta Chem. Scand., 1989, <u>43</u>, 813), 4--hydroxymethylcar-2-ene (E.N. Manukov *et al.*, Vestsi Akad. Navuk BSSR,

344

Ser. Khim. Nauvk, 1986, 67; U.S.S.R. SU 1,373,702, 1988; E. Kh. Kagakova *et al., ibid.*, 926,900, 1982), carane-2α,3ß- and -2ß,3ß-diol (B.A. Arbuzov *et al.*, Izv. Akad. Nauk SSSR, Ser. Khim., 1979, 1294), conformation of carane-2,3-diols (*idem, ibid.*, p2231), *cis*-carane-3,4-diol (B.A. Arbuzov, V.A. Shaikhutdinov, and Z.G. Isaeva, *ibid.*, 1972, 2124), (+)--carane-3ß,4α-diol (W. Cocker and D.H. Grayson, J. Chem. Soc., Perkin I, 1978, 155), (+)-carane-3ß,4ß-diol acetals and ketals (A. Hendrick, K. Piatkowski, and J. Gora, Perfum. Flavor., 1986, 11, 85; Pol. PL 141,357--141,364, 1987), 2-methylcarane-3,4-diols (Z.G. Isaeva *et al.*, Izv. Akad. Nauk SSSR, Ser. Khim., 1979, 1107), acetolysis of carane-3ß,4ß-diol 4-*p*--toluenesulphonate (B.A. Arbuzov, Z.G. Isaeva, and R.R. D'yakonova, *ibid.*, 1975, 972), oxidation products of isomeric carane-3,4-diols (A. Henrick and K. Paitkowski, Pol. J. Chem., 1984, 58, 73), carenadiol and derivatives useful in treatment of neurological disorders (A.R. Martin, P.F. Consroe, and V.J. Shah, U.S. Pat. US 4,758,597, 1988), enthalpies of combustion and formation of caran-3α- and -3ß-ol (M.P. Kozina *et al.*, Vestsi Akad. Navuk BSSR, Ser. Khim. Navuk, 1980, 31). The reaction of 4-hydroxymethylcar-2--ene tosylate with Et₃Al in CH₂Cl₂ gives a complex mixture of products (G.A. Tolstikov *et al.*, Izv. Akad. Nauk SSSR, Ser. Khim., 1985, 1814) and car-3--ene with PhSO₂N:S:O in dry THF at 0°, followed by reduction of the product with LiAlH₄ affords car-2-ene-4α-thiol (A. Gadras *et al.*, J. Org. Chem., 1984, 49, 442).

Epoxidation of 4α-acetoxycar-2-ene with perbenzoic acid gives in addition to 4α-acetoxy-2α,3α-epoxycarane, the rearranged hydroxyacetate (26) (R.H. Naik, G.D. Joshi, and G.H. Kulkarni, Indian J. Chem., 1986, 25B, 306).

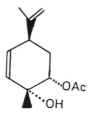

(26)

2,3-Epoxycarane is converted to *cis*-isolimonenol in the presence of metatitanic acid (J.O. Bledsoe, Jr., J.M. Derfer, and W.E. Johnson, Jr., U.S. Pat. 3,814,733, 1974). The isomerization of 2,3- and 3,4-epoxycaranes over solid acids and bases (K. Arata, J.O. Bledsoe, Jr., and K. Tanabe, Tetrahedron Letters, 1976, 3861; J. Org. Chem., 1978, 43, 1660); preparation of 2-acetyl- (N.D. Ibragimova *et al.*, Izv. Akad. Nauk SSSR, Ser. Khim., 1982, 351) and 4α-acetyl-2,3-epoxycarane (P.P. Pai *et al.*, Indian J. Chem., 1979, 18B, 549); Lewis acid rearrangement of 2,3-epoxycarane (B.C. Clark, Jr., *et al.*, J. Org. Chem., 1978, 43, 519); ring opening of isomeric acetyl-substituted 2,3- and 3,4-epoxycarane (B.A. Arbuzov *et al.*, Chem. Abs., 1978, 88, 191088j); and by-products from the epoxidation of (+)-car-3-ene with acetonitrile-hydrogen peroxide (W. Cocker and D.H. Grayson, J. Chem. Soc., Perkin I, 1976, 791) have been reported.

3α,4α-Epoxycarane isomerizes in boiling Me$_3$COK-pyridine to give a mixture containing car-2-en-4α-ol (27; R = α-OH) (35%) and car-4-en-3α-ol (28; R = α-OH) (65%). Similarly 3ß,4ß-epoxycarane yields car-2-en-4ß-ol (27; R = ß-OH) (72%), car-4-en-3ß-ol (28; R = ß-OH) (8%) and car-3(10)--en-4-ol (29) (8%). The alcohols have been converted into car-2-en-4-one, car-3-en-5-one, and car-3-ene-2,5-dione (Z.G. Isaeva, G.A. Bakaleinik, and A.N. Karaseva, Izv. Akad. Nauk SSSR, Ser. Khim., 1983, 1651; E.N. Manukov and T.R. Urbanovich, Vestsi Akad. Navuk BSSR, Ser. Khim. Navuk, 1985, 70). For the isomerization of 3,4-epoxycarane in presence of a Re$_2$O$_7$/Al$_2$O$_3$ catalyst see E.N. Manukov and G.N. Bazhina (Zh. Org. Khim., 1988, 24, 121).

(27)   (28)   (29)   (30)   (31)

The reaction of 3α,4α-epoxycarane with HCl (B.A. Arbuzov *et al.*, Doklady Akad. Nauk SSSR, 1972, <u>207</u>, 853; 1977, <u>237</u>, 98), halogen acids and acetic acid (W. Cocker and D.H. Grayson, J. Chem. Soc., Perkin I, 1975, 1217), cold concentrated H_2SO_4 (D.H. Grayson, Proc. R. Ir. Acad., 1983, <u>83B</u>, 85), aqueous $NaHSO_3$ (E. Myslinski and E. Michalek, Rocz. Chem., 1973, <u>47</u>, 285), AcOH-AcONa (B.A. Arbuzov, Z.G. Isaeva, and E.Kh. Kazakova, Izv. Akad. Nauk SSSR, Ser. Khim., 1975, 2554), $LiAlH_4$ (B.A. Arbuzov *et al.*, Doklady Akad. Nauk SSSR, 1977, <u>233</u>, 366), aqueous KOH (E.Kh. Kazakova and L.N. Surkova, Izv. Akad. Nauk SSSR, Ser. Khim., 1983, 2391), and N_2H_4 (*via* the Wharton reaction) (Z.G. Isaeva, A.N. Karaseva, and I.S. Ikhtonova, *ibid.*, 1985, 2628); of 3ß,4ß-epoxycarane with Ac_2O (B.A. Arbuzov, Z.G. Isaeva, and V.A. Shaikhutdinov, Doklady Akad. Nauk SSSR, 1973, <u>210</u>, 837), HCl (B.A. Arbuzov *et al.*, Izv. Akad. Nauk SSSR, Ser. Khim., 1975, 969; 1980, 2778), and alcohols in 0.1% H_2SO_4 (Z.G. Isaeva and G.A. Bakaleinik, *ibid.*, 1985, 648); and of 3α,4α- and 3ß,4ß-epoxycarane with $(PhSe)_2$ (A. Uzarewicz and E. Zientek, Rocz. Chem., 1977, <u>51</u>, 181), diborane and disiamylborane (*idem., ibid.*, p723; Pol. J. Chem., 1978, <u>52</u>, 389), and isothiuronium salts (N.P. Artemova *et al.*, Zh. Obshch. Khim., 1989, <u>59</u>, 2718) have been reported.

Epoxidation of 4α-chlorocar-3(10)-ene by peracetic acid gives a mixture of epoxides (30 and 31), the latter predominating (B.A. Arbuzov, Z.G. Isaeva, and G.A. Bakaleinik, Izv. Akad. Nauk SSSR, Ser. Khim., 1979, 832). 4ß--Chlorocar-3(10)-ene has been treated in a similar manner (Z.G. Isaeva *et al., ibid.*, 1985, 2808) and the interaction of stereoisomeric 4ß-chloro-3,10--epoxycaranes with methanol investigated (*idem, ibid.*, 1987, 2809). Also reported are the epoxidation of the diastereomeric 4-(α-hydroxyethyl)car-3--enes (*idem, ibid.*, 1979, 1299) and car-3-en-10-al and -10-ol (*idem, ibid.*, p1889), reactions of 4-acetyl-3α,4α-epoxycarane with $NaBH_4$ (B.A. Arbuzov *et al., ibid.*, p1156) and HCl (B.A. Arbuzov, A.N. Karaseva, and Z.G. Isaeva, Doklady Akad. Nauk SSSR, 1979, <u>247</u>, 364), hydrolysis of 4-acetyl-3,4--epoxycarane with H_2SO_4 in dioxane (B.A. Arbuzov, N.D. Ibragimova, and I.P. Povodyreva, Izv. Akad. Nauk SSSR, Ser. Khim., 1980, 1052), epoxidation of the dehydration products of caran-3ß-ol-4α-acetate (B.M. Mane, K.G. Gore and G.H. Kulkarni, Indian J. Chem., 1979, <u>18B</u>, 395), acid-catalyzed ring cleavage of 3,4-epoxycaranones (B.A. Arbuzov, Z.G. Isaeva, and A.N. Karaseva, Izv. Akad. Nauk SSSR, Ser. Khim., 1982, 698), reactions of epoxycaranones in the presence of bases (A.N. Karaseva *et al., ibid.*, 1984, 677), and isomerization of 3α,4α-epoxy-4-methylcarane (Z.G. Isaeva and G.Sh. Bikbulatova *ibid.*, 1982, 454).

The enthalpies of formation of 3α,4α- and 3ß,4ß-epoxycarane (M.P. Kozina *et al.*, Vestsi Akad. Navuk BSSR, Ser. Khim. Navuk, 1977, 94; Chem. Abs., 1980, <u>92</u>, 94582r) and their configurations and conformations (B.A. Arbuzov, Yu. Yu. Samitov, and Sh.S. Bikeev, Doklady Akad. Nauk SSSR, 1974, <u>216</u>, 550), and those of 4α-chloro-3,10-epoxycarane (G.I. Kovylyaeva *et al*, Izv. Akad. Nauk SSSR, Ser. Khim., 1988, 2018) have been discussed.

Heating 3α,4α-epoxycarane with (EtO)$_2$P(S)SH gives 3α,4α-thioepoxy-
carane, analogously obtained is 3ß,4ß-thioepoxycarane (O.N. Nuretdinova,
G.A. Bakaleinik, and B.A. Arbuzov, *ibid.*, 1975, 962). For their ^{1}H and ^{13}C
NMR spectral data see Yu.Yu. Samitov *et al.* (*ibid.*, 1976, 2696).

(c) Ketones

(i) Caranones, carenones, and acetylcarenes
The following have been reported, preparation of (+)- and (-)-car-3-en-2-
-one and (-)-car-2-en-4-one (D.D. Maas, M. Blagg, and D.F. Wiemer, J. Org.
Chem., 1984, 49, 853) and 5-hydroxymethylene-*cis*-caran-4-one derivatives
(F. Bondavalli, B. Schenone, and M. Lonobardin, Farmaco, Ed. Sci., 1974, 29,
48; J. Chem. Soc., Perkin I, 1976, 678), photochemical transformation of 3-
-methylcar-4-en-2-one into derivatives of 3,3,7-trimethylocta-4,6-dienoic acid
(A.J. Bellamy and W. Crilly, J. Chem. Soc., Perkin II, 1973, 122), Diels-Alder
reaction of (+)-car-3-en-2-one, (+)-car-4-en-3-one, and car-3-ene-2,5-dione
with 2,3-dimethylbuta-1,3-diene (L. Minuti *et al.*, J. Org. Chem., 1990, 55,
4261), Darzens reaction of (-)-*cis*-caran-4-one with ethyl chloroacetate and
chloroacetonitrile (Z. Kubica, Zesz.-Nauk-Wyzsza Szk. Pedagog. im.
Powstaneow Slask. Opolu. [Ser.]: Chem., 1984, 6, 121), oxidation and
reduction of some hydroxymethyl derivatives of caran-4-one, thujan-3-one
and pinan-3-one [C.H. Brieskorn and Chung Kyu Ryu, Arch. Pharm.
(Weinheim, Ger.), 1985, 318, 788], Wittig-Horner reaction of (-)-*cis*-caran-4-
-one (Z. Kubica, Z. Burski, and K. Piatkowski, Pol. J. Chem., 1985, 59, 827),
configuration of 3-aminocar-4-one (A.V. Tkachev *et al.*, Zh. Org. Khim.,
1990, 26, 1693), LiAlH$_4$ reduction of stereoisomeric (-)-5-hydroxymethylene-
-*cis*-caran-4-ones (L. Otorowska and K. Paitkowski, Pol. J. Chem., 1983, 57,
1187), microbiological reduction of car-3-ene-2,5-dione by *Rhodotorula
mucilaginosa* (A. Siewinski *et al.*, Tetrahedron, 1977, 33, 1139), and
irradiation of 4α-acetylcar-2-ene (H.R. Sonawane, V.G. Naik, and B.S.
Nanjundiah, Tetrahedron Letters, 1983, 24, 3025). 4-Formylcar-2-ene and
semicarbazone (H. Sadowska and J. Gora, Zesz. Nauk Akad. Roln.-Tech.
Olsztynie, Technol. Zywn., 1985, 39, 111) have been synthesized.
   4-Acetylcar-2-ene tolylsulphonylhydrazone when treated with BuLi in
(Me$_2$NCH$_2$)$_2$ at -78°, followed by the addition of DMF at 0° gives 2-(car-2-en-
-4-yl)prop-2-en-1-al (P.C. Traas, H. Boelens, and H.J. Takken, Tetrahedron
Letters, 1976, 2287).

(ii) Eucarvone
Evidence has been provided for an ionic intermediate in the
photoisomerization of eucarvone in acidic media (K.E. Hine and R.F. Childs, J.
Amer. Chem. Soc., 1973, 95, 6116). The UV irradiation of γ,δ-epoxy-
eucarvone (32) (A.P. Alder and H.R. Wolf, Helv., 1975, 58, 1048) and α,ß-
-epoxyeucarvone (33) (B. Frei and H.R. Wolf, *ibid.*, 1976, 59, 82), and the

photoisomerization of BF_3, BCl_3, and BBr_3 complexes of eucarvone (R.F. Childs and Yee-Chee Hor, Canad. J. Chem., 1977, 55, 3501) have been studied. Also reported are the Beckmann rearrangement of eucarvone oxime (A. Zabza et al., Bull. Acad. Pol. Sci., Ser. Sci. Chim., 1972, 20, 841) and of tetrahydroeucarvone (idem, ibid., 1973, 21, 1) and variable temperature NMR studies of eucarvone (E. Cuthbertson and D.D. MacNicol, Tetrahedron Letters, 1974, 2689). The NMR of the enolate ion (34) of eucarvone indicates that an equilibrium mixture containing the electrocyclized anion (35) (67%) is formed (A.J. Bellamy and W. Crilly, ibid., 1973, 1893).

| (32) | (33) | (34) | (35) |

(d) Carboxylic acids

The *trans* configuration of chamic acid (*trans*-car-4-en-10-oic acid) has been determined by chemical and NMR spectral correlation with (+)-*trans*-carane and (+)-*trans*-car-4-ene (T. Norin, S. Stroemberg, and M. Weber, Chem. Scr., 1982, 20, 49). (±)-Chaminic acid [(±)-car-3-en-10-oic acid] has been synthesized from dimethyl 3-hydroxycyclohexane-1,3-dicarboxylate (W.J. Gensler and P.H. Solomon, J. Org. Chem., 1973, 38, 1726). Oxidation of 4a-acetylcar-2-ene with $KMnO_4$ in acetone furnishes *cis*-caronic acid (36) (R.B. Mitra et al., Indian Pat. IN 150,470, 1982). (1R)-*cis*-Caronaldehyde (37) (R.B. Mitra, G.H. Kulkarni, and P.N. Khanna, Synth. Comm., 1987, 17, 1089), methyl (1R)-*trans*- and (1R)-*cis*-hemicaronaldehydes (A. Krief et al., Tetrahedron, 1989, 45, 3039), and (1R)-*cis*-caronaldehydric acid (S.A. Roman, Canad. Pat. CA 1,120,486, 1982) have been synthesized. (+)-Car-3-ene has been converted into (-)-cis-caronaldehydric acid hemiacetal (D. Bakshi et al., Tetrahedron, 1989, 45, 767), and *trans*-carane-4-carboxylic acid synthesized (F. Frinquelli, L. Minuti, and A. Taticchi, Synth. Comm., 1990, 20, 2507).

HO$_2$C   CO$_2$H

H·····   ····H

HO·····◯–O◯=O

H·····   ····H

(36)   (37)

Chrysanthemic acid and some derivatives and related pyrethroids, derivatives of cyclopropane, have been obtained from car-2- and -3-ene. Although there are many reported syntheses of the above compounds only those starting from the carenes are indicated; *trans*-chrysanthemic acid (38) from 4-acetylcar-2-ene (T.L. Ho, Synth. Comm., 1983, 13, 761) and (+)-car--3-ene (R. Sobti and S. Dev, Tetrahedron, 1974, 30, 2927; A.S. Khanra and R.B. Mitra, Indian J. Chem., 1976, 14B, 716), chrysanthemic acid analogues from car-2-ene (L. Chen, Huaxue Shijie, 1982, 23, 39), methyl (+)-*cis*--chrysanthemate (39) (T. Inokuchi et al., Chem. Express, 1988, 3, 623), 1R--cis- and 1R-*trans*-chrysanthemic acid (A.V. Rukavishikov et al., Zh. Org. Khim., 1989, 25, 1665), methyl (+)-*cis*-chrysanthemate and (+)-*cis*--homochrysanthemate (R.H. Naik and G.H. Kulkarni, Indian J. Chem., 1983, 22B, 859), (+)-dihydrochrysanthemolactone and methyl (+)-*cis*--chrysanthemate (B.M. Mane, K.G. Gore, and G.H. Kulkarni, ibid., 1980, 19B, 605), methyl 1R-(+)-*cis*-chrysanthemate and methyl 1S-(+)-*cis*-2,2-dimethyl--3-(2-phenylprop-1-enyl)cyclopropane-1-carboxylate (N.G. Bhat et al., ibid., 1981, 20B, 204), methyl 1R-(-)-*cis*-2,2-dimethyl-3-(2-phenylprop-1--enyl)cyclopropane-1-carboxylate (idem, ibid., p558), methyl 1S-*cis*-2,2--dimethyl-3-(2-chloro-2-phenylvinyl)cyclopropanecarboxylate (B.M. Mane et al., ibid., p1029), 1R-*cis*-2,2-dimethyl-3-(2,2-dibromovinyl)cyclopropane-carboxylic acid (G.A. Tolstikov et al., Zh. Org. Khim., 1989, 25, 2633), and (±)-dihydrochrysanthemolactone (40) (S. Lochynski et al., J. Prakt. Chem., 1988, 330, 284) from car-3-ene; (-)-*cis*-dihydrochrysanthemolactone via two routes from car-2-en-4-one obtained from car-3-ene (T.L. Ho and Z.U. Din, Synth. Comm., 1980, 10, 921); and some pyrethroids (B.G. Mahamulkar et al., Indian J. Chem., 1983, 22B, 355, 1261), 1R-(-)-*cis*-permethrin, 1R-(+)--cis-cypermethrin and 1R-*cis*-(+)-deltamethrin(decis) (A.K. Mandal et al., Tetrahedron, 1986, 42, 5715), γ-lactone of 1R-*cis*-2,2-dimethyl-3-(1-

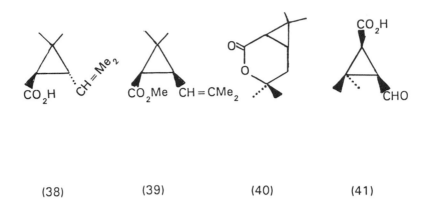

(38)   (39)   (40)   (41)

-hydroxy-2,2,2-tribromoethyl)cyclopropanecarboxylic acid a key intermediate for deltamethrin (G.H. Kulkarni and S.M. Toke, Synth. Comm., 1989, 19, 13), and methyl and *tert*-butyl (+)-1*R-trans*-2,2-methyl-3-[2-(4-chlorophenyl)-ethynyl]cyclopropanecarboxylate (S.S. Bhosale and Kulkarni, Curr. Sci., 1989, 58, 561) from car-3-ene. Caronic aldehyde (41) has been synthesized from (+)-4*a*-acetylcar-2-ene (G.A. Tolstikov *et al.*, Izv. Akad. Nauk SSSR, Ser. Khim., 1989, 2653).

(e) Halogeno and cyano derivatives

Conformational analysis of (-)-3*a*,4ß- and (+)-3ß,4*a*-dichlorocarane by electron diffraction methods shows that the former exists in an almost pure diequatorial conformation, whereas the latter occurs as both diaxial and diequatorial conformations (2:3) (B.A. Arbuzov *et al.*, Doklady Akad. Nauk SSSR, 1972, 207, 596). Dehydrochlorination of 4-chlorocar-3-ene and 4--chlorocar-3(10)-ene by Me₃COK in DMSO gives 3,7,7-trimethylcyclohepta--1,3,5-triene and cara-3(10),4-diene, respectively (E.N. Manukov *et al.*, Vestsi Akad. Navuk BSSR, Ser. Khim. Navuk, 1988, 55). Treatment of 4ß--chlorocaran-3*a*-ol with POCl₃ in pyridine yields 4-chlorocar-3(10)-ene (31.5%), 4-chlorocar-3-ene (3%), 3-chloromethylcar-3-ene (5%), and 3,4--dichlorocarane (3%) (Z.G. Isaeva *et al.*, Izv. Akad. Nauk SSSR, Ser. Khim., 1985, 919), and (-)-3,4-dibromocarane with aqueous Ag₂O affords ketone (42) as the main product and glycols (43; R = ß-,*a*-OH) (M. Walkowicz and J. Konopka, Pol. J. Chem., 1982, 56, 439).

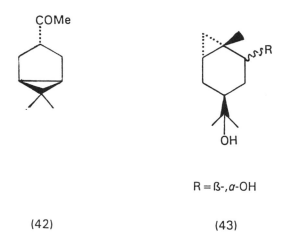

COMe

R = ß-,α-OH

(42)   (43)

Reaction of 3ß-bromocaran-4α-ol and its acetate with AgOAc gives a mixture of products (B.A. Arbuzov, Z.G. Isaeva, and R.R. D'yakonova, Izv. Akad. Nauk SSSR, Ser. Khim., 1980, 2141). For dipole moments, Kerr constants, and IR spectra of 3,4-dibromo-3,7-dichloro- and 3,4,7,7-tetrabromocarane see B.A. Arbuzov *et al.*, (*ibid.*, 1973, 2231), and for conformational analysis of some 3(4)-halogenocaran-4(3)-ols and 3-methylnorcaran-4(3)-ols, R.R. Shagidullin *et al.*, (*ibid.*, 1987, 591). An explosion of 4-iodo-3- -methoxycarane after standing 10 days in a sealed container has been reported (D.R. Dimmel, Chem. Eng. News, 1977, 55, 38).

Synthesis and chemical transformations of (+)-*cis*-4-cyano-*cis*-carane (Z. Kubica and K. Piatkowski, Pol. J. Chem., 1983, 57, 1197), the synthesis of 10-cyanocarane (T.F. Braish and P.L. Fuchs, Synth. Comm., 1985, 15, 549), and 4-(2-cyanoethyl)carane and related derivatives (D.K. Kettenes and W. Lenselink, Ger. Offen. 2,812,288, 1978) have been reported.

(f) Norcaranes and Norcarenes

The following norcarane and norcarene derivatives have been prepared; *endo*-7-phenylnorcarane (P.R. Morris and J.D. Woodyard, Org. Prep. Proced. Int., 1973, 5, 275), 1-aryl-7,7-dichloronorcaranes (D.A. Pisanenko and Yu. V. Balitskii, Vestn. Kiev Politekh. Inst., [Ser.]: Khim. Mashinostr. Tekhnol., 1983, 20, 22), 7,7-dibromonorcarane and hence *anti*-1,2-di(7'-norcaran-ylidene)cyclopropane (Y. Fukuda *et al.*, Tetrahedron Letters, 1979, 877) and other derivatives (A. Oku, Y. Yamaura, and T. Harada, J. Org. Chem., 1986, 51, 3730). The hydroxylation of 3-methylnorcar-2-ene (E. Kh. Kazakova and

L.N. Surkova, Izv. Akad. Nauk SSSR, Ser. Khim., 1981, 372), *cis*-2,3-epoxy-
-3-methylnorcarane (*idem, ibid.*, 1983, 2392), and 3-methylnorcar-3-ene
(E.Kh. Kazakova and S.V. Filippova, *ibid.*, 1980, 2791) has been reported.

4. The Wagner-Meerwein rearrangement

Camphene rearranges in boiling MeOH in the presence of $CuCl_2$ or
$Th(NO_3)_4.6H_2O$ to give isobornyl methyl ether (85 and 80%, respectively) and
in AcOH to yield isoboryl acetate (95%) (R.K. Thappa *et al.*, Indian Perfum.,
1979, 23, 172). The rearrangement to give isobornyl acetate without a Me
shift, proceeds presumably *via* a Cu(II) complexed non-delocalized carbenium
ion (1) (A. Heumann and B. Waegell, Nouv. J. Chim., 1977, 1, 275). 2-
-Chloro-3-nitroso-2,7,7-trimethylbicyclo[3.1.1]heptane reacts with KCN *via* a
Wagner-Meerwein rearrangement to furnish product (2) instead of the
expected substitution product (V.P. Papageorgios, Chem. Chron., 1974, 3,
149).

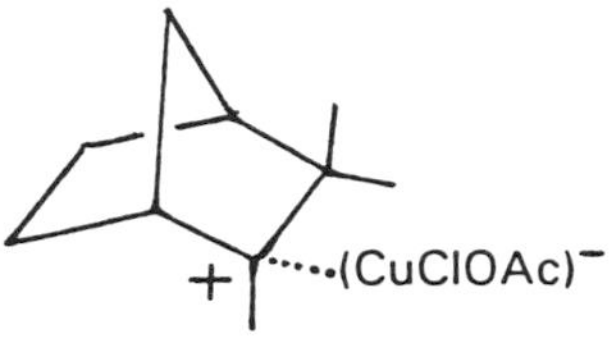

(1) 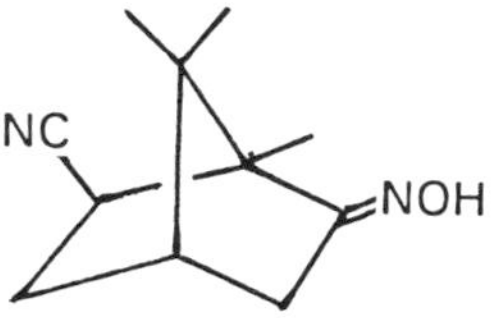 (2)

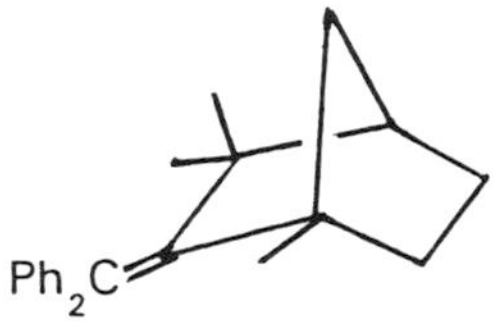

(3) 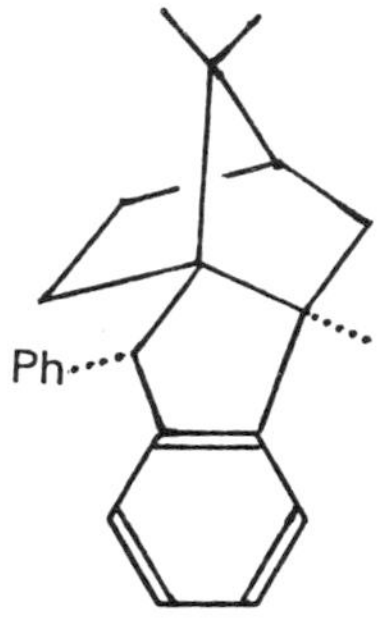 (4)

Photochemical rearrangement of 2-diphenylmethylenefenchane (3) affords
product (4) (61%) and probably a diastereoisomer of (4) (23%). Formation of
(4) involves a Wagner-Meerwein rearrangement and a twisted (*ca.* 90°)
zwitterionic excited singlet state of (3) is suggested as a reaction
intermediate (S.S. Hixson *et al.*, J. Amer. Chem. Soc., 1980, 102, 412).
Cases of other Wagner-Meerwein rearrangements are illustrated in the
appropriate subject section.

5. The pinane group

(a) Hydrocarbons

(i) Pinane

Thermal isomerization of (-)-pinane affords dihydromyrcene in ~ 60% yield (K. Suga, S. Watanabe, and T. Fujita, Yukagau, 1973, 22, 738; Y. Fujihara, C. Hata, and Y. Matubara, Yuki Gosei Kagaku Kyokai Shi, 1974, 32, 933). ¹³C NMR (J.M. Coxon and G.J. Hydes, J. Chem. Soc., Perkin II, 1984, 1351) and mass spectral data (Tse Lok Ho, Bull. Chem. Soc. Jpn., 1988, 61, 4127), ORD and CD curves (T. Hirata, ibid., 1972, 45, 3458) of pinane and derivatives, and some advances in pinane chemistry (Z. Chabudzinski, Wiad. Chem., 1975, 29 77) have been reported.

(ii) Pin-2-ene (a-pinene, pin-2,3-ene)

Treatment of (-)-pin-2(10)-ene with $KNHCH_2CH_2CH_2NH_2$ gives (-)-pin-2--ene (93-95%) (C.A. Brown, Synth., 1978, 754; C.A. Brown and P.K. Jadhav, Org. Synth., 1987, 65 224). 2-Azido-cis-pin-3-ene and 2-amino-cis--pin-3-ene on hydroboration afford pin-2-ene (I. Uzarewicz and A. Uzarewiez, Pol. J. Chem., 1978, 52, 1907). Hydride reduction of the tosylate of filipendulol gives cyclopinene (1) (66%) and (+)-pin-2-ene (A.D. Dembitskii, R.A. Yurina, and M.I. Goryaev, Khim. Prir, Soedin., 1982, 710). Total syntheses of racemic pin-2-ene and a-trans- and a-cis-bergamotene (S.D. Larsen and S.A. Monti, J. Amer. Chem. Soc., 1977, 99 8015) have been described. Pin-2-ene, pin-2(10)-ene, car-3-ene, and other products are contained in the seed oil of Bunium persicum Boiss (siah zira) (A. Karim, M. Pervez, and M.K Bhatty, Pak. J. Sci. Ind. Res., 1977, 20, 106). Pin-2- and -2(10)-ene have been separated by adsorption (A.J. De Rossef and R.W. Neuzil, U.S. Pat. 3,851,006, 1974; J.W. Priegnitz, ibid., 3,845,151), and azeotropic distillation methods (P.M. Koppel and W.I. Taylor, ibid., 3,987,226, 1976) and by gas-liquid chromatography based on formation of their a-cyclodextrin complexes (T. Koscielski, D. Sybilska, and J. Jurczak, J. Inclusion Phenom., 1987, 5, 69). For the upgrading of commercial (+)- and (-)-ⁱ pin-2-ene to material of high optical purity see H.C. Brown, P.K. Jadhav, and M.C. Desai (J. Org. Chem., 1982, 47, 4583). Pin-2-ene on pyrolysis at 500-600° gives a mixture of dipentene and ocimene (G. Rice and J.F. Pollock, U.S. Pat. 3,714,283, 1973).

Isomerization of pin-2- and -2(10)-ene to camphene, dipentene and other products has been effected by the following catalysts or reagents; zeolite 13X (C.B. Davis, ibid., 3,696,164, 1972; E.A. Takacs, ibid., 3,700,746; 3,700,747; Arizona Chemical Co., Fr. Demande 2,176,513, 1973), salicylic acid (I.I. Bardyshev, E.N. Manukov, and V.A. Chuiko, Vestsi Akad. Navuk BSSR, Ser. Khim. Navuk, 1977, 87; V.A. Chuiko, Tezisy Doklady-Resp.

Konf. Molodykh Uch. - Khim., 2nd, 1977, 1, 57, Akad. Nauk Est. SSR, Inst.
Khim. : Tallinn, USSR), and acid (G.A. Rudakov and L.S. Ivanova, Mezhdunar.

CH_2SnMe_3

(1)

(2)

Kongr. Efirnym Maslam, [Mater.], 4th, 1968 [Pub. 1971], 1, 285. Ed. P.V.
Naumenko, "Pishchevaya Promyshlennost" : Moscow, U.S.S.R.), including
formation of σ-menthenes by ring-opening of pin-2-ene derivatives (N. Lander
and R. Mechoulam, J. Chem. Soc., Perkin I, 1976, 484). Catalytic
isomerization of pin-2-ene in the presence of amorphous $FePO_4.2.5H_2O$ or
crystalline $FePO_4.2H_2O$ (at 180-560°) gives α-terpinene, dipentene,
camphene, and other products (V.V. Pechkovskii et al., Izv. Vyssh. Ucheb.
Zaved., Les. Zh., 1973, 16, 107); activated carbon (156-165°), camphene
(32.5%) and other products (C.B. Davis and J. McBride, Jr., U.S. Pat.
3,842,135, 1974); Cu and Zn systems (250 ± 5°), alloocimene and p-
-mentha-1,8-diene (Y. Fujihara, Yuki Gosei Kagaku Kyokai Shi, 1973, 31 928;
Y. Matsubara, Y. Fujihara, and C. Hata, Japan P. 75 32,104, 1975);
titanium, pinenes, tricyclene, and monocyclic terpenes (A.A. Popov and V.A.
Vyrodov, Gidroliz. Lesokhim. Prom-st., 1978, 19); TiO_2 (idem, ibid., 1979,
18); titanic acid (129.5°), camphene, tricyclene, limonene, and other
products (G.A. Rudakov et al., ibid., 1975, 7); $Pt-Al_2O_3$ (V. Krishnasamy, P.
Mathur, and K. Chandrasekharan, J. Chem. Technol. Biotechnol., 1982, 32,
454); 39.3% H_2SO_4, bicyclic and monocyclic hydrocarbons and alcohols,
1,4- and 1,8-cineole, and other products (I.I. Bardyshev, A.E.Sedel'nikov, and
T.S.Tikhonova, Vestsi Akad. Navuk BSSR, Ser. Khim. Navuk, 1975, 66; I.I.
Bardyshev, A.E. Sedel'nikov, and O.N. Druzhkov, ibid., 1976, 57);
vermiculite activated by H_2SO_4, camphene (53-58%), dipentene-limonene
(17-18%), tricyclene (7.3-8%), and other products (Sh. B. Battalova and T.R.
Mukitanova, Izv. Akad. Nauk Kaz. SSR, Ser. Khim., 1975, 25, 49; Vestn.
Akad. Nauk Kaz. SSR, Ser. Khim., 1977, 33); hydrobiotite activated with

H_2SO_4, camphene and tricyclene (M.I. Goryaev *et al.*, U.S.S.R., 321,085, 1976); sulphonated cation exchanger Ankalite (L.P. Petaline *et al.*, Izv. Akad. Nauk Kaz. SSR, Ser. Khim., 1976, <u>26</u>, 81); china clay treated with H_2SO_4 and activated at 350°, camphene (M. Nazir, M. Ahmad, and F.M. Chaudhary, Pak. J. Sci. Ind. Res., 1976, <u>19</u>, 175); Albanian clay, camphene and tricyclene (S. Kullaj, Bul. Shkencave Nat., 1985, <u>39</u>, 47); borophosphate (Sh.B. Battalova *et al.*, Izv. Akad. Nauk Kaz. SSR, Ser. Khim., 1986, 82); and mordenite, a mixture of 12 products (Y. Liu *et al.*, Linchan Huaxue Yu Gongye, 1988, <u>8</u>, 10).

Studies have been made of the following; the catalytic isomerization of pin-2-ene and car-3-ene in the vapour phase [V.V. Bazyl'chik *et al.*, Zh. Prikl. Khim. (Leningrad), 1975, <u>48</u> 582]; of pin-2-ene in the vapour phase by diatomite (*idem*, Izv. Vyssh. Uchebn. Zaved., Khim. Khim. Tekhol., 1976, <u>19</u>, 234) and to obtain maximum yield of camphene (S. Li, Linchan Huaxue Yu Gongye, 1983, <u>3</u>, 37); direct isomerization and esterification of pin-2-ene to bornyl and isobornyl acetates in the presence of AcOH and boracetic acid anhydride (M. Riaz *et al.*, Pak. J. Sci. Ind. Res., 1988, <u>31</u>, 541); isomerization of pin-2-ene in the presence of organic acids (R.M. Markevich, A.I. Lamotkin, and V.M. Reznikov, Khim. Drev., 1985, 103), eg. trifluoroacetic acid (*idem, ibid.*, p96); thermal rearrangement of pin-2-ene (J. De Pascual Teresa *et al.*, An. Quim., 1978, <u>74</u>, 301) and derivatives (Y. Bessiere, C. Grison, and G. Boussac, Tetrahedron, 1978, <u>34</u>, 1957); and different catalytic systems in the isomerization of pin-2-ene (Sh.B. Battalova, A.A. Likerova, and T.R. Mukitanova, Izv. Akad. Nauk SSR, Ser. Khim., 1975, <u>25</u>, 70). A number of catalysts have been used to facilitate the isomerization of pin-2-ene to pin-2(10)-ene and *vice versa*, e.g., CaO, MgO, SrO, and BaO (R. Ohnishi and K. Tanabe, Chem. Letters, 1974, 207), 1% Pd/Al_2O_3 (G.L. Kaiser, U.S. Pat. 3,974,103, 1976), and mixed metal catalysts (9:1 Pd-Au) (*idem, ibid.*, 3,974,102).

Treatment of (+)-pin-2-ene with $BuLi-(Me_2NCH_2)_2$ followed by Me_3SnCl gives the stannane derivative (2) (59%), which on hydrolysis with HCl in $MeOH-4\%H_2O$ yields (+)-pin-2(10)-ene (85%) (M. Andrianome and B. Delmond, Chem. Comm., 1985, 1203). Basic catalyzed isomerization of (+)--pin-2-ene and (-)-pin-2(10)-ene yields an equilibrium mixture (2-ene : 2(10)--ene = 96:4) (A. Ferro and Y.R. Naves, Helv., 1974, <u>57</u>, 1152). The D^+--catalyzed rearrangement of pin-2-ene using lanthanoid-assisted ^1H NMR spectroscopy (R. Muneyuki *et al.*, Chem. Letters., 1979, 49) and of pin-2-ene and apopinene (*idem*, J. Org. Chem., 1988, <u>53</u>, 358) have been investigated.

$Mn(OAc)_3$ oxidation of (+)-pin-2-ene affords *α*-terpineol acetate, *cis*-pin-3-en-3-ol acetate, myrtenol acetate, and compound (3) (K. Witkiewicz and Z. Chabudzinski, Rocz. Chem., 1977, <u>51</u>, 475). Oxidation of pin-2-ene by lead oxide in AcOH gives *cis*- and *trans*-carveol (Y. Fijihara and Y. Matsubara, Kinki Daigaku Kogakuba Kenkyu Hokoku, 1976, <u>10</u>, 7), other products including *p*-menth-4(8)-en-l-yl acetate, and *trans*-verbenyl acetate have also been obtained (Y. Fujihara, M. Nomura, and Y. Matsubara, *ibid.*, p1); CrO_3 in various solvents affords verbenone; and SeO_2, myrtenal (J.A. Retamar, Essenze Deriv. Agrum., 1989, 59, 159). Oxidation by palladium (II)

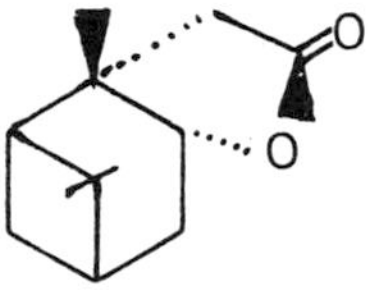

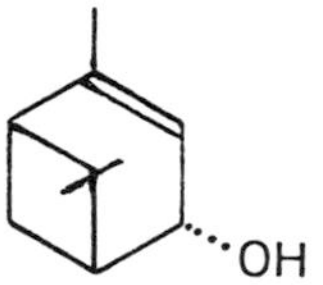

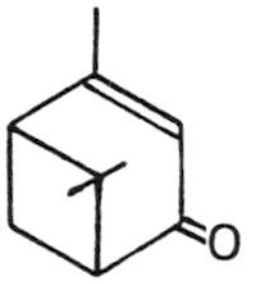

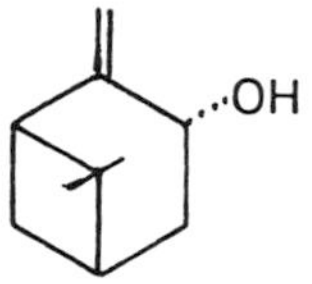

(3)   (4)   (5)   (6)

salts [V. Sirakova and D. Dimitrov, Khim. Ind. (Sofia), 1987, 59, 205], and OsO_4 catalyzed trimethylamine N-oxide (E. Erdik and D.S. Matteson, J. Org. Chem., 1989, 54, 2742) have been studied. (+)-Pin 2-ene oxidized by *Acetobacter methanolicus* gives (+)-*trans*-verbenol (4), verbenone (5), *trans*--pinocarveol (6), and *trans*-sobrerol (L. Weber and M. Doerre, Z. Chem., 1988, 28, 98).

Photooxygenation of pin-2-ene in protic media (MeCN:H_2O, 3:2) gives epidioxybornane (7) (18%), hydroperoxymenthenol (8) (5%) and acetamido-hydroperoxymenthene (9) (2%) (P. Capdevielle and M. Maumy, Tetrahedron Letters, 1980, 21, 2417) and the reaction of photogenerated singlet oxygen with pin-2-ene give *trans*-3-hydroperoxypin-2(10)-ene (99.3%) and with pin--2(10)-ene, 10-hydroperoxypin-2-ene (99.9%) (C.W. Jefford *et al.*, Helv., 1973, 56 2649). The oxidation of (-)-pin-2- and -2(10)-ene in the presence of microwave discharge oxygen yields (-)-*cis*-pin-3-en-2-ol and *trans*-verbenol, and (-)-myrtenol and (-)-myrtenal, respectively (M. Nomura and Y. Fujihara, Yukagaku, 1985, 34, 467). The anodic oxidation of (-)-pin-2-ene in aqueous THF containing $HClO_4$, $NaClO_4$, or NaOH affords racemic monoterpenes *trans*--sobrerol (21%), pinol (30%), *cis*- and *trans*-carveol (6 and 20%), and cymene (6%) [V. Montiel *et al.*, J. Chem. Res. (S), 1987, 27].

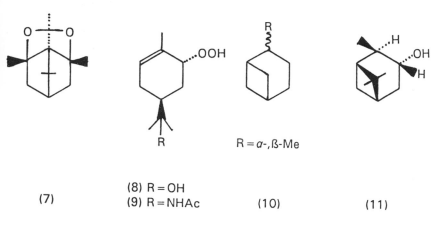

(7)

(8) R = OH
(9) R = NHAc

R = α-, ß-Me

(10)

(11)

The autoxidation of pin-2-ene has been carried out using a number of reagents and catalysts, eg. sulpholane (M. Nomura and Y. Fujihara, Kinki Daigaku Kogakubu Kenkyu Hokoku, 1980, 14, 7), cobalt acetylacetone (P.J. Martinez de la Cuesta, E.R. Martinez, and A.S. Carrillo, Afinidad, 1988, 45, 523; 1991, 48, 32), cobalt naphthenate (L.R. Bohe, Essenze Deriv. Agrum., 1983, 53, 148; S.C. Sethi, A.D. Natu, and M.S. Wadia, Indian J. Chem., 1986, 25B, 248) and (PrO)$_3$B [W.S. Angelow, D.I. Dimitrov, and I.N. Vuchkov, Chem. Tech. (Liepzig), 1975, 27, 167]. The ozonolysis of (-)-pin-2--ene furnishes a diastereoisomeric mixture of bis(6,6-dimethylnorpinane-2,2--diyl)diperoxide (K.H. Overton and P. Owen, J. Chem. Soc., Perkin I, 1973, 226).

The hydrogenation of pin-2-ene in the presence of Ru gives cis-pinane stereoselectively (M.S. Pavlin, U.S. Pat. U.S. 4,310,714, 1982) and the rate of hydrogenation of pin-2-ene to pinane over a Ni catalyst increases with temperature at 120-170° (A. Krasuska and J. Guberska, Prezm. Chem., 1981, 60, 259). Pin-2-ene has been hydrogenated over Pd/C, Pd/Amberlyst, Pd/Amberlyst 15, Pd/basic Sephadex QAE50, and Pd/alginic acid to give pinanes (10; R = α-, ß-Me) (C. Allandrieu et al., Bull. Soc. Chim. Fr., 1977, 519). Hydroboration of pin-2-ene with BH$_2$Cl in Et$_2$O at -5° affords isopinocampheol (11), and pin-2(10)-ene gives cis-myrtanol. Also reported are the hydroboration of camphene and car-3-ene (I. Uzarewicz and A. Uzarewicz, Rocz. Chem., 1976, 50, 1315; 1975, 49, 1113).

The adducts from pin-2- and -2(10)-ene and dibromocarbene have been isolated and characterized. Pin-2-ene-dibromocarbene adduct furnishes

3-bromo-8,8-dimethyl-2-methylenebicyclo[4.1.1]oct-3-ene (12) and 3-bromo-
-2,8,8-trimethylbicyclo[4.1.1]octa-2,4-diene (13) (G. Mehta and S.C. Narang,
Indian J. Chem., 1972, 10, 1057). Pin-2-ene reacts with dichlorocarbene to
give the tricyclooctane (14), which on LiAlH$_4$ reduction yields a mixture of

(12)   (13)   (14) R^1 = Me, R^2 = H   (16)
                        (15) R^1 = R^2 = H

bicyclo[4.1.1]octenes. Similarly apopinene affords (15), which is
dechlorinated to the hydrocarbon with Na/NH$_3$, and cis-pin-3-ene gives 2-
-dichloromethyl-2,6,6-trimethylbicyclo[3.1.1]hept-3-ene, trans-pin-3-ene does
not react (D. Joulain and F. Rouessac, Bull. Soc. Chim. Fr., 1973, 1428; S.
Julia and A. Ginebreda, Synth., 1977, 682). Carbenylation of pin-2-ene by
N$_2$CHCO$_2$Et catalyzed by CuSO$_4$ gives the tricyclic ester (16) (99%) [P.A.
Krasutskii et al., Ukr. Khim. Zh. (Russ. Ed.), 1989, 55, 842]. The addition of
IN$_3$ to pin-2-ene in MeCN affords (±)-cis-6-azido-8-(5-methyltetrazolyl)-p-
-menth-1-ene, whereas in HCONMe$_2$ the product is optically active endo-2-
-azido-6-iodobornane. Similar additions to pin-2(10)-ene have also been
investigated (B. Bochwic and B. Olejniczak, Rocz. Chem., 1973, 47, 315;
Zesz. Nauk, Politech. Lodz., Chem., 1973, 27, 157).

Electrophilic thiylation of pin-2-ene by H$_2$S in the presence of EtAlClBr
yields trans-pinane-2-thiol and p-menth-1-ene-8-thiol. Similarly pin-2(10)-ene
in the presence of AlBr$_3$ in CHCl$_3$ gives a mixture of products containing p-
-menth-1-ene-8-thiol and camphene yields diisobornyl sulphide. Also the
reaction between pin-2-ene and BuSH has been investigated (G.A. Tolstikov
et al., Zh. Org. Khim., 1983, 19, 2075). The addition of ClSO$_2$NCO to pin-2-,
-2(10)-ene, car-3-ene and camphene, and the rearrangement of some of the
products have been reported (T. Sasaki, S. Eguchi, and H. Yamada, J. Org.

Chem., 1973, <u>38</u>, 679; J.R. Malpass, Tetrahedron Letters, 1972, 4951; G.T. Furst *et al.*, Tetrahedron, 1973, <u>29</u>, 1675). The addition of *N,N*-dichloro-phosphoramide to pin-2-ene furnishes adduct (17), probably by a radical reaction, followed by a Wagner-Meerwein rearrangement. Norbornene has been treated in a similar fashion (B. Olejniczak, K. Osowska, and A. Zwierzak, *ibid.*, 1978, <u>34</u>, 2051).

Chlorination of pin-2-ene at room temperature with Me_3COCl in the presence of free radical initiators gives myrtenyl and verbenyl chlorides and 2,6-dichlorocamphane along with other products (I. Uzarewicz and A. Uzarewicz, Rocz. Chem., 1973, <u>47</u>, 921) and on bromination it affords *endo*--2-*endo*-6-dibromobornane, which on bromination gives 2,3-*endo*-6--tribromoborn-2-ene (R.M. Carman and G.J. Walker, Austral. J. Chem., 1977, <u>30</u>, 1393). Preparation of 9-iodopin-2-ene (F. Derguini, Y. Bessiere, and G. Linstrumella, Synth. Comm., 1981, <u>11</u>, 859).

R = $(EtO)_2P(O)$

(17) (18) (19) R = OH (20) R = OH
              (21) R = CH_2CH_2OH (22) R = CH_2CH_2OH

Treatment of pin-2-ene with Me_3COK and BuLi gives intermediate (18), which is oxidized with H_2O_2 to yield (1*S*,5*R*)-myrtenol (19) (42%) and (1*R*,3*R*,5*S*)-*trans*-pinocarveol (20) (~1%); (18) with ethylene oxide affords (21) (25%) and (22) (13%) (G. Rauchschwalbe and M. Schlosser, Helv., 1975, <u>58</u>, 1094). The hydration of pin-2- and -2(10)-ene, and car-3-ene with chlorinated acetic acid (Y. Matsubara, K. Tanaka, and M. Kenbou, Nippon Kagaku Kaishi, 1974, 1590) and along with camphene with formic acid in the presence of synthetic zeolites (M. Nomura and Y. Fujihara, Kinki Daigaku Kogakubu Kenkyu Hokoku, 1985, <u>19</u>, 1); and the effects of Na_2SO_4 on the hydration of pinene [A.I. Bibicheva, Z.R. Golovina, and L.I. Petukhova, Pishch. Prom-st. (Moscow), 1988, 41] have been investigated. Pin-2- and -2(10)-ene are methoxylated in the presence of zeolite catalysts (S.S. Koval'skaya, N.G.

360

Kozlov, and S.V. Shavyrin, Zh. Org. Khim., 1990, 26, 1947). The alkylation of phenol by pin-2-ene (V.I. Moskvichev and L.A. Kheifits, *ibid.*, 1973, 9, 2256; Zh. Vses. Khim. O-va., 1975, 20, 479) and the Friedel-Crafts acylation of pin-2-ene (S.K. Srivastava, A. Akhila, and M. C. Nigam, Indian J. Chem., 1984, 23B, 897) have been reported.

The reactions of pin-2-ene with NOCI (S.W. Markowicz, Rocz. Chem., 1975, 49, 2117), HOCl (J. Wolinsky and M.K. Vogel, J. Org. Chem., 1977, 42, 249), $(Me_2CHCH_2)_2AlH$ (V.P. Yur'ev *et al.*, Zh. Obshch. Khim., 1974, 44, 2084), and carboxylic acids (G.N. Valkanas, J. Org. Chem., 1976, 41, 1179); and pin-2- and -2(10)-ene with $NaNO_2$ (C.G. Francisco *et al.*, J. Chem. Soc., Perkin I, 1984, 459), and along with camphene with diethyl phosphonate (R.L. Kenney and G.S. Fisher, J. Org. Chem., 1974, 39, 682) have been discussed.

10-Phenylpin-2-ene (P.E. Peterson and G. Grant, J. Org. Chem., 1991, 56, 16), phenylpinene derivatives (R. Mechoulam, N. Lander, and S. Dikstein, Brit. UK Pat. Appl. 2,027,021, 1980), pin-2-ene-10-d$_3$ (Y. Stenstroem and L. Skatteboel, Acta Chem. Scand., 1980, B34, 131), 10-aminopin-2-ene and *trans*-3-aminopin-2(10)-ene (I. Wyzlic, I. Uzarewicz, and A. Uzarewicz, Pol. J. Chem., 1990, 64, 113) and 10-alkylaminopin-2-enes (S.W. Markowicz *et al.*, Pol. PL 112,750, 1982) have been prepared. The following have been synthesized from pin-2-ene, juvenile hormone analogues (R. Sterzycki, W. Sobotka, and M. Kocor, Rocz. Chem., 1977, 51, 735; L. Borowiecki and E. Reca, Ann., 1982, 1775), 4,4,6,6-tetramethylbicyclo[3.1.1]hept-2-yl propionate and 6,6-dimethyl-4-ethylbicyclo[3.1.1]hept-3-en-2-yl propionate used as insecticides against *Periplaneta americana* (Mitsubishi Chemical Industries Co. Ltd., Japan P. 81 87,536, 1981), 1,7,7-trimethyl[2.2.1]-heptane-2-thiol, 8-*N*-acylamino-*p*-menth-6-en-2-one oximes (S.S. Koval'skaya, N.G. Kozlov, and S.V. Shavyrin, Zh. Org. Khim., 1990, 26, 1947), (+)--carvone (W.-L. Liu and Y.-S. Cheng, Proc. Natl. Sci. Counc. Repub. China, 1981, 5, 21; H.J. Liu and J.M. Nyangulu, Tetrahedron Letters, 1989, 30, 5097), and (3,3-dimethylallyl)diisopinocampenylborane a novel reagent for chiral isoprenylation of aldehydes (H.C. Brown and P.K. Jadhav, *ibid.*, 1984, 25, 1215). The C-3 atom of pin-2-ene has been exchanged for nitrogen to yield 3-azapin-2-ene (23) [M. Gannon, A. Postlewhite, and R.S. McElhinney, J. Chem. Res. (S), 1979, 393).

(23)

¹H NMR spectral data of pin-2-ene and apopinene (C.A.N. Catalan *et al.*, Riv. Ital. EPPOS, 1981, 63, 289), ¹³C NMR spectral data (M.C. Hall, M. Kinns, and E. J. Wells, Org. Magn. Reson., 1983, 21, 108) and enthalpies of vaporization (X. An *et al.*, Wuli Huaxue Xuebao, 1987, 3, 668) of pin-2- and -2(10)-ene, and derivatization and separation by gas chromatography of sub--milligram quantities of the enantiomers of pin-2-, and -2(10)-ene, camphene, sabinene, thuj-3-ene, and car-3-ene (D.M. Satterwhite and R.B. Croteau, J. Chromatogr., 1987, 407, 243) have been reported.

(iii) *Pin-2(10)-ene (ß-pinene, nopinene, pin-2,10-ene)*

Pinane is catalytically converted into pin-2(10)-ene (24) by IrH₅-[P(CHMe₂)₃]₂ at 100° in the presence of Me₃CCH = CH₂ as hydrogen acceptor (Y.Lin, D. Ma, and X. Lu, J. Organometallic Chem., 1987, 323, 407). (+)-Pin-2(10)-ene has been synthesized from (+)-pin-2-ene (L.M. Harwood and M. Julia, Synth., 1980, 45; H.C. Brown, M. Zaidlewicz, and K.S. Bhat, J. Org. Chem., 1989, 54, 1764) and (-)-pin-2(10)-ene from (-)--pin-2-ene (Y.F. Min, B.W. Zhang, and Y. Coo, Synth., 1982, 875). The reaction of linalool with BF₃ or iodine does not give pin-2(10)-ene has previously reported (Y. Fujita, S. Fujita, and H. Okura, Nippon Kagaku Kaishi, 1974, 132).

(24) 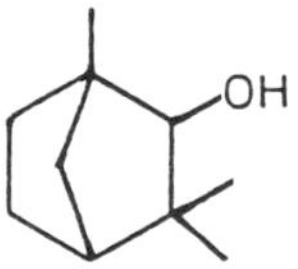   (25) 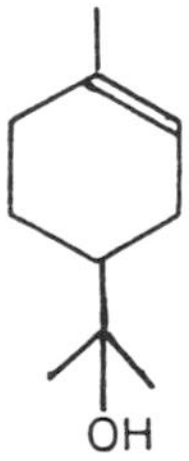   (26) 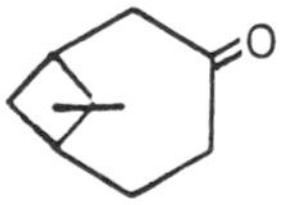   (27)

Isomerization of pin-2- and -2(10)-ene in the presence of TiO₂-H₂SO₄ at 155-180° gives mainly camphene (~60%) (M. Dul and M. Bukala, Chem. Stosow, 1973, 17, 19) and the thermal rearrangement of pin-2(10)-ene affords myrcene and other products (J. De Pascual Teresa *et al.*, An. Quim., 1978, 74, 305; P.J.R. O'Malley, Chem. Abs., 1984, 100, 175074n; Y.

362

Yuan and Z. Cheng, Linchan Huaxue Yu Gongye, 1989, 9, 10). For the catalytic hydrogenation of olefins homologous to pin-2(10)-ene see M. Barthelemy, A. Gianfermi, and Y. Bessiere (Bull. Soc. Chim. Fr., 1976, 182) and for the dehydrogenation of pin-2(10)-ene to *p*-cymene over Pt-Al$_2$O$_3$ catalyst, V. Krishnasamy, P. Mathur, and K. Chandrasekharan (J. Indian Chem. Soc., 1983, 60, 49).

Oxidation of pin-2(10)-ene by Pb(OAc)$_4$ in aqueous solution yields fenchyl alcohol (25), *a*-terpineol (26), and 7,7-dimethylbicyclo[4.1.1]octan-3-one (27), in addition to *trans*-pinocarveol, myrtenol, and perillyl alcohol (H. Ohue *et al.*, Osaka Kogyo Daigaku Kiyo, Rikohen, 1978, 22, 175; K. Yokoi and Y. Matsubara, Nippon Kagaku Kaishi, 1979, 641) and by SeO$_2$ in pyridine, myrtenal (28) and pinocarvone (29) (H. Ohue *et al.*, Osaka Kogyo Daigaku Kiyo, Rikohen, 1978, 22, 183). For oxidation using PbO$_2$ in AcOH see Y. Fujihara and Y. Matsubara (Yuki Gosei Kagaku Kyokai Shi, 1976, 34, 243).

| (28) | (29) | (30) | (31) | (32) |

Bromination of pin-2(10)-ene with *N*-bromosuccinimide furnishes myrtenyl bromide (30) (C.A.N. Catalan, D.J. Merep, and J.A. Retamar, An. Soc. Cient. Argent., 1973, 196, 35) and hydrosilylation by HSiCl$_3$ and Me$_2$CHCH$_2$PCl$_2$ in the presence of Ni(acac)$_2$ in an autoclave at 120° yields products (31 and 32) (V.V. Kaverin *et al.*, Izv. Akad. Nauk SSSR, Ser. Khim., 1980, 2657). Addition of small amounts of HBr to pin-2(10)-ene results in isomerization to pin-2-ene, whereas with excess HBr bornyl and fenchyl bromide are formed (M. Barthelemy and Y. Bessiere-Chretin, Bull. Soc. Chim. Fr., 1974, 1703).

The thermal ene addition reaction of (1S,5S)-(-)-pin-2(10)-ene to Cl$_3$CCHO yields a 17:83 mixture of adduct (33) and its diastereoisomer (G.B. Gill and B. Wallace, Chem. Comm., 1977, 382) and addition reactions with

MeCOCO$_2$Me, maleic anhydride, HC≡CCO$_2$Et, and CH$_2$=CClCN give [34; R = MeC(OH)CO$_2$Me (100%), 2,5-dioxo-3-furyl (74%), E-CH=CHCO$_2$Et (80%), and CH$_2$CHClCN (85%), respectively] (J.A. Gladysz and Y.S. Yu, *ibid.*, 1978, 599). Also for the addition of maleic anhydride see R.K. Hill, J.W. Morgan, and R.V. Shetty (J. Amer. Chem. Soc., 1974, 96 4201).

(34) R = MeC(OH)CO$_2$Me,
2,5-dioxo-3-furyl,
E-CH = CHCO$_2$Et,
CH$_2$CHClCN
(35) R = SO$_2$NHSO$_2$Ph
(36) R = CH$_2$SiMe$_3$, Bu,
CH$_2$CH$_2$CHMe$_2$, Ph

(33)

Aromatic aldehydes add to pin-2(10)-ene under Lewis acid catalyzed Prins reaction conditions to afford the corresponding homoallylic alcohols (M. Majewski and G.W. Bantle, Synth. Comm., 1990, 20, 2549). Reaction of pin-2(10)-ene with PhSO$_2$NSO gives derivative (35), which on treatment with a Grignard reagent derived from RCl (R = Me$_3$SiCH$_2$, Bu, Me$_2$CHCH$_2$CH$_2$, Ph) yields (36). Similarly related compounds were prepared from pin-2-ene and car-2-ene (G. Deleris, J. Dunogues, and A. Gadras, Tetrahedron Letters, 1984, 25, 2135). For the hydration of pin-2(10)-ene in the presence of mordenite catalyst see Y. Liu, Q. Chen, and Y. Li (Linchan Huaxue Yu Gongye, 1989, 9, 35).

Pin-2(10)-ene reacts with AcOCH$_2$CH$_2$OAc in the presence of (Me$_3$C)$_2$O$_2$ to give the menthene derivatives (37 and 39), which on hydrolysis yield (ρ-menth-1-en-7-yl)ethanediol (38) and (ρ-menth-1-en-7-yl)propanoic acid (40) (M. Cazaux, B. Maillard, and R. Lalande, Compt. Rend., 1972, 275, 1133) and with (NH)$_2$Ce(NO$_3$)$_6$ and MeOH absorbed on silica gel to furnish methyl α-terpinyl ether (T. Kurata, N. Kobayashi, and M. Kawarada, Yukagaku, 1989, 38, 553). 10-Trimethylsilylpin-2-ene, 2-dimethylmethoxysilylpinane (D. Wang and T.H. Chan, Canad. J. Chem., 1987, 65, 2727), (+)-hinesol (41) and its

10-epimer (D.A. Chass, D. Buddhasukh, and P.D. Magnus, J. Org. Chem., 1978, 43, 1750), (+)-nootkatone (42) (G. Asanuma and S. Torii, *ibid.*, 1982, 47, 4622), and (+)-ß-selinene (L. Moore, D. Gooding, and J. Wolinsky, *ibid.*, 1983, 48, 3750) have been synthesized from (-)-pin-2(10)-ene. 3-Acylpin--2(10)-enes (J.P. Pillot *et al.*, *ibid.*, 1979, 44, 3397).

(37) R = Ac
(38) R = H

(39) R = CH$_2$CH$_2$OAc
(40) R = H

(41)

(42)

^{1}H NMR spectral data and conformation of pin-2(10)-ene, pinocarvone, and *cis*- and *trans*-pinocarveols (R.J. Abraham *et al.*, Org. Magn. Reson., 1973, 5, 373), structure of pin-2(10)-ene by electron diffraction methods (V.A. Naumov and V.M. Bezzubov, Zh. Strukt. Khim., 1972, 13, 977), its molecular geometry as deduced from the crystal and molecular structure of *cis*-pinocarvyl-*p*-nitrobenzoate (G.F. Richards *et al.*, J. Org. Chem., 1974, 39, 86), and irradiation in MeCN-MeOH (D.R. Arnold and X. Du, J. Amer. Chem. Soc., 1989, 111, 7666) have been discussed.

(iv) Pin-3-ene (δ-pinene, pin-3,4-ene)

The elimination reaction of (-)-isopinocampheol tosylate in the presence of Me$_3$COK at 105° affords *cis*-pin-3-ene (43) (63%) (Z. Rykowski, H. Orszanska, and Z. Chabudzinski, Bull. Acad. Pol. Sci., Ser. Sci. Chim., 1976, 24, 681). Treatment of (+)-4-trimethylstannylpin-2(10)-ene (*cis:trans* = 1:1.8) with HCl-dioxane gives a mixture (85%) containing *cis*-pin-3-ene (43) (77%), *trans*-pin-3-ene (44) (16%), and pin-2-ene (7%). Similar treatment of *trans*-(+)-4-trimethylstannylpin-2(10)-ene yields a mixture (87%) containing *cis*-pin-3-ene (83%) and *trans*-pin-3-ene (17%) (A.N. Kashin *et al.*, Izv. Akad. Nauk SSSR, Ser. Khim., 1981, 1180; Zh. Org. Khim., 1982, 18, 2233).

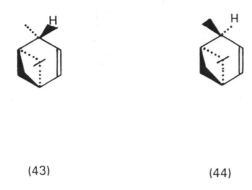

(43)   (44)

(v) Miscellaneous hydrocarbons

Reaction of *trans*-4-bromoapopin-2-ene (45) with RMgX (R = Me, Et, CH_2 = CMeCH$_2$, PhCH$_2$; X = I, Br, Cl) affords the corresponding 4-alkylapopin- -2-ene (46) (D. Joulain, C. Moreau, and M. Pfau, Tetrahedron, 1973, 29, 143). Apopinene undergoes ring cleavage at the C(1)-C(7) bond on treatment with PdCl$_2$ in AcOH to give 1,2,3-trimethylbenzene (R.M. Giddings and D. Whittaker, Tetrahedron Letters, 1978, 4077).

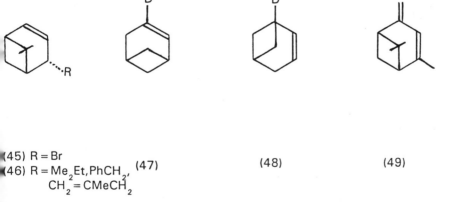

(45) R = Br
(46) R = Me, Et, PhCH$_2$,   (47)   (48)   (49)
    CH_2 = CMeCH$_2$

366

Hydrolysis of norpin-2-one *p*-tolylsulphonylhydrazone with DCl or D_3PO_4
gives the bicycloheptene (47), which is isomerized at 226-236° to the
valence isomer (48) (K. Dietrich and H. Musso, Ber., 1974, <u>107</u>, 731). *Cis*-
and *trans*-6,6-dimethyl-2-vinylnorpinane (M. Ukita and Y. Hashioka, Yuki
Gosei Kagaku Kyokai Shi, 1975, <u>33</u>, 698) have been prepared, and the
photochemistry of 4-methylverbene (49) (P.S. Mariano and D. Watson, J.
Org. Chem., 1974, <u>39</u>, 2774) and the addition of HCl to nopadiene and
homoverbenene (B. Bochwic and S.W. Markowicz, Rocz. Chem., 1976, <u>50</u>,
87) have been reported.

(b) Epoxides, alcohols, aldehydes, ketones, and acids

(i) Epoxides
Epoxidation of pin-2- and -2(10)-ene by Me_3CCH_2OOH in the presence of
$MoCl_5$ or $Mo(CO)_6$ gives 2,3- and 2,10-epoxypinane (50 and 51),
respectively, in 85-95% yields (V.P. Yur'ev *et al.*, Izv. Akad. Nauk SSSR, Ser.
Khim., 1974, 919). Similarly 2,3- and 3,4-epoxycarane are obtained from
car-2- and -3-ene, respectively. 2,3-Epoxypinane is also prepared by the
oxidation of pin-2-ene with aqueous AcOOH in the presence of Na_2CO_3 (A.M.
Romanikhin and N.I. Popova, Izv. Vyssh. Uchebn. Zaved., Khim. Khim.
Tekhnol., 1975, <u>18</u>, 1967) and 2,10-epoxypinane (88%) by treating pin-
-2(10)-ene with AcOOH, Na_2CO_3, and AcONa in CH_2Cl_2 (M. Guo *et al.*,
Huaxue Shijie, 1987, <u>28</u>, 495). Epoxidation of pin-2-ene by pin-2-ene
hydroperoxide in the presence of molybdenyl acetylacetonate affords 2,3-
-epoxypinane (50%) and a mixture of alcohols (A.M. Romanikhin, Izv. Vyssh.
Uchebn. Zaved., Khim. Khim. Tekhnol., 1977, <u>20</u>, 1807).

(50)   (51)

Hydrogenolysis of 2*a*,3*a*-epoxypinane in EtOH yields pinol and sobrerol. In hexane at 60° and 60 atmospheres pressure, and with the use also of Raney-Ni, isocarvomenthol and isopinocampheol are obtained in high yield (Z. Rykowski, K. Burak, and Z. Chabudzinski, Rocz. Chem., 1975, 49, 1335). Addition of a solution of 2,10-epoxypinane in THF to a solution of B_2H_6-LiBH$_4$ in THF affords a mixture of monohydric alcohols (~90%) [containing myrtenol (74%)] and a diol (~10%). If the addition is reversed the products are monohydric alcohols (~10%) and diol (~90%) (E. Segiet-Kujawa and A. Uzarewicz, Rocz. Chem., 1974, 48, 2303).

Isomerization of 2,3-epoxypinane with H_2CrO_4/SiO_2 and H_2CrO_4/Al_2O_3 affords *a*-campholenaldehyde (52) and carvone (53), respectively (T. Kurata and T. Koshiyama, Yukagaku, 1988, 37, 130); over SiO_2-Al_2O_3, SiO_2-TiO_2,

$R = a\text{-}CH_2OH, \text{ ß-}CH_2OH$

(52)　　　　(53)　　　　(54)　　　　(55)

solid H_3PO_4 and $FeSO_4$, preferentially 5-membered ring aldehydes (K. Arata and K. Tanabe, Chem. Letters, 1979, 1017); in the presence of Pt, Pd, Ru, and Rh at 100-200°, carvacrol and other products (T. Kurata, Yukagaku, 1981, 30, 562); with $Al(OCHMe_2)_3$ in an organic solvent, pinocarveol (F. Scheidl, Ger. Offen. DE 3,143,227, 1983); and with Me_3COK in DMF or pyridine, *trans*-pinocarveol (54) (77 and 91%, respectively) (Z. Rykowski, K. Burak, and Z. Chabudzinski, Rocz. Chem., 1974, 48, 1619). In the presence of synthetic zeolites, 2,3-epoxypinane in DMF gives *trans*-carveol, and 2,10--epoxypinane in sulpholane yields myrtenol and in Cl_3CCO_2H, perillyl alcohol (M. Nomura and Y. Fujihara, Nippon Kagaku Kaishi, 1985, 990; 1987, 883; Yukagaku, 1986, 35, 454). Also described are products obtained under different reaction conditions.

Reactions of 2,3-epoxypinane with $KMnO_4$ (T. Kurata, T. Koshiyama, and H. Kawashima, *ibid.*, 1987, <u>36</u>, 206), Na_2SO_3 (E. Myslinski, Rocz. Chem., 1973, <u>47</u>, 1755), trimethylamine hydrofluoride (M.M. El Gaied and A. Selmi, J. Soc. Chim. Tunis., 1987, <u>2</u>, 3), RCN (R = Me, $MeOCH_2CH_2$, Ph) in the presence of H_2SO_4 (S.S. Koval'skaya, N.G. Kozlov, and S.V. Shavyrin, Zh. Obshch. Khim., 1989, <u>59</u>, 1356), and Li in ethylenediamine (K.N. Gurudutt, S. Rao, and A.K. Shaw, Indian J. Chem., 1991, <u>30B</u>, 345) have been reported. (+)-2a,3a-Epoxypinan-4-one on treatment with 0.5 or 2M amounts of borane gives 2a,3a-epoxypinan-4-ol and the triols (55), respectively (A. Uzarewicz and E. Segiet-Kujawa, Pol. J. Chem., 1978, <u>52</u>, 63).

Both 2,3- and 2,10-epoxypinanes are oligomerized in the presence of PF_5 or BF_3 (E.R. Ruckel, R.T. Wojcik, and H.G. Arlt, Jr., J. Macromol. Sci., Chem., 1976, <u>A10</u>, 1365). Oleuropeic acid (56) has been obtained from (-)-2,10-epoxypinane (R. Pellegata *et al.*, Synth. Comm., 1985, <u>15</u>, 165) and some 4-substituted 2,3-epoxypinanes and pin-2-enes synthesized (I. Ribas *et al.*, An. Quim., Ser. C, 1982, <u>78</u>, 31).

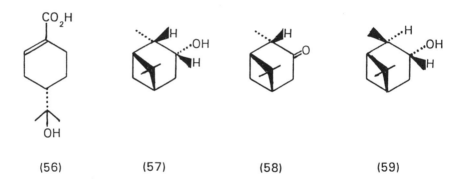

(56)   (57)   (58)   (59)

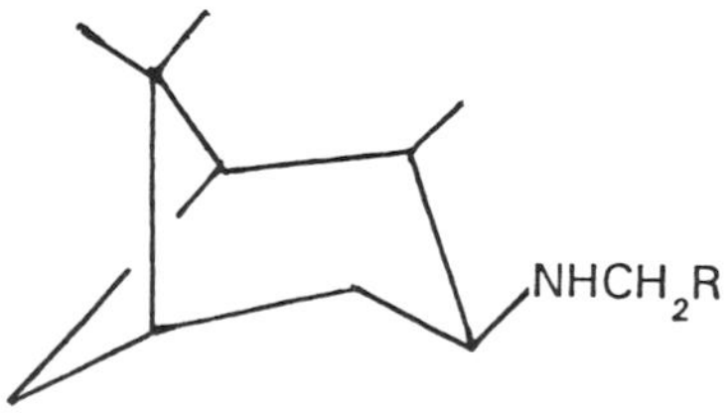

R = Me, Et, Pr

(60)

(ii) Pinocampheol (pinan-3-ol) (57) and pinocamphone (pinan-3-one) (58)

Hydroamination of isopinocampheol (59) with the appropriate nitrile yields the (1*S*,2*S*,3*R*)-bicycloheptylamine (60) (G.V. Kalechits, N.G. Kozlov, and T. Valimae, Zh. Org. Khim., 1987, 23, 2377).

Pinocamphone (58) practically free of pinocarveol is prepared by heating 2,3-epoxypinane with 0.03-0.3 mol. Al alkoxide in xylene at 140-180° (F. Scheidl, Ger. Offen. DE 3,143,226, 1983). The oximes of pinocamphone (58) and isopinocamphone (61) undergo the Beckmann rearrangement in aqueous Me$_2$CO-NaOH containing *p*-MeC$_6$H$_4$SO$_2$Cl to yield lactams (62 and 63), respectively (A. Zabra, C. Wawrzenczyk, and H. Kuczynski, Bull. Acad. Pol. Sci., Ser. Sci. Chim., 19, 22, 855). The acid-catalyzed reaction between

(61)

(62) R^1 =H, R^2 =Me
(63) R^1 =Me, R^2 =H

(64)

pinocamphone and HCHO (C.K. Ryu and S.H. Park, Nonchong-Han'guk Saenghwal Kwahak Yonguwon, 1988, 42, 129) has been studied. A condensation product with diethylmalonate (Z. Rykowski and Z. Chabudzinski, Pol. J. Chem., 1980, 54, 741), some Schiff bases (S.W. Markowicz, J. Karolak-Wojciechowska, and W. Kwiatkowski, J. Crystallogr. Spectrosc. Res., 1989, 19, 535), an oxime and a 4-bromo derivative (A. Gilis, Tezisy Doklady-Resp. Konf. Molodykh Uch-Khim., 2nd, 1977, 1, 6, Akad. Nauk Est. SSR. Inst. Khim: Tallin, USSR) have been obtained from 2*α*--hydroxypinocamphone. The structure of (+)-2-hydroxypinocamphone (64) obtained by the KMnO$_4$ oxidation of (-)-pin-2-ene and the ^{1}H NMR spectral data of its *O*-benzyl derivative (A. Solladi-Cavallo *et al.*, Bull. Soc. Chim. Fr., 1989, 544) have been determined. The stereoisomerism and crystal structure of (1*R*,2*S*,5*R*)-(+)-4-[benzoyloxy)methylene]pinan-3-one (N. Azani *et al.*, Bull. Chem. Soc. Jpn., 1989, 62, 3403) have been discussed.

Epimeric 4-amino-3ß-hydroxy-*cis*-pinanes have been obtained from 2-hydroxy-isopinocamphone (K. Burak and Z. Chabudzinski, Pol. J. Chem., 1981, 55, 387).

(iii) Myrtanol, myrtenol, myrtenal, and myrtenic acid

Hydroboration-oxidation of (-)-pin-2(10)-ene and (±)-pin-2-ene with BH$_3$--Me$_2$S in hexane, followed by H$_2$O$_2$ affords (-)-*cis*-myrtanol (85%) and (±)--isopinocampheol (92%), respectively (C.F. Lane, J. Org. Chem., 1974, 39, 1437). Treatment of pin-2(10)-ene with (Me$_2$CHCH$_2$)$_3$Al, followed by oxidation with oxygen and hydrolysis gives *trans*-myrtanol (62.5%) and *trans*--pinane-1,10-diol (2.2%), the tosylate of which on reduction with LiAlH$_4$ yields (-)-*trans*-pinan-1-ol (H. Benn, J. Brandt, and G. Wilke, Ann., 1974, 189). Photosensitized oxidation of pin-2(10)-ene using natural chlorophyll as the sensitizer yields a hydroperoxide, which on reduction with NaBH$_4$ affords myrtenol. Similarly, *trans*-pinocarveol is obtained from pin-2-ene (R.L. Kenney and G.S. Fischer, Ind. Eng. Chem., Prod. Res. Develop., 1973, 12, 317). Treatment of (+)-2α,10-epoxypinane with BH$_3$ in THF in the presence of LiBH$_4$ gives a mixture of mono alcohols containing (-)-myrtenol (74%) (A. Uzarewicz and E. Segiet-Kujawa, Rocz. Chem., 1977, 51. 2147). Myrtenyl and perillyl esters (H. Miyawaki, Japan P. 79 59,253, 1979), methyl myrtenyl thioether (F. Chatopoulos-Ouar and G. Descotes, J. Org. Chem., 1985, 50, 118). Methylepoxymyrtanol (65) reacts with LiAlH$_4$ to yield diol (66) and unsaturated alcohol (67), whereas dimethylepoxymyrtanol (68) affords only diol (69) (F. Chatzopoulos and Y. Bessiere, Bull. Soc. Chim. Fr., 1982, 362).

(65) R = H   (66) R = H
(68) R = Me   (69) R = Me   (67)   (70)   (71)

Oxidation of pin-2-ene with activated SeO_2 in EtOH furnishes myrtenal
(70) (61%) (Q. Lin, X. Li, and S. Deng, Huaxue Shijie, 1990, 31, 299).
Treatment of 2,3-epoxymyrtanol with $LiAlH_4$ in Et_3N gives myrtenal (70)
(20%) and pinocarvone (75%) (M.M. El Gaied and A. Selmi, J. Soc. Chim.
Tunis., 1985, 2, 27). Myrtenic acid (71) has been converted to (-)-pinane-
-3,10-diol (L. Borowiecki and E. Reca, Rocz. Chem., 1976, 50, 1689). For
the ^1H NMR spectral data of methyl myrtenate and trans-apoverbenyl bromide
see C.A.N. Catalan and D.I. Acosta De Iglesias [Arch. Bioquim., Quim. Farm.,
1977 (Pub. 1978), 20, 133].

(iv) Verbanol, verbanone, verbenol, and verbenone
(-)-Neoverbanol (72) is prepared by the $LiAlH_4$ reduction of (+)-trans-
-verbanone, obtained by treating 6,6-dimethylnorpin-3-en-2-one with CuI and
MeLi (P.D. Hobbs and P.D. Magnus, J. Chem. Soc., Perkin I, 1973, 2879).
The irradiation of (-)-2,3-epoxyverbanone results in ring contraction to a
mixture of epimeric bicyclo[2.1.1]hexane ß-diketones (T. Gibson, J. Org.
Chem., 1974, 39, 845). Alkylation of cis-verbanone with
$CH_2 = CH(CH_2)_8CO_2Me$ affords ketone (73) (M. Chatzopoulos, B. Boinon, and
J.P. Montheard, Compt. Rend., 1975, 281, 1015).

$$R = (CH_2)_{10}CO_2Me$$

(72)   (73)   (74)   (75)

(+)-trans-Verbenol and its antipode, the pheromone of *Dendroctonus* bark
beetles have been synthesized from (+)- and (-)-pin-2-ene, respectively (K.
Mori, Agric. Biol. Chem., 1976, 40, 415) and the enantiomers of cis-verbenol
from (+)- and (-)-verbenone, respectively, (1S,4S,5S)-pin-2-en-4-ol is the
pheromone of the *Ips* beetle (K. Mori, N. Mizumachi, and M. Matsui, *ibid.*,
p1611). (+)-trans-Verbenol is obtained by oxidation of pin-2-ene with
$Pb(OAc)_4$, followed by conversion to its 3,5-dinitrobenzoate and fractional
crystallization of the (±)-, (-)-mixture (Z.G. Isaeva and V.V. Karlin, Doklady

Akad. Nauk SSSR, 1984, <u>279</u> 113). The conformations of some verbenols and methylnorpinols have been deduced from the ¹H NMR spectral data obtained using the shift reagent tris(dipivaloylmethane)europium (C. Nishino and H. Takayanagi, Agric. Biol. Chem., 1979, <u>43</u>, 2323). Studies on the sex pheromone mimic (+)-*trans*-verbenyl acetate of the American cockroach (*idem, ibid.*, 1980, <u>44</u>, 2877) and the conformation of some related pinan-3--ol derivatives (*idem, ibid.*, 1979, <u>43</u>, 2399) have been reported. The dichlorocarbene adduct of pin-2-ene on reduction with Na/NH₃, followed by ozone oxidation gives the verbenone analogue (74), which on reduction with NaBH₄ yields the related verbenol (75) (J. Wu *et al.*, Youji Huaxue, 1986, 456). *Trans*- and *cis-p*-mentha-1,8-dien-5-ol have been synthesized from *trans*-verbenol (M. Bulliard, G. Balme, and J. Gore, Synth., 1988, 972). Both antipodes of ipsdienol have been obtained by the pyrolysis of the respective pinenols, resulting from the LiAlH₄ reduction of (+)- and (-)-verbenone (G. Ohloff and W. Giersch, Helv., 1977, <u>60</u>, 1496).

Reaction of apopinene with *N*-bromosuccinimide gives bromo derivative (76), which on treatment with KOH affords *trans*- and *cis*-apoverbenol [77 (88%) and 78 (12%), respectively], converted to apoverbenone (79) on oxidation with Jones reagent or MnO₂. Hydrogenation of *trans*-apoverbenol over PtO₂ gives *trans*-nopinol (80), whereas *cis*-apoverbenol affords a mixture of nopinone (81) and *cis*-nopinol (82) (D.J. Merep *et al.*, Riv. Ital. Essenze, Profumi, Piante Off., Aromat., Syndets, Saponi, Cosmet., Areosols, 1978, <u>60</u>, 613).

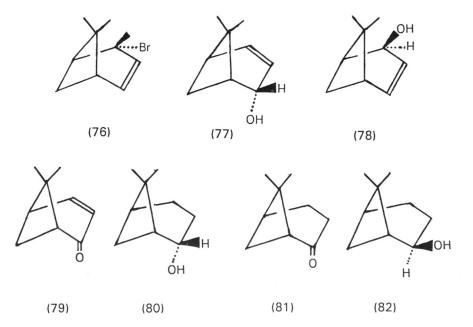

(76)   (77)   (78)

(79)   (80)   (81)   (82)

(v) Chrysanthenone

Reduction of chrysanthenone (83), prepared by photoisomerization of verbenone, with LiAlH₄ gives chrysanthenol and hydrogenation over a PtO₂ or Pd-C catalyst yields chrysanthanone, which on LiAlH₄, reduction affords chrysanthanol (D.J. Merep, C.A.N. Catalan, and J.A. Retamar, *ibid.*, 1975, 57, 197). Treatment of chrysanthenone with HCl causes rearrangement, probably to compound (84) and oxidation with SeO₂ yields aldehyde (85) [D.J. Merep and J.A. Retamar, An. Acad. Bras. Cienc., 1972, 44 (Suppl.), 355].

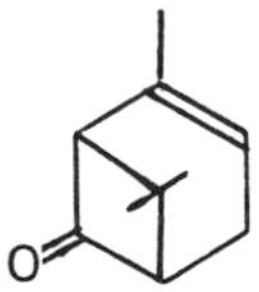

(83)	(84)	(85)	(86)

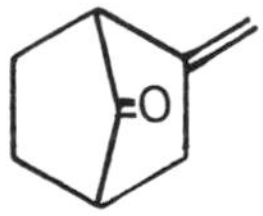

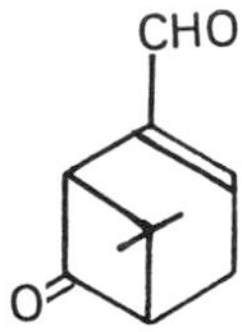

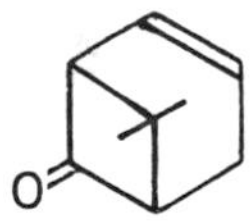

Irradiation of 9-trideuterioverbenone in cyclohexane or AcOH affords a 1:1 mixture of 8- and 9-trideuteriochrysanthenone (G.W. Shaffer and M. Pesaro, J. Org. Chem., 1974, 39, 2489). UV irradiation of (±)-apoverbenone gives (±)-apochrysanthenone (86), (±)-apoisopiperitenone, and Me₂C=CHCH₂CH=CHCH=CO. Apochrysanthenol, apochrysanthanone, and apochrysanthanol are obtained from apochrysanthenone (C.A.N. Catalan *et al.*, Riv. Ital. EPPOS, 1981, 63, 209).

(vi) Pinocarveol and pinocarvone

Oxidation of pin-2(10)-ene with 50% aqueous H₂O₂ in the presence of SeO₂ in Me₃COH gives *trans*-pinocarveol (87) (J.M. Coxon, E. Dansted, and M.P. Hartshorn, Org. Synth., 1977, 56, 25), also obtained on boiling 2,3--epoxypinane (88) in Et₂O in the presence of LiNEt₂ (J.K. Crandall and L.C. Crawley, *ibid.*, 1973, 53, 17). ¹H NMR spectral data of *cis*- and *trans*--pinocarveol (C.C. Hinckley and W.C. Brumley, J. Magn. Reson., 1976, 24, 239; H.T. Velez, Rev. Cienc. Quim., 1983, 14, 51) and photolyis of pinocarvone (89) (T.D.R. Manning, Tetrahedron Letters, 1974, 2669) have been reported.

374

(87)   (88)   (89)   (90)   (91)

(vii)Miscellaneous alcohols, ketones, and derivatives

The preparation of pinan-2-ols (90) (R.R. Risco and S. Lemberg, U.S. Pat. 3,723,542, 1973; L.A. Shutikova *et al.*, Maslo-Zhir. Prom., 1973, 23; Stepan Chemical Co., Fr. Demande 2,216,252, 1974), and their conformations (J. Texter and E.S. Stevens, J. Org. Chem., 1979, 44, 3222), ^{13}C NMR spectral data (J.W. Blunt and P.J. Steel, Austral. J. Chem., 1982, 35, 2561), and rearrangements (H. Indyk and D. Whittaker, J. Chem. Soc., Perkin II, 1974, 313) have been discussed. Grandisol has been synthesized from pinan-2ß-ol (P.D. Hobbs and P.D. Magnus, Chem. Comm., 1974, 856).

3*a*-Tosyloxy-pin-2*a*-ol (Z. Rykowski, O. Gubrynowicz, and J. Wrzesien, Pol. J. Chem., 1983, 57, 1237), 3-aminopinan-2-ol (K. Burak and Z. Chabudzinski, *ibid.*, 1978, 52, 1721), 3-aminopinan-4-ol (*idem, ibid.*, 1981, 55, 2015), and 10-(dimethylaminomethyl)pin-2-en-4-one (H. Krieger, A. Kojo, and A. Oikarinen, Finn. Chem. Letters, 1978, 185), have been prepared and the gas phase pyrolysis of some *endo*-pin-2-en-7-ols (P.A.E. Cant, J.M. Coxon, and M.P. Hartshorn, Austral. J. Chem., 1975, 28, 621), and the pyrolysis of *endo*-10ß-pinan-7-yl acetate (*idem, ibid.*, p391), have been reported. Pure (-)-3-formylpinane (W. Himmele *et al.*, Tetrahedron Letters, 1976, 911) has been prepared and tricyclic spiroketone (91) obtained from (+)-3-formylpinane (W. Giersch and K.H. Schulte-Elte, Eur. Pat. Appl. EP 382,934, 1990). 2-Methylpinan-4-one, 4-acetoxy-2-methyl- and 4-acetoxy--10-methylpinane, and 4-acetoxy-10-methylpin-2-ene (Mitsubishi Chemical Industries Co., Ltd. Japan P. 81 30,940, 1981) have been prepared.

The structure of filipendulol (pin-2-en-7-ol) isolated from *Achillea filipendulin* has been confirmed by chemical and spectral methods (A.D. Dembitskii, R.A. Yurina, and M.I. Goryaev, Izv. Akad. Nauk kaz. SSR, Ser. Khim., 1980, 55).

The reaction between nopyl tosylate (92) and AlEt$_3$ (G.A. Tolstikov et al., Izv. Akad. Nauk SSSR, Ser. Khim., 1985, 1814) and the fluorination of nopol (93) with CF$_3$CHFCF$_2$NEt$_2$ to give nopyl fluoride (94) and nopyl tetrafluoropropionate (95) (S. Watanabe et al., Yukagaku, 1984, 33, 58) have been reported. Acetolysis of tosylate ester of nopol (92) gives the acetate of 8,8--dimethyltricyclo[5.1.1.0^{2,5}]nonan-2ß-ol (96) (R.M. Giddings et al., J. Chem. Soc., Perkin II, 1986, 1525).

CH$_2$CH$_2$O$_3$SC$_6$H$_4$Me-4

CH$_2$CH$_2$R

HO

(92)

(93) R = OH
(94) R = F
(95) R = CF$_3$CHFCO$_2$

(96)

(-)-cis-Pinane-trans-2,3-diol, prepared by oxidation of (-)-pin-2-ene or reduction of (+)-2-hydroxypinocamphone, on dehydration with p--MeC$_6$H$_4$SO$_3$H gives (+)-pinol, (-)-isopinocamphone, and (+)-α--campholenaldehyde. The latter product is the result of a Wagner-Meerwein rearrangement, followed by a bond cleavage (J. De Pascual Teresa et al., An. Quim., 1976, 72, 560; 1978, 74, 1012). The dehydration of cis-pinane-cis--2,3-diol has also been investigated (idem, ibid., p950). Detosylation of 3α--tosyloxypinan-2α-ol in aprotic solvents gives mainly pinocamphone and isopinocamphone (K. Witkiewicz and Z. Chabudzinski, Rocz. Chem., 1977, 51, 475). For the reaction of pinane-2α,3α-diol with Ac$_2$O see Z. Rykowski and M. Skwarek (ibid., 1974, 47, 1555) and for direct chiral syntheses using pinanediol boronic esters, D.S. Matteson and R. Ray (J. Amer. Chem. Soc., 1980, 102, 7590).

(viii) Nopinone
Oxidation of (-)-pin-2(10)-ene with aqueous alkaline KIO$_4$ in the presence of KMnO$_4$ affords in addition to the expected nopinone (97), a ketol (98) (C.W. Jefford, A. Roussel, and S.M. Evans, Helv., 1975, 58, 2151).

376

(97)   (98)   (99)

The following have been reported, a 10 step synthesis of nopinone (G.S.S.
Murthi and Alok Mazumder, Indian J. Chem., 1971, 20B, 339), the
conversion of (1R,5R)-(+)-pin-2-ene to (1S,5R)-(-)-nopinone (P. Lavallee and
G. Bouthillier, J. Org. Chem., 1986, 51, 1362), preparation of 2-
-methylnopinone (A. Yoshikoshi, Y. Takagi, and T. Akiyama, Japan P. 73
91,047, 1973) and 3-methylnopinone (M. Barthelemy and Y. Bessiere,
Tetrahedron, 1976, 32, 1665), the acid-catalyzed ring opening of a number
of substituted nopinones to yield 4-(2-propyl)cyclohex-2-enones (J.M. Coxon,
G.J. Hydes, and P.J. Steel, ibid., 1985, 41 5213), cyclobutane ring cleavage
of (+)-cis-3-methylnopinone (M. Kato et al., J. Org. Chem., 1989, 54,
1536), crystal structure and circular dichroism of (-)-3-[(o-chlorobenzoyloxy)-
(o-chlorophenyl)-α-methylene]nopinone (99) (N. El Batouti et al., Z.
Kristallogr., 1989, 187, 85) and (-)-3-(p-bromobenzylidene)nopinone [R.
Roques et al., J. Chem. Res. (S), 1980, 370], and the conformation and
electronic structures of nopinone, isonopinone and related ketones (J.
Fournier, ibid., 1977, 320).

Acetonylnopinone (100) is obtained on reacting nopinone with
$CH_2 = CMeOAc$ in the presence of $Mn(OAc)_3$. Similarly cis-verbanone and
isonopinone yield acetonyl derivatives (101 and 102, respectively) (M.
Chatzopoulos and J.P. Montheard, Compt. Rend., Ser. C, 1977, 284, 133).
Nopinylamines (103) are prepared by catalytic hydrogenation of nopinone in
the presence of NH_3 or amines (W. Hoffmann, N. Mueller, and J. Paust, Ger.
Offen. 2,545,657, 1977). Treatment of 3ß-bromonopinone with 1,8-
-diazabicyclo[5.4.0]undec-7-ene in DMF affords apoverbenone (Mitsubishi

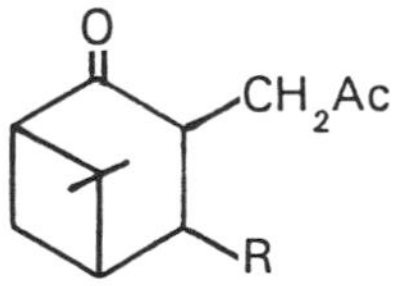

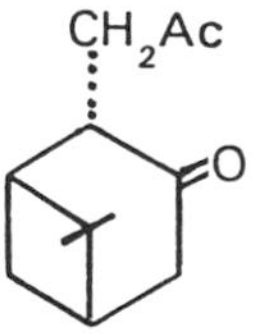

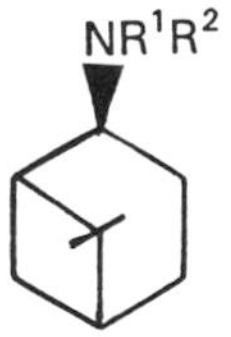

O

CH₂Ac

CH₂Ac

NR¹R²

O

R¹ = R² = H,Me

R

(100)R = H
(101)R = Me   (102)   (103)

Chemical Industries Co., Ltd. Japan P. 81 79,640, 1981). (+)-3-*trans*-
-Ethylidenenopinone (T. Yanami *et al., ibid.*, 80 45,649, 1980), (-)-nopinone
enol acetate [R.A. Archer and W.A. Day, Pat. Specif. (Aust.) AU 517,594,
1981].

(ix) Acids

Stereoisomeric pinonic acid (104) is obtained in optimum yield by the
ozonization of pin-2-ene in 98% AcOH (S.V. Chudinov, E.T. Nesterova, and
L.B. Kuprina, Lesokhim. Podsochka, 1976, 10). ^{1}H NMR spectral data of
methyl (±)-*cis*-pinonate have been recorded (E. Liepins, R. Kampare, and F.
Avotins, Latv. PSR Zinat. Akad. Vestis, Kim. Ser., 1975, 89). Bromination of
(±)-*cis*-pinonic acid by dioxane dibromide in Et$_2$O gives compounds (105 and
106) (F. Avotins and E. Liepins, *ibid.*, 1976, 220). (±)-*cis*-2,2-Dimethyl-3-
-(bromoacetyl)cyclobutylacetic acid (105) has been converted to
2,2-dimethyl-3-(2-aminothiazol-4-yl)cyclobutylacetic acids (F. Avotins and I.
Mikelsone, *ibid.*, 1984, 347). Capture of the radical from the decarboxylation
of (+)-*cis*-pinonic acid has been investigated (D.H.R. Barton, N. Ozbalik, and
M. Schmitt, Tetrahedron Letters, 1989, 30, 3263). Pin-2-ene-4-carboxylic
acids (107) analogues of cockroach pheromones have been synthesised (H.
Takahashi and C. Nishino, Koen Yoshishu-Koryo, Terupen oyobi Seiyu Kagaku
ni Kansuru Toronkai, 23rd, 1979, 80. Chem. Soc. Japan: Tokyo, Japan).

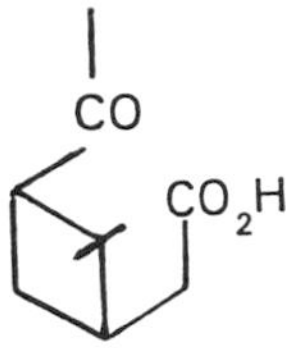

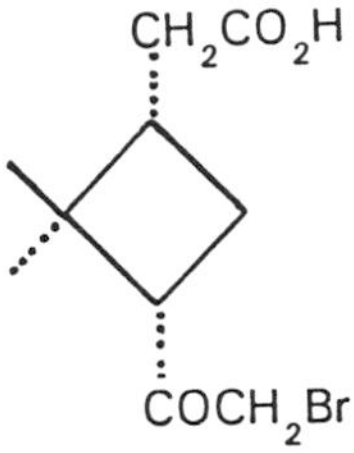

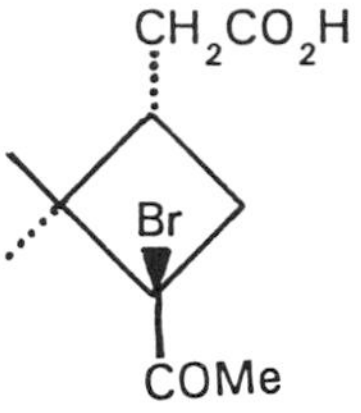

378

$Z = \alpha$-H, ß-CO$_2$H;
α-CO$_2$H, ß-H

(104)   (105)   (106)   (107)

(c) Halogeno, amino, and other derivatives

The *syn*-addition of DCl to the less hindered site of pin-2- and -2(10)-ene gives chlorides (108 and 109), respectively. Rapid rearrangement of the tertiary chlorides yields the corresponding bornyl chlorides (E.F. Weigand and H.J. Schneider, Ber., 1979, 112, 3031). Stereochemistry of the reaction of (+)-*trans*-4-chloropin-2-ene with Me$_3$SnLi (A.N. Kashin *et al.*, Izv. Akad. Nauk SSSR, Ser. Khim., 1980, 1950), cleavage of some verbenylstannanes (D. Young and W. Kitching, J. Org. Chem., 1985, 50, 4098), rearrangement of 2-chloro(or bromo)-3-nitrosopinane in C$_6$H$_6$ or CCl$_4$ in the presence of SiO$_2$ to give 6-halogenocamphor oxime (C.H. Brieskorn and E. Hemmer, Ber., 1976, 109, 1418), deamination of 10-aminopinanes (E. Chong-Sen, R.A. Jones, and T.C. Webb, J. Chem. Soc., Perkin II, 1974, 38) and 2α(H),3α-aminopinane (R.M. Giddings *et al.*, *ibid.*, 1982, 725), and reaction of 2α(H),10- -aminopinane (*cis*-myrtanylamine) with nitrous acid (P.I. Meikle and D. Whittaker *ibid.*, 1974, 318) have been investigated.

The reaction of pin-2-ene with Me$_3$SiN$_3$ and Pb(OAc)$_4$ in CH$_2$Cl$_2$ yields *cis*-pin-3-ene-2-yl azide (110), which on heating in AcOH rearranges to give azide (111) (A. Stuetz and E. Zbiral, Ann., 1972, 765, 34). The pinenyl phenyl selenide (112) on treatment with *N*-chlorosuccinimide is transformed to the rearranged allylic chloride (113) (T. Hori and K.B. Sharpless, J. Org. Chem., 1979, 44, 4208).

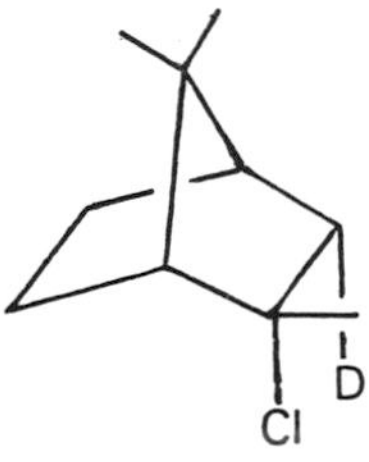

(108) 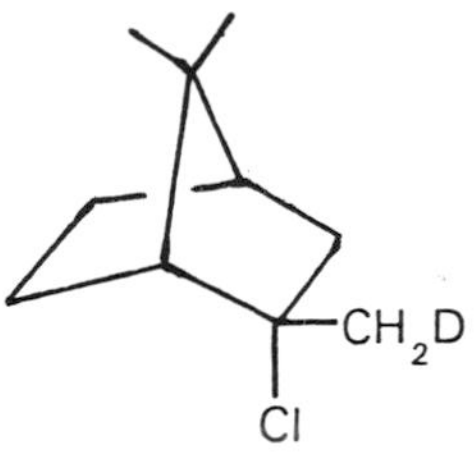   (109)    (110) 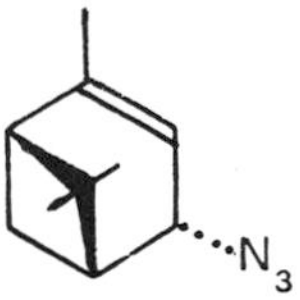   (111)

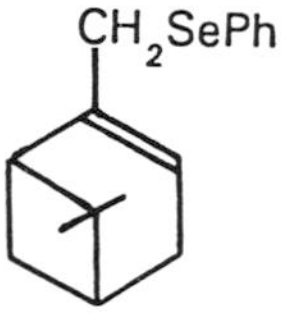

(112) 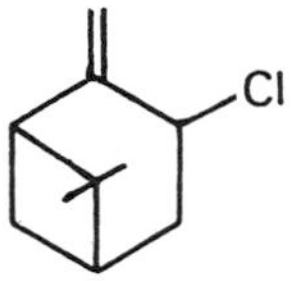   (113)

6. Tricyclene

Although tricyclene does not occur naturally it is a product or an intermediate of many reactions of bicyclic monoterpenes and its chemistry is reported in section 6 (the santene group) of Volumn IIC of the second edition.

The oxidation of tricyclene (1) with Pb(OAc)$_4$ gives 2,2-dimethyl-5--hydroxynorbornane (2), tricyclenyl alcohol (3,3-dimethyl-2-hydroxymethyltricyclo[2.2.1.0^{2,6}]heptane) (3), and camphor (H. Iwamuro et al, Nippon Nogei Kagaku Kaishi, 1983, 57, 1097).

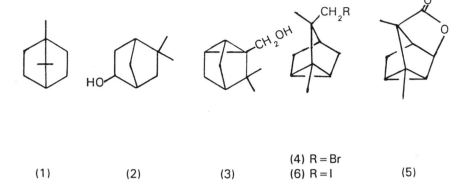

(1)　　　　　　(2)　　　　　　(3)

(4) R = Br
(6) R = I

(5)

8-Bromotricyclene (7-bromomethyl-1,7-dimethyltricyclo[2.2.1.0²,⁶]heptane)
(4) has been synthesized from methyl 3-methyl-2-methylenenorborn-5-ene-3-
-carboxylate (H. Monti and D. Raffin, Compt. Rend., Ser. 2, 1984, 299, 539).
Carbocationic intramolecular cyclocondensation of 3-methyl-2-methylene-
norborn-5-ene-3-carboxylic acid gives the oxatetracyclononanone (5), which
on reduction with Et₃SiH and then with Me₂SBH₃, followed by tosylation and
iodination yields 8-iodotricyclene (6), a key intermediate for terpene synthesis
(J. Kang, W.K. Lee, and H.T. Shin, Bull. Korean Chem. Soc., 1987, 8, 264).
Tricyclene does not react with MeCN in aqueous H₂SO₄ at 30°, but
undergoes addition and cleavage at 50° to yield amide (7) (N.G. Kozlov and
T.E. Kozlova, Zh. Obshch. Khim., 1986, 56, 233).

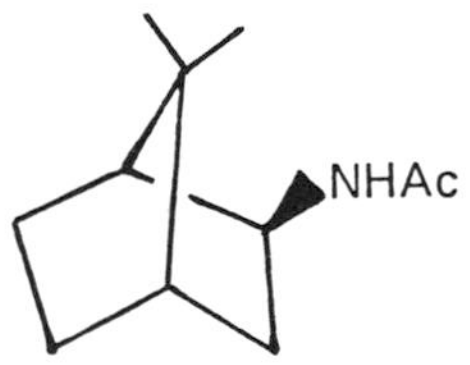

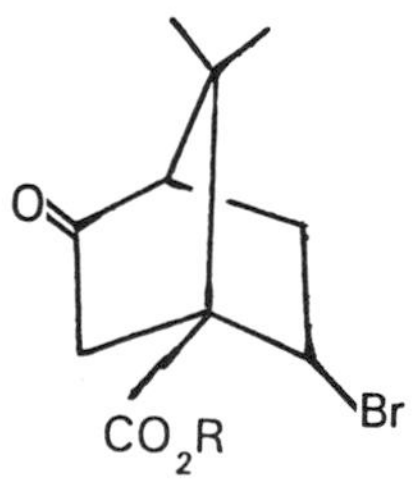

R = H, Me

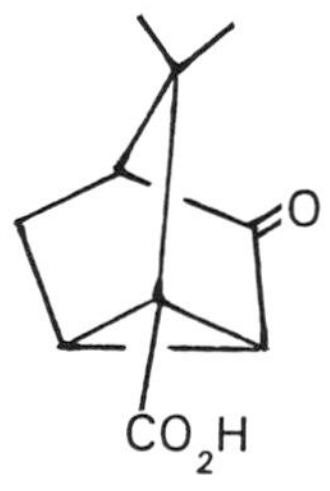

(7)　　　　　　(8)　　　　　　(9)　　　　　　(10)

3-Oxotricylenic acid (8) is prepared by the KMnO$_4$ oxidation of tricyclenic acid (9). Hydrobromination of (8) and its methyl ester gives *exo*-(10) (L. Borowiecki, B. Makowshi, and W. Wodzki, Pol. J. Chem., 1979, 53, 2267). The decomposition of tricyclenone tosylhydrazone by the Bamford-Stevens method has been investigated (L. Borowiecki and M. Welniak, *ibid.*, 1978, 52, 2173) and a synthesis of tricyclenone reported (L. Borowiecki and M. Welniak, Rocz. Chem., 1977 51, 1751).

Reaction of isotricyclen-2-one (1,7,7-trimethyltricyclo[2.2.1.0^{3,5}]heptan--2-one) (11) with (F$_3$CSO$_2$)$_2$O in the presence of F$_3$CSO$_3$H affords compound (12), which on cyclization in MeOH-H$_2$O containing Et$_3$N at 80° yields the tricycloheptyl ester (13) (A. Garcia Martinez and A. Garcia Fraile, An. Quim., Ser. C, 1980, 76, 127). Ketone (11) is obtained on reacting 3-diazocamphor in THF with catalytic CF$_3$CF$_2$CF$_2$CO$_2$Ag (F.C. Brown, D.G. Morris, and A.M. Murray, Synth. Comm., 1975, 5, 477).

| (11) | (12) | (13) |

7. The norbornane group: camphor and related compounds

(a) Camphene and related compounds

Conditions for obtaining the maximum yield of camphene (1) by heating pin-2-ene over titanic acid have been reported (N.M. Lien, P.T. Phuong, and N.T. Thuy, Tap Chi Hoa Hoc, 1977, 15, 13; Yasuhara Yushi Kogyo Co., Ltd., Japan P. 58 26,826 [83 26,826], 1983). Camphene is obtained in good yield by direct isomerization of Chinese turpentine in the presence of Ti catalyst (S. Li, Linchan Huaxue Yu Gongye, 1985, 5, 29) and (-)-camphene has been prepared from pure (+)-isocamphenilanic acid (G.W. Hana and H. Koch, Ber., 1978, 111, 2527). Camphene-8-[14]C (J.E. Oliver, J. Labelled Compd. Radiopharm., 1977, 13, 349; M. Stock and R. Horst, Z. Chem., 1987, 27, 65).

382

(3) R^1R^2 = CHI

(1)   (2)   (4) $R^1 = N_3$, $R^2 = CH_2I$ (5)

Camphene in aqueous $HClO_4$ at 80° gives a dimer (48%) and at 0°
diisobornyl ether (56%) (E.B. Starostina, A.M. Chashchin, and N.P.
Polyakova, Gidroliz. Lesokhim. Prom-st., 1990, 19). Camphene dimer (2)
along with other dimeric products is obtained on treating camphene with an
alumino-silicate catalyst (E.B. Starostina et al., Khim. Drev., 1989, 107).
$Pb(OAc)_4$ oxidation of camphene gives bicycloheptane and bicyclooctane
derivatives (H. Iwamuro et al., Nippon Nogei Kagaku Kaishi, 1983, 57, 1097)
and $PhC(O)O_2CMe_3$ catalyzed by Cu(I) salts gives a mixture of alcohols,
ethers, and hydrocarbons (M. Julia, D. Mansuy, and P. Detraz, Tetrahedron
Letters, 1976, 2141). Also reported is the oxidation of camphene with
$Mn(OAc)_3$ in the presence of cyanoacetic acid (K. Witkiewicz and Z.
Chabudzinski, Bull. Acad. Pol. Sci., Ser. Sci. Chim., 1978, 26 753).

The reaction between camphene and $ClCH_2CO_2H$, followed by $LiAlH_4$
reduction gives a hydroxydimethylnorbornylpropanol (T. Kishimoto, H.
Ishihara, and Y. Matsubara, Bull. Chem. Soc., Jpn., 1977, 50, 1897). With
HOCl a mixture of compounds including 10-syn- and 10-anti-chlorocamphene
are obtained (G. Buchbauer et al., Monatsh., 1984, 115, 509). The products
from reactions with IN_3 are compounds (3, 4, and 5) (B. Bochwic, G. Kuswik,
and B. Olejniczak, Tetrahedron, 1975, 31, 1607), with N_3H-BF_3-Et_2O azides
(6 and 7) (A. Pancrazi, I. Kabore, and K.H. Qui, Bull. Soc. Chim. Fr., 1977,
162), and with $Me_3CN:OsO_3$ an adduct, which on reduction gives derivative
(8) [G. Ruecker, M. Neugebauer, and B. Daldrup, Arch. Pharm. (Weinheim,
Ger.), 1988, 321 831]. The last reaction has been applied to pin-2(10)-ene
and the one with IN_3 to bornene.

(6)   (7)   (8)

The rearrangements during the interactions between camphene or α-
-fenchene and phenol in the presence of acid catalysts (I.S. Aul'chenko *et al.*,
Zh. Org. Khim., 1975, 11, 738), the reaction between camphene and α-
-cresol (W. Minemathu and Y. Mathubara, Nippon Kagaku Kaishi, 1974,
2361) and benzyl 3,4-dimethylphenyl ether (J.M. Espinos Taya *et al.*, Span.
461,473, 1978), and the prenylation of camphene (Azizur-Rahman *et al.*,
Tetrahedron, 1987, 43, 4119) have been investigated.

The preparation and/or reactions of the following camphene derivatives
have been reported; 8-, 9-, and 10-methylcamphene (J. Wolinsky and E.J.
Eustace, J. Org. Chem., 1975, 40, 3654); rearrangement of 8-alkyl-
camphenes (E.J. Eustace, Diss. Abstr. Int. B, 1972, 33, 2527); 1,4-
-dimethylcamphene (V. Tamminen, Finn. Chem. Letters, 1984, 49); 10-
-methyl- and 10,10-dimethyl-camphene (J.B. Lamture and U.R. Nayak, Indian
J. Chem., 1983, 22B, 853); phenylation of camphene with benzyne (G.
Mehta and B.P. Singh, Tetrahedron Letters, 1974, 4297); 8,8-dideuterio-
camphene (G. Buchbauer and H. Koch, Ber., 1978, 111, 2533); 7-bromo-,
7,8-dibromo-camphene, 2,3-dibromobornane, and bornyl bromides (L.
Borowiecki and M. Welniak, Rocz. Chem., 1975, 49, 559); 6-bromo-
camphene (T. Onishi *et al.*, Japan P. 60 181,037 [85 181,037], 1985); 4-
-nitrocamphene from 2-bromo-2-nitrobornane [J.P. Begue, C. Pardo, and J.
Sansoulet, J. Chem. Res. (S), 1978, 52]; 10-nitro- and 10-nitroso-camphene
(S. Ranganathan, B.B. Singh, and C.S. Panda, Tetrahedron, 1977, 33, 2415);
1-chloromercuriocamphene (V.G. Andrianov *et al.*, Izv. Akad. Nauk SSSR,
Ser. Khim., 1979, 2021); 1- and 4-chloromercuriocamphene (I.V. Shchirina-
Eingorn *et al.*, *ibid.*, 1983, 374); 1-camphenyl-lithium, -isobutylmercury, and
-triethyltin (D. Schenke and K.H. Thiele, J. Prakt. Chem., 1990, 332, 496);
camphenyl derivatives of chromium and uranium (A.N. Nesmeyanov *et al.*,
Izv. Akad. Nauk SSSR, Ser. Khim., 1979, 2826); camphene sultone,
sulphonation of 8-methylcamphene and pin-2-ene (J. Wolinsky, R. Marhenke,
and E.J. Eustace, J. Org. Chem., 1973, 38, 1428); LiAlH$_4$ reduction of

camphene sultone (J. Wolinsky and R. Marhenke, *ibid.*, 1975, 40 1766); 1- and 4-camphenyl triflates (A. Garcia Martinez, M. Gomez Marin, and L.R. Subramanian, An. Quim., 1978, 74, 972); aminomethylation of camphene (H. Krieger, A. Arstila, and L. Koskenniska, Oulun Yliopiston Kem. Laitoksen Raporttisar., 1983, 10); photochemistry (H.R. Sonawane, B.S. Nanjundiah, and M.D. Panse, Tetrahedron Letters, 1985, 26, 3507) of vinyl halides based on camphene (H.R. Sonawane, B.S. Nanjundiah, and S.I. Rajput, Indian J. Chem., 1984, 23B, 331); 1- and 4-(carbethoxyethoxymethyl)camphene (H.J. Liu, J. Org. Chem., 1975, 40, 2252); potential juvenoids with the camphene system in the terminal position (M. Welniak, J. Prakt. Chem., 1989, 331, 1002; L. Borowiecki, M. Welniak, and W. Wodzki, Pol. J. Chem., 1988, 62, 739); bishomocamphene, bishomocamphenilone, and bishomocamphenilanic acid (G. Buchbauer *et al.*, Monatsh., 1989, 120, 299); and 7-oxacamphene, 7-oxacamphenilone, and 7-oxacamphenilanic acid (G. Buchbauer and H. Holbik, Heterocycles, 1988, 27, 1217).

Hydroboration and hydroalumination of camphene, followed by oxidation affords *endo*- and *exo*-isocamphan-10-ol (A.V. Kuchin *et al.*, Zh. Obshch. Khim., 1979, 49, 1567) and Bertram-Walbaum hydration (AcOH-8%H_2SO_4) and hydrolysis gives 6-hydroxyisocamphane (6-hydroxy-2,2,3-trimethyl-norbornane) (K. Yokoi and Y. Matsubara, Nippon Kagaku Kaishi, 1979, 1121). 6-Hydroxycamphene (9), has been synthesized from tricyclene (10) (D.G. Patil *et al.*, Indian J. Chem., 1983, 22B, 189) and its ring cleavage by electrophilic reagents investigated (J.S. Yadav *et al.*, Tetrahedron, 1982, 38, 1003). Treatment of 6-hydroxycamphene with Cl_2 in CCl_4 affords cyclo-pentene (11) (J.S. Yadav, H.P.S. Chawla, and S. Dev, Tetrahedron Letters, 1977, 201). The reaction of camphene with alcohols in the presence of H_2SO_4 yields the corresponding bornyl ether (F. Scheidl and M. Gscheidmeier, Ger. Offen. DE 3,327,014, 1985).

(9)

(10)

(11)

(12)

(13)

R = CH(OAc)CH=CH$_2$

Optimal conditions (L.P. Meteshkina *et al.*, Gidroliz. Lesokhim. Prom-st., 1977, 21; I.N. Klabukova, G.V. Nesterov and B.A. Radbil, *ibid.*, 1980, 14), the use of a number of catalysts (Y. Kou and Y. Yin, Fenzi Cuihua, 1989, 3, 262), and a continuous process (I.N. Klabukova *et al.*, Gidroliz. Lesokhim. Prom-st., 1983, 15) for the formylation of camphene to give isobornyl formate have been reported. The vapour-liquid equilibrium of camphene--isobornyl acetate (P. Qian and A. Wang, Nanjing Linxueyuan Xuebao, 1984, 68) has been investigated. 10-Acetoxy-10-vinylcamphene (12) on treatment with BF_3 or $HClO_4$ rearranges to a bicyclo[3.2.1]octane derivative (13) (N. Lamb *et al.*, Canad. J. Chem., 1982, 60, 1055).

A helium photoelectron study of functionally substituted camphene derivatives (A.N. Nesmeyanov *et al.*, Docklady Akad. Nauk SSSR, 1981, 256, 121), the photolysis of 10-bromo-10-phenylcamphene (H.R. Sonawane *et al.*, Indian J. Chem., 1985, 24B, 202), ^{13}C NMR spectral data of camphene and pin-2-ene (W. Offermann, Org. Magn. Reson., 1982, 20, 203) and of some camphene derivatives (A. Garcia Martinez and M. Gomez Marin, An. Quim., 1978, 74, 339), ^{13}C and ^{1}H NMR and UV spectral data of some camphene derivatives (X. Mao *et al.*, Bopuxue Zazhi, 1988, 5, 117), and a NMR study of camphene under high hydrostatic pressure (K. Holderna-Matuszkiewicz, Acta Phys. Pol., 1987, A72, 637) have been reported.

Reaction of bornene with chloramine T and OsO_4 in Me_3COH furnishes compound (14), which on reduction with Na-liq.NH_3 gives (-)-3-*endo*-amino-bornan-2-*endo*-ol (15) (J. Wrzesien and Z. Rykowski, Pol. J. Chem., 1981, 55, 2629). The mercuration of bornene under varying conditions has been investigated (E.V. Skorobogatova, L.N. Povelikina, and V.R. Kartashov, Zh. Org. Khim., 1980, 16, 2318).

(14) R = SOC$_6$H$_4$Me-4
(15) R = H

(16)

6-*endo*-Hydroxy-*exo*-isocamphane and acetate (Y. Matsubara, K. Yokoi, and Y. Sato, Japan P. 79 30,144, 1979), the isocamphane analogue of mandelic acid (G. Buchbauer *et al.*, Monatsh., 1981, 112, 517), and isocamphylguaiacol ester (T. Sato, H. Tsuruta, and T. Yoshida, Japan P. 79 148,763, 1979; K. Bauer and G.K. Lange, Ger. Offen. 2,917,360, 1980) have been prepared and the Ritter reaction of isocamphan-6-ol with some nitriles investigated (N.G. Kozlov, L.A. Popova, and Nesterov, Zh. Obshch. Khim., 1986, 56, 1562). Also reported are, an approach to *endo*--isocamphane derivatives (G. Buchbauer *et al.*, Monatsh., 1982, 113, 1433), the syntheses of some *endo*-isocamphane derivatives (16; e.g., R = CO_2Et, CN, acyl, hydroxymethyl) (R. Vitek and G. Buchbauer, *ibid.*, 1985, 116, 801), an isocamphane analogous to ephedrine (G. Buchbauer *et al.*, *ibid.*, p.1209), isocamphanyl-3-hydroxypentanoic acid ethyl ester and isocamphanyl alkenes (G. Buchbauer, J. Zehetner, and R. Reidinger, Sci. Pharm., 1987, 55, 139), and isocamphanylboronic acid (17) (M. Tokles and J.K. Snyder, Tetrahedron Letters, 1988, 29, 6063). Treatment of isoborneol with PCl_5 and $CaCO_3$ in $CHCl_3$ at 0° gives *exo*-2-chloroisocamphane (*exo*-2-chloro-2,3,3-trimethyl-norbornane, camphene hydrochloride) (18) (R.M. Carman and I.M. Shaw, Austral. J. Chem., 1980, 33, 1631). Attempts to synthesize *endo*-2-chloro-isocamphane have been described (D.J. Brecknell, R.M. Carman, and K.L. Greenfield, *ibid.*, 1984, 37, 1075). 10-Bromo- and -chloro-isocamphane [2,2--dimethyl-3-(bromo and chloromethyl)bicyclo[2.2.1]heptane], 2--(bromomethyl)bornane, and 2-(chloromethyl)bornane have been prepared from (+)-camphor (M. Falorni, L. Lardicci, and G. Giacomelli, J. Org. Chem., 1986, 51, 5291).

CH₂B(OH)₂

(17)   (18)   (19)

The reactions of 2,10-dibromobornane with aqueous $AgNO_3$, bromine, and Mg (G. Mehta, Indian J. Chem., 1973, 11, 843), pyrolysis of 2-bromo-2-
-nitrobornane (S. Ranganathan and H. Raman, Tetrahedron, 1974, 30, 63), irradiation of polychlorobornane (H. Parlar and F. Korte, Chemosphere, 1983, 12, 927), preparation of bornane-2-thiol (1,7,7-trimethylbicyclo[2.2.1]-heptane-2-thiol) (19) (V.B. Baltrushaitis and K.V. Sadauskas, Fr. Demande 2,442,834, 1980), 2-chlorobornane-4-carboxamide and 4-aminomethyl-2-
-chlorobornane (G. Minardi, E. Bottini, and A. Gallazzi, Farmaco, Ed. Sci., 1977, 32, 147), 2-bornanylidenebornanes (H. Wynberg, K. Lammertsma, and L.A. Hulshof, Tetrahedron Letters, 1975, 3749) and their photochemical oxidation (H. Takeshita, T. Hatsui, and O. Jinnai, Chem. Letters, 1976, 1059), 2-(ß-hydroxyethoxy)bornane (J. Podlejski et al., Pol. PL 136,059, 1988), N-bornyl urea (M.E. Spiridonova, O.I. Korobkova, and L.I. Olishevets, Khim. Prir. Soedin., 1985, 841), and the bornane spirooxetanes (20) (E.F. Marchik et al., U.S.S.R. SU 1,225,840, 1986) have been reported.

R = Lower alkyl

(20)

(21)

6-Isobornyl-3,4-dimethylphenol (21), prepared by the reaction of camphene with 3,4-xylenol in the presence of $SnCl_4$, has bacteriostatic activity against Staphylococcus, Streptococcus, Pneumococcus, and Enterococcus (J.M. Gazave, Ger. 1,668,427, 1976). Some bornanes as potential juvenoids (L. Borowiecki and M. Welniak, Pol. J. Chem., 1989, 63, 149) and some stereoisomeric bornane and norbornane derivatives with spasmolytic activity [V. Vasilev, R. Marev, and B. Blagoev, Arch. Pharm. (Weinheim, Ger.), 1984, 317, 967] have been synthesized, and the action of red light on solid (-)-2-chloro-2-nitrosocamphane [(-)-2-chloro-2-nitroso-bornane] (N.N. Majeed et al., J. Chem. Soc., Perkin II, 1988, 1027) investigated. ^{1}H NMR spectral data of some 2-(trialkylsiloxy)camphane [2--(trialkylsiloxy)bornane] (I.P. Biryukov *et al.*, Zh. Fiz. Khim., 1974, 48, 2371), the chemical shifts of the C-8 and C-9 methyl groups of some bornane derivatives [L. Lacombe and L. Mamlok, J. Chem. Res. (S), 1977, 111], and ^{13}C NMR spectral data of some bornane (H. Duddeck *et al.*, Org. Magn. Reson., 1983, 21, 122) and isocamphane derivatives (V.I. Lysenkov, Zh. Org. Khim., 1988, 24, 1677) have been recorded.

(b) Camphor, epicamphor and related compounds

(i) Camphor, bornan-2-one, 2-oxobornane

Camphor has been synthesized from (+)-2-acetoxy-*p*-mentha-1,8-diene (J.C. Fairlie, G.L. Hodgson, and T. Money, J. Chem. Soc., Perkin I, 1973, 2109), pin-2-ene (Y. Matsubara *et al.*, Kinki Daigaku Rikogakubu Kenkyu Hokoku, 1974, 9, 11) and dihydrocarvone (G.L. Lange and J.M. Conia, Nouv. J. Chim., 1977, 1, 189), and by the reaction between dibromo ketone (22) and $Fe_2(CO)_9$ in benzene at 100-110° (R. Noyori *et al.*, J. Amer. Chem. Soc., 1979, 101, 220). The optimum conditions for the oxidation of camphene to camphor by $Na_2Cr_2O_7$ and $K_2Cr_2O_7$ in aqueous H_2SO_4 (N.M. Lien, P.T. Phuong, and T.T. Van, Tap San, Hoa- -Hoc, 1975, 13, 27), methods for the purification of synthetic camphor (L.G. Slivkin *et al.*, Gidroliz. Lesokhim. Prom-st., 1982, 9; 1983, 10), and reviews on the nomenclature, reactions and synthetic production of camphor [P.A. Verbrugge, Chem. Tech. (Amsterdam), 1974, 29, 229; 1975, 30, 4] have been reported.

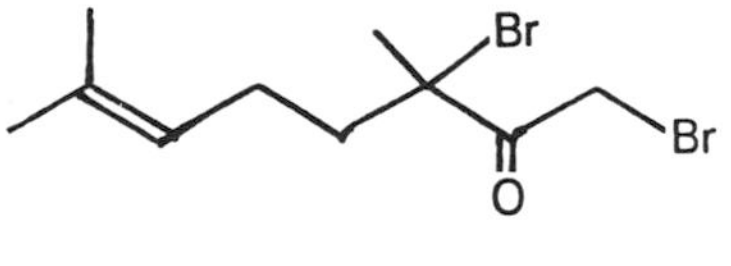

(22)

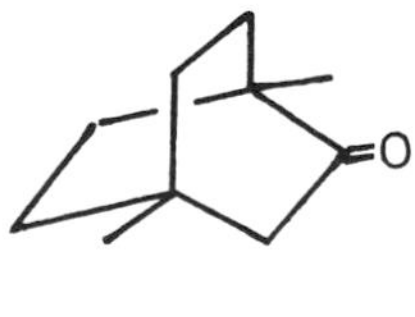

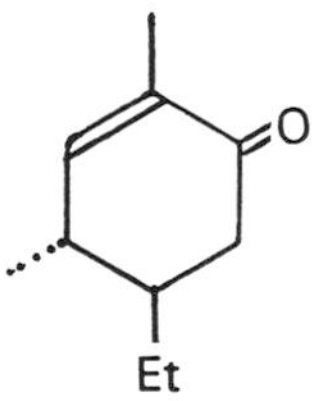

(24) 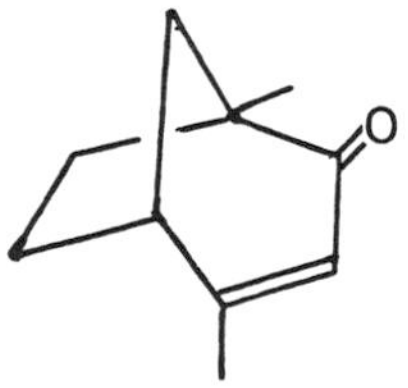   (25)

(23)

In HF-SbF$_5$ camphor isomerizes to give 1,4-dimethylbicyclo[2.2.2]octan-2-
-one (23) (12%), 2,4-dimethyl-5-ethylcyclohex-2-en-1-one (24) (27%), and
1,4-dimethylbicyclo[3.2.1]oct-3-en-2-one (25) (20%) (J.C. Jacquesy, R.
Jacquesy, and J.F. Patoiseau, Tetrahedron, 1976, 32, 1699). Camphor is
stereoselectively reduced by PhSiH$_3$, PhMeSiH$_2$, or Et$_2$SiH$_2$ at 0-80° in the
presence of (Ph$_3$P)$_3$RhCl to give isoborneol (26) (73-90%) and lesser amounts
of borneol (27). Et$_3$SiH yields isoborneol (30%) but PhMe$_2$SiH does not
reduce camphor (I. Ojima, M. Nihonyanagi, and Y. Nagai, Bull. Chem. Soc.
Jap., 1972, 45, 3722). Reduction of (+)-camphor with TiCl$_4$-Mg-Me$_3$COH
gives isoborneol (58%) and borneol (32%) (M. Pierrot, J.M. Pons, and M.
Santelli, Tetrahedron Letters, 1988, 29, 5925) and the alcohols are obtained

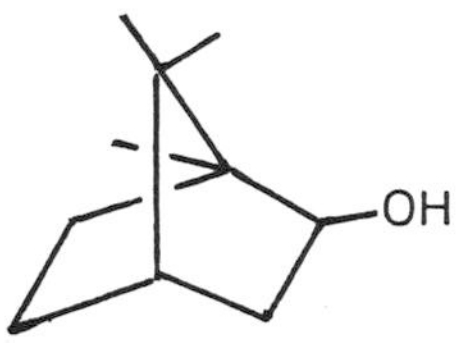

(26) 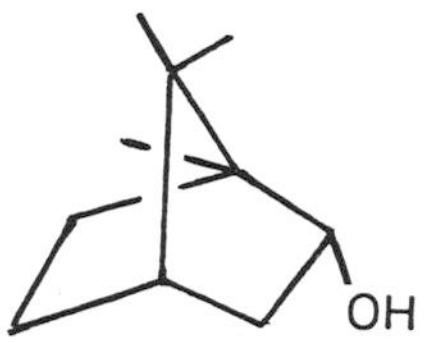   (27) 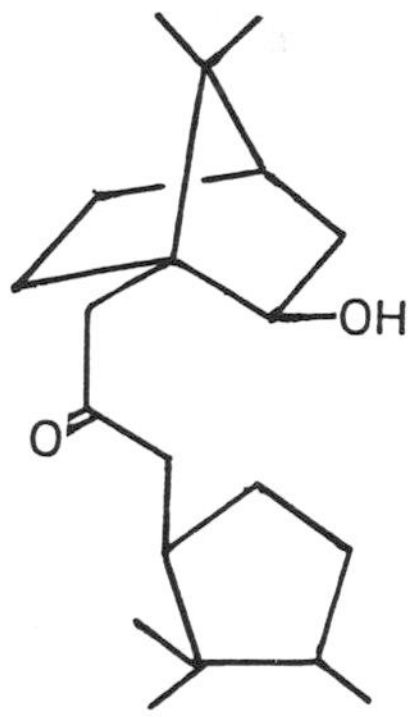   (28)

on using Li, Na, or K in the presence of hexamethylphosphorotriamide (HMPT)
and a protic solvent, such as Me$_3$COH or MeOH (M. Larcheveque and T.
Cuvigny, Bull. Soc. Chim. Fr., 1973, 1445). The hydrogen-transfer step in
the dissolving metal reduction in the absence of an added proton donor of
(+)- and (±)-3-exo- and 3-endo-deuteriocamphor has been investigated and a
detailed study made of the dimeric reduction products from (+)- and (±)-
camphor, which besides the pinacols also contained ketone (28) (J.W.
Huffmann and R.H. Wallace, J. Amer. Chem. Soc., 1989, 111, 8691). For
the reduction of (+)-[3,3-D$_2$]camphor with Li, Na, and K in NH$_3$ and THF,
H$_2$O, or NH$_4$Cl see V. Rautenstrauch et al. (Helv., 1981, 64, 2109).

(29)

(30)

(31)

(32)

$R = CH_2CO_2Me$

$R^1 = O_2CCH = CHMe$
$R^2 = H, Me, Ph, PhCH_2$

Anodic oxidation of camphor in dilute aqueous Na_2CO_3-EtOH or dilute aqueous Na_2CO_3-dioxane gives lactones (29 and 30), (29) on treatment with acid isomerizes to (30) (F. Barba *et al.*, An. Quim., 1979, 75, 967). The kinetics of the chromic acid oxidition of camphor (P.S. Subramanian and D.R. Nagarajan, Indian J. Chem., 1979, 17A, 170) and the liquid-phase oxidation of camphor in the presence of variable-valence metal salts (Yu. A. Vasil'ev *et al.*, Deposited Doc., 1982, VINITI 367) has been studied. Baeyer-Villiger oxidation of camphor affords ß-dihydrocampholenolactone, which on methanolysis gives methyl ß-campholenoate (31) (K. Schulze and H. Trauer, Z. Chem., 1989, 29, 59).

Camphor is methylated with MeI in the presence of $NaNH_2$ to give 3-methylcamphor (57.2%) and 3,3-dimethylcamphor (13.8%) [E.E Aringoli and L.E. Vottero, Rev. Fac. Ing. Quim. Univ. Nac. Litoral, 1971-1972 (Pub. 1973), 40-41, 27]. A number of arylidenecamphors with the *E*-configuration have been prepared and their UV and IR spectra discussed (N. El Batouti and J. Sotiropoulos, Compt. Rend., Ser. C, 1974, 278, 1109). Some 3-Benzyl-idenecamphors are useful in cosmetics and pharmaceutical preparations (R. Welters and H. Russmann, Ger. Offen. 2,336,219, 1975; C. Bouillon and C. Vayssie, *ibid.*, 2,811,041, 1978; Fr. Demande 2,421,878, 1979; Can. CA 1,113,480, 1981; G. Lang *et al.*, Ger. Offen. DE 3,321,679, 1983; 3,324,735, 1984; 3,445,365, 3,445,712, 1985; S. Forestier *et al.*, Eur. Pat. Appl. EP 390,682, 1990). The reaction of (+)-camphor with $HCONMe_2$ in the presence of KH stereoselectively gives bis[(1*R*,3*S*,4*R*)-2-oxo-3-bornyl]methane (S. Kiyooka *et al.*, Bull. Chem. Soc. Jpn., 1989, 62, 1364; A.M. Reinbol'd, G.S. Pasechnik, and D.P. Popa, Khim. Prir. Soedin., 1983,

389). The chiral enoates (32) readily available from (+)-camphor by Grignard alkylation and esterifiction with crotonoyl chloride have been prepared (P. Somfai, D. Tanner, and T. Olsson, Tetrahedron, 1985, 41, 5973).

(+)-Camphor with diMe phosphite in the presence of NaOMe gives *endo*--2-dimethylphosphone-*exo*-2-hydroxy-(-)-camphane (G. Adiwidjaja, B. Meyer, and J. Thiem, Z. Naturforsch., Anorg. Chem., Org. Chem., 1979, 34B, 1547), and with PhLi, 2-phenyl-1,7,7-trimethylbicyclo[2.2.1]heptan-2-*exo*-ol (N.G. Kozlov *et al.*, Zh. Org. Khim., 1989, 25, 783). Treatment of camphor with HC(OEt)$_3$ in the presence of 95% H$_2$SO$_4$ or 56% oleum affords acetal (33) [V.V. Belogorodski, M.V. Korsakov, and T.N. Nevinskaya, Zh. Prikl. Khim. (Leningrad), 1980, 53, 713) and with (F$_3$CSO$_2$)$_2$O in CH$_2$Cl$_2$ in the presence of Na$_2$CO$_3$, camphen-4-yl triflate (65%) and camphen-1-yl triflate (35%) (H. Bentz *et al.*, Tetrahedron Letters, 1977, 9). It has also been reported that the products are camphen-1-yl triflate, carvenone, and carvacryl triflate (W. Kraus and G. Zartner, *ibid.*, p13). Fenchone has been treated in a similar manner. The above reaction in the absence of base yields a mixture of bornane and isocamphane derivatives, which are hydrolyzed during final aqueous treatment to give camphenyl triflates (E. Teso Vilar, M. Gomez Marin, and C. Ruano Franco, Ber., 1985, 118, 1282). The Reformatski reaction of (+)-camphor with BrCH$_2$C(:CH$_2$)CO$_2$Et affords the spirofuranone (34) (G. Ruecker and W. Gajewski, Eur. J. Med. Chem. - Chim. Ther., 1985, 20, 87).

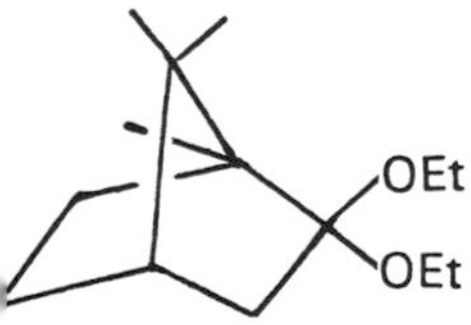

(33) 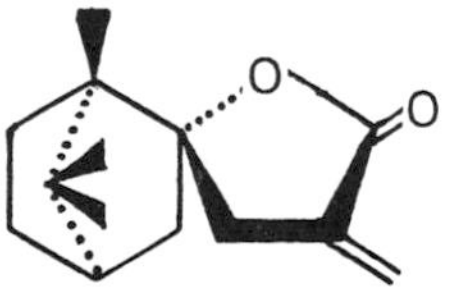   (34) 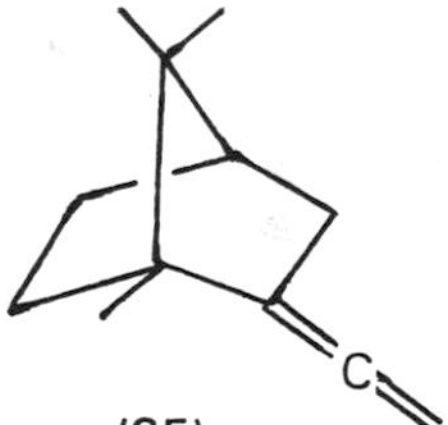   (35)

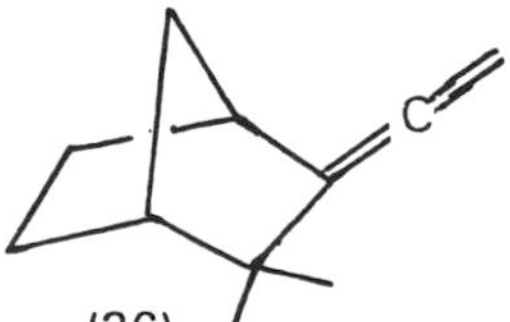

(36)

Decarbonylation of camphor furnishes 1,6,6-trimethylbicyclo[2.1.1]hexane (G. Kruppa and H. Suhr, Ann., 1980, 677), and ethenylidenebicycloheptanes (35 and 36) have been prepared from (+)-camphor and camphene, respectively (J. Mattay, M. Conrads, and J. Runsink, Synth., 1988, 595).

Camphor has been converted to nopinone (J.V. Paukstelis and B.W. Macharia, Tetrahedron, 1973, 29, 1955); and the following have been synthesized (3R)-1-vinyl-(3R)-1-hydroxypropenyl- and (3R)-1-epoxyethyl-5- -methoxy-1,2,2-trimethylcyclopentane derivatives from (+)-camphor via (+)- -5-oxobornyl acetate (H. Shibuya et al., Chem. Pharm. Bull., 1982, 30, 1271), 3-diphenylphosphinocamphor (S.D. Perera and B.L. Shaw, J. Organometal. Chem., 1991, 402, 133), bis[3-(trifluoroacetyl)camphorato]- barium (II) (M. Elian, G. Deleanu, and S. Rosca, Rev. Roum. Chim., 1990, 35, 299), 3-aminomethylcamphors (E. Occelli et al., Farmaco, Ed. Sci., 1985, 40 86), (+)-exo-3-(dimethylaminomethyl)camphor (N.L. McClure, G.Y. Dai, and H.S. Mosher, J. Org. Chem., 1988, 53, 2617), and 9,9,9-trideuteriocamphor (R.N. McCarty, Diss. Abs. Int. B, 1973, 33, 3558). Alcohols have been obtained by the addition of MeCH(OBu)OXC≡CH (X = CH$_2$, CHMe) to camphor (T.S. Kuznetsova et al., Uch. Zap., Yarosl. Gos. Pedagog. Inst., 1973, 122, 50).

The Li enolate of (+)-camphor reacts with SO$_2$ and 3-methylbut-2-enyl bromide to give ß-keto sulphone (37) (N.P. Singh and J.F. Biellmann, Synth. Comm., 1988, 18, 1061; N.P. Singh, B. Metz, and J.F. Biellmann, Bull. Soc. Chim. Fr., 1990, 98). The stereochemical aspects of the [2,3]-Wittig rearrangement of optically active tertiary allyl ethers (38 and 39) derived from (+)-camphor and (-)-fenchone have been investigated (D.S. Keegan et al., J. Org. Chem., 1991, 56, 1185).

$$R^1 = OCH = CH_2$$
$$R^2 = OCH_2CH = CH_2$$

(37) (38) (39)

A number of spectral investigations have been carried out and the resulting data reported; coupling constants between 3-exo H and 5-exo H for a number of 4-substituted bornan-2-ones (D.G. Morris and A.M. Murray, Org. Magn. Reson., 1974, 6, 510), ^{1}H NMR of (+)-camphor under high hydrostatic pressure [J. Klimowski et al., Mater. Ogolnopol. Semin. "Magn. Rezon. Jad. Jego Zastosow.", 13th 1980 (Pub. 1981), 119. Inst. Fiz. Jad.:

Krakow, Pol.], ^{13}C NMR of solid (+)-camphor (R.E. Wasylishen and M.R. Graham, Mol. Cryst. Liq. Cryst., 1979, 49, 225), EPR of the cation radicals generated from dehydrocamphor, dehydronorcamphor, and dehydroepi-camphor in H_2SO_4 and in CF_3CO_2H (J. Eloranta, E. Salo, and P. Malkonen, Fin. Chem. Letters, 1977, 217), and the mass spectrum of camphor-5,6-d$_2$ (J. Korvola and P.J. Malkonen, *ibid.*, 1974, 25). Heats of solution of camphor, borneol, and isoborneol in a number of solvents [V.G. Tsvetkov and B.A. Radbil, Zh. Prikl. Khim. (Leningrad), 1982, 55, 669] and the base-catalized D exchange rates of 3-*exo* H and 3-*endo* H for camphor and related bornanones [F.C. Brown *et al.*, J. Chem. Res (S), 1977, 335] have been determined.

^{1}H NMR spectral data of camphor oxime with reference to the three methyl groups and the hydroxyl group have been reported (A.K. Singh and S.M. Verma, Indian J. Chem., 1981, 20B, 33). Reaction of camphor oxime with 3 moles of PhMgBr in PhMe affords 4-cyano-3,3,4-trimethylcyclohex-1--ene (R. Chaabouni and A. Laurent, Tetrahedron Letters, 1973, 1061). Hydrosilylation of camphor oxime with $HSiR^1R_2^2$ (R^1 = H, R^2 = Ph; R^1 = Me, R^2 = Cl) in PhMe containing RhCl(PPh$_3$)$_3$, K[PtCl$_3$C$_2$H$_4$], or PtO$_2$-H$_2$O, followed by acid hydrolysis gives in addition to the expected amines (40), the (3S)-cyclopentylethanamine (41) stereoselectively (H. Brunner and R. Becker, Angew. Chem., 1985, 97, 713). For reductive alkylation of camphor oxime, including formation of *N*-cycloalkyl-1,7,7-trimethylbicyclo-[2.2.1]heptyl-2-amines see N.G. Kozlov, T. Pehk, and T. Valimae (Khim. Prir. Soedin., 1983, 41).

$R^1 = NH_2, R^2 = H$
$R^1 = H, R^2 = NH_2$

(40)   (41)

Beckmann rearrangement of (*E*)-camphor oxime (42) in boiling concentrated HCl at 110° (10h) yields lactam (43) and nitriles (44 and 45) in a 1:3:1 ratio (N.G. Kozlov and T. Pehk, Zh. Org. Khim., 1982, 18, 1118).

Heating camphor oxime with polyphosphoric acid yields besides the known fragmentation products four isomeric ketones, camphen-5-one, camphen-6--one, tricyclenone, and endo-2,4-dimethylbicyclo[3.2.1]oct-2-en-7-one (R.K. Hill *et al.*, Tetrahedron, 1988, 44, 3405). Camphor oxime undergoes Beckmann fragmentation in the presence of tosyl chloride in pyridine generating olefinic nitriles (N. Satyanarayana, H.R. Shitole, and U.R. Nayat, Indian J. Chem., 1985, 24B, 997). Camphor on boiling (20h) with H_2NOSO_3H in HCO_2H gives *a*-camphidone (46) by a Beckmann rearrangement involving methylene group migration (G.R. Krow and S. Szezepanski,

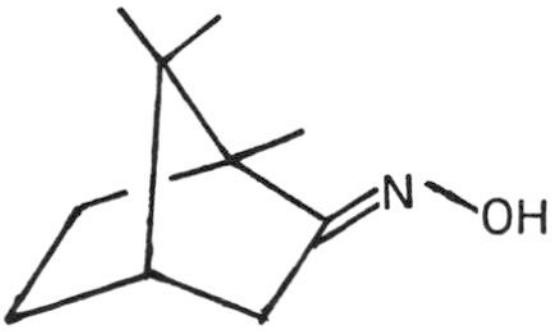

(42) 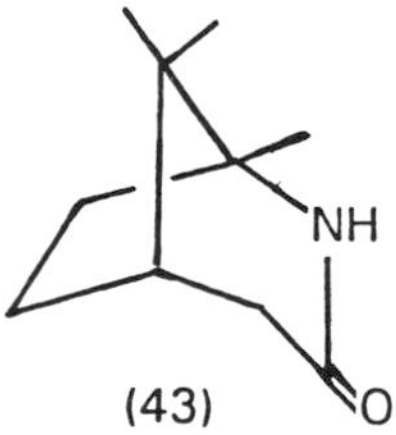   (43) 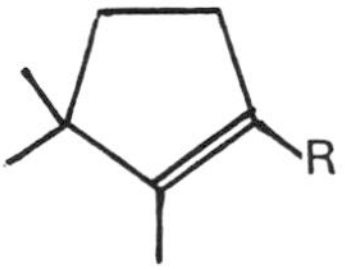   (44)

(45) $R = CH_2CN$

Tetrahedron Letters, 1980, 21, 4593). IR and UV spectral data and X-ray crystallography indicate that (*E*)-3-isonitrosocamphor exists in the oxime (47) form in solution and in the crystal (A.B. Zolotoi *et al.*, Izv. Akad. Nauk SSSR, Ser. Khim., 1987, 791). For some transformations involving camphor semithiocarbazone see B.V. Thachuk and N.M. Turkevich [Farm. Zh. (Kiev), 1981, 24].

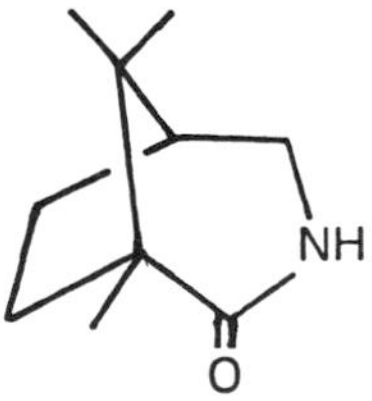

(46) 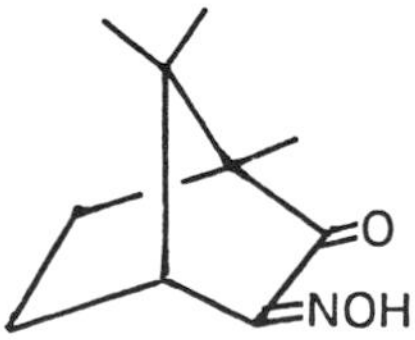   (47)

(ii) Substitution products of camphor

Halogeno derivatives of camphor

Direct bromination of (+)-3,3-dibromocamphor affords a mixture of (+)--3,8,8-tribromocamphor, 1,7-dibromo-3,3,4-trimethylnorbornan-2-one, and 1,7-dibromo-4-dibromomethyl-3,3-dimethylnorbornan-2-one. It has also been shown that (+)-3,9,9-tribromocamphor is a minor product in the conversion of (+)-3-bromocamphor to (+)-3,9-dibromocamphor. (+)-3,9-Dibromo-camphor has been isomerized to (-)-6,9-dibromocamphor (P. Cachia et al., J. Chem. Soc., Perkin I, 1976, 359). 3,3-Dibromo- or endo-3-bromo-camphor on treatment with ZnEt$_2$ in boiling benzene gives 4,7,7-trimethyltricyclo-[2.2.1.0^{2,6}]heptan-3-one (48) (L.T. Scott and W.D. Cotton, J. Amer. Chem. Soc., 1973, 95, 2708). The mechanism of dehalogenation of 3-bromo-camphor to camphor by Me$_2$NPh (A.G. Giumanini and M.M. Musiani, Z. Naturforsch., Anorg. Chem., Org. Chem., 1977, 32B, 1314) and to camphene, tricyclene, and bornylaniline with MeNHPh (idem, J. Prakt. Chem., 1980, 322, 423) has been discussed. Dehalogenation of endo-3--bromocamphor by AgSbF$_6$-CH$_2$Cl$_2$ yields 6-methyl-6-isopropenylcyclohex-2--enone (J.P. Begue et al., Tetrahedron, 1978, 34, 293). The reaction

Z = NNHPh

(48)　　　　　　　　(49)　　　(50)Z^1 = NHNHPh, H; Z^2 = NNHPh
　　　　　　　　　　　　　　(51)Z^1 = Z^2 = NNHPh

between 3-bromocamphor and phenylhydrazine under nitrogen at 100° gives bis(phenylhydrazone) (49) and derivative (50), and at 150° derivative (51) (A.G. Giumanini, L. Caglioti, and W. Nardini, Bull. Chem. Soc. Jap., 1973, 46, 3319). Reduction of 10-bromo- and 10-chloro-camphor with Al(OCHMe$_2$)$_3$ yields both the exo and endo alcohols. endo-10-Bromoborneol readily loses HBr to give α-campholenaldehyde (N. Proth, Rev. Tech. Luxemb., 1976, 68, 195).

Bromination of 3,3-dibromocamphor by Br_2-$ClSO_3H$, followed by reaction with Zn-HBr affords 8-bromocamphor (52) (75%). 1,7-Dibromo-3,3,4-tri-methylbornan-2-one (53) is a by-product resulting from a Wagner-Meerwein rearrangement (C.R. Eck, R.W. Mills, and T. Money, J. Chem. Soc., Perkin I, 1975, 251) and D-labelling shows that the mechanism of the C-8 bromination of 3,3-dibromocamphor involves two 2,3-*endo*-Me shifts (W.M. Dadson and T. Money, Chem. Comm., 1982, 112). Reduction of 8-bromocamphor by Na--K alloy in ether or hexane gives *p*-menth-8-en-2-one, and in $OP(NMe_2)_3$,

| (52) | (53) | (54) |

camphor. Electrochemical and Bu_3SnH reductions have also been discussed (D.P.G. Hamon and K.R. Richards, Austral. J. Chem., 1983, 36, 109). Treatment of 8-bromocamphor with Mg in THF causes reductive cyclization to camphor-1,4-homoenol (54) and reaction with Bu_3SnD in benzene yields 8-deuteriocamphor (W.M. Dadson and T. Money, Canad. J. Chem., 1980, 58, 2524). The ring cleavage of 8,10- and 9,10-dibromo-camphor to provide intermediates for the synthesis of natural products (J.H. Hutchinson, T. Money, and S.E. Piper, *ibid.*, 1986, 64, 854) and the preparation of (-)-(1S)-6-*endo*-bromo-2-oxobornane-8-sulphonic acid (J. Degnbol and A. Hammershoei, Acta Chem. Scand., 1988, B42, 390), 3-bromocamphor-8--sulphonyl chlorides (R. Cremlyn, M. Bartlett, and L. Wu, Phosphorus Sulphur, 1988, 39, 173), 10-chlorocamphor-10-sulphine (M.F. Haslanger and J. Heikes, Synth., 1981, 801), and europium and praseodymium chelates of fluorinated camphor derivatives (H.L. Goering, J.N. Eikenberry, and G.S. Koermer, U.S. Pat. 3,789,060, 1974) have been reported. Kinetic methylation of camphor, 9-bromo-, and 9,10-dibromocamphor furnishes 3-*exo*-Me derivatives as the major product and small quantities of 3-*endo*-Me derivatives (J.H. Hutchinson and T. Money, Canad. J. Chem., 1984, <u>62</u>, 1899). Regiospecific bromination and debromination reactions provide a convenient synthetic route from camphor to optically active 8,10- and 9,10-disubstituted camphor derivatives (W.M. Dadson et al., ibid., 1983, <u>61</u>, 343).

Camphorsulphonic acids

Positions of sulphonation in camphor have been determined by ^1H and ^{13}C NMR spectroscopy (P.D. Stanley et al., Magn. Reson. Chem., 1988, <u>26</u>, 14). Camphor-10-sulphonic acid has been used to provide an efficient new route to enantiomerically pure halogenohydrins and epoxides (W. Oppolzer and P. Dudfield, Tetrahedron Letters, 1985, <u>26</u>, 5037) and epimeric 2-[(1S)-7,7--dimethyl-2-oxonorbornan-1-yl]thiirane dioxides (55 and 56) have been prepared from (1S)-camphor-10-sulphonyl chloride (T. Kempe and T. Norin, Acta Chem. Scand., 1973, <u>27</u>, 1452). Camphor-10-sulphonamide derivatives (57) are prepared by the reaction between camphor-10-sulphonyl chloride and the appropriate amine (A.K. Shubber and S.Y. Kazandji, Iraqi J. Sci., 1990, <u>31</u>, 529). Also reported are the preparation of camphor and

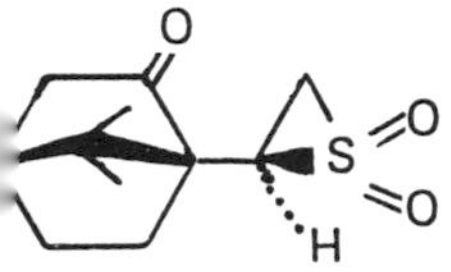

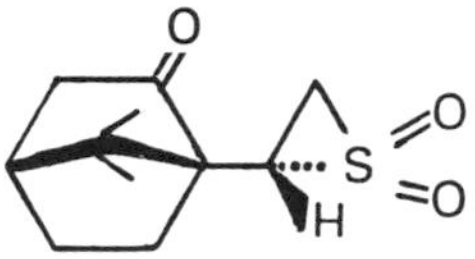

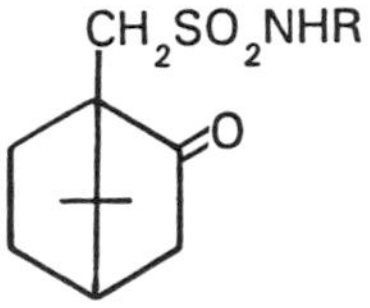

R = 2-,3-,4-pyridyl,
2-pyrimidinyl, 2-picolyl,
2-thiazolyl, c-pentyl,
c-hexyl, CMe_3

(55)   (56)   (57)

camphor-10-sulphonamide sulphonohydrazones (R. Cremlyn, M. Bartlett, and J. Lloyd, Phosphorus Sulphur, 1988, <u>40</u>, 91) and of camphor-3-sulphonic acid (R. Cremlyn and L. Wu, ibid., <u>39</u>, 165).

Sultam (58) prepared from (-)-camphor-10-sulphonic acid on treatment with (E)-crotonoyl chloride gives sultam (59), which undergoes a Diels-Alder reaction with cyclopentadiene (M. Vandewalle et al., Tetrahedron, 1986, <u>42</u>,

4035). For the cycloaddition reaction between (-)-camphor sultam acrylate (60) and methylenecyclopropane see P. Binger and B. Schaefer (Tetrahedron Letters, 1988, 29, 529), for reactions of Oppolzer's camphor sultam, D.P. Curran *et al.* (J. Amer. Chem. Soc., 1990, 112, 6738), and for the reactivity of [hydroxy(((+)-camphor-10-sulphonyl)oxy)iodo]benzene towards carbonyl compounds, E. Hatzigrigoriou, A. Varvoglis, and M. Bakola-Christianopoulou (J. Org. Chem., 1990, 55, 315).

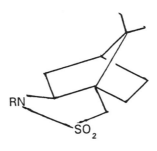

(58) R = H
(59) R = COCH = CHMe
(60) R = COCH = CH$_2$

Oxocamphors, bornanediones
(-)-Camphorquinone. Ozonolysis of 3-hydroxymethylenecamphor (61) gives, in addition to the expected camphorquinone (62), a surprisingly large amount of camphoric anhydride (56%) *via* Baeyer-Villager reaction (D.T.C. Yang *et al.*, Proc. Arkansas Acad. Sci., 1976, 30, 97). Condensation of camphorquinone with excess BzNHNH$_2$ yields only camphorquinone-3-
-benzoylhydrazone (63) (E.S.H. El Ashry *et al.*, Indian J. Chem., 1980, 19B, 716). Details are provided for the preparations of camphorquinone semicarbazones and thiosemicarbazones (geometric isomers) (64) (A.B. Tomchin and U. Lepp, Zh. Org. Khim., 1981, 17, 1262), camphorquinone arylhydrazones (65) (K.N. Zelenin, Tomchin, and U. Lepp, *ibid.*, p.1253), 3-
-(hydroxyimino)camphornitrimine and camphorquinonedinitrimine [A. Ranise *et al.*, J. Chem. Res. (S), 1988, 282; 1989, 96], and (-)-camphorquinone α- and ß-hydrazones (D.L. Cullen *et al.*, Tetrahedron, 1983, 39, 733).

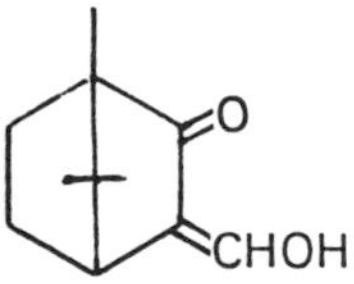

(61)

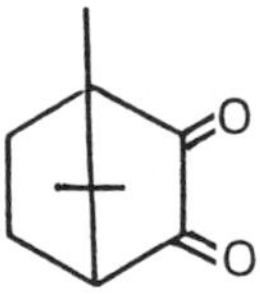

(62)

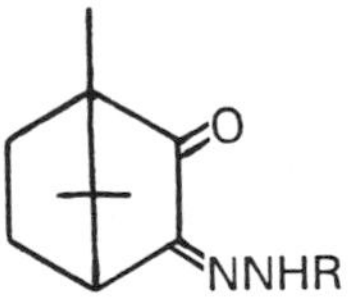

(63) R = Bz
(64) R = CXNH$_2$
X = O,S
(65) R = C$_6$H$_4$R^1-4
R^1 = H,NO$_2$,CO$_2$Et,
Br,I,Me,MeO,NMe$_2$

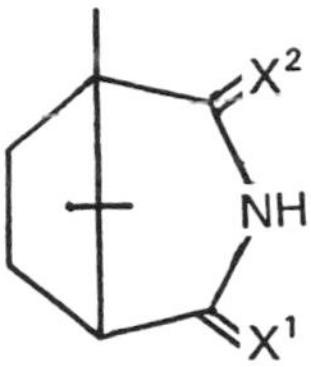

X^1 = NOH, X^2 = O
X^1 = O, X^2 = NOH

(66)

An inseparable mixture of camphorimide oximes (66) is obtained on heating camphorquinone dioxime with KOH in ethylene glycol (V.K. Kapoor et al., Indian J. Chem., 1990, 29B, 971). For a preparation of 3-benzoyl-(+)-camphor and (3-benzoyl-(+)-camphorato)diphenylboron see W.R. Cullen et al., (Canad. J. Chem., 1988, 66, 2007) and for ^{13}C NMR chemical shifts of bornanediones (67), F.C. Brown, D.G. Morris, and A.M. Murray (Tetrahedron, 1978, 34, 1845). Rotary dispersion and CD data have been reported for Cu chelates based on (-)-camphor ß-carbonyl derivatives (V.M. Potapov, G. V. Panova, and N.K. Vikulova, Zh. Obshch. Khim., 1973, 43, 939).

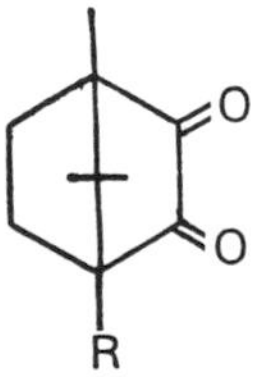

R = H,Me,CO$_2$H,Br,Cl,NO$_2$

(67)

Camphorcarboxylic acids

An equilibrium mixture of camphor-3-carboxylic acid contains 76:24 of the *endo* and *exo* diastereomers by analysis of the ^{1}H NMR spectral data. The pure *endo* diastereomer is obtained only as the oxime (68) (R. Antkowiak and W.Z. Antkowiak, Bull. Acad. Pol. Sci., Ser. Sci. Chim., 1975, 23, 717). Methyl camphor-3-carboxylate on bromination affords a mixture of two 3-bromo diastereomers, from which only methyl 3-*endo*-bromocamphorcarboxylate (69) is isolated (*idem, ibid.*, p.723). Reaction of derivative (69) with strong bases (cyanide or *tert*-butanolate ion) yields a mixture of decarboxylated and debrominated products, while reaction with NaBH$_4$ in MeOH results in debromination, but gives decarboxylation and carbonyl reduction when carried out in diglyme furnishing 3-bromoborneol (*idem, ibid.*, 1978, 26, 933).

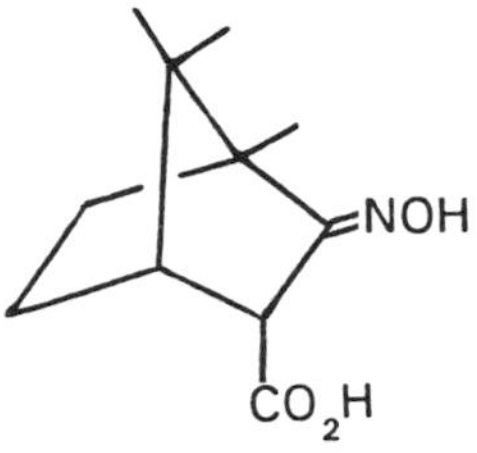

(68) 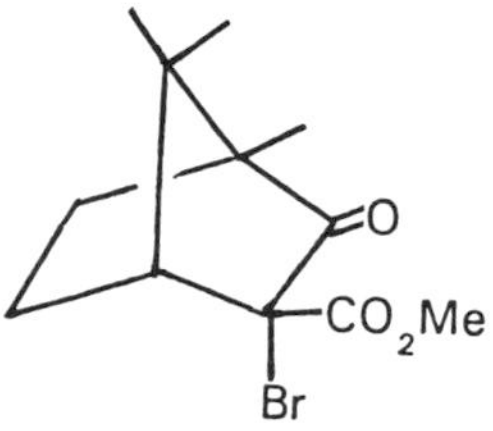　　　　　　　　(69) 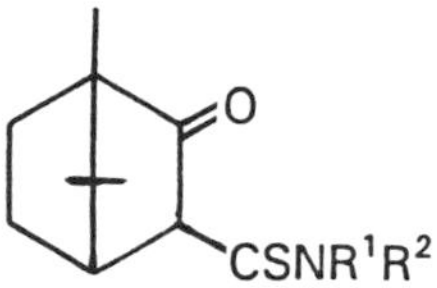　　　　　　　(70) $R^1 = H; R^2 = H, Me$
　　　　　　　　　　　　　　　　　　　　　　　Et, PhCH$_2$, c-hexyl
　　　　　　　　　　　　　　　　　　　　　　　Ph, C$_6$H$_4$Me-4
　　　　　　　　　　　　　　　　　　　　　(71) $R^1 = R^2 = Me$
　　　　　　　　　　　　　　　　　　　　　(72) $R^1 = Ph, R^2 = Me$

The conformations of *N*-monosubstituted camphor-3-carbothioamides (70) (A.M. Lamazouere and J. Sotiropoulos, Bull. Soc. Chim. Fr., 1974, 2989), and of *endo-N,N*-dimethylcamphor-3-carbothioamide (71) and *endo-N*-methyl-*N*-phenylcamphor-3-carbothioamide (72) (*idem, ibid.*, p2995), and the preparation of a number of *N,N*-disubstituted camphor-3-carbothioamides (72;

NR^{1}R^2 = piperidino, pyrrolidino, morpholino, indolino, 4-ethoxycarbonyl-piperidino) (*idem, ibid.*, 1976, 1851) have been discussed. Camphor-3-carbo-thioamide reacts with H$_2$N(CH$_2$)$_n$NH$_2$ (n = 2,3) to give the bisoxothioamides (73) or the heterocyclic compounds (74) *via* cleavage of both C-S bonds and transamination (S. Seube, A.M. Lamazouere, and J. Sotiropoulos, Heterocycl. Chem., 1978, 15, 343).

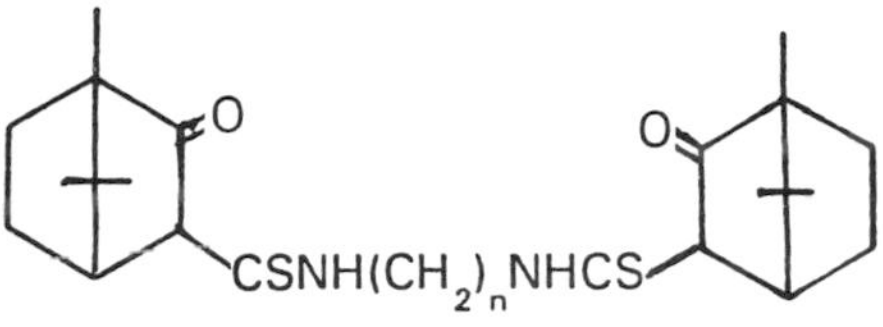

(73) 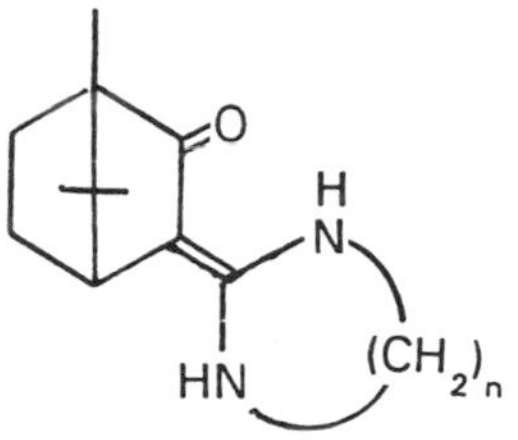　　　　　　　　　　　　　(74)

(iii) Epicamphor, bornan-3-one

Condensation of sodioepicamphor with 2- and 4-chlorobenzaldehyde gives 2-(2- and 4-chlorobenzylidene)epicamphor, respectively, in both *E* and *Z* isomers (F. Labruyere and C. Bertrand, Compt. Rend., Ser. C, 1972, 275, 673). 2-*endo*-Phenyl-2-*exo*-hydroxyepicamphor (75) in the presence of H$_2$SO$_4$ at 0° rearranges to give 3-hydroxy-4-phenyl-2,2,3-trimethylcyclo-hexanecarboxylic acid lactone (76) (P. Wilder, Jr. and W.-C. Hsieh, J. Org. Chem., 1975, 40, 717). The mutarotation of several amines derived from 2-hydroxymethylenebornan-3-one has been investigated (F. Labruyere and C. Bertrand, Compt. Rend., Ser. C, 1980, 290, 73).

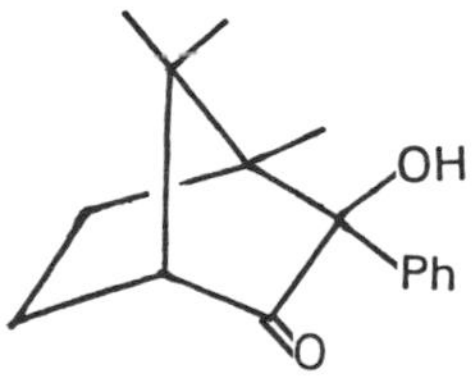

(75) 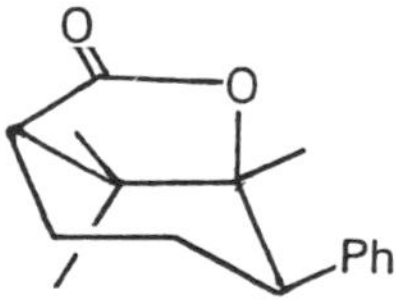　　　　　　　　　　　　　(76)

(iv) Compounds related to camphor
Isocamphan-5-one (77) on treatment with PhLi affords alcohol (78),
which undergoes a Ritter reaction with RCN to give the corresponding *exo-*
-amide (79). Hydrolysis of amide (79) by alcoholic HCl yields 2-*exo*-amino-6-
-phenylbornane (80) (N.G. Kozlov *et al.*, Zh. Obshch. Khim., 1990, 60, 198).

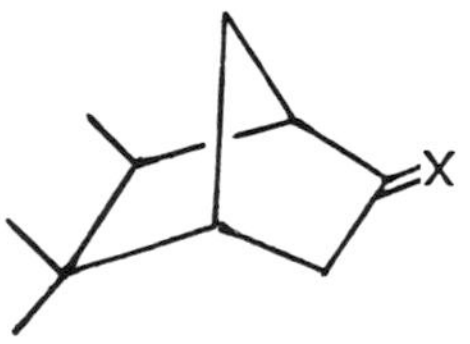

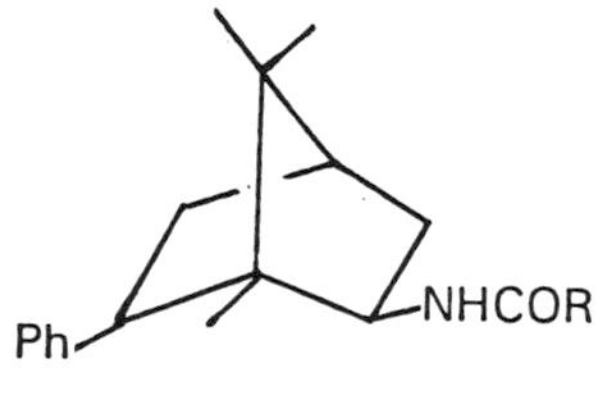

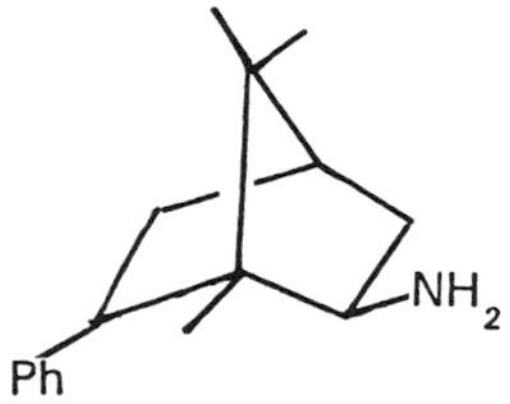

$$R = Me, Me_2CHCH_2, MeOCH_2CH_2$$

(77) X = O
(78) X = OH,Ph   (79)   (80)

Reacting isocamphan-5-one with MeCN at 20°, followed by concentrated
H_2SO_4 gives bisacetamide (81), which on reduction with $LiAlH_4$ in ether
furnishes amine (82) (*idem, ibid.*, 1988, 58, 2593). Oximation of

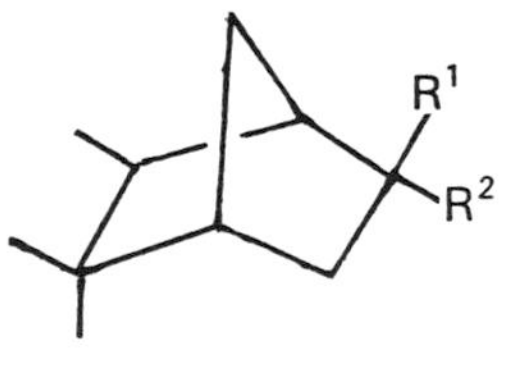

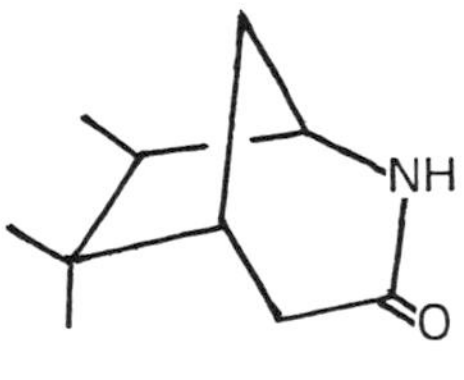

(81) $R^1 = R^2 = AcNH$
(82) $R^1 = EtNH, R^2 = H$ (83)

isocamphan-5-one gives the *E*-oxime (84%) which undergoes Beckmann rearrangement in boiling concentrated HCl to give *a*- and ß-campholenitrile (10% combined) and lactam (83) (36%) (*idem, ibid.*, p.821).

Camphenilone derivatives (84) are ß-enolized by Me_3CO^- in Me_3COH at 180° to give mixtures of ketones (84 and 85) (S. Peiris, A.J. Ragauskas, and J.B. Stothers, Canad. J. Chem., 1987, 65, 789). Substitutent effects on ^{13}C NMR chemical shifts in 1-substituted camphenilones and some derived *N*-nitro-imines (F.C. Brown and D.G. Morris, J. Chem. Soc., Perkin II, 1977, 125) have been reported and carbocamphenilone, 6,7-dehydrocarbocamphenilone (R. Nayori, T. Souchi, and Y. Hayakawa, J. Org. Chem., 1975, 40, 2681) and *exo*-3-formyl-3-methylcamphor (K. Suzuki *et al.*, Nippon Kagaku Kaishi, 1987, 257) have been synthesized.

$R = Me, R_2 = CH_2CH_2, Me_2C =$

(84) (85) (86)

The reaction between thiocamphor and Grignard reagents affords bornane-2-thiol, 2-alkylthiobornanes (*endo-exo* mixtures) and 2-alkylthioborn--2-enes. Grignard reactions with thiofenchone give only fenchane-2-thiol and its alkyl derivatives (M. Dagonneau, D. Paquer, and J. Vialle, Bull. Soc. Chim. Fr., 1973, 1699). Condensation of thiocamphor with N_2CHCO_2Et in THF yields thienol ether (86) instead of the expected thiazolidine (J.M. McIntosh, K.C. Cassidy, and P.A. Seewald, J. Org. Chem., 1989, 54, 2457). *a*--Arylmethylenethiocamphor *S*-oxides have been synthesized (S. Motoki *et al.*, Chem. Letters, 1988, 319) and the rates of base-catalyzed proton-deuterium exchange (*a*-thioenolization) of the *exo* and *endo* protons of thiocamphor in dioxane-D_2O (2:1) determined (N.H. Werstiuk and P. Andrews, Canad. J. Chem., 1978, 56, 2605).

(c) Alcohols

Borneol and isoborneol. Borneol (87) is obtained by hydration of pin-2--ene or camphene with $ClCH_2CO_2H$ with 96% selectivity at 10% conversion. Epimerization of isoborneol (88) yields borneol and isoborneol in the ratio 4:1 (K. Tanaka and Y. Matsubara, Kinki Daigaku Rikogakubu Kenkyu Hokoku, 1978, 13, 33; S. Li, S. Gu, and C. Chen, Linchan Huaxue Yu Gongye, 1984, 4, 10). Hydrolysis of isobornyl acetate with NaOH in MeOH, followed by thermal isomerization of isoborneol causes conversion to borneol (T. Sen *et al.*, Riv. Ital. Essenze, Profumi, Piante Off., Aromat., Syndets, Saponi, Cosmet., Aerosols, 1979, 61, 112). Isobornyl acetate has been prepared in 82-95% yields by treatment of camphene with AcOH in the presence of H_2SO_4, askanite, KU23, or KU2-8 catalyst (G.I. Moldovanskaya *et al.*, Maslo-Zhir. Prom-st., 1976, 29) and by reaction with AcOH over a natural zeolite catalyst (Q. Chen *et al.*, Faming Zhuanli Shenqing Gongkai Shuomingshu CN 1,036,950, 1989) or with D72 resin as catalyst (W. Ma, Linchan Huaxue Yu Gongye, 1989, 9, 18). The production of borneol by direct hydration of pin-2-ene in the presence of GC-82 type solid acid catalyst (R. Tan and Q. Chen, Linchan Huaxue Yu Gongye, 1988, 8, 1) and a method of purifying crude borneol containing isoborneol (M. Kasano and H. Okitsu, Japan P. 62 19,546 [87 19,546], 1987) have been reported. The following preparations or chemical/physical properties have been described, *endo*-bornyl esters (T.-H.L. Perng, L.-L. Liu, and K.-K. Lee, T'ai-wan Yao Hsuch Tsa Chih, 1979, 31, 119), mass spectrometry of naturally occurring esters (A.M. Bambagiotti *et al.*, Biomed. Mass Spectrom., 1982, 9, 495), *N*-[2-(2-*exo*--bornyloxy)ethyl]ethylenediamine as a chelating agent for potassium chloride (V.I. Lysenkov *et al.*, U.S.S.R. SU 1,216,178, 1986), 3-mercaptoborneol (S.M. Hung, D.S. Lee, and T.K. Yang, Tetrahedron, Asymmetry, 1990, 1, 873), phosphorylated derivatives of borneol (R.J. Cremlyn, R.M. Ellam, and N. Akhtar, Phosphorus Sulphur, 1979, 7, 109), and borneol-derived chiral silyl enol ether (89) (P.T. Kaye and R.A. Learmonth, Synth. Comm., 1989, 19, 2337). The mass spectrometric fragmentation of the acetyl and other alcohol derivatives of 9-bromoborneol are dependent on both the orientation and identity of the leaving group at the 2-position (U.N. Sundram and K.F. Albizati, J. Org. Chem., 1991, 56, 2622).

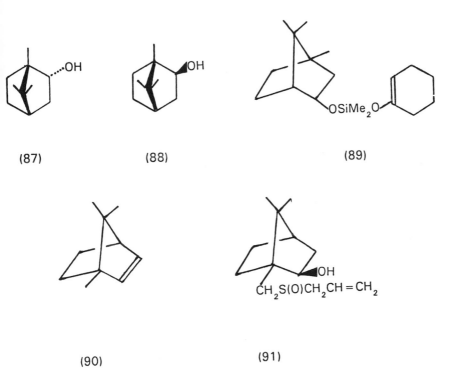

(87)　　　　　　(88)　　　　　　　　　(89)

(90)　　　　　　　　　(91)

Treatment of bornyl *p*-toluenesulphonate with Me₃COK in C₆H₆ at 80°
(17h) affords elimination product (90) (13%), tricyclene (8%), and camphene
(79%), whereas with Me₃COK in C₆H₆ in the presence of 18-crown-6 at 80°
(3h) only (90) (100%) is obtained (R.A. Bartsch, J.R. Allaway, and J.G. Lee,
Tetrahedron Letters, 1977, 779). Vapour-phase dehydrogenation of technical
isoborneol, containing ~ 4% H₂O at 240-260° affords camphene and not the
expected camphor (I.N. Kuzyukova and G.V. Nesterov, Chem. Abs., 1976,
85, 177618p). For the evaluation and selectivity of catalysts for the vapour-
-phase dehydrogenation of borneols see G.V. Nesterov, B.A. Radbil, and G.A.
Rudakov (Sb. Tr.-Tsentr. Nauchno-Issled. Proektn. Inst. Lesokhim. Prom-st.,
1975, 24, 113). The preparations of the following isoborneol derivatives
have been described, isoborneol aliphatic acid esters (Y. Matsubara and H.
Yada, Japan P. 74 13,158, 1974), isoborneol amino esters (T.S. Raikova *et
al.*, Chem. Abs., 1983, 98, 215811z), stable isoborneol allylic sulphoxide
(91) (R. Annunziata *et al.*, Tetrahedron, 1987, 43, 1013), chiral (*E*)- and (*Z*)-
-(*Rs*)-10-isobornyl vinyl sulphoxides (S.G. Pyne, P. Bloem, and R. Griffith,
ibid., 1989, 45, 7013), isoborneol ethers from camphene and the appropriate

alcohol, with simultaneous camphene rearrangement in the presence of 72- -85% $HClO_4$ (S.P. Pavlova and N.A. Kamannov, U.S.S.R. 352,870, 1972), isobornyl alkyl ethers (S. Xiao et al., Huaxue Xuebao, 1988, 46, 693), isobornyl phenoxyalkyl ethers (L. Borowiecki and M. Welniak, Pol. PL 149,264, 1990), isobornyl cresyloxyalkyl ethers (idem, ibid., 149,263, 1990), isobornyl trichlorophenoxyalkyl ethers (idem, ibid., 149,262, 1990), and isobornyl aryl ethers (idem, Pol. J. Chem., 1988, 62, 235).

The rearrangement of 3,3-ethylenedioxyisobornyl tosylate leads to an efficient synthesis of camphen-7-one (D.G. Patil, H.P.S. Chawla, and Sukh Dev, Tetrahedron, 1979, 35, 527). Also, the dehydration of 3,3-ethylene-dioxyisoborneol (92) with PCl_5 in ligroin-pyridine yields the camphen-7-one ketal (93) and product (94), whereas decomposition of the tosylhydrazone (95) under Bamford-Stevens reaction conditions affords products (93 and 96) and tricyclenone ketal (97) (L. Borowiecki and M. Welniak, Pol. J. Chem., 1983, 57, 99).

(93)

(92) $R^1 = H, R^2 = OH$
(95) $R^1R^2 = NNHSO_2C_6H_4Me-4$

(94) $R^1 = Cl, R^2 = H, R^3 = Me$
(96) $R^1-R^3 = OH, Me, H;$
       OH, H, Me

(97)

The oxidation of 10-(alkylsulphenyl)- and 10-(alkenylsulphenyl)-isoborneol (98 and 99) with $3-ClC_6H_4CO_2OH$ affords the S-oxide in good yields and high selectivity (Y. Arai, M. Matsui, and T. Koizumi, Synth., 1990, 320). Asymmetric synthesis using Eliel's $(1R)-(+)$-isoborneol-10-thiol (100) as a template has been reported, e.g., the synthesis of (+)-citronellal (M. Isobe et al., Tetrahedron Letters, 1988, 29, 4773). Enantiomerically pure N- -substituted a-(2-exo-hydroxy-10-bornylsulphinyl)maleimides (101) have been synthesized diastereoselectively from isoborneol-10-thiol (10-mercaptoiso-

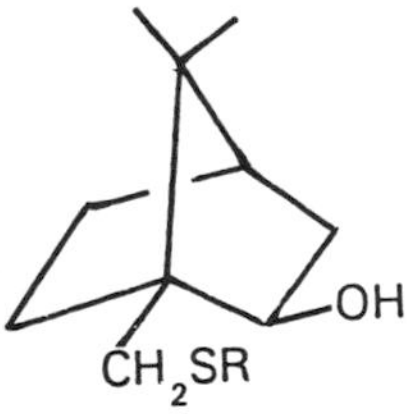

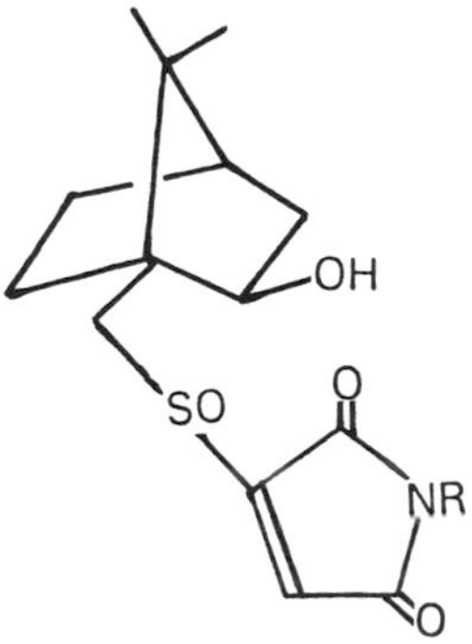

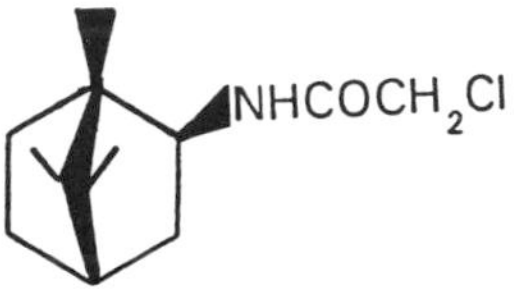

(98) R = Me,Et,CHMe$_2$
(99) R = CH$_2$CH = CH$_2$
(100)R = H

(101) R = Me,Ph,PhCH$_2$

(102)

(103)

borneol) (Y. Arai et al., J. Org. Chem., 1991, 56, 1983). The Ritter reaction of isoborneol with ClCH$_2$CN in H$_2$SO$_4$-AcOH at 20° (24h) gives a mixture of amides [102 (90%) and 103 (10%)] (V.I. Lysenkov, Zh. Org. Khim., 1989, 25, 1570).

The following reactions and/or properties of both borneol and isoborneol have been described, determination of kinetic parameters for the dehydrogenation of (±)-isoborneol and (-)-borneol to camphor in presence of Cu-Ni catalyst [G.V. Nesterov, B.A. Radbil, and G.A. Rudakov, Gridroliz. Lesokhim. Prom., 1973, 6; Zh. Prikl. Khim. (Leningrad), 1975, 48, 2334], treatment of (±)-borneol and (±)-isoborneol with BF$_3$.OEt$_2$ to give bornyl isobornyl ether and diisobornyl ether, respectively (K. Nagai et al., Bull. Chem. Soc. Jpn., 1974, 47, 1193), preparation of esters of hydroxyalkyl ethers of bornane (C.R. Gorman, J.M. Evans, and S.G. Traynor, U.S. Pat. US 4,620,028, 1986), pyrolysis of 4-nitrobenzoates and 3,5-dinitrobenzoates of borneol and isoborneol to yield mainly hydrocarbons, e.g., camphene, tricyclene, and bornene (J. Korvola and P. Malkonen, Finn. Chem. Letters, 1974, 23), and of phenyl- and 1-naphthyl-urethanes of borneol and isoborneol to give hydrocarbons including pin-2-ene (idem, ibid., p.19), and the catalytic deuterization of dehydroborneol, dehydroisoborneol, and 2-methyl-dehydroiso-borneol (idem, ibid., p.14).

The reduction of 3-bromocamphor with LiAlH$_4$, NaBH$_4$, and Al(OCHMe$_2$)$_3$ gives varying amounts of 3-bromoborneol and 3-bromoisoborneol, both of which with PCl$_5$ yield 6-bromopin-2-ene (N. Proth, Rev. Tech. Luxemb.,

408

1978, <u>70</u>, 23). The structures of 6-*endo*-bromoisoborneol and 6-*endo*--bromoborneol have been confirmed by the lanthanide induced shifts in their ¹H NMR spectra (L. Hietaniemi and P. Malkonen, Finn. Chem. Letters, 1982, 41). Azobenzoates (104 and 105) thermally rearrange to afford lactam (106) and camphor, respectively (T. Tezuka, T. Otsuka, and H. Kasuga, Tetrahedron Letters, 1990, <u>31</u>, 7633).

(107)

(104) R¹ = N = NC₆H₄Br-4
       R² = O₂CC₆H₄Cl-3
(105) R¹ = O₂CC₆H₄Cl-3
       N² = N = NC₆H₄Br-4

(106)

Asymmetric Diels-Alder reactions of acrylates derived from 3-alkyl--borneols and -isoborneols (W. Oppolzer and C. Chapuis, *ibid.*, 1984, <u>25</u>, 5383), stereochemistry of some diastereoisomeric 2-substituted borneols (B. Blagoev *et al.*, J. Mol. Struct., 1984, <u>116</u>, 189), synthesis of (-)-(1*R*, 2*S*,-3*R*,4*S*)-2-benzyloxybornan-3-ol (H. Herzog and H.D. Scharf, Synth., 1986, 788) and of a single isoborneol allyl sulphoxide derivative (107) by a stereospecific conversion of (+)-camphor (M.R. Binns *et al.*, Tetrahedron Letters, 1985, <u>26</u>, 6381), CD and absorption spectra of (-)-borneol in the vapour phase and in solution (P.A. Synder and W.C. Johnson, Jr., J. Amer. Chem. Soc., 1978, <u>100</u>, 2939), mass spectra of borneol and isoborneol (R. Robbiani *et al.*, Org. Mass. Spectrom., 1978, <u>13</u>, 275), and internal dynamics of (±)-borneol under hydrostatic pressure by NMR methods (K. Holderna-Matuszkiewicz, Acta Phys. Pol., 1988, <u>A74</u>, 3) have been discussed.

Nojigiku alcohol [camphen-6-ol, (5*R*)-2,2-dimethyl-3-methylenebicyclo-[2.2.1]heptan-5-ol] (108), m.p. 52-53°, [α]$_D$ + 12° (CHCl₃), together with its acetate (109) and a number of monoterpenoids and sesquiterpenoids has been isolated from the essential oil of *Chrysanthemum japonese* and its structure and absolute configuration and that of its acetate determined from chemical and spectral data and by synthesis from (±)-camphene (A. Matsuo *et al.*, Tetrahedron Letters, 1974, 4219; Y. Uchio, Bull. Chem. Soc. Jpn., 1978, <u>51</u>, 2342). It has also been synthesized from (-)-isobornyl acetate (N. Darby, N. Lamb, and T. Money, Canad. J. Chem., 1979, <u>57</u>, 742) and from ether (110) obtained by the photo-induced decomposition of *endo-* and *exo--isocamphanyl* hypobromites in *n*-pentane (Z. Liu, Q. Chen, and Y. Liu, Huaxue Xuebao, 1989, <u>47</u>, 609).

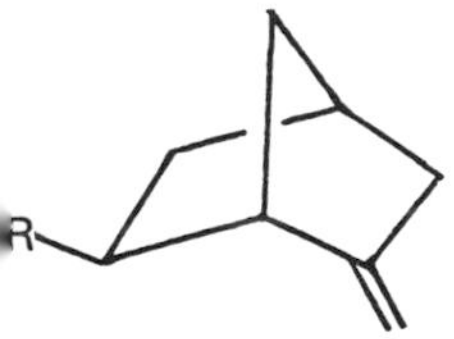

(108) R = OH
(109) R = OAc   (110)    (111)

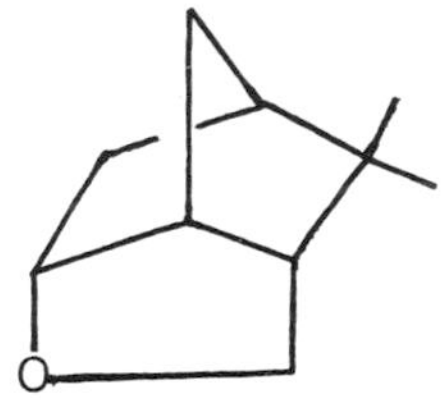

Bornane-2,3-diol. ^{1}H NMR spectra (M.A. Johnson and M.P. Fleming, Canad. J. Chem., 1979, <u>57</u>, 318), ^{1}H and ^{13}C NMR and mass spectra (T. Partanen *et al.*, J. Chem. Soc., Perkin II, 1990, 777), ^{13}C NMR and crystal structure (S.J. Angyal, D.C. Craig, and T.Q. Tran, Austral. J. Chem., 1984, <u>37</u>, 661) of the isomeric bornane-2,3-diols and the synthesis of 2-OH, 3-ester and 2-ester, 3-OH derivatives of bornane (R. Gamboni *et al.*, Tetrahedron Letters, 1985, <u>26</u>, 203) have been reported.

Platydiol (111) a bornane-2,6-diol has been isolated from the pericarp of *Platycladus orientalis* and its structure elucidated by spectroscopic and chemical methods (Y.H. Kuo, W.C. Chen, and K.S. Shih, Chem. Expres, 1989, <u>4</u>, 511).

(+)-Camphor-10-thiol has been prepared by the reduction of (+)-camphor-10-sulphenyl chloride with red P-iodine in AcOH (N. Proth, Rev. Tech. Luxemb., 1979, <u>71</u>, 129).

(d) Bases of bornane series

2-*N,N*-Diethylbornylamine is prepared by boiling bornyl chloride with Et$_2$NH in dry C$_6$H$_6$ (M.E. Spiridonova *et al.*, Khim. Prir. Soedin., 1981, 673) and *endo*-2-chlorobornane (112) on treatment with RNH$_2$ (R = Ph, 2-ClC$_6$H$_4$, 3-MeC$_6$H$_4$) at 185° (5h) yields the corresponding *exo*-aminobornane derivative (113) (31.3, 7.9, and 36.6%, respectively) (V.I. Lysenkov *et al.*, Zh. Org. Khim., 1982, 18, 2617). Syntheses have been recorded of the following: *N*-acetylisobornylamine (J. Caram *et al.*, Rev. Latinoam. Quim., 1986, 17, 39), *exo*- and *endo*-bornyl-*tert*-octylamines (E.J. Corey and A.W. Gross, J. Org. Chem., 1985, 50, 5391), *N*-bornyl- and *N*-isobornyl- -tryptamines (S.A. Kozhin, P.R. Aier, and S.V. Vitman, Zh. Org. Khim., 1975, 11,1640), *N*-substituted isocamphylamines (N.G. Kozlov, T. Valimae, and G.V. Nesterov, Zh. Org. Khim., 1984, 20, 295).

(112)

(113) R = Ph, 2-ClC$_6$H$_4$, 3-MeC$_6$H$_4$
(115) R = Me, Et

(114) R = Me, Et

Reductive amination of (-)-camphor by MeCN and H$_2$C = CHCN at 15 atmospheres (H$_2$) over a Cu-Al$_2$O$_3$ catalyst affords stereoisomeric mixtures of (114 and 115) (I.I. Bardyshev, N.G. Kozlov, and T. Valimae, Vestsi Akad. Navuk BSSR, Ser. Khim. Navuk, 1980, 71). For the reductive amination of (-)camphor with HCONH$_2$ see I.I. Bardyshev and N.G. Kozlov (Doklady Akad. Nauk BSSR, 1979, 23, 630) and for the hydrogenation of camphor enamines, F. Bondavalli, P. Schenone, and A. Ranise [J. Chem. Res. (S), 1980, 257].

(1*R*)-3-*endo*-Aminocamphor (116) is stereospecifically reduced with Me$_2$CHCH$_2$AlCl$_2$ to furnish (1*R*)-3-*endo*-aminoborneol (117) (H. Pauling, Helv., 1975, 58, 1781). Treatment of 3-aminocamphor with p-MeC$_6$H$_4$N$_2^+$Cl$^-$ yields mainly compound (118), which is in slight equilibrium with the 3-diazoamino-camphor derivative (119) (S. Treppendahl and P. Jakobsen, Acta Chem. Scand., 1980, B34, 303).

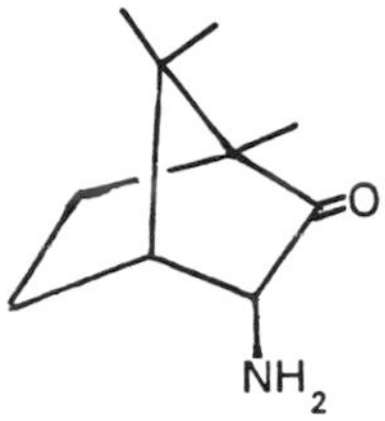

(116)

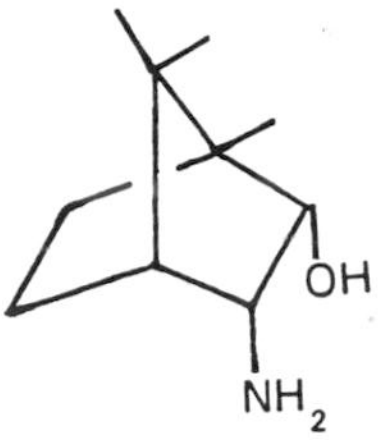

(117)

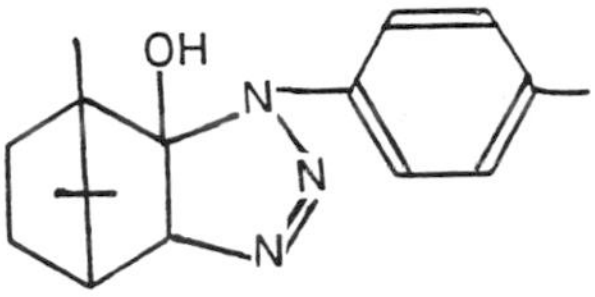

(118)

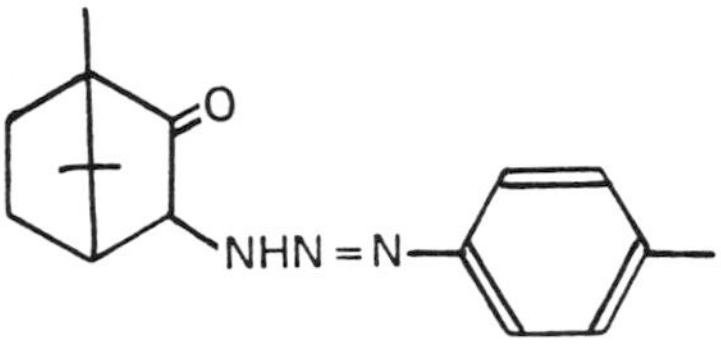

(119)

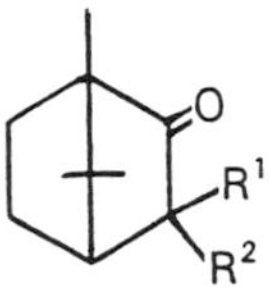

(120)

R^1 = Ac,Bz,1-naphthoyl
R^2 = 2-,3-,4-$O_2NC_6H_4N=N$

Camphorimines (I. Shahak and Y. Sasson, Synth., 1973, 535; P. Guo *et al.*, Huaxue Xuebao, 1990, 48, 78), 4,N-dinitrobornan-2-imine (4,N-dinitro-camphorimine) (T.S. Cameron *et al.*, J. Chem. Soc., Perkin II, 1979, 300), alkylation of camphorimines of glycinates (J.M. McIntosh and P. Mishra, Canad. J. Chem., 1986, 64, 726). The asymmetric alkylation of D-camphor-imine of 1-aminomethylnaphthalene (Y. Jiang *et al.*, Synth. Comm., 1989, 19, 1297) and the aldol reaction of D-camphorimine derived from benzylamine (*idem, ibid.*, 1988, 18, 1291) have been investigated.

Reactions between 3-acylcamphor and nitrobenzenediazonium salts give the 3-acyl-3-nitrophenylazocamphor (120) (J.C. Gullaumon *et al.*, Compt.

Rend., Ser. C, 1973, <u>276</u>, 1111). ^{13}C NMR spectra, chemical shifts and substituent effects of 4-substituted camphors, *N*-nitrocamphorimines, and diazocamphors have been discussed (D.G. Morris and A.M. Murray, J. Chem. Soc., Perkin II, 1976, 1579).

8. The norbornane group: fenchone and related compounds

(a) Ketones and alcohols

(i) Fenchone(1,3,3-trimethylnorbornan-2-one) and epifenchone(1,2,2-tri-methylnorbornan-3-one)

The Diels-Alder reaction between methylcyclopentadiene and $CH_2=CCICN$, catalyzed by $Cu(BF_4)_2$, gives 1-methylnorborn-5-en-2-one, which on hydrogenation yields 1-methylnorbornan-2-one. Methylation of the latter ketone by MeI, catalyzed by lithium cyclohexylisopropylamide affords fenchone (1) (G. Buchbauer and H.C. Rohner, Ann., 1981, 2093). For absolute configuration of (+)-fenchone and (+)-dehydrofenchone [(+)-1,3,3--trimethylnorborn-5-en-2-one] see J. Korvola and P.J. Malkonen (Suom. Kemistilehti, B, 1972, <u>45</u>, 381).

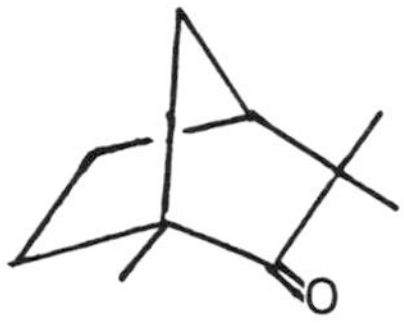

(1) 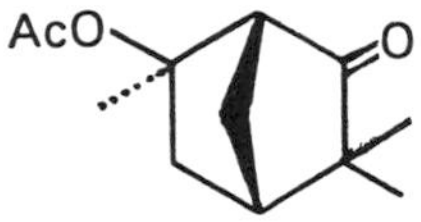   (2) 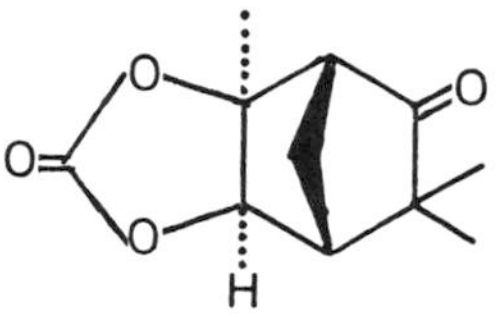   (3)

Oxidation of (-)-fenchone with $CrO_2(OAc)_2$ in AcOH yields (-)-fenchane--2,5-dione, (+)-fenchane-2,6-dione, acetate (2) and carbonate (3). Similar oxidation of (+)-2-*endo*-fenchyl acetate furnishes a mixture of products. Bromination of (+)-fenchone with Br_2 in presence of Cu affords (+)-1-bromo-methylfenchone [(+)-10-bromofenchone], (-)-(1*R*)-7-*anti*-bromo-1,3,3--trimethylnorbornan-2-one, and (+)-3-*endo*-bromocamphor [W. Cocker, R.L.

413

Gordon, and P.V.R. Shannon, J. Chem. Res. (S), 1985, 172]. The Grignard
reduction of fenchone with various alkylmagnesium halides gives α-fenchol as
the main product and ß-fenchol as a minor product (J. Korvola, Suom.
Kemistilehti B, 1973, 46, 212).

The photo-Beckmann rearrangement of (+)-fenchone and (+)-camphor
oximes (H. Suginome, F. Furukawa, and K. Orito, Chem. Comm., 1987,
1004), and their treatment with Me_3COCl to give the corresponding
chloronitroso derivatives (5 and 6) (T. Bosch, G. Kresze, and J. Winkler,
Ann., 1975, 1009) have been reported. Also reported are the catalytic
hydroamination of (+)-fenchone (N.G. Kozlov, G.V. Kalechits, and T.
Valimae, Khim. Prir. Soedin., 1983, 480), synthesis of (±) homofenchone
[(±)-1,4,4-trimethylbicyclo[3.2.1]octan-3-one] (7) (W. Kreiser, P. Below, and
L. Ernst, Ann., 1985, 194) and (+)-homofenchone (W. Kreiser and P. Below,
ibid., p.203), and high-field ^2H NMR spectra of deuterated fenchone (N.H.
Werstiuk, G. Timmins, and B. Sayer, Canad. J. Chem., 1986, 64, 1465).

(5)   (6)   (7)

Irradiation of thiofenchone and thiocamphor affords cyclofenchyl thiol and
tricyclyl thiol, respectively, which on heating revert to the thiocarbonyl
compounds (D.S.L. Blackwell and P. De Mayo, Chem. Comm., 1973, 130).
For Rydberg states in thiofenchone and thiocamphor see K.J. Falk and R.P.
Steer (Canad. J. Chem., 1988, 66, 575) and for natural and magnetic circular
dichroism spectra of selenofenchone, W.M.D. Wijekoon et al., (J. Phys.
Chem., 1983, 87, 3034).

Epifenchone (8) has been synthesized by the cycloaddition reaction between $H_2C=CMeCMe=CH_2$ and $H_2C=CHCHO$ to give product (9), which was
reductively cyclized and methylated to furnish (8). Dehydrocamphenilone
(10) has also been synthesized (E. Dworan and G. Buchbauer, Ber., 1981,
114, 2357).

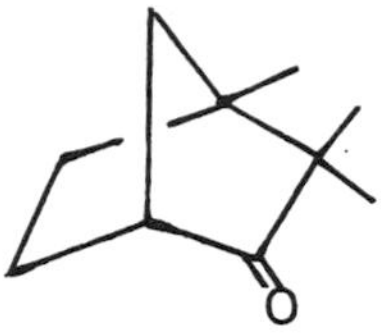

414

(8) 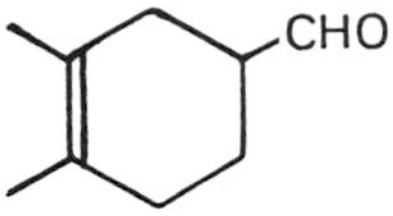   (9) 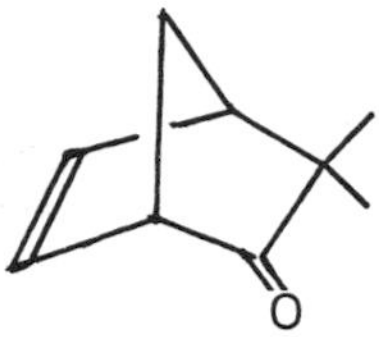   (10)

(ii) Isofenchone(1,5,5-trimethylbornan-2-one)

Beckmann rearrangement of isofenchone oxime (11) gives the azabicyclo derivative (12), which on reduction with LiALH$_4$ yields (13). Reduction of oxime (11) by LiAlH$_4$ in THF affords a mixture containing (13) (5%), *endo*--(14) (45%), *exo*-(14) (10%), and azabicyclo derivative (15) (40%) (N.G. Kozlov and L.A. Popova, Zh. Org. Khim., 1987, <u>23</u>, 990).

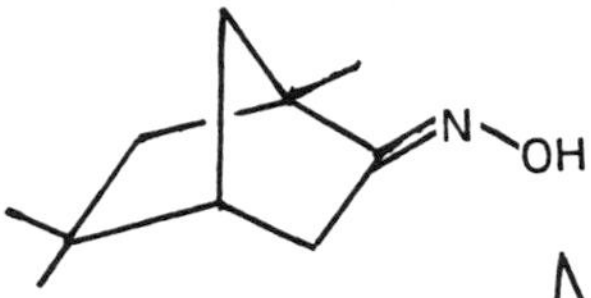

(11)

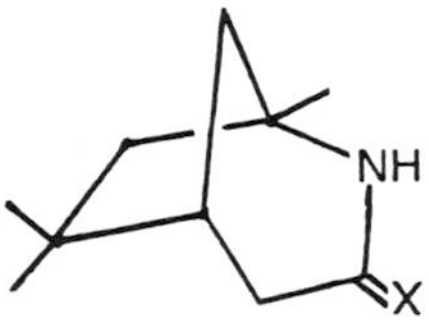

(12) X = O
(13) X = H$_2$

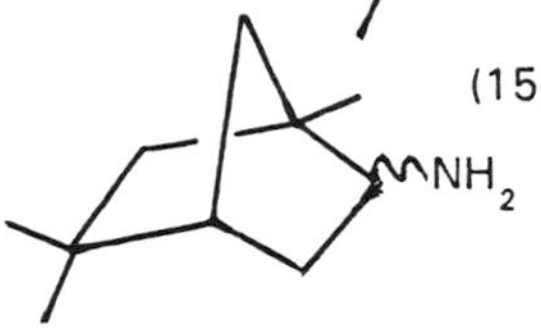

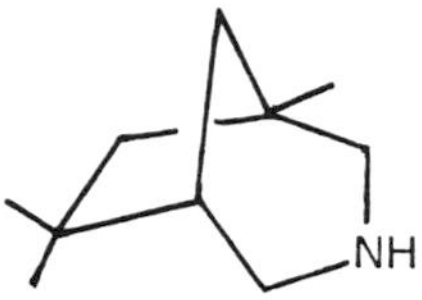

(15)

(14)

(iii) Fenchyl alcohols

Catalytic hydrogenation of (+)-fenchone over Raney Ni gives (+)-*α*- and (-)-ß-fenchol (1,3,3-trimethylnorbornan-2*α*- and -2ß-ol) (16 and 17) (P. Teisseire and A. Galfre, Recherches, 1974, <u>19</u>, 232). Thermal decomposition of fenchyl acetate in a fused mixed salt of NaNO$_2$, NaNO$_3$, and KNO$_3$ (40:7:53 wt.%) gives cyclofenchene and *α*-fenchene. Similarly isobornyl acetate yields tricyclene, camphene, and bornene (M. Nomura, Y. Fujhara, and Y. Matsubara, Nippon Kagaku Kaishi, 1979, 1282). Reaction of (+)--fenchol with $ClCH_2COCl$, followed by the addition of the appropriate amine affords the related amino ester (18) of fenchol (T.S. Raikova et al., Khim. Prir. Soedin., 1983, 305). 10,10-Dinitrofenchol (19) on treatment with mild base gives the epimer (20), presumably by participation of the 10,10-dinitro group (C.G. Franisco et al., Tetrahedron Letters, 1983, 24, 3907).

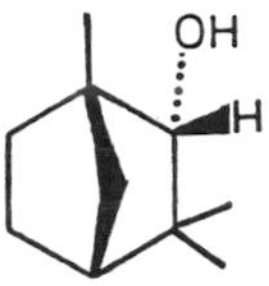

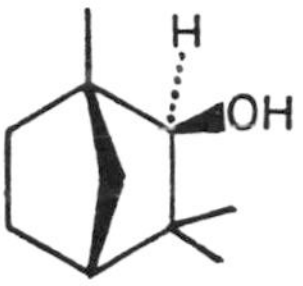

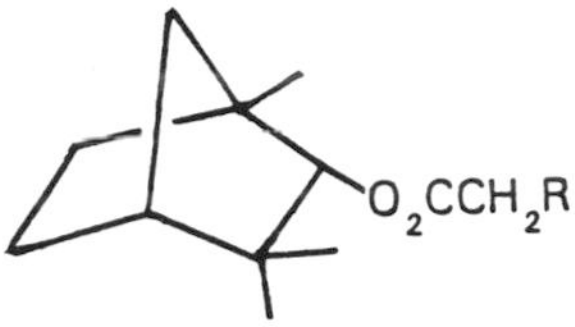

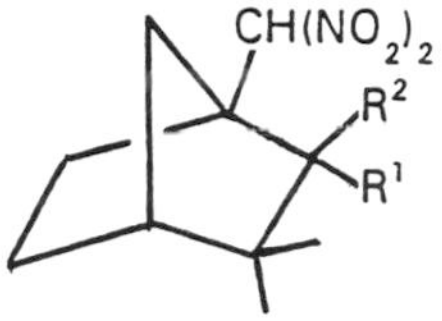

$R = Me_2N, Et_2N, Bu_2N,$
morpholino, piperidino

(19) $R^1 = OH, R^2 = H$
(20) $R^1 = H, R^2 = OH$

(16)   (17)   (18)

Synthesis of exo- and endo-isofenchol (exo- and endo-1,5,5-trimethylnorbornan-2-ol) (21 and 22) (Y. Matsubara and K. Yokoi, Nippon Kagaku Kaishi, 1979, 955) has been described. Catalytic hydroamination of isofenchol (21) by RCN gives a mixture of bicycloheptyl-2-endo- and -2-exo-amine (23 and 24) in a 2:1 ratio (N.G. Kozlov, T. Valimae, and G.V. Nesterov, Zh. Obshch. Khim., 1984, 54, 958). Amino esters of isofenchol (T.S. Raikova et al., Vestsi Akad. Navuk BSSR, Ser. Khim. Navuk, 1984, 85).

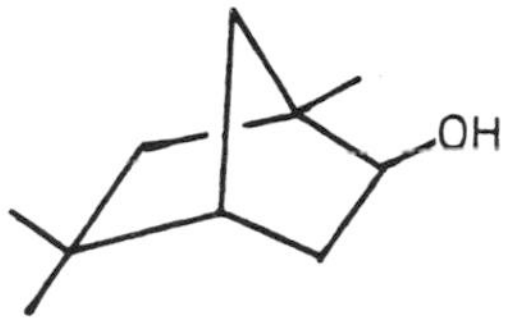

(21)

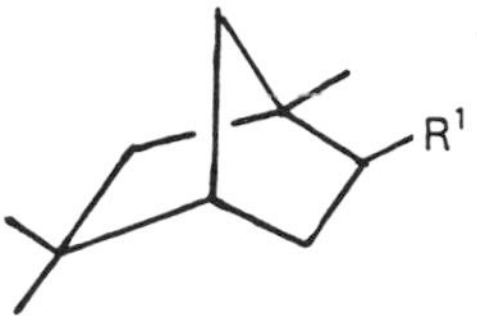

(24)

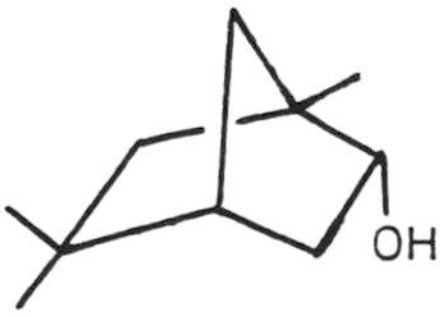

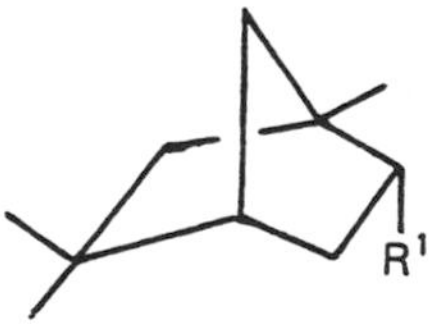

$R^1 = NHCH_2R^2; R^2 = Me, Et, Pr$

(22)   (23)

416

6-Bromoisofenchol on boiling with AgNO$_3$ and MeOH gives acetal (25) [K. Schulze, H. Uhlig and M. Muehlstaedt, Ger. (East) DD 245,660,1987]. Other acetals and fencholene aldehyde (26) have also been prepared (H. Uhlig and K. Schulge, Z. Chem., 1988, 28, 97).

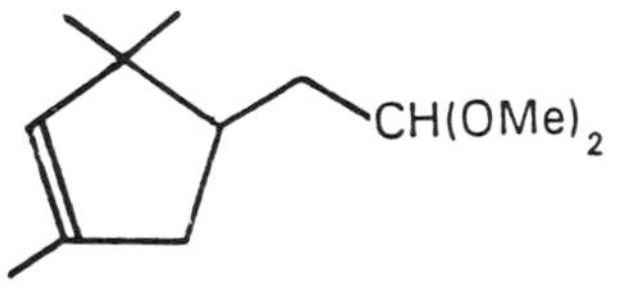

(25) 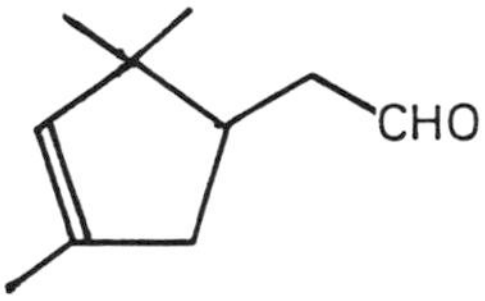   (26)

(b) Hydrocarbons

Fenchenes

Equilibration of a-, β-, γ-, δ-, ζ-, and cyclo-fenchenes (27, 28, 29, 30, 31 and 32) is accomplished by acetoxylation and deacetoxylation. The most stable fenchenes are a, β, and cyclo (Y. Castanet, F. Petit, and M. Evrard, Bull. Soc. Chim. Fr., 1976, 1583).

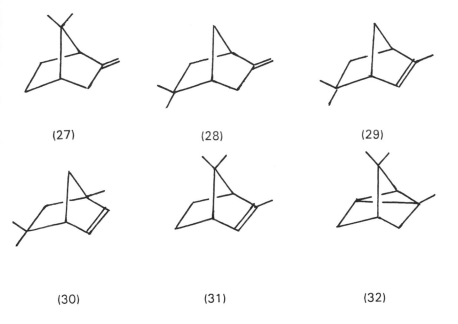

(27)   (28)   (29)

(30)   (31)   (32)

Heating pin-2-ene with titanic acid at 130° yields a mixture containing α-
(27%), ß- (45%), cyclo- (13%), ς,α- (13%), and δ-fenchene (2%). A
mechanism based on 2,6-hydride shifts and Wagner-Meerwein rearrangements has been proposed (G.A. Rudakov, T.N. Pisareva, and N.F. Ovsyukova,
Zh. Org. Khim., 1977, 13, 332). α-Fenchen-9-ol has been synthesized (C.
Grison and Y. Bessiere-Chretien, Bull. Soc. Chim. Fr., 1972, 4570).
Alkylation of phenol by α-fenchene catalyzed by 35% BF₃-AcOH gives a
40:60 mixture of σ- and p-isofenchylphenol (T.F. Gavrilova, I.S. Aul'chenko,
and L.A. Kheifits, Zh. Org. Khim., 1973, 9, 89) and in the presence of
(PhO)₃Al the above two compounds and two other isomers (T.F. Gavrilova *et
al.*, *ibid.*, p.2260). For reactions of cyclo- and α-fenchene see J. Paasivirta
and P. Hirsjarvi (Acta Chem. Scand., 1973, 27, 1098).

Oxidation of cyclo- and ß-fenchene with Pb(OAc)₄, followed by hydrolysis
gives *exo*-ß-fenchol, *endo*-5,5-dimethyl-2-methylenenorbornan-6-ol, 5,5-
-dimethyl-2-(hydroxymethyl)norborn-2-ene, (*R*)-homo-ß-fenchenilone,
endo-(*R*)-homo-ß-fenchenilol, isofenchone, 2,5,5-trimethylnorbornan-6-one,
endo-isofenchol, cyclofenchenyl alcohol, *exo*-5,5-dimethyl-2-methylenenor-
bornan-6-ol, and *exo*-7,7-dimethyl-2-methylenenorbornan-6-ol (T. Uchida *et
al.*, Nippon Nogei Kagaku Kaishi, 1986, 60, 443).

*Second Supplements to the 2nd Edition of Rodd's Chemistry
of Carbon Compounds, Vol. II B(Partial), C, D and E,* edited by M. Sainsbury
© 1994 Elsevier Science B.V. All rights reserved.

Chapter 13

THE SESQUITERPENOIDS

A.T.HEWSON

1. Introduction.

The period covered by this review has seen an enormous increase in our knowledge in the field of organic chemistry and this is reflected in terms of published work in the sesquiterpene area, as indeed it is in all areas of the subject. New structures have continued to be isolated from natural sources, but perhaps the major developments have been in synthetic methodology, with recent attention being directed towards target molecules in chirally pure form. Biosynthetic studies have also resulted in a more detailed understanding of nature's routes to these compounds.

Clearly, in the limited space available here, only a small fraction of the work published in the last twenty years can be considered . Emphasis will be placed largely on those areas which are not merely a development of work previously covered. More detailed reviews containing this information are available in Chem. Soc. Annual Reviews and more recently in Natural Products Reports. A comprehensive review of synthetic work up to 1979 is also available ("The Total Synthesis of Natural Products" (Vol 5), Ed. J. ApSimon, Wiley Interscience, New York, 1983).

In describing synthetic approaches to sesquiterpenoids, only the key step(s) will be given.

2. Acyclic compounds

The unusual sesquiterpene derivative of taurocyamine, agelasidine A (1), has been isolated from the sea sponge *Agelas nakamurai* (H. Nakamura et al., Tetrahedron Letters, 1983, 24, 4105), and a synthesis based on the proposed biosynthesis has been described involving a [2,3]sigmatropic rearrangement of the allylic sulphinate (2) followed by substitution of the acetate in (3) (Y. Ichikawa, T. Kashiwagi and N. Urano, J. Chem. Soc. Perkin Trans I, 1992, 1497).

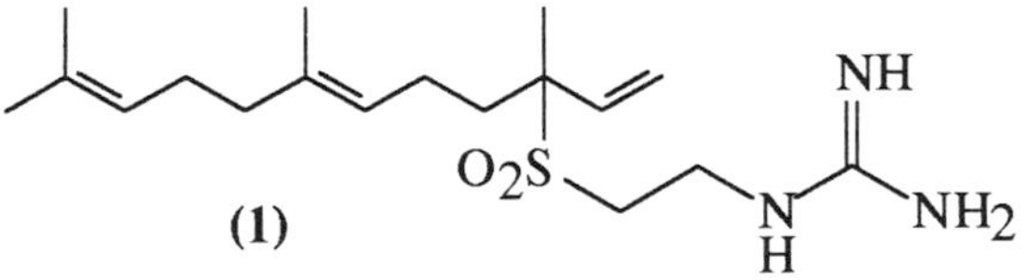

(1)

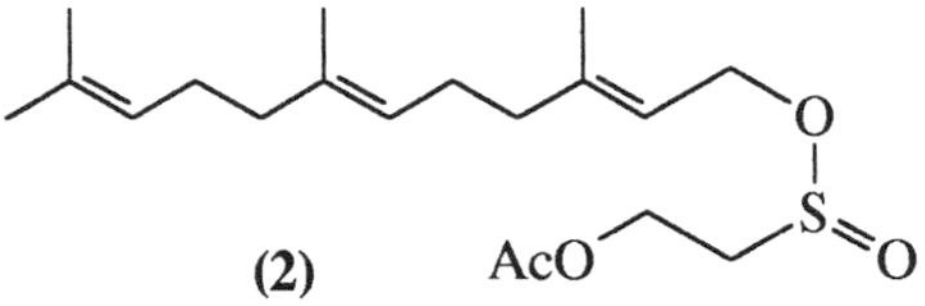

(2)

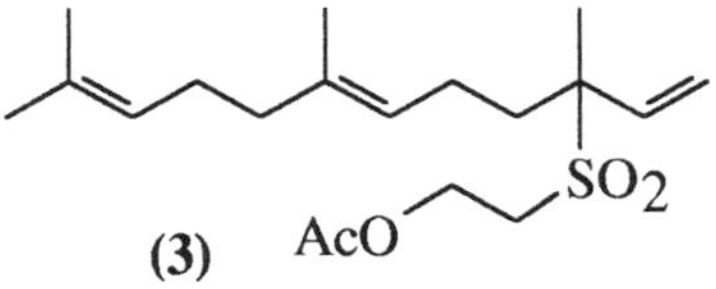

(3)

3. Monocyclic compounds

The bisabolenes paniculides A-C, (4), (5) and (6) respectively, have been isolated from *Andrographis paniculata*. Paniculide A has been synthesised via intramolecular cycloaddition of the acetylenic oxazole (7) which gives the intermediate (8) (P. A. Jacobi, C. S. R. Kaczmarek and U. E. Udodong, Tetrahedron, 1987, 43, 5475).

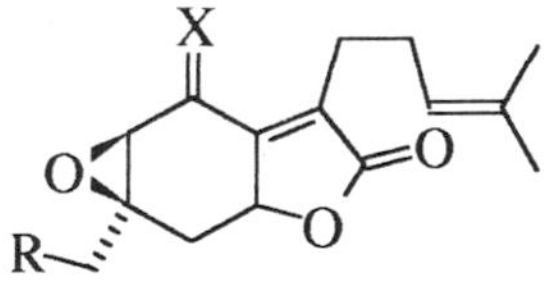

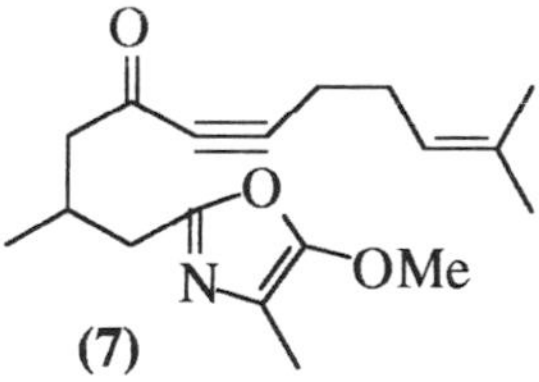

(4) R=H, X=O

(5) R=OH, X=β-OH

(6) R=OH, X=O

(7)

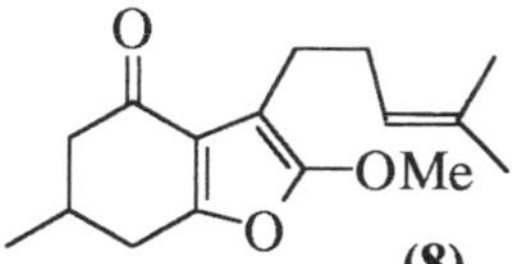

(8)

The important antineoplastic bisabolane glycone phyllanthoside has been isolated from the central American tree *Phyllanthus acuminatus* and its structure elucidated as (9) (G. R. Pettit et al., Can. J. Chem., 1982, 60, 544 and 939). A formidable synthetic effort has resulted in the synthesis of (9) by first preparing the aglycone phyllanthocin (10), followed by glycosidation. The spiro acetal ring system in (10) was set up by acid treatment of the diketo alcohol (11) to give (12) (A. B. Smith et al., J. Amer. Chem. Soc., 1991, 113, 2092).

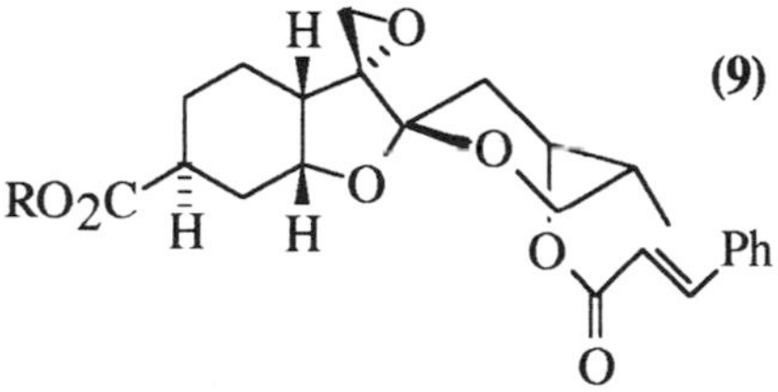

(9) R=

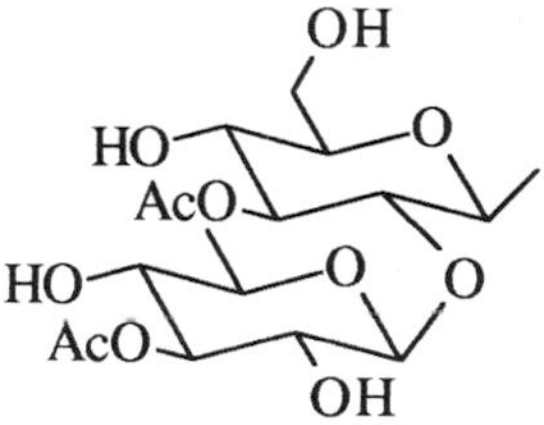

(10) R=Me

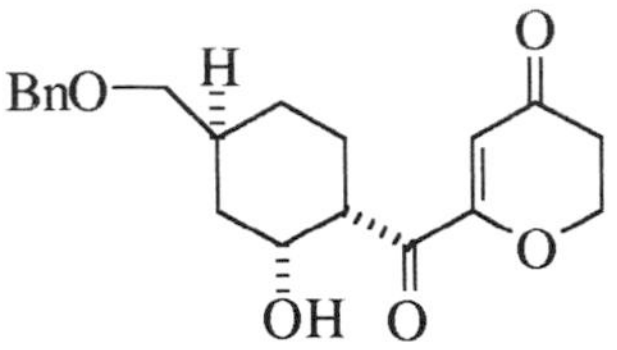

(11) 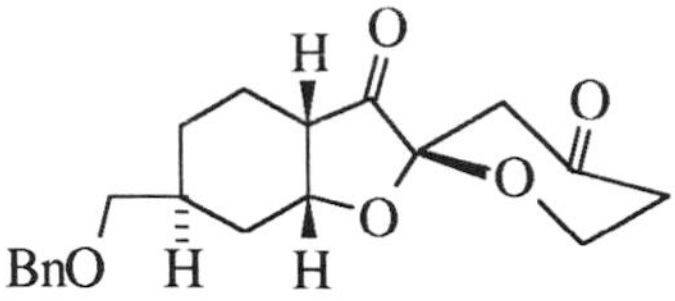   **(12)**

The bisabolane derivative (+)-hernandulcin (**13**), which is more than a thousand times sweeter than sucrose, has been isolated from *Lippia dulcis* (C. M. Compadre et al., J. Agric. Food Chem, 1987, 35, 273). It has been synthesised from (R)-(+)-limonene, proving its absolute stereochemistry to be as shown (K. Mori and M. Kato, Tetrahedron Letters, 1986, 27, 981). Cassiolide (**14**) isolated from *Cinnamonum cassia* shows serotonin-induced antiulcerogenic activity (Y. Shigara et al., Tetrahedron, 1988, 44, 4703). The aglycone cassiol (**15**), which is a stronger inhibitor in rats, has been synthesised from the commercially available optically active keto ester (**16**) (T. Takemoto, C. Fukaya and K. Yokoyama, Tetrahedron Letters, 1989, 30, 723).

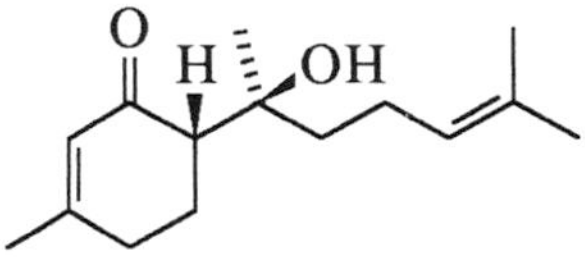

(13)

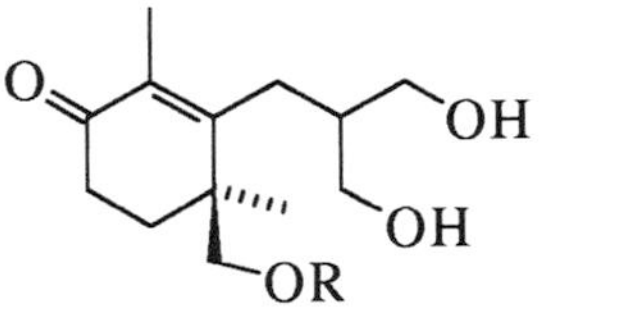

(14)   R= β-D-glucopyranosyl

(15)   R=H

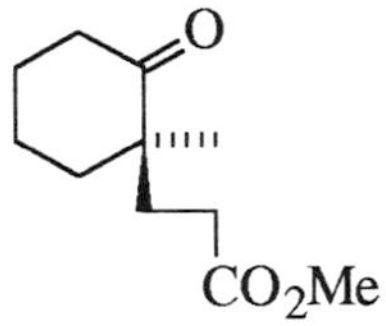

(16)

A number of nitrogen containing sesquiterpenes have been found in marine organisms, exemplified by the bisabolanes (**17-20**) isolated from two sponges of the genus *Theonella* (I. Kitigawa et al., Chem. Pharm. Bull., 1987, 35, 928).

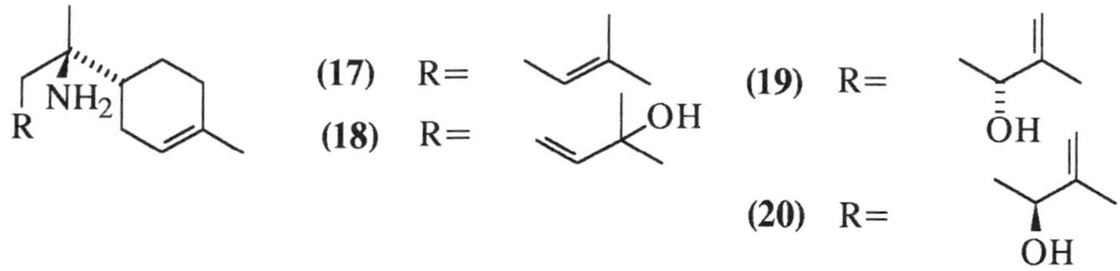

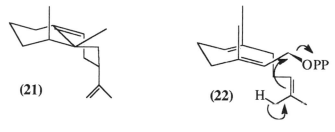

(17) R=

(18) R=

(19) R=

(20) R=

4. Bicyclo[4.4.0]decanes.

The (-) enantiomer of the eremophilane aristolchene (**21**) was reported in the earlier supplement, and now the (+) enantiomer has been obtained from the fungus *Aspergillus terreus* (D. E. Cane, E. J. Salaski and P. C. Prabhakaran, Tetrahedron Letters, 1990, <u>31</u>, 1943). Using the aristolchene synthase isolated from *A. terreus*, and appropriate isotopically labelled precursors, detailed information on the biosynthesis has been obtained (D. E. Cane et al., J. Amer. Chem. Soc., 1990, <u>112</u>, 3209). Farnesyl pyrophosphate (**22**) cyclises to germacrane A (**23**) with inversion of configuration at C-1 and deprotonation of the C-12 (cis) methyl group. Further cyclisation gives the cation (**24**) which undergoes methyl migration and then loss of H-8*si* in double bond formation.

(21)

(22)

424

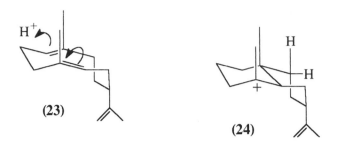

(23)

(24)

The drimane derivative warburganal (25) has attracted much attention due to its insect anti-feedant activity. One route to the (-)-isomer involves controlled oxidative degradation of the readily available (-)-abietic acid (26) to give the intermediate (27) which can be further degraded to (25) (H. Okawara, H. Nakai and M. Ohno, Tetrahedron Lett., 1982, <u>23</u>, 1087). Another approach uses a Diels Alder reaction to produce the key intermediate (28) (S. V. Ley et al., J. Chem. Soc., Perkin Trans I, 1983, 1579). A third approach, which is very efficient, uses sequential oxidations of the diene (29) to afford the diol (30), which undergoes base catalysed isomerisation to the triol (31) (T. Nakano and A. Martin, J. Chem. Research (S), 1989, 52).

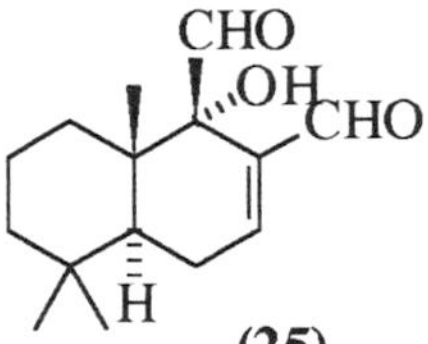

(25)

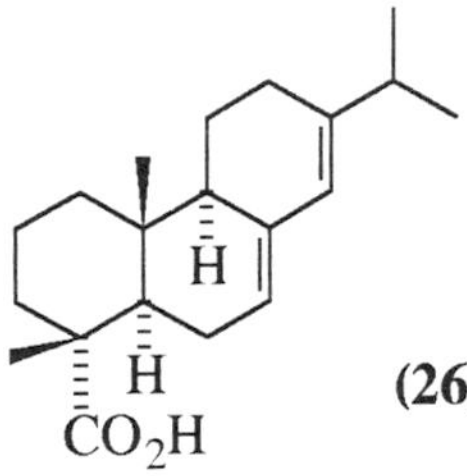

(26)

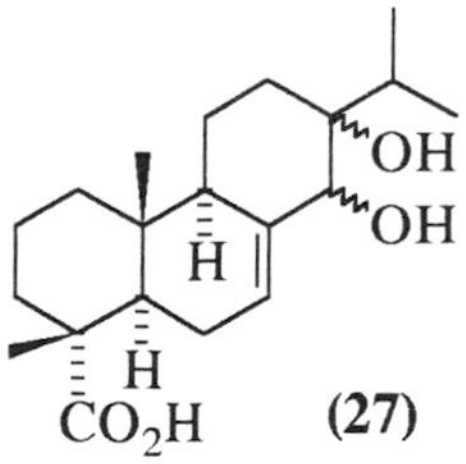

(27)

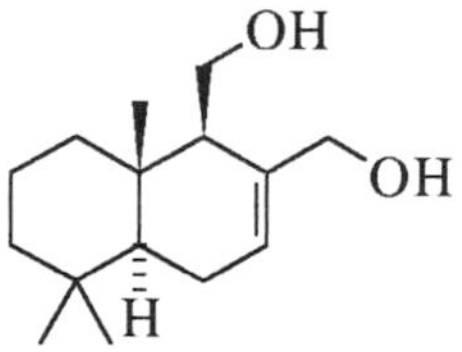

(28)

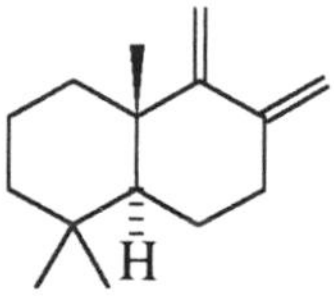

(29)

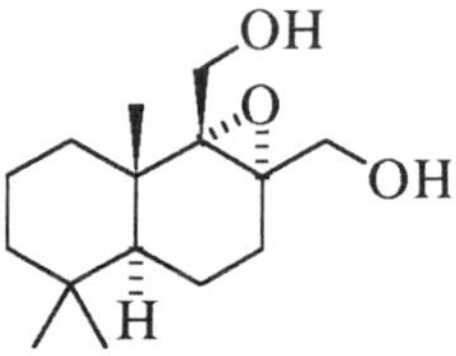

(30)

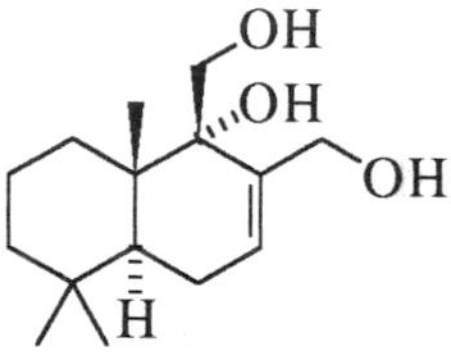

(31)

The fascinating cadinane derivative artemisinin (also known as qinghaosu) (32) has potent plasmocidal properties. It has undergone trials in China as an antimalarial, and the derivative (33) is nearing commercial development. (A. R. Butler and Yu-Lin Wu, Chem. Soc. Rev., 1992, 21, 85). The peroxide bridge is crucial to its biological activity. Artemisinin has been synthesised by photooxidation of (34), itself derived from citronellal, followed by acidification (W. Zhou et al., Tetrahedron, 1986, 42, 819).

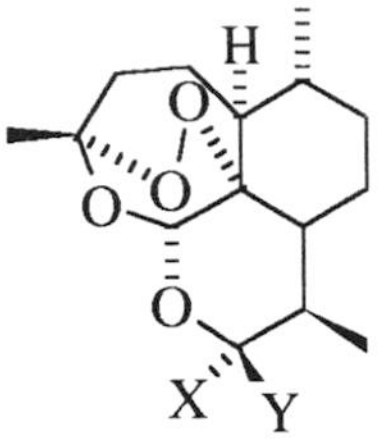

(32) X+Y=O

(33) X=H,Y=OEt

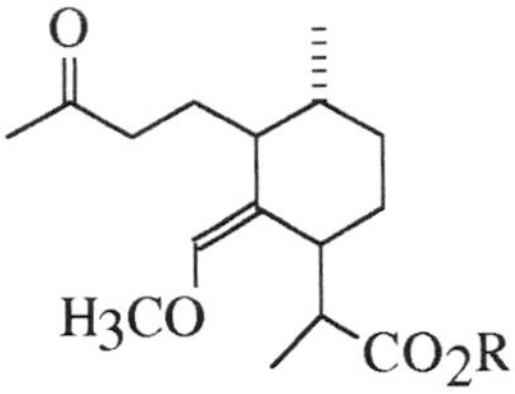

(34)

5. Bicyclo[4.3.0]nonanes.

The highly functionalised compound picrotoxin (35) has been synthesised in an elegant manner from carvone, with a key step being the conversion of (36) to (37) by means of a remarkable oxidative lactonisation process using $Pb(OAc)_4$ (E. J. Corey and H. L. Pearce, J. Amer. Chem. Soc., 1979, 101, 5841).

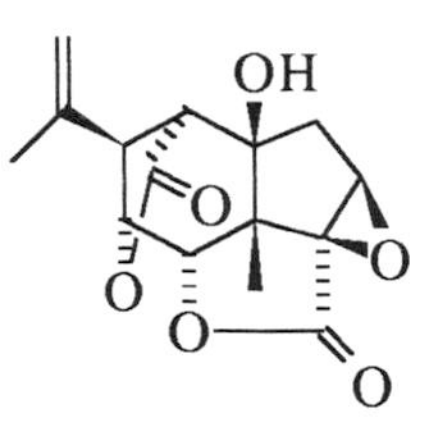

(35) 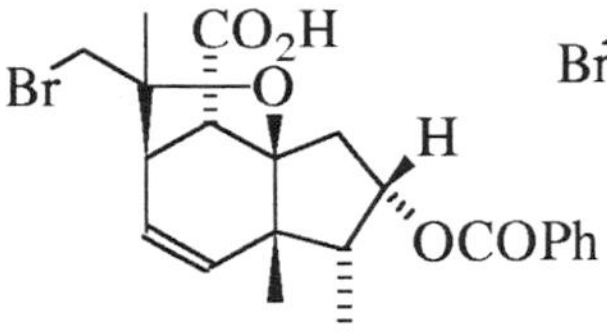   (36) 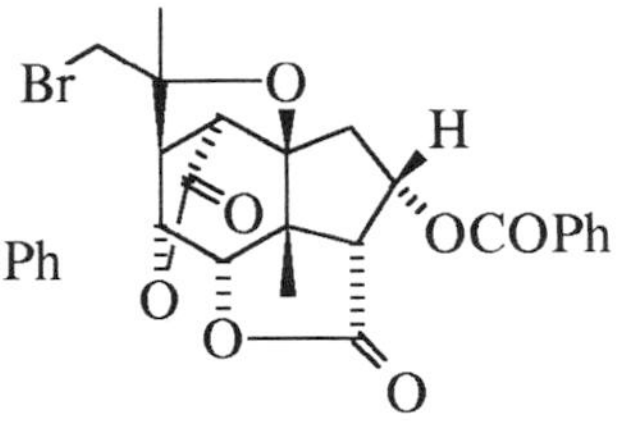   (37)

6. Cyclopentyl cyclohexanes.

Herbertene (38) has been obtained in a fascinating acid catalysed rearrangement of the drimane-type alcohol (39) (G. Frater, J. Chem. Soc. Chem. Commun., 1982, 521). The conversion probably involves a series of Wagner-Meerwein and cyclopropyl-carbinyl rearrangements.

Further details of the biosynthesis of trichodiene (**40**) have been elucidated using the enzyme isolated from a variety of fungal sources and appropriately labelled farnesyl pyrophosphate(**41**). The intermediacy of nerolidyl pyrophosphate has also been demonstrated (D. E. Cane and H-J Ha, J. Amer. Chem. Soc., 1989, 110, 6855).

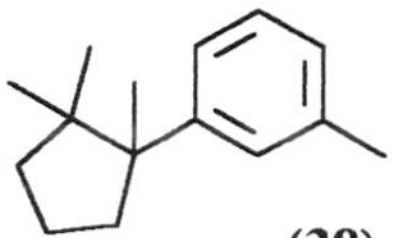

(38)

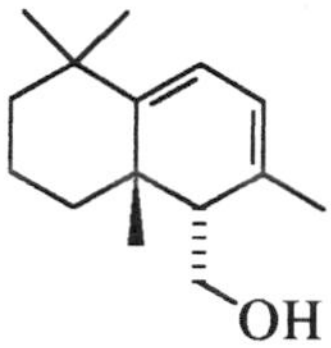

(39)

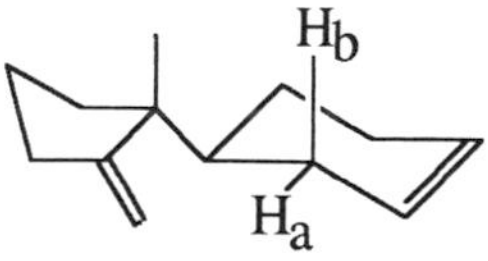

(40)

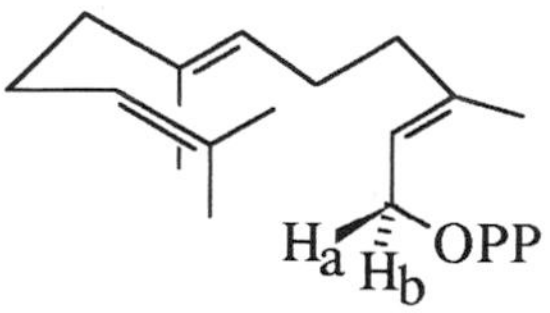

(41)

Trichodiene (**40**) has been synthesised in a stereoselective manner in which the enelactone (**42**) undergoes regio and stereoselective Diels-Alder reaction to give (**43**) which is readily converted to the key precursor (**44**) (R. H. Schlessinger and J. A. Schultz, J. Org. Chem., 1983, 48, 407).

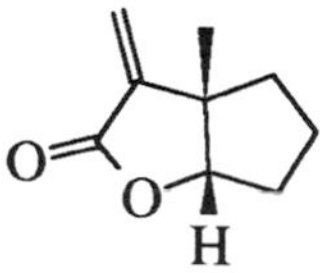

(42)

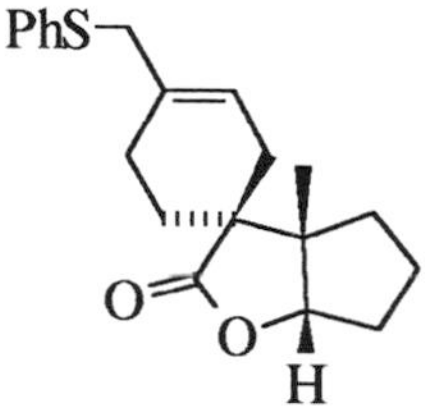

(43)

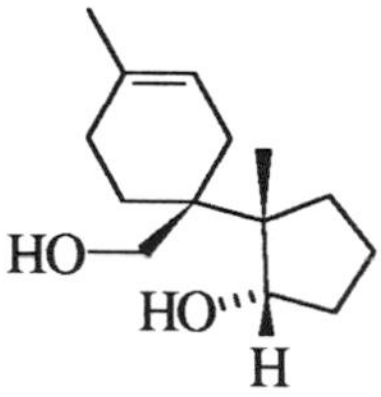

(44)

A number of syntheses of verrucarol (**45**) have been reported with one involving a Diels Alder reaction on the ene-lactone (**46**) to give the spiro compound (47), which on reduction, acid catalysed rearrangement and functional group manipulation gives (45) (W. R. Roush and T. E. D'Ambra, J. Amer. Chem. Soc., 1983, 105, 1058).

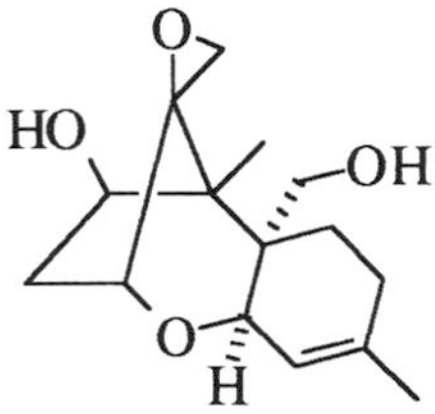

(45) 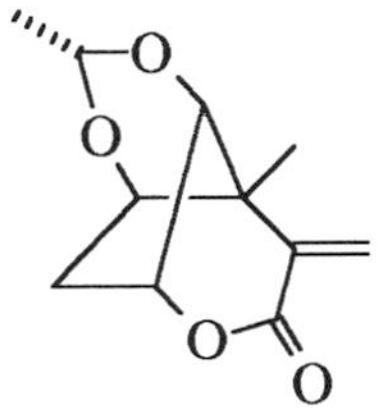   (46) 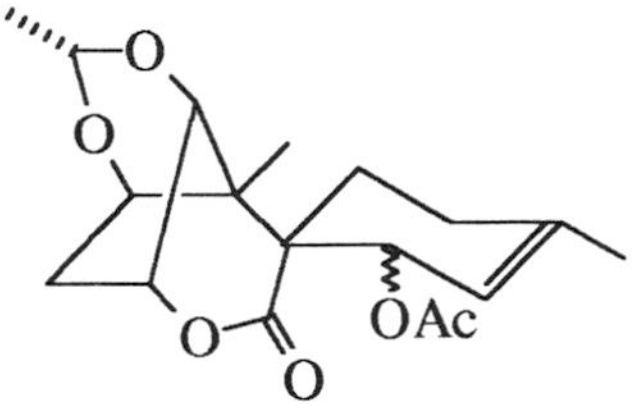   (47)

7. Spiro compounds.

A large number of halogenated chamigranes has been obtained from marine organisms. For example a Sengalese variety of *Laurencia intricata* contains the novel compound almadioxide (48) (M. Aknin et al., Tetrahedron Letters, 1989, 30, 559), while the related (49), isolated from *Laurencia nipponica*, has been shown to exist as a mixture of two stable conformational isomers (M. Segawa et al., J. Chem. Soc., Perkin Trans. II, 1989, 335). The latter organism has also yielded laurenones A (50) and B (51) which possess a novel skeleton (A. Fukuzawa et al., Chem. Lett., 1984, 1349).

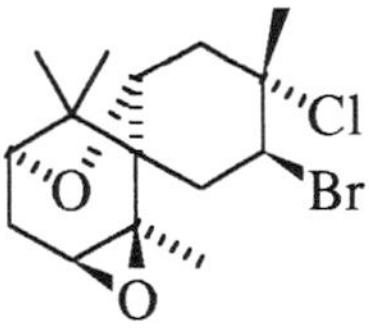

(48) 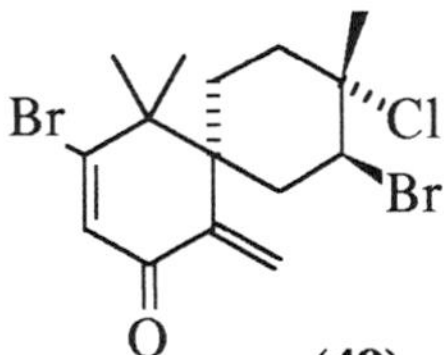   (49)

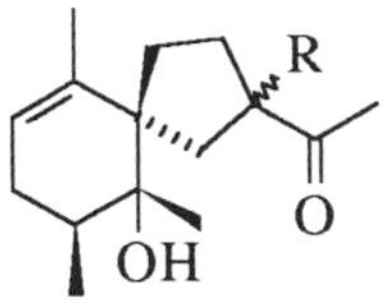

(50) R=β-H

(51) R=α-H

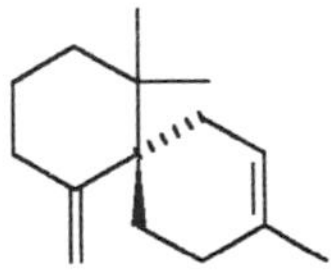

(52)

(+)-β-Chamigrene (52) has been isolated from *Marchantia polymorphia*. This compound is enantiomeric to that isolated from higher plants, suggesting a special position for liverworts in the plant kingdom (A. Matsuo, N. Nakayama and M. Nakayama, Phytochemistry, 1985, 24, 777).

8. Bicyclo[5.3.0]decanes.

A versatile synthetic route to the guaiane and pseudoguaiane systems relies on the [2+2] photocycloaddition of cyclopentenones to 1,2-bistrimethylsilyloxycyclopentenes to give intermediates of general structure (53). Hydrolysis, diol cleavage and simple functional group manipulation afford the important intermediates (54) or (55). (F. Audenart, D. De Keukeleire and M. Wandewalle, Tetrahedron, 1987, 43, 5593).

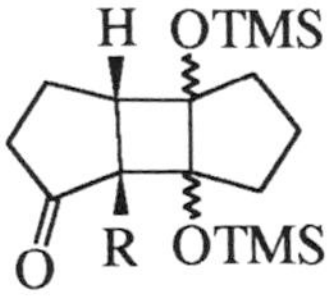

(53) 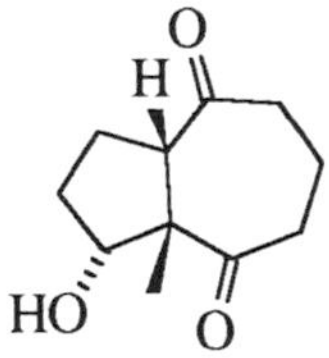   (54) 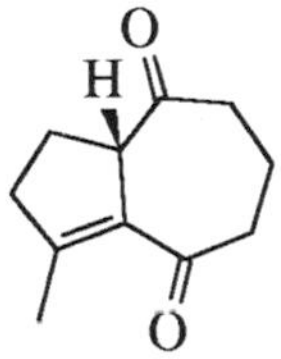   (55)

The method is very versatile since it is also applicable to eudesmane synthesis by varying the ring sizes of the starting materials.
Another useful route to the the hydroazulene skeleton is illustrated by the rhodium(II) catalysed ring expansion of the diazoketone (56) to afford (57)

(M. Kennedy and M. A. McKervey, J. Chem. Soc., Chem. Commun., 1988, 1028).

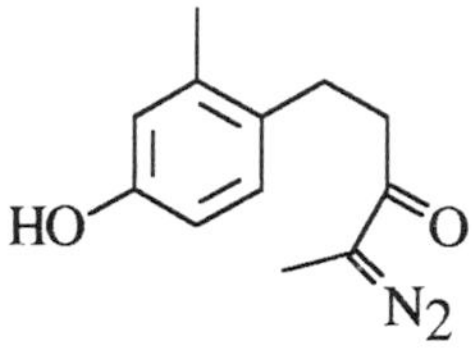

(56)

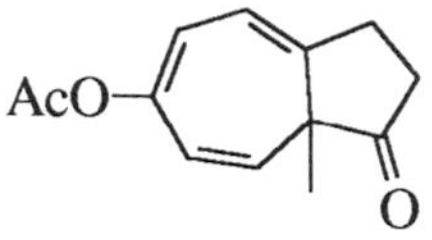

(57)

9. Macrocyclic compounds.

A very short synthesis of humulene (**58**) has been described which illustrate a number of developments made in organometallic chemistry (J. E. McMurry, J. R. Matz and K. L. Kees, Tetrahedron, 1987, 43, 5489). The π-allyl palladium complex (**59**) reacts with an organozirconium complex to give the aldehyde (**60**), which cyclises by treatment with the low valent titanium species generated from TiCl$_3$/Zn/Cu to give humulene .

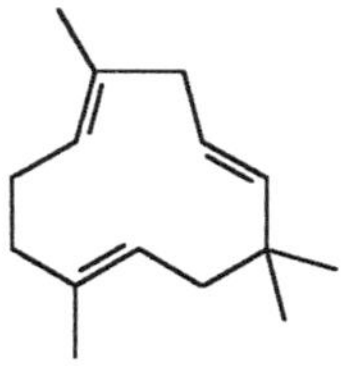

(58)

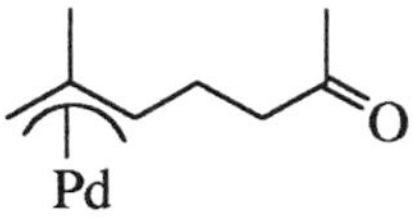

(59)

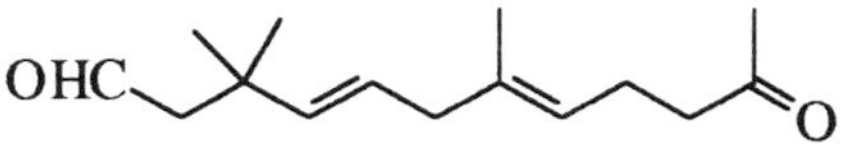

(60)

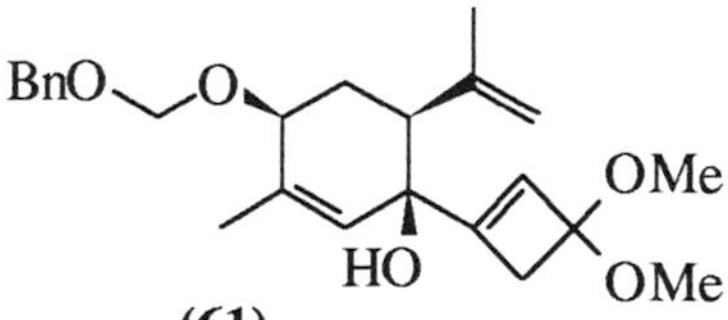

(61)

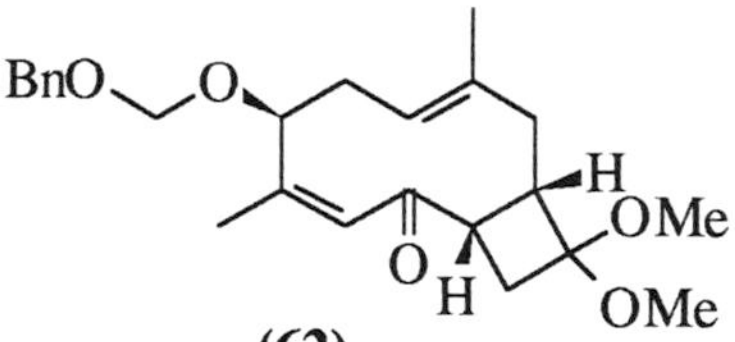

(62)

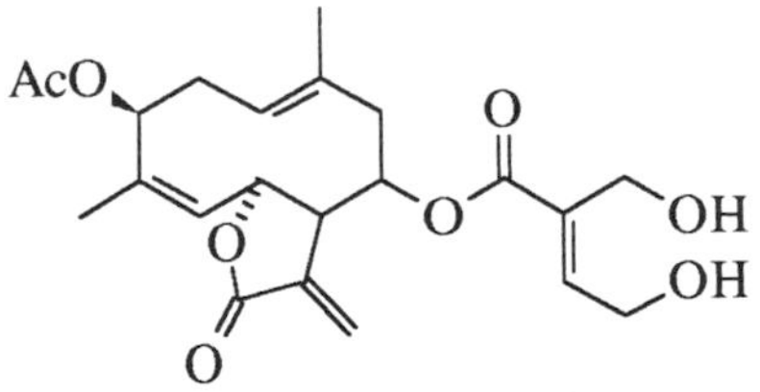

(63)

The oxy-Cope rearrangement has been used in the synthesis of germacrane derivatives as illustrated by the rearrangement of (**61**) to (**62**) in the synthesis of eucannabinolide (**63**) (W. C. Still et al, J. Amer. Chem. Soc., 1983, <u>105</u>, 625).

10. Triquinanes

A family of compounds containing three five-membered rings has been the focus of much work. The hydrocarbons hirsutene (**64**) and capnellene (**65**) have a "linear" skeleton, whilst pentalenene (**66**), isocomene (**67**), silphinene (**68**) and silphiperfolene (**69**) are "angular" triquinanes and modhepene (**70**) has the [3.3.3]propellane skeleton. These compounds are all derived from humulene via a series of cyclisations accompanied by carbocation rearrangements. Detailed studies on the biosynthesis of pentalenene, for example, using a cyclase isolated from *Streptomyces* UC5319, have suggested a route in which farnesyl pyrophosphate (**71**) cyclises to humulene (**72**) which is protonated leading to cyclisation and formation of the protoilludyl cation (**73**). This can undergo a hydride shift followed by further cyclisation and deprotonation to give pentalenene (**74**), labelled as shown (D. E. Cane et al., J. Amer. Chem. Soc., 1988, <u>110,</u> 4081).

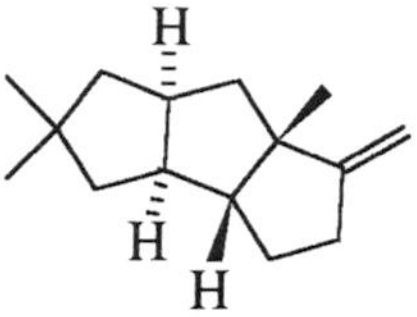

(64)

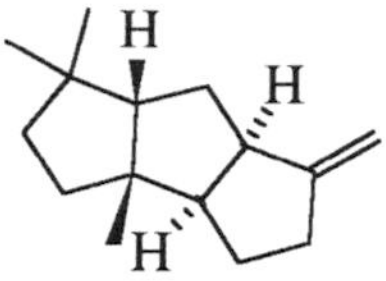

(65)

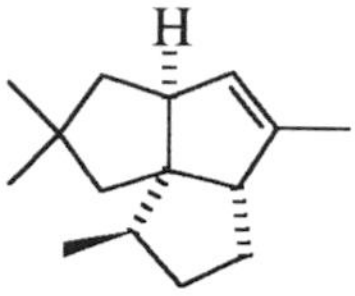

(66)

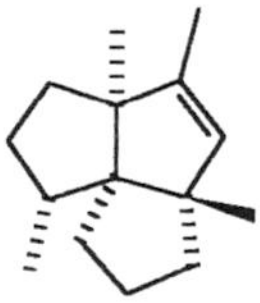

(67)

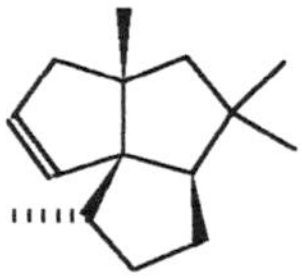

(68)

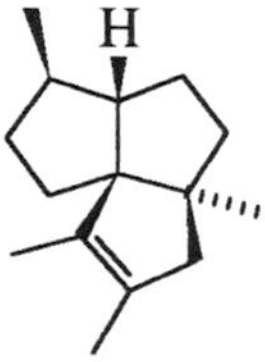

(69)

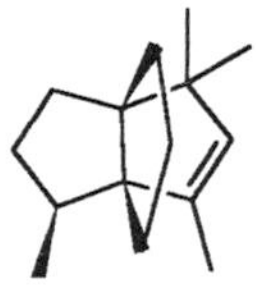

(70)

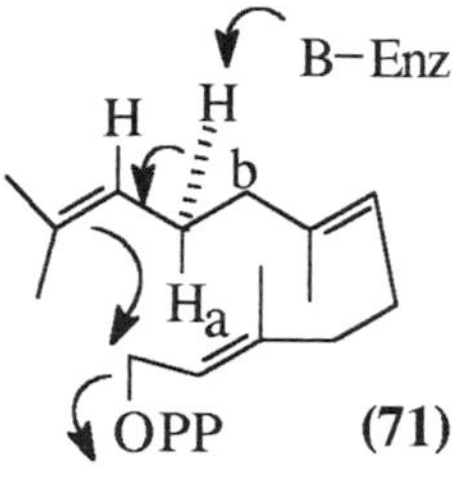

(71)

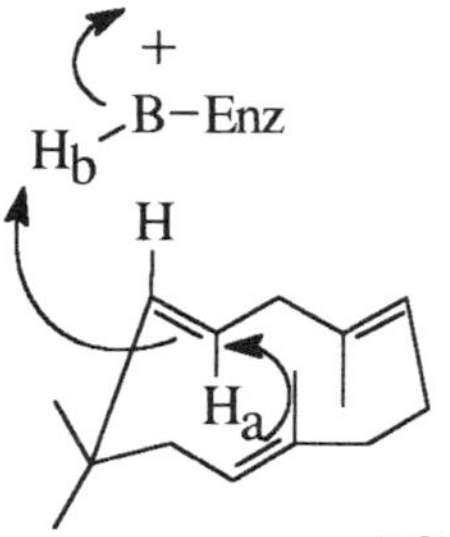

(72)

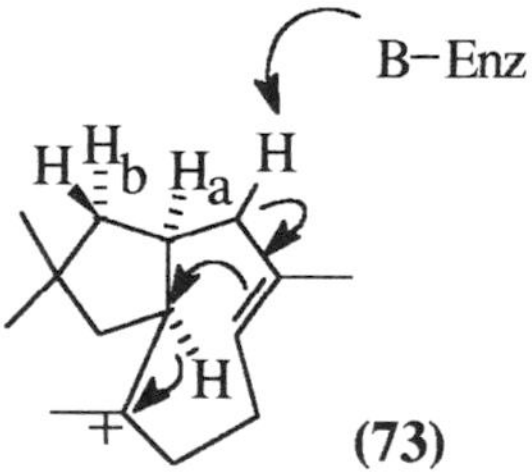

(73)

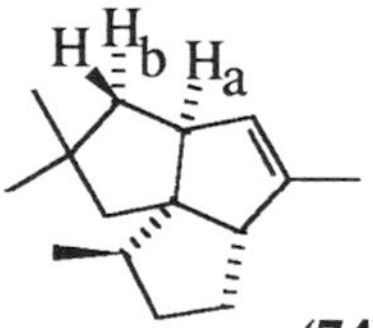

(74)

These triquinanes, both as the hydrocarbons and as various oxygenated derivatives, have been the targets of a vast amount of synthetic effort some of which is described below. Early work in the area has been reviewed (L. A. Paquette, Top. Curr. Chem, 1984, 84, 1).

Hirsutene (64) has been obtained via the trimethylsilyl iodide promoted rearrangement of the dione (75) which provides the intermediate (76) (M. Iyoda et al., J. Chem. Soc., Chem. Commun., 1986, 1049). Isocomene (67) has been synthesised by a route which uses alkynone cyclisations as the key ring-forming reactions, as illustrated by the step which converts (77) to (78) by heating at 540° (G.G.G. Menzardo, M. Karpf and A. S. Dreiding, Helv. Chim. Acta., 1986, 108, 2090).

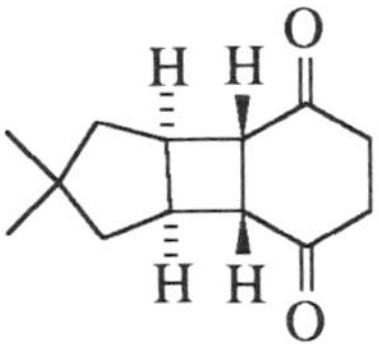

(75) 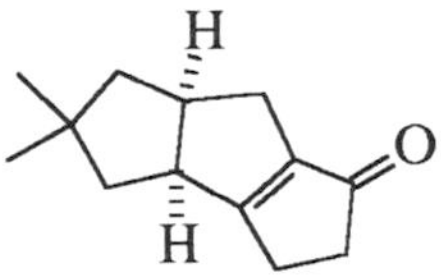   (76) 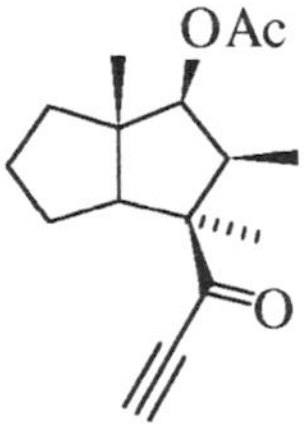   (77) 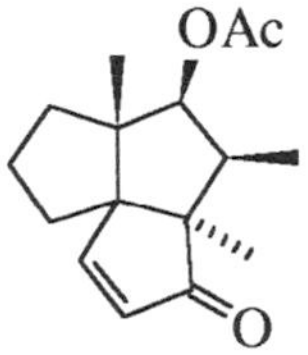   (78)

An asymmetric route to (+)-pentalenene (66) has been described which makes use of the asymmetric induction in the reaction of a chiral sulphinylallyl anion with an enone. Thus the anion derived from the sulphoxide (79) reacts with the enone (80) to give the intermediate (81) which undergoes acid catalysed cyclisation with formic acid to give (82) readily converted to the target molecule (D. H. Hua, J. Amer. Chem. Soc., 1986, 108, 3835).

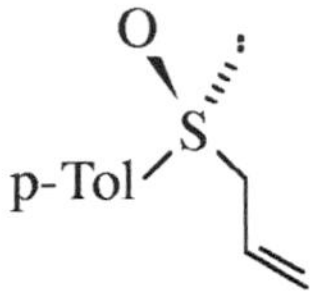

(79)

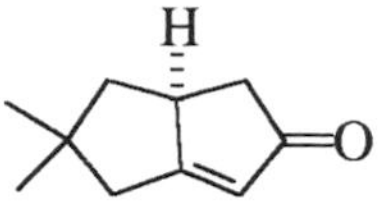

(80)

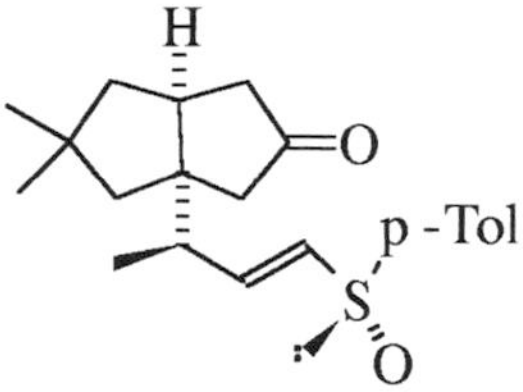

(81)

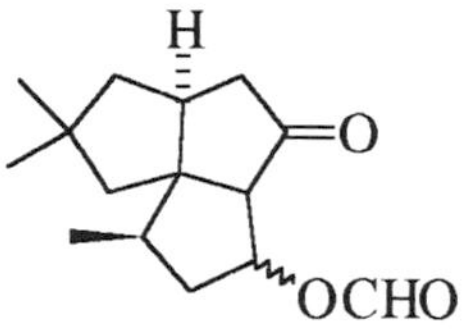

(82)

A concise route to silphinene (**68**) involves an intramolecular photochemical cyclisation of the dienone (**83**) to give the ketone (**84**) which is converted to the iodide (**85**) before treatment with tributyltin hydride which causes a radical induced fragmentation to give silphinene (M. T. Crimmins and S. W. Mascarella, Tetrahedron Letters, 1987, <u>28</u>, 5063).

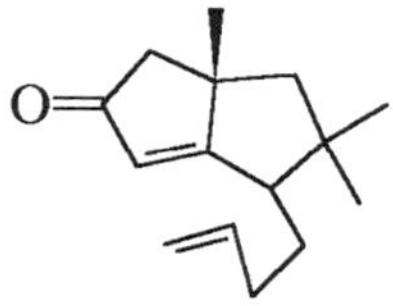

(**83**)

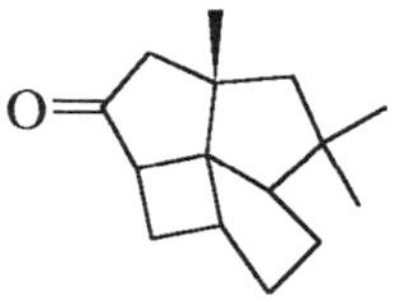

(**84**)

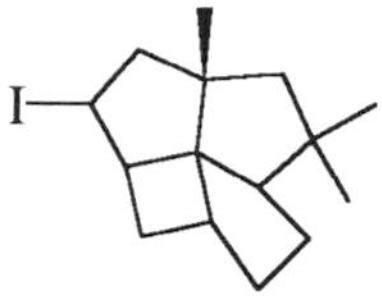

(**85**)

The current interest in radical chemistry is nicely illustrated by an approach to racemic silphiperfolene (**69**) in which the radical generated by Bu$_3$SnH reduction of the vinyl bromide (**86**) cyclises to give (**87**) which is easily converted into (**69**) (D. P. Curran and S-C Kuo, Tetrahedron, 1987, <u>43</u>, 5653). A synthesis of the bromide (**86**) in chiral form has allowed the method to be used to obtain (-)-silphiperfolene (A. I. Meyers and B. A. Lefker, Tetrahedron, 1987, <u>43</u>, 5663).

The photochemical oxa-di-π-methane in the conversion of (**88**) to (**89**) provides a neat route to an intermediate which is readily converted into modhepene (**70**) (G. Metha and D. Subrahmanyam, J. Chem. Soc., Chem. Commun., 1985, 768).

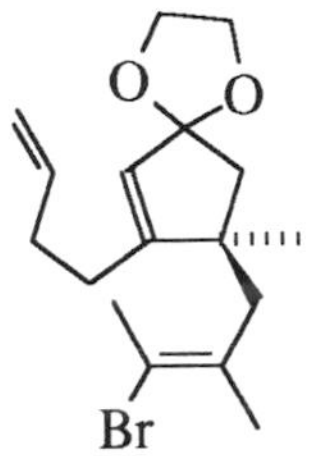

(**86**) 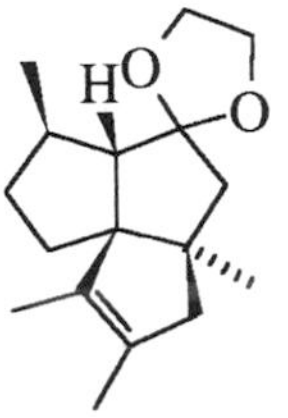 (**87**) 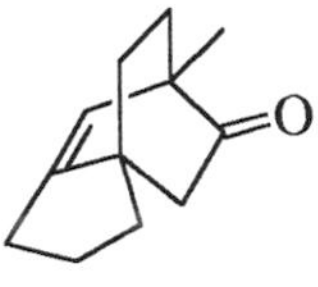 (**88**) 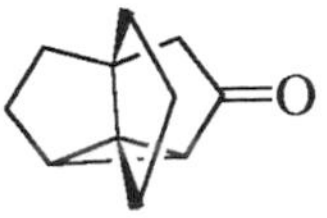 (**89**)

The pentalenolactones are a series of metabolites of *Streptomyces* related to pentalenene, with examples of the family being pentalenolactone (**90**), pentalenolactone E (**91**), pentalenolactone G (**92**) and pentalenolactone H (**93**).

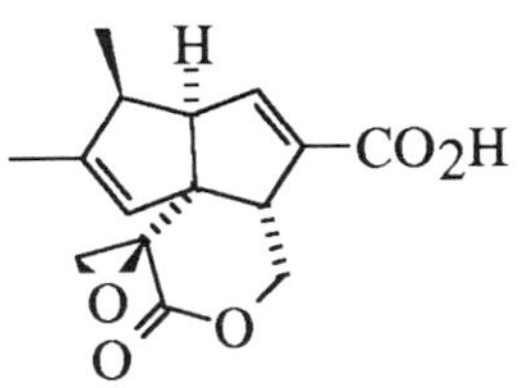

(**90**) 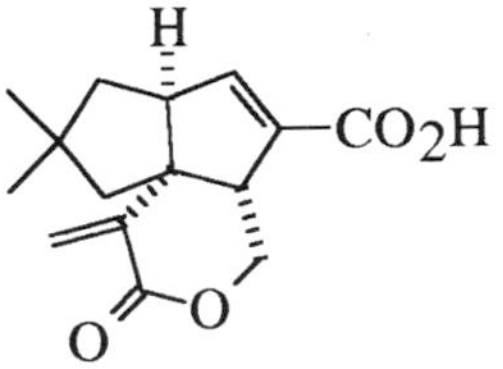 (**91**) 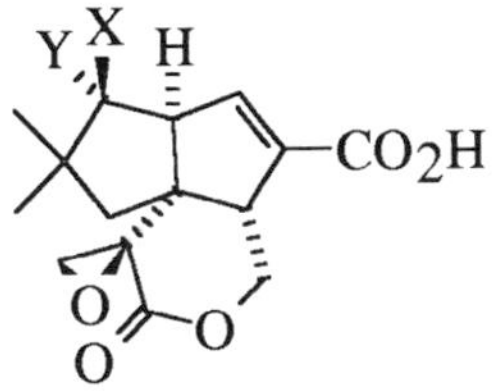 (**92**) X+Y=O

(**93**) X=H, Y=OH

An efficient route to pentalenolactone E involves as the key step a rhodium mediated intramolecular C-H insertion which converts the diazoketoester (**94**) to the tricyclic ketoester (**95**) (D. F. Taber and J. L. Schuchardt, J. Amer. Chem. Soc., 1985, 107, 5289).

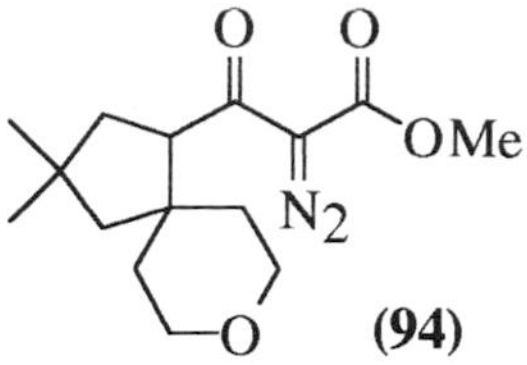

(94) 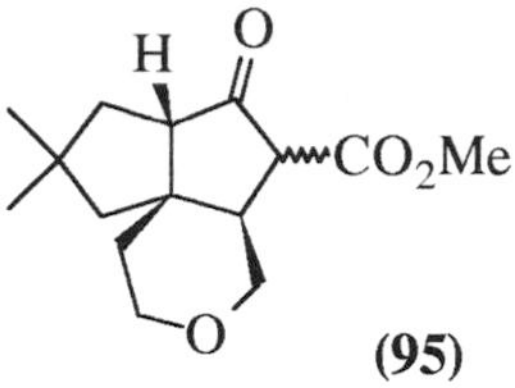　　　　　　　　**(95)**

11. Other polycyclic systems.

The versatile "magnesium-ene" reaction is illustrated in a synthesis of racemic khusimone (**96**). Thus the allylic Grignard reagent (**97**) is converted in a regio- and stereocontrolled manner into a cyclised product, which is trapped by CO_2 to form the precursor (**98**) (W. Oppolzer and R. Pitteloud, J. Amer. Chem. Soc., 1982, <u>104</u>, 6478). The process has also been reported using the optically pure Grignard reagent, hence providing (-)-khumisone (W. Oppolzer et al., Tetrahedron Letters, 1983, <u>24</u>, 4975).

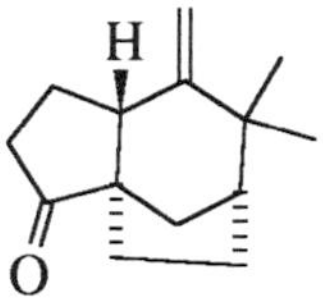

(96) 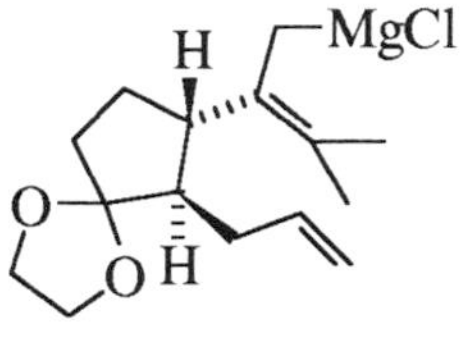　　　　　　　　**(97)** 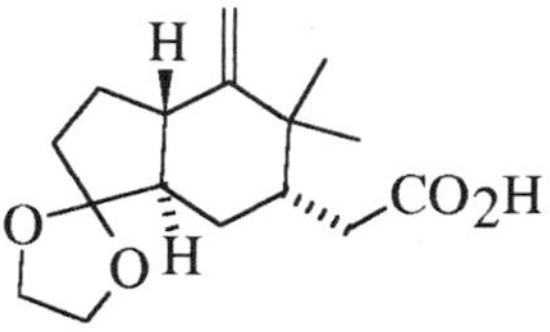　　　　　　　　**(98)**

Punctatin A (**99**), also known as antibiotic 95464, isolated from the dung fungus *Poronia punctata*, has been synthesised by a route in which the key step is a Norrish type II photoreaction on the ketone (**100**) to afford the alcohol (**101**) (L. A. Paquette and T. Sugimura, J. Amer. Chem. Soc., 1986, <u>108</u>, 3841).

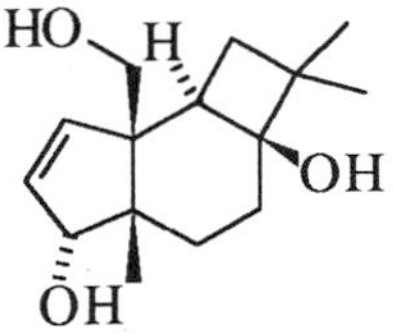

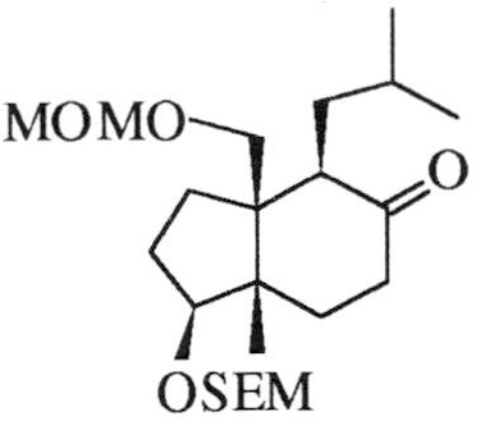

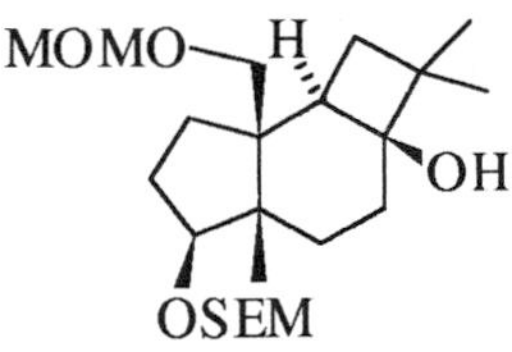

(99)   **(100)**   **(101)**

The unstable nor-sesquiterpene glucoside ptaquiloside (**102**), with the illudane skeleton, has been isolated from the bracken fern *Pteridium aquilinum*. It is a potent carcinogen since it is readily converted to the highly reactive dienone (**103**) which reacts with nucleophiles to give (**104**) (M. Ojika et al., Tetrahedron, 1987, 43, 5261).

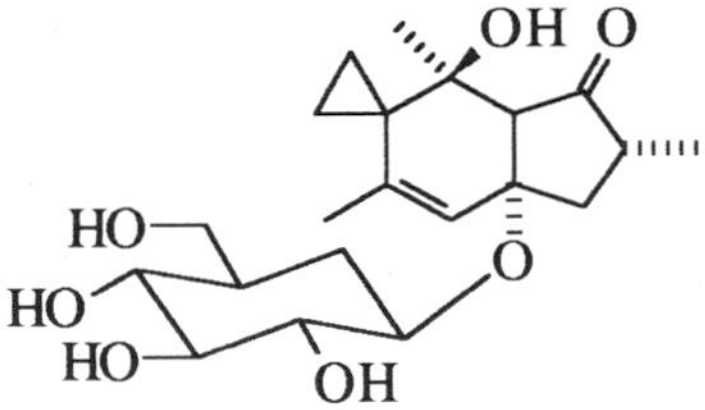

(102) 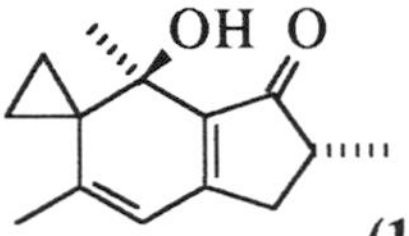   **(103)** 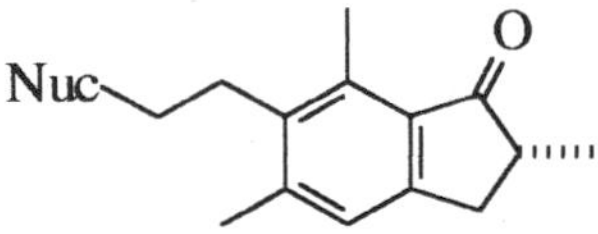   **(104)**

The tricyclic sterpurene skeleton has been shown to be derived from the humulyl cation by a route involving a 1,2-hydride shift and two 1,2-alkyl shifts. Labelling experiments have demonstated the cleavage of the farnesyl 5-6 bond in (**105**) to give (**106**). The latter cation undergoes the second alkyl shift to the intermediate (**107**), which loses a proton to give sterpurene (**108**) (C. Abell and A. P. Leech, Tetrahedron Letters, 1987, 28, 4887).

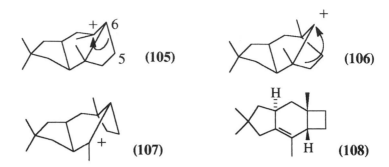

(105)

(106)

(107)

(108)

An interesting synthesis of (+)-sterpurene has been achieved by phenyl sulphenyl mediated ring closure of the dienynol (**109**) which provides sulphoxide (**110**). The latter is converted into sterpurene (**108**) in several steps (R. A. Gibbs and W. H. Okamura, J. Amer. Chem. Soc., 1988, <u>110</u>, 1028).

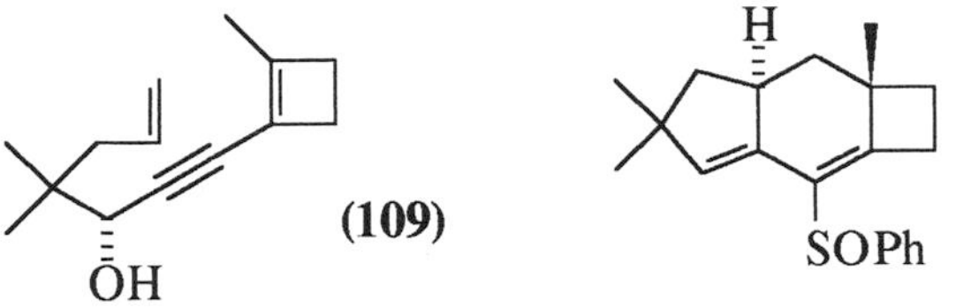

(109)

(110)

The reported anti-tumour activity of the fungal metabolite quadrone (**111**), isolated from *Aspergillus terreus*, together with its unusual structure have attracted much attention. In one route the diester (**112**), readily obtained by an intramolecular Diels-Alder reaction, is converted to the chloride (**113**) which undergoes a silver(I) assisted rearrangement to give the enone (**114**) which can be converted into quadrone (P. A. Wender and D. J. Wolanin, J. Org. Chem., 1985, <u>50</u>, 4418).

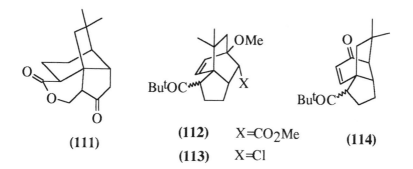

(111)

(112)   X=CO₂Me

(113)   X=Cl

(114)

Second Supplements to the 2nd Edition of Rodd's Chemistry of Carbon Compounds, Vol. II B(Partial), C, D and E, edited by M. Sainsbury
© 1994 Elsevier Science B.V. All rights reserved.

Chapter 14

DITERPENOIDS AND SESTERTERPENOIDS

J.R. HANSON

1. Introduction

The classification of the diterpenoids in this chapter follows the biogenetic pattern used   in the first supplement . The isolation of novel diterpenoids has been reviewed annually in the Royal Society of Chemistry Specialist Periodical Report Series *Terpenoids and Steroids* (1983, **12**, 185 and previous volumes) and subsequently in *Natural Product Reports* (1991, **8**, 1 and previous reports). Some more specialist reviews will be mentioned in the body of the text.

Major phytochemical surveys of the Compositae and of the Labiatae have led to the isolation of a large number of novel diterpenoids. Some of these are highly oxidized compounds in which the parent diterpenoid carbon skeleton has undergone rearrangement or oxidative cleavage. Furthermore not only are diterpenoids of both enantiomeric series known but some of the novel skeleta are close stereochemical relatives of each other. Consequently it is now rarely a valid approach to structural work in this series just to assume a diterpenoid skeleton and place the functional groups on it in accordance with their spectroscopic properties. There are an increasing number of X-ray crystal structures that have been determined and there is now a need to establish inter-relationships between novel natural products and these key compounds.

Marine organisms have provided a wide range of diverse diterpenoid structures, many based on unusual cyclizations of geranylgeranyl pyrophosphate and this area is now in need of systematization.

During the period since the last supplement, diterpenoids have attracted considerable attention in the context of their biological activity. Amongst the diterpenoids are tumour inhibitory substances, antibiotics, bitter principles, sweet tasting compounds, psychotropic substances, abortiofacients, feeding deterrents, insect defensive secretions, plant hormones and phytotoxins. Despite this range of activity, the systematic investigation of structure:activity relationships has been limited to relatively few compounds.

442

The diterpenoids differ from the triterpenoids and steroids in that many of their functional groups lie on pendant groups attached to the ring system. In many cases the compounds are also more highly functionalized. Hence their chemistry is dominated by interactions between the functional groups and by neighbouring group participation in reactions. Furthermore interactions within the bridged polycyclic ring systems provide the driving force for some skeletal rearrangements several of which have been investigated from the chemical and biogenetic standpoint. The combination of skeletal and stereochemical variety with biological activity has made these compounds attractive targets for synthesis. A number of significant synthetic achievements have recently been recorded in this area. Although not germane to this chapter, nevertheless it is also worth noting that the biosynthesis and bio-transformation of these compounds has also attracted interest.

2. Acyclic and related diterpenoids

(1)

A number of hydroxylation products of geranylgeraniol (e.g. 1, oleaxillaric acid from *Olearia axillaris*; U. Warning *et al., J. Nat. Prod.*, 1988, **51**, 513) have been isolated during the course of phytochemical surveys particularly of the Compositae. It has been suggested that some of these, particularly with butenolide end groups, may have taxonomic significance in the classification of the tribe, *Astereae*. There has been considerable interest in the oxepan, zoapatanol (2) which is an abortiofacient that has been obtained (S.D. Levine *et al., J. Am. Chem. Soc.*, 1979, **101**, 3404) from a Mexican plant, *Montanoa tomentosa*, (zoapatle) that is used in folk medicine. The synthesis of this compound has been reported (R.C. Cookson and N.J. Liverton, *J. Chem. Soc., Perkin Trans. 1*, 1985, 1589). A number of these compounds may arise through the opening of epoxides of geranylgeraniol.

(2)

3. Bicyclic diterpenoids

(a) Labdanes
Since the previous supplement was published a large number of hydroxylated labdanes of both enantiomeric series have been described. Although

not confined to the Compositae, they are particularly common in these plants. The range of nuclear hydroxylation is also accompanied by a variety of side chains which include not only simple alcohols, as in scoparic acid (3) (M. Kawasaki *et al., Chem. Pharm. Bull.*, 1987, **35**, 3963), aldehydes and acids but also furans, spirolactones and butenolides (e.g. 4; J.M. Fang *et al., Phytochemistry*, 1989, **28**, 1173). A hot-tasting 15,16-dialdehyde has been isolated from *Aframomum daniellii* which is used as a condiment. Plants of the genus Grindelia (Asteraceae) often produce an abundant resinous exudate which contains diterpenoids (e.g. 5; B.M. Timmermann *et al., Phytochemistry*, 1987, **26**, 467). A number of these have insect antifeedant activity. Examination of the plant *Halimium viscosum* (Cistaceae) has afforded a range of labdane diterpenoids and rearrangement products known as the ent-halimenes and exemplified by (6; J.G. Urones *et al., Phytochemistry*, 1990, **29**, 1247). This plant also contains diterpenoids with the tormesane skeleton.

Coleus forskohlii (Labiatae) has played an important role in Hindu and Ayurvedic traditional medicine. A group of 11-oxomanoyl oxide derivatives including forskohlin (7; S.V. Bhat *et al., Tetrahedron Lett.*, 1977, 1669) have been isolated from this plant. Forskohlin has attracted considerable interest because of its anti-hypertensive activity and its action on adenylate cyclase (Y. Khandelwal *et al., J. Med. Chem.*, 1988, **31**, 1872). Consequently its chemistry has been explored including the α-ketol rearrangement to afford (8). It has also been a target for synthesis (S. Hashimoto *et al., J. Am. Chem. Soc.*, 1988, **110**, 3670; E.J. Corey *et al., J. Am. Chem. Soc.*, 1988, **110**, 3672).

The preparation of potential perfumery compounds from manool and other readily accessible labdanes, has involved examination of various aspects of the oxidative chemistry in this area and in particular the

444

(7)   (8)

formation of cyclic ketals involving the side chain (P.K. Grant *et al., Aust. J. Chem.*, 1988, **41**, 711; 1991, **44**, 433). Another aspect of the chemistry of these compounds which has been studied is the composition of the mixtures obtained from the acid-catalysed cyclization of manool. This interest stems from the role of the labdanes in the biosynthesis of tri- and tetracyclic diterpenoids.

Examination of *Ballota* species (Labiatae), some of which are medicinal herbs, has afforded a number of labdanes related to marrubiin. These are exemplified by 18-hydroxyballonigrin (9; G. Savona *et al., J. Chem. Soc., Perkin Trans.* 1, 1978, 1271). Some prefuran-9:13-ethers have also been isolated. The diterpenoid lactones of *Leonotis leonurus* (Labiatae) (e.g. 10, D.E.A. Rivett, *S. Afr. J. Chem.*, 1988, **41**, 124) also reveal this structural feature.

(9)   (10)

(b) Clerodanes

Clerodanes possessing both a *cis* and a *trans* ring-junction continue to be isolated in large numbers. In some plants, e.g. *Solidago* species, both ring fusions occur together. *Ajuga* and *Teucrium* (Labiatae) species have been major sources of clerodanes (for a review see F. Piozzi, *Heterocycles*, 1987, **25**, 807). Many of these compounds possess a 4:18-epoxide and variously oxygenated side chains. Typical examples the structures of which have been established by X-ray crystallography are auropolin (11; from *Ajuga chamaepitys*, F. Camps *et al., Phytochemistry*, 1987, **26**, 1475) and tafricanin (12, from *Teucrium africanum*, J.R. Hanson *et al., J. Chem. Soc., Perkin Trans.* 1, 1982, 1005). These compounds have attracted interest because of their insect antifeedant activity. Jodrellin B (13, M.D. Cole *et al., Phytochemistry*, 1990, **29**, 1793) is the most active compound against *Spodoptera littorales* to have been detected so far. A consequence of this has been synthetic activity which has included a total synthesis of ajugarin 1 (S.V. Ley *et al., J. Chem. Soc., Chem. Commun.*, 1983, 503).

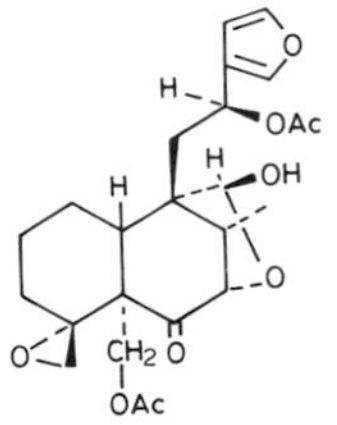

(11)

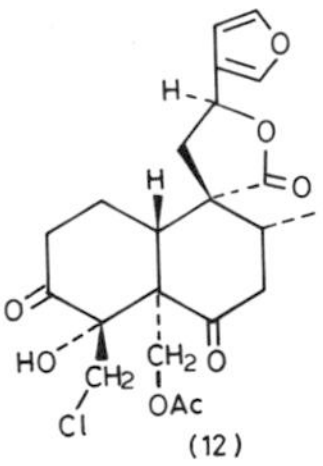

(12)

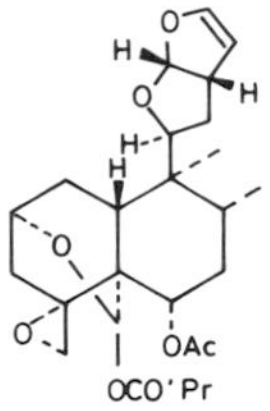

(13)

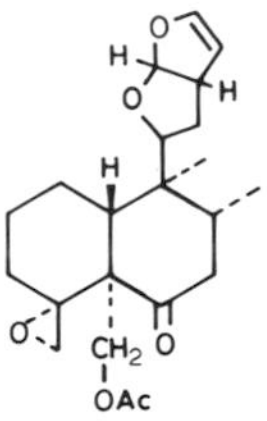

(14)

Although the structure of clerodin was originally established by X-ray crystallography, the absolute stereochemistry had to be revised to (14) (D. Rogers *et al., J. Chem. Soc. Chem. Commun.*, 1979, 97). Subsequently the absolute stereochemistry of a number of other clerodanes has been clarified (M. Martinez-Ripoll *et al., J. Chem. Soc., Perkin Trans.* 1, 1981, 1186). Where they are related to the clerodin structure they are known as neo-clerodanes.

Plants of the genus *Salvia* (Labiatae) include a number of medicinal herbs. These have yielded clerodanes such as salvarin (15; G. Savona *et al., J. Chem. Soc., Perkin Trans.* 1, 1978, 643) and salvinorin A (divinorin) (16). The latter is the psychotropic constituent of the Mexican hallucinogenic plant, *S. divinorum* (A. Ortega *et al., J. Chem. Soc., Perkin Trans.* 1, 1982, 2505; L.J. Valdes *et al., J. Org. Chem.*, 1984, **49**, 4716). Clerocidin and terpentecin (17) are antibiotics obtained from Actinomycetes and possess this skeleton.

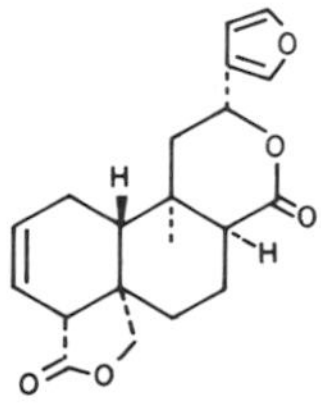

(15)

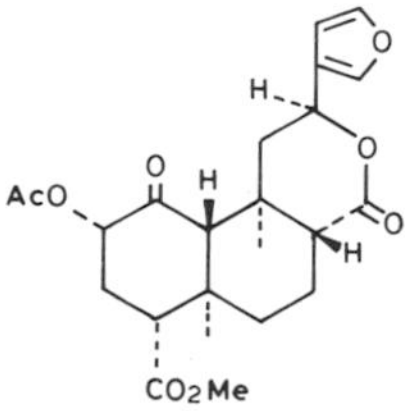

(16)

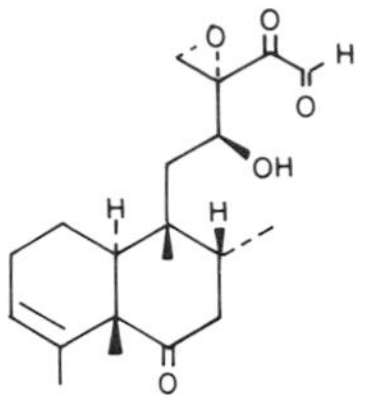

(17) 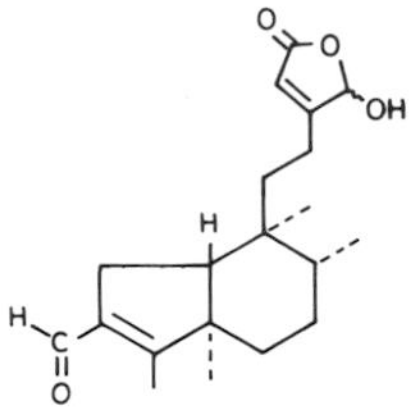   (18)

Modifications of the clerodane skeleton include some rearrangements of ring A as in the (4-2)-abeoclerodane (18) which was isolated from *Polyalthia viridis* (Annonaceae) (A. Kijjoa *et al., Phytochemistry*, 1990, **29**, 653). Another modification is found in the nor-clerodane, teuflin (19) obtained from *Teucrium flavum* (G. Savona *et al., J. Chem. Soc., Perkin Trans.* 1, 1979, 1915). This compound may be formed by the loss of an 18-hydroxymethyl group.

Clerodanes possessing the *cis*-ring junction have been obtained from various sources. Their structure permits the formation of bridges between the rings which have facilitated structure determination. They can be exemplified by the zuelanin (20) series of compounds (M. Khan *et al., Phytochemistry*, 1990, **29**, 1609). The X-ray crystal structures have been reported for a number of *cis*-clerodanes, including tinosporide and columbin (K. Swaminathan *et al., Acta Crystallogr.*, Sect. C, 1989, **45**, 300).

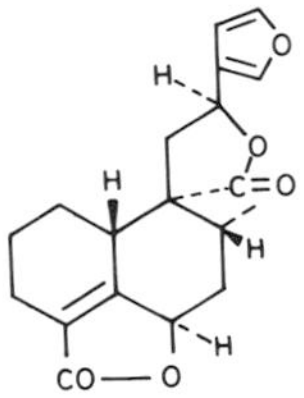

(19) 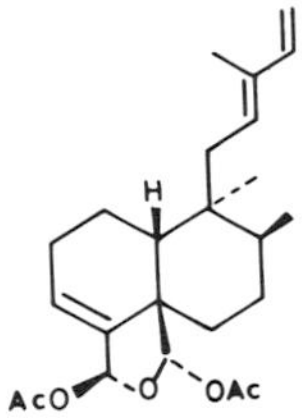   (20)

4. Tricyclic diterpenoids

For many years it was accepted that the bi-, tri-, and tetracyclic diterpenoids with the labdane, pimarane and kaurane skeleta, all possessed a *trans* relationship between the C-10 methyl and the C-9 hydrogen atom which arose because of a chair:chair cyclization of geranylgeranyl pyrophosphate. However a small but increasing group of diterpenoids has been found in which there is a *cis* relationship arising from a chair:boat cyclization. Although it is now generally believed that the tetracyclic diterpene hydrocarbons are formed directly from the bicyclic labdanes without the intervention of discrete free tricyclic hydrocarbons, nevertheless it is likely that some common tricyclic carbocationic intermediates exist. The tricyclic carbocation which is formed first from a labdadienol pyrophosphate may either initiate further cyclization, be discharged by loss of a proton from C-7, C-9 or C-14 or migrate with an accompanying hydride shift to C-9 where the new carbocation may in turn initiate cyclization, further back-bone rearrangement by methyl group migration from C-10 or be dis-charged by proton loss from C-11. An increasing number of compounds have been described at the tricyclic level arising from these different processes. Since many of these different types have spectroscopic features which at first sight are rather similar, e.g. a trisubstituted double bond, spectroscopic evidence for their structure needs careful evaluation. One example are compounds with the 9(11)-double bond. Thus the tree *Dacry-dium biforme*, which is endemic to New Zealand, possesses contrasting forms of juvenile and adult foliage. Whereas 8α-isopimara-9(11),15-diene is a characteristic hydrocarbon of the juvenile foliage, the tetracyclic diterpene, phyllocladene, is found in the adult foliage (A.R. Hayman *et al.*, *Phytochemistry*, 1986, **25**, 649). Several $\Delta^{9(11)}$-pimaradiene derivatives including (21) have been isolated from *Mikania triangularis* (Compositae) (F.S. Knudsen *et al.*, *Phytochemistry*, 1986, **26**, 1240), whilst the structure and stereochemistry of viguiepinol (22), obtained from *Viguiera pinnatilo-bata* (Compositae), was established by X-ray crystallography (C. Guerrero *et al.*, *Acta Crystallogr.* Sect. C., 1986, **42**, 729). Maximol (23) and the corresponding acid are unusual ent-nor-rosane derivatives in which ring A is aromatic. These were found in the fronds of the fern, *Arachniodes maximowiczii* (N. Tanaka *et al.*, *Chem. Pharm. Bull.*, 1986, **34**, 1015).

(21) R=H, R'=OH
(22) R=OH, R'=H

(23)

Diterpenoids with an aromatic ring C are much more common. The medicinal and other properties of the Neem tree (*Azadirachta indica*) are well known. A number of aromatic diterpenoids (e.g. 24) have been isolated from this source (S. Siddiqui *et al.*, *Phytochemistry*, 1990, **29**, 911). Some of the phenols may be associated with antibacterial activity.

Plants of the genus *Salvia* (Labiatae) are a rich source of highly oxidized abietanes and their rearrangement products. Thus rosmariquinone (25) has been obtained from *S. canariensis* (A.G. Gonzalez *et al.*, *Can. J. Chem.*, 1989, **67**, 208). This plant also afforded salvicanol which possesses a

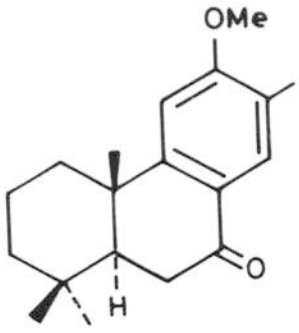

(24)

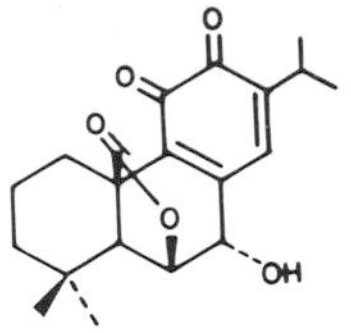

(25)

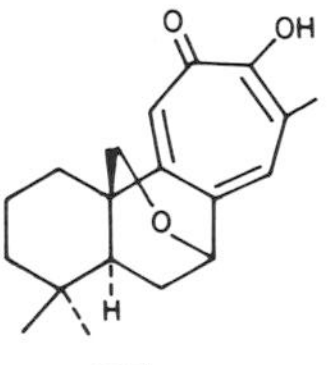

(26)

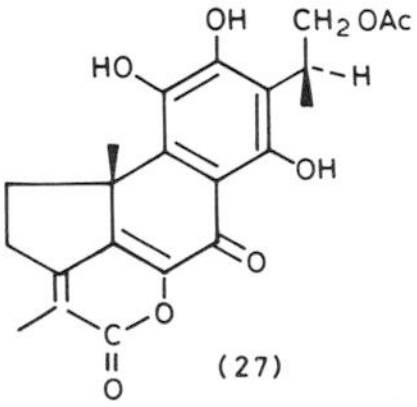

(27)

seven-membered ring B (B.M. Fraga *et al., Phytochemistry*, 1986, **25**, 269). The Chinese medicinal drug 'Tan-Shen' is the dried roots of *S. miltiorrhiza* and these have been the source of many diterpenoids including the tropolone miltipolone (26) (G. Haro *et al., Chem. Lett.*, 1990, 1599).

The glandular pigments obtained from the leaves of plants of the genera *Coleus* and *Plectranthus* (Labiatae) include a wide variety of oxidized abietanes in which rearrangements of both the side-chain and ring A have taken place. These are exemplified by edulon A (27), obtained from *Plectranthus edulis* (J.M. Kuenzle *et al., Helv. Chim. Acta*, 1987, **70**, 1911) and the 16-R-plectrinone (28) (P. Ruedi *et al., Helv. Chim. Acta*, 1986, **69**, 972). The diterpenoid barbatusin (29), from *Coleus barbatus*, has attracted interest because of its tumour-inhibitory activity (R. Zelnik *et al., Tetrahedron*, 1977, **33**, 1457). Caudicifolin (30) has been obtained from *Euphorbia acaulis* (Euphorbiaceae) (N.K. Satti, *Phytochemistry*, 1986, **25**, 1411). This plant has been used in folk medicine to alleviate inflammatory disease. There have been further reports of podolactones which have been isolated from *Podocarpus nagi* (I. Kubo *et al., Phytochemistry*, 1991, **30**, 1967). Many of these show cytotoxic and insect antifeedant activity.

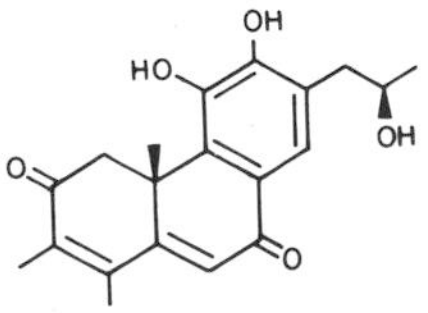

(28)

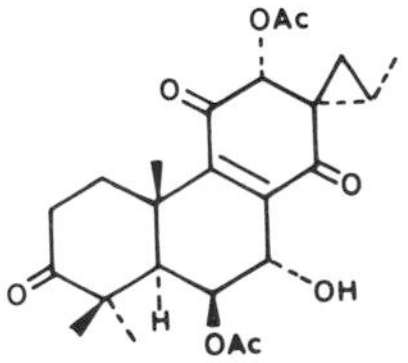

(29)

Tricyclic diterpenoids and their relatives have been detected in marine organisms. For example the structure of the bromoparguerene derivative (31), which was obtained from the red alga *Laurencia obtusa*, was established by X-ray crystallography (T. Suzuki *et al., Chem. Lett.*, 1989, 969). If the hydrogen atom at C-8 is of mevalonoid origin, its stereochemistry suggests that this diterpenoid may have arisen via a chair:boat cyclization of geranylgeranyl pyrophosphate.

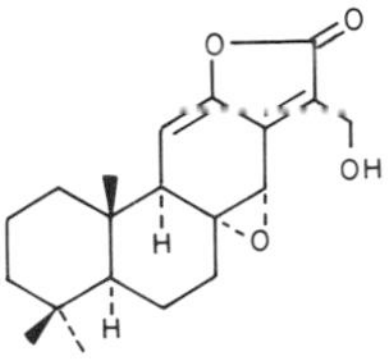

(30) 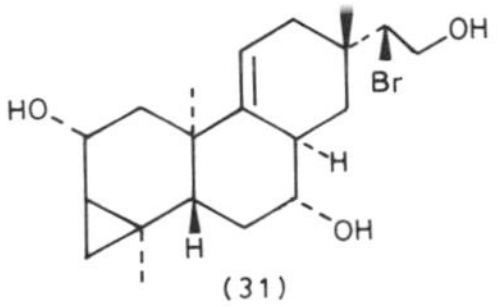 (31)

The ready availability of a number of tricyclic diterpenoids has led to their utilization as starting materials for various synthetic endeavours. These studies have included the selective modification of the double bonds of the pimaric acid series and an extensive series of modifications of podocarpic acid (32). The interaction of the aromatic ring with organometallic reagents has been investigated in connection with studies aimed at adding the cyclopentane ring characteristic of the steroids. Other studies have been directed at the introduction of functional groups on ring A via, for example, the epoxide (33) (R.C. Cambie *et al., Aust. J. Chem.*, 1990, **43**, 883).

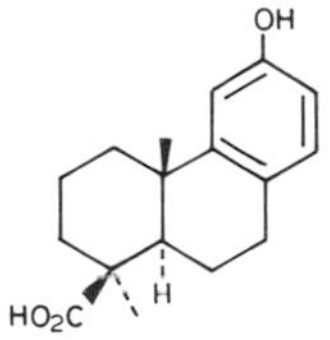

(32) 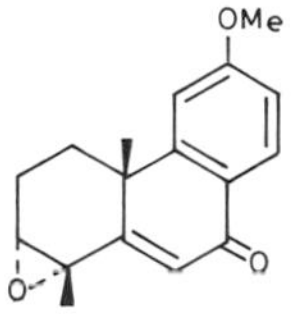 (33)

5. Tetracyclic diterpenoids

(a) Kaurenes

The phytochemical investigations of the Compositae have led to the isolation of many ent-kaurenoid diterpenes. Although C-19 is the most common site of oxygenation, hydroxylation has been observed at almost all the other sites on the skeleton. ent-Kaurenes are not restricted to the Compositae. Indeed they are widespread in the Labiatae particularly *Sideritis* species, exemplified by epicandicandiol and the corresponding C-18 carboxylic acid,

450

sventenic acid (34) (B.M. Fraga *et al., Phytochemistry*, 1990, **29**, 591). They are also common in the liverworts such as *Jungermannia infusca* which contains the infuscicides e.g. (35) (F. Nagashima *et al., Phytochemistry*, 1990, **29**, 1619; for review see Y. Asakawa, *Fortschr. Chem. Org. Naturst.*, 1982, **42**, 1).

(34)   (35)

Biological activity is common in these diterpenoids. For example a number of the ent-kaurenoic and trachylobanic acids obtained from *Helianthus* (sunflower) species exert an antifeedant activity against the sunflower moth. Many tetracyclic diterpenoids occur as their glycosides. Thus stevioside, the very sweet glycoside obtained from *Stevia rebaudiana*, is now commercially available as a sweetner. It has been the subject of a number of investigations aimed at enhancing its sweetness (S. Kitahata *et al., Agric. Biol. Chem.*, 1989, 2923). Its chemistry is also of interest, particularly the rearrangement of the aglycone, steviol (36) to isosteviol (37) (A.G. Avent *et al., J. Chem. Soc. Perkin Trans.* 1, 1990, 2661). A number of other glycosides are bitter tasting. Recently mozambioside (38) has been obtained as a bitter component from *Coffea pseudozanguebariae*, a caffeine-free coffee (R. Prewo *et al., Phytochemistry*, 1990, **29**, 990). Some of these glycosides are toxic and a number related to atractyloside have been obtained from livestock poisons such as the 'alfumbrilla' plant, *Drymaria avenariodes*, and various 'cockleburr' poisons (D. Vargas *et al., Phytochemistry*, 1988, **27**, 1532).

(36)   (37)

Medicinal plants of the genus *Rabdosia* (Labiatae), particularly those of Chinese and Japanese origin, have been the source of many highly hydroxylated kaurenes, several of which have anti-tumour activity (for a review see E. Fujita and M. Node, *Fortschr. Chem. Org. Naturst.*, 1984, **46**, 77). Structure:activity relationships have associated this with the α-methylenecyclopentanone moiety on ring D although other structural features enhance the activity, (Y. Fuji *et al., Chem. Pharm. Bull.*, 1989, **37**, 1472). In several of these compounds ring B has been cleaved.

Rotation about the C-9:C-10 bond then produces natural products with lactone rings involving C-7:C-20 or C-7:C-1 oxygen bridges (see for example Y. Takeda *et al., Chem. Pharm. Bull.*, 1990, **38**, 439). A number of partial syntheses in this area have been directed at studying the structure:activity relationships, (M.S. Ali *et al., J. Chem. Soc., Perkin Trans.* 1, 1991, 2679).

The role of ent-kaurene and the oxidative modification of ring B in the biosynthesis of the gibberellin plant hormones, has led to studies on its stereospecific labelling, (S.J. Castellaro *et al., Phytochemistry*, 1990, **29**, 1823) The key gibberellin biosynthetic intermediate, gibberellin A_{12} 7-aldehyde, is prepared for biosynthetic studies, by the chemical ring contraction of the kaurenolide lactones. The stereochemistry of this rearrangement has been studied (M. Alam *et al., J. Chem. Soc., Perkin Trans.* 1, 1990, 2577). The unique biosynthetic step involving the ring contraction of ent-7α-hydroxykaurenoic acid to form gibberellin A_{12} 7-aldehyde is the target of some active-site directed inhibitors of gibberellin biosynthesis (M.K. Baynham *et al., Phytochemistry*, 1988, **27**, 761), Some of these inhibitors have been prepared from the interesting aldehyde:anhydride, fujenal (39).

(38)   (39)

The flexibility of a biosynthetic pathway in accepting analogues of natural intermediates has been the object of a number of studies with ent-kaurenoids and the kaurenolide:gibberellin pathway in *Gibberella fujikuroi*. These biotransformations have shed light on the constraints of the pathway and provided access to many substituted ent-kaurenes and gibberellins (for a review see J.R. Hanson, *Nat. Prod. Rep.,*, 1992, **9**, 139).

(b) Gibberellins

There are now over 80 gibberellin plant hormones known (for a review of their isolation see M.H. Beale and C.L. Willis, Gibberellins, in *Methods in Plant Biochemistry*, eds. B.V. Charlwood and D.V. Banthorpe, Academic Press, London, 1991, Vol. 7 p. 289). Gibberellin A_{78} (40) is an example of a gibberellin that has been isolated from wheat (P.S. Kirkwood and J. MacMillan, *J. Chem. Soc.*, Perkin Trans. 1, 1982, 689). An interesting feature of recent work has been the demonstration that the antheridium inducing factors of ferns possess a gibberellin-like structure exemplified

(40)　　　　　　　　　　　　(41)

by antheridic acid (41) which was obtained from *Anemia phyllitidis* (E.J. Corey *et al.*, *Tetrahedron Lett.*, 1986, **27**, 5083).

One of the milestones of total synthesis has been the total synthesis of the gibberellins (E.J. Corey *et al.*, *J. Am. Chem. Soc.*, 1978, **100**, 8034; L.N. Mander *et al.*, *J. Am. Chem. Soc.*, 1979, **101**, 3373). Because many of the gibberellins that have been obtained from higher plants are relatively rare, numerous partial syntheses have been developed starting from the readily available gibberellic acid or gibberellin A_{13}. Progress in this area has been reviewed (L.N. Mander, *Nat. Prod. Rep.* , 1988, **5**, 541). The chemistry of the gibberellins has been explored not only in the context of partial synthesis and labelling for metabolic studies, but also because these molecules possess a juxtaposition of functionality that leads to neighbouring group participation in many of their reactions. Studies on the chemistry of ring A of the gibberellins have been concerned, *inter alia*, with the labelling of ring A by the conjugate reduction of unsaturated ketones, the alkylation of this ring and the rearrangement of esters of Δ^1-3-hydroxygibberellins to form 1-hydroxy-Δ^2-isomers. A number of transformations have been facilitated by the formation of ring A dienes and the subsequent reconstruction of the 19-10-lactone ring by iodo-lactonization reactions involving a $\Delta^{1(10)}$-alkene. Various mechanistic studies have been reported on the rearrangements of this ring and on some of the more deep-seated skeletal rearrangements which occur in concentrated acid. On ring B epimerization and labelling reactions at C-6 have been studied and various methods have been developed for the hydrolysis of C-7 esters. The study of the chemistry of rings C and D has revealed a number of aspects of neighbouring group participation. Various methods have been explored for the removal of the bridge-head hydroxyl group in the partial synthesis of 13-desoxygibberellins. The chemistry of the gibberellins has been reviewed on a number of occasions (see for example, J.R. Hanson, *Nat. Prod. Rep.*, 1990, **7**, 41).

Although biosynthesis, metabolism and mode of action are outside the scope of this chapter, nevertheless it is important to record that these have been very active areas of investigation.

(c) Atiserenes, Beyerenes and Trachylobanes

New compounds of these skeletal types have continued to be isolated. They may be exemplified by antiquorin (42) obtained from *Euphorbia fidjiana* (A.R. Lal *et al.*, *Acta Crystallogr.*, Sect. C, 1990, **46**, 2387).

(d) Aphidicolin and Stemaranes

Aphidicolin (43) has attracted considerable biological interest in the light of its selective action on DNA polymerase α. A number of total syntheses have been reported (for refs. see C.J. Rizzo and A.B.Smith III, *J. Chem. Soc., Perkin Trans.* 1, 1991, 969). Some aspects of the chemistry of aphidicolin have been explored both in the context of structure:activity studies and in the preparation of intermediates for biosynthetic work. A number of rearrangements have been uncovered (J.R. Hanson *et al., J. Chem. Soc., Perkin Trans.* 1, 1992, 41) New naturally-occurring aphidicolanes such as (44) have been reported (J.R. Hanson *et al., Phytochemistry,* 1992, **31**, 799). Compounds possessing related skeleta, for example scopadulcic acid (45), have been isolated from the medicinal plant, *Scoparia dulcis* (T. Hayashi *et al., Chem. Pharm. Bull.,* 1990, **38**, 945) whilst some new stemodane diterpenoids have been isolated from *Stemodia maritima* (C.D. Hufford *et al., J. Nat. Prod.,* 1992, **55**, 48).

(42)

(43) R = H
(44) R = OH

(45)

6. Cembranes

Many cembranoids have been isolated from tobacco particularly from those hybrid varieties of *Nicotiana tabacum* derived from *N. sylvestris*. These compounds are prone to undergo biodegradation and consequently they give rise to a variety of volatile compounds which impart an aroma to cured tobacco. This area of tobacco chemistry has been reviewed (T. Wahlberg and C.R. Enzell, *Nat. Prod. Rep.,* 1987, **4**, 237) as has that of the naturally occurring cembranes (see A.J. Weinheimer, *Fortschr. Chem. Org. Naturst.,* 1979, **36**, 285). Cembranoid diterpenoids are also common constituents of corals. For example sarcophytol A (46) and the lactone sarcophytonin A (47) were obtained from *Sarcophyton glaucum*, (M. Kobayashi *et al., Chem. Pharm. Bull.,* 1979, **27**, 2382).

(46)   (47)

The total synthesis of the cembranes and the cembranolides has also been reviewed (M.A. Tius, *Chem. Rev.*, 1988, 88, 719). Since a number of polycyclic diterpenoid skeleta may be formally derived from the further biosynthetic cyclization of cembrenes, the conformation, selective epoxidation and cyclizations of this macrocyclic system have been examined on a number of occasions.

Amongst the polycyclic diterpenoids derived from this pathway are the taxane series which are obtained from various species of yew tree. There is considerable interest in this area because of the anti-cancer activity of taxol (48) which is obtained from the bark of the Pacific yew tree, *Taxus brevifolia*. Not only have a number of novel derivatives been isolated (for a review see F. Gueritte-Voegelein *et al., J. Nat. Prod.*, 1987, **50**, 9) but other potent analogues have been prepared (L. Mangatal *et al., Tetrahedron*, 1989, **45**, 4177; *J. Med. Chem.*, 1991, 34, 992).

(48)   (49)

Ingol is a key derivative in the lathyrane series and biogenetically plausible cyclizations of this may lead to the tigliane, daphnane and ingenane series. An X-ray crystal structure of ingol tetra-acetate (49) has been reported (H. Lotter, H.J. Opferkuch and E. Hecker, *Tetrahedron Lett.*, 1979, 77). Considerable attention has been devoted to the pro-inflammatory tumour-promoting and anti-tumour diterpenoids of the plant families of the Euphorbiaceae and Thymelaceae, many of whose biologically-active constituents are esters and ortho-esters of this series. This topic has been reviewed (F.J. Evans and S.E. Taylor, *Fortschr. Chem. Org. Naturst.*, 1983, .44, 1; *Naturally-Occurring Phorbol Esters*, ed. F.J.Evans, CRC Press, Boca Raton, USA, 1986). New compounds continue to be isolated. For example, esulone C (50) is a jatraphane diterpenoid which has been found in the roots of the leafy spurge, *Euphorbia esula*. This plant is a toxic weed that poses a threat to livestock (G.D. Manners and D.G. Davies, *Phytochemistry*, 1987, **26**, 727). C-13 Esters of 5-deoxy-13-hydroxyingenol from the latex of the chechum tree, *Mabea excelsa*, produce a debilitating inflam-

(50)　　　　　(51)

matory skin reaction and this led to a loss of military manpower with troops
on patrol in Belize (G. Brooks *et al.*, *Phytochemistry*, 1990, **29**, 1615).

7. Miscellaneous diterpenoids

Marine organisms have furnished a wide variety of different diterpenoid
skeleta. Some such as the eunicellin series (e.g. 51, obtained from a
Cladiella species; Y. Uchio *et al.*, *Tetrahedron Lett.*, 1989, **30**, 3331) and
the briarane diterpenoids (e.g. the juncellolide 52, obtained from *Junceella
fragilis*, J. Shin *et al.*, *Tetrahedron*, 1989, **45**, 1633) have skeleta which
may be derived from cembrene. Others such as the xenicane diterpenoids
(e.g. 53 from *Xenia garciae*; G.M. Konig *et al.*, *J. Nat. Prod.*, 1989, **52**, 294)
have skeleta which represent a different cyclization of geranylgeranyl
pyrophosphate. Brown algae, particularly *Dictyota* species, have been a
fruitful source of diterpenoids.

(52)　　　　　(53)

The dolestane (54) obtained from *Dictyota furcellata* (R.W. Dunlop *et al.*,
Aust. J. Chem., 1989, **42**, 315) is a typical example. Other tricyclic diterpenes obtained from marine organisms are exemplified by the spongiane
series (e.g. 55, obtained from *Hyatella intestinalis*; R.C. Cambie *et al.*, *J.
Nat. Prod.*, 1988, **51**, 293) and the bromo-diterpene (56), obtained from
Sphaerococcus coronopifolius (F. Cafieri *et al.*, *Gazz. Chim. Ital.*, 1990, **120**,
139). Some skeleta take the form of prenylated sesquiterpenoids exemplified by the dictyotins (e.g. 57) obtained from *Dictyota dichotoma*,
(M.O. Ishitsuka *et al.*, *Phytochemistry*, 1990, **29**, 2605). The biological
activity of these compounds particularly as feeding deterrents, has attracted attention. A number of compounds also have antibiotic and cytotoxic activity. The chemistry of marine organisms is regularly reviewed

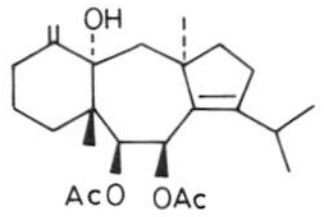

456

(54)

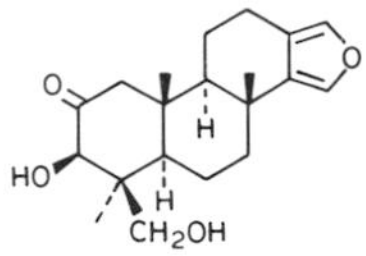

(55)

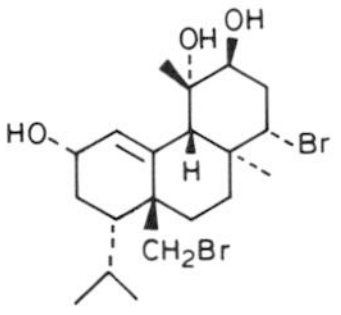

(56)

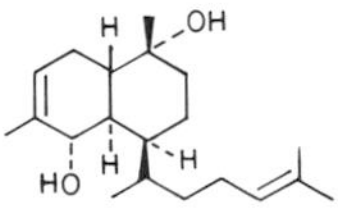

(57)

(see D.J. Faulkner, *Nat. Prod. Rep.*, 1991, **8**, 97).

The diversity of diterpenoid structures is not restricted to marine organisms. Plants of the genus *Eremophila* have afforded a number of novel diterpenoids exemplified by the serrulatane diterpenes (58), obtained from *E. rotundifolia* (A.D. Abell *et al., Aust. J. Chem.*, 1985. **38**. 1263) and the eremane (59), obtained from *E. cuneifolia* (K.D. Croft *et al., Aust.J.Chem.*, 1984, **37**, 785).

The plant *Halimium viscosum* has been shown to contain labdanes, ent-halimanes and more recently diterpenoids of the tormesane skeleton (e.g. 60) (J.G. Urones *et al., Phytochemistry*, 1990, **29**, 2585). Diterpenoids are also found quite often in the liverworts. Thus the verrucosane (61) was obtained from *Gyrothyra underwoodiana* (I. Kubo *et al., J. Org. Chem.*, 1984, **49**, 4644).

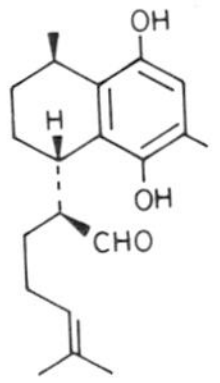

(58)

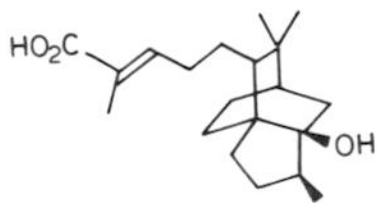

(59)

Termites produce a number of defensive substances that are diterpenoids including trinervitane alcohols such as (62) (G.D. Prestwich *et al., Tetrahedron Lett.*, 1981, **22**, 1563) and secotrinervitanes (e.g. 63) (R. Baker *et al., Tetrahedron Lett.*, 1984, **25**, 579).

Some fungal metabolites are also diterpenoids with somewhat unusual structures. Those from the *Basidiomycetes* have been reviewed (W.A. Ayer

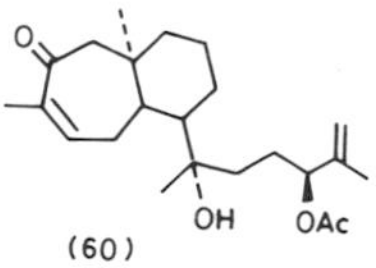

(60)

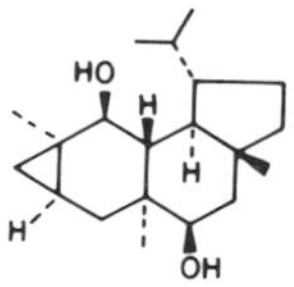

(61)

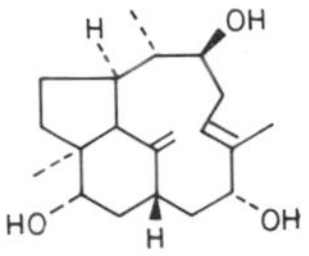

(62)

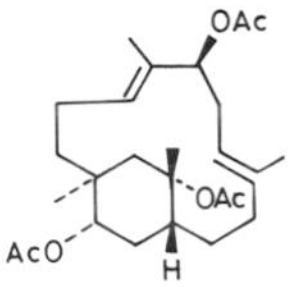

(63)

and L.M. Browne, *Tetrahedron*, 1981, **37**, 2199). More recent examples of fungal metabolites that have been isolated include traversiadiene (64) and its relatives (A. Stoessl, *Can. J. Chem.*, 1989, **67**, 1302).

Pleuromutilin has continued to attract interest not only because of its biological activity but also as a synthetic target (L.A. Paquette *et al.*, *J. Org. Chem.*, 1988, **53**, 1441).

The strained ring system of the hydrocarbon lauren-1-ene (65) has provided a number of interesting rearrangements (R.T. Weavers *et al.*, *Aust. J. Chem.*, 1983, **36**, 2588; 1990, **43**, 719).

8. Sesterterpenoids

Although the occurrence of sesterterpenes in nature is relatively uncommon, an increasing number have been reported. The occurrence of sesterterpenes has been reviewed (P. Crews and S. Naylor, *Fortschr. Chem. Org. Naturst.*, 1985, **48**, 203; J.R. Hanson, *Nat. Prod. Rep.*, 1986, **3**, 123).

Linear sesterterpenes containing terminal furan rings and tetronic acid units have been found in a number of sponges of the genera *Ircinia* and its relatives (see for example M.R. Kernan *et al.*, *J. Nat. Prod.*, 1991, **54**, 265).

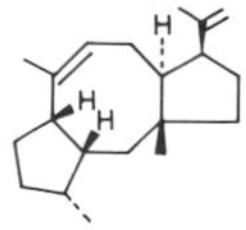

(64)

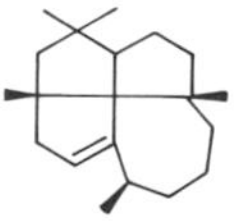

(65)

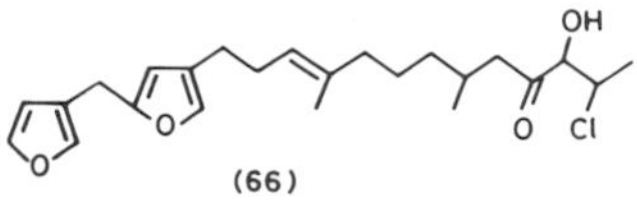

(66)

458

In a number of instances these have undergone partial degradation to nor-sesterterpenes (for example, 66) (A. De Giulio *et al., J. Nat. Prod.*, 1990, **53**, 1503).

A number of bicyclic sesterterpenes have also been obtained from marine sponges and these include some unusual sulfates such as sulfiricin (67) which was obtained as its *N,N*-dimethylguadinium salt (A.E. Wright *et al., J. Org. Chem.*, 1989, 54, 3472), cacospongionolide (68) which possesses tumour-inhibitory properties, (S. De Rosa *et al., J. Org. Chem.*, 1988, **53**, 5020) and a number of cyclic peroxides such as (69) which possess antibiotic properties (R.J. Capon and J.K. Macleod, *J. Nat. Prod.*, 1987, **50**, 225).

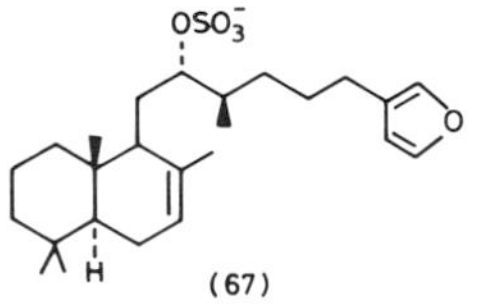

(67)

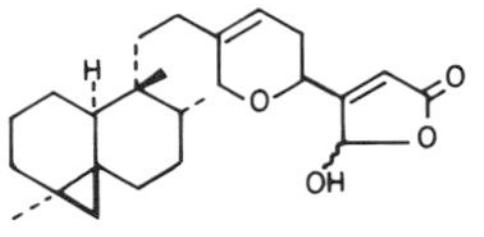

(68) 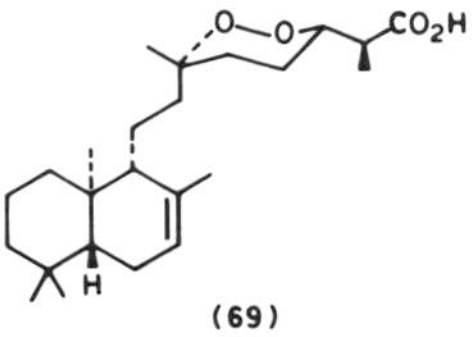   (69)

Sesterterpenes with structures based on the scalarane skeleton have also been reported from many marine sponges. In a number of cases these possess additional methyl groups at C-24 and C-19 (or C-20). They are exemplified by (70) (P.F. Barron *et al., Aust. J. Chem.*, 1991, **44**, 995) and (71) (C.B. Rao *et al., J. Nat. Prod.*, 1991, **54**, 364).

Novel sesterterpenoids have also been reported as fungal metabolites. Ophiobolin K (72) is a nematocidal agent from *Aspergillus ustus* (S.B. Singh *et al., Tetrahedron*, 1991, **47**, 6931) whilst others have been obtained as phytotoxic agents from *Drechslera* species (F. Sugawara *et al., J. Org.*

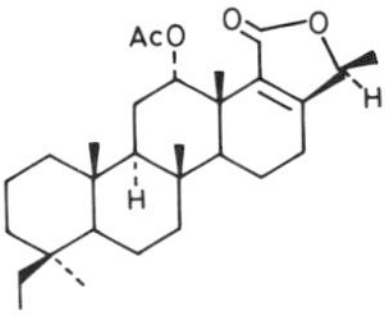

(70) 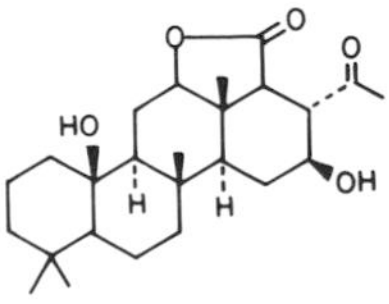   (71)

Chem., 1988, **53**, 2170). The synthesis of the ophiobolins has been reported (see, *inter alia*, Y. Kishi *et al., J. Am. Chem. Soc.*, 1989, **111**, 2735; R.K. Boeckman *et al., J. Am. Chem. Soc.*, 1989, **111**, 2737).

Variculanol (73) possessing a novel 5/12/5 tricyclic skeleton (S.B. Singh *et al., J. Org. Chem.*, 1991, **56**, 5618) and variecolin (74) (O.D. Hensens *et al., J. Org. Chem.*, 1991, **56**, 3399) are two biologically-active metabolites of *Aspergillus variecolor*. The citreohybridones A (75) and B possess a novel sesterterpenoid skeleton and were obtained from the mycelium of *Penicillium citreo-viride* (S. Kosemura et al., *Tetrahedron Lett.*, 1991, **32**, 3543). They may however be meroterpenoids rather than genuine sesterterpenes.

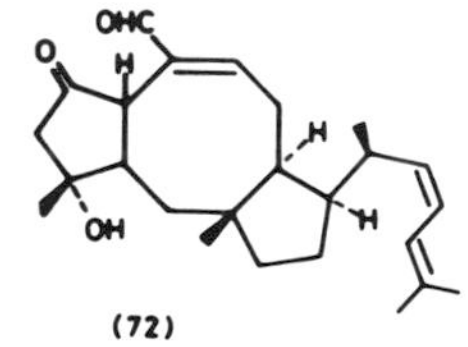

(72)

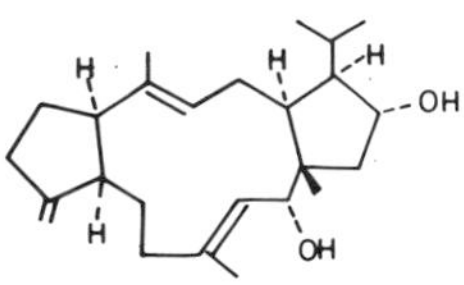

(73)

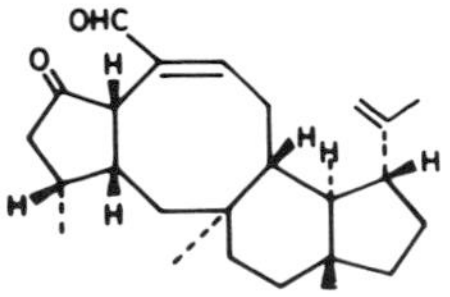

(74)

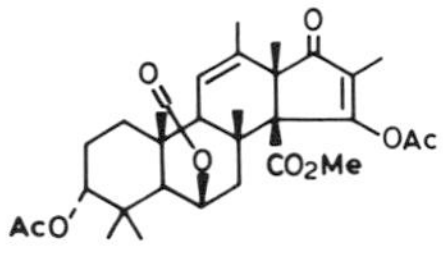

(75)

Second Supplements to the 2nd Edition of Rodd's Chemistry of Carbon Compounds, Vol. II B(Partial), C, D and E, edited by M. Sainsbury
© 1994 Elsevier Science B.V. All rights reserved.

Chapter 15

STEROIDS

F.J.ZEELEN

1. Introduction

The steroids were reviewed in volumes IIE and IIF of the second edition of Rodd's Chemistry of Carbon Compounds (1971) and again in the supplement to this edition (1974).

For the pharmaceutical industry, the early 1970's marked the end of a period where large research groups worked on the development of novel steroid drugs. Vast numbers of derivatives of

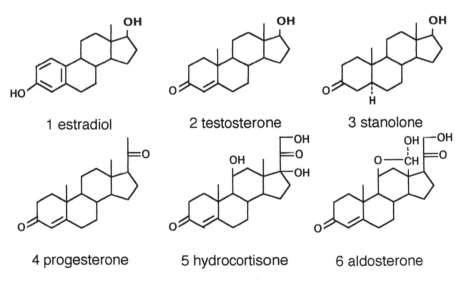

1 estradiol   2 testosterone   3 stanolone

4 progesterone   5 hydrocortisone   6 aldosterone

Fig.1 Steroid hormones

the female hormones estradiol **1** and progesterone **4**, of the male hormones testosterone **2** and stanolone **3**, and of the adreno-cortical hormones hydrocortisone **5** and aldosterone **6**, had been synthesized and tested. This had yielded an impressive number of novel drugs (F.J.Zeelen, Medicinal Chemistry of Steroids, Elsevier, Amsterdam 1990).

The resulting synthetic methodology had been collected in an interesting series of books, such as: J.Fried and J.A.Edwards Organic Reactions in Steroid Chemistry (Van Nostrand Reinhold Company, New York 1972), D.N.Kirk and M.P.Hartshorns Steroid Reaction Mechanisms (Elsevier, Amsterdam 1968) and, as was mentioned already, in Rodd's Chemistry of Carbon Compounds.

Steroid research continued, though at smaller scale and with stronger emphasis on biochemistry. Next to the well known aldosterone antagonist spironolactone **7** (A.Karim, Drug Metab. Rev.,1978,**8**,151), other hormone antagonists were developed. For example mifeprostone **8**, an antagonist of progesterone and

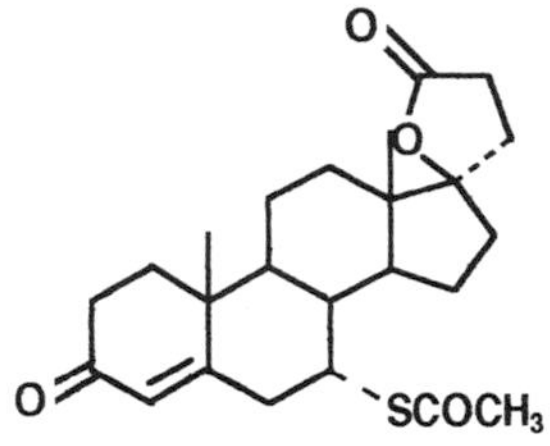

7 spironolactone 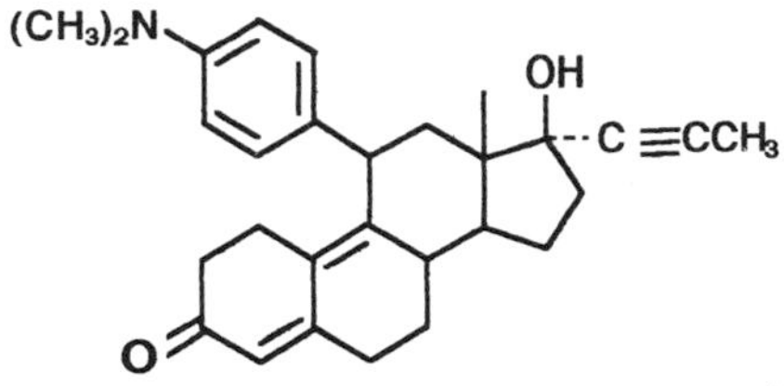   8 mifeprostone

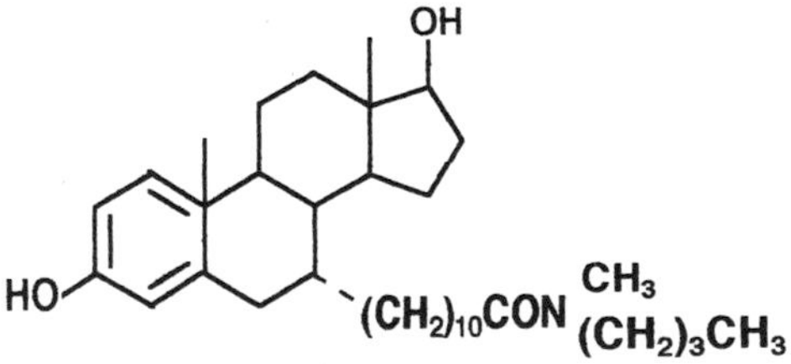

9 ICI 164 386

Fig.2 Hormone antagonists

hydrocortisone (R.C.Henshaw and A.A.Templeton, Drugs 1992, **44**,531) and the anti-estrogen ICI 164 386 **9** (A.E.Wakeling in Regulatory Mechanisms in Breast Cancer, M.Lippman and R.Dickson eds. Kluwer Academic Publishers, Boston, 1990, 237). These compounds form interesting tools for biological studies on the mechanism of action of the steroid hormones.

More recently, interest focussed on the development of inhibitors of the biosynthesis of the hormones. Trilostan **10** is an inhibitor of adrenal 3ß-hydroxy-5-ene-steroid dehydrogenase and so of the biosynthesis of hydrocortisone. The drug is used to treat Cushing's syndrome (P.Komanicky, R.S.Spark and J.C.Melby, J.Endocrinol. Metab. ,1978, **47**,1042; G.O.Potts *et al.*, Steroids, 1978, **32**, 257). Finasteride **11** is used in the treatment of prostatic hyperplasia. This compound inhibits the conversion of testosterone into stanolone (M.Ohtawa, H.Morikawa and J.Shimazaki, Eur. J. Drug. Metab. Pharmacokinet., 1991, **16**, 15; J.Imperato-McGinley *et al.*, Endocrinology, 1992, **131**, 1149; J.D.McConnell *et al.*, J. Clin. Endocrinol. Metab., 1992, **74**, 505). Hydroxyandrostenedione **12**, an inhibitor of the enzyme aromatase, is used for the treatment of estrogen dependent breast tumours (D.A.Marsh *et al.*, J. Med. Chem. 1985, **28**, 788; R.D.Koos*et al.*, Steroids 1985, **45**, 143).

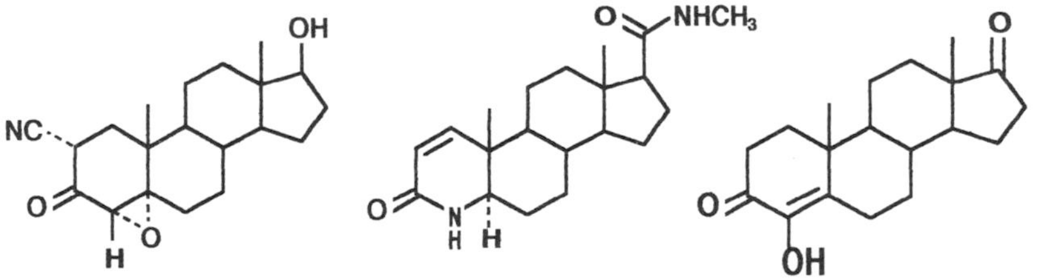

10 trilostan   11 finasteride 12 hydroxyandrostene
dione

Fig.3 Inhibitors of steroid biosynthesis

The synthetic chemistry of the steroids has been reviewed regularly, from 1971-1983 in a series of year books (Terpenoids and steroids, vol 1-12, The Chemical Society, London) and since 1984 in the Journal Natural Product Reports. In this update the

464

major developments during the period 1974 till mid 1992 will be reviewed.

2. Technical steroid synthesis

In the 1970's, major changes in technical steroid production were introduced. Diosgenin **13**, stigmasterol **14**, hecogenin **15** and cholic acid **16** had been the classical starting materials for the production of the corticosteroids, sex hormones, contraceptives and spironolactone **7**. These were to a great extent replaced by starting materials derived from sterols by microbiological oxidation.

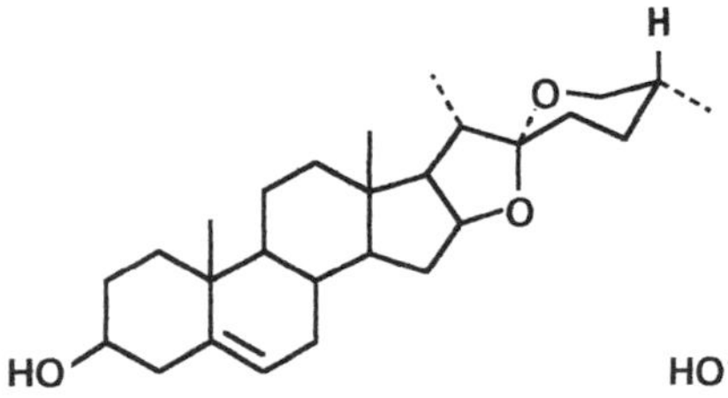

13 diosgenin 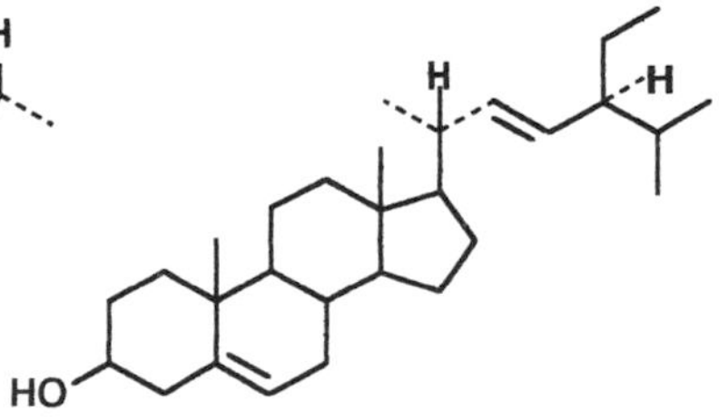   14 stigmasterol

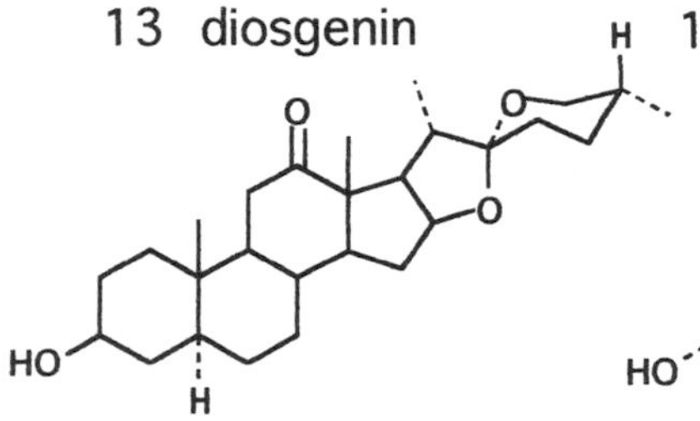

15 hecogenin 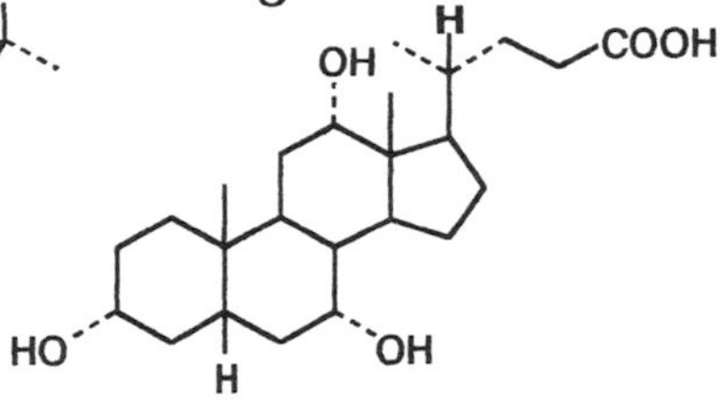   16 cholic acid

Fig.4 The classical starting materials

In the 1960's, basic studies had unravelled the sequence of reactions involved in the microbiological oxidation of sterols (C.J.Sih *et al.*, J.Biol. Chem. 1966, **241**, 540; D.T.Gibson *et al.*, J.Biol. Chem. 1966, **241**, 551; M.A.Rahim and C.J.Sih, J.Biol. Chem. 1966, **241**, 3615) (fig. 5). Biotechnological development using selected organisms and mutated strains made androstadienedione **17**, 9α-hydroxyandrosta-1,4-diene-3,17-dione **18** and the lactone **19** attractive intermediates for steroid synthesis (M.Wovcha *et al.*, Biochem. Biophys. Acta 1978, **531**, 308; T.Nakamutso, T.Beppu and K.Arima, Agric. Biol. Chem. 1983, **47**, 1449; N.P.Ferreira *et al.*, Biotechnology Lett. 1984, **6**, 517). In 1976, large-scale production of these intermediates started. Meanwhile organic chemists had developed routes to convert these new starting materials into the usual end-products.

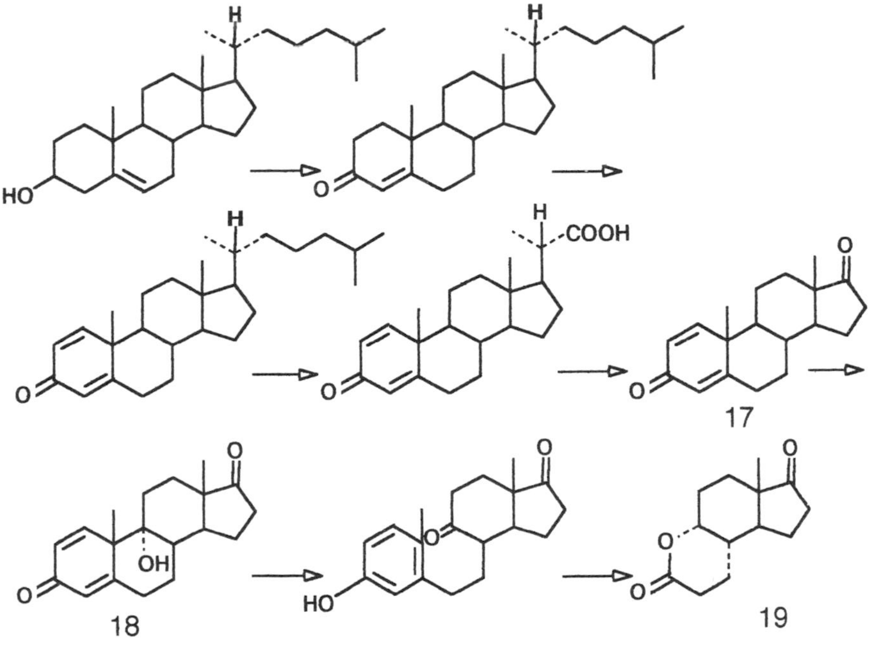

Fig.5 Microbiological oxidation of cholesterol

(a) Corticosteroid synthesis

For the synthesis of corticosteroids or progesterone from these 17-ketosteroid starting materials, a two-carbon 17ß side-chain has to be built up. A large number of syntheses has been reported. The early work in this field was reviewed by E.P.Oliveto (in 'Organic Reactions in Steroid Chemistry' , J.Fried and J.A.Edwards, eds.,

Van Nostrand Reinhold Company, New York 1972, Vol. II, chapter 11), D.M.Piatak and J.Wicha (Chem.Rev. 1978, **78**, 199) and J.Redpath and F.J.Zeelen (Chem. Soc. Rev. 1983, **12**, 75). In this update only a few representative examples can be discussed.

Selective ethynylation of androsta-1,4-diene-3,17-dione **17** gives the 17α-ethynyl derivative **20** in high yield. Epimerization of the side-chain was achieved through silver or copper (I) catalyzed solvolysis of the nitrate ester **21** (H.Hofmeister *et al.*, Chem. Ber. 1978, **111**, 3086; I.Nitta *et al.*, Bull. Chem. Soc. Jpn 1985, **58**, 981; H.Hofmeister *et al.*, Liebigs Ann. Chem. 1987, 423). The 17ß acetyl side-chain is then formed by mercury acetate catalyzed hydration. The dione **22** can be converted into corticosteroids in the classical manner via introduction of the 21-hydroxyl group and microbiological 11ß-hydroxylation. Several variations on this theme have been described.

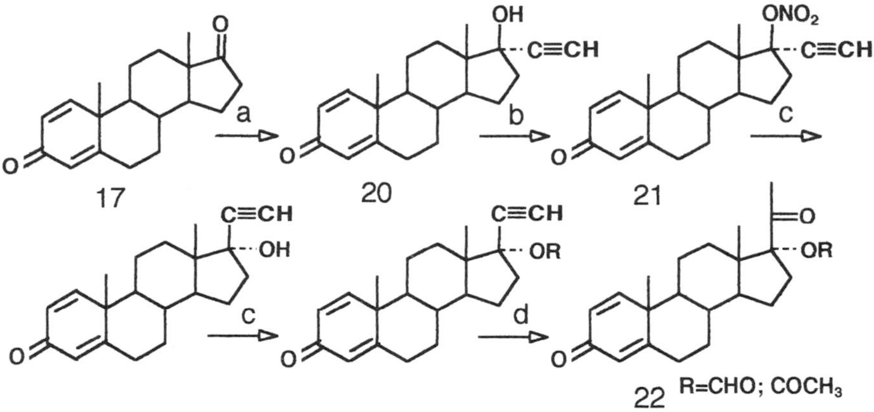

Scheme 1 Introduction of the 17ß-acetyl side-chain
Reagents: a.K-acetylide; b.HNO_3/ $(CH_3CO)_2O$;
c.$AgOCOCH_3$/ HCOOH; d.$Hg(OCOCH_3)_2$/ HCOOH

The additional microbiological 11ß-hydroxylation can be avoided by starting from 9α-hydroxyandrost-4-ene-3,17-dione **23** (J.C.Kapur, A.F.Marx and J.Verwey, Steroids 1988, **52**, 181; J.N.M.Batist, A.F.M.Slobbe and A.F.Marx, Steroids 1989, **54**, 321). This starting material is available through oxidation of soya bean derived sterols using a mutant of *Micobacterium fortuitum*

(M.G.Wovcha*et al.*, Biochem. Biophys. Acta 1978, **539**, 308).

V.V.VanRheenen and K.P.Shephard (J.Org.Chem., 1979, **44**, 1582) used the allylic sulphoxide-sulphenate rearrangement for the epimerization of the side-chain. The sulphenate ester **26**,

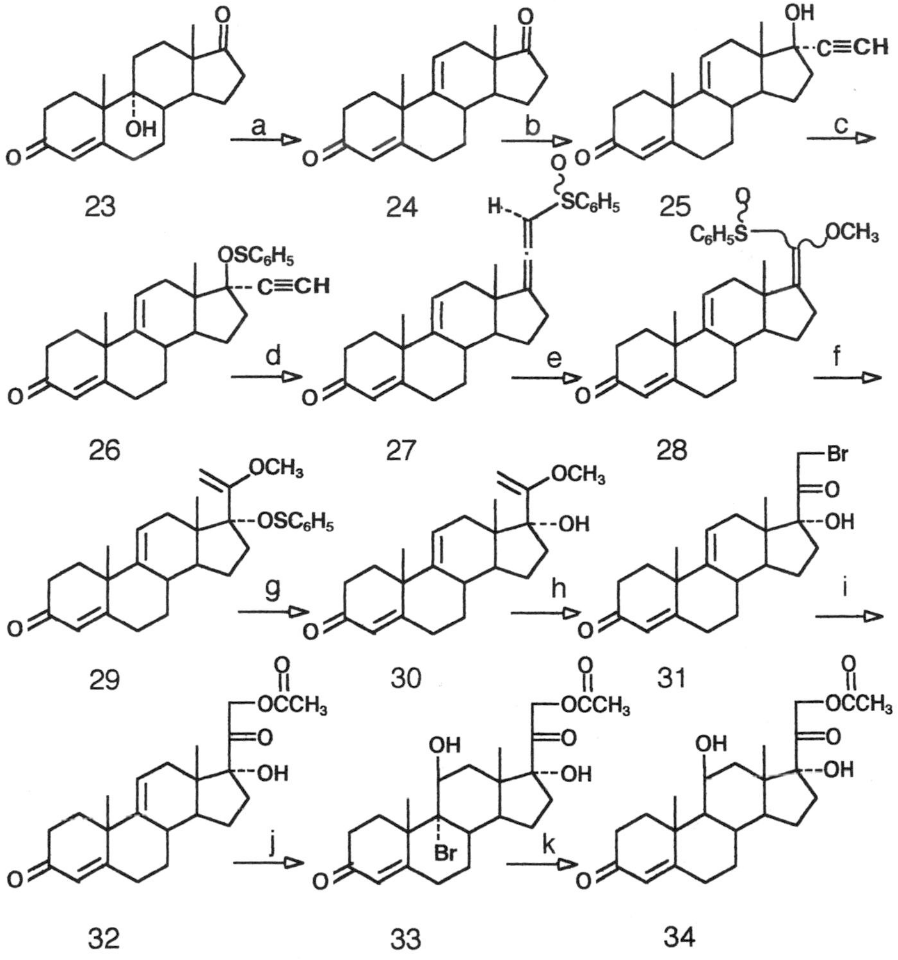

Scheme 2 Synthesis of hydrocortisone acetate
Reagents: a.C_6H_5SOCl/pyridine; p-TSA; b.K-acetylide;
c.C_6H_5SCl/$(C_2H_5)_3$NH; d. heat; e. NaOCH$_3$; f.,g. P(OCH$_3$)$_3$;
h.Br$_2$/pyridine; i.KOCOCH$_3$; j. NBS, H$^+$; k. Cr^{++}.

prepared by reaction of the alcohol **25** with phenyl sulphenyl chloride, is rearranged to the sulphoxide **27** which, upon treatment with sodium ethoxide in methanol, gives the enolether **28**. In refluxing methanol the sulphoxide **28** and the sulphenate **29** are in equilibrium and treatment with trimethyl phosphite provides the 17-hydroxy-20-methylenolether **30**. Bromination, followed by displacement of the bromide with acetate completes the synthesis of the side-chain. The synthesis of hydrocortisone acetate **34** is finished by reaction with HOBr and reduction of the 9-bromide with chromous ions. The reported overall yield is 46%.

The use of cyanohydrins for the synthesis of corticosteroids has been explored by several groups (I.Nitta, S.Fujimori and H.Ueno, Bull. Chem. Soc. Jpn, 1985, **58**, 978; J.G.Reid and T.Debiak-Krook, Tetrahedron Lett., 1990, **31**, 3669; J.N.M.Batist,

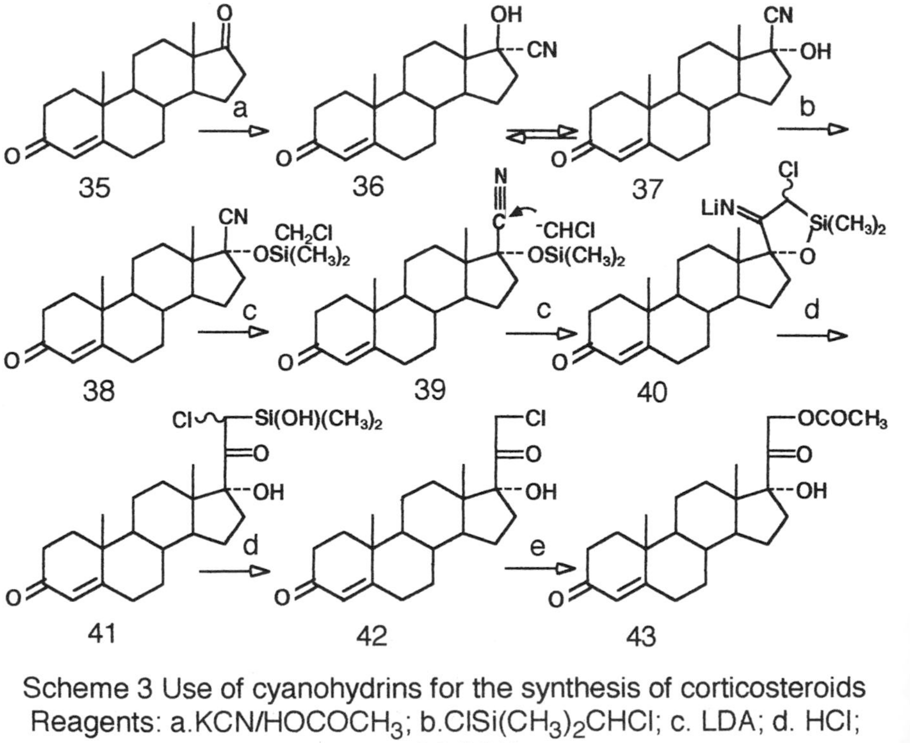

Scheme 3 Use of cyanohydrins for the synthesis of corticosteroids
Reagents: a.KCN/HOCOCH$_3$; b.ClSi(CH$_3$)$_2$CHCl; c. LDA; d. HCl;
e. KOCOCH$_3$

N.C.M.E.Barendse and A.F.Marx, Steroids 1990, **55**, 109; N.I.Carruthers, S.Garshasb and A.T.McPhail, J.Org.Chem., 1992, **57**, 961; N.I.Carruthers *et al.*, J. Chem. Soc. Perkin Trans. 1, 1992, 1195). The addition of cyanide anion to 17-ketosteroids gives a mixture of stereoisomers. However, if conditions of equilibration and selective crystallization are established, the 17ß-cyanohydrins can be obtained in high yield. Livingstone *et al.* (J. Am. Chem. Soc., 1990, **112**, 6449) converted the 17-hydroxyl group of the cyanohydrin **37** into the (chloromethyl)dimethylsilyl ether **38**. Treatment with the strong base lithium diisopropylamide resulted in cyclization unto the nitrile with formation of the imine **40**. The 21-acetate **43** was formed by displacement with potassium acetate.

Ester condensations have also been used for the construction of the corticosteroid side-chain. In contrast with the syntheses described above, this approach usually requires protection of the 3-keto-4-ene system (G.Neef *et al.*, Chem. Ber. 1980, **113**, 1184; A.R.Daniewski and W.Wojciechowska, J. Org. Chem. 1982, **47**,

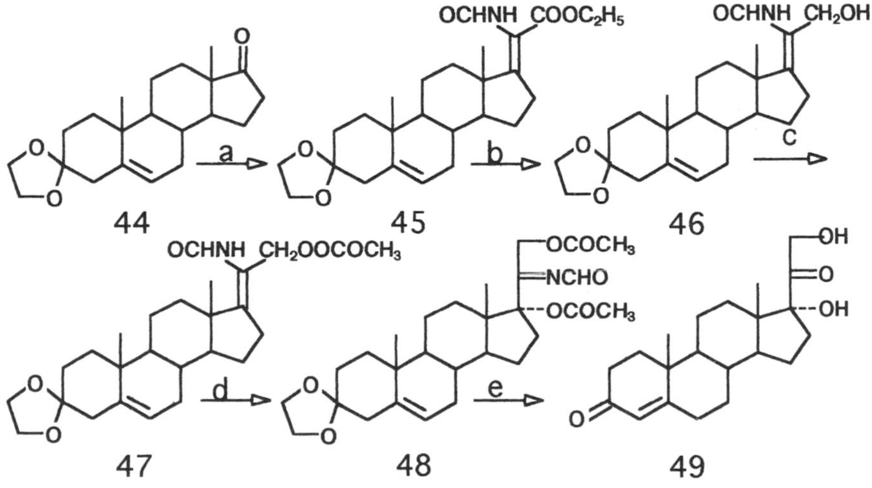

Scheme 4 Use of ester condensation for the construction of corticosteroid side-chain
Reagents: a. $C_2H_5OOCCH_2NC$; b. $LiAlH_4$; c. $(CH_3CO)_2O$; d. $Pb(OCOCH_3)_4$; e. KOH

2993; A.R.Daniewski and W.Wojciechowska, Synthesis, 1984, 132). L.Nédélec, V.Torelli and M.Hardy (J. Chem. Soc. Chem. Commun. 1981, 775) used the α-formylaminoacrylic ester **45**, formed by condensation of ethyl isocyanoacetate with the 17-ketone **35**. The ester was reduced selectivily with lithium aluminium hydride to the alcohol **46** and protected as the acetate. Oxidation with lead tetraacetate and hydrolysis gave cortexolone **49** .

A Knoevenagel-type condensation of 4-tolylsulfonylmethyl-isocyanide with 17-ketosteroids forms the basis of a synthesis

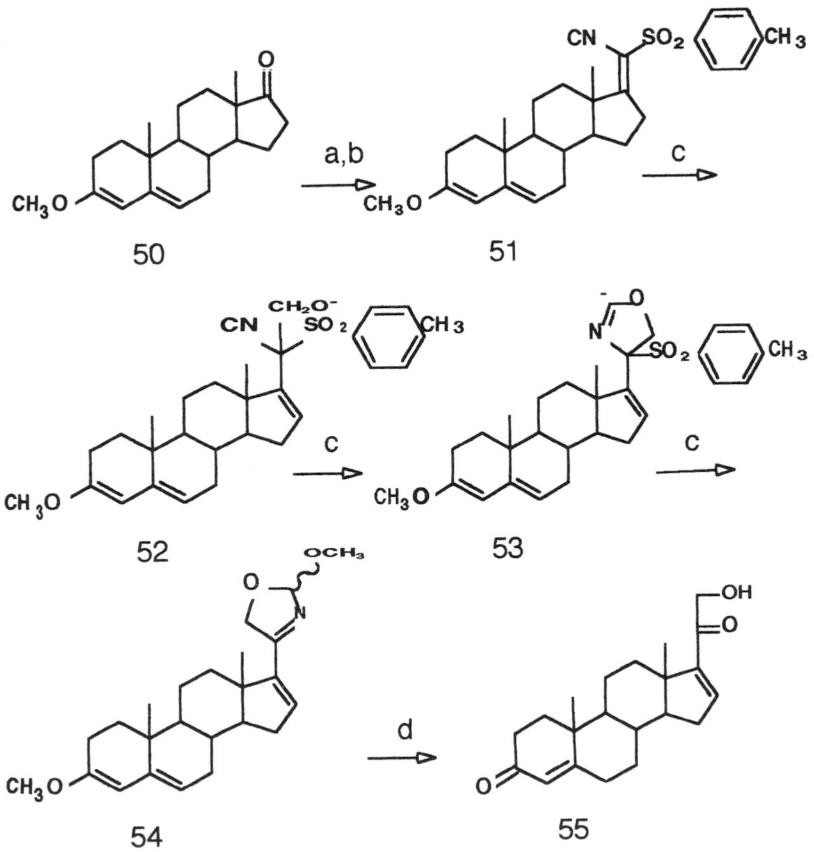

Scheme 5 Use of 17-(isocyanotosylmethylene)-steroid
Reagents: a. $TosCH_2N=C$; b.$POCl_3/(C_2H_5)_3N$; c.CH_2O;
d.H_2SO_4

developed by D. and A.M.van Leusen (Tetrahedron Lett., 1984, **25**, 2581; Recl.Trav. Chim. Pays-Bas, 1991, **110**, 393). Specific reaction conditions were worked out for the addition of form-aldehyde and the intramolecular cyclization to the oxazoline **54**. Acid hydrolysis gave 21-hydroxypregna-4,16-diene-3,20-dione **55**. The 16,17-double bond makes this compound a suitable interme-diate for the synthesis of 16,17-substituted derivatives.

Related approaches have been described by D.H.R.Barton, W.B. Motherwell and S.Z.Zard (J. Chem. Soc. Chem. Commun., 1981, 774; Bull. Soc. Chim. France II, 1983, 61); D.H.R.Barton and S.Z. Zard (J.Chem. Soc. Perkin Trans. I, 1985, 2191) and by S.Sólyom, K.Szilágyi and L.Toldy (Liebigs Ann. Chem., 1987, 153).

(b) Synthesis of 19-norsteroids

New routes to 19-norsteroids were opened up when the lactone **19** became available as starting material. Most contraceptive ster-oids belong to this class of compounds. As illustration, the synthe-

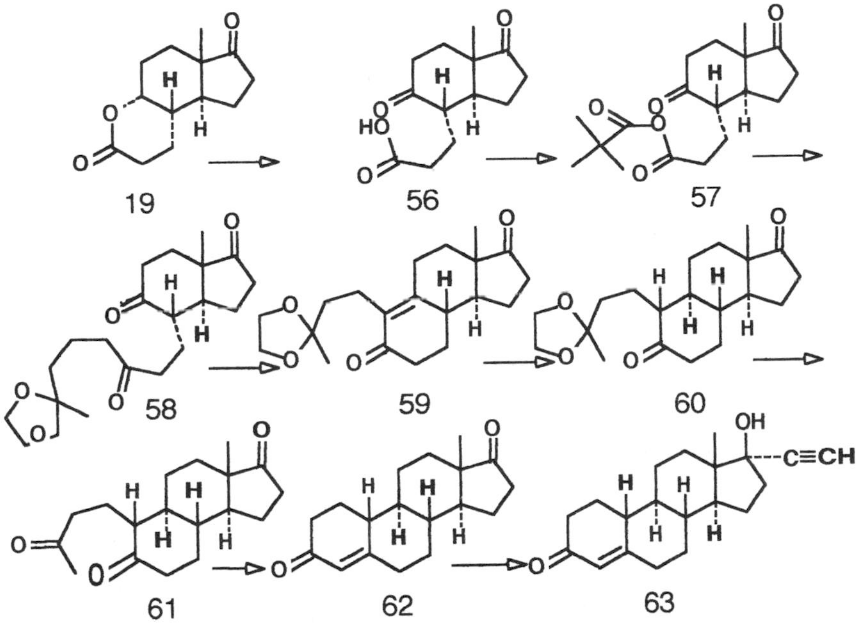

Scheme 6 Synthesis of norethisterone

sis of norethisterone **63**, as descibed by G.F.Cooper and A.R. Van Horn (Tetrahedron Lett., 1981, **22**, 1479), is given. It is related to an earlier described total synthesis (G.Nomine, G. Amiard and V.Torelli, Bull. Soc. Chim. France, 1968, 3664).

The ketocarboxylic acid **56**, prepared by oxidation of the lactone **19** was converted into the mixed anhydride **57** by condensation with pivaloyl chloride in the presence of triethylamine. At -70⁰ C, the anhydride was reacted with the Grignard reagent, derived from 5-chloro-2-pentanone ethyleneketal, followed by treatment with base to form the tricyclic enedione **59**. Catalytic hydrogenation and isomerization afforded the dione **60** with the desired 9α,10β-configuration. Deprotection and cyclization to the tetracyclic estr-4-ene-3,17-dione **62** followed by selective ethynylation gave norethisterone **63**.

3. Steroid total synthesis

The total synthesis of steroids remains a challenge for the organic chemist. A recent review (M.B.Groen and F.J.Zeelen, Recl. Trav. Chim. Pays Bas, 1986, **105**, 465), covering the period 1977-1985, discussed more than 200 publications. In this section, the major developments are discussed.

(a) Intramolecular Diels-Alder cycloadditions

Several groups have explored intramolecular cycloadditions for their utility in steroid total synthesis. In most cases, highly reactive o-quinodimethanes were used as dienes. The first example of such

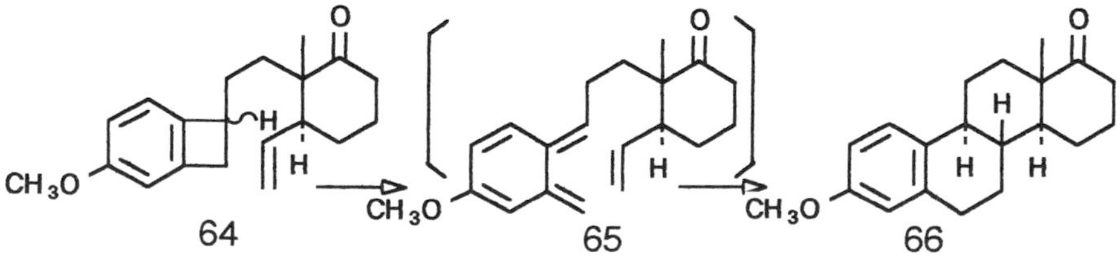

Scheme 7 Generation and intramolecular
cycloaddition of a o-quinodimethane

an approach was the synthesis of D-homoestrone **66** (T.Kametani *et al.*, J. Am. Chem. Soc., 1976, **98**, 3378; 1977, **99**, 3461), where the cyclobutene derivative **64** was used to generate the o-quinodi-methane. The yield of the cycloaddition was 95% and the stereo-selectivity was high.

The enantioselective total synthesis of (+)-cortisone by H.Nemoto *et al.* (J.Org.Chem., 1990, **55**, 5625) forms an inter-esting example of this approach. Intramolecular cycloadditions for the formation of ring C of the steroids had already been used suc-cessfully for the synthesis of *racemic* estrone and adrenosterone (H.Nemoto *et al.*, J. Chem. Soc. Perkin Trans. I, 1989, 1639) and the *racemic* 19-norderivative of canrenone (H.Nemoto *et al.*, J. Am. Chem. Soc., 1988, **110**, 2931). In the synthesis of cortisone a chiral stereodirecting auxiliary group was used to induce the desired configuration of the natural steroids. The cycloaddition pro-ceeded quantitatively and no other isomers could be detected in the A-norsteroid **68**. After protection of the 17-hydroxyl group as its methoxymethyl ether, the auxiliary group was removed by oxida-tive hydrolysis. The resulting aldehyde **69** was reduced to the al-cohol **70**.

Birch reduction of the alcohol **70**, followed by acid-catalyzed isomerization afforded the thermodynamically favoured α,β-unsat-

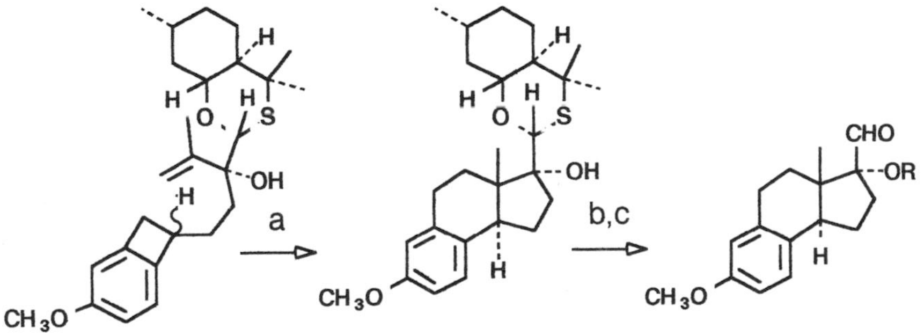

67 68 69

Scheme 8 Cortisone synthesis, formation of ring C
Reagents: a.Reflux in o-dichlorobenzene; b.CH_3OCH_2Cl;
c.N-chlorosuccinimide/$AgNO_3$

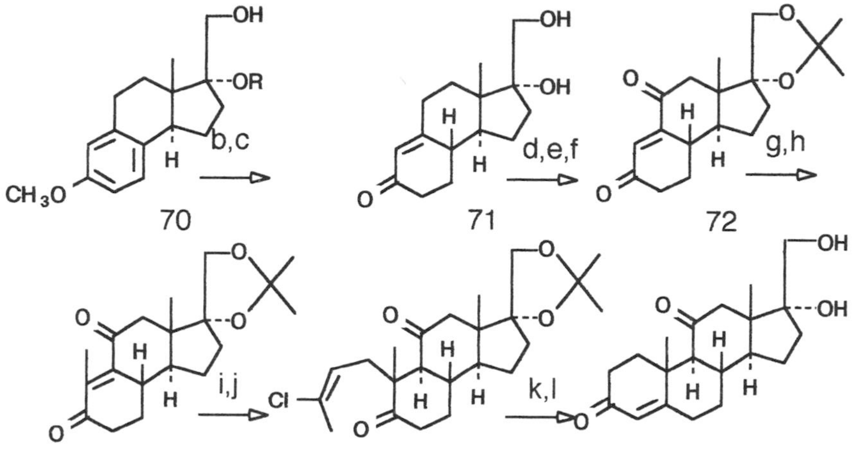

Scheme 9 Cortisone synthesis,
introduction of 11-ketogroup and addition of ring A
Reagents: a.NaBH$_4$; b.Li/ liq. NH$_3$; c.HCl; d. (CH$_3$O)$_2$C(CH$_3$)$_2$;
e.SeO$_2$; f. pyridinium chlorochromate; g.CH$_2$N$_2$; h.reflux in
o-dichlorobenzene; i.Li/ liq. NH$_3$; j.BrCH$_2$CH=CClCH$_3$;
k.Hg(OCOCF$_3$)$_2$ then HCl; l. KOH

urated ketone **71**. After protection of the vicinal hydroxyl groups as
an acetonide, the 11ß-ketogroup was formed by allylic oxidation.
The 19-methyl group was introduced via regioselective 1,3-dipolar
addition of diazomethane to the enedione **72** and thermolysis of
the resulting mixture of pyrazolines.

Stereoselective A-ring formation was carried out using a procedure developed by G.Stork and E.W.Logush (J. Am. Chem. Soc.,
1980, **102**, 1219). The enedione **73** was reduced with lithium in
liquid ammonia-tetrahydrofuran and the enolate anion, thus generated, was alkylated with CH3CCl=CHCH2Br to give the dione **74**,
which was hydrolyzed to a triketone. Base catalyzed cyclization
furnished the steroid **75**, which was converted into (+)-cortisone. A
similar approach has been followed for the synthesis of (+)-19-nor-
deoxycorticosterone (H. Nemoto et al., J. Chem. Soc. Chem.
Commun., 1990, 1001).

Several groups have explored the use of dienes less reactive than the o-quinodimethanes. 1,2-Naphtoquinodimethanes were used, for example, for the formation of ring-C of 16-oxasteroids (K.Kobayashi, M.Itoh and H.Suginome, J. Chem. Soc. Perkin Trans. I, 1991,2135).

Transannular Diels-Alder reactions of 14-membered macrocyclic trienes can form three rings in one step. The (E,E,E)-triene **76** gave 5ß-androst-6-ene-3,17-dione **77** in 84% yield. No diastereo-isomers could be detected by HPLC or NMR analysis (T.Takahasi *et al.*, J. Am. Chem. Soc., 1988, **110**, 2674).

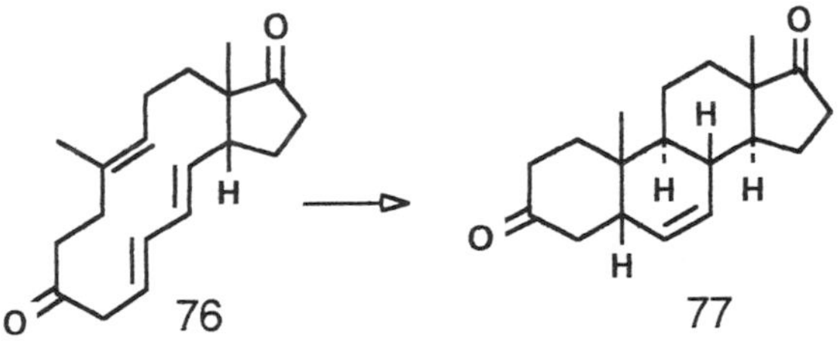

Scheme10 Transannular Diels-Alder reaction

It is of interest to note that this work was guided by conformational analysis of the macrocyclic trienes (S.Lamothe, A.Ndibwami and P.Deslongchamps, Tetrahedron Lett., 1988, **29**, 1639 and 1641) and molecular mechanics calculations on transition state models (T.Takahashi, Y.Sakamoto and T.Doi, Tetrahedron Lett., 1991, **33**, 1992).

An example of creative use of an intramolecular Diels-Alder reaction is provided by a steroid total synthesis developed by Stork and coworkers (G.Stork and E.W.Logush, J. Am. Chem. Soc., 1980, **102**, 1218, 1219; G.Stork, G.Clark and C.S.Shiner, J. Am. Chem. Soc., 1981, **103**, 4948; G.Stork, J.D.Winkler and C.S.Shiner, J. Am. Chem. Soc., 1982, **104**, 3767; G.Stork and D.H.Sherman, J. Am. Chem. Soc., 1982, **104**, 3758). The enedi-ene **80** was formed by reaction of the ketoacid **78** with isopropenyl lithium followed by dehydration. Intramolecular Diels-Alder cyclo-addition provided the cyclohexene **81** which does not resemble a steroid at all. However, when subjected to ozonolysis a tetraket-one **82** was formed, which could be cyclized to the steroid **83**. An improved method for the conversion of pregna-4,16-dien-3,11,20-

trione **83** into racemic cortisone was described by Y.Horiguchi, E.Nakamura and I.Kuwajima (J. Am. Chem. Soc., 1989, **111**, 6257).

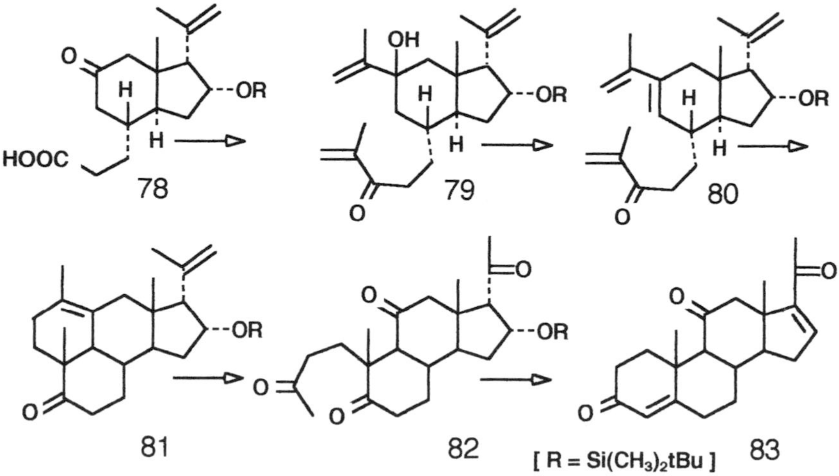

Scheme 11 Stork's total synthesis

The Stork methodology has also been used for the synthesis of deuterium-labelled corticosteroids (K.Minagawa *et al*, J. Chem. Soc. Perkin Trans. I, 1988, 587; H.Shibasaki, T.Furuta and Y.Kasuya, J. Labelled Compounds Radiopharm., 1991, **29**, 1033; Steroids, 1992, **57**, 325).

(b) Intermolecular Diels-Alder cycloadditions

Intermolecular Diels-Alder reactions played an important role in early total syntheses such as that described by R.B.Woodward *et al.* (J. Am. Chem. Soc., 1951, **73**, 2403 and 4057) or by L.H.Sarett *et al.* (J. Am. Chem. Soc., 1952, **74**, 4974). In later years interest shifted to the intramolecular Diels-Alder reactions. Recently, however, when it was demonstrated that intermolecular Diels-Alder reactions can be catalyzed with Lewis acids, interest in this approach revived. The catalysts not only influence the reaction rate, but also the stereoselectivity (G.Quinkert *et al.*, Tetrahedron Lett., 1991, **32**, 3357; S.Takano, M.Moriya and K.Ogasawara,

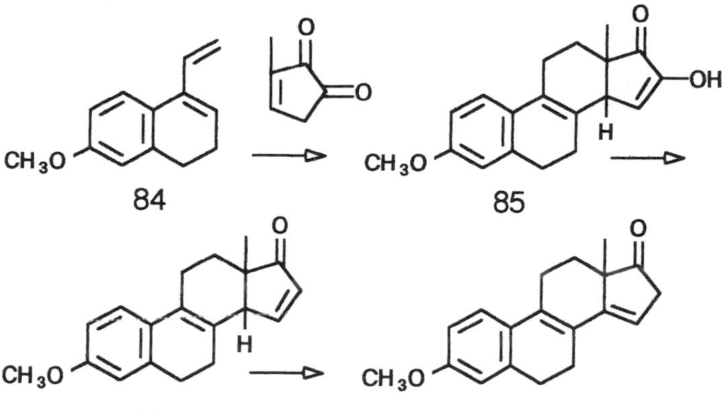

Scheme 12 Diels-Alder cyclization using a chiral catalyst

Tetrahedron Lett. 1992, **33**,1909). This may be illustrated by a recently described synthesis of 3-methoxyestra-1,3,5(10),8,14-pentaen-17-one **87** where a chiral complex of $TiCl_2(OiC_3H_7)_2$ and the diol **88** was used as catalyst (G.Quinkert *et al.*, Tetrahedron Lett., 1992, **33**, 3617). On a preparative scale, the adduct **85** was formed in 64% yield. The enantiomeric purity (e.e. 73%) is still too low to be of practical use but it illustrates the potential of this type of approach. Reduction and isomerization provided the pentaene **87**, which is also an intermediate in Torgov's total synthesis (S.N.Ananchenko and I.V.Torgov, Tetrahedron Lett., 1963, 1553).

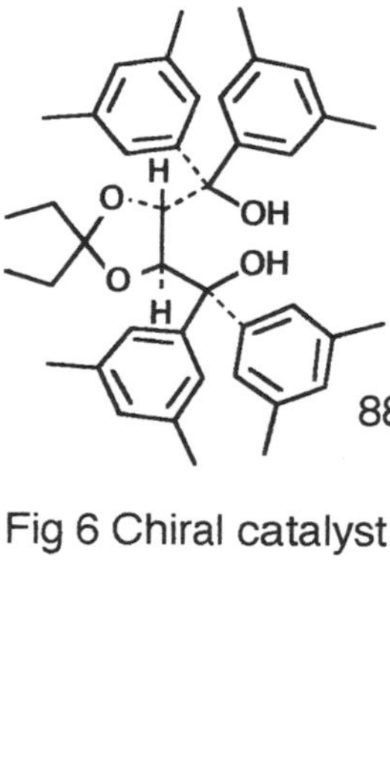

Fig 6 Chiral catalyst

(c) Asymmetric cyclizations

The power of using chiral catalysts in steroid total syntheses had already been demonstrated by the efficient asymmetric cyclization of the trione **89** (U.Eder, G.Sauer and R.Wiechert, Angew. Chem., 1971, **83**, 492; Z.G.Hajos and D.R.Parrish, J. Org. Chem., 1974, **39**, 1615). Chiral amino acids such as proline or phenylalanine are used as catalysts. This discovery made use of the indenedione **90,** now a readily availible synthon, often employed in total synthesis. A recent example is a synthesis of androst-4-ene-3,11,17-trione **96**, a starting material for synthesis of glucocorticoids (A.R.Daniewski, E.Piotrowska and W.Wojciechowska, Liebigs Ann. Chem., 1989, 1061; A.R.Daniewski and E.Piotrowska, Liebigs Ann. Chem., 1989, 571).

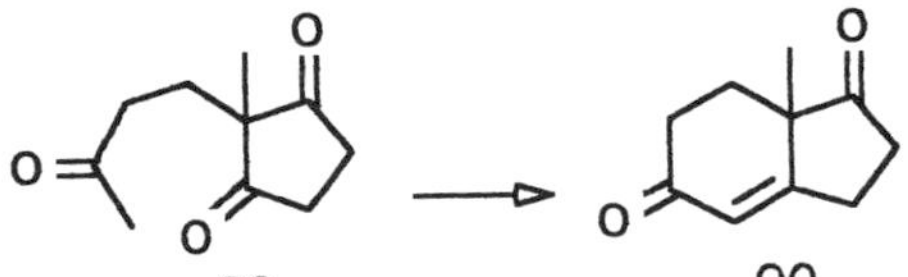

Scheme 13 Asymmetric cyclization

The enolate **91**, formed by cuprous iodide catalyzed diisobutyl-aluminiumhydride reduction of the indenedione **90** (A.R.Daniewski *et al.*, Liebigs Ann. Chem., 1988, 593), was converted into the silyl ether **92** and coupled with the alkene **93**. For this coupling tin tetrachloride was used as a catalyst. After removal of the ketal, the resulting trione **94** was cyclized to the dione **95**. The conversion of this dione into androst-4-ene-3,11,17-trione **96** followed the route as described above (scheme 9).

(d) Free-radical cyclizations

Free-radical cyclizations form an interesting new approach to steroid total synthesis (G.Stork and R.Mah, Tetrahedron Lett., 1989, **30**, 3609; V.H.Rawal, R.C.Newton and V.Krishnamurthy, J. Org. Chem., 1990, **55**, 5181; P.A.Zoretic, M.Ramchandani and M.L.Caspar, Synthetic Commun., 1991, **21**, 923). The power of the present methodology is illustrated by the free-radical cyclization of

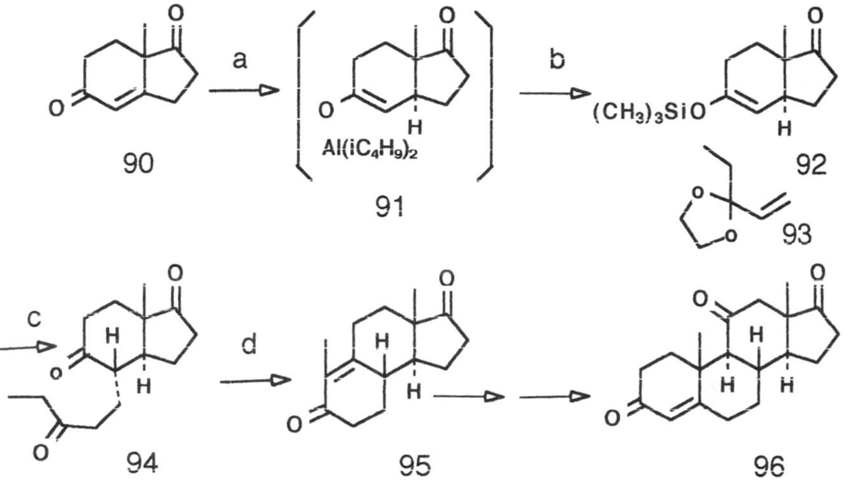

Scheme 14 Use of chiral synthon for glucocorticosteroid synthesis
Reagents: a. t-C_4H_9Cu/(i-C_4H_9)$_2$AlH; b. (CH_3)$_3$SiCl; c. $SnCl_4$;
d. CH_3ONa

the tetraene ß-ketoester **97** to the D-homosteroid **98**. Manganese (III) acetate was used to generate the ß-ketoester radical and copper (II) acetate was used to terminate the cyclization. The yield of this step was 31% (P.A.Zoretic *et al.*, Tetrahedron Lett., 1991, **32**, 4819).

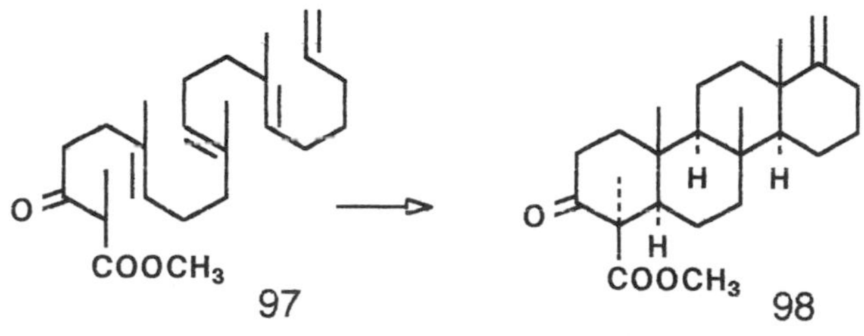

Scheme 15 Free-radical cyclization

(e) Cationic cyclizations

It is of interest to compare the above result with those of the cationic cyclizations, explored by W.S.Johnson and coworkers. Also

480

here, success depends strongly upon the careful selection of initiating and terminating groups, but impressive results have been
obtained. The tetraene **99** was cyclized, hydrolyzed and acetylated, to facilitate purification, to give in 58% yield the 11α-hydroxy-
A-norsteroid **101**. This compound was converted by ozonolysis
and cyclization into 11α-hydroxypreg-4-ene-3,20-dione **102**
(W.S.Johnson, T.A.Lyle and G.W.Daub, J. Org. Chem., 1982, **47**,
163).

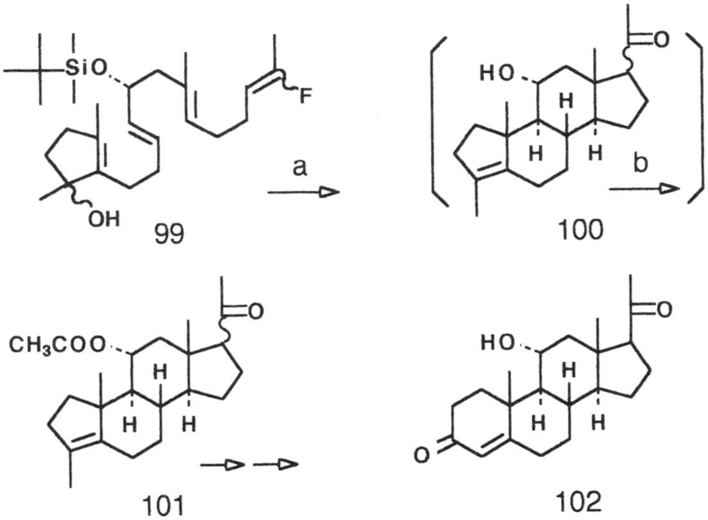

Scheme 16 Cationic cyclization
Reagents: a.CF$_3$COOH/CF$_3$CH$_2$OH; b.(CH$_3$CO)$_2$O/pyridine

4. Remote functionalization of non-activated positions in steroids

In a long and systematic series of investigations R.Breslow and
coworkers developed template-catalyzed and directed chlorination
methods for the functionalization of non-activated positions in
steroids (R.Breslow, Chem.Soc. Rev., 1972, **1**, 553; R.Breslow,
B.B.Snider and R.J.Corcoran, J. Am. Chem. Soc., 1974, **96**, 6792;
R.Breslow et al., J. Am. Chem. Soc., 1977, **99**, 905). These reactions depend on careful geometric control and the choice of proper
reaction conditions, but efficient syntheses can be developed. This

was demonstrated with the conversion of 5α-cholestan-3ß-ol **103** into 9α-fluoro-3α-hydroxy-5α-androsta-11,17-dione **111** in 36% overall yield (R.Breslow and T.Link, Tetrahedron Lett., 1992, **33**, 4145). This dione is a potential intermediate in the synthesis of 9α-fluorocorticosteroids.

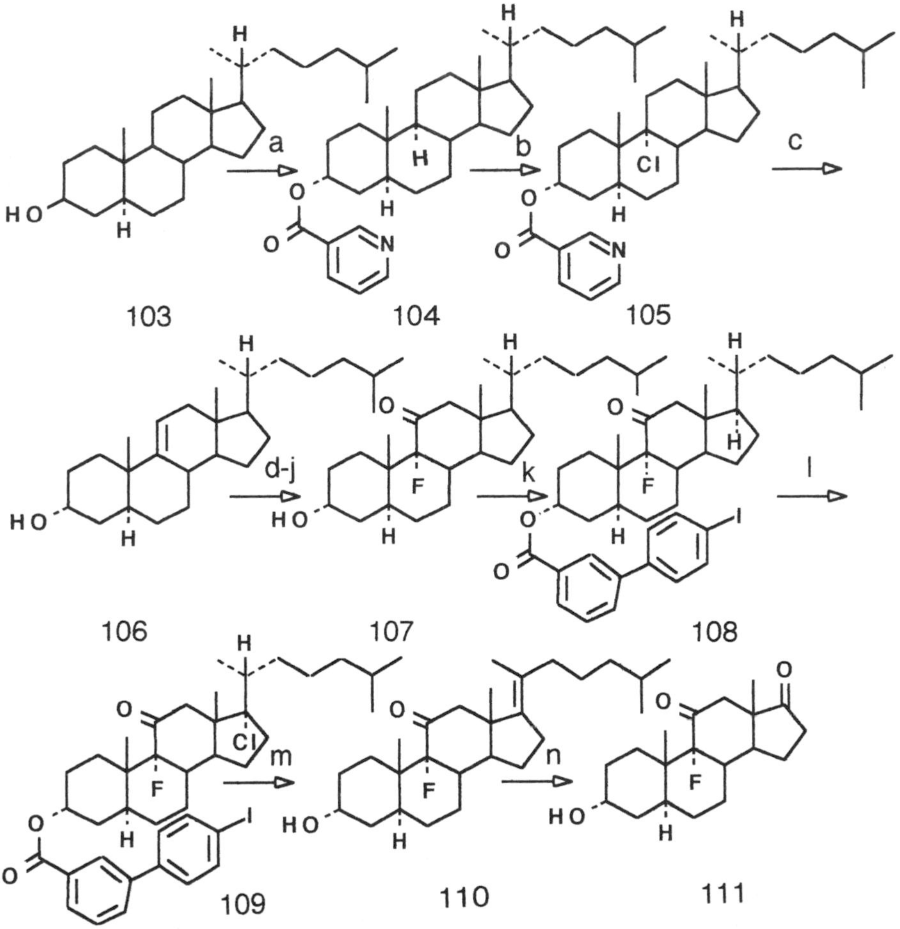

Scheme 17 Template-catalyzed and directed chlorination
Reagents: a.Ph₃P/diethyl azodicarboxylate/nicotinic acid; b.PhICl₂, hv; c.KOH; d.(CH₃CO)₂O; e. NBS; f.KOH; g.(CH₃CO)₂O; h. HF; i. PCC; j. KOH; k.acid chloride; l. PhICl₂ , hv; m. DBU; n. O₃/Ph₃P

In the first photochemical step the pyridine ring of the nicotinate ester **104** serves as relay for the transfer of a chlorine atom (R.Breslow et al., J. Am. Chem. Soc., 1987, **109**, 3799). In that way free-radical chlorination with phenyliodosochloride gave the 9α-chloroderivative **105** which upon treatment with KOH afforded the Δ9(11)-unsaturated sterol **106**. In order to obtain the optimal geometry for the next photochemical step and subsequent dehydrohalogenation (R.Breslow and U.Maitra, Tetrahedron Lett., 1984, **25**, 5843), the Δ9(11)-double bond was converted in the classical manner via the bromohydrin, epoxide, fluorohydrin into the 9α-fluoro-11-ketone **107**. In the following photochemical reaction, the iodobenzene group of the ester **108** served as relay for the chlorination. The resulting 17α-chloroderivative **109** was converted into the exocyclic Δ17(20)-unsaturated sterol **110** upon treatment with DBU (1,8-diazabicyclo[5,4,0]undec-7-ene). Ozonolysis gave the 17-ketone **111**.

5. The brassinosteroids

(a) Introduction

In 1979, M.D.Grove *et al.* (Nature,1979, **281**. 216) reported isolation and structure determination of brassinolide **112**, a highly active plant growth promotor from the pollen of rape, *Brassica napus*, where it occurs in a concentration of 200 μg/kg. This plant growth-promoting steroid stimulates both cell elongation and cell division. Later, many related steroids have been identified and today some 27 natural brassinosteroids are known. Examples are castasterone **113** from insect gall of the chestnut tree, *Castanea spp.* (T.Yokota, M.Arima and N.Takahashi, Tetrahedron Lett., 1982, **24**, 1275), typhasterol **114** from cat-tail, *Typha latifolia*, pollen (J.A.Schneider *et al.*, Tetrahedron Lett., 1983, **24**,3859) and teasterone **115** from leaves of green tea, *Thea sinensis* (H.Abe *et al.*, Agric. Biol. Chem., 1983, **47**, 2419). Brassinosteroids are widely distributed in plants, but in very low concentrations (N.B.Mandava, Ann. Rev. Plant Physiol. Plant Mol. Biol., 1988, **29**, 23). For this reason, synthetic reference samples have been very important for the structure elucidation of the natural products.

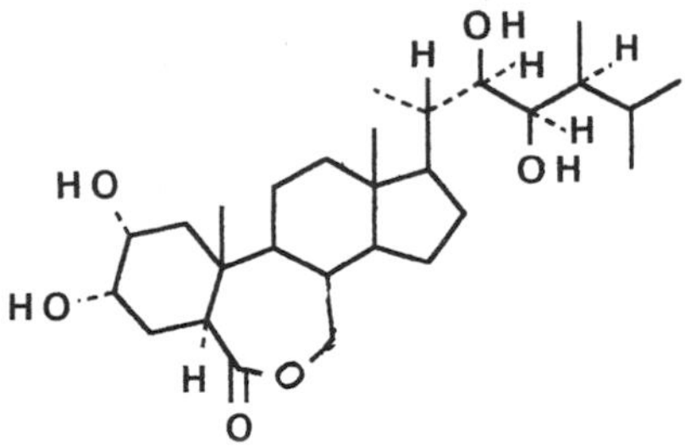

112 Brassinolide

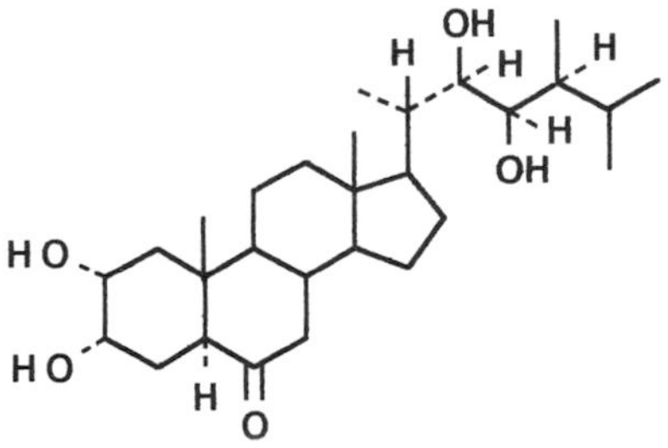

113 Castasterone

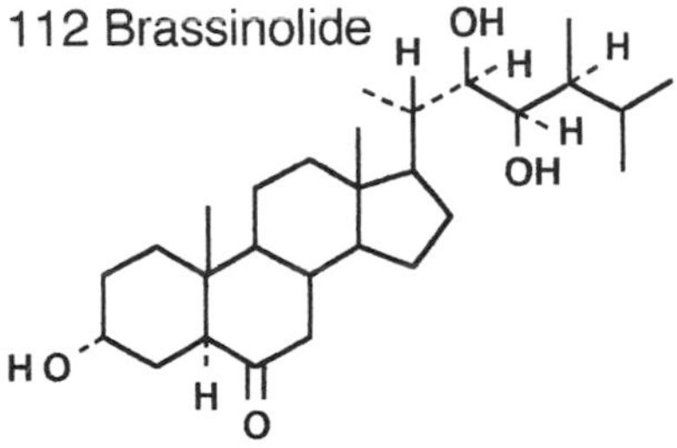

114 Typhasterol

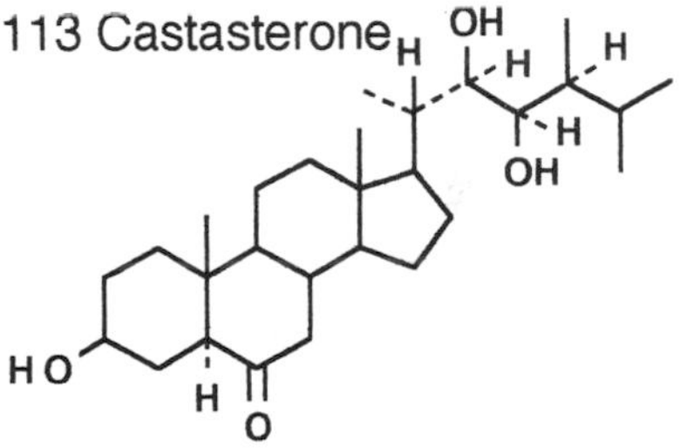

115 Teasterone

Fig. 7 Brassinosteroids

(b) Synthesis of brassinosteroids

The aldehyde **116**, easily obtainable from stigmasterol, proved to be an attractive starting material for the synthesis of brassinosteroids. In S.Fung and J.B.Siddall's synthesis of brassinolide **127** (J.Am. Chem. Soc., 1980, **102**, 6580), the aldehyde **116** was alkylated with lithium butyldimethyl-(E)-2,3-dimethylbutenyl alanate **117**. The reaction was not fully stereospecific, the desired 22(S)-allylic alcohol **118** was the major product and could be isolated in 46% yield. Hydroxyl-directed epoxidation with m-chloroperbenzoic acid gave the epoxide **119** and the synthesis of the side-chain was completed by reduction of the epoxide with LiBH4-BH3.THF. The regioselectivity for formation of the vicinal diol **120** was 3:1. The 3ß-hydroxy-5-ene system was regenerated by treatment with acid.

To complete the brassinolide synthesis, the vicinal hydroxyl groups of the side-chain were protected as an acetonide and the 3ß-hydroxyl group was converted into a tosylate. A 6α-hydroxyl

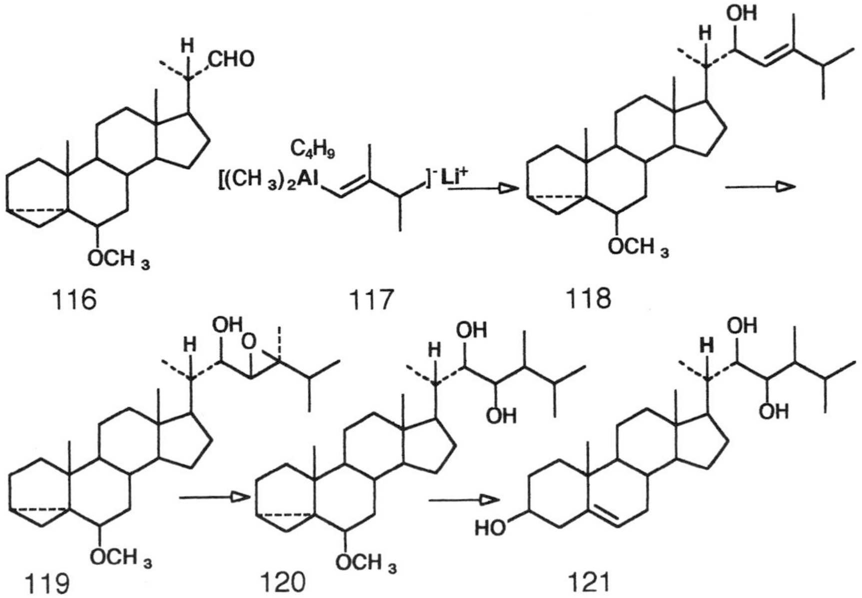

116 117 118

119 120 121

Scheme 18 Construction of the brassinolide side-chain

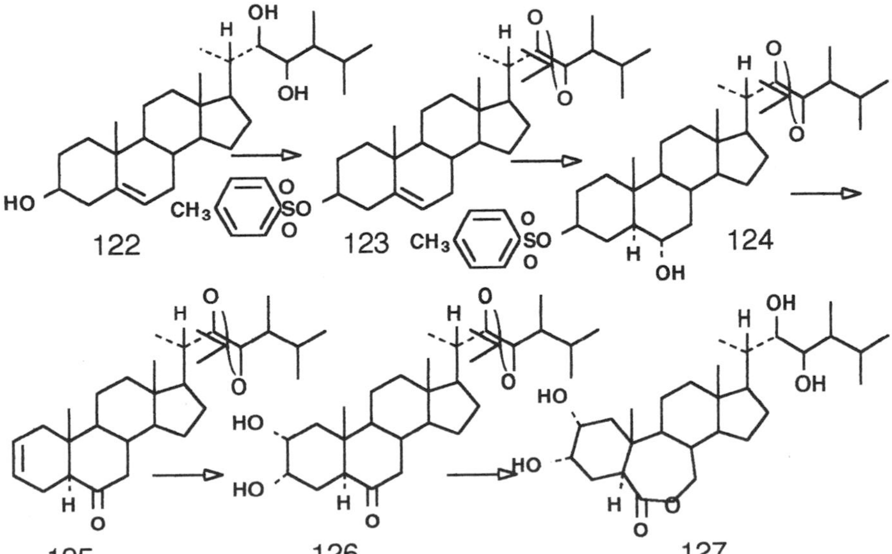

122 123 124

125 126 127

Scheme 19 Formation of the brassinolide nucleus

group was introduced by hydroboration-oxidation and the 2,3-ene function was formed by elimination of the tosylate group with Li2CO3 in dry dimethylacetamide. This sequence was followed by Jones oxidation to give the 6-ketone **125**. Stereospecific α-hydroxylation with OsO4 in pyridine gave the 2α,3α-diol **126** which was simultaneously deprotected and oxidized under Bayer-Villiger conditions to yield brassinolide **127**.

T.G.Back, P.G.Blazecka and M.V.Krishna (Tetrahedron Lett., 1991, **32**, 4817) followed a related approach for the synthesis of the side-chain and encountered comparable problems with stereoselectivity. Alkylation of the aldehyde **116** with (E)-propenyl lithium gave a mixture of the desired alcohol **129** and its epimer in a ratio 72:28. Sharpless oxidation of the major product **129** afforded a mixture of the *threo* epoxide **130** and the *erythro* diastereoisomer in a ratio 7:3 Finally, reaction of the unprotected epoxide **130** with excess of the higher order mixed cuprate **131** produced the diol **120** in 61% yield. Variations of this approach have been described by H.Hayami *et al.* (J. Am. Chem. Soc., 1983, **105**, 4491) ; M.Ishiguri *et al.* (J. Chem. Soc. Chem. Commun., 1980, 963;

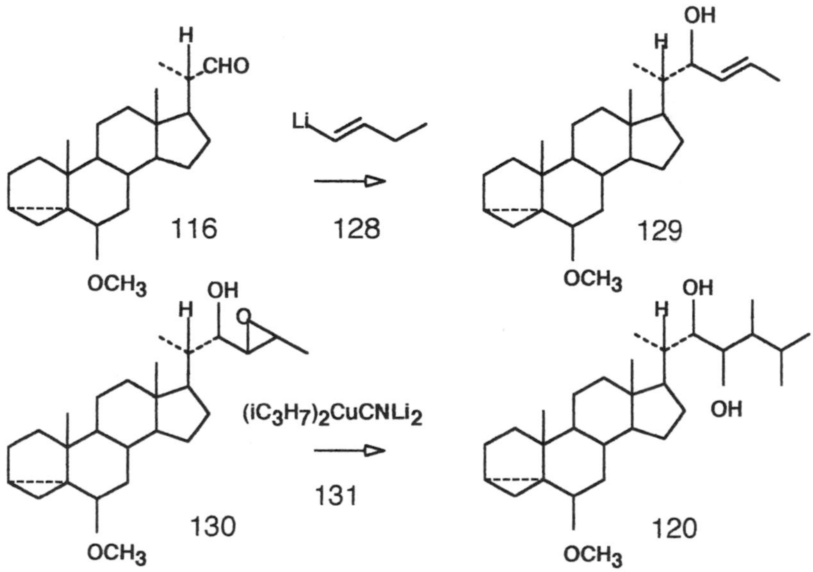

Scheme 20 Synthesis of side-chain of the brassinosteroids

S.Takatsuto *et al.* (J. Chem. Soc. Perkin Trans. I, 1984, 139) and by T.G.Back and M.V.Krishna (J. Org. Chem., 1991, **56**, 454).

With the aim of improving the stereoselectivity, the use of more rigid cyclic side-chain intermediates was investigated. Using kinetic reaction conditions, the butenolide **133** could be obtained in 65% yield by aldol condensation of the aldehyde **116** and the anion from 3-isopropylbut-2-enolide **132** (J.R.Donaubauer, A.M.Greaves, and T.C.McMorris, J. Org. Chem., 1984, **49**, 2833; W.-S.Zhou and W.-S. Tian, Tetrahedron, 1987, **43**, 3705; M.Tsubuki, K,Keino and T.Honda, J.Chem.Soc. Perkin Trans. I, 1992, 2643). Catalytic hydrogenation gave 78:22 mixture of the desired trans lactone **134** and its cis isomer. Reduction with lithium aluminum hydride afforde the triol **135**. After protection of the vicinal 22- and 23-hydroxyl groups as an acetonide, the 29-hydroxyl group was oxidized to aldehyde. This aldehyde was decarbonylated with tris(triphenylphosphine)rhodium chloride to give the

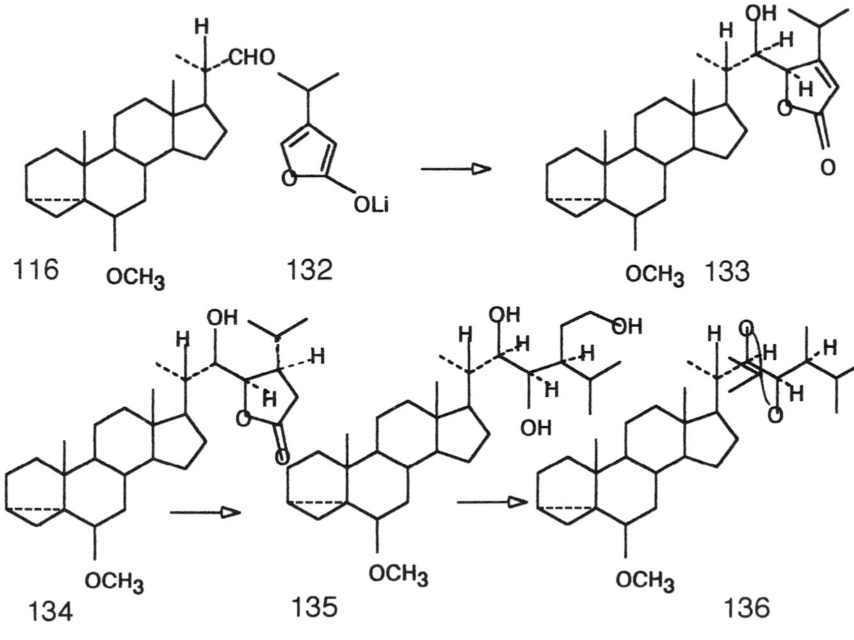

Scheme 21 Use of cyclic side-chain intermediates

24(S)-methyl derivative **126**, which was converted to brassinolide in the classical way. Several variations of this scheme have been reported (T.Kametani *et al.*, J.Org. Chem., 1986, **51**, 2932; Tetrahedron Lett., 1989, **30**, 3141).

W.-S.Zhou, B,Jiang and X.-F.Pan (J. Chem. Soc., Chem. Commun., 1989, 612; Tetrahedron, 1990, **46**, 3173; J.Chem. Soc. Perkin Trans. I, 1990, 2281) used hydrogen ation of the α,β-

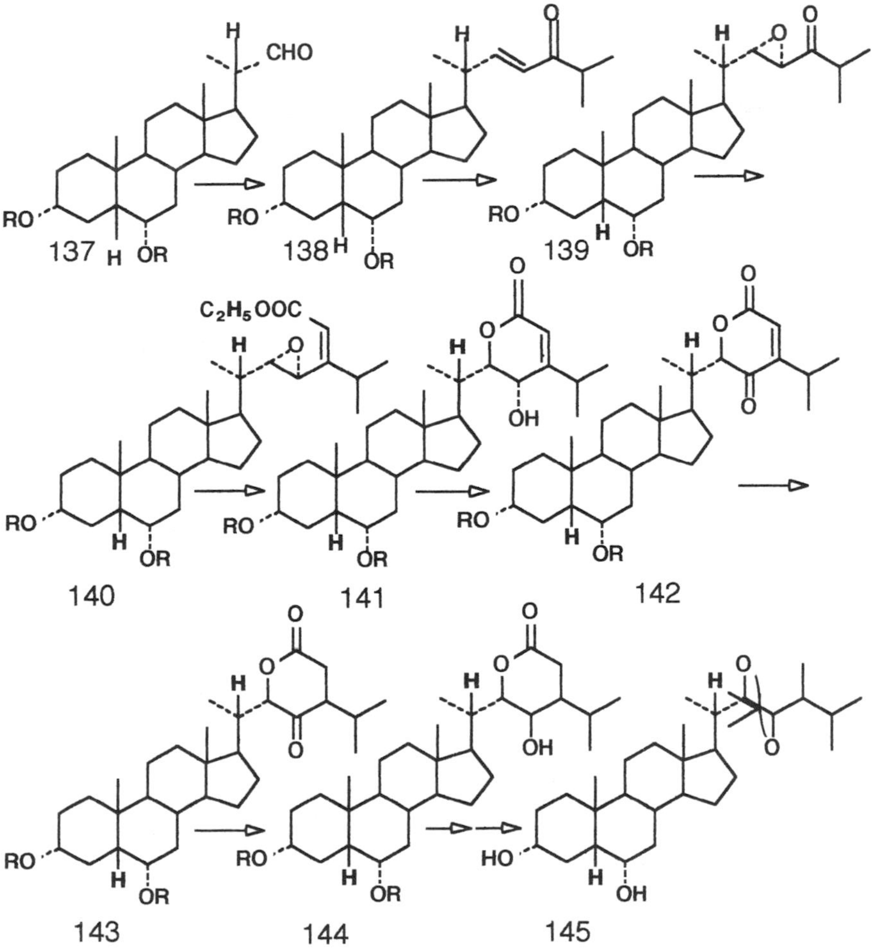

Scheme 22 Brassinosteroid synthesis

unsaturated lactone **142** for the introduction of the chiral centre at C-24. The aldehyde **137**, prepared from hyodeoxycholic acid, was reacted with isobutylcarbonyl arsonium ylide to form the α,β-unsaturated ketone **138**. Epoxidation with H_2O_2-NaOH gave the epoxyketone **139** as a single isomer. Wittig-Horner reaction of the ketone with ethyldimethylphosphono acetate afforded a mixture of the (z)-α,β-unsaturated acid ester **140** and the (E)-isomer in a ratio of 10:1 Acid treatment of the ester **140** gave the α,β-unsaturated δ-lactone **141**.

For the synthesis of brassinosteroids the hydroxyl group at C-23 has to be inverted. For that reason it is oxidized first to the ketone **142**. (An alternative, but less stereoselective, route to steroids with such an α,β-unsaturated δ-lactone in the side-chain has been reported by T.Kametani et al. (J. Chem. Soc. Perkin Trans. I, 1988, 1503) and by M.Tsubuki, K.Keino and T.Honda (J.Chem.Soc. Perkin Trans. I, 1992,2643). Catalytic hydrogenation of the α,β-unsaturated δ-lactone **142** established the wanted stereochemistry at C-24 and reduction with lithium borohydride provided the C-23 alcohol with the configuration of the brassinosteroids. The δ-lactone can be isomerized to the more stable γ-lactone and the synthesis of the brassinosteroid side-chain completed in an analogous manner to the synthesis described previously (scheme 21). Alternatively the δ-lactone **144** could be reduced with DIBAH to a hemiacetal, which after protection of the 22- and 23-hydroxyl groups as an acetonide, was decarbonylated with tris(trimethylphosphine)-rhodium chloride to give the 24(S)-methyl derivative **145**. A similar approach has been followed for the synthesis of castasterone (T.Honda, K.Keino and M.Tsubuki, J. Chem. Soc., Chem.Commun., 1990, 65)).

T.Kametani et al (J.Am.Chem.Soc., 1986, **108**, 7055; J. Org. Chem., 1988, **53**, 1982) reported an approach whereby four chiral centres on the steroid side-chain were formed in one step through β-face hydrogenation of a 5-ylidenetetronate derivative **149**. The saturated lactone **150**, formed in 90% yield, upon reduction with lithium aluminum hydride gave the diol **151**, which could be converted to brassinolide.

The 5-ylidenetetronate **149** was synthesized as follows. The ketone **146** was coupled with the dianion **147** and gave the tetronate **148** and its diastereoisomer at C-22 in a ratio of 91:9.

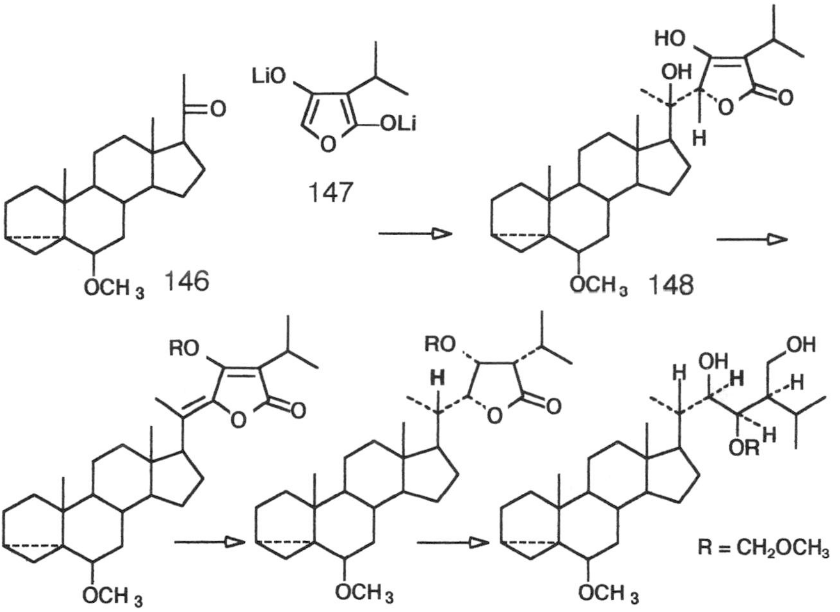

Scheme 23 Formation of four chiral centres in one step

The 23-hydroxyl group was protected as methoxymethyl ether and the 20-hydroxyl group was converted into the trifluoroacetate. This compound upon treatment with 1,8-diazabicyclo[5,4,0]undec-7-ene gave the (Z)-5-ylidenetetronate **149** and its (E)-isomer in 73% yield in a ratio of 82:18.

The concept of chirality transfer has also been applied in the construction of brassinosteroid side-chains. The Δ^{22}-unsaturated ester **154** was obtained via stereospecific Claisen rearrangement of the 22(R)-alcohol **152** (M.Anastasia et al., J. Chem. Soc. Perkin Trans. I, 1983, 383). Reduction of the ester to aldehyde and decarbonylation gave the methyl derivative **155** with the desired stereochemistry. Oxidation to the 6-ketone, opening of the cyclopropan ring with HCl and dehydrohalogenation produced the diene **156**, which upon treatment with OsO_4 afforded a mixture of castasterone **157** and its 22,23-isomer in a ratio of 2:13. In related

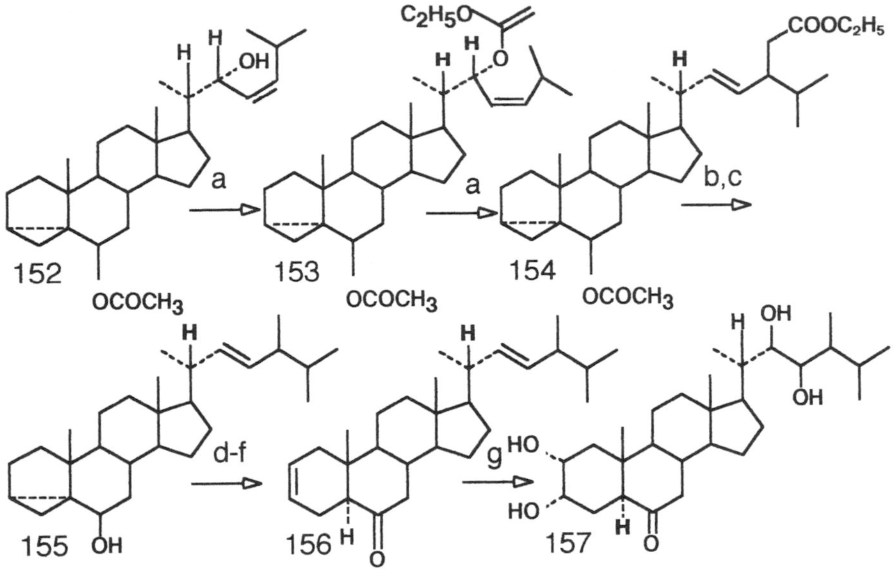

490

Scheme 24 Use of chirality transfer
Reagents: a.$CH_3C(OC_2H_5)_3$; b.$(i-C_4H_9)_2AlH$; c. $[(C_6H_5)_3P]RhCl$;
d. Jones reagent; e. HCl; f.LiBr/DMF; g.OsO_4

syntheses, the stereoselectivity of the conversion of a Δ^{22}-double bond in the side-chain to the desired 22(R),23(R)-diol, using epoxidation or osmylation, proved ineffective (M.Anastasia et al., J.Org.Chem., 1985, **50**, 321; K.Mori et al., Tetrahedron, 1982, **38**, 2099; S.Takatsuto, J. Chem. Soc. Perkin Trans. I, 1986, 2833; S.Takatsuto and N.Ikekawa, J. Chem. Soc. Perkin Trans. I, 1984, 439 and 1986, 2269).

Claisen rearrangement also formed a key step in the synthesis of deuterated brassinosteroids. The ester group formed in this way, was converted into a CD3 group by successive treatment with LiAlD4, CH3SO2Cl-pyridine and again LiAlD4 (S.Takatsuto and N.Ikekawa, Chem. Pharm. Bull., 1986, **34**, 1415, 4045; J. Chem. Soc. Perkin Trans. I, 1986, 591). In this example, the 22(R)-alcohol 158 had to be used as starting compound. It remains a problem that the conversion of C-22 aldehydes into hydroxyethynyl derivatives proceeds with hardly any stereoselectivity (Y.Yamamoto, S.Nishii and K.Maruyama, J.Chem. Soc. Chem. Commun.,

1986,102). The unwanted alcohol may, however, be recycled via oxidation to the 22-ketone and asymmetric reduction with L-Selectride (T.Takahashi *et al.*, Tetrahedron Lett., 1985, **26**, 69) or (R)-Alpine-Borane (Y.M.Midland and Y.C.Kwon, Tetrahedron Lett., 1984, **25**, 5981).

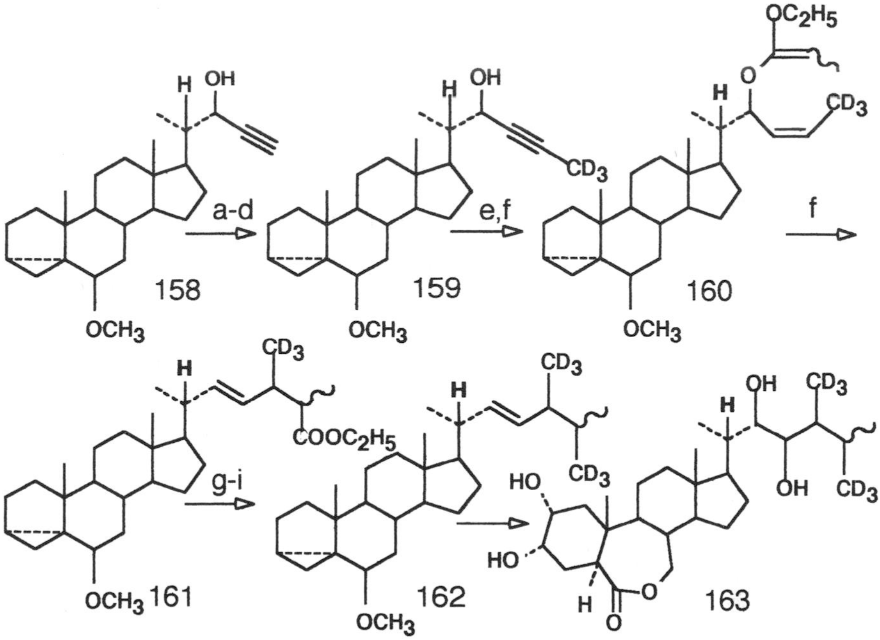

Scheme 25 Synthesis of deuterated brassinosteroids
Reagents: a. $(C_2H_5)_3SiCl$/pyridine; b.C_4H_9Li; c.CD_3I; d. $(C_4H_9)_4NF$; e. H_2/Lindlar catalyst; f. $C_2H_5C(OC_2H_5)_3$; g.$LiAlD_4$; h.CH_3SO_2Cl; i. $LiAlD_4$

As an alternative approach to stereoselectivity, the use of chiral acetal templates has been reported recently (Y.Yamamoto, S.Nishii and J.-I Yamada, J.Am. Chem. Soc., 1986, **108**, 7116; Y.Yamamoto *et al.*, J.Chem. Soc. Perkin Trans. I, 1991,3253). This illustrates that the stereoselective synthesis of these complicated steroids remains a challenge for the organic chemist.

6. Synthesis of vitamin D and analogues

(a) Introduction

The discovery in the late 1960's, that vitamin D is only a prodrug, which has to be activated first in the liver by hydroxylation at C-25 then in the kidney by 1α-hydroxylation (H.F.DeLuca, Endocrinology, 1992, **13**, 1763), stimulated the development of new synthetic routes to the active steroids. Subsequently it was found that calcitriol not only stimulates intestinal calcium transport and bone calcium mobilization, but also plays a role in cell differentiation, the immune system and reproduction (L.Binderup, Biochem. Pharmacol., 1992, **43**, 1885). In view of this wide range of activities hundreds of analogues have been synthesized and tested in order to find selective compounds, which may be used therapeutically.

(b) Partial syntheses starting with cholesterol or ergosterol

D.H.R.Barton et al. (J. Am. Chem. Soc., 1973, **95**, 2748) developed

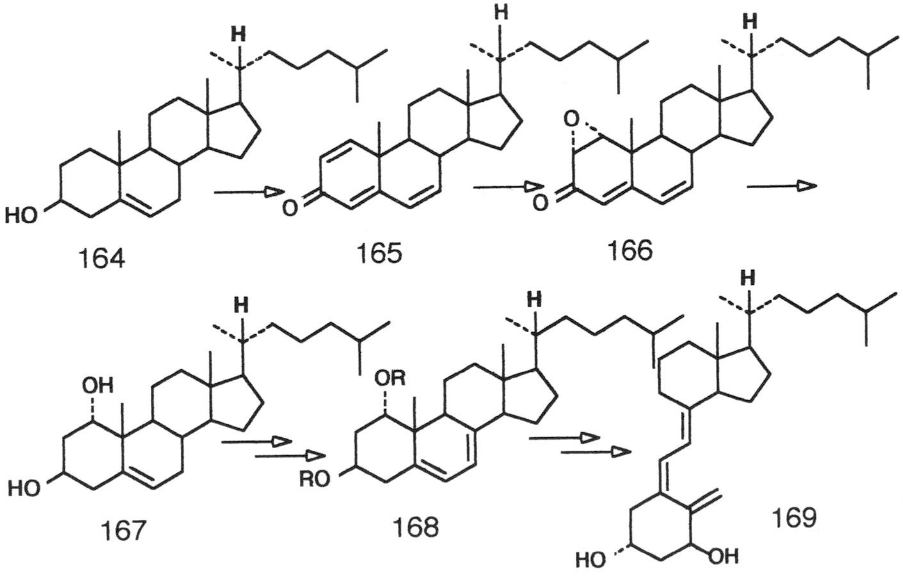

Scheme 26 Synthesis of 1-hydroxy metabolite of vitamin D

an interesting method for the introduction of the 1α-hydroxyl group. The trienone **165**, prepared by oxidation of cholesterol **164** with dichlorodicyanoquinone, was converted into the 1α,2α-epoxide **166**. Treatment of this epoxide with large excesses each of lithium metal and ammonium chloride in ammonia-tetrahydro-furan under reflux, led to the 1α-hydroxy derivative **167**, which was converted into the vitamin D metabolite **169**. The overall yield for the introduction of the 1α-hydroxyl group was 27%.

An alternative route is based upon the deconjugation of the trienone **170** either via kinetic protonation of the enolate (D.W.Guest and D.H.Williams, J. Chem. Soc. Perkin I, 1979, 1695) or via enolacetylation (A.Emke et al., J. Chem. Soc. Perkin I, 1977, 821; S.S.Toh et al., Steroids, 1991, **56**, 30). Reduction gives the 3ß-alcohol **173**. Its 5,7-diene system is protected by formation of

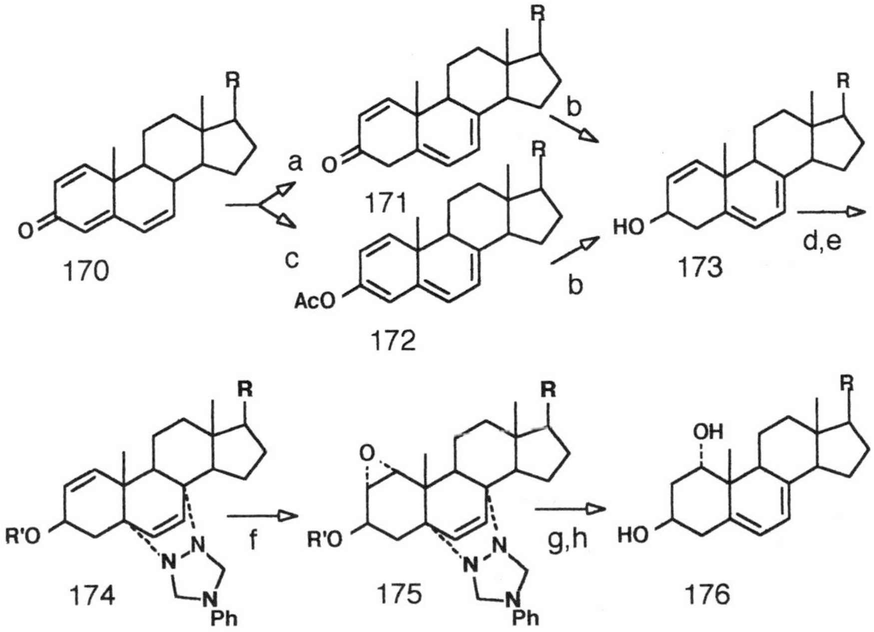

Scheme 27 Alternative route to 1-hydroxy metabolites
Reagents: a.NaOCH₃/ DMSO then CH₃COOH; b. CaBH₄;
c. CH₂=C(CH₃)OCOCH₃; d.4-phenyl-1,2,4-triazoline-3,5-dione;
e.t-C₄H₉(CH₃)₂SiCl; f.m-ClC₆H₄CO₃H; g. CH₃COOH; h.LiAlH₄

the Diels-Alder adduct with 1-phenyl-1,2,4-triazoline-3,5-dione. In order to achieve the desired stereoselectivity of the epoxidation, the 3ß-hydroxy group is first converted into the dimethyl-t-butylsilyl ether. After the epoxidation the protecting group is removed. Reduction with lithium aluminium hydride gives the 1α-hydroxyl group with regeneration of the 5,7-diene system.

Y.Tachibana (Bull. Chem. Soc. Jpn., 1988, **61**, 3915) used this approach for the synthesis of the 1α-hydroxy metabolite of ergocalciferol [R = CH(CH3)CH=CHCH(CH3)CH(CH3)2] (vitamin D2). It was then found that the 22,23-double bond of the silyl ether **174** is more reactive than the 1,2-double bond so that epoxidation leads to an isomeric mixture of diepoxides. At the end of the reaction sequence the 22,23-double bond was regenerated by deoxygenation of the 22,23-epoxides with sodium iodide and trifluoroacetic anhydride.

The 25-hydroxyl group may be introduced directly into cholest-

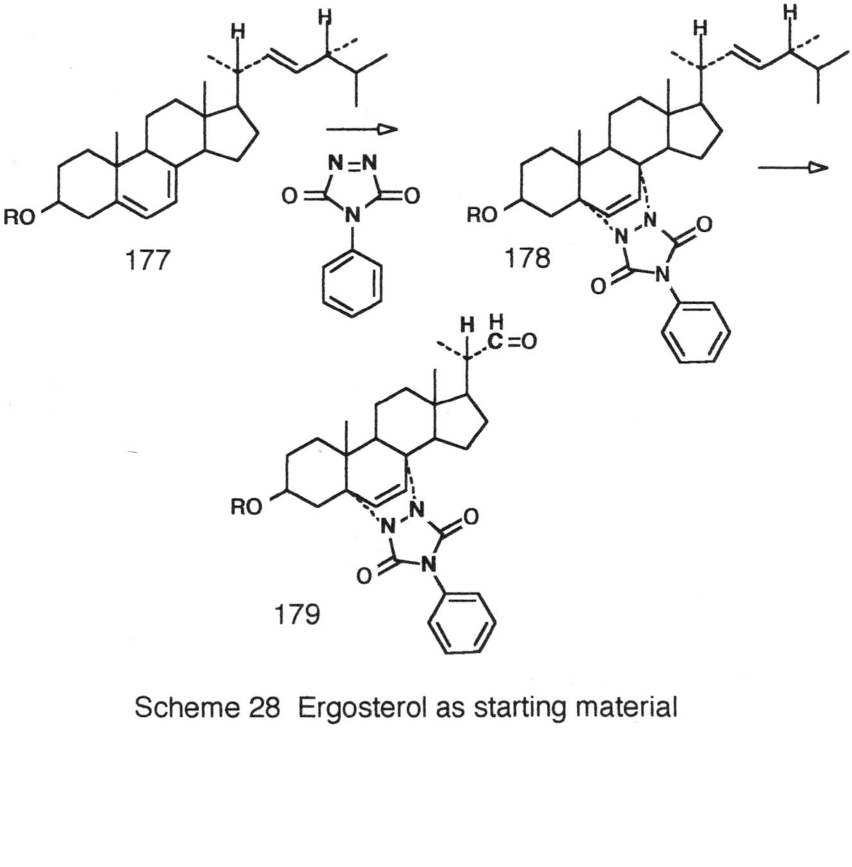

Scheme 28 Ergosterol as starting material

anes through oxidation with dimethyldioxirane (P.Bovicelli *et al.*, J. Org. Chem., 1992, **57**, 5052) but usually the 25-hydroxy side-chain is introduced separately.Ergosterol **177** (R = H) is often used as starting material for the synthesis of vitamin D derivatives with a modified side-chain. Protection of the 5,7-diene as the Diels-Alder adduct **178** and ozonolysis gives the aldehyde **179**, which is a good starting material for the construction of the desired side-chains. Regeneration of the 5,7-diene system, irradiation and thermal isomerization provides the vitamin D derivatives (K.Katsumi *et al.*, Chem.Pharm. Bull., 1984, **32**, 3744; 1987, **35**, 970). Similarly, the 1α-hydroxy derivative of ergosterol has been used for the synthesis of analogues (A,Kutner *et al.*, Tetrahedron Lett., 1987, **28**, 6129). M.Tsuji *et al.* (Bull. Chem. Soc. Jpn., 1990, **63**, 2233) prepared the iodide **181** (R = COCH$_3$) by reduction of the aldehyde **180** followed by tosylation and iodide exchange. The acetate groups were removed by saponification and replaced by

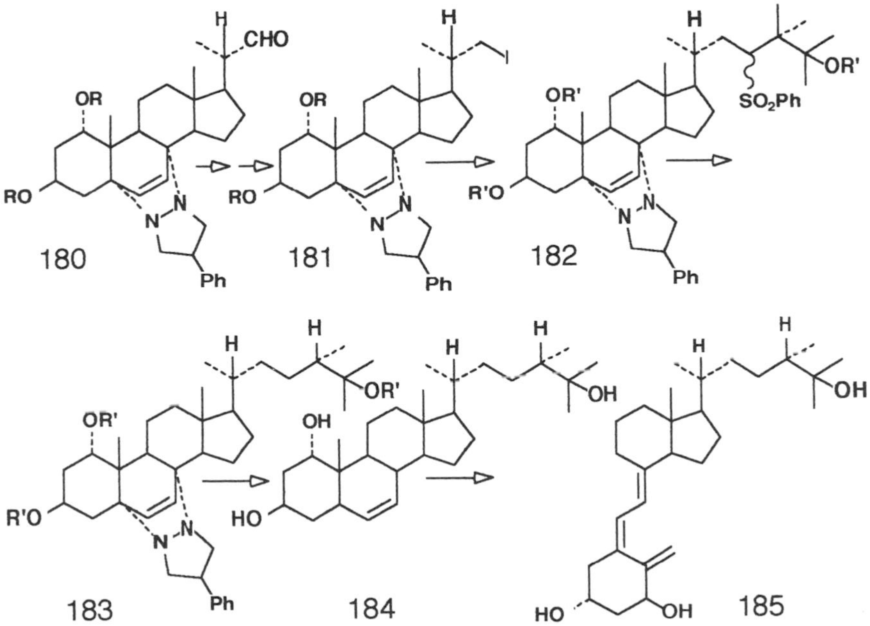

Scheme 29 Synthesis of the 1,25-dihydroxy-22,23-dihydro-derivative of vitamin D$_2$

tetra-hydropyranyl protecting groups. The resulting iodide was coupled with the anion of (3R)-3-phenylsulfonyl-2,3-dimethylbutan-2-ol tetrahydropyranylether to give the sulfone **182** (R' = THP). Reductive desulfonylation with sodium amalgam in buffered methanol provided the desired side-chain. Acid treatment removed the tetrahydropyranyl protecting groups and reduction with lithium aluminium hydride regenerated the 5,7-diene system. Finally the vitamin D derivative was obtained by irradiation of the triol **184** and thermal isomerization of the resulting previtamin.

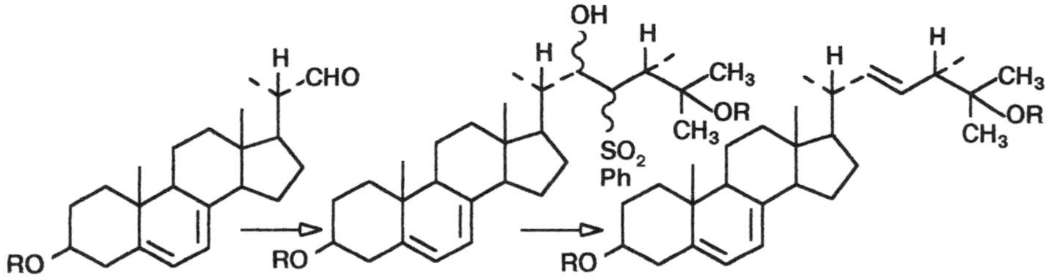

Scheme 30 Synthesis of derivatives with a 22,23-double bond

Derivatives with a 22,23-double bond in the side-chain may be prepared by coupling of a C-22 aldehyde with a sulfone followed by reduction with sodium amalgam to generate the double bond (T.Taguchi *et al.*, Tetrahedron Lett., 1988, **29**, 227; M.Tsuji, S.Yokoyama and Y.Tachibana, Bull. Chem. Soc, Jpn., 1989, **62**, 3132; M.Chodynski and A.Kutner, Steroids, 1991, **56**, 311; K.L.Perlman, H.K.Schnoes and H.F.DeLuca, J. Chem. Soc. Chem. Commun., 1989, 1113). For example S.Yamada *et al.* (Tetrahedron Lett., 1984, **25**, 3347) coupled the aldehyde **186** (R = THP) with the anion of (3R)-3-phenylsulfonyl-2,3-dimethylbutan-2-ol tetrahydropyranyl ether whereby a mixture of ß-hydroxysulfones **187** was formed. Reduction of these ß-hydroxysulfones yielded the desired side-chain derivative **188** with a 22E-double bond. Similarly, side-chain homologues have been synthesized starting from a C-24 aldehyde prepared from hyodeoxycholic acid (Y.Kobayashi *et al.*, Chem. Pharm. Bull., 1982, **30**, 4297; 1988, **36**, 4144).

When discussing the synthesis of brassinosteroids, the use of chirality transfer for the synthesis of steroids with a 22,23-double bond and a chiral centre at C-24 of the side-chain was discussed (scheme 24). This same approach has also been used for the synthesis of vitamin D derivatives (F.J.Sardina, A.Mouriño and L.Castedo, Tetrahedron Lett., 1983, **24**, 4477; M.M.Midland and Y.C.Kwon, Tetrahedron Lett., 1985, **26**, 5017; I.Horibe et al., J. Chem. Soc. Perkin Trans. I, 1989, 1957).

(c) Microbiological route to 1α-hydroxy derivatives

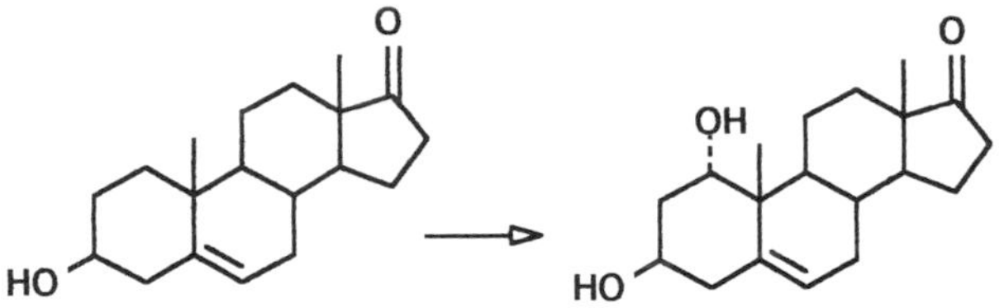

189   190

Scheme 31 Microbiological introduction of a 1-hydroxyl group with *Penicillium sp*

1α,3β-Dihydroxyandrost-5-en-17-one **190**, prepared by microbiological hydroxylation of 3β-hydroxyandrost-5-en-17-one **189** (R.M.Dodson, A.H.Goldcamp and R.D.Muir, J.Am.Chem.Soc., 1960, **82**, 4026), proved to be a versatile intermediate for the synthesis of metabolites and analogues of vitamin D. The desired side-chain can be built up first, followed by the introduction of the 5,7-diene system and irradiation (A.D.Batcho, D.E.Berger and M.R.Uskokovic, J. Am. Chem. Soc., 1981, **103**, 1293; M.Ohmori et al., Tetrahedron Lett., 1982, **23**, 4709; 1986, **27**, 71) but the diene system can also be introduced first (E.Murayama et al., Chem. Pharm. Bull. 1986, **34**, 4410). For example, Wittig reaction of the 17-ketone **191** (R = t-butyl dimethylsilyl) produced the (Z)-ethylidene derivative **192**, which in an ene reaction with paraformaldehyde in the presence of borontrifluoride-etherate stereoselectivily gave the 20(S)-alcohol **193**. Selective hydrogenation of the Δ16-double bond yielded the alcohol **194** with the correct stereochemistry at C-17. Introduction of the 7,8-double bond via allylic bromination and dehydrobromination produced the diene **195,** which may be used for the introduction of the desired side-chain

(K.Konno *et al.*, Chem. Pharm. Bull., 1992, **40**, 1120). The oxa deri-vative **196**, for example, was formed via the mesylate ester (N.Kubodera *et al.*, Chem. Pharm. Bull., 1991, **39**, 3221).

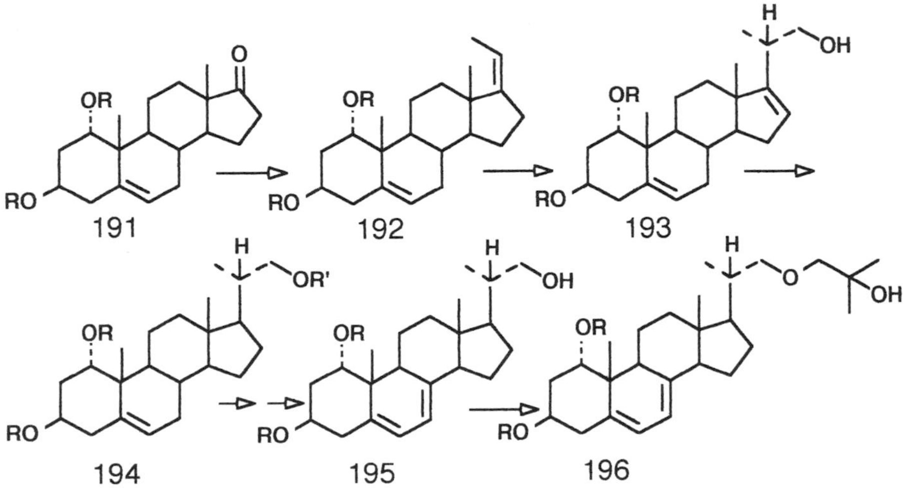

Scheme 32 1,3-Dihydroxyandrost-5-en-17-one as starting material for the synthesis of vitamin D derivatives

(d) Introduction of the 1α-hydroxyl group in the secotriene stage

Several groups have explored a completely different synthetic scheme, whereby the introduction of the 1 α-hydroxyl group comes after the difficult irradiation-isomerization sequence. H.E.Paaren, H.F.DeLuca and H.K.Schnoes (J.Org.Chem., 1980, **45**, 3253) con-verted the triene **198** into the 3,5-cycloderivative. The 1α-hydroxyl group was introduced by allylic oxidation with selenium dioxide and t-butylhydroperoxide. The triene group was regenerated by treatment with glacial acetic acid, followed by hydrolysis of the acetate thus formed. Reported overall yields for the introduction of the 1α-hydroxyl group are 15-20%. This reaction sequence has been used with a variety of side-chains and has also been applied for the synthesis of 19-nor- and thio-derivatives (R.P.Esvelt *et al.*, J. Org. Chem., 1981, **46**, 456; A.Kutner *et al.*, J. Org. Chem., 1988, **53**, 3450; R.M.Moriarty, J.Kim and R.Penmasta, Tetrahedron Lett., 1992, **33**, 3741; K.L.Perlman *et al.*, Tetrahedron Lett., 1990, **31**,

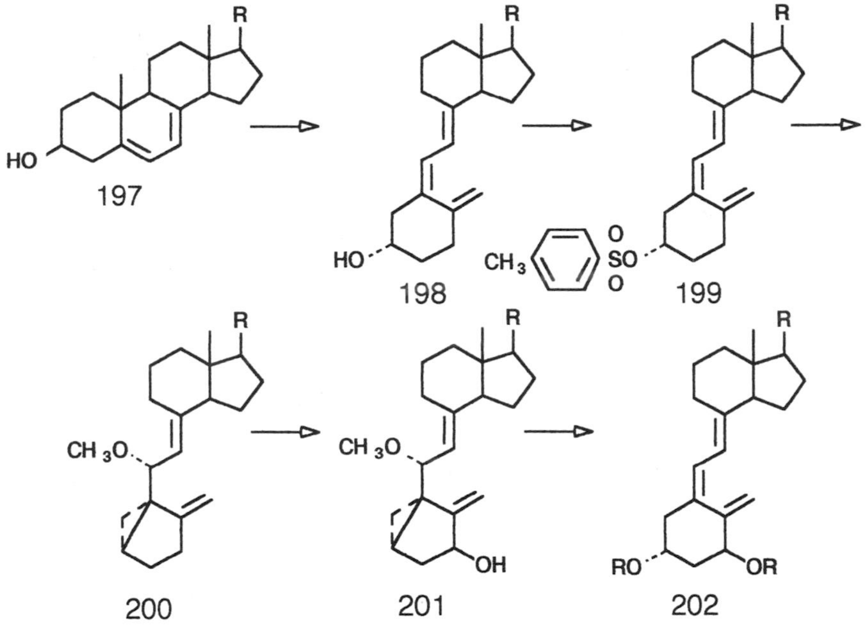

Scheme 33 Introduction of the 1-hydroxyl group via allylic oxidation

1823; B.R.de Costa, S.A.Holick and M.F.Holick, J. Chem. Soc. Chem. Commun., 1989, 325).

Direct introduction of a 1α-hydroxyl group is possible by allylic oxidation of the 5,6-cis isomers. This forms the basis for a versatile approach to vitamin D derivatives (D.R.Andrews *et al.*, J. Org. Chem., 1986, **51**, 4819). Conversion of ergocalciferol **203** with sulphur dioxide gives an isomeric mixture of adducts **204**. Due to the strong electron-withdrawing properties of the sulphur dioxide moiety, the 7,8- and 5,10-double bonds are inert to ozonolysis, so that selective cleavage of the side-chain is possible. The aldehyde **205** formed may be used for the introduction of a modified side-chain (M.J.Calverley, Tetrahedron Lett., 1987, **28**, 1337; Tetrahedron 1987, **43**, 4609) or may be reduced to the corresponding alcohol and protected (K.Ando *et al.*, Chem. Pharm. Bull., 1992, **40**, 1662). Thermolysis gives the 5,6-cis isomer **207**, which is oxidized to the 1α-hydroxyl derivative **208**. The 5,6-trans isomer **210** is formed by photoisomerization. This approach has also

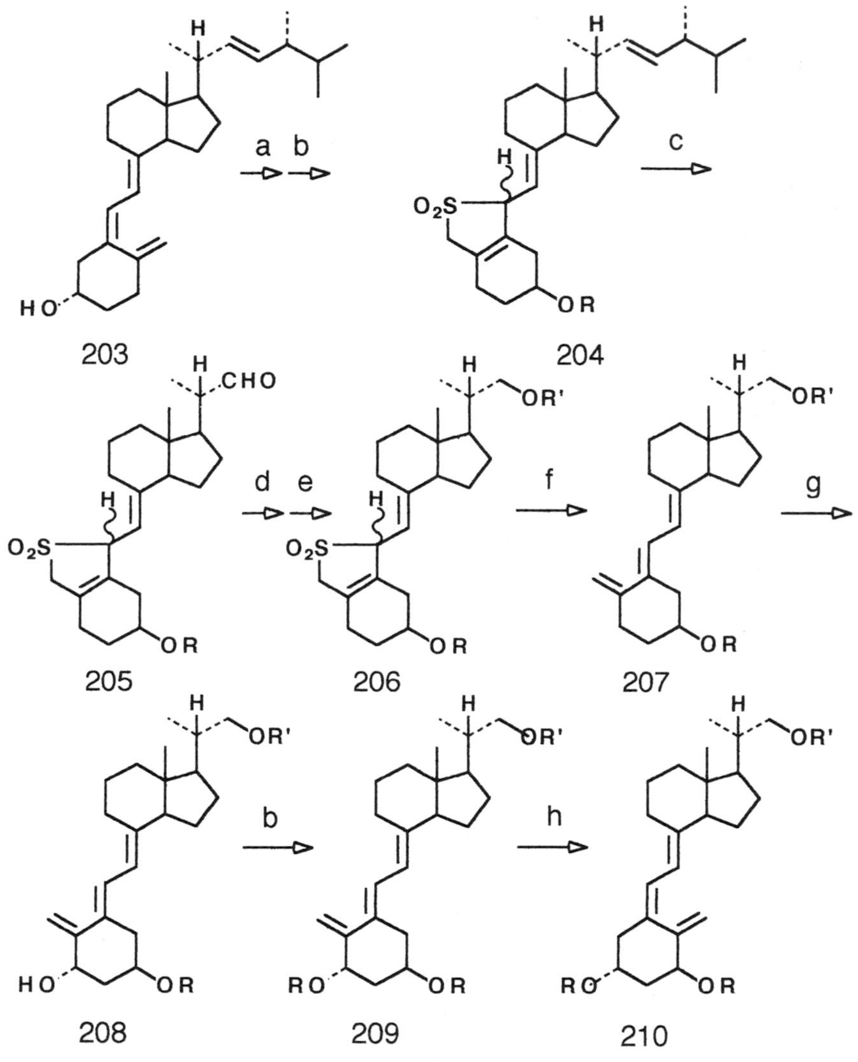

Scheme 34 Use of 5,6-cis derivatives for introduction of the
1-hydroxyl group. Reagents: a.SO_2; b. $(C_2H_5)_3SiCl$/imidazole;
c. O_3-$(C_6H_5)_3P$; d. $NaBH_4$; e. p-$CH_3C_6H_4SO_2Cl$/pyridine;
f. thermolysis; g. SeO_2/N-methylmorpholine N-oxide;
h. photoisomerization

been used for the synthesis of deuterated vitamin D derivatives (S.Yamada et al., Steroids, 1989, **54**, 145; R.Ray et al., Steroids, 1992, **57**, 142).

The following scheme illustrates the use of Ziegler bromination for the introduction of the 1α-hydroxyl group (W.Nerinckx *et al.*, Tetrahedron, 1991, **47**, 9419). Low temperature irradiation of provitamin D3 **211** gave the previtamin **212**, which was converted into the Diels-Alder adduct **213**. Allylic bromination gave in >90% yield the pure 1α-bromide. After protection of the 3β-hydroxyl group as a t-butyldimethylsilylether, treatment with mercuric acetate in methylene dichloride gave the acetate **215** with retention of configuration. Deprotection and isomerization then gave the 1α-hydroxyderivative of vitamin D3.

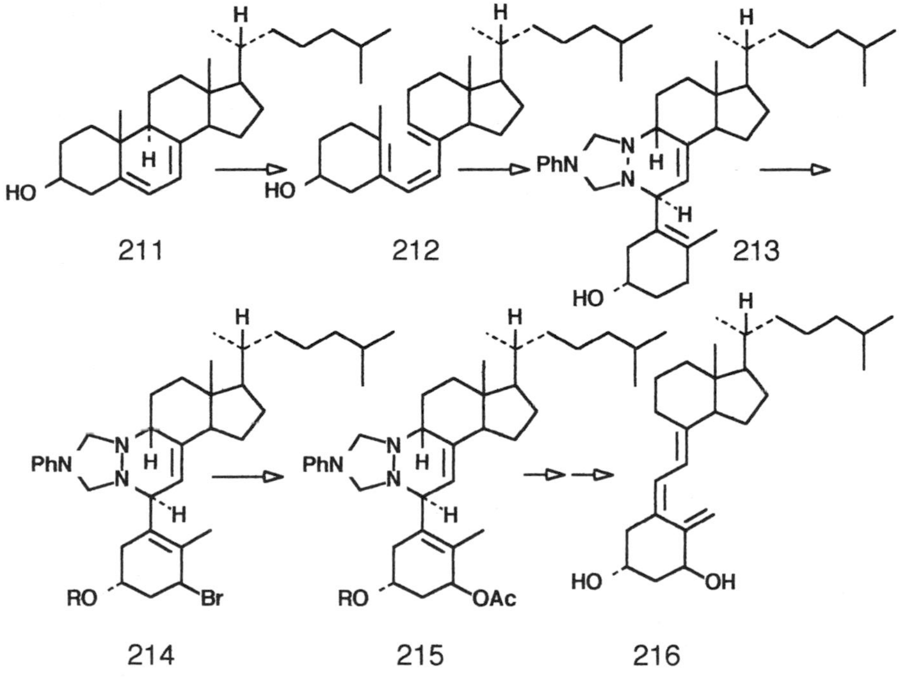

Scheme 35 Use of Ziegler bromination for the introduction
of a l-hydroxyl group

(e) Convergent syntheses of vitamin D and analogues

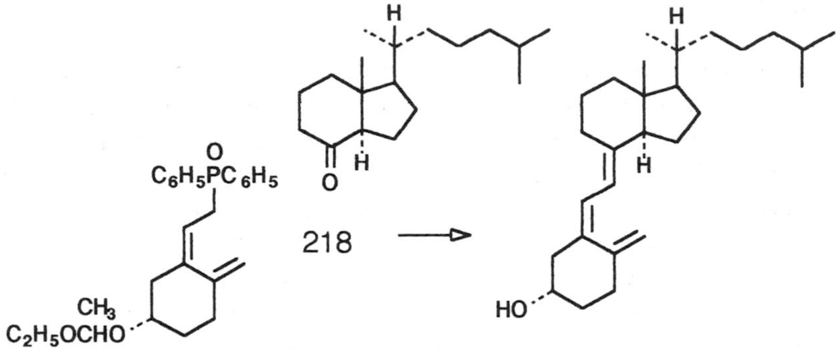

Scheme 36 Lythgoe synthesis of calciferol

A review (B.Lythgoe, Chem. Soc. Rev., 1980, **9**, 449) describes the pioneering work of Lythgoe and coworkers in this area of research. Many routes had to be explored in order to find a stereoselective synthesis, which gives vitamin D in a reasonable yield. The first route is based upon a Wittig-Horner coupling of the A-ring synthon with the C/D-ring frament. Coupling of the allylic phosphine oxide **217** with the ketone **218** gave, after hydrolysis of the protecting group, calciferol **219** in *ca* 60% yield (B.Lythgoe *et al.*,Tetrahedron Lett., 1975, 3863; J. Chem. Soc. Perkin I, 1978, 591). The original (Z)-allyl geometry of the phosphine oxide was completely preserved in the 5,6-double bond of the product, and the new 7,8-double bond was formed with exclusive (E)-geometry.This route has been used succesfully for the synthesis of a large number of vitamin D derivatives, such as the 25-hydroxy- and 1,25-dihydroxy-derivatives of calcifereol and ergocalciferol (F.J.Sardina, A.Mouriño and L.Castedo, J. Org. Chem., 1986, **51**, 1264; E.G.Baggiolini *et al.*, J. Org. Chem., 1986, **51**, 3098; J.Kiegiel, P.M.Wovkulich and R.Uskokovic, Tetrahedron Lett., 1991, **43**, 6057; J.L.Mascareñas *et al.*, J. Org. Chem., 1986, **51**, 1269, Tetrahedron Lett., 1991, **32**, 2813), the 19-nor- and 8(14)a-homo-derivatives of calcitriol (K.L.Perlman *et al.*, Tetrahedron Lett., 1991, **32**, 7663; 1992, **33**, 2937; A.Steinmeyer *et al.*, Steroids, 1992, **57**, 447), a series of 1,

9, 11 or 18-substuted analogues (G.H.Posner *et al.*, J. Med. Chem., 1992, **35**, 3280;W.G.Dauben and L.J.Greenfield, J. Org. Chem., 1992, **57**, 1597; R.Bouillon *et al.*, J. Biol. Chem., 1992, **267**, 3044; D.F.Maynard, A.W.Norman and W.H.Okamura, J. Org. Chem., 1992, **57**, 3214), 25-hydroxydihydrotachysterol (M.A.Maestro, L.Castedo and A.Mouriño, J. Org. Chem., 1992, **57**, 5208; J.C.Hanekamp *et al.*, Tetrahedron, 1992, **48**, 9283), phospha- and fluoro-derivatives of calcitriol (W.G.Dauben *et al.*, Tetrahedron Lett., 1989, **30**, 677; K.Iseki, T.Nagai and Y.Kobayashi, Chem. Pharm. Bull., 1992, **40**, 1346; Y.Ohira *et al.*, Chem. Pharm. Bull., 1992, **40**, 1647).

Several interesting syntheses for the A-ring synthons have been developed, mainly starting from natural products such as (R)-(-)-carvone (S.Hatakeyama *et al.*, J. Org. Chem., 1989, **54**, 3515), (-)-quinic acid (D.Desmaele and S.Tanier, Tetrahedron Lett., 1985, **26**, 4941) or (S)-(+)-carvone (E.G.Baggiolini et al., Tetrahedron Lett., 1987, **28**, 2095; L.Castedo, J.L.Mascareñas and A,Moriño, Tetrahedron Lett., 1987, **28**, 2099; J.M.Aurrecoechea and W.H.Okamura, Tetrahedron Lett., 1987, **28**, 4947). Asymmetric reaction conditions (K.Nagasawa *et al.*, Tetrahedron Lett., 1991, **32**, 4937; B.Fernández *et al.*, J. Org. Chem., 1992, **57**, 3173) or enzymatic resolution (S.Kobayashi *et al.*, Tetrahedron Lett., 1990, **31**, 1577) have also been used to obtain the desired chiral synthons.

Similar approaches have been used for the synthesis of the C/D-ring synthons (T.Takahashi, *et al.*, Tetrahedron Lett., 1985, **26**, 4463; W.S.Johnson, J.D.Elliott and G.J.Hanson, J. Am. Chem. Soc., 1984, **106**, 1138; W.J.Johnson and M.F.Chan, J. Org. Chem., 1985, **50**, 2598; R.V.Stevens and D.J.Lawrence, Tetrahedron, 1985, **41**, 93; J.H.Hutchinson and T.Money, J. Chem. Soc. Chem. Commun., 1986, 288; L.Castedo *et al.*, Tetrahedron Lett., 1987, **28**, 4589; S.Hatakeyama *et al.*, J. Chem. Soc.Chem. Commun., 1989, 1893; H.Nemeto, M.Ando and K.Fukumoto, Tetrahedron Lett., 1990, **31**, 6205; S.R.Wilson, A.E.Davey and M.E.Guazzaroni, J. Org. Chem., 1992, **57**, 2007; Y.Fall *et al.*, Tetrahedron Lett., 1992, **33**, 6683).

Another route to vitamin D and derivatives proceeds via the previtamins (R.G.Harrison, B.Lythgoe and P.W.Wright, J. Chem.

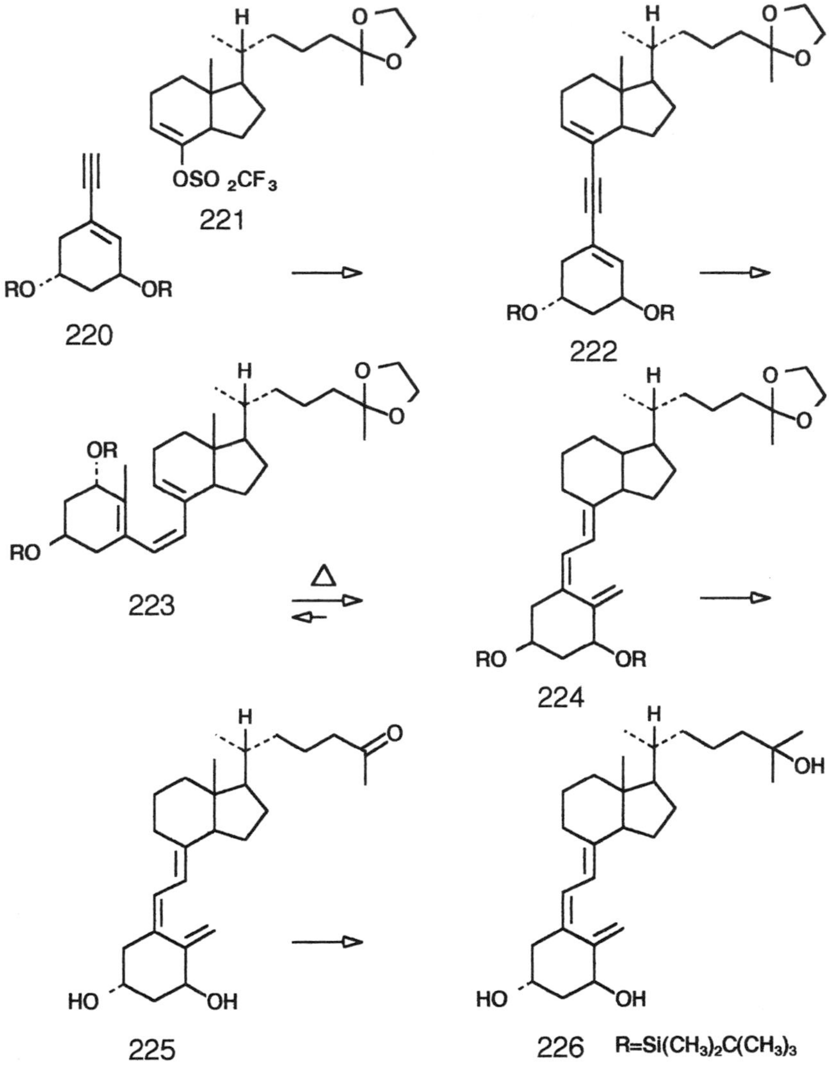

Scheme 37 Convergent synthesis of a previtamin and
conversion to calcitriol

Soc. Perkin I, 1974, 2654). Palladium-catalyzed coupling
(L.Castedo, L.Mouriño and L.A.Sarandeses, Tetrahedron Lett.,
1986, **27**, 1523) of the enyne **220** with the vinyl triflate **221** gave

the dienyne **222** in 91% yield. Semi-hydrogenation and thermal isomerization gave a mixture of the previtamin **223** and the vitamin **224** in a ratio of approximately 12:88 (combined yield 94%). Removal of the protecting group and reaction with methyl lithium gave calcitriol **226** in 65% yield (J.L.Mascareñas *et al.*, Tetrahedron, 1991, **47**, 3485).

This route has been used for the synthesis of a number of vitamin D derivatives, such as derivatives with a heterocyclic-, nor- or homo-A-ring (S.A.Barrack, R.A.Gibbs and W.H.Okamura, J. Org. Chem., 1988, **53**, 1790; A.S.Lee, A.W.Norman and W.H.Okamura, J. Org. Chem., 1992, **57**, 3846; J.D.Enas, G.-Y.Shen and W.H.Okamura, J. Am. Chem. Soc., 1991, **113**, 3873; J.D.Enas, J.A.Palenzuela and W.H.Okamura, J. Am. Chem. Soc., 1991, **113**, 1355), 19-nor-, C-11- or C-18-substituted derivatives (L.A.Sarandeses *et al.*, Tetrahedron Lett., 1992, **33**, 5445; M.Torneiro *et al.*, Tetrahedron Lett., 1992, **33**, 105; M.J.Vallés, L.Castedo and A.Mouriño, Tetrahedron Lett., 1992, **33**, 1503) and analogues with a modified side-chain (B.Figadère *et al.*, J. Med. Chem., 1991, **34**, 2452; A.S.Craig, A.W.Norman and W.H.Okamura., J. Org. Chem., 1992, **57**, 4374).

7. Biosynthesis of the steroids

In 1971 (vol IIE), the biosynthesis of the steroids was reviewed and the overall scheme seemed well established. Recently the bio- cyclization of 2,3-oxidosqualene **227** to lanosterol was reinvestigated. In the classical scheme it had been assumed that the enzymatic cyclization of 2,3-oxidosqualene **227** first gave the cation **228**, followed by a 120⁰ rotation of the side-chain to give the cation **229**. Lanosterol **230** was then formed through a series of migrations and the release of a proton. It remained hard to visualize this large rotation of the side-chain in the active site of the enzyme. A.Krief *et al.* (J. Am. Chem. 1987, **109**, 7910) prepared the 2,3-oxidosqualene analogue **231** with unnatural (Z)-stereo- chemistry at the 18,19-double bond, where cyclization should have led directly to the cation **229**, but enzymatic cyclization of that analogue yielded tricyclic products only. E.J.Corey, S.C.Virgil and S.Sarshar (J. Am. Chem. Soc., 1991, **113**, 8171) studied the enzymatic cyclization of the analogue **232**, where the cyclized

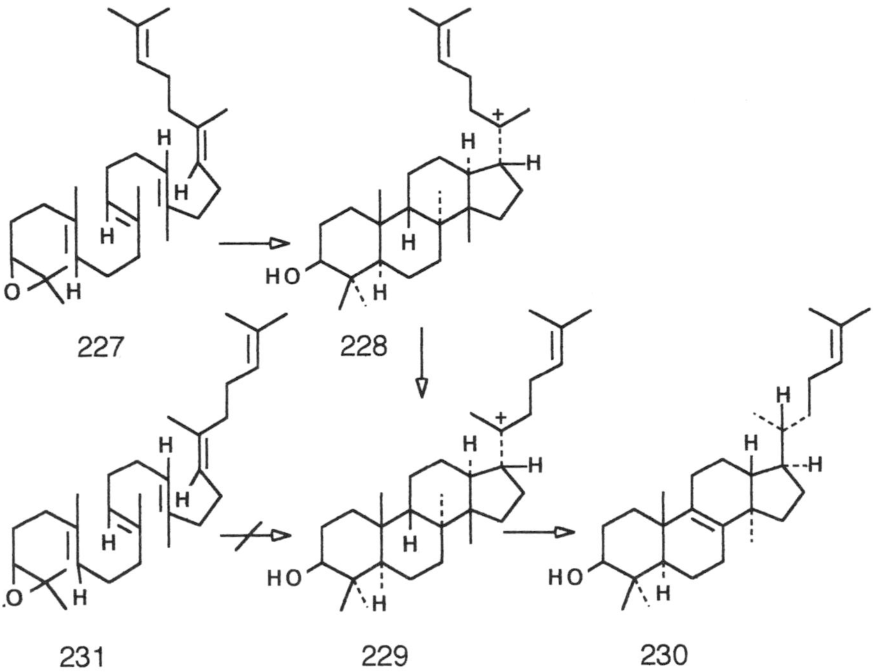

Scheme 38 The classical scheme of sterol biosynthesis

cation **233** is stabilized by conjugation. It was then found that the first product in the enzymatic cyclization has a 17β-side-chain and not a 17α-side-chain, as had been assumed previously. Cyclization of the 20-oxa analogue of 2,3-oxidosqualene gave a similar result (E.J.Corey and S.C.Virgil, J. Am. Chem. Soc., 1991, **113**, 4025). These results demonstrate that the substrate is folded on the enzyme in such a manner that the side-chain occupies a 17β-like position. In the following series of 1,2 shifts the leaving 17α-hydrogen and the incoming 13α-hydrogen are in a cis relationship. Steric factors seem to play a role in this folding. The methyl group at C-10 of 2,3-oxidosqualene , for example, proved to be crucial for correct folding of the substrate (E.J.Corey et al., J. Am. Chem. Soc., 1992, **114**, 1524). However, recent work of W.S.Johnson and his group (W.S.Johnson, Tetrahedron, 1991, **47**, xi) demonstrated that electronic factors are equilly important. These interesting results demonstrate that study of steroid biosynthesis is still an active

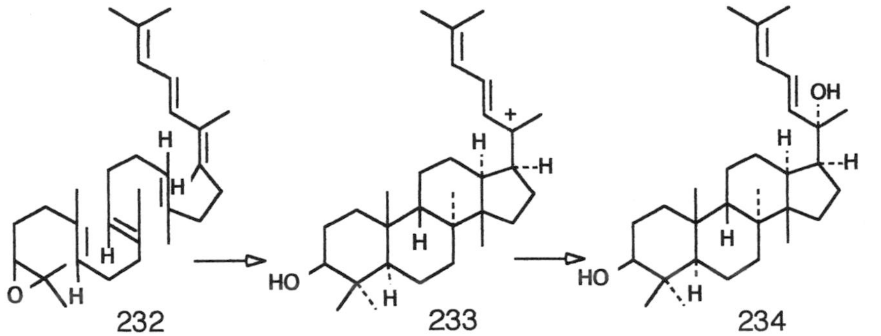

Scheme 39 Enzymatic cyclization of a dehydro analogue
of 2,3-oxidosqualene

research field. ^{13}C-NMR spectroscopic analysis of [^{13}C]- and
[^{13}C^{2}H]- labelling patterns of the products formed, has become a
very powerful tool for these studies (S.Seo *et al.*, J. Chem. Soc.
Perkin Trans. I, 1991, 2065).

Second Supplements to the 2nd Edition of Rodd's Chemistry of Carbon Compounds, Vol. II B(Partial), C, D and E, edited by M. Sainsbury   509

Chapter 16

GLYCOSIDES, SAPONINS AND SAPOGENINS

S. B. MAHATO

Glycosides are sugar conjugates of aglycones of various types and these are classified according to the nature of the aglycone. The saponins are naturally occurring glycosides which have the property of forming a soapy lather when shaken with water and producing haemolysis when water solutions are injected into the blood stream. Several types of the glycosides, e.g. cardiac glycosides, cucurbitacins etc. possess these properties, but they are classified separately because of their specific biological activities. The glycoalkaloids possessing a basic steroidal aglycone share many physical, physicochemical and physiological properties with saponins, but these are also usually considered separately. The steroid saponins and sapogenins have been reviewed by Elks in this series (Vol.IIE, 1971, P.1; supplement to part IIE, 1974, P.205). The present article covers literature from 1973 to 1991 and describes recent developments with regard to isolation, chemistry and biosynthesis of steroid and triterpenoid saponins and sapogenins. The glycone part of these compounds are generally oligosaccharides, linear or branched, attached to a hydroxy or a carboxyl group or both. The sites of attachment may be one (monodesmosides), two (bisdesmosides) or three (tridesmosides).

Excellent general reviews on saponins have been published during the period by Kawasaki (Method. Chim., 1978, **11**, 87), Tschesche and Wulff (Fortschritte der chemie organischer Naturstoffe, 1973, **30**, 461), Tanaka and Kasai (ibid., 1984, **46**, 1), Rastogi et al. (Phytochemistry, 1967, **6**, 1249; ibid., 1974, **13**, 2623; ibid., 1980, **19**, 1889) and Mahato et al. (Phytochemistry, 1982, **21**, 959; ibid., 1988,

27, 3037; ibid, 1991, **30**, 1357). The chemistry and biological significance of saponins in food and feeding-stuffs have been reviewed by Price et al. (CRC Crit. Rev. Food Sci. Nutr. 1987, **26**, 27). Ginseng saponins have been reviewed by Pleinard (Planta Med. Phytother. 1979, **13**, 4) and Shibata et al. (Economic and Med. Plant Res. 1985, **1**, 155 , Academic Press). The biological activity of triterpenoid saponins have been described by Adler and Hiller (Pharmazie, 1985, **40**, 676; ibid, 1982, **37**, 619).

1. ISOLATION

The classical methods of isolation of saponins are found to be inadequate for the separation of complex saponin mixtures into individual components. The advent of recent chromatographic techniques have provided valuable means for the isolation of pure saponins or their derivatives. Although the integration of silica gel column chromatography, semi-preparative HPLC and repeated preparative TLC may yield expected separation of saponins in most cases, special isolation techniques have been adopted in particular situations(M. Adinolfi et al. Can. J. Chem. 1988, **66**, 2787). The technique of droplet counter-current chromato-graphy (DCCC) has been developed and described by Hostettmann et al. (J. Chromatogr. 1979, **170**, 355; Planta Med. 1980, **39**, 1; Adv. Med. Plant Res. 1984, 225; Nat. Prod. Rep. 1984, **1**, 471). The isolation of a new bisdesmoside of quinovic acid from the roots of Guettarda platypoda was achieved by HPLC and DCCC (R. Aquino et al. Phytochemistry, 1988, **27**, 2927). The mixture of steroid saponins from Tribulus terrestris was separated by HPLC on a column of μ-Bondapak C_{18} using methanol as the mobile phase (S.B. Mahato et al. J. Chem. Soc. Perkin. I 1981, 2405). A reversed phase S-10-ODS column was used with solvent system methanol-water (7:3) for the separation of a number of saiko-saponins-like triterpenoid glycosides, e.g. corchorusins from Corchorus acutangulas (S.B. Mahato et al. J. Chem. Soc. Perkin I, 1987, 629; Phytochemistry, 1988, **27**, 1433).

The isolation of genuine saponins sometimes necessitates pretreatment of plant material before subjecting it to the usual extraction and isolation procedures. For example, mormordin I and mormordin II have been isolated from the roots of <u>Momordica Cochinchinensis</u> (N. Kawamura et al. Phytochemistry, 1988, **27**, 3585). It was found that mormordin II which is a bisdesmoside is converted to the monodesmoside momordin I by intracellular esterases present in the root during the drying process. This was proved by the observation that the bisdesmoside alone could be isolated when the fresh root was subjected to hydrochloric acid treatment followed by solvent extraction and chromato-graphic purification of the saponins. During the extraction of hebevinosides, the tetracyclic triterpenoid saponins from <u>Hebeloma vinosophyllum</u> by Fujimoto et al. (Chem. Pharm. Bull. 1986, **34**, 88; ibid, 1987, **35**, 2254) with 90% methanol, it was observed that methylation of the allylic 7-hydroxy group occurred leading to confusion regarding the identity of the aglycone. This ambiguity was resolved by maintaining the pH of the extracting solution at 7 by occasional addition of a small amount of pyridine during extraction. Now only the genuine saponins containing a 7-hydroxy group could be exclusively isolated. Acylated triterpenoid saponins occur in some plants as rather inseparable mixtures; however, if the acyl groups are cleaved by treatment of the saponin mixture with $NaHCO_3$ solution, it is usually possible to separate them by standard procedures. Thus, Higuchi et al. (Phytochemistry, 1987, **26**, 229) isolated two desacyl saponins by treatment of the saponin mixture obtained from the bark of <u>Quillaza saponaria</u> with 6% $NaHCO_3$ solution and subsequent chromatographic separation. The following method proved to be useful for isolation and separation of Ginseng saponins : column chro-matography of the aquous suspension of the McOH extract on Amberlite XAD-2, Diaion MCl-gel HP-20 or Kogel B-G 4600 using water and methanol (or aquous methanol) as subsequent developing solvents. This furnished mainly saponins which could then be separated into individual constituents by HPLC or MPLC (O.Tanaka et al., Abstracts of Chinese Medicines, 1986, **1**, 130).

(a) Structure elucidation

The conventional method of structure elucidation of

512

saponins starts with acid hydrolysis which yields the aglycone and the sugar moieties which are investigated separately. In the classical method, the structure of the
sugar moieties of the saponins is determined by
identification of the monosaccharides obtained on acid
hydrolysis by PC and GLC, quantitative determination of
monosaccharides by GLC, partial hydrolysis followed by
isolation and characterization of prosapogenins and also,
where possible, by characterization of oligosaccharides.
The points of attachment of different sugar units are
revealed by permethylation of the saponins followed by
hydrolysis or methanolysis and identification of the
methylated sugars by PC or GLC. The mode of sugar linkage
in saponins is determined by enzymic hydrolysis with α-and
β-glycosidases or by the application of Klyne's rule (W.
Klyne, Biochem. J. 1950, 47, XLI) on molecular rotation
difference. However, both of these methods are not always
applicable, particularly in the case of complex glycosides.
A micromethod based upon split circular dichroism (CD)
curves of pyranose polybenzoates has been proposed by Liu
and Nakanishi (J. Amer. Chem. Soc. 1981, 103, 7005) for
determination of the glycosidic linkages at the branching
points. This method which has been applied to molluscidal
saponins, e.g. balanitin-1 (1) consists of permethylation

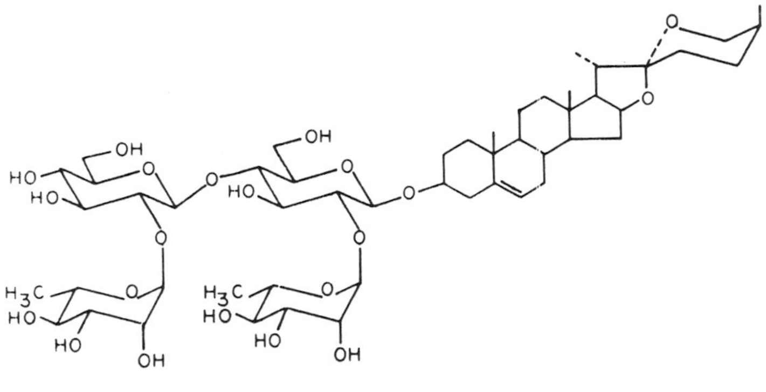

of the oligosaccharide, methanolysis and p-bromobenzoylation which converts the nonanomeric hydroxyl groups involved in glycosidic linkages into p-bromobenzoyl derivatives. The difference in CD values of the two extrema of split CD curves of the di- and tribenzoates derived from branched pyranose groups then establishes the spatial arrangement of benzoate groups and hence of the hydroxyl groups which were involved in the branching.

(b) Mass spectroscopy

The determination of structures of saponins by spectroscopic methods has the advantage that it examines the intact saponin prior to any treatment which might produce artifacts. There has been a significant increase over the last two decades in the application of NMR and mass spectroscopy. The newer techniques of mass spectroscopy particularly FD-MS, and FAB-MS are also of immense help in the determination of M_rs of saponins and the sequence of their sugar units. FD-MS of underivatized steroid and triterpenoid saponins have been reported (H.R. Schulten et al. Tetrahedron, 1977, **33**, 2595; ibid. 1978, **34**, 1003; T.Komori et al. Z. Naturforsch, Teil C 1979, **3/4**, 1094). The FD-MS of the saponins not only gives the M_r but also clear information with regard to the sequence of the sugar units in the molecule and their individual chemical structures by the fragment ions formed by the direct bond cleavages in the oligosaccharide moiety of the saponins. The new technique of plasma desorption mass spectrometry (PDMS)(R.D.Macfarlane and D.F.Torgerson, Science, 1976, **191**, 920; H. Kasai et al. J. Am. Chem. Soc. 1976, **98**, 5044) has also been successfully employed for M_r determination of underivatized steroidal saponins (K. Hostettmann et al. Helv. Chim. Acta 1978, **61**, 1990). The rapid sample heating (flash desorption) technique has been used for the structural analysis of α-hederin, a triterpenoid saponin (W.R. Anderson, Jr. et al. J. Am. Chem. Soc. 1978, **100**, 1974). The technique of fast atom bombardment (FAB) mass spectrometry has of late become one of the most powerful tools in the structural investigation of saponins (D.H. Williams et al. J. Am. Chem. Soc. 1981, **103**, 5700; C.Fenselau, J. Nat. Prod. 1984, **47**, 215). The FAB-MS and their

514

FAB-MIKE (mass analysed ion kinetic energy) spectra (R.G.Cooks and R.W. Kondat Analyt. Chem. 1978, **50**, 881) have been employed for the structural investigation of arvenososide A and arvenososide B, two new triterpenoid glycosides from the aerial parts of Calendula arvensis (R. Chemli et al. Phytochemistry, 1987, **26**, 1785). Secondary ion mass spectrometry (SI-MS) (H.Kambara and S.Hishida Org. Mass Spectrom. 1981, **16**, 167) has recently been proposed for the M_r determination and structural investigation of complex saponins. Kitagawa et al. (Chem. Pharm. Bull. 1988, **36**, 2819) successfully demonstrated the utility of this method in the structural investigation of three new bitter and astringent bisdesmosides acetyl-soyasaponins A_1, A_2 and A_3 isolated from American soyabean seeds (Glycine max). Other techniques which have potential for structural analysis of saponins include laser desorption-MS (R.B. Van Breemen et al. Biomed. Mass spectrom. 1984, **11**, 278), diethanolamine-assisted SI-MS (K.I.Harada et al. Org. Mass Spectrum. 1982,**17**, 386), HPLC-MS, liquid-secondary-ion mass spectrometry (LSI-MS) (A. Tsarbopoulos et al. Analytica Chemica Acta 1990, **241**, 315; S.B. Mahato et al. Tetrahedron, 1992, **48**, 2483).

(c) **NMR Spectroscopy**

NMR spectroscopy is now widely used for the structure determination of saponins. Assignments of the signals of various carbons of the saponins are generally made by comparison with the c.m.r. data of the aglycone and methyl sugars using known chemical shift rules and glycosylation shifts (R. Kasai et al. Tetrahedron, 1979, **35**, 1427; S. Seo et al. J. Am. Chem. Soc. 1978, **100**, 3331). However, in complex cases, utilization of certain other techniques is necessary for complete signal assignments. Some recent assignment techniques including 2D-spectroscopy and selective single frequency proton decoupling have been reviewed for the structure elucidation of steroid saponins (P.K. Agrawal et al. Phytochemistry, 1985, **24**, 2479). These techniques are equally applicable to triterpenoid saponins.

The spin-lattice relaxation time (T_1) (A. Neszmelyi et al. J. Chem. Soc. Chem. Commun. 1977, 613; I. Kitagawa et al. Chem. Pharm. Bull. 1981, **29**, 1951) has also been used to determine sugar sequences of the echinocystic acid bisglycosides (T.Konoshima and T.Sawada Chem. Pharm. Bull. 1982 **30**, 2747). For a carbon atom, the product of T_1 and N, the number of directly bonded hydrogen atoms is a function of the flexibility of the molecule at that carbon atom. The values of NT_1 are shortest for aglycone and inner sugar carbon atoms and longest for those of the terminal sugar. This technique has been used by Ishii et al. (Tet. Lett. 1981, 1529) and Wagner et al. (Planta Med. 1983, **48**, 136) for determination of sugar sequences in saponins.

2D FT NMR techniques have greatly enhanced the power of NMR to unravel complex chemical structures including saponins. Homonuclear 2DJ-resolved experiments are helpful in the elucidatin of complex overlapping coupling patterns. Homonuclear chemical shift correlation experiments may be used to determine the homonuclear coupling networks. Heteronuclear chemical shift correlation experiments can be used to relate the proton spectrum to the carbon spectrum. Carbon-carbon connectivity experiments reflect carbon homonuclear coupling. Application of 2D NMR in carbohydrate structural analysis has been reviewed (S.L. Patt, J. Carbohydr. Chem. 1984, 3, 493). The structure of medicoside I (**2**) from the roots of <u>Medicago</u> <u>sativa</u> has been determined, including sequencing of its sugars, using p.m.r. spectroscopy alone (G. Massiot et al. J. Chem. Soc. Chem. Commun. 1986, 1485). A 500 MHz p.m.r. spectrum of the peracetate of medicoside I showed that the sugar protons resonances split into two zones; one between $\delta 4.75$ and 5.4 assigned to CHOAc, the other between $\delta 3.0$ and 4.3 assigned to CH_2OAc, CHOR and CH_2OR. The anomeric protons were located between these two zones in the case of ether linkages and at a higher frequency than $\delta 5.5$ for the ester linkage. A single COSY experiment (A. Bax et al. J. Carbohydr. Chem. 1984, **3**, 593) revealed the interproton connectivities and thus the location of the branching points of each sugar. The nature of individual sugars was determined by counting their protons and examination of their coupling constants. A long range COSY experiment (A. Bax and R. Freeman, J. Magn. Reson. 1981, **44**, 542) helped in the location of a 4_J intersugar coupling between the anomeric proton of the inner glucose and H-2 of the inner arabinose. Thus three NMR

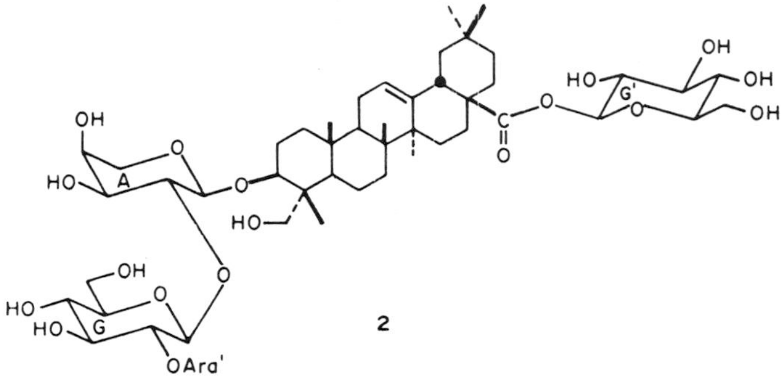

2

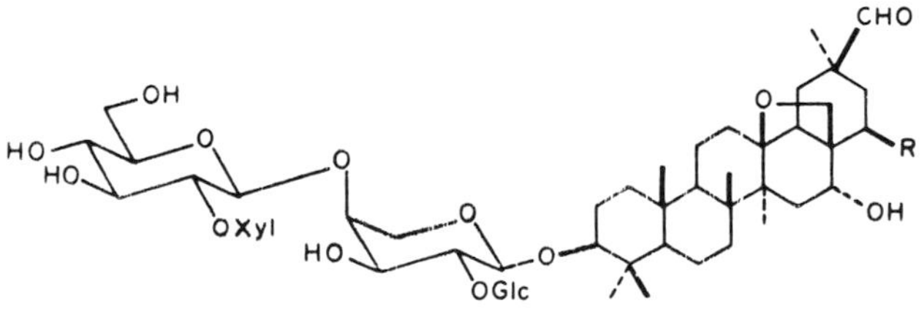

3 R = OAc
4 R = H

experiments : a 1D normal spectrum, 2D COSY and 2D long range COSY were sufficient to identify and sequence the sugars of the saponin. All the available chemical methods, p.m.r. and c.m.r. spectroscopy were found to be inadequate for unambiguous determination of the positions of the terminal xylose and glucose of two triterpenoid tetrasaccharides saxifragifolin A (**3**) and saxifragifolin B (**4**) from Androsace saxifragifolia (J.P. Waltho et al. J. Chem. Soc. Perkin I, 1986, 1527). The problem was successfully overcome by the use of two-dimensional COSY-45 and two-dimensional NOESY experiments. The complete structures of the tetrasaccharide moiety could be elucidated including conformations of the monosaccharide units. The structure of Camellidin II (**5**) isolated from Camellia japonica (A. Numata et al. Chem. Pharm. Bull. 1987, 35, 3948) has been determined mainly by NMR experiment on its methyl ester acetate. The

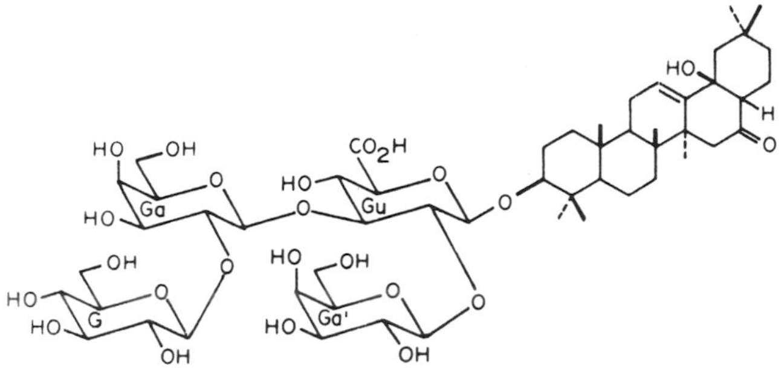

5

signals of four anomeric protons of Gu, Ga, G and Ga', two 4'-protons of Ga and Ga' and H-5 of G were easily assigned by their characteristic splitting patterns, chemical shifts and ^{1}H-^{13}C heteronuclear shift correlated 2D NMR (HETCOR) spectrum. The multiple-relayed homonuclear shift correlated 2D (COSY) spectrum showed the protons correlated with four anomeric protons. Thus H-2 of Gu was observed in the COSY, H-2 and H-3 of Gu in the single-relayed COSY and H-2, H-3 and H-4 of Gu in the double-relayed COSY. The protons H-2, H-3 and H-4 of Ga, G and Ga' were detected in the multiple-relayed COSY in the same manner. Assignment of the overlapping signals in the multiple-relayed COSY and H-5 and H-6 of all the four sugars were made by 1H-1H homonuclear decoupling. Assignment of the c.m.r. signals in the sugar moieties were achieved by correlation with the fully assigned proton signals in the HETCOR spectrum. The anomeric configurations of the four sugar moieties were deduced from the J_{H-H} and J_{C-H} values of the anomeric protons and carbons of the methyl ester acetate.

(d) X-ray crystallography

The application of X-ray crystallography for the determination of the structure and stereochemistry of saponins is not very common. The limited application of this technique to saponins may be attributed to the difficulty in obtaining a single crystal of a suitable size. The molecular geometry of asiaticoside, the major triterpenoid trisaccharide isolated from Centella asiatica reputed for its antileprotic activity has been determined by a single crystal X-ray analysis (S.B. Mahato et al. J. Chem. Soc. Perkin II, 1987, 1509).

2. BIOLOGICAL ACTIVITY

The wide occurrence in nature and varied biological activities of saponins have attracted the attention of mankind from the early stage of civilization. The use of saponins as natural detergents and fish poison were known to the primitive people. Earlier works on biological activities of saponins were mainly conducted on saponin-rich extracts or crude saponin mixtures containing not necessarily only saponins but also other constituents. Consequently, the results of such studies are non-specific and should be accepted with reservation. The advent of various efficient sophisticated techniques of isolation and structure elucidation and sporadic reports on interesting biological activities of crude saponins have prompted many workers to study the biological activities of pure saponin constituents or pure saponin mixtures and a considerable literature has already been accumulated. An excellent review on the occurrence and biological activities of triterpene glycosides in plants and animals has been published by Anisimov and Chirva (Pharmazie 1980, **35**, 731). The effects of triterpenoid saponins derived from several plants on the secretion of ACTH and corticosterone have been discussed in detail by Hiai (Adv. Chin. Med. Mater. Res. Int. Symp. 1986, 49 Chem. Abs. **104**, 161363) and a review on the pharmacology of ginseng saponins, Bupleurum falcatum saponins and glycyrrihizin from Glycyrrihiza glabra has been published by Ohura (Mod. Med. Jpn. 1981, **10**, 21, Chem. Abs. **94**, 149797). The antibiotic properties of triterpene glycosides of the Holothuroidea class of marine echinoderm animals have been reviewed by Somolilov and Girshovich (Antibiotiki 1980, **25**, 307).

(a) Action on metabolism

The hypocholesterolaemic effect of some saponins has prompted considerable clinical interest. Malinow et al. (J. Clin. Invest. 1981, **67**, 156) and Oakenfull (Food Tech. Aust.

1981,33, 432) have suggested that consumption of saponins may provide a useful means of controlling plasma cholesterol in man. Saponins (10g/kg) in the high cholesterol diet of rats reversed the hypercholesteromia and increased both the rate of bile acid secretion and the fecal excretion of bile acids and neutral sterols (D.G. Oakenfull et al. Br. J. Nutr. 1979, **42**, 209). Ivanov et al. (Khim-Farm. Zh. 1987, **21**, 1091) studied hypolipidemic properties of synthetically prepared betulin glycosides. Both monodesmosidic and bisdesmosidic betulin glycosides were prepared from betulin derivatives for this study. Liposomes containing the glycosides were prepared using lecithin, cholesterol, stearic acid and the glycosides. In in vitro experiments, these authors found that the cholesterol content was decreased 187–211 mg% by the glycosides. In a study in cats with experimentally induced hypercholesteromia, the glycosides decreased the cholesterol levels. The highest activity was observed in glycosides with a free hydroxyl group, i.e. in monodesmosidic betulin glycosides.

(b) Antitumour and antileukemic activity

Triterpenoid glycosides foetoside C and cyclofoetoside B from Thalictrum foetidum and thalicoside A from T. minus were studied for their antitumour activity in rats with implanted tumours (K.D. Rakhimov et al. Khim-Farm. Zh. 1987, **21**, 1434). Each agent given in dose of 30–50 mg kg^{-1} i.p. daily for 10 days, had an appreciable antitumour activity. Foetoside C was found to be most effective, especially in the treatment-resistant types of tumours. According to the authors, the Thalictrum glycosides are promising agents in the development of new neoplasm inhibitors. Two oleanolic acid glycosides giganteaside D and flaccidin B isolated from Anemone flaccida showed an inhibitory effect on reverse transcriptase from RNA tumour virus (Y. Shen. et al. Shoyakugaku Zasshi 1988, **42**, 35; Chem. Abst. **109**, 222012). Bioassay-directed fractionation of the cytotoxic antileukemic extracts of Prunella vulgaris, Psychotria serpens and Hyptis capitata led to the isolation of ursolic acid as one of the active principles (K.H.Lee et al. Planta Med. 1988, **54**, 308). Ursolic acid showed significant cytotoxicity in the lymphocytic leukemia cells P-388 and L-1210 as well as the human lung carcinoma cell A-549. It also demonstrated marginal cytotoxicity in the KB and human colon (HCT-8) and mammary (MCF-7) tumour cells. Esterification of the hydroxyl group at C-3 and the carboxyl group at C-17 led to compounds with decreased cytotoxicity in human tumour cell lines, but with equivalent or slightly increased activity against the growth of L-1210 and P-388 leukemic cells. Tubeimoside I, a cyclic bisdesmoside isolated from the bulb of Bolbostemma paniculatum showed moderate antitumour activity in primary in vivo pharmacological study (F.H. Kong et al. Tetrahedron Letters 1986, 27, 5765). Afromontoside, a steroidal saponin isolated from Dracaena afromontana, its aglycone diosgenin, as well as dihydrodiosgenin and several structurally related compounds have been shown by Reddy et al. (J. Chem. Soc. Perkin I 1984, 987) to be cytotoxic to cultured KB cells. Cytotoxic activities of triterpene glycosides of holothurian A and B from Holothuria mexicana and stichoposide A and C from Cucumaria frandatrix were compared. Their cytotoxic activity depended on the number of monosaccharide units attached to the hydroxy group at C-3. Holothurian A, having four units are more toxic than holothurian B, with only two sugar units. On the other hand, the increase in the length of saccharide chains to six (stichoposide C with A) had little effect on activity. The changes noted indicated the general tendency of these glycosides to show increased physiological activity in response to an increase in length of the chains to 4-6 monosaccharide residue (M.M. Anisimov et al. Toxicon 1980, 18, 221).

(c) Antimicrobial activity

Saponins are generally good antifungal and antibacterial agents. The antifungal activity is found to be more effective with saponins than the sapogenins and the acylated saponins. Digitonin has a considerable fungistatic activity (Y. Assa et al. Life Sci. 1972, 11, 637). A glucoside of Δ^{12} oleanene -23,28-dioic acid isolated from alfalfa root demonstrated considerable activity against Cryptococcus neoformans suggesting that it might be a useful active agent in the treatment of cryptococcosis (R. Zhang et al. Yaoxue

Xuebao 1986, **21**, 510). Alfalfa root saponins and sapogenins reduce the fungal population, but did not affect the bacterial population of peat (D. Levanon et al. Soil Boil Biochem. 1982, **14**, 501).

(d) Molluscicidal activity

Molluscicidal or snail-killing activities of saponins are of special importance for the control of schistosomiasis. Balanitin-1, 2 or 3 in dilute (5-10 ppm) solution kill schistosomiasis transmitting snails Biomphalaria glabrata within 24h, and hence they are potent molluscicides (H. Liu, and K.Nakanishi, Tetrahedron 1982, **38**, 513). Two monodesmosides of oleanolic acid isolated from Xeromphis spinosa were found to be lethal against the snails at concentration of 15-20 ppm (O.P. Sati et al. Planta Med. 1987, **53**, 530).

(e) Miscellaneous

The glycosidic fraction isolated from Maesa chisia var. angustifolia showed antiinflammatory, analgesic and antipyretic activities in experimental animals (A. Gomes et al. Indian J. Exp. Biol. 1987, **25**, 826). Saponins isolated from Luffa cylindrica have been shown to be effective antiobesity agents and have the ability to control the unspecified side effects of steroid drugs (Japanese patent JP 59, 203, 451, 1984; Chem. Abst. 1985, **102**, 165552 u). The spermicidal activities of the metabolites of Cusonia spicata (J. Gunzinger et al. Phytochemistry 1986, **25**, 2501), Phytolacca dodecandra (A.C. Dorsaz and K. Hostettmann, Helv. Chim. Acta 1980, **69**, 2038) and Acacia auriculiformis (A. Pakrashi et al. Contraception 1991, **43**, 475) are shown to be due to the water soluble saponins. They are potentially valuable as contraceptives.

The report on biological activities of saponins presented here is not essentially comprehensive However, the recognised pharmacological activities of saponins are antitussive, expectorant, antiinflammatory, antiulcer, antiallergic and CNS-effective (S. Shibata, Advances in Medicinal Phytochem., Barton and Ollis eds, John Libbey,

1986, p.159). The beneficial actions of ginseng saponins, Bupleurum falcatum saponins and saponins of Glycyrrihiza glabra have been established. Chinese, Korean and Japanese pharmaceutical research groups have devoted a great deal of attention to ginseng, a very expensive traditional medicine, in recent years, particularly, to its constituti- onal saponins which appear to be responsible for much of the biological activity of the plant. The most active constituents of the ginseng saponins have been shown to be dammarane-type triterpenoid saponins. Considerable research activities are underway on saponin constituents of other recognised medicinal plants and more useful results are expected to be forthcoming.

3. BIOGENESIS

A detailed coverage of the biogenesis of the steroidal and triterpenoidal sapogenins and saponins is beyond the scope of this presentation. The biochemistry of the steroidal saponins has been reviewed by Heftmann (Lipids 1974, **9**, 626), Tschesche (Proc. Roy. Soc. Ser.B 1972, **180**, 187) and Russo (Fitoterapia 1971, No.2, 61). Most saponins possess an oxygen function at C-3 which has been shown to result from an intermediate, 2,3-epoxysqualene. Labelled 10α- cucurbita-5,24-dien 3β-ol was obtained from [3-^{3}H] squalene 2,3-epoxide incubated with microsomes of Cucurbita maxima seedlings. By contrast the lanostane triterpenoids [2-^{3}H], cycloartenol, [2-^{3}H] parkeol and their corresponding derivatives [2,12-^{3}H] 11-ketocycloartenol and [2-^{3}H] 24,25- dihydro-9α,11α -epoxyparkeol incubated under the same conditions, gave no rearranged products with a cucurbitane skeleton (G. Balliano et al. Phytochemistry 1983, **22**, 915). Biosynthesis of triterpenoid sapogenols in soybean and alfalfa seedlings was investigated by Peri et al. (Phytochemistry 1979, **18**, 1671). By incubation of germinating soybeans with mevalonate-[2-^{14}C](MVA), radioactivity was incorporated into four soyasapogenols A,B,C and E. When alfalfa seedlings were incubated with MVA-[2-^{14}C], about two-thirds of the radioactivity incorporated into the sapogenols was associated with medicagenic acid. It has been proposed (W.R. Nes Advances in Lipid Research, eds., R.Paoletti and D. Kritchevsky 1977,

Vol.15, p.233, Academic Press, New York) that there are essentially two pathways leading from epoxysqualene to the sterols. In photosynthetic organisms the pathway passes through cycloartenol whereas in nonphotosynthetic organisms it proceeds via lanosterol. However, although the transformation of cycloartenol to cholesterol has not been observed in nonphotosynthetic organisms, lanosterol does act as a precursor of sterols in photosynthetic organisms, e.g. in Chlorella ellipsoidea (L.B.Tsai and G.W.Patterson, Phytochemistry 1976, 15, 1131). On the other hand lanosterol, but no cycloartenol was found in Euphorbia pulcherrima (B.C.Sekula and W.R.Nes, Phytochemistry 1980, 19, 1509). According to Heftmann (Isopentenoids in Plants 1984, W.D.Nes et al. eds. p.487 Marcel Dekker, Inc., New York) this phenomenon may be explained by assuming cycloartenol as a precursor of lanosterol at least in Euphorbia (G.Ponsinet and G.Ourisson, Phytochemistry 1968, 7, 757). Cholesterol is converted to diosgenin in tissue culture of Dioscorea deltoidea (S.J.Strohs et al. Phytochemistry 1969, 8, 1679; A.R. Chowdhuri and H.C. Chaturvedi, Curr. Sci. 1980, 49, 237) and D. tokoro (Y.Tomita and A.Uomori, J. Chem. Soc.,D 1971, 284). When $[26-^{14}C]-(25R)-26$-hydroxycholesterol was administered to D. floribunda plants it was converted to diosgenin, but not to yamogenin (R.D.Bennett et al. Phytochemistry 1970, 9, 349). Apparently the C-25 stereochemistry is determined by the location of the oxygen function at the terminal carbon atom C-26 or C-27 of cholesterol. Convallamerogenin, a sapogenin containing a methylene instead of a methyl group at C-25 is not formed from cholesterol, but from the Δ^{24} precursor of cholesterol, desmosterol (R. Tschesche et al. Phytochemistry 1975, 14, 129). Seo et al. (J. Chem. Soc. Perkin I 1984, 869) administered $[1,2-^{13}C_2]$ acetate in cell cultures of D. tokoro for the study of the biosynthesis of (25S)- and (25R)-furostanol glycosides, protoneotokorin and prototokoronin. The ^{13}C labelling patterns of neotokorogenin and tokorogenin obtained from protoneotokorin and prototokorinin respectively indicated that the hydrogen atom at C-25 is introduced on the 25-si face of the Δ^{24} intermediate followed by oxidation of the pro-R(C-26) and the pro-S(C-27) methyl groups at C-25 leading to (25S)-and (25R)-furostanol glycosides, respectively. The results were confirmed from the labelling patterns of yamogenin and diosgenin isolated by hydrolysis of crude furostanol glycosides. The oxidation of the pro-R methyl group was accelerated by increasing the concentration of sodium acetate without any effect on the ^{13}C-labelling patterns.

Kintia et al. (Bull. Acad. Pol. Sci., Ser. Sci. Biol. 1974, **22**, 73) and Wojciechowski (Phytochemistry 1975, **14**, 1749) investigated the metabolic interrelationships of oleanolic acid saponins, calendulosides A-D, D_2 and F from Calendula officinalis.

Labelling studies, the use of cell free enzyme preparations as well as chemical analysis revealed oleanolic acid to be first converted to its 3β - glucuronoside (calenduloside F) via UDP-glucuronate. The necessary glucuronosyl transferase appeared to be highly specific for oleanolic acid, since other structurally related triterpenes were not utilized. Calenduloside D_2 was then formed by glucose transfer to C-28 from UDP-glucose, the enzyme involved again being specific. Elongation of the sugar chain at C-3 occurred with UDP-galactose/galactosyl transferase to form calenduloside C and D and UDP-glucose/ glucosyl transferase to yield calenduloside A and B. The authors concluded that the consecutive steps of sugar chain elongation in the oleanolic acid glycoside are catalysed by specific transferases, localized in three cellular compartments. The biosynthesis of the 3-glucuronoside takes place in the microsomes, the elongation of the sugar chain at C-3 of the aglycone proceeds in heavy membrane structures which are probably fragments of the Golgi complex while a cytosol enzyme(s) is involved in glucosylation of C-17 carboxyl group of oleanolic acid.

4. STEROID SAPONINS AND SAPOGENINS

(a) Saponins

A large number of steroid saponins have been isolated during the period and instead of giving a comprehensive list of the saponins, some well characterized saponins of interest are mentioned. Extensive chemical studies on the aglycones revealed that they are almost exclusively spirostane derivatives. But furastanol glycosides have also been isolated and characterized. The furostanol glycosides with some exceptions (T. Nohara et al. Chem. Pharm. Bull. 1975, 23, 872) show a characteristic red colour on a TLC plate when sprayed with p-dimethylaminobenzaldehyde and hydrochloric acid (Ehrlich reagent). Moreover, the furostane skeleton does not exhibit the characteristic i.r. absorptions of spirostane derivatives. Confirmatory evidence for the furostane structure is obtained by examination of the products of Marker's degradation or Baeyer-Villiger oxidation followed by hydrolysis (S.B. Mahato et al. Indian J. Chem. 1978, 16B, 350). Some glycosides have been isolated whose aglycones are not spirostanols but a modification. In general, the sugar moieties contain 2-4 kinds of sugar units, e.g. D-glucose, D-galactose, D-xylose and L-rhamnose. D-Xylose and L-rhamnose generally occur at the terminal positions. Arabinose containing steroidal saponins are also known. In a very few cases quinovose occurs as the carbohydrate moiety. Trillenoside A (6), a novel 18-norspirostanol glycoside, contains xylose, rhmnose, arabinose and apiose as the sugar constituents (T. Nohara et al. Tetrahedron Letters 1975, 4381; Chem. Pharm. Bull. 1980, 28, 1437). Another unusual saponin consisting of kammogenin and five molecules of 2-deoxyribose has been reported by Backer et al. (J. Pharm. Sci. 1972, 61, 1665). The furostanol bisglycosides generally contain a glucose unit attached to the C-26 hydroxyl.

However, there has been one recent example of a spirostanol saponin having L-rhamnose attached to ring F (K. Nakano et al. J. Chem. Soc. Chem. Commun. 1982, 789). Afromontoside (7), a new cytotoxic principle from Dracaena afromontana is the first example of a furostanol glycoside having α-L-rhamnose at C-26 (K.S. Reddy et al. J. Chem. Soc. Perkin I, 1984, 987). Convallamaroside (8) is the first example of a tridesmoside isolated from Convallaria majalis. Monodesmosides containing sugar moiety attached to 5β-hydroxyl group and bis-desmoside, in which the sugar moieties are attached to 5β-and 26-hydroxyls have also been isolated (R. Tschesche et al. Chem. Ber. 1973, 106, 3010). A new glucuronide, (22R)-3β,16β,22,26-tetrahydroxycholest-5-

6

7

8

ene-3-0-α-L-rhamnopyranosyl-(1→2)-β-D-glucuronopyranoside
(**9**) has been isolated from the aerial parts of Chinese
Solanum lyratum (S. Yahara et al. Chem. Pharm. Bull. 1989,
37, 1802). This compound appears to be an important
biogenetic precursor of spirostanol and furostanol
glycosides. Novel steroidal saponins, pardarinosides A-G
have been isolated as the bitter ingredients from the fresh
bulbs of Lilium pardarinum. The structure of pardarinoside A
is 22-0-methyl-26-0-acetyl-(25R)-5α-furost-3β, 14α,17α,22α,
26-pentaol 3-0-[α-L-rhamnopyranosyl-(1→2)-β-D-glucopyrano-
side (**10**). Naturally occurring 22,26-hydroxyl furostanol
saponins exist in the form of bisdesmosides, bearing sugars
at both C-3 and C-26 hydroxyl positions without any
exceptions. During partial hydrolysis of the sugar linkage

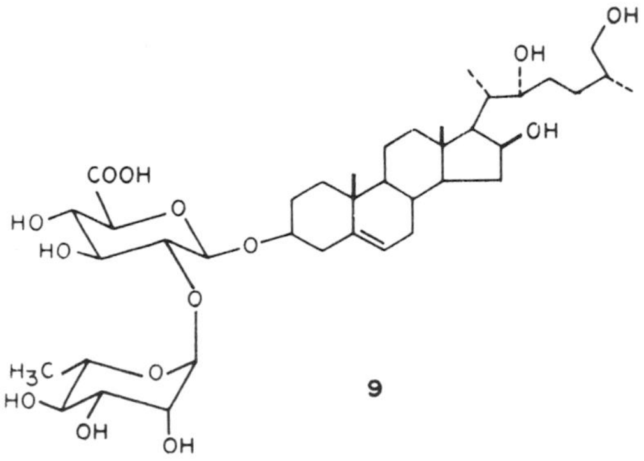

9

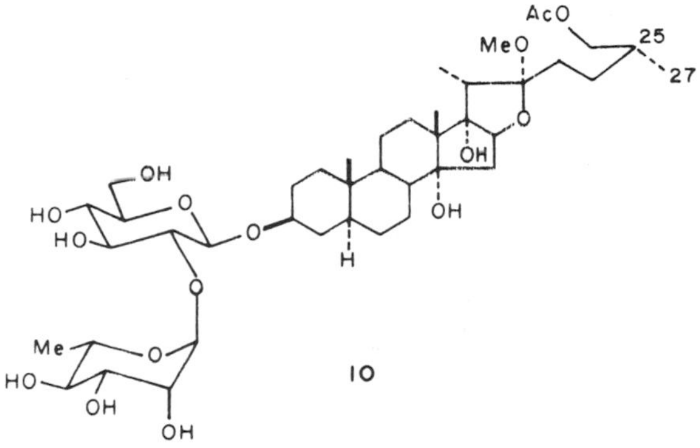

10

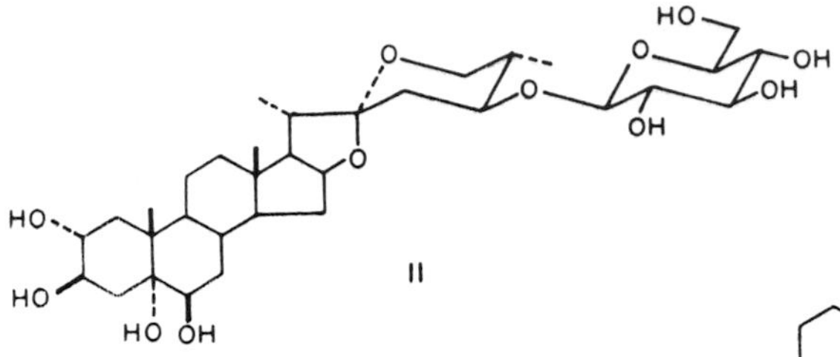

11

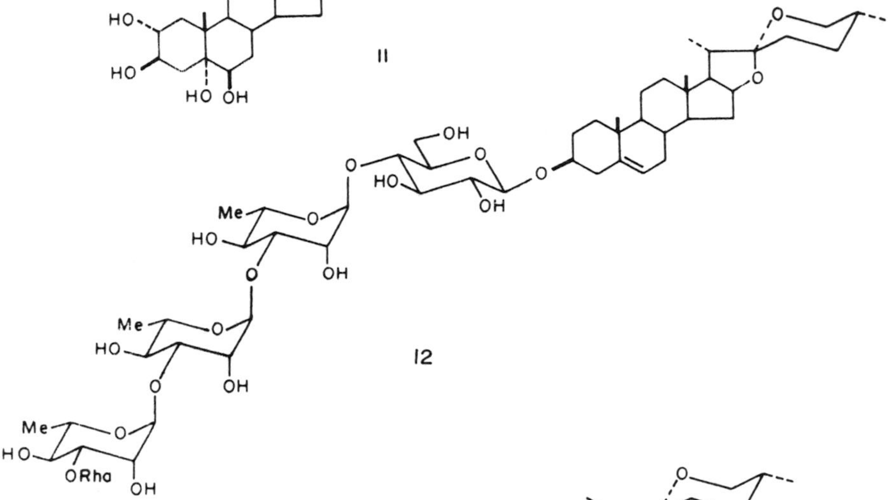

12

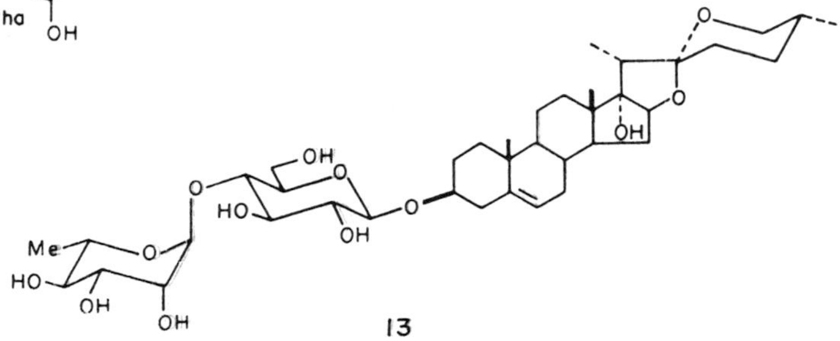

13

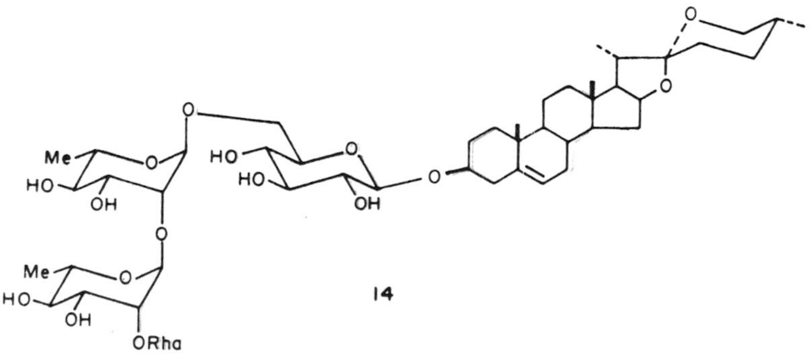

14

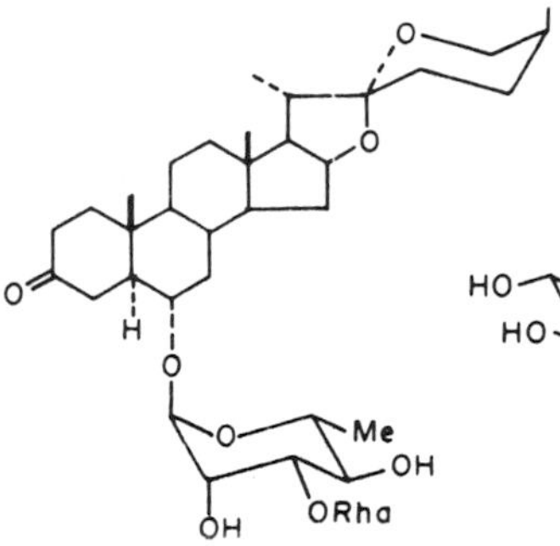

15

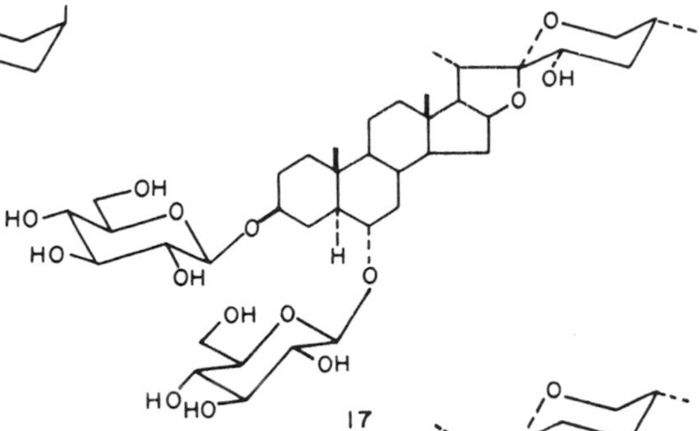

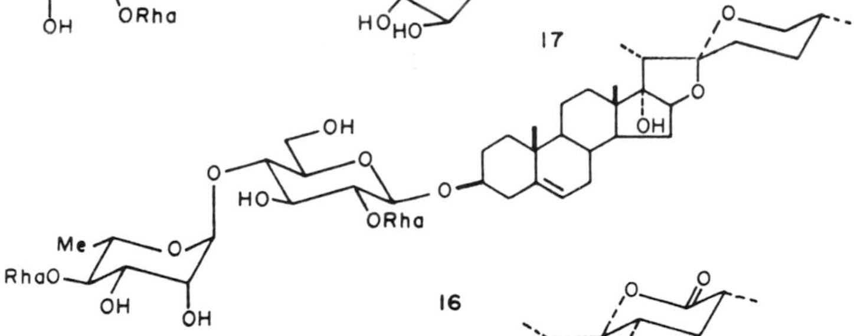

16

17

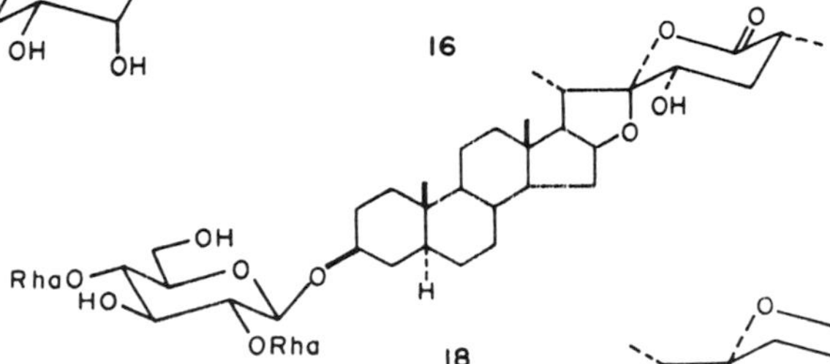

18

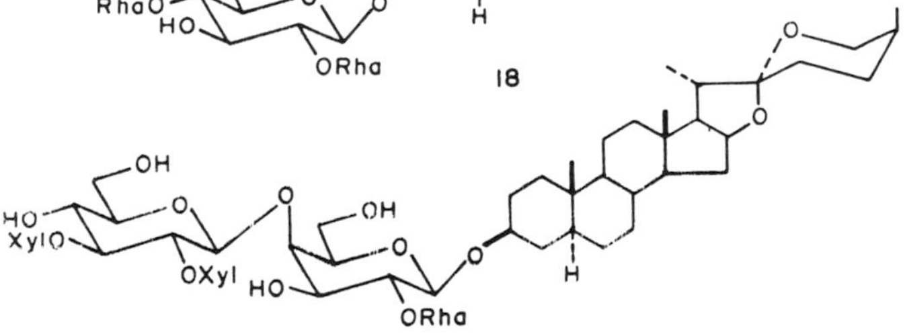

19

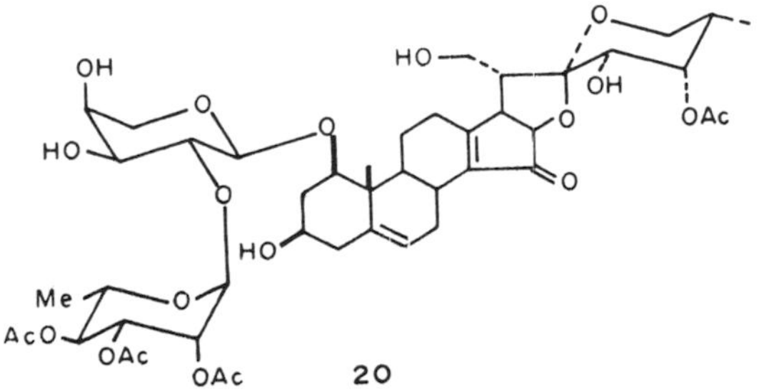

20

to the C-26 hydroxyl position, they are readily cyclized to give the corresponding spirostanol glycosides. Pardarinosides, except for pardarinoside E, are 22,26-hydroxyl furostanol derivatives, and it is noteworthy that they are distinctive in carrying an acyl substitution in place of sugar to the C-26 hydroxyl position. A novel polyhydroxylated steroidal saponin has been isolated from the bulbs of Allium giganteum (K. Kawashima et al. Phytochemistry 1991, 30, 3063). The structure of the saponin has been established as (24S,25R)-5α-spirostane-2α,3β,5α,6β,24-pentaol 24-0-β-D-glucopyranoside (11). The structure is distinctive in carrying a glucose moiety at C-24 position.

The yams of Dioscorea floribunda are used as raw material for extraction of diosgenin. A number of saponins have been isolated from the yams (S.B. Mahato et al. Indian J. Chem 1978, 16B, 350; Phytochemistry 1981, 20, 1943). Floribundasaponin D, a diosgenin pentaglycoside (12) was obtained as the major saponin and floribundasaponin B (13), a pennogenin diglycoside was isolated as more abundant of the two glycosides containing pennogenin as the aglycone. Kallstroemin D (14) is the major saponin isolated from Kallstroemia pubescens, a new potential source of diosgenin (S.B. Mahato et al. Indian J. Chem. 1977, 15B, 445). A few other saponins e.g. hispinin B (15) (A.K. Chakrabarty et al. Phytochemistry 1979 18, 902), pennogenin tetraglycoside (16) (T. Nohara et al. Chem. Pharm. Bull. 1975, 23, 872), spirostanol bisdesmoside (17) (O.P. Sati et al. J. Nat. Prod. 1985, 48, 395), soladulcoside B (18) (T. Yamashita et al. Chem. Pharm. Bull. 1991, 39, 1626), tribulosin (19) (S.B. Mahato et al. J. Chem. Soc. Perkin I 1981, 2405) and Ts-d (20) (K. Nakano et al. Phytochemistry 1983, 22, 1047) are illustrated as representatives of various types of steroidal saponins.

(b) Sapogenins

Steroid sapogenins that have been isolated since the earlier chapter was written are listed in table 1. Most of the sapogenins are variants on known types and their structures have been determined by spectroscopic methods e.g. u.v., i.r., proton and carbon n.m.r. and mass spectral analysis. Simple chemical transformations have also been made sometimes for confirmation of the structures determined by the spectroscopic methods.

Table 1

Sapogenin[a] (Structure)	Source	mp($°$C),$[\alpha]_D$(degree)
1	2	3
Sapogenin(65)	Cordyline rubra[1]	151-153,
Epidiosgenin(46)	Gynura japonica[2]	244-246, -145
Crabbogenin(58)	Cordyline stricta[3]	203-204
Episceptrumgenin(59)	Gynura Japonica[2]	197-198, -136
22-Epirhodeasapo- genin(53)	Rhodea japonica[4]	-
Brisbagenin(21)	Cordyline canifolia[5]	203-204, -76 c
Cannigenin(22)	Cordyline canifolia[6]	215.5-217, -58.03 c
Barbourgenin(23)	Agave sisalana[7]	228-230
Cordylagenin(31)	Cordyline stricta[8]	216, -50 c
Sapogenin(66)	Cordyline rubra[1]	-
Solagenin(32)	Solanum hispidum[9]	218-220, -51.6 c
Prazerigenin A(47)	Dioscorea prazeri[10]	- , -84.9
Sapogenin(60)	Reineckia carnea[11]	-
Sapogenin(61)	Cordyline stricta[3]	-
Sapogenin(62)	Helleborous multifidus[12]	223-225,
Chenogenin(26)	Cordyline rubra[13]	278-280,
Agigenin(24)	Allium ampeloprasum[14]	267-269,-75.3 c+m
Sapogenin(40)	Yucca gloriosa[15]	204-205
Paniculogenin(34)	Solanum hispidum[16]	236-240,-70.8 c+e
Solaspigenin(25)	Solanum hispidum[16]	>300, -70.5 c+e
Pompeygenin(33)	Cordyline stricta[3]	260
Strictagenin(35)	Cordyline stricta[3]	271-273, -25
Hispigenin(52)	Solanum hispidum[17]	258-260, +23.3 p
Epimetagenin(39)	Metanarthecium luteo- viride[18]	231-232, -62 p

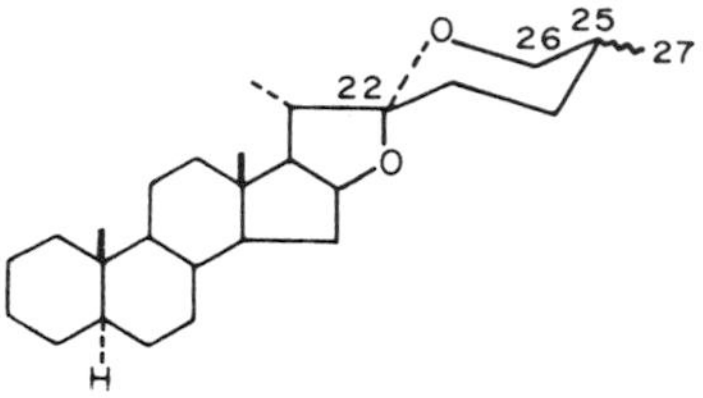

26 25 27 22 O

O 26 CH₃ 22 25 27 CH₃

21 Brisbagenin,25(R)-,1β,3β-OH
22 Cannigenin,25(R)-,1β,3α-OH
23 Barbourgenin,25(S)-,3β,27-OH
24 Agigenin,25(R)-,2α,3β,6β-OH
25 Solaspigenin,25(R)-,3β,6α,23β-OH
26 Chenogenin,25(R)-,1β,3α,25-OH
27 Sapogenin,25(R)-,3β,23α-OH,26-oxo
28 Sapogenin,25(R)-,3β,15α,23α-OH, 26-oxo
29 Alliogenin,25(R)-,2α,3β,5α,6α-OH
30 Sapogenin,25(R)-,2α,3β,5α,6β, 24-OH
31 Cordylagenin,25(S)-,1β,3α-OH
32 Solagenin,25(S)-,6α-OH,3-oxo
33 Pompeygenin,25(S)-,1β,3α,25-OH
34 Paniculogenin,25(S)-,3β,6α,23β-OH

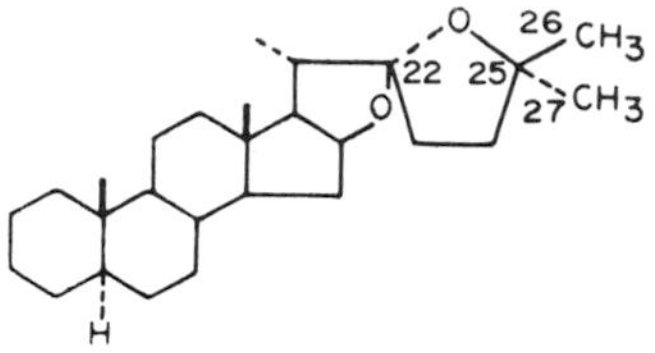

35 Strictagenin,22(S)-,25(S)-,1β,3α,26-OH
36 Rubragenin,22(S)-,25(R)-,1β,3α,26-OH
37 Wallogenin,22(R)-,25(R)-,1β,3α,26-OH
38 Taccagenin,22(R)-,3β,26,27-OH, 5-ene

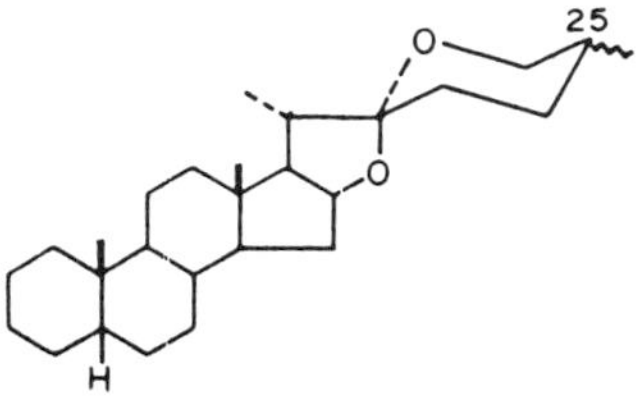

25 O O

39 Epimetagenin,25(R)-,2β,3α,11α-OH
40 Sapogenin,25(R)-,2β,3β,12β-OH
41 Protometeogenin,25(R)-,2β,3β,11α-OH, 4-ene
42 Sapogenin,25(R)-,1β,3β,5β-OH,4β-OSO₃Mg
43 Sapogenin,1β,3β-OH,25-oxo
44 1β-Hydroxydiotigenin,25(S)-,1β,2β,3α,4β-OH
45 Neopentologenin,25(S)-,1β,2β,3β,4β,5β-OH

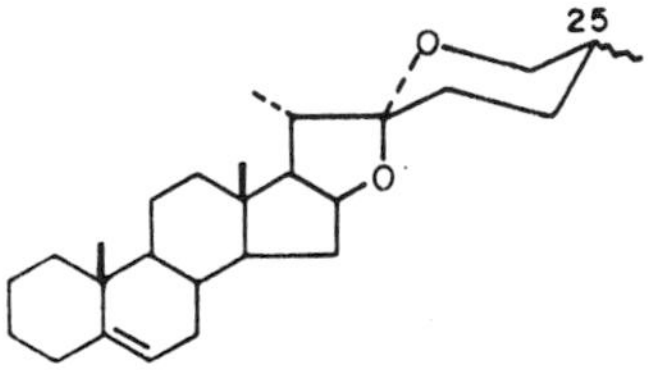

46 Epidiosgenin,25(R)-,3α-OH
47 Prazerigenin A,25(R)-,3β,
   14α-OH
48 Prazerigenin D,25(R)-,3β,
   14α,27-OH
49 Bahamgenin,25(R)-,3β,12β,
   15α-OH
50 Sapogenin,25(R)-,3β,17α,
   24β-OH
51 Spirotaccagenin,25(R)-,3β,
   25,27-OH

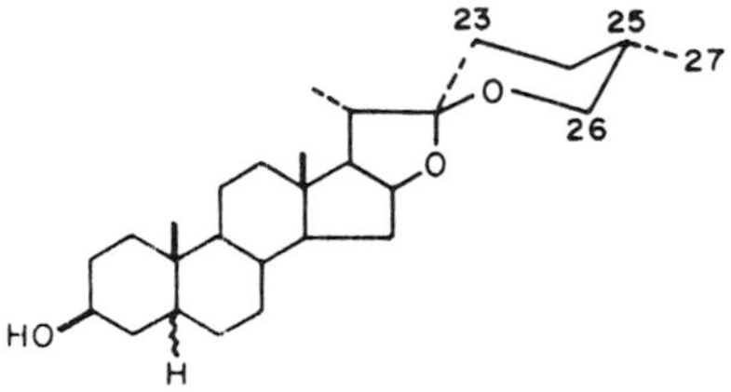

52 Hispigenin,6β,23α-OH,5α-H
53 22-Epirhodeasapogenin,1β-
   OH,5β-H

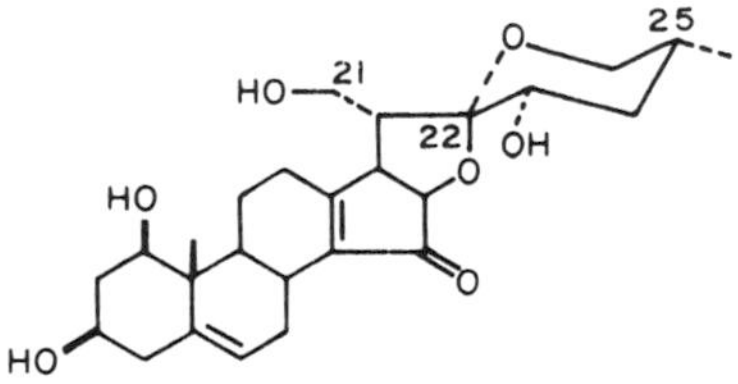

54 Trillenogenin,24β-OH
55 Epitrillenogenin,24α-OH
56 Epitrillenogenin 21-O-
   acetate
57 Epitrillenogenin 24-O-
   acetate

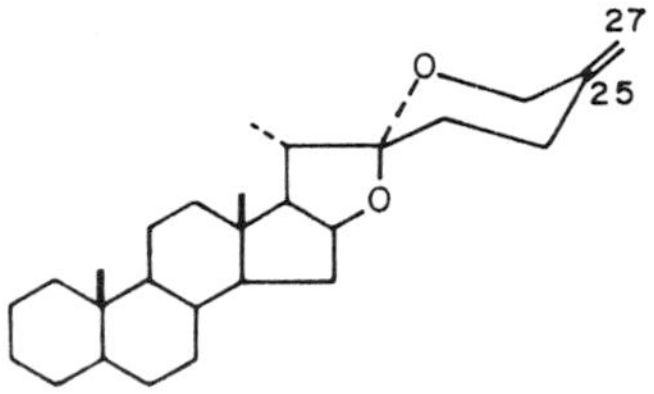

58 Crabbogenin,3α-OH,5α-H
59 Episceptrumgenin,3α-OH,5-
   ene
60 Sapogenin,1β,3β-OH,5β-H
61 Sapogenin,1β,3α-OH,5α-H
62 Sapogenin,3β,11α-OH,5-ene
63 Sapogenin,1β,2β,3β,4β,
   5β,7α-OH,6-oxo
64 Sapogenin,1β,2β,3β,4β,
   5β,6β,7α-OH

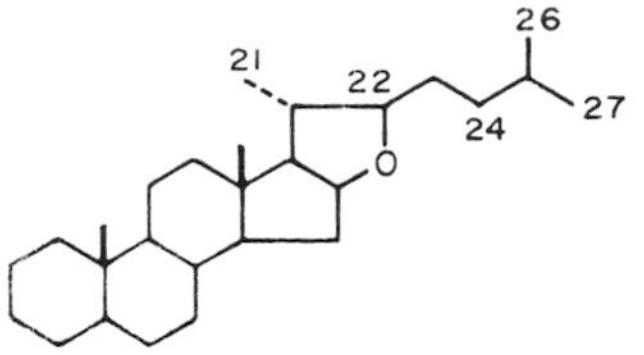

534

65 1β,3α-Dihydroxyfurost-5-ene
66 1β,3α,26-Trihydroxy-5α-
   furostane

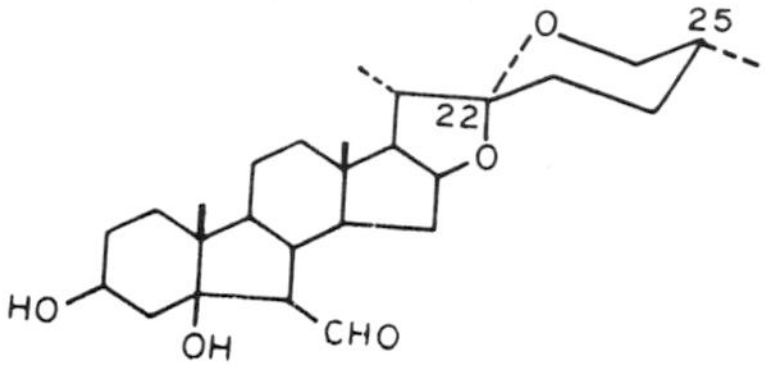

67 Leontogenin

Table 1 (Contd.)

1	2	3
Rubragenin(36)	Cordyline rubra[13]	248-250
Wallogenin(37)	Cordyline rubra[13]	-
Leontogenin(67)	Tacca leontopetaloides[19]	-
Sapogenin(43)	Reineckia carnea[11]	-
Sapogenin(27)	Solanum dulcamara[20]	-
Protometeogenin(41)	Metanarthecium luteo-viride[21]	-
Bahamgenin(49)	Solanum bahamense[22]	257-258,-86.2 c+m
Taccagenin(38)	Tacca leontopetaloides[23]	246-248
Prazerigenin D(48)	Dioscorea prazeri[24]	280-282,-116.6 p
Sapogenin(50)	Trillium kamtschaticum[25]	-
Spirotacca-genin(51)	Tacca leontopetaloides[23]	225-227
Alliogenin(29)	Allium aflatunense[26]	- , -57 p
Sapogenin(44)	Dioscorea tenuipes[27]	308-310,-28 m
Sapogenin(28)	Solanum dulcamora[20]	-
Sapogenin(30)	Allium giganteum[26]	-

Table 1 (Contd.)

1	2	3
Neopentologenin(45)	Aspidistra elatior[28]	288-293(dec),-62 p
Trillenogenin(54)	Trillium kamtschati-cum[29]	250-251,-198
24-Epitrilleno-genin(55)	Trillium tschonoskii[25]	-
24-Epitrillenoge-nin-21-0-acetate(56)	Trillium tschonoskii[30]	-
24-Epitrillenogenin-24-0-acetate(57)	Trillium tschonoskii[31]	-
Sapogenin(63)	Rhodea japonica[32]	244-247,86.9 p
Sapogenin(64)	Rhodea japonica[32]	273-275, -95.5 p
Sapogenin(42)	Reineckia carnea[11]	332, -38 d

a   The compounds are listed in order of increasing number of oxygen atom; c in chloroform;   m in methanol; p in pyridine;   d in DMSO; c+m in chloroform and methanol; c+e   in chloroform and ethanol.

(c) References

1. M. Yang et al. Phytochemistry 1990, **29**, 1332.
2. M. Takahira et al. Tetrahedron Letters 1977, 3647.
3. G. Blunden et al. Tetrahedron 1981, **37**, 2911.
4. K. Kudo et al. Chem. Pharm. Bull. 1984, **32**, 4229.
5. K. Jewers et al. Steroids 1974, **24**, 203.
6. W.J. Griffin et al. Phytochemistry 1976, **15**, 1271.
7. G. Blunden et al. J. Nat. Prod. 1986, **49**, 687.
8. K. Jewers et al. Tetrahedron Letters 1974, 1475.
9. A.K. Chakravarty et al. Phytochemistry 1979, **18**, 902.
10. K. Rajaraman et al. Indian J. Chem. 1976, **14B**, 735.
11. K. Iwagoe et al. Yakugaku Zasshi 1987, **107**, 140.
12. S.M. Vladimirov et al. Phytochemistry 1991, **30**, 1724.
13. M. Yang et al. Phytochemistry 1989, **28**, 3171.
14. T.Morita et al. Chem. Pharm. Bull. 1988, **36**, 3480.
15. K. Nakano et al. Phytochemistry 1991, **30**, 1993.
16. A.K. Chakravarty et al. Phytochemistry 1980, **19**, 1249.

536

17. A.K. Chakravaarty et al. Tetrahedron Letters 1978, 3875.
18. I. Yosioka et al. Tetrahedron 1974, **30**, 2283.
19. A. Abdel-Aziz et al. Phytochemistry 1990, **29**, 2623.
20. T. Yamashita et al. Chem. Pharm. Bull. 1991, **39**, 1626.
21. I. Kitagawa et al. Tetrahedron Letters 1976, 1885.
22. F. Coll et al. Phytochemistry 1983, **22**, 787.
23. A. Abdel-Aziz et al. Phytochemistry 1990, **29**, 1643.
24. K. Rajaraman and S. Rangaswami, Indian J. Chem. 1982, **21B**, 832.
25. N. Fukuda et al. Chem. Pharm. Bull. 1981, **29**, 325.
26. K. Kawashima et al. Phytochemistry 1991, **30**, 3063.
27. S. Kiyosawa et al. Phytochemistry 1982, **21**, 2913.
28. T. Konishi et al. Chem. Pharm. Bull. 1984, **32**, 1451.
29. T. Nohara et al. Tetrahedron Letters 1975, 4381.
30. M. Ono et al. Phytochemistry 1986, **25**, 544.
31. K. Nakano et al. Phytochemistry 1983, **22**, 1047.
32. K. Miyahara et al. Tetrahedron Letters 1980, **21**, 83.

Some points of special interest regarding steroid sapogenins are described here. Trillenogenin (54) and epitrillenogenin (55) are 18-norspirostane derivatives possessing the following structural peculiarities : (i) an enone system in the D-ring, (ii) a hydroxylated 21-methyl group and (iii) an α-glycol system in the F-ring. The p.m.r. and c.m.r. spectral studies of trillenogenin revealed the presence of one secondary methyl, one tertiary methyl, four secondary hydroxyls, one primary hydroxyl, one tetrasubstituted double bond, one trisubstituted double bond and one carbonyl function. The structure of trillenogenin was determined by single crystal X-ray crystallographic analysis of the tetraacetyl mono brosylate. The circular dichroism of the sapogenin showed positive cotton effect curve at 326 nm due to n $\longrightarrow \pi^*$ transition of an enone system (T. Nohara et al. Chem. Pharm. Bull. 1980, **28**, 1437). Thus the absolute configuration was deduced to be that of a normal steroid form, a conclusion based upon the Snatzke rule for transoid cyclopentenones.

The u.v.,c.d,,i.r. and electron impact ionisation mass spectra of epitrillenogenin (55) were very similar to those of trillenogenin (N. Fukuda et al. Chem. Pharm. Bull. 1981, **29**, 325) suggesting a close structural resemblance of the compound. Treatment of **55** with acetone and p-toluenesulfonic acid yielded a product the acetate of which was shown to be a monoacetonide triacetate by p.m.r. spectroscopy

Moreover, the coupling patterns of 23-H and 24-H of **55** indicate that of 24-H has β (equatorial) configuration and the 24-acetoxy group has α-configuration. Thus the 23- and 24-hydroxy groups in sapogenin (**55**) should have the same configuration (α) and epitrillenogenin is the 24-diastereo-isomer of trillenogenin.

$1\beta,2\beta,3\beta,4\beta,5\beta,7\alpha$-Hexahydroxy spirost-25(27)-en-6-one (**63**) and $1\beta,2\beta,3\beta,4\beta,5\beta,6\beta,7\alpha$-heptahydroxy spirost-25(27)-ene(**6 4**), the two spirostane derivatives highly oxygenated at A- and B-rings have been isolated and structures elucidated by Miyahara et al. (Tetrahedron Lett. 1980, 21, 83). The i.r. spectrum of compound (**63**) indicate the presence of spirost-25(27)-ene together with hydroxyl and 6-membered ring ketone. The p.m.r. spectrum of its acetate exhibits the signals of 18-, 19- and 21-methyl groups and five acetoxyls and one hydroxyl proton. This spectrum also showed a singlet (2H, 4.75 ppm) and a double doublet (2H, 3.48 and 4.26 ppm) respectively ascribable to the terminal 27-methylene and the equatorial and the axial proton at C-26. An attempted regeneration of the sapogenin (**63**) from the acetate was not successful and yielded a new compound (**68**) the structure of which was proposed to be 5-hydroxy-1,2,3,4-tetranor spirost-5(10),25(27)-dien-6-one on the basis of its molecular formula and spectral data. The structure of the sapogenin (**63**) was established by X-ray crystallography and that of sapogenin (**64**) was suggested from the study of its i.r. and n.m.r. spectral data in relation to those of sapogenin (**63**).

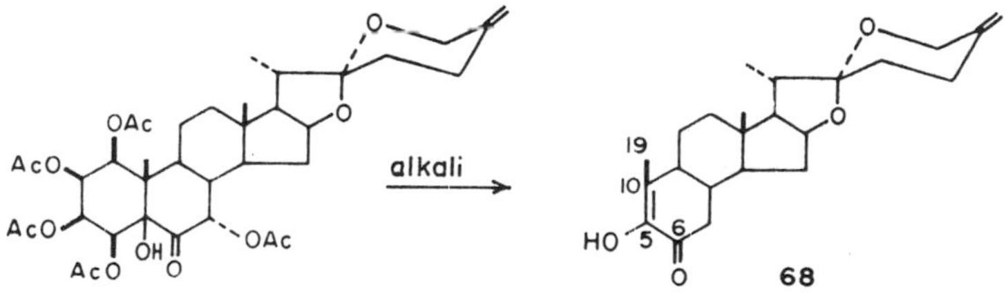

Acetate of **63**   alkali ⟶   **68**

The furostane sapogenins (**65,66**) have been isolated from Cordyline rubra (M. Yang et al. Phytochemistry 1990, **29**, 1332) which may be regarded as intermediates in the biosynthetic pathway for the formation of the spirostanol sapogenins. The biosynthetic origin of the spirostane nucleus from cholesterol requires hydroxylation at C-16, C-22 and C-26/27. 22-Hydroxylation is the final step as 22,26-dihydroxy furospirostanes readily cyclize unless glycosylated. Ronchetti et al. (Phytochemistry 1975, **14**, 2423) demonstrated stereospecific incorporation of [2,4,2',4'-3H$_4$][25S]-5α -furostane - 3β ,26-diol into neotigogenin in Lycopersicon esculentum. Thus isolation of these two sapogenins is significant from the point of view of biogenesis of the spirostanol sapogenins.

Wallogenin (**37**), (22R, 25R)-5α -furostan-22,25-epoxy-1β ,3α,26-triol is epimeric at C-22 to all the previously reported furanose F-ring sapogenins and its structure has been determined by mass spectral analysis as well as by comparison of its c.m.r. data, particularly of the absorptions of C-15, C-16, C-21, C-23, C-24 and C-26 to those of rubragenin, (22S, 25R), 5α-furostane-22,25-epoxy-1β ,3α,26-triol and strictagenin,(22S, 25S), 5α-furostan-22,25-epoxy - 1β ,3α ,26-triol (M. Yang et al. Phytochemistry 1989, **28**, 3171).

(22S, 25S)-5β-Spirostan-22,26-epoxy-1β,3β-diol (**53**) (22-epirhodeasapogenin) and (22R, 25S)-5α-spirostan-22,26-epoxy-3β,6β,23α-triol (**52**) (Hispigenin) are epimeric at C-22 to rhodeasapogenin and penniculogenin respectively (cf table-1).

A B-ring contracted spirostane, leontogenin has been isolated and characterized by Abdel-Aziz et al. (Phytochemistry 1990, **29**, 2623). Its structure has been determined to be that of (25R)-B-nor(7)-6β-formyl-spirostane-3β,5β-diol (**67**) by spectroscopic methods. The formation of leontogenin from a Δ^5-precursor possessing a 3β-OH group has been suggested to have occurred by oxidation of the 5-ene precursor producing the keto aldehyde which undergoes condensation to give the sapogenin.

The roots of Gynura japonica are used in Chinese medicine for the purpose of the coagulation of blood. Four new steroidal compounds, 3-epidiosgenin-3β -D-glucopyranoside (**69**), 3-episceptrumgenin-3β-D-glucopyrano-

side (**70**), 3-epiruscogenin (**71**) and 3-epineoruscogenin (**72**)
were isolated as the major constituents from the roots. The
compounds are the first examples of naturally occurring 3α-
hydroxyspirost-5-enes. Their structures were established by
chemical transformations (M.Takahira et al.Tetrahedron Lett.
1977, 3647).

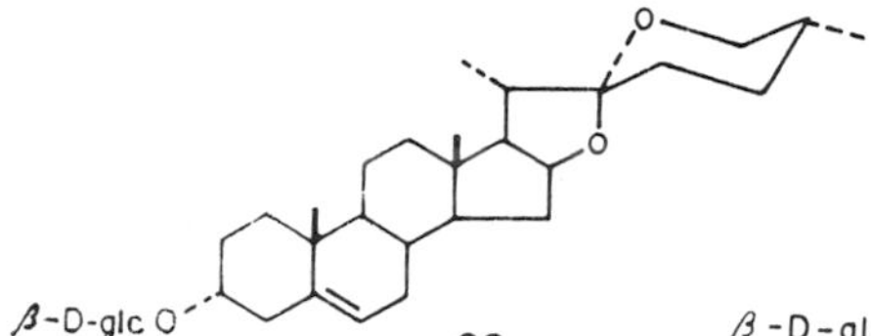

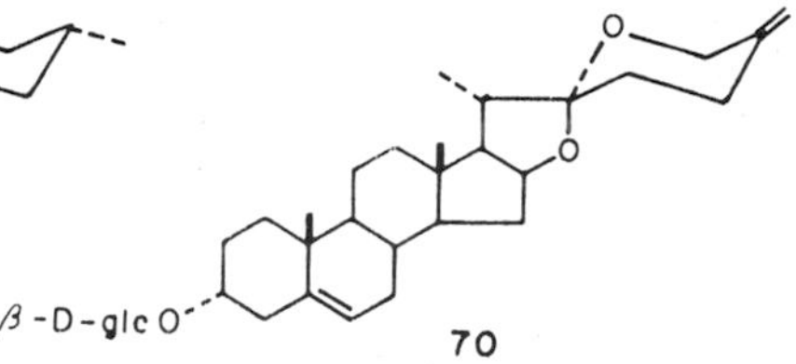

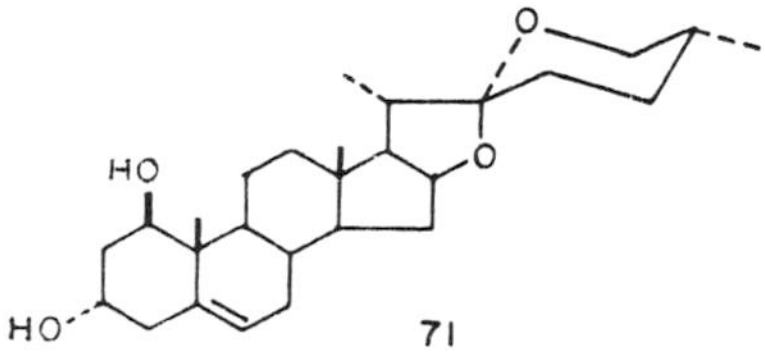

(d) Nomenclature and configuration

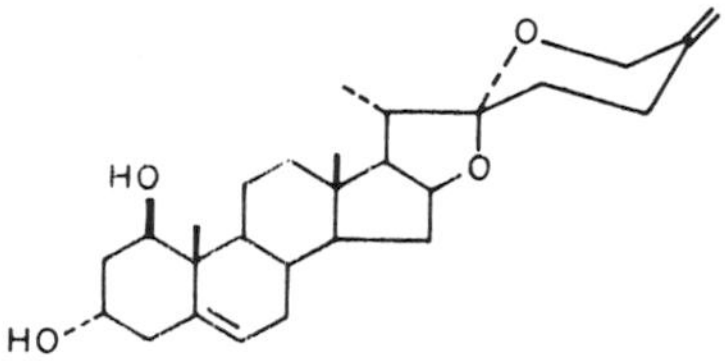

The name spirostan is given to the skeleton **73** and
furostane to the skeleton **74** which specify configuration of
chiral centres other than C-5 and C-25. Configurations at 5-,
22- and 25- positions of spirostans and furostans are

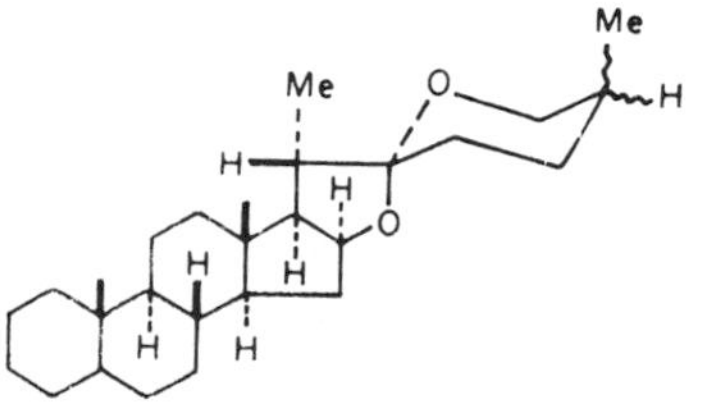

expressed in the modifications if known. For example
pompeygenin (**33**) 22-epirhodeasapogenin (**53**), rubragenin
(**36**) and wallogenin (**37**) are named as $(22R,25S)-5\alpha$-
spirostane-1β,3α,25-triol;(22S,25S)-5β-spirostane-1β,3β-diol;

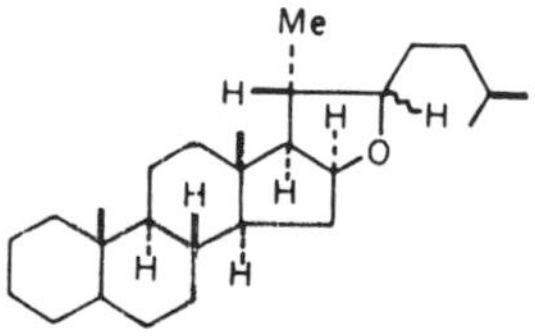

(22S,25R)-5α -furostan-22,25-epoxy-1β,3α,26-triol and (22R, 25R)-5α-furostan-22,25-epoxy-1β ,3α ,26-triol respectively using sequence rule procedure (Cahn, Ingold and Prelog Angrew Chem. Int. Edn. Engl. 1966, 5, 385)

A convenient method has been suggested by Eliel (J. Chem. Edunc. 1985, 62, 223) and elaborated by Nasipuri (Stereochemistry of Org. Compds. Principles and Applications, Wiley Eastern Ltd, New Delhi, 1991) for determination of absolute configuration of each of the chiral centres in the cyclic molecules e.g. steroids and terpenes which are usually projected on the plane of the paper and hydrogen (or substituent) located below and above the plane are assigned α -and β -descriptors respectively (α is represented by dotted and β by thick lines) as shown in trillenogenin (54).

54

At any particular chiral centre one ligand must be clearly in front (F) or clearly in back (B), this would be regarded as the reference ligand. The order (clockwise or anticlockwise) of the remaining three can be easily determined, all three being in the same plane of the paper. If this reference ligand is 4/B (lowest locant and in the back), the sequence of the remaining three ligand would give the correct descriptor. So 4/B (+) may be used as a mnemonic [(+) stands for correct]. For other combination the numbers with alternate signs, e.g. 4/B(+), 3/B(-), 2/B(+) and 1/B(-) and 4/F(-), 3/F(+), 2/F(-) and 1/F(+). Thus the C-25 chiral centre of trillenogenin corresponds to 4/F and the order 1 → 2 → 3 is anticlockwise and so the configuration is R.

(e) Triterpenoid saponins and sapogenins

Triterpenoid saponins and sapogenins are perhaps the most ubiquitous non-steroidal secondary metabolites in terrestrial and marine flora and fauna. The methods for isolation and structure elucidation of both steroidal and triterpenoid saponins are similar. However, some specific techniques have been developed in relation to the structure elucidation of triterpenoid saponins because of their special structural features. Chapter 17 of this volume deals with triterpenes and instead of listing the triterpenoid saponins and sapogenins some salient points of interest are described here.

(i) Saponins

Generally acid hydrolysis is used for liberating the aglycones and sugar moieties of saponins. However, if a saponin contains an acid labile aglycone milder hydrolysis techniques are needed. The method of photochemical degradation has been developed by Kitagawa et al. (Chem. Pharm. Bull. 1974, **22**, 1339) for selective cleavage of the glucuronide linkage. Various other techniques have also been reported for specific cleavage of glucuronide linkages, lead acetate and alkali (I.Kitagawa et al. Tetrahedron Lett. 1976, 549), anodic oxidation (I.Kitagawa et al. Chem. Pharm. Bull. 1980, **28**, 3078) or acetic anhydride and pyridine (I. Kitagawa et al. Chem. Pharm. Bull. 1977, **25**, 1408) have been used for the purpose. The method of diazomethane degradation of the sugar aglycone linkage is desacylsaponins obtained from the bark of Quillaja saponaria has been used by Higuchi et al. (Phytochemistry 1987, **26**, 229). The mechanism of degradation has been studied using gypsogenin 3-O-glycoside (R. Higuchi et al. Leibigs Ann. Chem. 1988, **3**, 249) which contains a 4α-aldehyde group in the aglycone, and a carboxy and 4'-hydroxy group in the aglycone-bound glucuronic acid of the carbohydrate moiety. The role of each of these functional groups was examined using model compounds and it was observed that the presence of 4α-aldehyde group was essential. By taking into account this evidence and the reaction of the aldehyde group with diazomethane, the degradation is thought to proceed through an epoxide intermediate. Smith degradation (I.J.Goldstein et al. Methods Carbohydr. Chem. 1965, **5**, 361) is sometimes employed for the determination of the sugar sequence in saponins. Thus, the structure of the sugar moiety of guaianin, a new nor-triterpenoid saponin isolated from Guaiacum officinale has been elucidated by this technique (V.U.Ahmed et al. J. Nat. Prod. 1986, **49**, 784). Alkaline hydrolysis is generally adopted for the cleavage of the ester glycosidic bond.

Bellericoside, a new pentacyclic triterpenoid ester-glucoside isolated from Terminalia bellerica (A.K.Nandy et al. Phytochemistry 1989, **28**, 2769) was subjected to alkaline hydrolysis with 5% methanolic KOH (aq.). This method was successfully applied for the selective cleavage of ester glycosidic bond of guaianin E, a nortriterpenoid bisdesmoside from Guaiacum officinale (V.U. Ahmed et al. Tetrahedron 1988, **44**, 247).

Triterpenoid saponins often contain acyl functions either in the aglycone or the sugar moieties. Mild alkaline hydrolysis with 2% $NaHCO_3$ in ethanol was found to be suitable for partial deacylation of gymnocladus saponin G, a complex triterpenoid saponin isolated from the fruits of Gymnocladus chinensis (T.Konoshima et al. Chem. Pharm. Bull. 1987, **35**, 46). A newer technique of hydrolysis using alcoholic-alkali-metal solution containing a trace of water has proved to be useful for isolation of acid labile aglycones (Y.Ogihara and M.Nose, J. Chem. Soc. Chem. Commun. 1986, 1417; Y. Chen et al. Chem. Pharm. Bull. 1987, **35**, 1653). The use of n-BuOH-Na-metal at water bath temperature (95°C) for 48 hours afforded the genuine triterpenoid anagalligenin B (S.B.Mahato et al. Tetrahedron 1991, **47**, 5215) as the major product which, however, could not be isolated by the usual acid hydrolysis.

The strategies employed for structure elucidation of two complex triterpenoid saponins of interest are briefly described. The structure of tubemoside I (**75**) a potent solubilizer and an antitumour agent from the bulb of Bolbostemma paniculatum (F.H. Kong et al. Tetrahedron Lett. 1986, **27**, 5765; R. Kasai et al. Chem. Pharm. Bull. 1986, **34**, 3974) was determined mainly by comparative n.m.r studies of 75 and its degradation products, 76 and 77. Compound 75 has an intersaccharide chain bridged by a dicrotalic acid to form an unique macrocyclic structure. The c.m.r.spectrum of 75 showed five anomeric sugar carbons. The presence of a dicrotalic acid moiety was also suggested from the n.m.r.data. The product 76 was obtained by enzymatic hydrolysis of 75 with cellulase which, on mild alkaline hydrolysis with 5% K_2CO_3, afforded the dicrotalic acid and the deacylated derivative 77. All proton signals of the sugar moieties of 76 and 77 were assigned using the COSY technique. Significance downfield shift of H-4 signals of rhamnose and the terminal arabinose indicated that the dicrotalic acid is linked to the 0-4 of rhamnose and the 0-4 of the

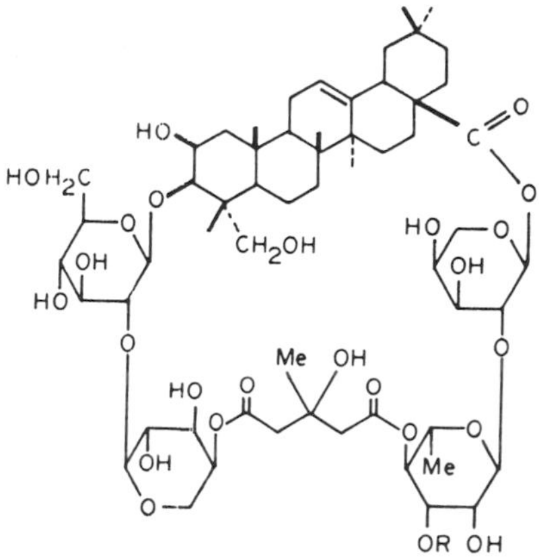

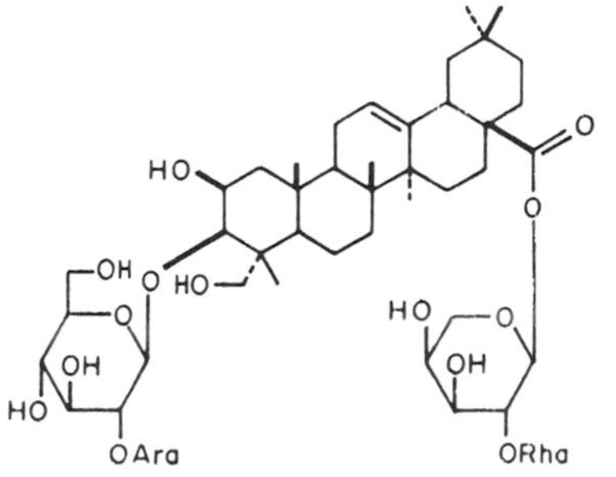

75 R = Xyl,   76 R = H

77

terminal arabinose. This was also corroborated by a set of acylation shifts (T.Konishima and T.Sawada, Chem. Pharm. Bull. 1982, **30**, 2747).

The complex structures of two acylated triterpenoid bisglycosides, acaciasides A (**78**) and B (**79**) were elucidated by a combination of fast atom bombardment mass

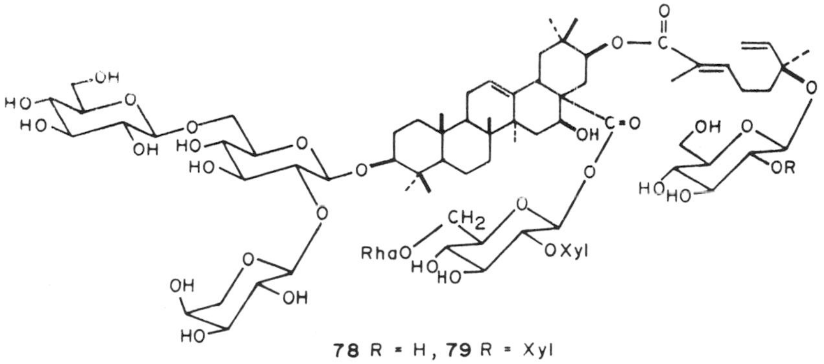

78 R = H, 79 R = Xyl

spectrometry, strategic chemical degradation, and n.m.r. spectroscopy (S.B. Mahato et al. Tetrahedron,1992,**48**, 6717) Acid hydrolysis of acaciaside A (**78**) yielded aglycones (**80**) and (**81**) and monosaccharides, D-glucose, D-xylose, L-arabinose and L-rhamnose. Hydrolysis of aglycone (**81**) with

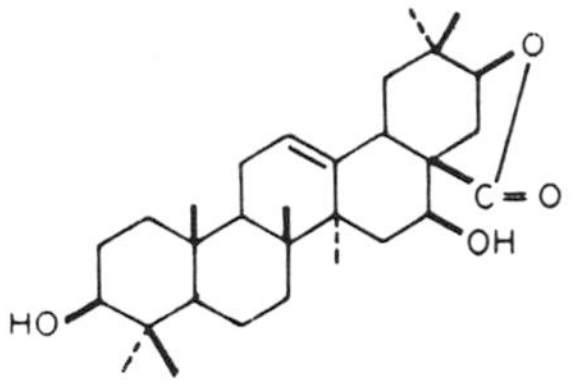

80

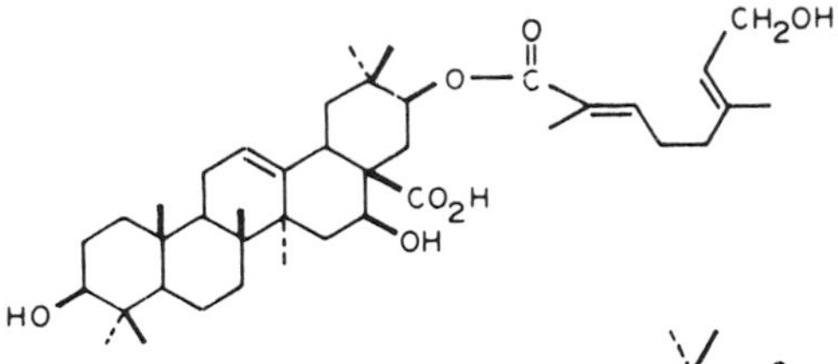

81

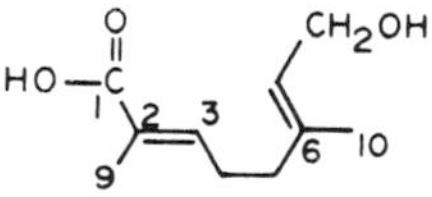

82

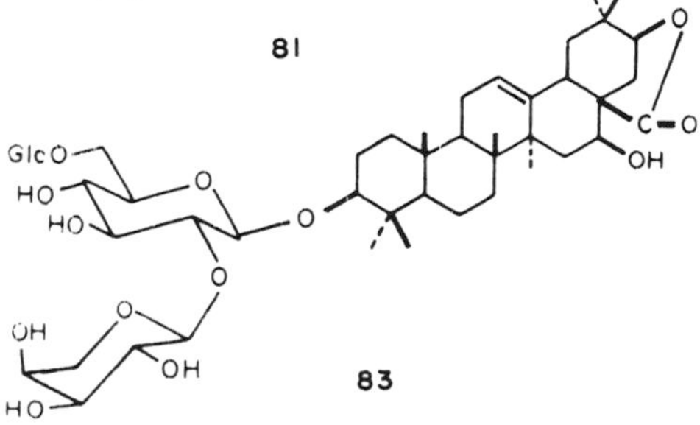

83

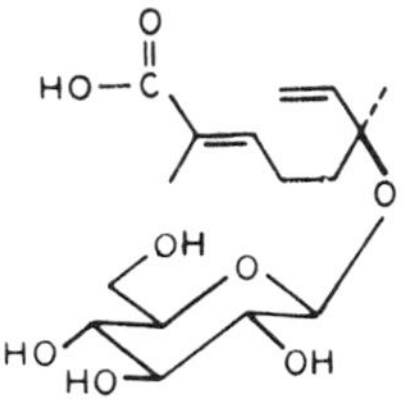

84

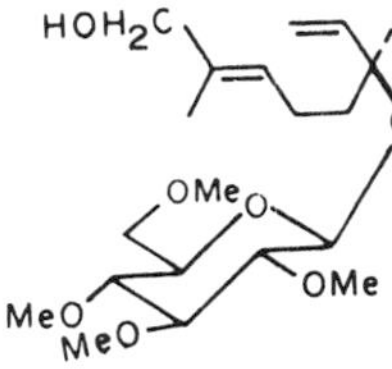

85

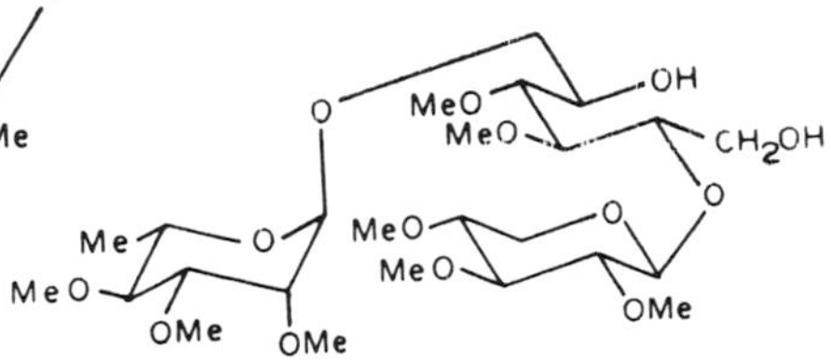

87

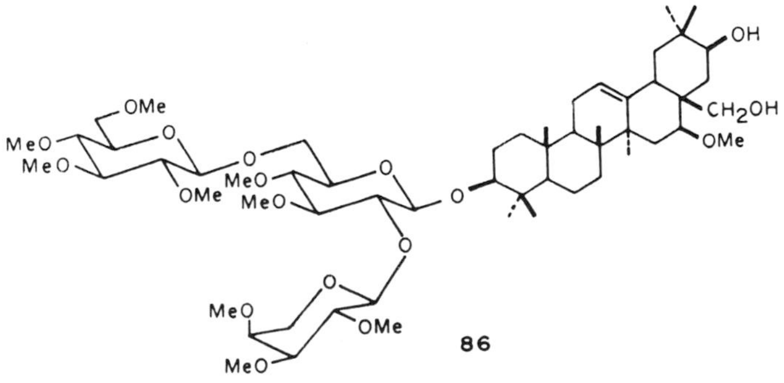

86

6% KOH in aqueous ethanol afforded the aglycone (**80**) and a monoterpene acid (**82**). However, alkaline hydrolysis of saponin (**78**) yielded acaciaside (**83**) and a monoterpene glycoside (**84**). Acaciaside A permethylate on reduction with LiAIH$_4$ furnished three products, a methylated monoterpene glycoside (**85**), a methylated triterpene glycoside (**86**) and a methylated oligosaccharide (**87**). These results together with spectroscopic data of the saponin disclosed its chemical structure. The structure of acaciaside B (**79**) was elucidated similarly.

(ii) Sapogenins

Unlike steroidal sapogenins which are mainly spirostane and furostane derivatives, triterpenoid sapogenins possess varied carbon skeletons (Chart-1). Despite the remarkable diversity that is already known to exist among the carbon skeletons of triterpenes, new variants continue to emerge. The triterpenoids including the new carbon skeleton triterpenoids have been discussed in a recent review (S.B. Mahato et al. Phytochemistry 1992, **31**, 2199) which covers literature from 1982 to 1989. Earlier comprehensive reviews on triterpenoids (M.C. Das and S.B. Mahato, Phytochemistry 1983, **22**, 1071; N. Basu and R.P. Rastogi, ibid., 1967, **6**, 1249; S.K. Agarwal and R.P. Rastogi, ibid 1974, **13**, 2623; P. Pant and R.P. Rastogi, ibid 1979, **18**, 1095; T.G. Halsall and R.T. Aplin , Progress in the Chemistry of Organic Natural Products, L. Zechmeister ed. 1964, **22**, 153, Springer, New York) covered literature upto 1981. Moreover, the triterpenes are described in chapter 17 of this volume.

"Friedo", "neo" and "abeo" system of nomenclature

"Friedo" preceded by D, D:A, D:B or D:C as prefixed to pentacyclic triterpene stereoparents indicates that angular methyl groups have shifted from their normal positions as follows :

D	14 to 13	D:B	14 to 13	D:A	14 to 13
D:C	14 to 13		8 to 14		8 to 14
	8 to 14		10 to 9		10 to 9
					4 to 5

The changes are cumulative in proceeding from the normal arrangement of methyl groups to the D:A-friedo arrangement (Chem. Abst. Index Guide 1972, **76**, 212)

546

Chart-1

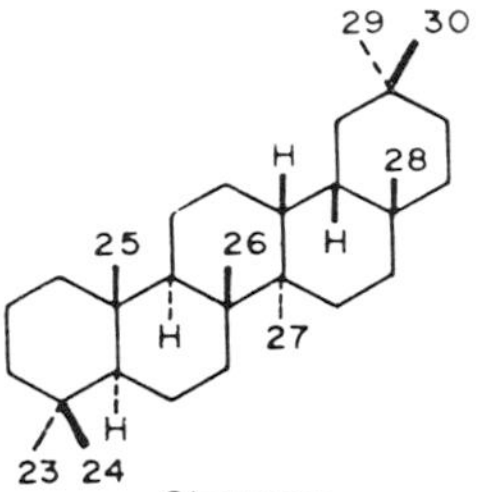

Oleanane

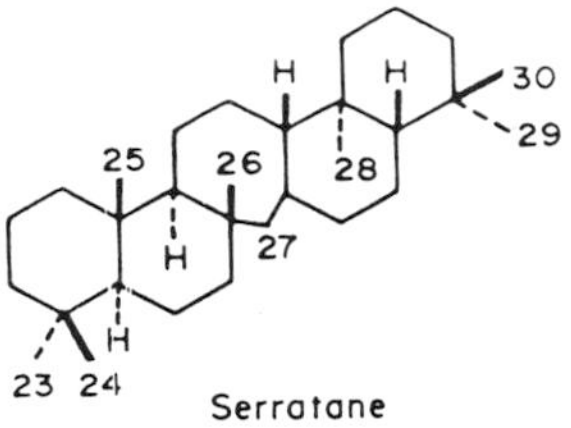

Serratane

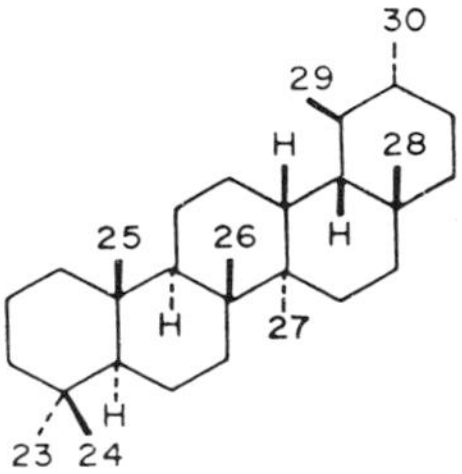

Urasane

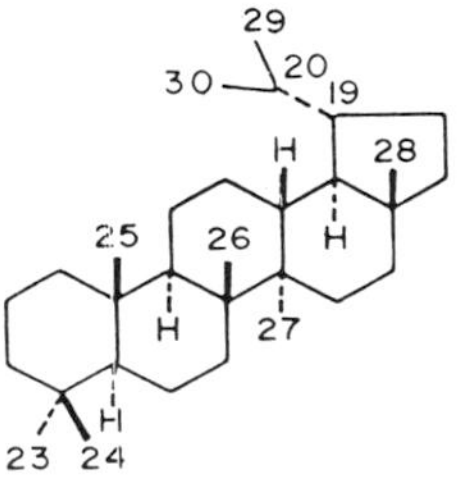

Lupane

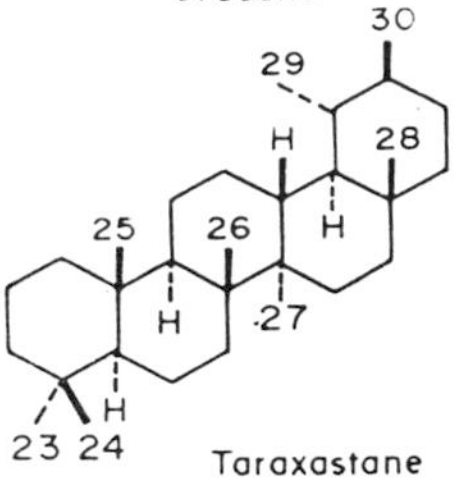

Taraxastane

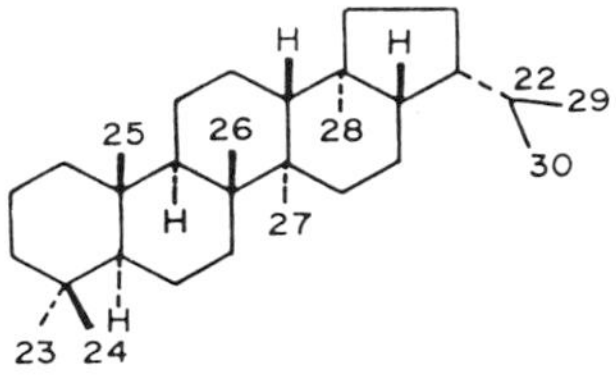

Hopane

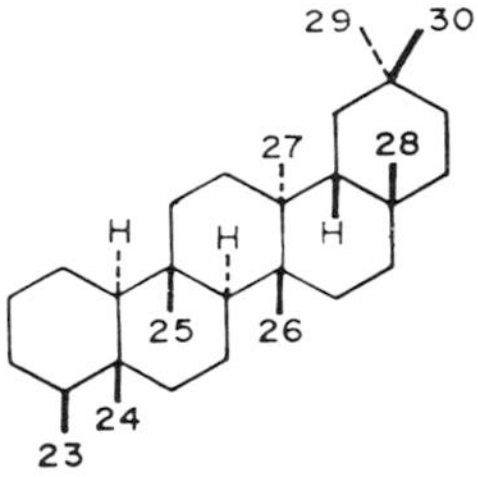

Friedelane

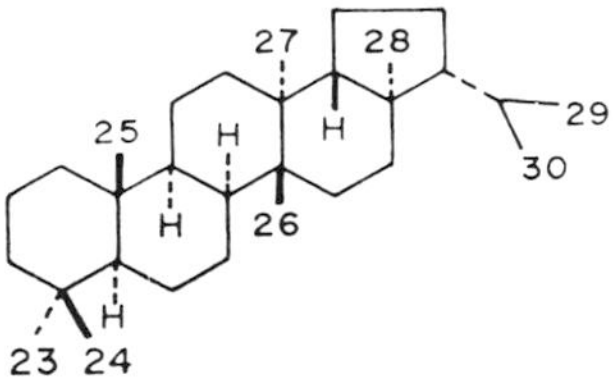

Fernane

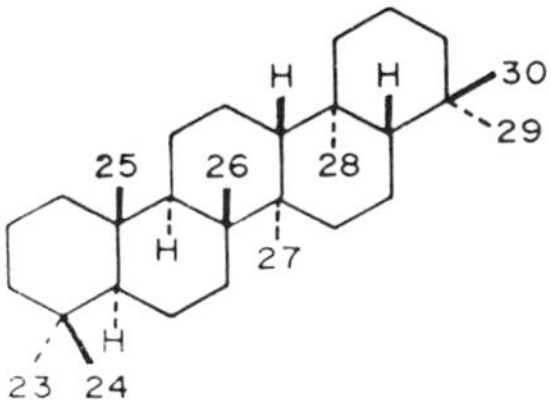

Gammacerane

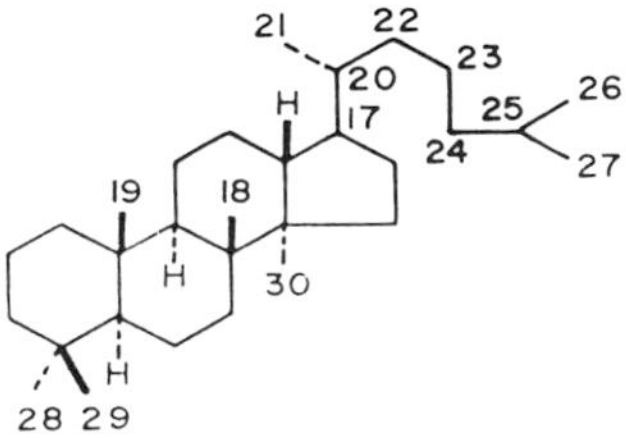

Dammarane

Chart 1 continued

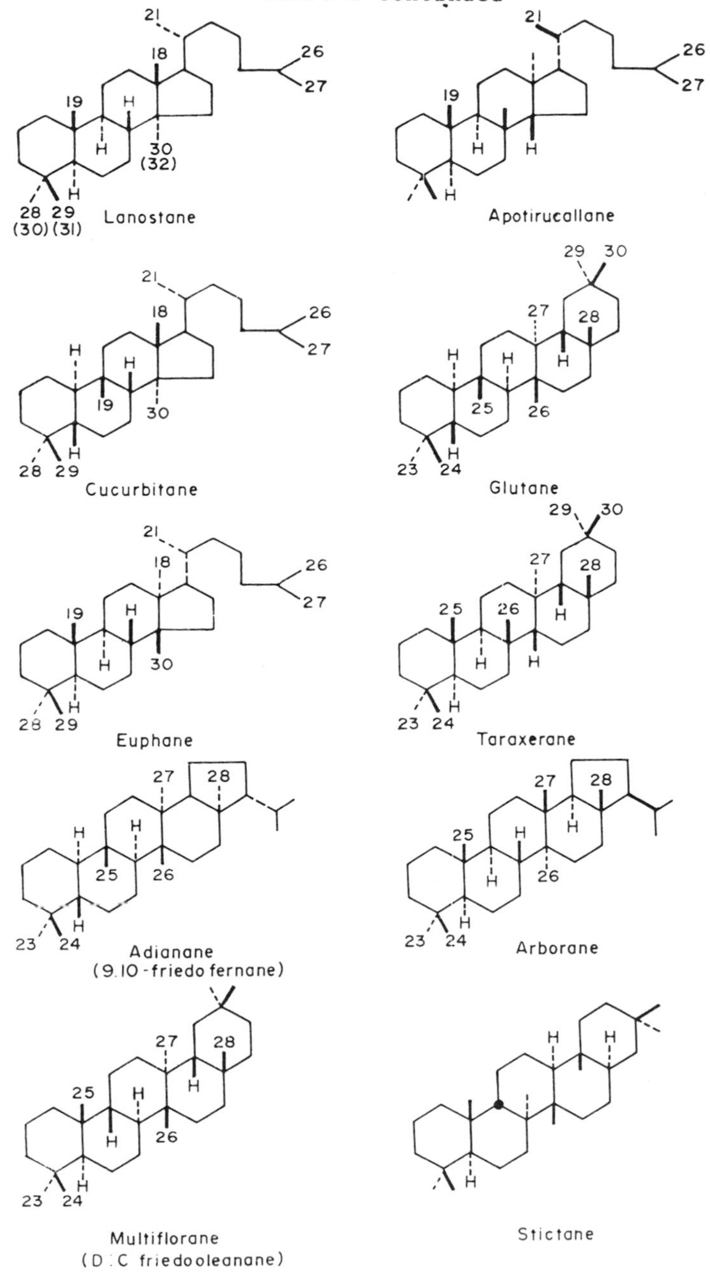

Lanostane

Apotirucallane

Cucurbitane

Glutane

Euphane

Taraxerane

Adianane
(9.10-friedo fernane)

Arborane

Multiflorane
(D:C friedooleanane)

Stictane

Chart 1 continued

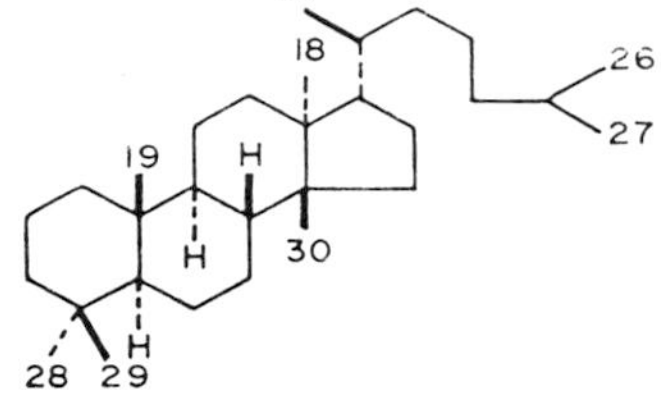

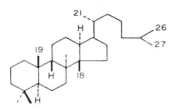

Tirucallane

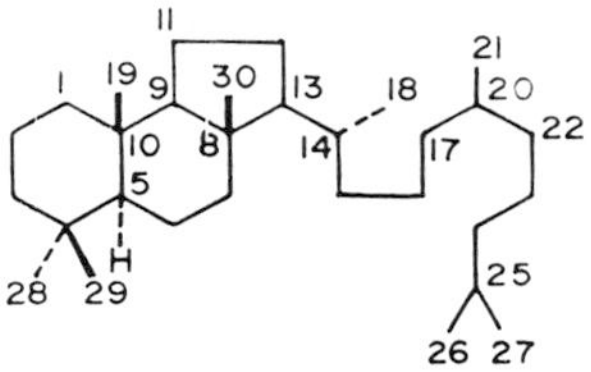

Malabaricane

Protostane

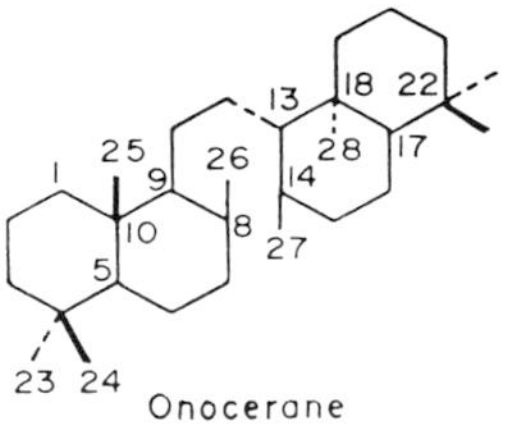

Onocerane

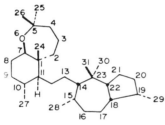

Sipholane

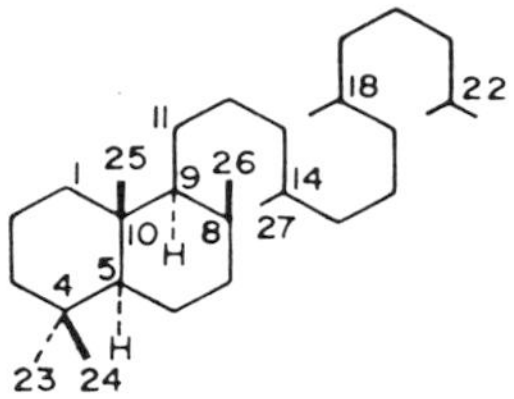

Polypodane

Siphonellane

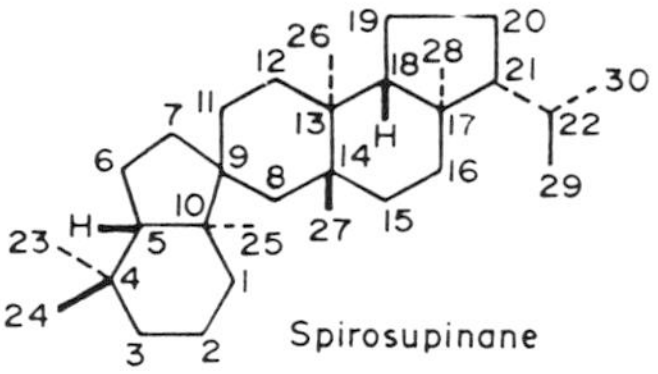

Spirosupinane

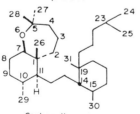

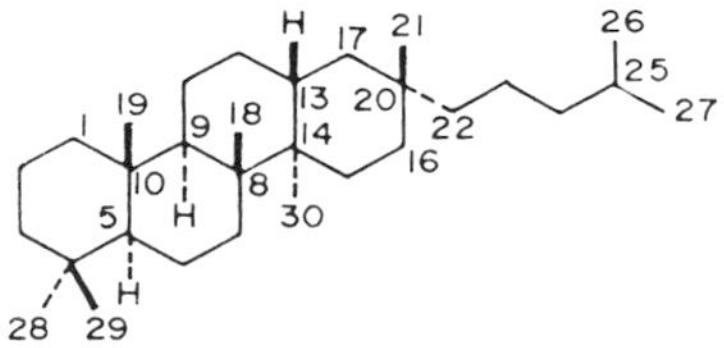

Baccharane

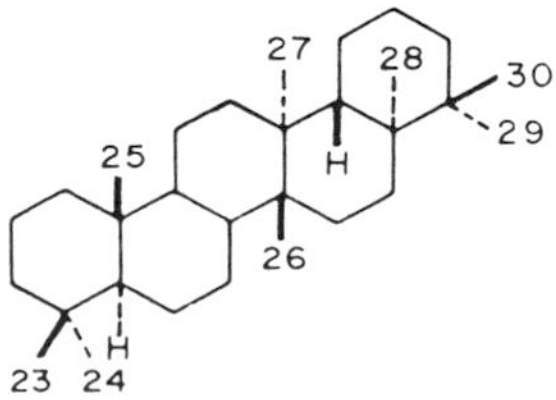

Swertane

Chart 1 continued

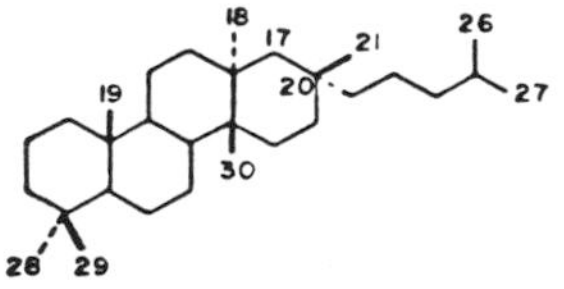

Lemmaphyllone

Pfaffone

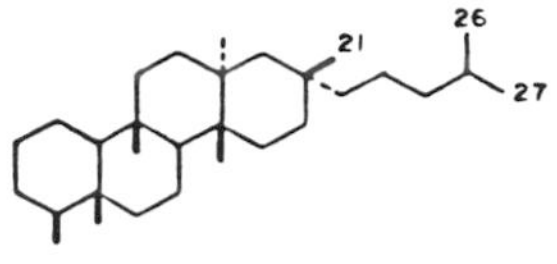

Shionone

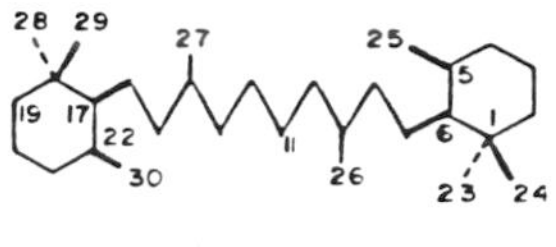

Carotenoid-like skeleton

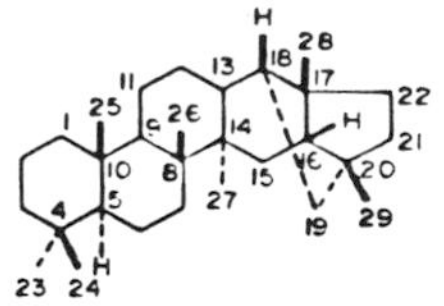

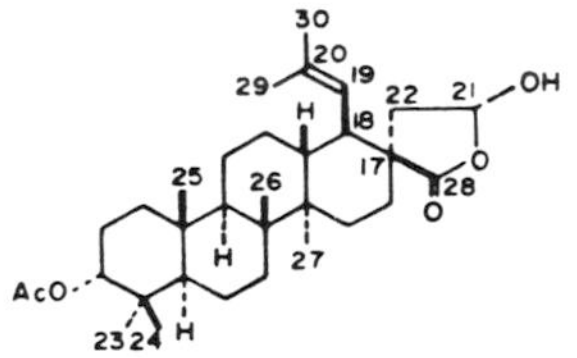

Radermasinin

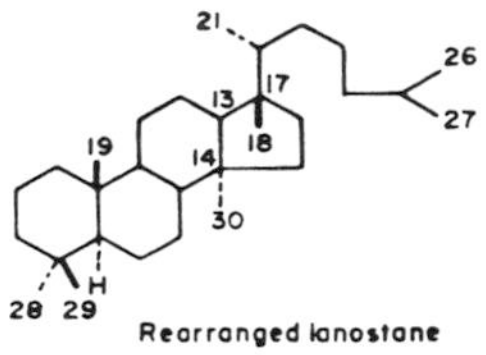

Rearranged lanostane

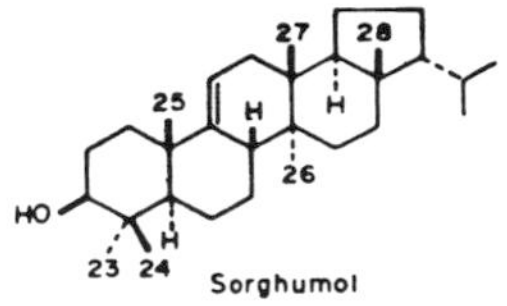

Sorghumol

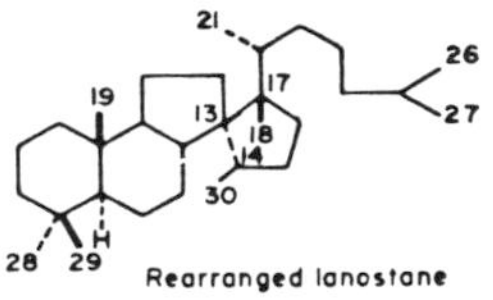

Rearranged lanostane

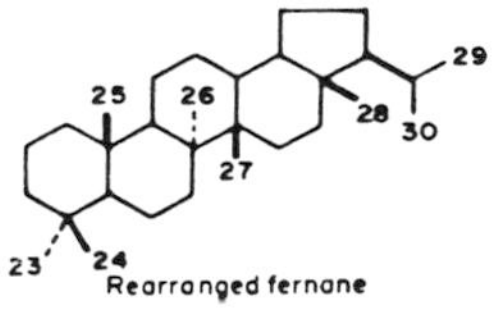

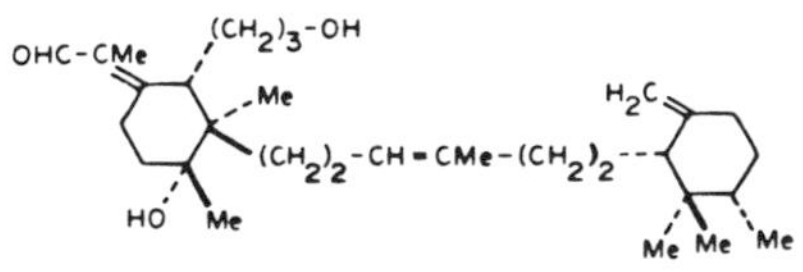

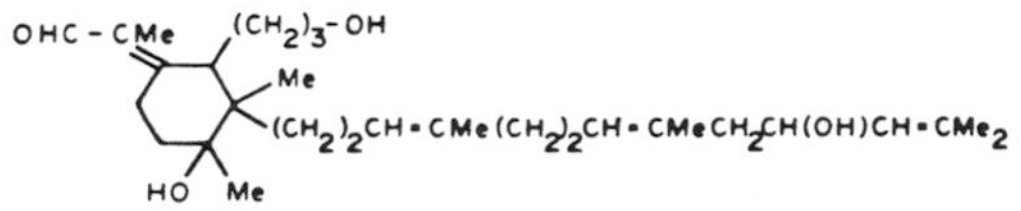

Rearranged fernane

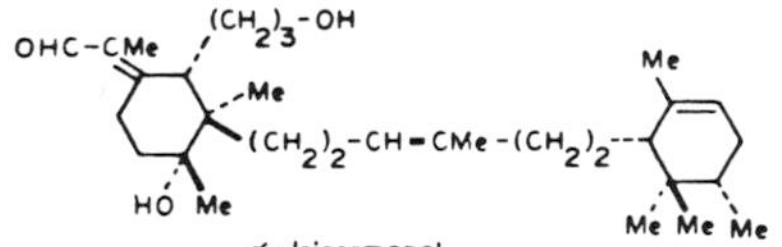

α-Irigermanal

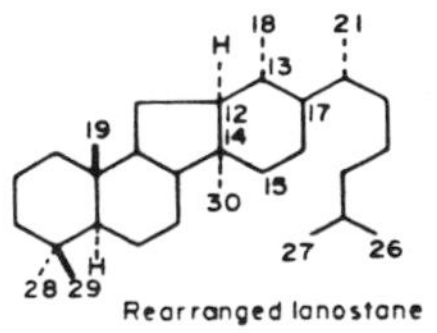

OHC–CMe (CH$_2$)$_3$–OH Me (CH$_2$)$_2$–CH=CMe–(CH$_2$)$_2$– HO Me Me Me Me

Rearranged lanostane

γ-Irigermanal

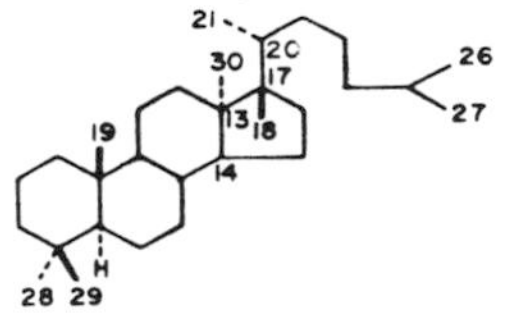

OHC–CMe (CH$_2$)$_3$–OH Me (CH$_2$)$_2$–CH=CMe–(CH$_2$)$_2$– HO Me Me Me Me

Rearranged lanostane

iridogermanal

OHC–CMe (CH$_2$)$_3$–OH Me (CH$_2$)$_2$CH=CMe(CH$_2$)$_2$CH=CMeCH$_2$CH(OH)CH=CMe$_2$ HO Me

Chart 1 continued

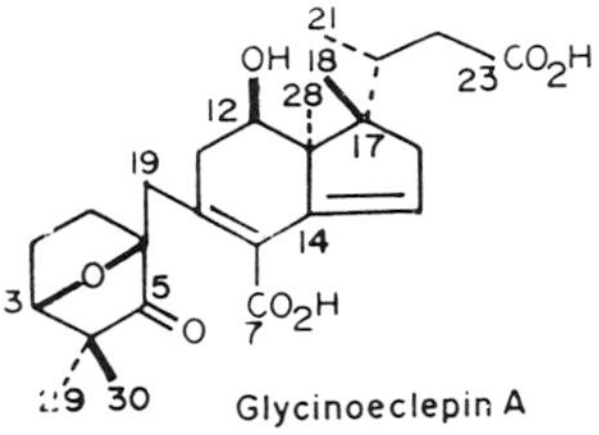

Glycinoeclepin A

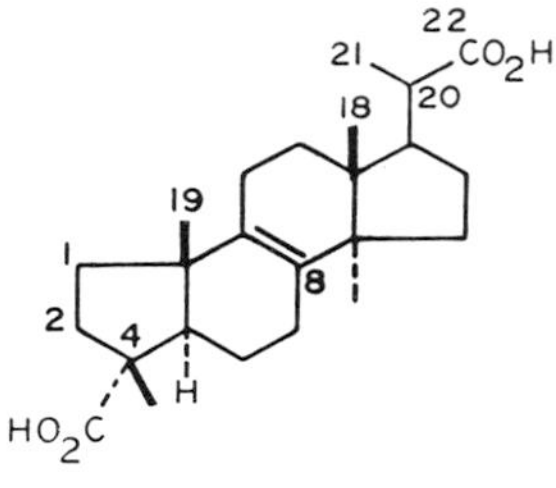

Javeroic acid

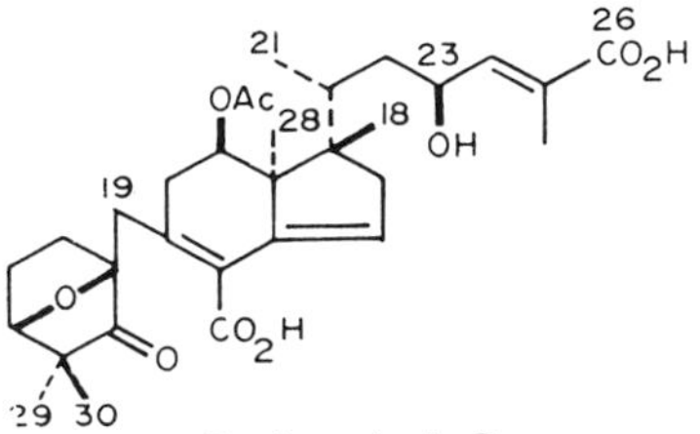

Glycinoeclepin B

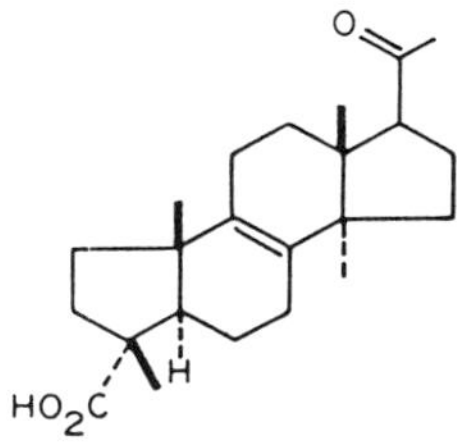

Phellinic acid (45)

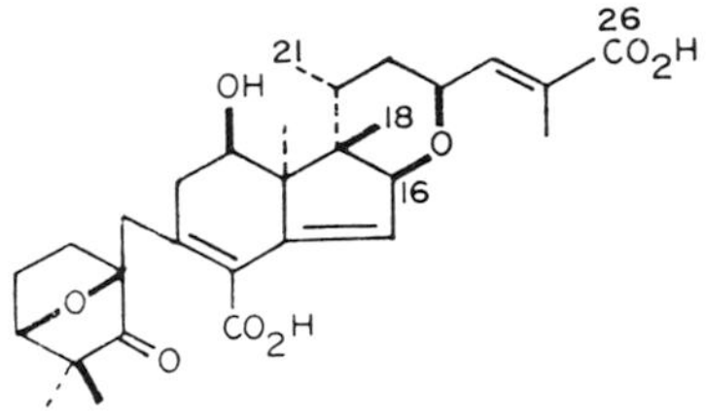

Glycinoeclepin C

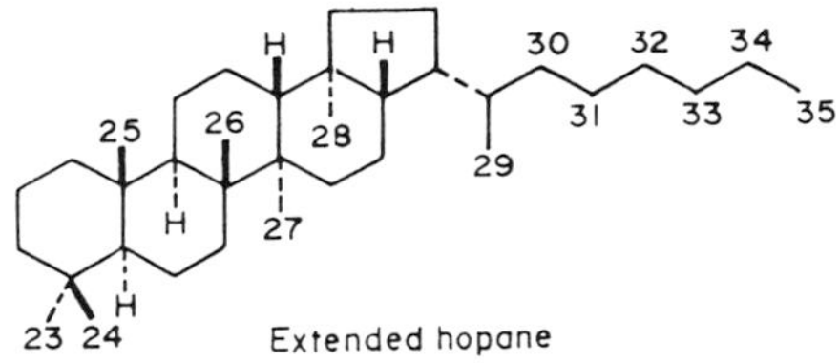

Extended hopane

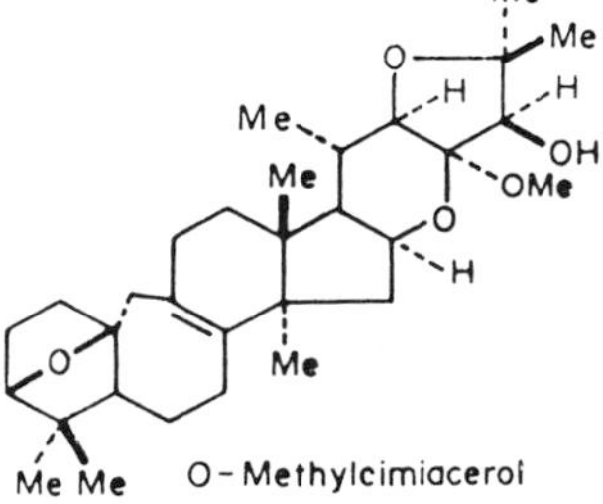

O-Methylcimiacerol

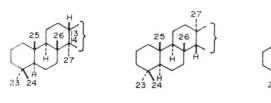

Normal arrangement   [D- Friedo-]   [D:A - Friedo-]

"Friedo" preceded by numbers indicates that angular methyl groups have shifted from their normal positions as follows.

17,13-friedolanostane	13 to 17
17,14-friedolanostane	13 to 17
	14 to 13

The prefix "neo" preceded by an italicized capital **A** indicates that ring **A** has undergone the following rearrangement.

The prefixes "A:B-Neo-", "A:C-Neo-" and "A:D-Neo-" indicate "A-Neo-" ring alteration plus migration of angular methyl groups located in positions 10,8 and 14 respectively For example

A:B-Neo-	A-Neo-alteration	A:D-Neo-	A-Neo-alteration
	10 to 5		10 to 5
A:C-Neo-	A-Neo-alteration		8 to 9
	10 to 5		14 to 8
	8 to 9		

Normal arrangement   A - Neo-   A:C - Neo   A:D - Neo

In the gammacerane and onocerane skeletons there is additional modifications A'-Neo- and B':A'-Neo- as follows.

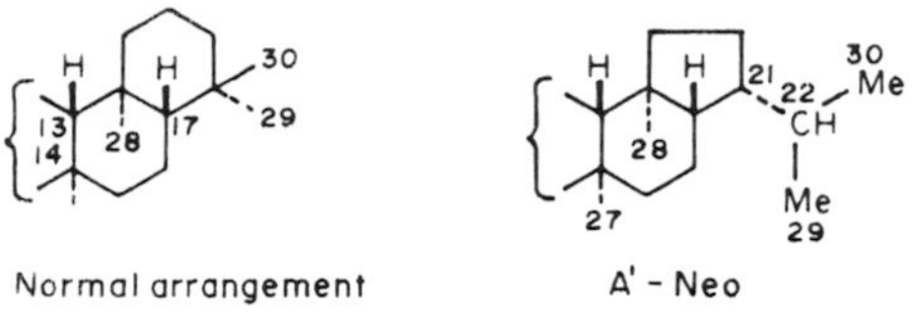

Normal arrangement   A' - Neo   B': A' - Neo -

"Abeo" terpenes are those in which a bond has migrated. Thus the skeleton **88** represents 17,13-friedolanostane and the skeleton **89** is 8(14 ⟶ 13R) abeo, 17,13-friedolanostane in which 8,14-bond has been replaced by a 8,13-bond, contracting ring C from six to five members (S. Hasegawa et al. Tetrahedron 1987, **43**, 1775).

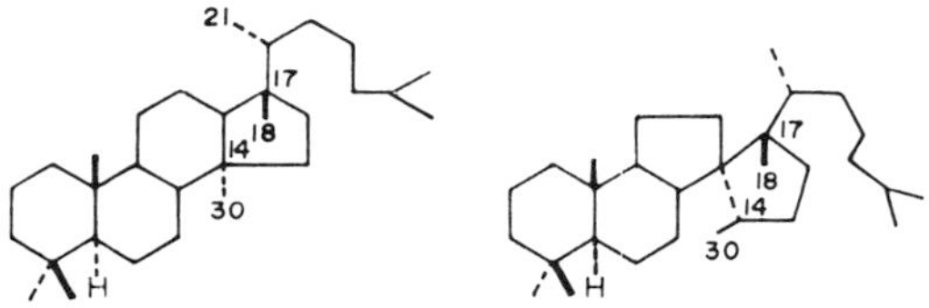

88   **89**

(f) Synthesis

It has now been established that saponins are important bioactive ingredients of various Oriental and Western plant drugs. However, complex mixtures of saponins are usually present in plants and chemical synthesis of saponins has been seriously hampered by their complex structural features. Nevertheless, it is of interest to synthesize simpler bioactive saponins. The first synthesis of mosesin-4 (**90**) a naturally occurring steroid saponin with shark repellent activity and its analogue (**91**) has been reported by Gargiulo et al. (Tetrahedron 1989, **45**, 5423). Both of these compounds possess a free galactose residue attached axially at the 7α-position of the steroidal moiety. Me-cholate 3-cathylate was used as a model for exploring various methods to glycosylate the severely hindered 7α-position. Best results were obtained with β-galactose pentaacetate, using trimethylsilyl triflate as a promoter in 1,2-dichloroethane at -20°C for 14h

Saikosaponins are important bioactive principles of Bupleuri Radix (roots of <u>Bupleurum</u> spp., Umbelliferae) which have been shown to exhibit antiinflammatory activity. Among

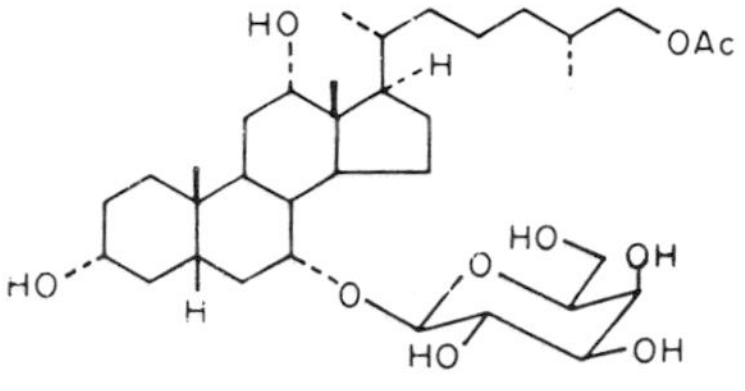

90

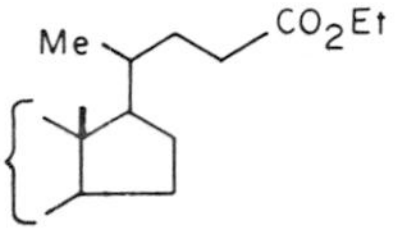

91

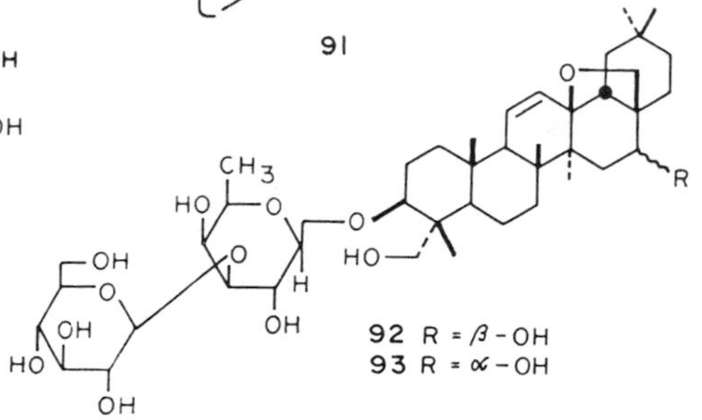

92 R = β-OH
93 R = α-OH

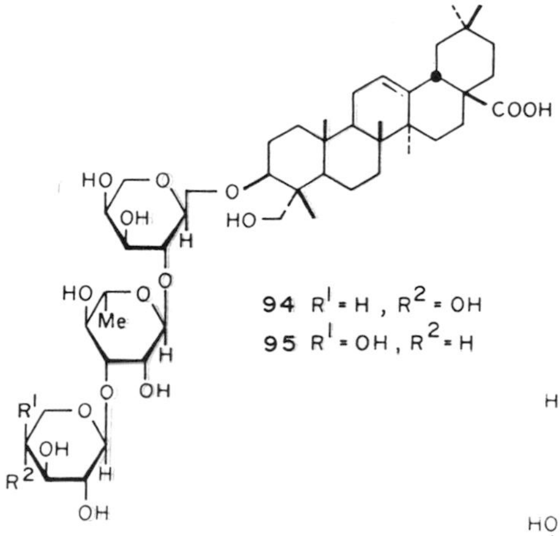

94 R¹ = H , R² = OH
95 R¹ = OH , R² = H

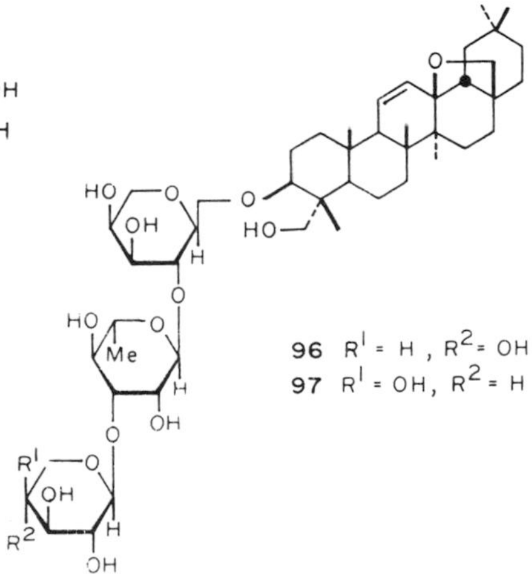

96 R¹ = H , R² = OH
97 R¹ = OH , R² = H

these saponins, saikosaponin – a (**92**) and – d (**93**), which possess the C-4 hydroxymethylfunction and the 11-en-13β,28-oxide moiety exhibit significant antiinflammatory activity (M. Yamamoto et al. Arzneim Forsch. 1975, **25**, 1021, 1240). Kitagawa (Isopentenoids in plants, W.D. Nes et al. eds Marcel Dekker Inc. New York 1984) has described the synthesis of 11-en-13β,28-oxide triglycosides (**96,97**) from readily available hederagenin triglycosides (**94,95**) involving the reduction of the methyl esters with LiAlH$_4$ in tetrahydrofuran followed by constant current electrolyis.

(g) Production of saponins by cell culture

In recent years there has been considerable interest in plant cell cultures as a potential alternative to traditional agriculture for large scale production of secondary plant metabolites and considerable effort in this direction is being made in the laboratories in Japan, Germany, Canada and to a lesser extent in the U.S.A. (M.F. Balandrin et al. Science 1985, **228**, 1154). However, cost analyses indicate that production of a secondary metabolite in plant cell culture is economical for cultures producing more than 1 gram of compound per litre of cell culture for compounds with a market value of at least $1000 per kg (M.E. Curtin Bio/Technology 1983, 1, 649). Mitsui Petrochemical Industries (Tokyo) is already marketing tissue culture-derived shikonin, a naphthaquinone dye and an astringent.

Extensive studies have been made on the production of saponins by cell suspension culture of Ginseng (T. Furuya et al. Planta Med. 1983, **47**, 183; ibid 1983, **48**, 83). It was observed that thiosemicarbazide promoted the biosynthesis of saponins in the presence of mevalonic acid, and the saponin biosynthesis was inhibited by the corresponding end products (T. Furuya et al. Planta Med. 1983, **47**, 200). The total saponin content in the cell suspension culture based on these studies was about 4 times higher than that in the parent plant. The production of Ginseng saponins in an industrial scale by cell culture technology is now being explored in Japan.

Chapter 17

TRITERPENOIDS

J.D. CONNOLLY AND R.A. HILL

In common with other classes of natural products there has been a huge increase in the number of triterpenoids isolated and identified over the past twenty years. This review concentrates on new triterpenoids that have novel skeletons or interesting structures and is therefore highly selective. Comprehensive coverage of triterpenoids can be found in recent compilations of data (J.D. Connolly and R.A. Hill, "Dictionary of Terpenoids", Chapman and Hall, London, 1991; S. Dev and B.A. Nagasampagi (vol. 1); S. Dev, A.S. Gupta and S.A. Parwardhan (vol. 2), "Handbook of Terpenoids, Triterpenoids", CRC Press, 1989). The isolation of new triterpenoids is covered in regular reviews (see J.D. Connolly and R.A. Hill, *Nat. Prod. Rep.*, 1989, 6, 475 and previous reviews; S.B. Mahato, A.K. Nandy and G. Roy, *Phytochemistry*, 1992, 31,2199).

1. The Squalene Group

The green alga *Botryococcus braunii* has attracted much attention since the report of the isolation of the rearranged tetramethylsqualene derivative botryococcene (**1**) in 1973 (R.E. Cox *et al., J. Chem. Soc., Chem. Commun.*, 1973, 284). The absolute configuration of botryococcene has been established as in (**1**) (J.D. White, T.C. Somers and G.N. Reddy, *J. Am. Chem. Soc.*, 1986, 108, 5352). Related botryococcenes e.g. (**2**) without additional methylation (P. Metzger *et al., Phytochemistry*, 1985, 24, 2995) and containing one [C-3 or C-20 (Z. Huang and C.D. Poulter, *Phytochemistry*, 1989, 28, 3043)] and two [C-3 and C-20 (P. Metzger and E. Casadevall, *Tetrahedron Lett.*, 1983, 24, 4013)] methylations have also been isolated. Biosynthetic studies reveal impressive incorporations of acetate (11.76%) and L-leucine (5.23%), with mevalonate being less well utilised (P. Metzger, E. Casadevall and M.-P. Peuch, *Tetrahedron Lett.*, 1984, 25, 4123) while additional methyl groups are derived from methionine (P. Metzger, M. David and E. Casadevall, *Phytochemistry*, 1987, 26, 129). Interestingly the tetramethylsqualene (**3**) occurs in *B. Braunii* var. *showa* (Z. Huang and C.D. Poulter, *Phytochemistry*, 1989, 28, 1467). Monocyclic botryococcenes, exemplified by braunicene (**4**), have recently been reported (Z. Huang *et al., J. Am. Chem. Soc.*, 1988, 110, 3959).

The alga *Caulerpa prolifera*, a natural source of squalene 2,3S-epoxide, also contains the 6S,7S- (**5**) and 10S,11S- (**6**) epoxides (L. De Napoli *et al., Phytochemistry*, 1982, 21, 782). The red alga *Laurencia okamura* produces the enantiomer of (**6**), the 10R,11R-epoxide the

556

structure of which was confirmed by chiral synthesis (H. Kigoshi *et al., Tetrahedron Lett.*, 1982, 23, 5413). Several squalene polyether derivatives have been found in *Laurencia* species. These are exemplified by thyrsiferol (**7**) from *L. thyrsifera* (J.W. Blunt *et al., Tetrahedron Lett.*, 1978, 69), its diastereoisomer venustatriol (**8**) from *L. venusta* (S. Sakami *et al., Tetrahedron Lett.*, 1986, 27, 4287) and magireol A (**9**) from *L. obtusa* (T. Suzuki *et al., Chem. Lett*, 1987, 361). Teurilene (**10**) from *L. obtusa* lacks bromine and represents a different mode of cyclisation (T. Suzuki *et al., Tetrahedron Lett.*, 1985, 26, 1329). The structures of thyrsiferol (**7**), venustatriol (**8**) and teurilene (**10**) were all established by X-ray

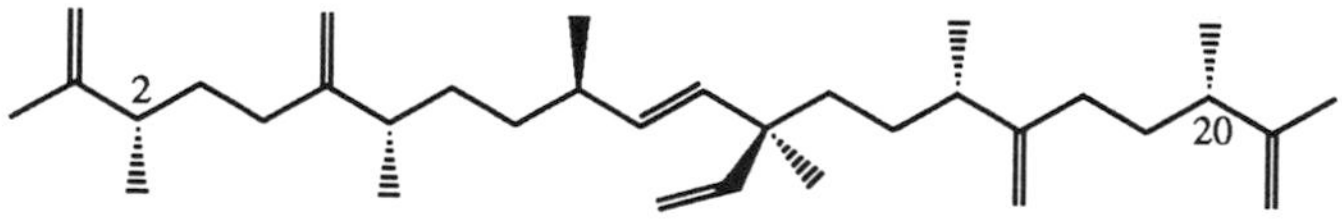

(1)

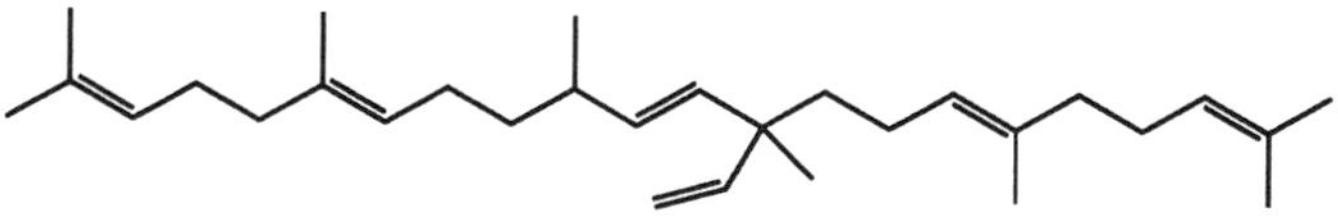

(2)

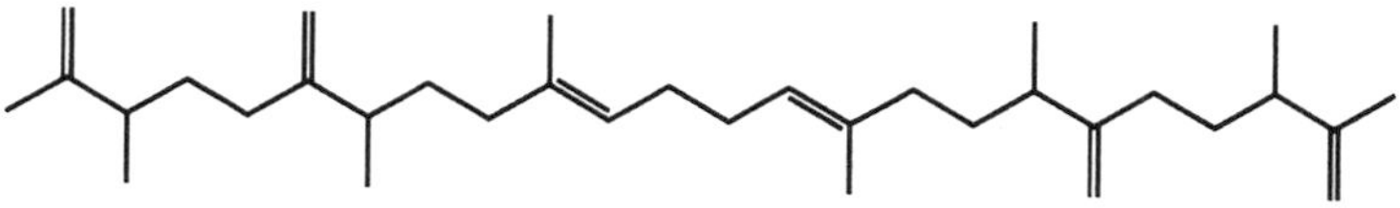

(3)

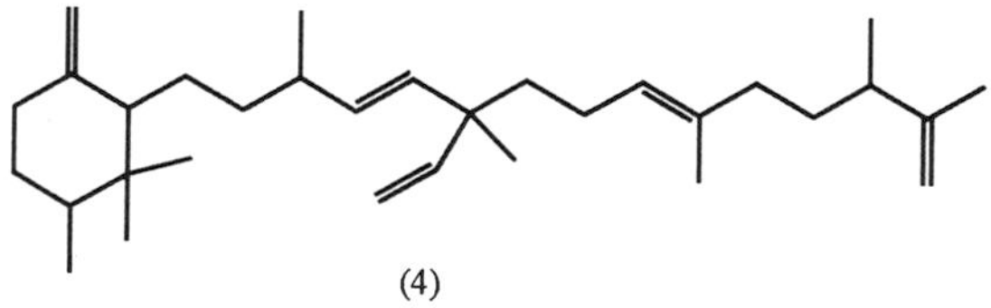

(4)

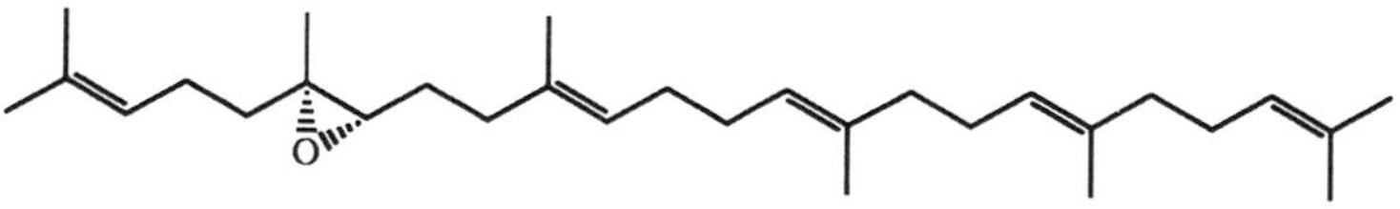

(5)

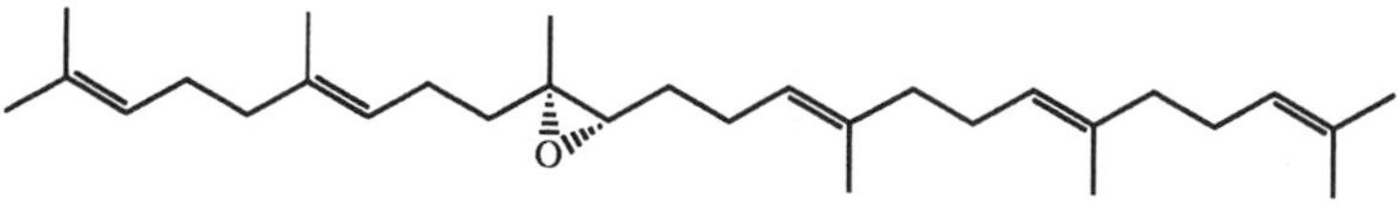

(6)

analysis and have all been recently synthesised (E.J. Corey and D.-C. Ha, *Tetrahedron Lett.*, 1988, 29, 3171; M. Hashimoto *et al., J. Org. Chem.*, 56, 2299; M. Hashimoto *et al., J. Org. Chem.*, 1991, 56, 2299). Quassiol A (**11**), from *Quassia multiflora* (W.F. Tinto *et al., Tetrahedron Lett.*, 1993, 34, 1705), eurylene (**12**) and longilene peroxide (**13**) from *Eurycoma longifolia* (H. Itokawa *et al., Tetrahedron Lett.*, 1991, 32, 1803; *Chem. Lett*, 1991, 2221) are examples of this type of compound from plant sources. Crystal structure analyses were performed on eurylene (**12**) and longilene epoxide (**13**).

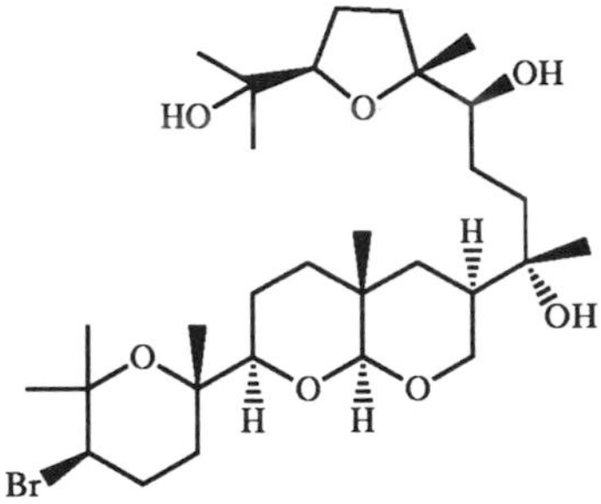

(7)

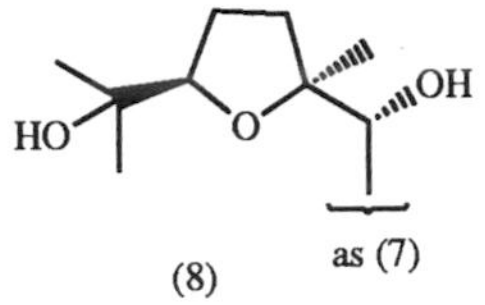

(8)   as (7)

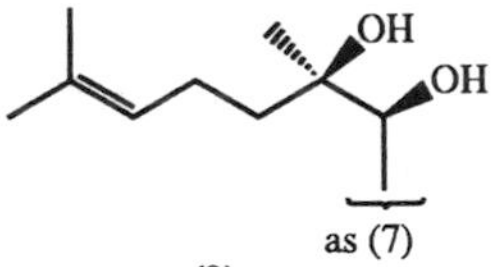

(9)   as (7)

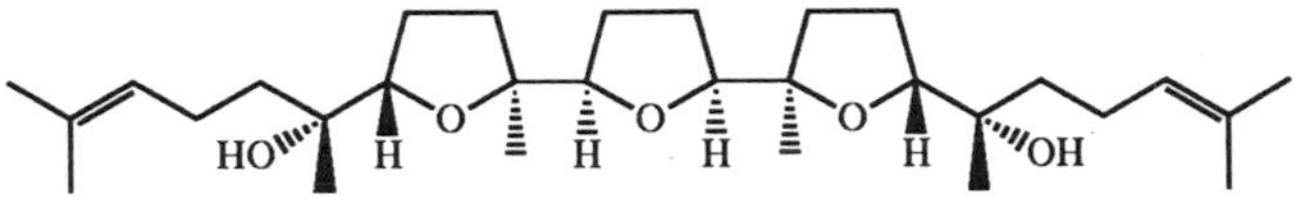

(10)

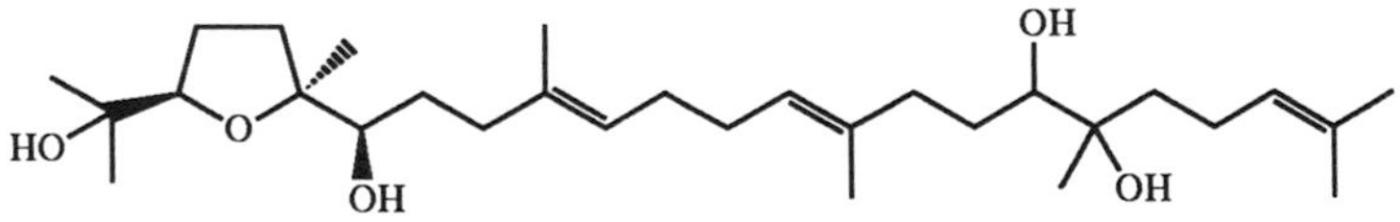

(11)

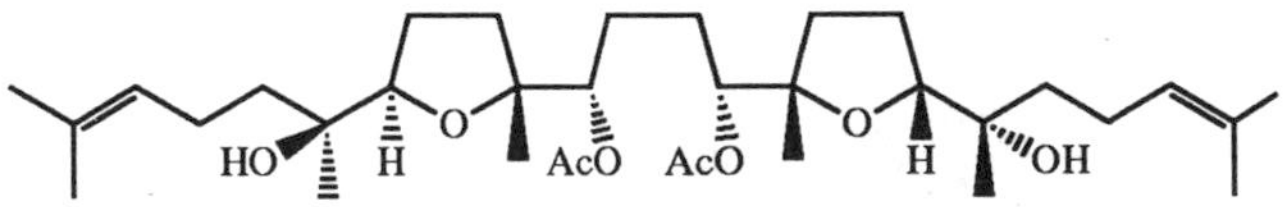

(12)

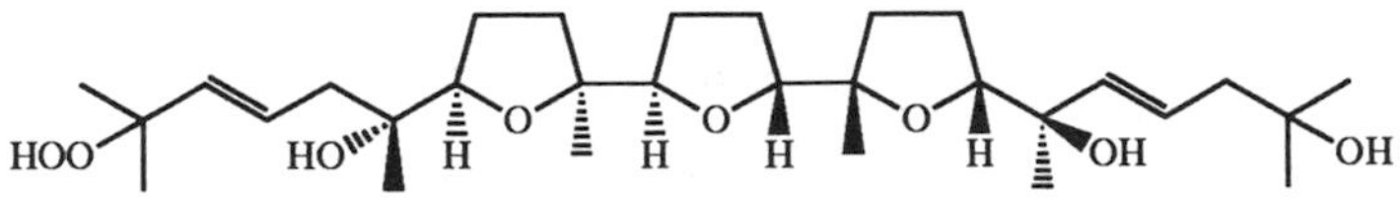

(13)

Squalene 2,3S-epoxide has been synthesised with an enantiomeric excess of 92% (E.J. Corey K.Y. Yi and S.P.T. Matsuda, *Tetrahedron Lett.*, 1992, 33, 2319). Considerable recent effort has gone into the preparation and characterisation of the 2,3S-oxidosqualene-lanosterol cyclase from yeast and liver (E.J. Corey and S.P.T. Masuda, *J. Am. Chem. Soc.*, 1991, 113, 8172; T. Hoshuno *et al.*, *Tetrahedron*, 1991, 47, 5925; M. Kusano, *Chem. Pharm. Bull.*, 1991, 39, 2397). Interest continues in the enzymic cyclisation of modified squalene epoxides, an area which owes much to the monumental contributions of van Tamelen and his colleagues (E.E. van Tamelen *et al.*, *J. Am. Chem. Soc.*, 1982, 104, 6479). Oxidosqualene cyclase controlled cyclisation of appropriately hydroxylated squalene epoxides yields lanosterols hydroxylated on C-28, C-19 or C-21 (X.-Y. Xiao and Prestwich *Tetrahedron Lett.*, 1991, 32, 6843; J.C. Medina and Kyler *J. Am. Chem. Soc.*, 1988, 110, 4818). The homosqualene oxides (14) and (15) were succesfully cyclised under enzymatic conditions to the homolanosterols (16) (J.C. Medina, R. Guajardo and K.S. Kyler, *J. Am. Chem. Soc.*, 1989, 111, 2310) and (17) (X.-Y. Xiao, S.E. Sen and G.D. Prestwich, *Tetrahedron Lett.*, 1990, 31 2097). The formation of (16) necessitates the migration of the vinyl group during the biosynthesis.

Bacterial squalene epoxide cyclase appears to be less specific than that present in typical eukaryotes and is able to utilise the 3R-enantiomer of squalene epoxide. A cell-free preparation from *Methylococcus capsulatus* converted the 3R-enantiomer into 3-*epi*lanosterol and hop-22(29)-en-3α-ol and the 3S-epoxide into the expected lanosterol and hop-22(29)-en-3β-ol (Ourisson *Eur. J. Biochem.*, 1980, 557; M. Rohmer, C. Anding and G. Ourisson, *Eur. J. Biochem.*, 1980, 112, 541). Presqualene alcohol has the absolute configuration 1R,2R,3R (L.J. Altman, R.C. Kowerski and D.R. Laungani, *J. Am. Chem. Soc.*, 1978, 100, 6174).

(14) R=Me ; R'=

(15) R= ; R'=Me

(16) R=Me ; R'=

(17) R= ; R'= Me

2. The Lanostane Group

Several minor co-metabolites of the important antibiotic fusidic acid have been identified from the fungus *Fusidium coccineum* (W.O. Godtfredsen *et al.*, *Tetrahedron*, 1979, 35, 2419).

The mushroom *Ganoderma lucida* has proved to be a most prolific source of lanostane triterpenoids. The lanostanes derivatives are often highly oxygenated. In the last decade well in excess of a hundred new compounds have been isolated from this source and characterised. The trivial nomenclature used for these compounds is often confused and confusing with several names given to one compound and the same name being used for two or more compounds. The lanostanes range from the relatively simple ganoderic acid Z (**18**) (J.O. Toth *et al.*, *J. Chem. Res (M)*, 1983, 299) through the trihydroxydione ganoderic acid C (or D or E) (**19**) (H. Kohda *et al.*, *Chem. Pharm. Bull.*, 1985, 24, 1367; T. Kikuchi *et al.*, *Chem. Pharm. Bull.*, 1985, 33, 2624; M. Hirotani and T. Furuya, *Phytochemistry*, 1986, 25, 1189) to the trisnorlanostane derivative lucidenic acid L (**20**) (T. Nishitoba, H. Sato and S.Sakamura, *Phytochemistry*, 1987, 26, 1777).

Related *Ganoderma* species also produce similar lanostanes. Furanoganoderic acid AP (**21**), from *G. applanatum*, is of interest because of its furan-containing side-chain (L.-J. Lin, M.-S. Shiao, and S.-F. Yeh, *J. Nat. Prod.*, 1988, 51, 918). Other furan side-chain arrangements occurs in pomacerone (**22**) from *Phellinus pomaceus* (A.G. González *et al.*, *Heterocycles*, 1990, 31, 841) and in pseudolarifuroic acid (**23**) from *Pseudolarix kaempferi* (K. Chen *et al.*, *Huaxue Xuebao*, 1990, 48, 591; *Chem. Abstr.* 114, 3442d).

(18)

(19)

(20)

(21)

560

The interesting rearranged lanostane spiroveitchionolide (24) has been reported from *Abies veitchii*. Its structure was established by X-ray analysis (R. Tanaka *et al.*, *J. Chem. Soc., Chem. Commun.*, 1992, 1351). *Abies* species produce several more rearranged lanostanes with mariesane and abiesane skeletons. Mariesiic acids A (25), B (26) and C (27) form a representative group from *A. mariesii* (S. Hasegawa *et al.*, *Tetrahedron*, 1987, <u>43</u>, 1775). A 3,4-secoabiesane (28) has been isolated from *A. alba* (T.V. Leibyuk *et al.*, *Chem. Nat. Compd. (Engl. Transl.)*, 1990, <u>26</u>, 651) while 3,4-secomariesanes such as *cis*-sibiric acid (29) occurs in *A. sibirica* (S.A. Shevtsov and V.A. Raldulgin, *Chem. Nat. Compd. (Engl. Transl.)*, 1989, <u>25</u>, 182).

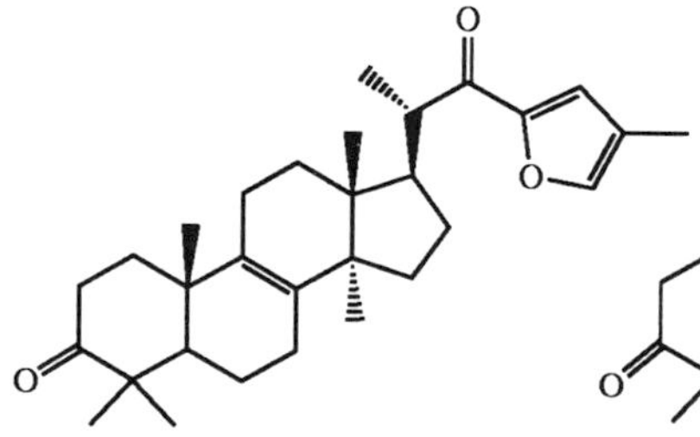

(22)

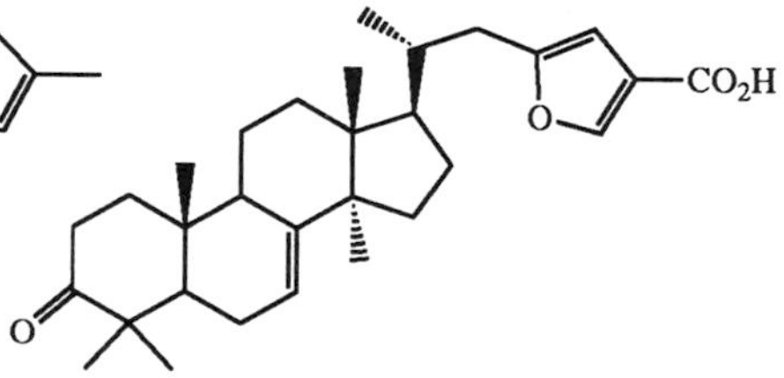

(23)

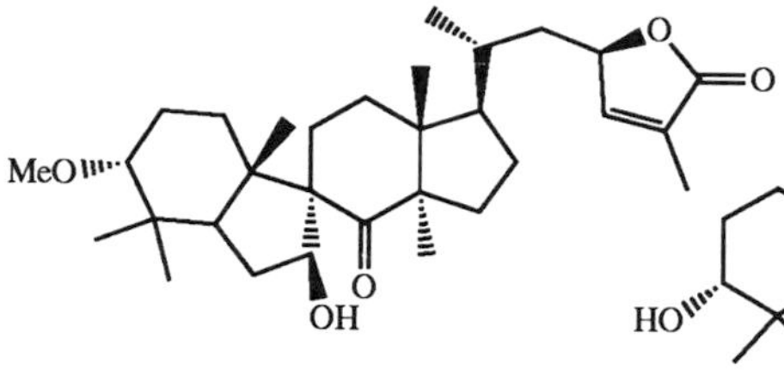

(24)

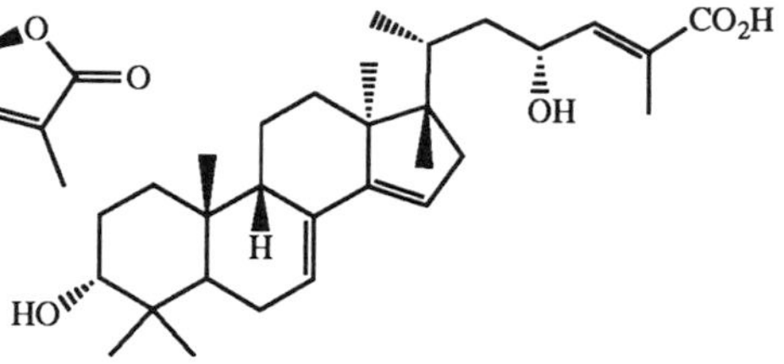

(25)

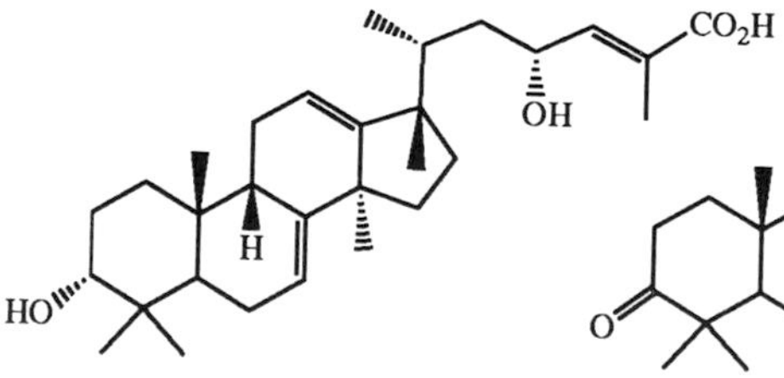

(26)

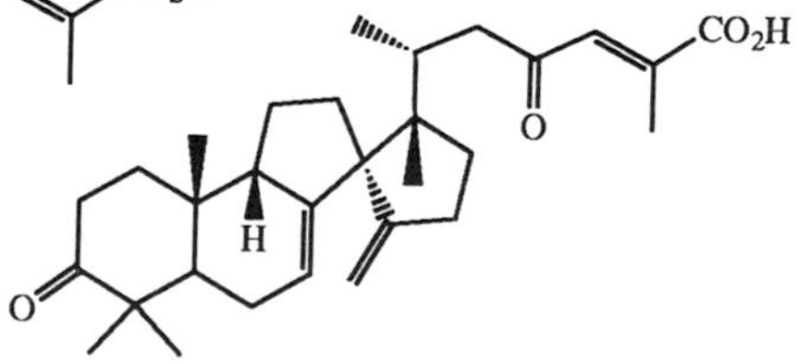

(27)

The 12α-hydroxylanostane derivative (30) is the probable precursor of several 14(13→12)-abeolanostanes including neokadsuranic acids A (31) and C (32) in *Kadsura heteroclita* (L. Lian-niang *et al., Planta Med.*, 1989, 55, 294). The fungus *Phellinus panaceus*, which grows on laurisilva trees, produces the degraded lanostanes javeroic acid (33) and phellinic acid (34) with contracted A-rings (A.G. González *et al., J. Chem. Soc., Perkin Trans. 1*, 1986, 551).

Pisolactone (35) from the fungus *Pisolithus tinctorius* has an unusual pattern of side-chain oxidation (A.M. Lobo *et al.,, Tetrahedron Lett.*, 1983, 24, 2205). The metabolites of toxic mushrooms *Neametaloma fasciculare*, e.g. fasciculol C (36), have plant growth inhibitory activity. They occur as depsipeptides (A. Takahashi *et al., Chem. Pharm. Bull.*, 1989, 37, 3247). Eucosterol (37) is a novel spirocyclic norlanostane from *Eurycoma* species (R. Ziegler and Ch. Tamm, *Helv. Chim. Acta*, 1976, 59, 1997).

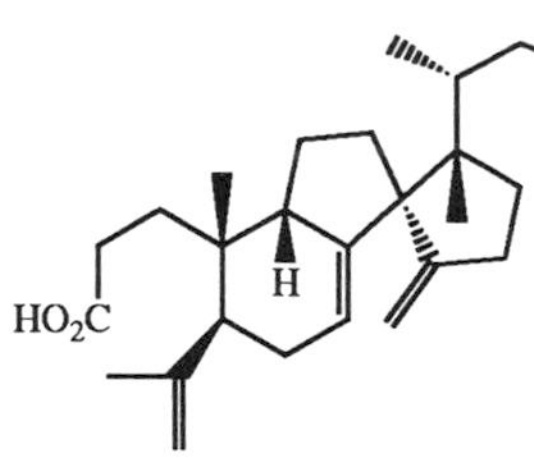

(28)

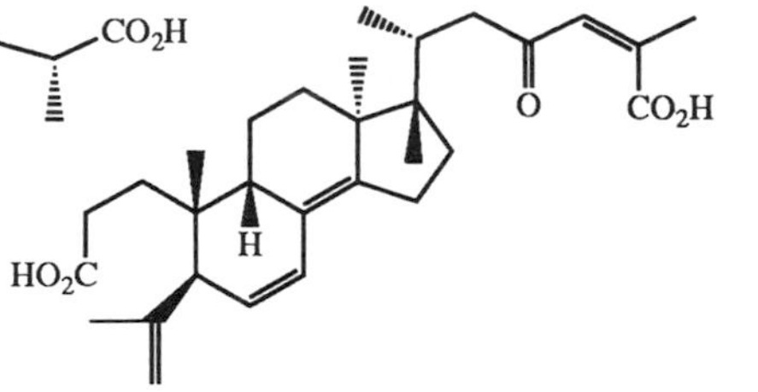

(29)

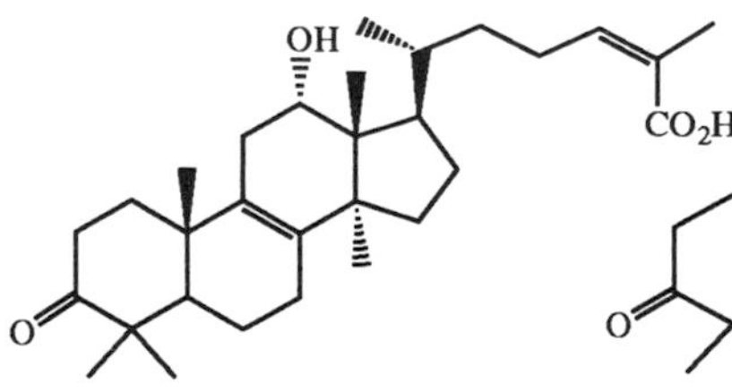

(30)

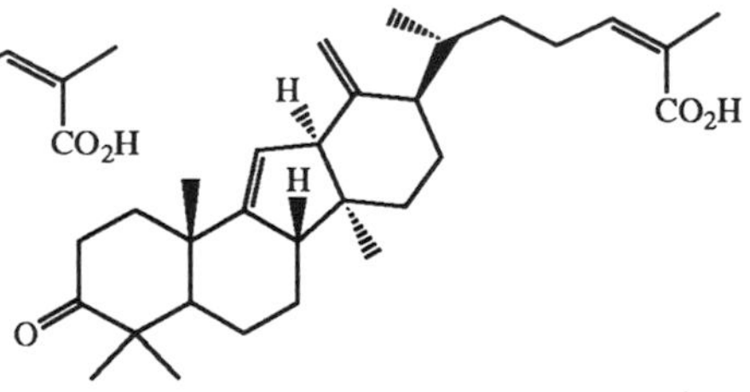

(31)

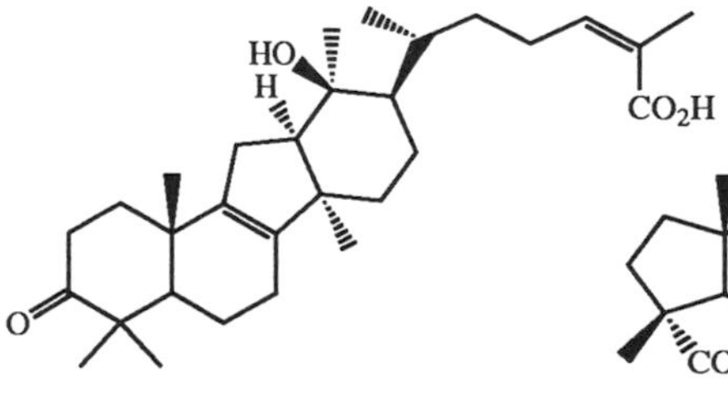

(32)

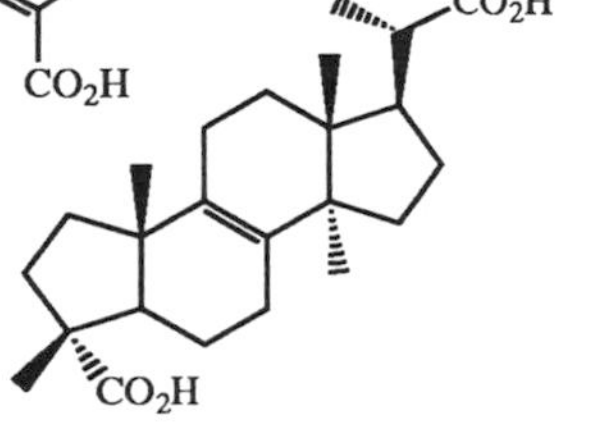

(33)

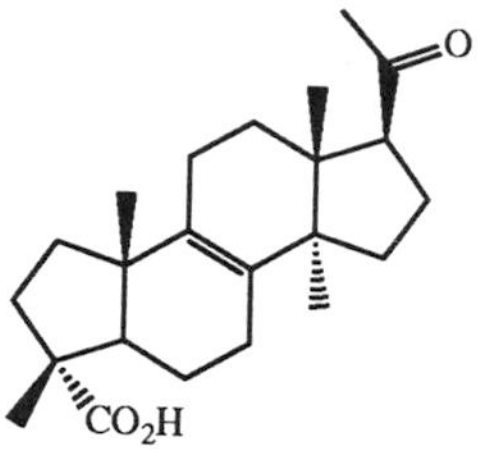

(34)

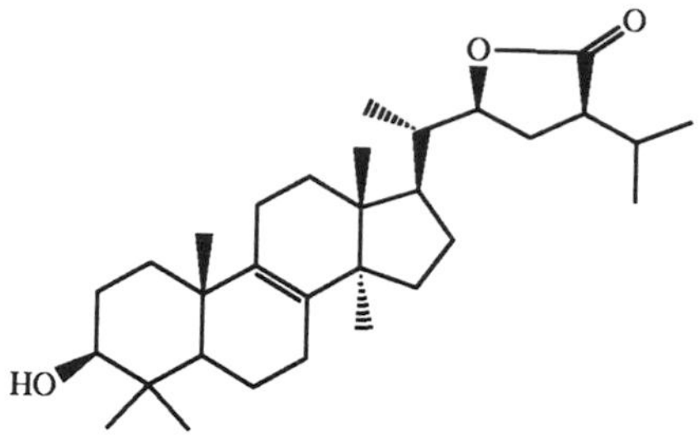

(35)

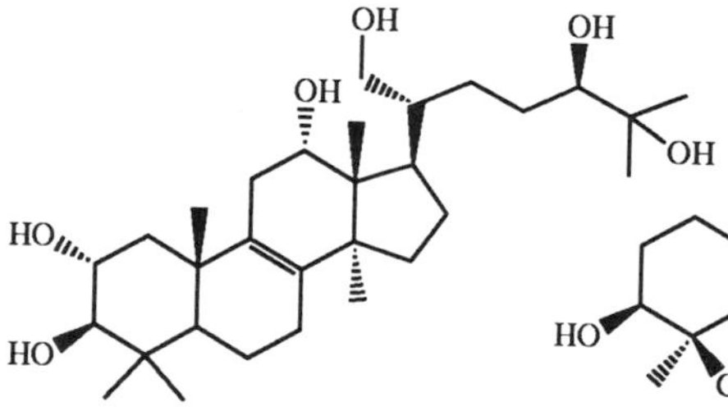

(36)

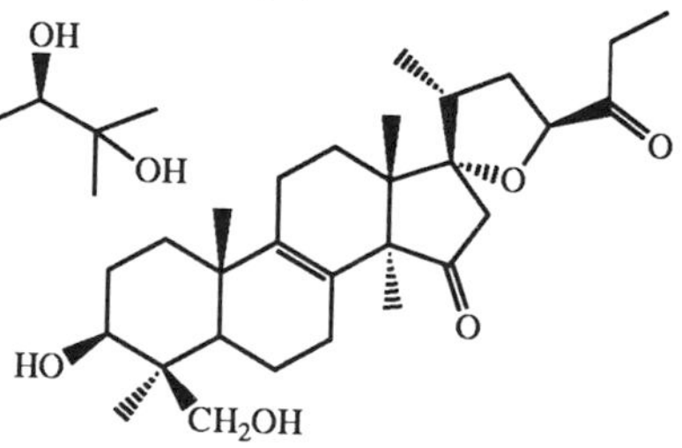

(37)

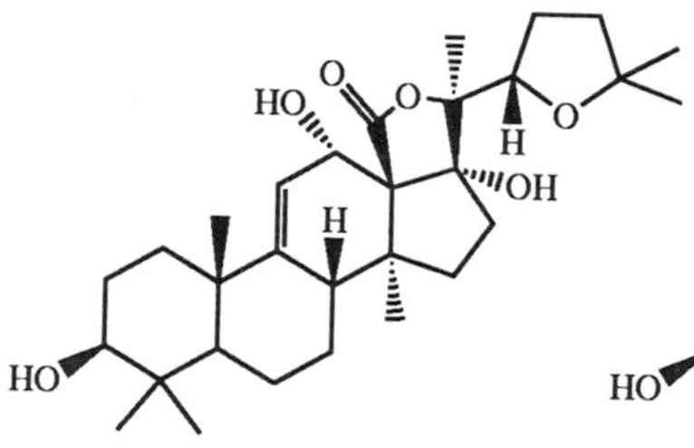

(38)

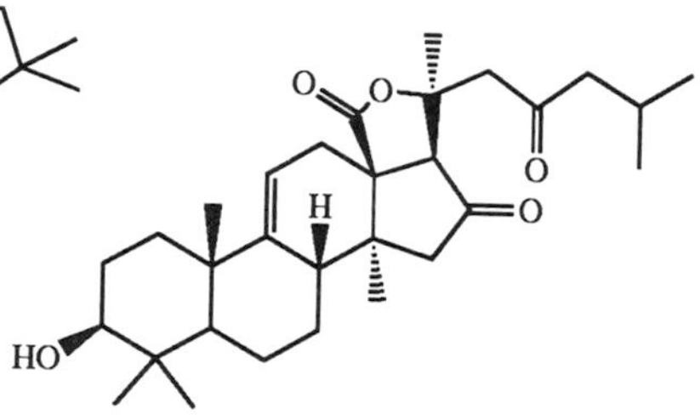

(39)

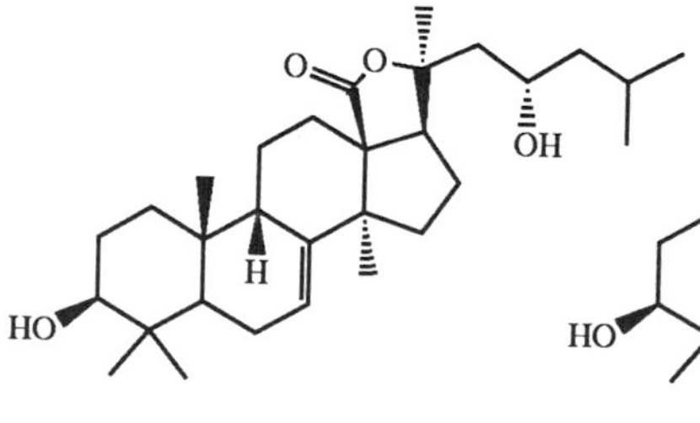

(40)

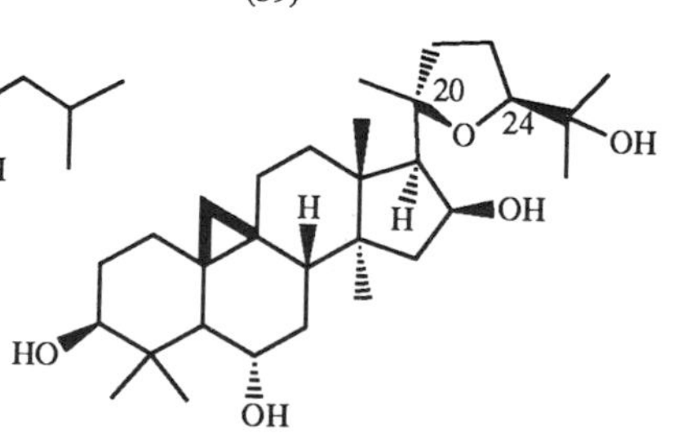

(41)
(42) 20,24-diepi

There is much interest in the antifungal oligoglycosides of sea cucumber species. The aglycones are illustrated by holothurigenol (**38**) from the saponins holothurins A and B of *Holothuria leucospilota* (I. Kitagawa *et al., Chem. Pharm. Bull.*, 1981, 29, 1942 and 1951) and cucumechinol A (**39**) from saponins of *Cucumaria echinata* (T. Miyamoto *et al., Liebigs Ann. Chem.*, 1990, 39). The structure of stichlorogenol (**40**) from *Stichopus chloronotus* was confirmed by X-ray analysis (I. Kitagawa *et al., Chem. Pharm. Bull.*, 1981, 29, 1189; 2387).

The confusion over the names and relative stereochemistry of the two main sapogenins, cyclogagaligenin (**41**) and cyclosieversigenin (**42**), of *Astragalus* species appears to have been resolved by an X-ray analysis of the former (V.K. Kravtsov*et al., Chem. Nat. Compd. (Engl. Transl.)*, 1988, 24, 458). *Nervilia purpurea* contains several new cycloartane derivatives including cyclohomonervilol (**43**) with a 24S-isopropenyl group and the rearranged cyclonervilasterol (**44**) (S. Kadota *et al., Chem. Pharm. Bull.*, 1987, 35, 200; T. Kikuchi *et al., Chem. Pharm. Bull.*, 1986, 34, 3183). Other side-chain variations are to be found in skimmiwallichin (**45**) from *Skimmia wallichi* (I.N. Kostova, N. Pardeshi and S. Rangaswami, *Indian J. Chem.*, 1977, 15B, 811), cycloswietenol (**46**) from *Swietenia mahogani* (A.S.R. Anjaneyulu, Y.L.N. Murty and L.R. Row, *Indiian J. Chem.*, 1978, 16B, 650), cyclopholidonol (**47**) from *Pholidota chinensis* (W. Lin *et al., Planta Med.*, 1986, 4), cyclopterospermol (**48**) from *Pterospermum heyneanum* (A.S.R. Anjaneyulu and S.N. Raju, *Phytochemistry*, 1987, 26, 2805), cyclopodmenyl acetate (**49**) from *Polypodium vulgare* (Y. Arai, K. Shiojima and H. Ageta, *Chem. Pharm. Bull.*, 1989, 37, 560), the 26,26-dimethylcycloartane derivative (**50**) from *Euphorbia soonigarica* (Y. Ding, Z. Jia and T. Chu, *Gaodeng Xuexiao Huaxue Xuebao*, 1989, 10, 1129; *Chem. Abstr.*, 113, 3235t), uniflorin (**51**) from *Coelogyne uniflora* (P.J. Majumder and S. Pal, *Phytochemistry*, 1990, 29, 2717)) and 25-methyl-24-methylenecycloartan-3β-ol derivative *from Neolitsea servicea* (K. Yano *et al., Phytochemistry*, 1992, 31, 1741).

The structure of abietospiran (**52**), the principal compound of the crystalline coating responsible for the silver-grey colour of the bark of *Abies alba* has been established by X-ray analysis (W. Steglich *et al., Angew. Chem., Int. Ed. Engl.*, 1979, 18, 698). Uvariastrol (**53**) from *Uvariastrum zenkeri* (P.G. Waterman *et al., Phytochemistry*, 1984, 23, 2077) and the peroxide (**54**) from *Lindheimera texana* (W. Herz, *et al., Phytochemistry*, 1985, 24, 2645) are cycloartan-3-one derivatives with unusual ketal side-chains.

The structure of glycinoeclepin A (**55**), a hatching stimulus for the soybean cyst nematode, was established by X-ray analysis (*cf* mariesiic acid A) (A. Fukuzawa *et al., J. Chem. Soc., Chem. Commun.*, 1985, 222). It was isolated from the roots of kidney beans (1.25 mg from *ca.* 1 tonne) and has recently been synthesised (E.J. Corey *et al., J. Am. Chem. Soc.*, 1990, 112, 8997; H. Watanabe and K. Mori, *J. Chem. Soc., Perkin Trans. 1*, 1991, 2919). Cimigenol (**56**) and *O*-methylcimiacerol (**57**) are acid rearrangement products of the unstable genins of *Cimifuga* xylosides (S. Sakurai, T. Inoue and M. Nagai, *Chem. Pharm. Bull.*, 1976, 24, 3220; N. Sakurai *et al., Chem. Pharm. Bull.*, 1981, 29, 955; G. Kusano *et al., Heterocycles*, 1983, 20, 1951). Passiflorin (**58**) from the leaves of *Passiflora edulis* (E. Bombardelli *et al., Phytochemistry*, 1975, 14, 2661) undergoes an interesting cyclopropane transposition to (**59**) on acid treatment. Abrusoside A, a highly sweet

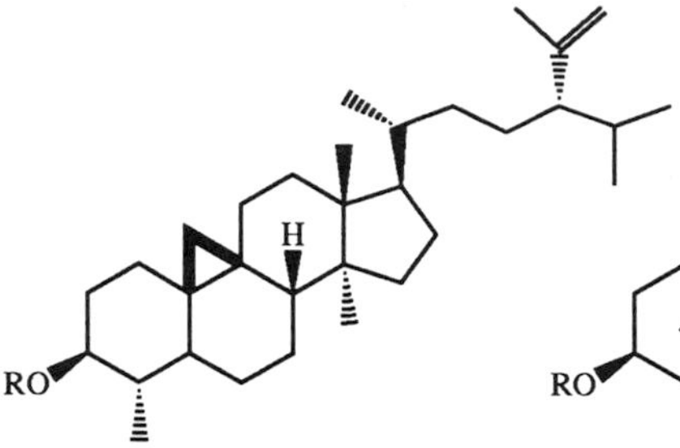

(43) R=H

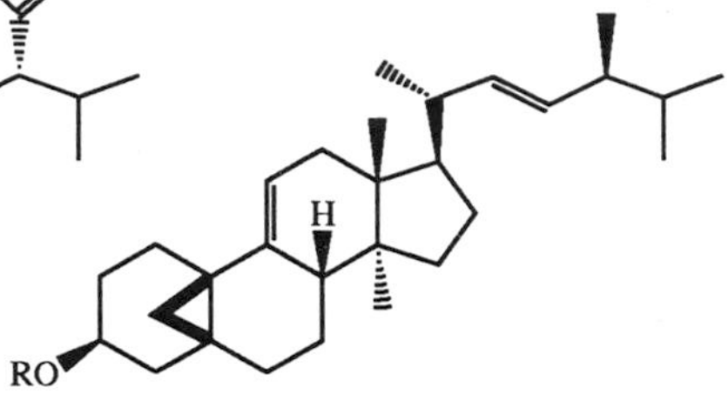

(44)

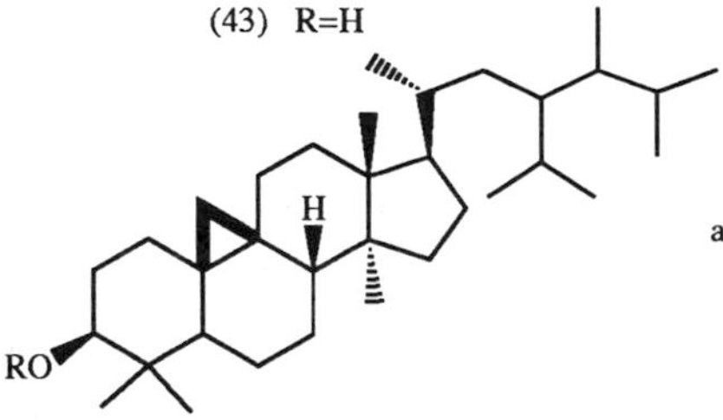

(45) R=Me

as (45)

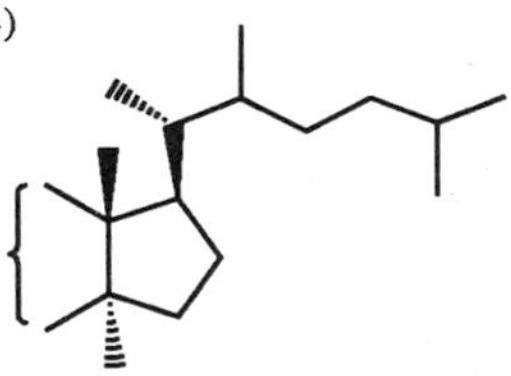

(46) R=H

as (43)

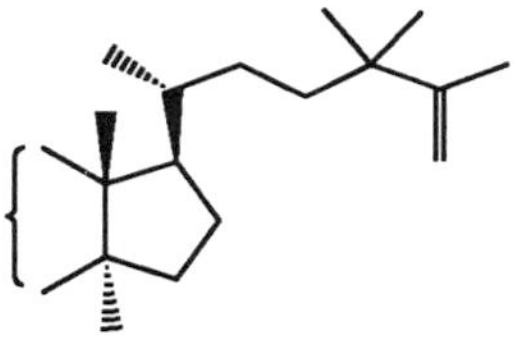

(47)

as (45)

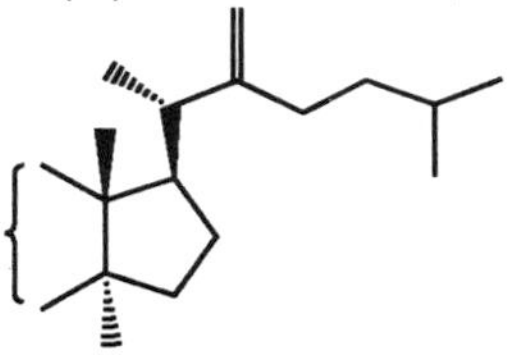

(48)

as (45)

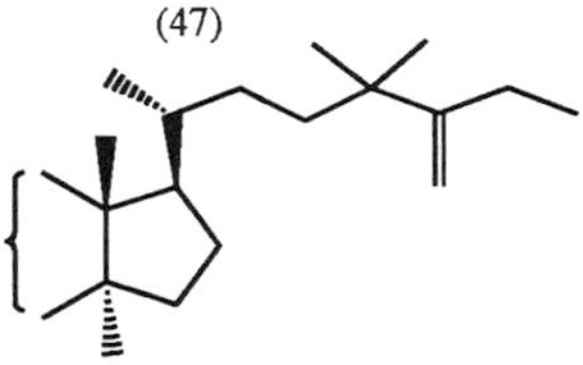

(49) R=Ac

as (45)

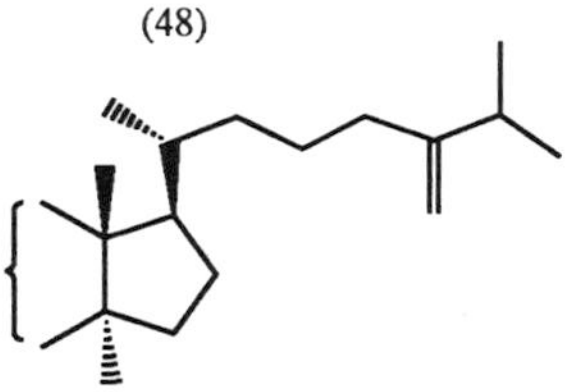

(50)

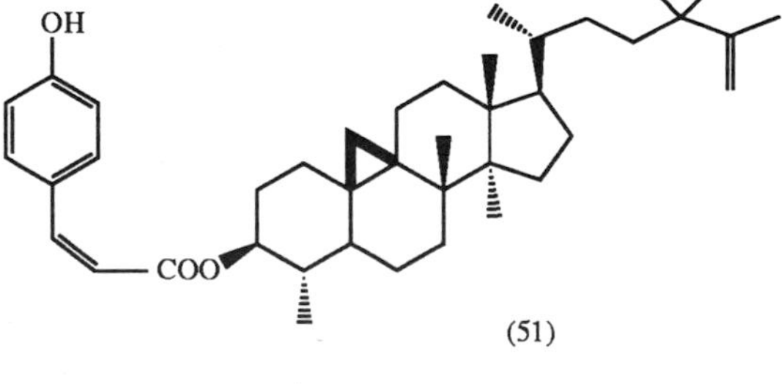

(51)

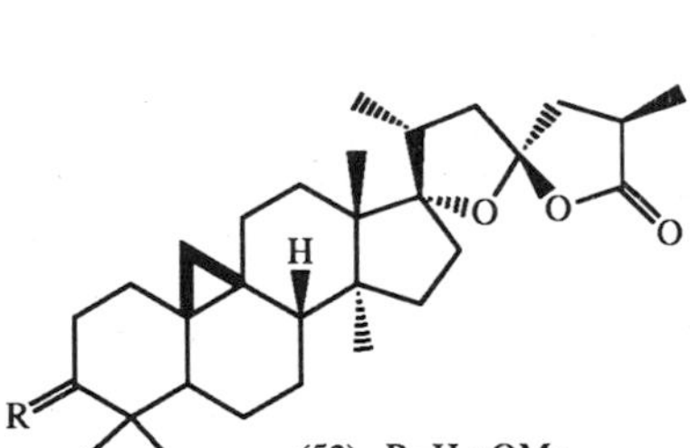

(52) R=H,αOMe

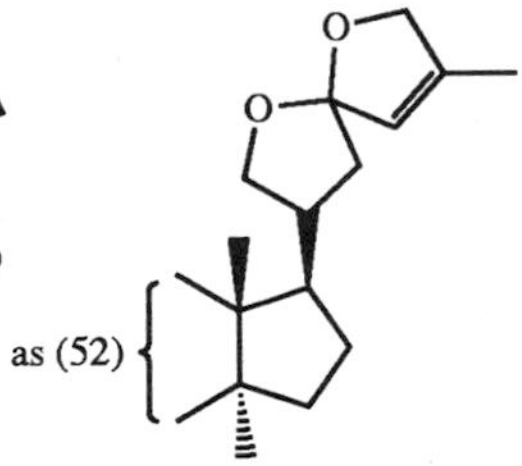

as (52) {

(53) R=H,βOH

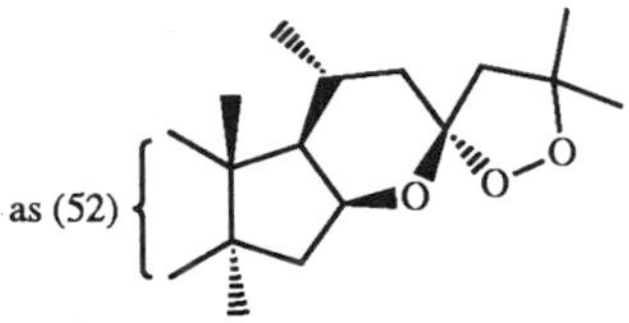

as (52) {

(54) R=O

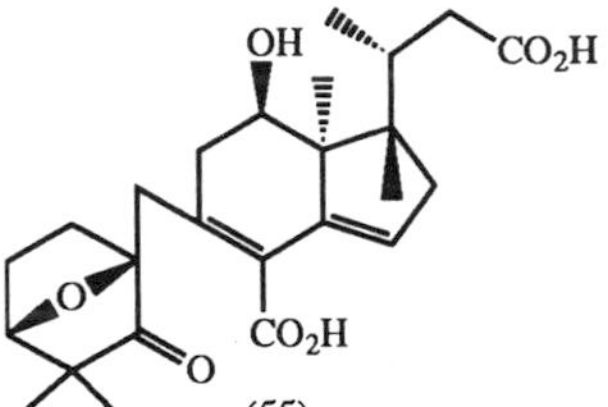

(55)

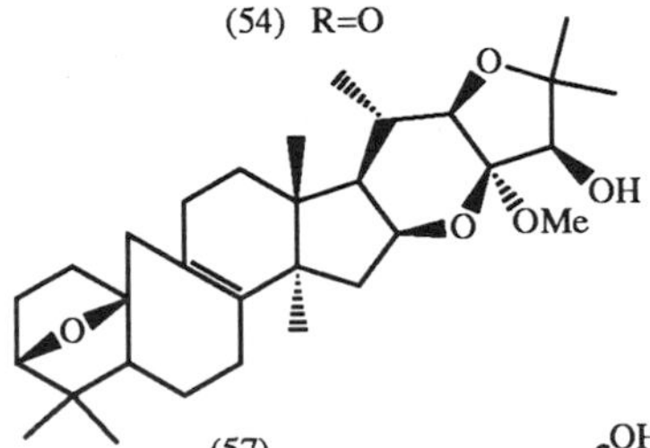

(57)

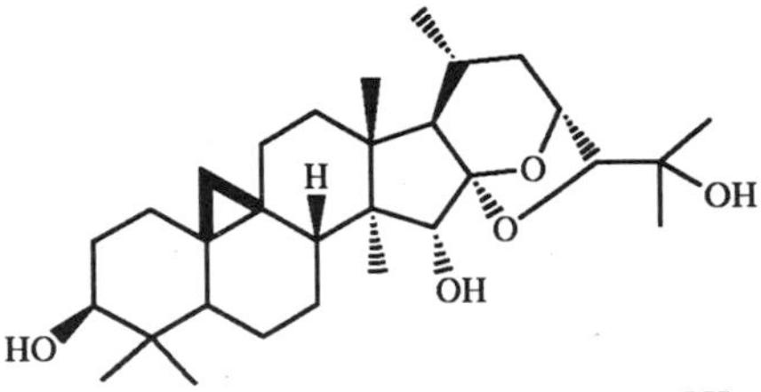

(56)

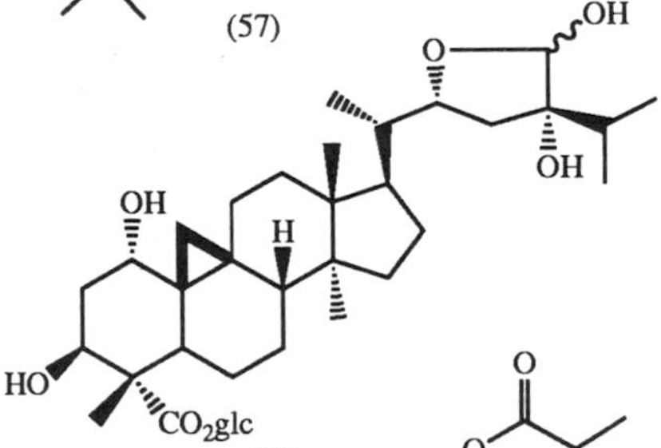

(58)

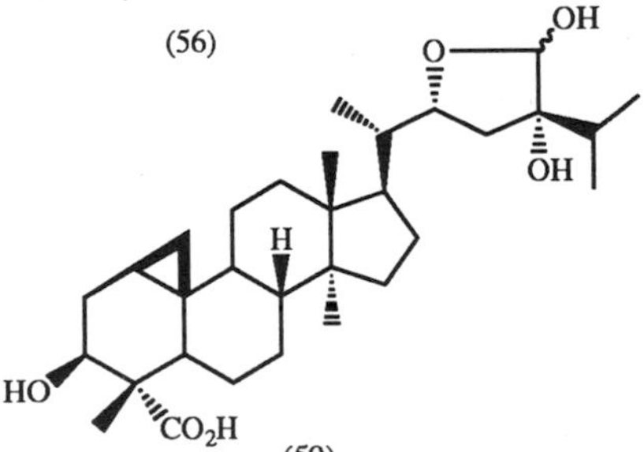

(59)

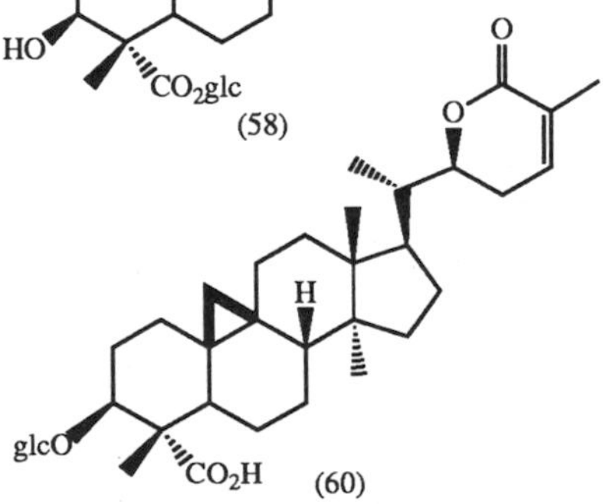

(60)

566

constituent of *Abrus precatorium*, has structure **(60)** (Y.-H. Choi *et al.*, *J. Chem. Soc., Chem. Commun.*, 1989, 887).

Several interesting cleaved cycloartane structures, based on X-ray analyses, have been reported recently. These include the peroxides pseudolarolides H **(61)** (G.-F. Chen *et al.*, *Tetrahedron Lett.*, 1990, 31, 3413) and I **(62)** (G.-F. Chen *et al.*, *Heterocycles*, 1990, 31, 1903 and the 8,9:9,10-cleaved derivative pseudolarolide E **(63)** (G.-F. Chen *et al.*, *J. Chem. Soc., Chem. Commun.*, 1990, 113) from *Pseudolarix kaempferi* and buxapentalactone **(64)** from *Buxux papillosa* (Atta-ur-Rahman *et al., Tetrahedron*, 1992, 48, 3577). Cycloartenol has been synthesised from 3β-acetoxylanostan-24-en-11β-yl nitrite (R.B. Boar and D.B. Copsey, *J. Chem. Soc., Perkin Trans. 1*, 1979, 563).

Various acid catalysed transformations of lanostanes into cucurbitacins have been achieved (O.E. Edwards and Z. Paryzek, *Can. J. Chem.*, 1983, 61, 1973; Z. Paryzek and R. Wydra, *Can. J. Chem.*, 1985, 63, 1280). Among the cucurbitacin aglycones of the saponins of *Momordica charantia* is the pentol **(65)** the structure of which was confirmed by X-ray analysis. (H. Okabe *et al.*, *Chem. Pharm. Bull.*, 1980, 28, 2753; Y. Miyahara *et al.*, *Chem. Pharm. Bull.*, 1981, 29, 1561). Aromatic cucurbitacins fevicordin A glucoside **(66)** from *Fevillea cordifolia* (H. Achenbach, U. Hefter-Bübl and M.A. Constela, *J. Chem. Soc., Chem. Commun.*, 1987, 441) and **(67)** from a *Wilbrandia* species (M.E.O. Matos *et al.*, *Phytochemistry*, 1991, 30, 1020) have been reported.

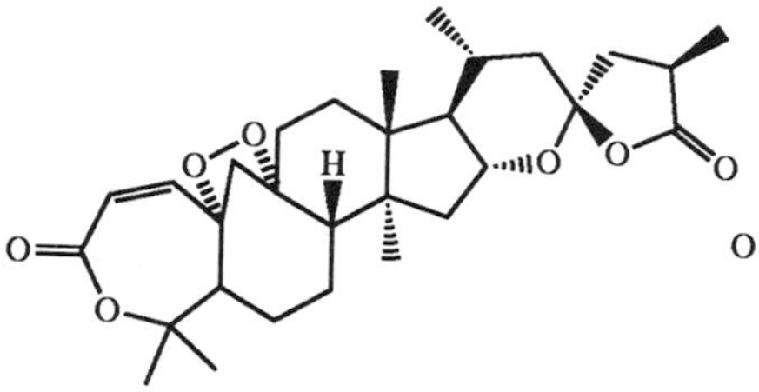

(61)

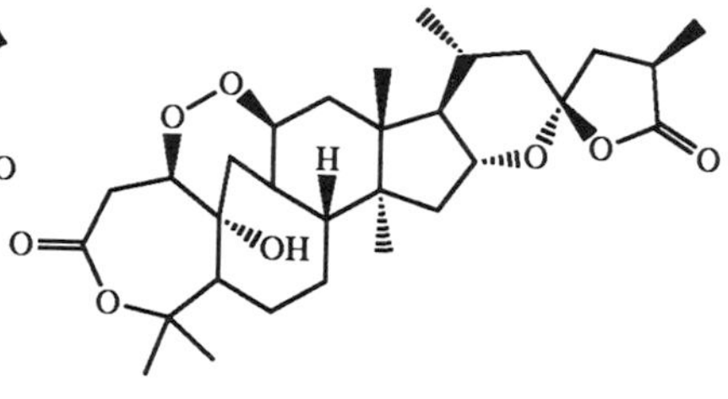

(62)

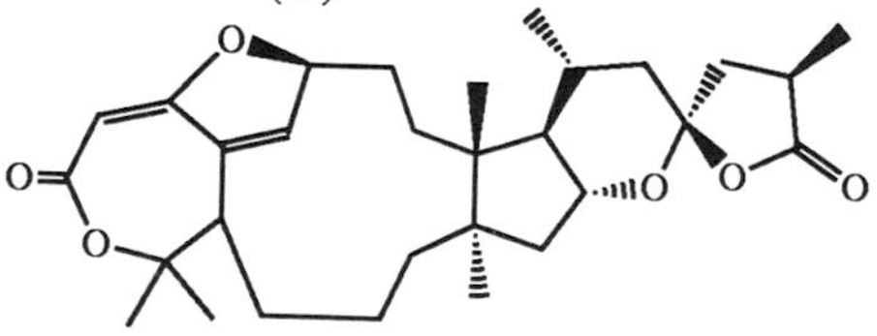

(63)

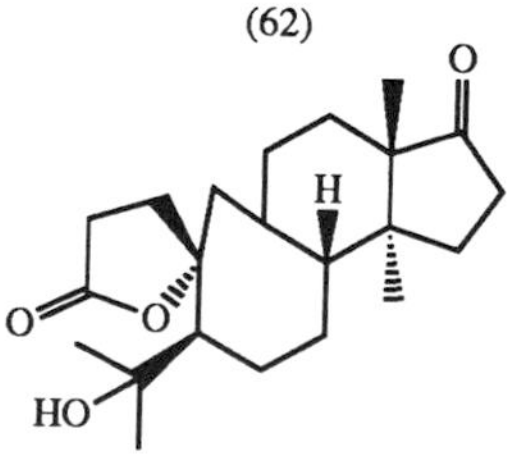

(64)

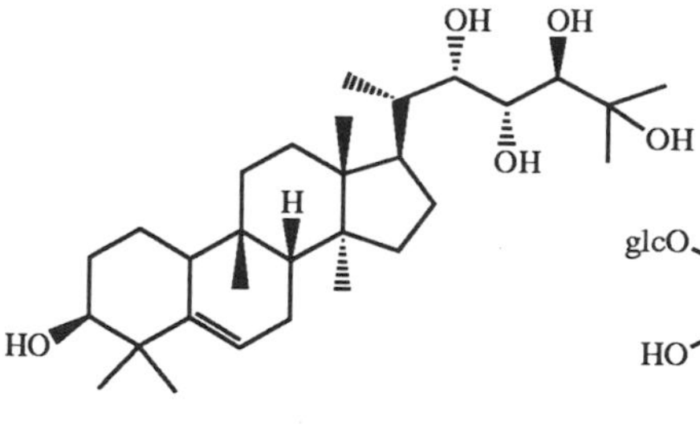

(65)

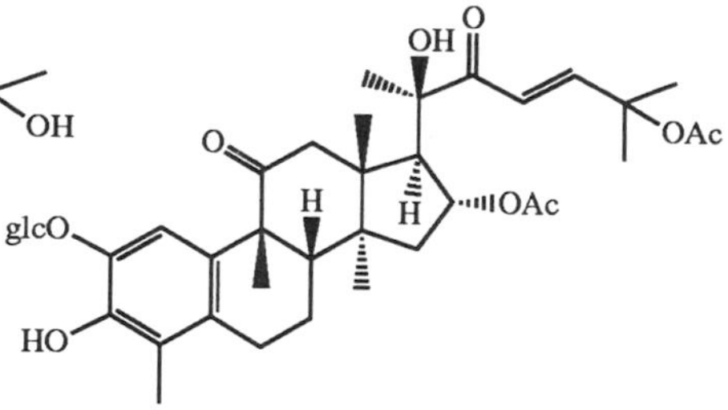

(66)

(67) Δ^6

3. The Dammarane Group

The novel reissantane skeleton found in reissantenol oxide (68) from *Reissantia indica* may arise by backbone rearrangement of either dammarane or euphane cations (C.B. Gamlath, A.A.L. Gunatilaka and E.O. Schlemper, *J. Chem. Soc., Perkin Trans. 1*, 1989, 2259). The similarly rearranged tirucallanes euferol (69) and its $\Delta^{1(10)}$-isomer, melliferol, have been found in *Euphorbia mellifera* (M.J.U. Ferreira *et al.*, *J. Chem. Soc., Perkin Trans. 1*, 1990, 185). Their structures were confirmed by X-ray analysis. Aonena-3,24-diene (70) represents complete backbone rearrangement of a dammarane intermediate. It is found in fresh rhizomes of *Polypodioides niponica* (Y. Arai, M. Hirohara and H. Ageta, *Tetrahedron Lett.*, 1989, 30, 7209). Euphol and tirucallol have been synthesised using a route based on acid catalysed cyclisation of a monocyclic-dienediyne precursor (W.R. Bartlett *et al.*, *J. Org. Chem.*, 1990, 55, 2215). 9β,19-Cyclotirucall-24-en-3β-ol (cycloroylenol) (71) the tirucallane equivalent of cycloartenol, has been isolated from *Euphorbia royleana* (V.S. Bhat, V.S. Joshi and D.D. Nanavati, *Tetrahedron Lett.*, 1982, 23, 5207).

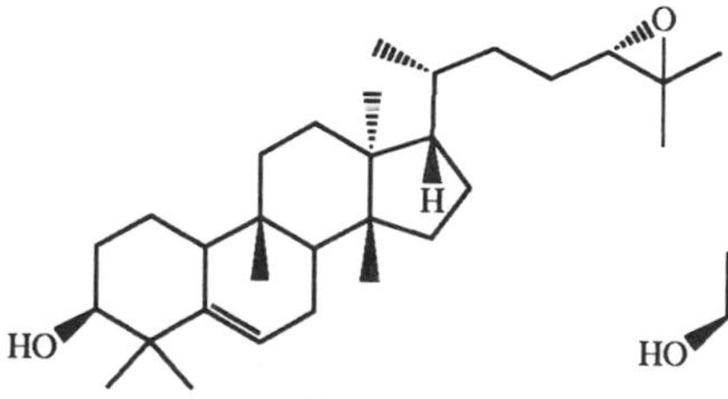

(68)

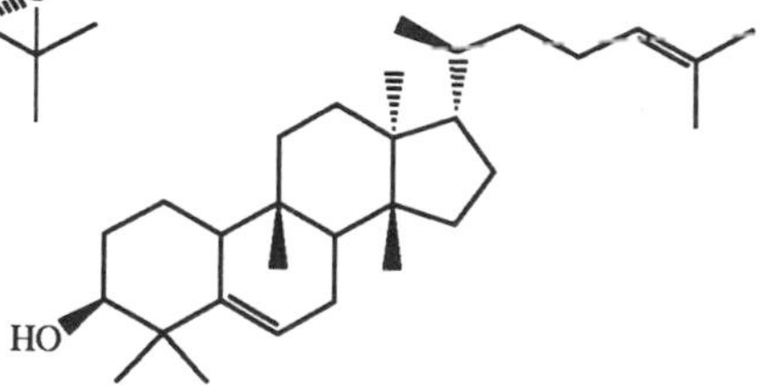

(69)

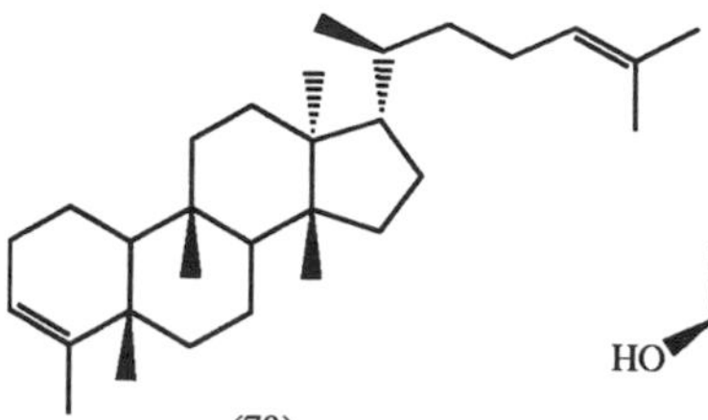

(70)

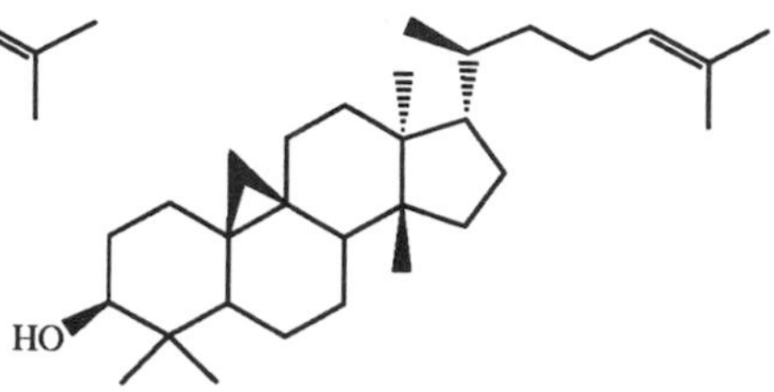

(71)

Mabioside A (**72**) is an unusual ring D cleaved dammarane saponin from *Colubrina elliptica* (C.E. Seaforth *et al., Tetrahedron Lett.*, 1992, 33, 4111). The authentic sapogenin of dammarane saponins from several species, for example *Zizyphus jujuba* (K.-I. Kawai *et al., Phytochemistry*, 1974, 13, 2829; K. Yoshikawa, N. Shimono and S. Arihara, *Tetrahedron Lett.*, 1991, 32, 7059) and *Hovenia dulcis* (O. Inoue, T. Takeda and Y. Ogihara, *J. Chem. Soc., Perkin Trans. 1*, 1978, 1289), is jujubogenin (**73**) which is extremely acid labile and on exposure to acid forms ebelin lactone (**74**). Hovenolactone (**75**) is a related sapogenin from *H. dulcis* (Y. Kobayashi, *J. Chem. Soc., Perkin Trans. 1*, 1982, 2795). A crystal structure analysis revealed that gynogenin (**76**), a sapogenin from *Gymnostemma pentaphyllum*, has an unusual cyclopentenone side-chain (M.F. MacKay, J.-X. Wei and Y.-G. Chen, *Acta Crystallogr., Sect. C*, 1991, 47, 790).

Many tirucallane, apotirucallane and tetranortriterpenoid derivatives have been isolated from *Azadirachta inica* (S. Siddiqui *et al., Phytochemistry*, 1991, 30, 1615; B.S. Siddiqui *et al., Phytochemistry*, 1992, 31, 4275). There are some recent additions (B.S. Joshi *et al., Tetrahedron Lett.*, 1985, 26, 1273; Y. Niimi *et al., Chem. Pharm. Bull.*, 1989. 37, 57) to the glabretal (**77**) (G. Ferguson *et al., J. Chem. Soc., Perkin Trans. 1*, 1975, 491) series of cyclopropyl apotirucallanes.

In the tetranortriterpenoid (limonoid) area intriguing new structural types continue to appear (D.A.H. Taylor, *Prog. Chem. Org. Nat. Prod.*, 1984, 45, 1). A series of dukunolides, for example dukunolide A (**78**), has been isolated from *Lansium domesticum* (M. Nishizawa *et al., J. Org. Chem.*, 1985, 50, 5487; *Phytochemistry*, 1988, 27, 237). The dukunolide framework probably arises by cleavage of a C9-C10 bond of a bicyclononanolide precursor

(72)   (73)

(74)   (75)   (76)

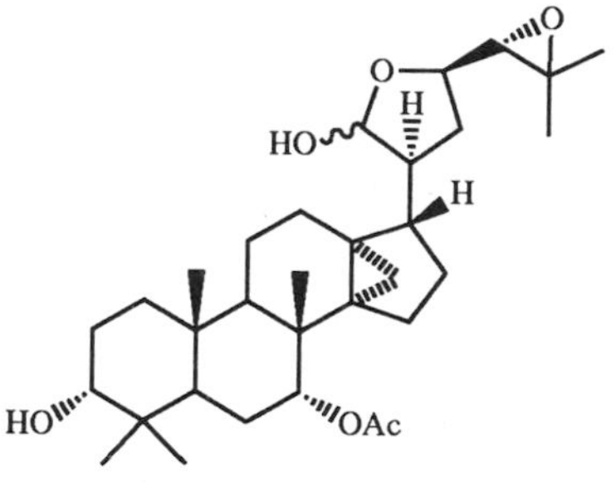

(77)

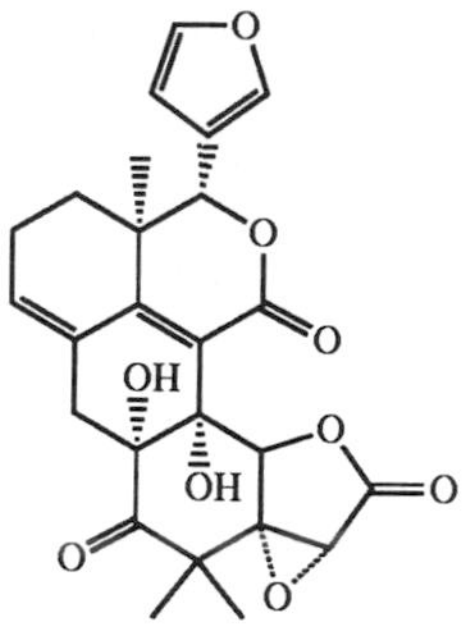

(78)

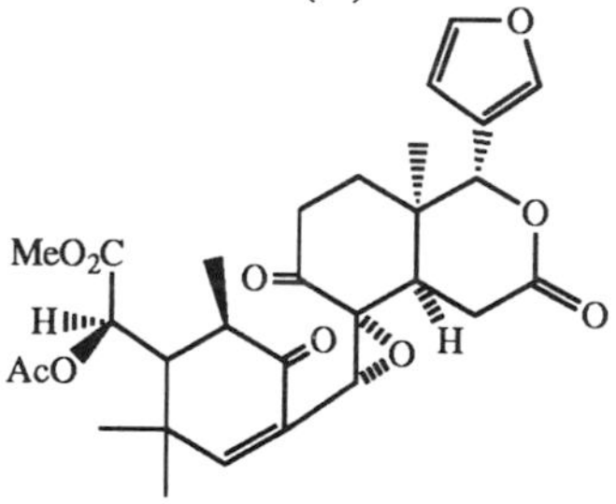

(79)

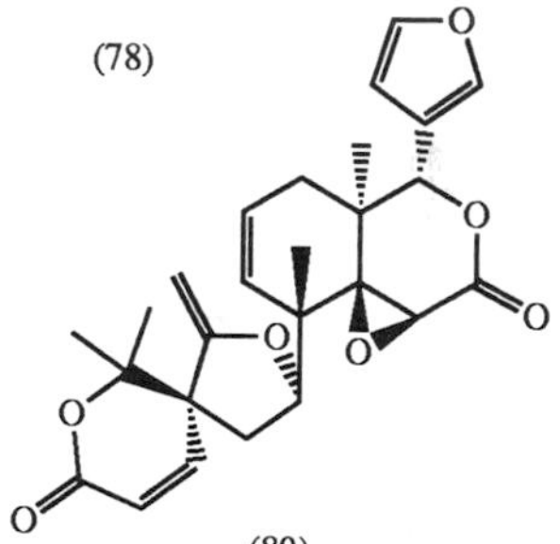

(80)

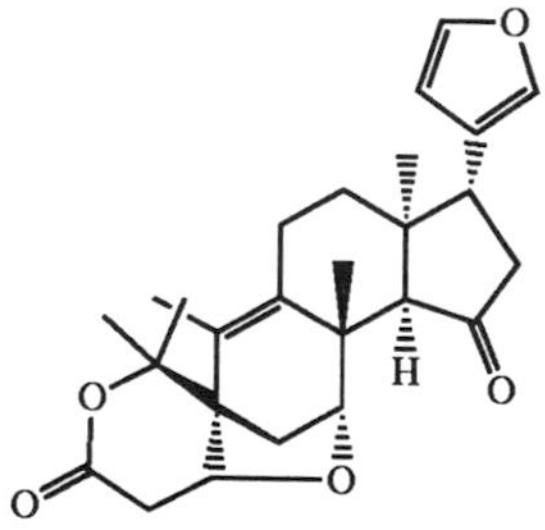

(81)

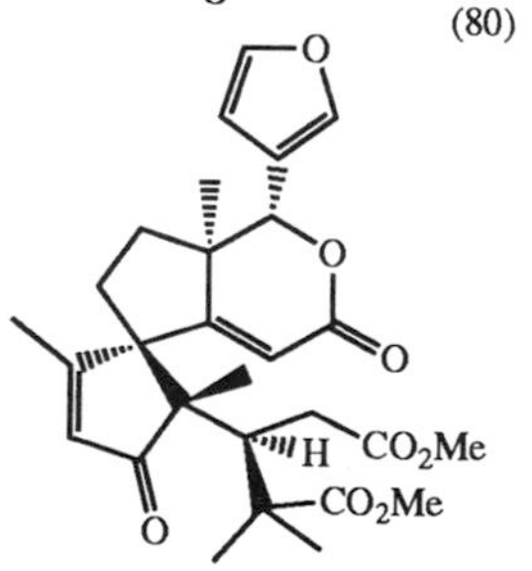

(82)

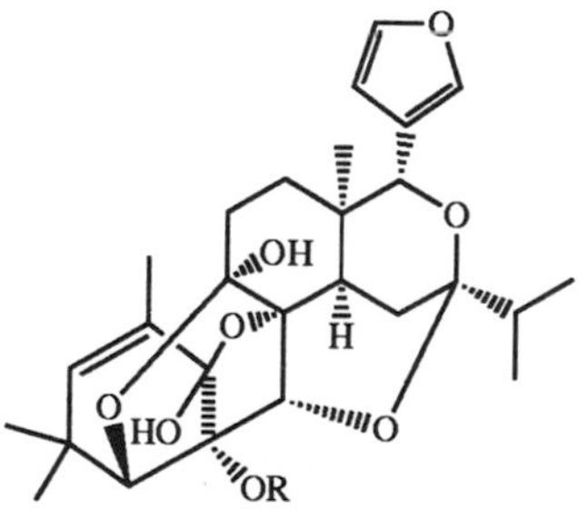

(83) R=H

(84) R=Ac

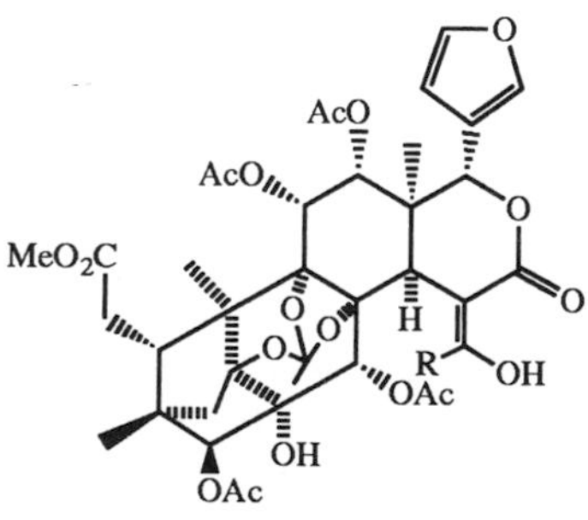

(85) R=ⁱPr

followed by bond formation between C1 and C14. Ecadorin (**79**) from *Guarea kuntthiana* (B.S. Mootoo *et al.*, *Can. J. Chem.*, 1992, 70, 1260) represents simple retro-aldol C9-C10 cleavage of a bicyclononane skeleton. The biogenesis of carapolides, for example carapolide D (**80**), from the seeds of *Carapa procera* and *C. grandiflora* probably involves C9-C10 cleavage in a spiro-precursor (S.F. Kimbu *et al.*, *Tetrahedron Lett.*, 1984, 25, 1613; 1617). The spirolactone (**81**) was isolated from the bark of *C. procera* (A.F. Cameron *et al.*, *Tetrahedron Lett.*, 1979, 967) and its structure was established by X-ray analysis. The structures of the highly cleaved derivatives guyanin (**82**) from *Hortia regia* (S. McLean *et al.*, *J. Am. Chem. Soc.*, 1988, 110, 5339) and entilins A (**83**) and B (**84**) from the stem bark of *Entandrophragma utile* (J.C. Tchouankeu *et al.*, *Tetrahedron Lett.*, 1990, 31, 4505) were solved using modern nmr methods. The structure of guyanin has been confirmed by X-ray analysis. The entilins are formally heptanortriterpenoids which have gained four carbons by acylation with isobutyric acid. Such acylation was originally observed in the busseins, for example bussein H (**85**), highly complex bicyclononanolide derivatives from *Entandrophragma bussei* (M. Geux and Ch. Tamm, *Helv. Chim. Acta*, 1984, 67, 885.

There is considerable interest in the limonoids of *Citrus* species and many limonoids have been isolated as glycosides (R.D. Bennett *et al.*, *Phytochemistry*, 1991, 30, 3803). Two unusual limonoids (**86**) and (**87**), with C-19 incorporated into ring A, have been isolated from a *Citrus* hybrid species (R.D. Bennett and S. Hasegawa, *Phytochemistry*, 1982, 21, 2349). Glaucin B (**88**), from *Evodia glauca*, has a rare *cis* AB junction (M. Nakatani *et al.*, *Phytochemistry*, 1988, 27, 1429). A unique arrangement of ring A is found in dumsin (**89**) from *Croton jatrophioides* (I. Kubo *et al.*, *Tetrahedron*, 1990, 46, 1515). Limonoids do not normally occur in the Euphorbiaceae.

The tetranortriterpenoid which has attracted most attention is undoudtedly azadirachtin (**90**) from *Azadirachta indica*. Many tetranortriterpenoids have insect antifeedant activity but azadirachtin is by far the most potent. Detailed accounts of the work of the various groups who contributed to its structural elucidation have appeared (W. Kraus *et al.*, *Tetrahedron*, 1987, 43, 2817; J.N. Bilton *et al.*, *Tetrahedron*, 1987, 43, 2805; C.J. Turner *et al.*, *Tetrahedron*, 1987, 43, 2789). The absolute configuration of azadirachtin has also been established (S.V. Ley, H. Lovell and D.J. Williams, *J. Chem. Soc., Chem. Commun.*, 1992, 1304). Azadirachtin H (**91**) and I (**92**) represent recent structural variants in this series (T.R. Govindachari, G. Sandhya and S.P.G. Raj, *J. Nat. Prod.*, 1992, 55, 596). A recent comprehensive review describes the history, the chemistry, the biological activity of azadirachtin and related compounds and details the considerable synthetic efforts of the Ley group in this area (S.V. Ley, A.A. Denholm and A. Wood, *Nat. Prod. Rep.*, 1993, 10, 109).

The cneorins and tricoccins form a fascinating group of C_{25} highly cleaved triterpenoid derivatives from the Cneoraceae family (A. Mondon and B. Epi, *Prog. Chem. Org. Nat. Prod.*, 1983, 44, 101). The fact that they co-occur with normal tetranortriterpenoids suggests that they share a common origin with them and are not conventional sesterterpenoids. The structural variations may be illustrated by cneorin B (**93**) and tricoccin S_{42} (**94**) (A. Mondon *et al.*, *Liebigs Ann. Chem.*, 1983, 1760; 1798; B. Epi *et al.*, *Tetrahedron Lett.*, 1979, 1365; 4045).

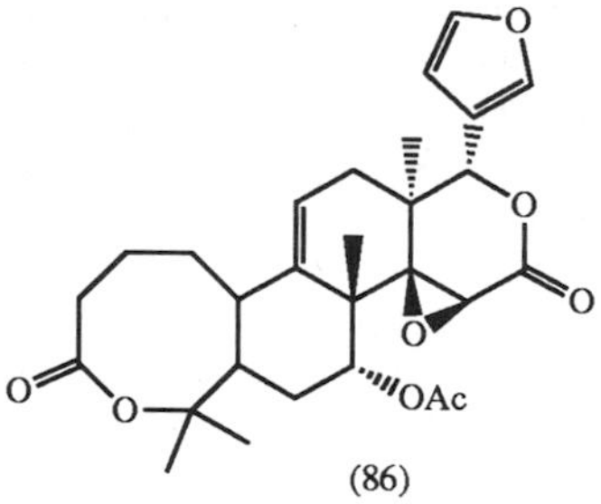

(86)

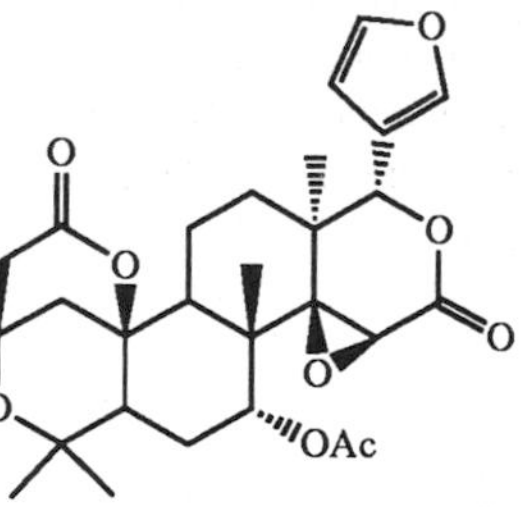

(87)

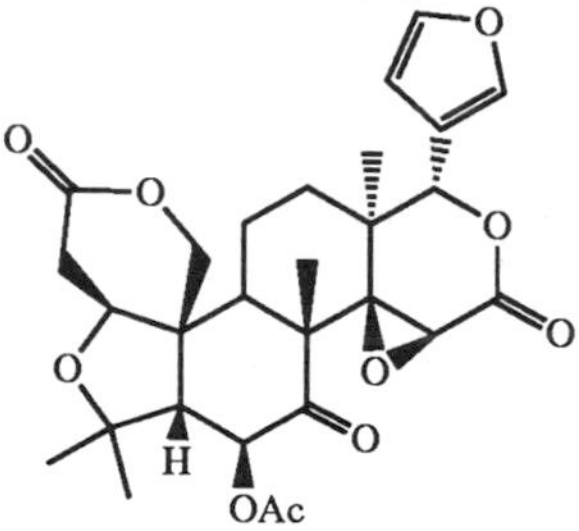

(88)

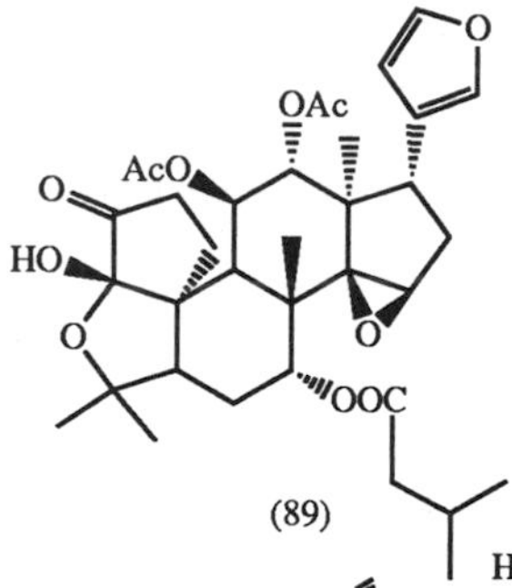

(89)

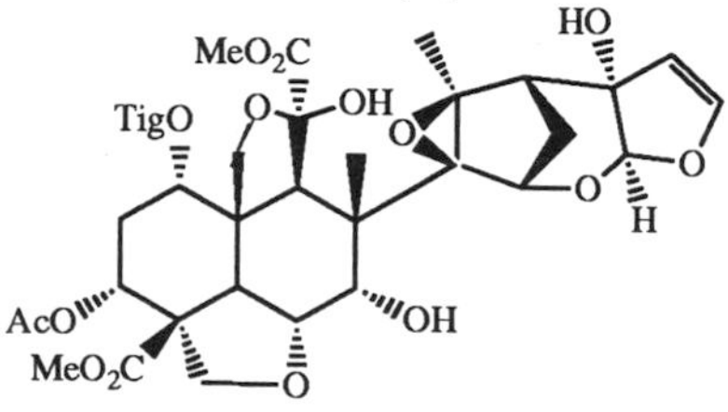

(90)

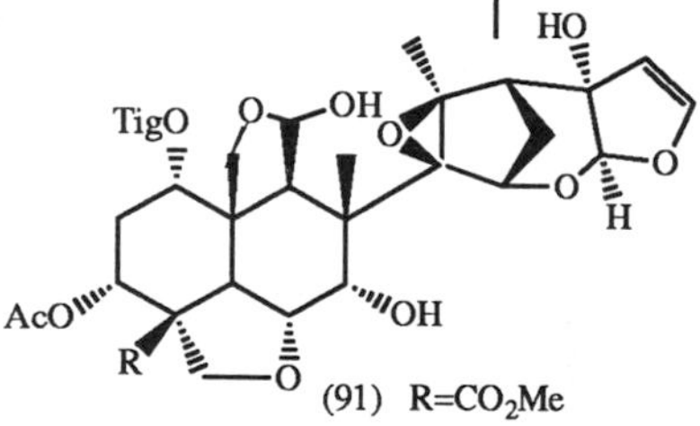

(91) R=CO$_2$Me

(92) R=CH$_3$

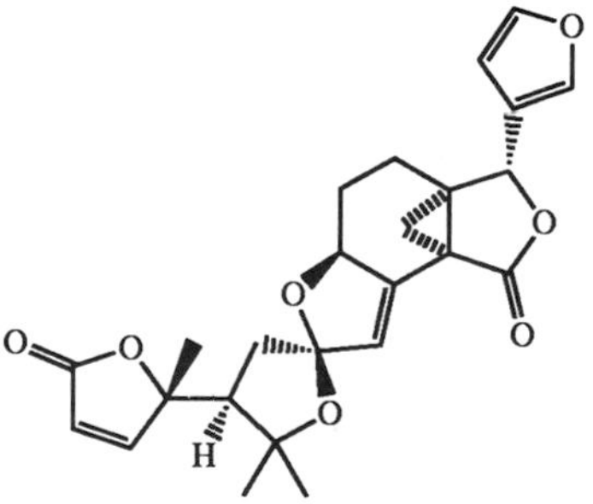

(93)

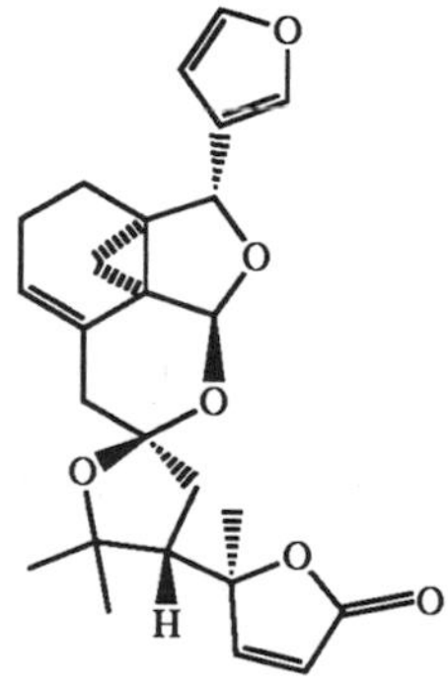

(94)

The biological activity of the quassinoids has ensured an ongoing structural (J. Polonsky, *Prog. Chem. Org. Nat. Prod.*, 1985, 47, 221) and synthetic interest in this group. There has been considerable synthetic effort in this area, some of which has been recently reviewed (K. Kawada, M. Kim and D.S. Watt, *Org. Prep. Proc. Int.*, 1989, 21, 521). Successful total syntheses of quassin (**95**) (G. Vidari, S. Ferrino and P.A. Grieco, *J. Am. Chem. Soc.*, 1984, 106, 3539), castelanolide (**96**) (P.A. Grieco *et al.*, *J. Org. Chem.*, 1984, 49, 2342), klaineanone (**97**) (P.A. Grieco, R.P. Nargund and D.T. Parker, *J. Am. Chem. Soc.*, 1989, 111, 6287), (+)-picrasin B (**98**) (M. Kim *et al.*, *J. Org. Chem.*, 1990, 55, 504), chaparrinone (**99**) (R.S. Gross, P.A. Grieco and J.L. Collins, *J. Am. Chem. Soc.*, 1990, 112, 9436), bruceantin (**100**) (M. Sasaki, T. Murae and T. Takahashi, *J. Org. Chem.*, 1990, 55, 528) and amarolide (**101**) (H. Hirota *et al.*, *J. Org. Chem.*, 1991, 56, 1119) have been achieved.

The group of C_{25}-quassinoid derivatives which represent the biogenetic link between tetranortriterpenoids and C_{20}-quassinoids has increased considerably. All members of this group lack oxygenation at C-12. Soulameolide (**102**) from *Soulamea tomentosa* (J. Polonsky *et al.*, *J. Chem. Soc.*, *Chem. Commun.*, 1979, 641) and klaineanolide A (**103**) from *Hannoa klaineana* (R. Vanhaelen-Fastré *et al.*, *Phytochemistry*, 1987, 26, 317) are representative examples.

Several unusual quassinoid structures have appeared. The structure of samaderine A (**104**), a C_{18} ring A contracted derivative from *Samadera indica*, was established by X-ray analysis (M.C. Wani *et al.*, *J. Chem. Soc.*, *Chem. Commun.*, 1977, 295). Lauricolactones A and B from *Eurycoma longifolia* are further additions to this C_{18} group (Nguyen-Ngoc-Suong *et al.*, *Tetrahedron Lett.*, 1982, 23, 5159). The novel carbon frameworks of shinjulactones B (**105**) (T. Furono *et al.*, *Bull. Chem. Soc. Jpn.*, 1984, 57, 2484) and C (**106**) and shinjudilactone (**107**) (M. Ishibashi *et al.*, *Bull. Chem. Soc. Jpn.*, 1983, 56, 3683), minor constituents of *Ailanthus altissima*, were all revealed by X-ray analysis. Shinjulactone F (**108**) from the same source has a *cis* AB ring junction (M. Ishibashi *et al.*, *Chem. Lett.*, 1984, 555). The biogenesis of the novel structure (**109**) of sergeolide from *Picrolemma pseudocoffea* may involve rearrangement of an acetylated precursor (C. Moretti *et al.*, *Tetrahedron Lett.*, 1982, 23, 647).

Many quassinoid glycosides have been isolated (T. Sakai *et al.*, *Bull. Chem. Soc. Jpn.*, 1985, 58, 2680; M. Okano *et al.*, *Bull. Chem. Soc. Jpn.*, 1985, 58, 1793). Of particular interest are the javanicinosides, such as javanicinoside K (**110**) (K. Koike and T. Ohmoto, *J. Nat. Prod.*, 1992, 55 482) which co-occur in *Picrasma javanica* with a series of aglycones such as javanicin J (**111**) (K. Koike *et al.*, *Phytochemistry*, 1991, 30, 933). Most of the compounds from *P. javanica* lack a methyl group at C-4. A ring A cleaved compound (**112**) from *Brucea javanica* has also been called javanicin (L.-Z. Lin *et al.*, *Phytochemistry*, 1990, 29, 2720).

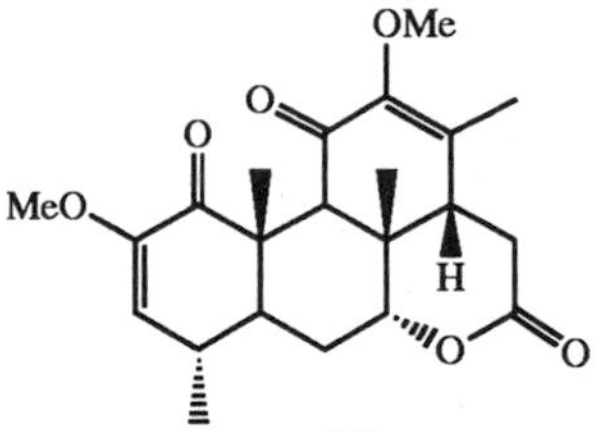

(95)

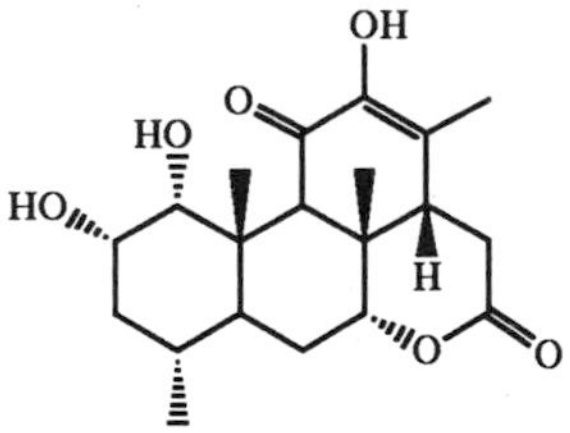

(96)

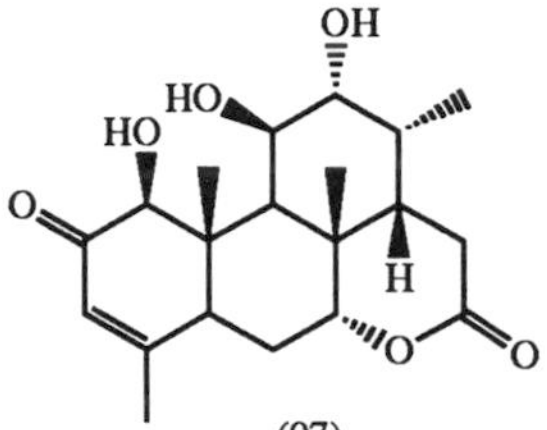

(97)

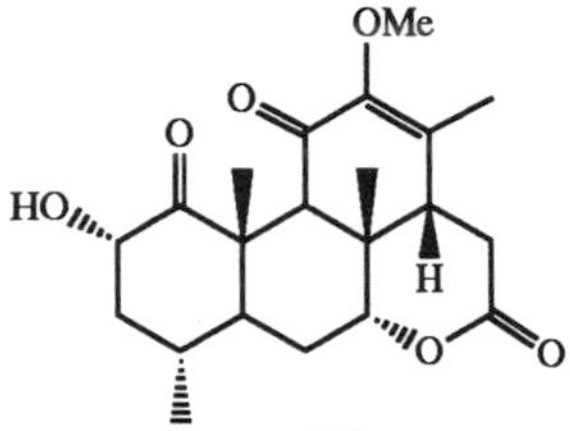

(98)

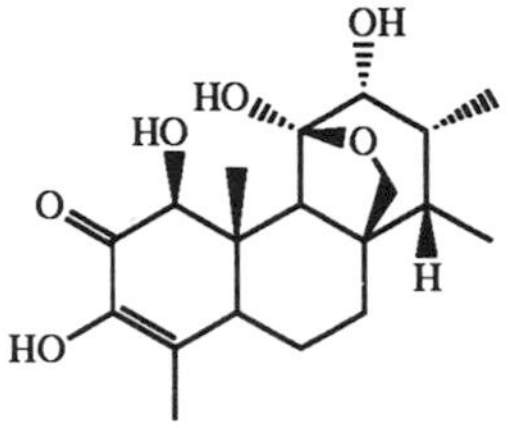

(99)

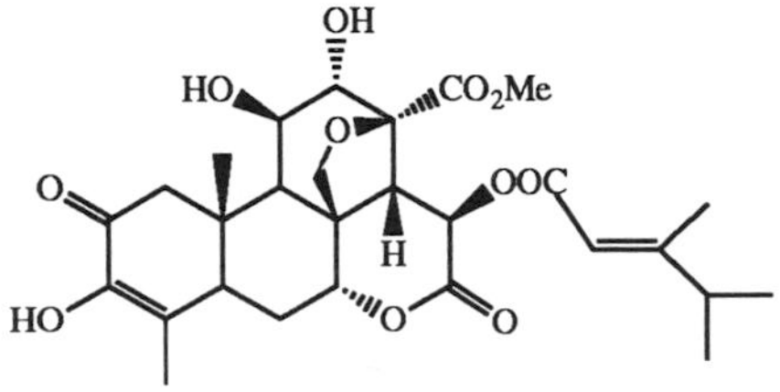

(100)

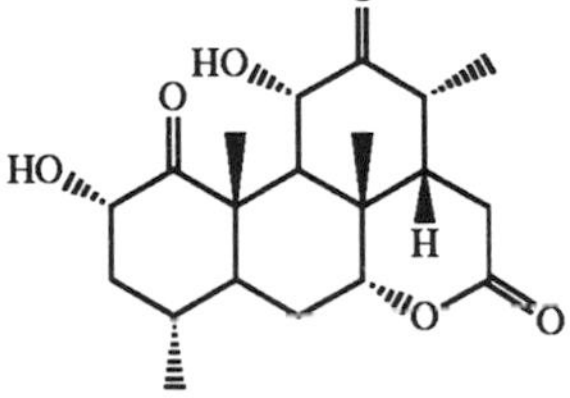

(101)

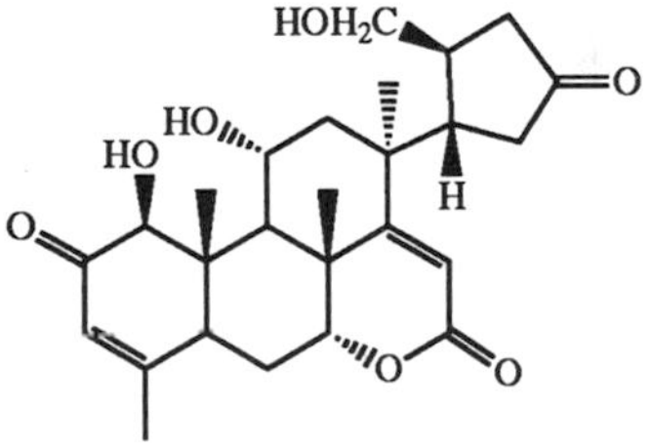

(102)

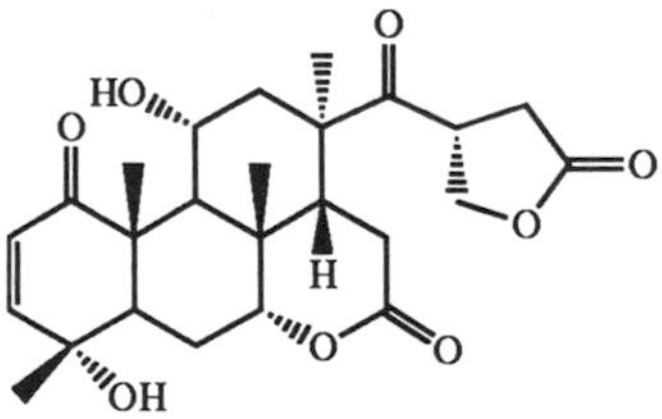

(103)

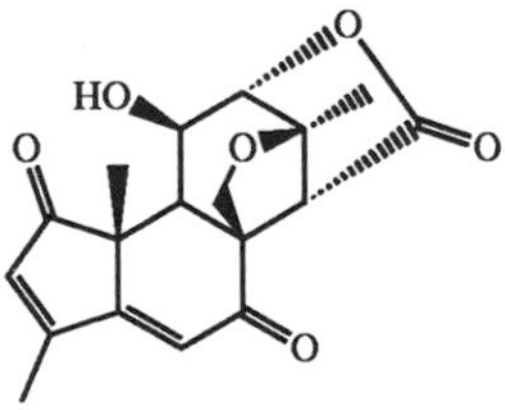

(104)

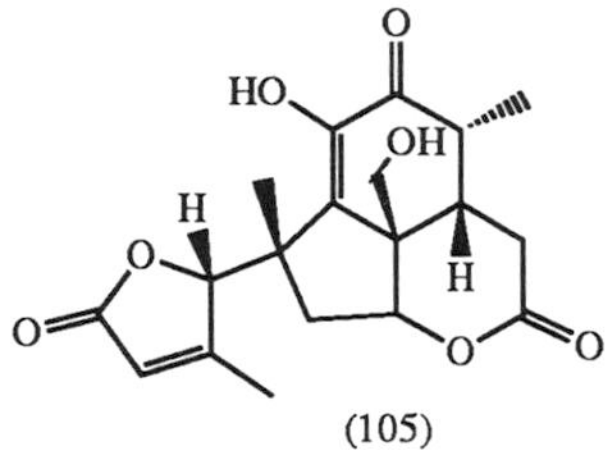

574

(105)

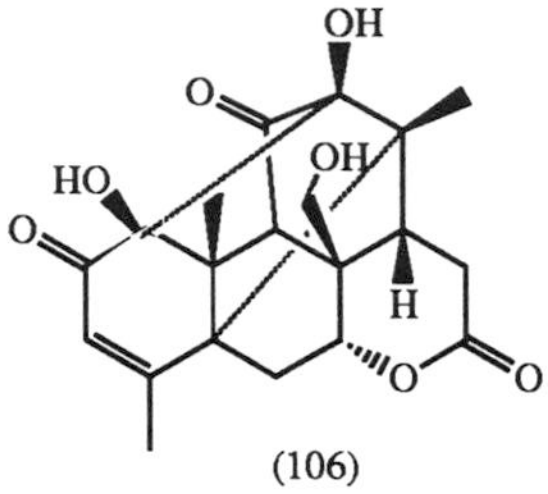

(106)

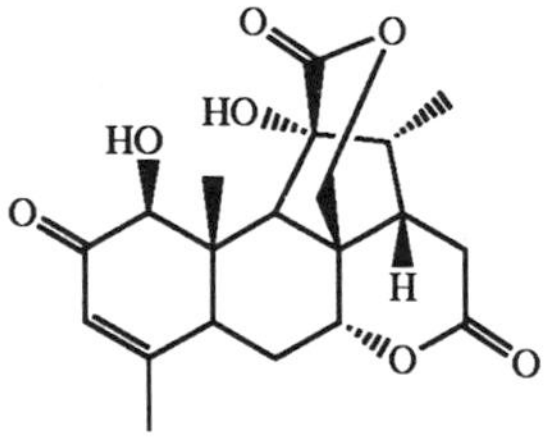

(107)

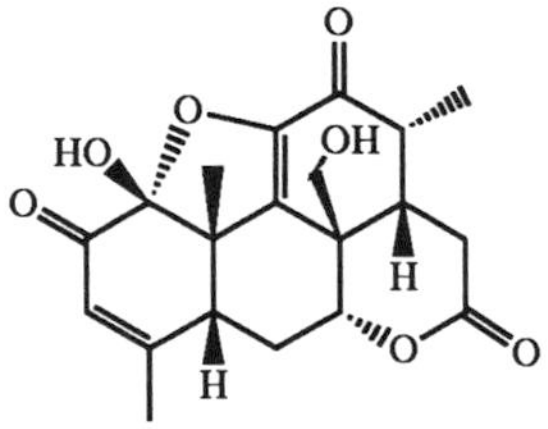

(108)

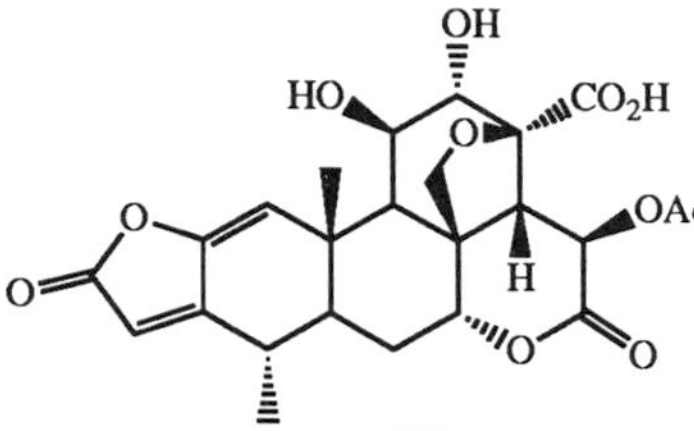

(109)

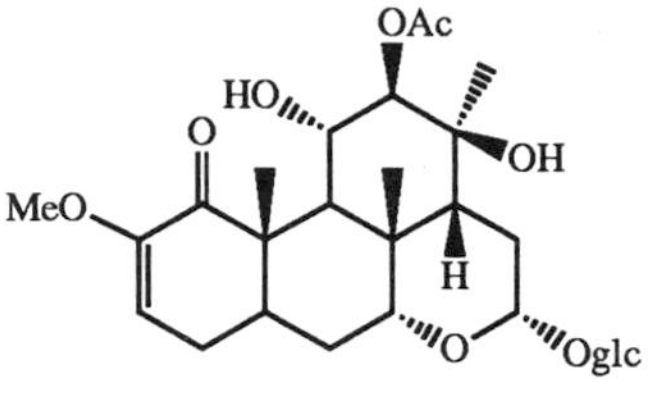

(110)

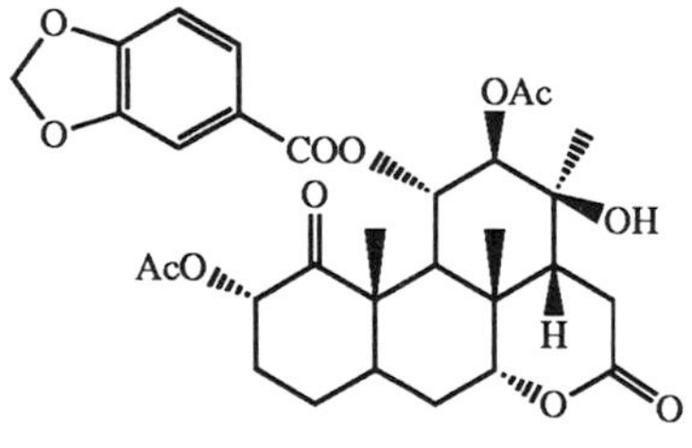

(111)

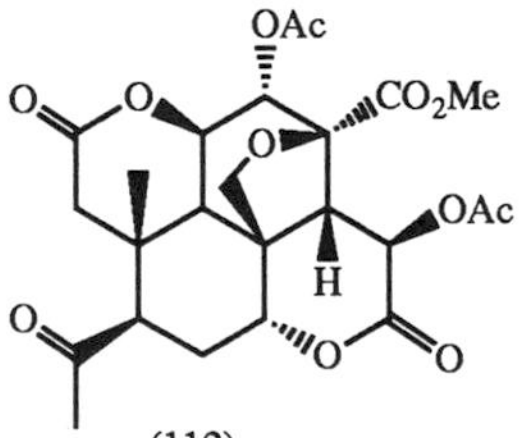

(112)

4. The Baccharane/Shionane Group

A total synthesis of shionone (113) has been achieved (R.E. Ireland *et al., J. Org. Chem.,* 1975, 40, 990). Backbone rearrangement of 3α,4α-epoxyshionane affords bacchar-12-en-3α-ol (114) (S. Yamada *et al., Bull. Chem. Soc. Jpn.,* 1976, 49, 1134). Natural compounds in this group include hosenkol A (115) from *Impatiens balsamina* (N. Shoji *et al., J. Chem. Soc., Chem. Commun.,* 1983, 871) and three related hydrocarbons bacchara-12(21)-diene, shiona-3,21-diene and lemnaphylla-7,21-diene (116) from *Lemnaphyllum microphyllum* (K. Masuda, K. Shiojima and H. Ageta, *Chem. Pharm. Bull.,* 1983, 31, 2530).

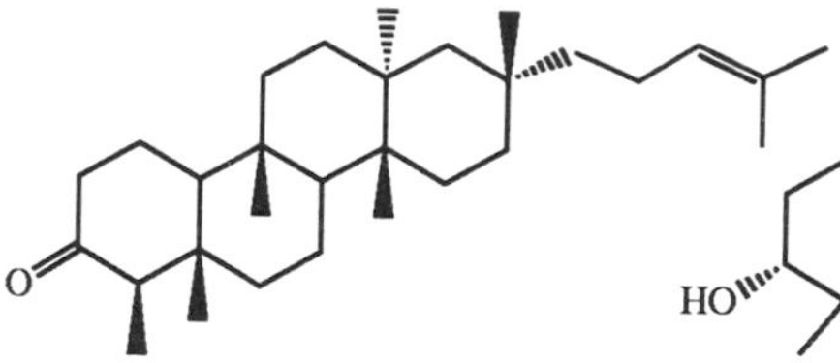

(113) 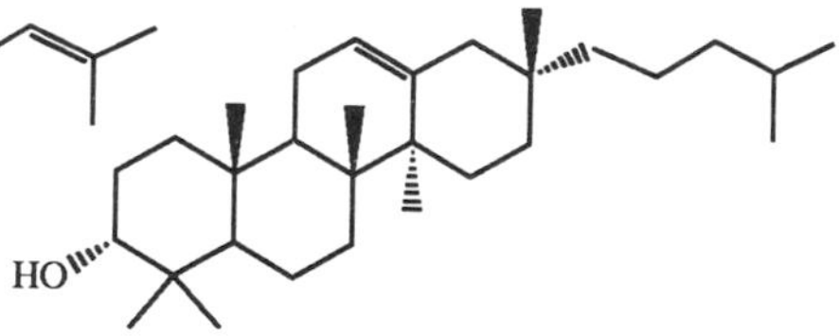　　　　　　　　　　(114)

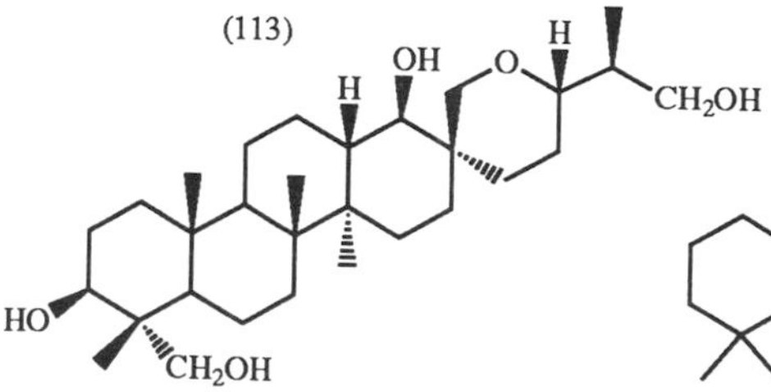

(115) 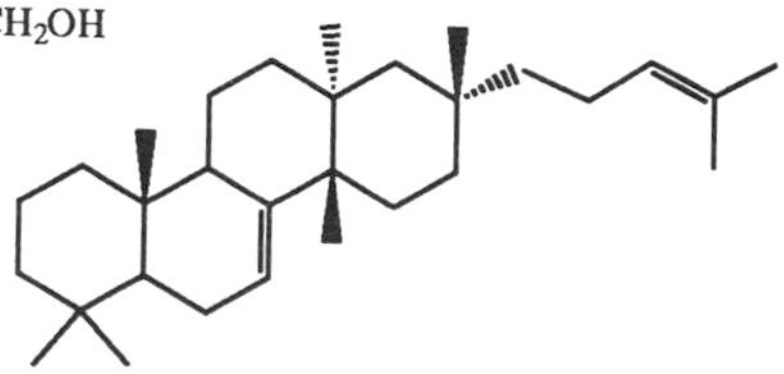　　　　　　　　　　(116)

5. The Lupane Group

Hancolupenol (117) and the corresponding 3-ketone, from *Cynanchum hancokianum* are representatives of a new skeletal type of triterpenoid. Hancokinol (118), from the same source, has a rearranged version of this new skeleton (Y. Konda *et al., Chem. Pharm. Bull.,* 1990, 38, 2899; H. Takayanagi *et al., Chem. Pharm. Bull.,* 1991, 39, 1234; H. Lou *et al., Chem. Pharm. Bull.,* 1991, 39, 2271). Neolupenol (119), from *Taraxacum japonicum* (H. Ageta *et al., Tetrahedron Lett.,* 1981, 22, 2289), and tylolupenols A (120) and B (121) from *Tylophora kerii* (K. Kawanishi *et al., Phytochemistry,* 1985, 24, 2051) represent various stages of backbone rearrangement of a lupane skeleton. Other friedolupanes include cymbopogone (122) and cymbopogonol (123) from lemongrass, *Cymbopogon citralis* (Y. Yokoyama *et al., Tetrahedron Lett.,* 1980, 21, 3701). The rearranged skeleton of opigenin (124) from *Opilia celtidifolia* is unusual (D. Druet, L. Comeau and J.-P. Zahra, *Can. J. Chem.,* 1986, 64, 295).

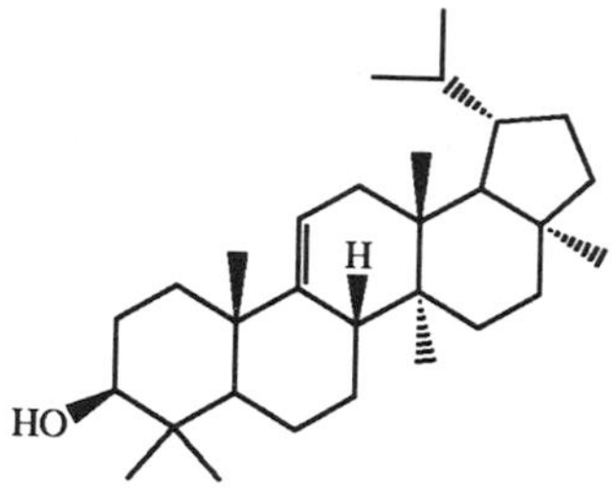

(117)

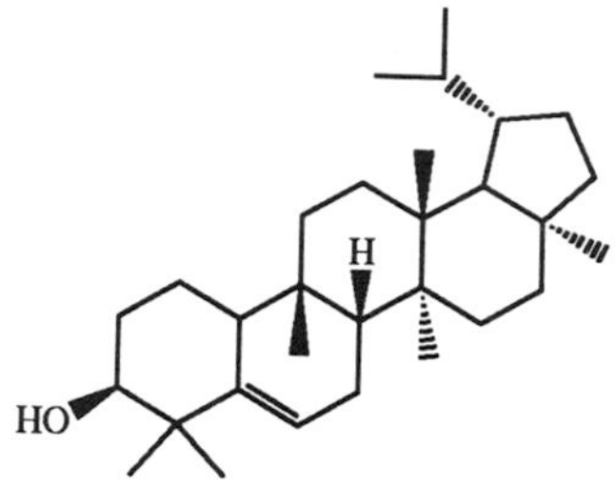

(118)

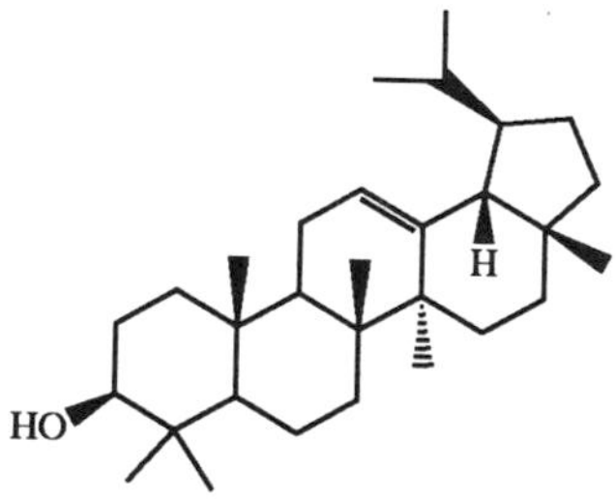

(119)

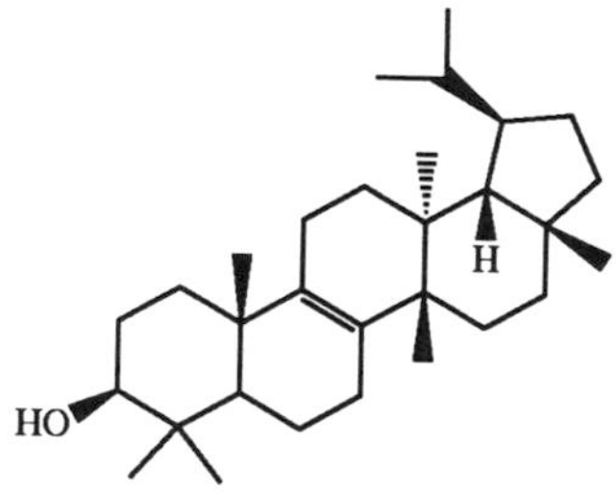

(120)

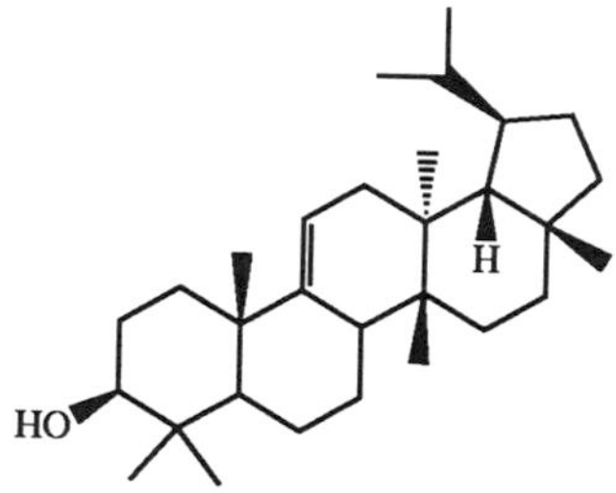

(121)

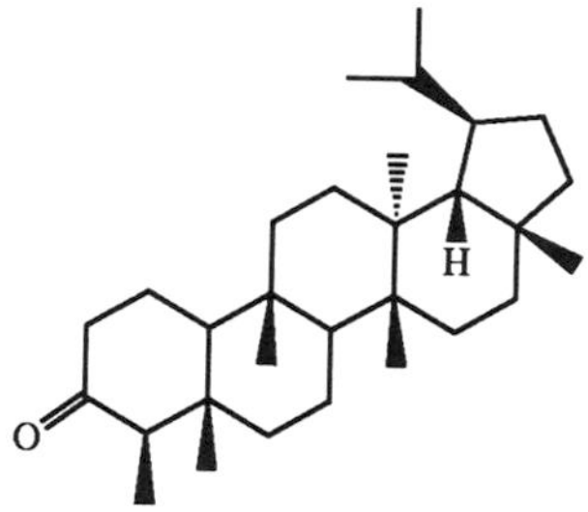

(122)

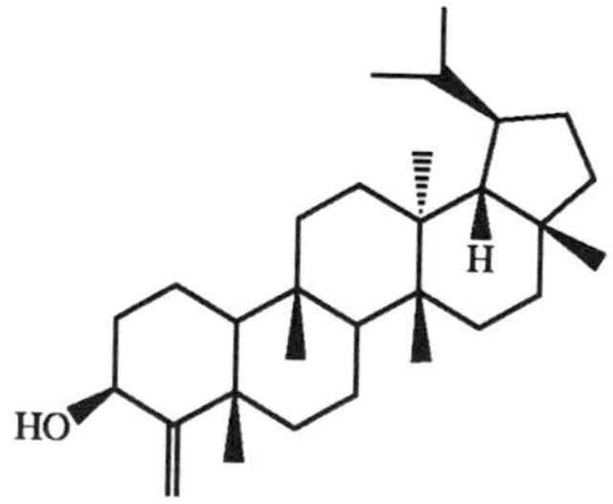

(123)

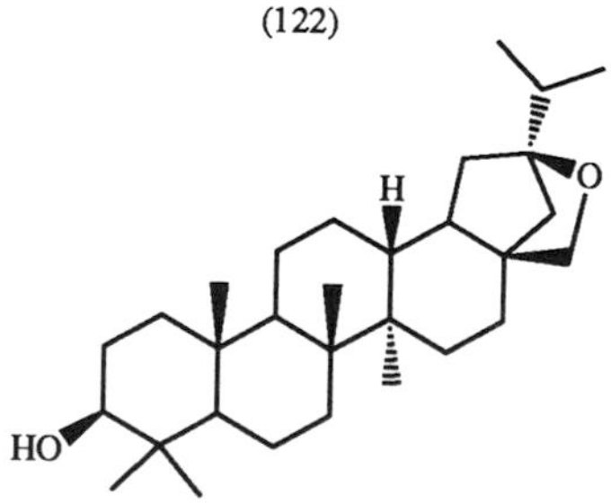

(124)

X-Ray analyses have revealed that in both emmolactone (**125**) from *Emmospermum alphitonioides* and the lactone (**126**) from *E. pancherianum*, the isopropenyl group has an α-orientation (G.V. Baddeley, J.J.H. Simes and T.-H. Ai, *Aust. J. Chem.*, 1980, 33, 2071). The stereochemistry of another ring A contracted lupane, granulosic acid from *Columbina granulosa* (D.K. Kulshreshtha, *Phytochemistry*, 1977, 16, 1783) and *Paliurus ramosissimus* (S.-S. Lee, W.-C. Lu, and K.C. Liu, *J. Nat. Prod.*, 1991, 54, 615) has been established as (**127**).

In common with other groups of triterpenoids, lupanes are found in geological samples. The hydrocarbon (**128**) and the corresponding Δ^9- and $\Delta^{5(10)}$-analogues have been detected in pond mud (J.M. Trendal *et al.*, *Tetrahedron*, 1989, 45, 4457). The aromatic derivatives (**129**) and (**130**) have been synthesised and identified in various samples (G.A. Wolff, J.M. Trendel and P. Albrecht, *Tetrahedron*, 1989, 45, 6721).

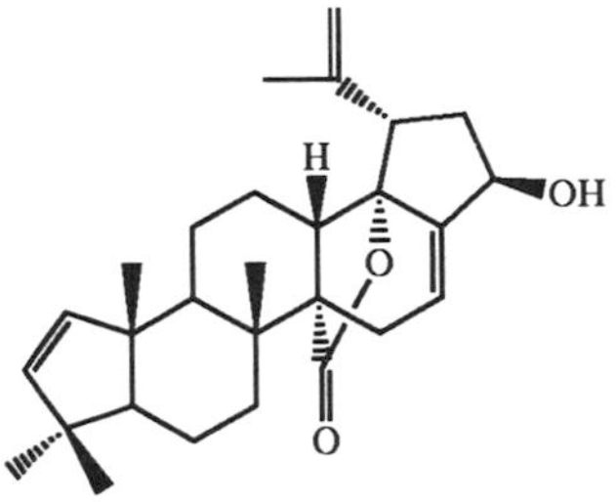

(125)

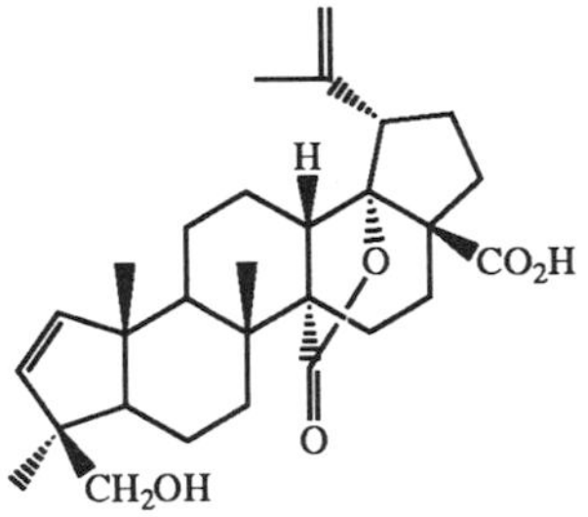

(126)

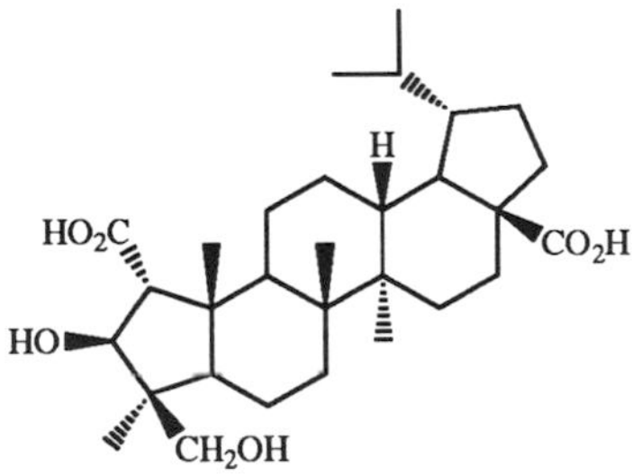

(127)

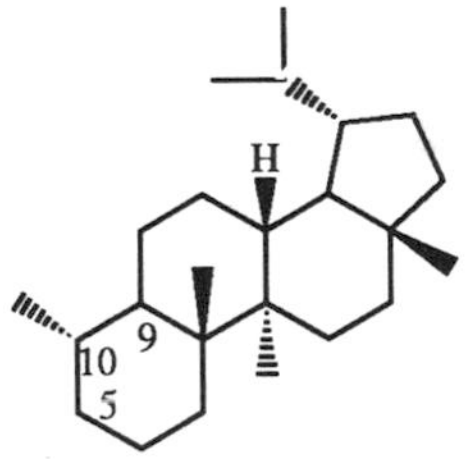

(128)

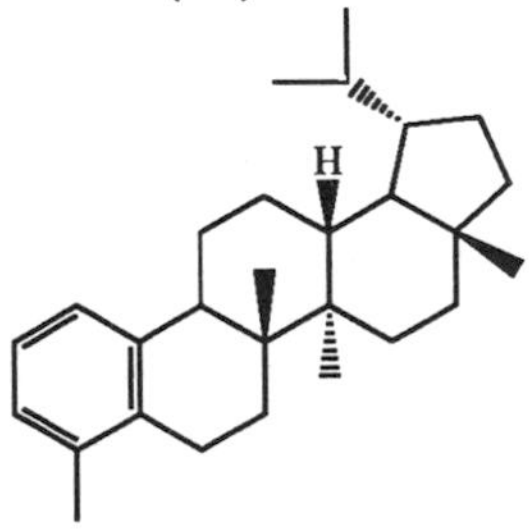

(129)

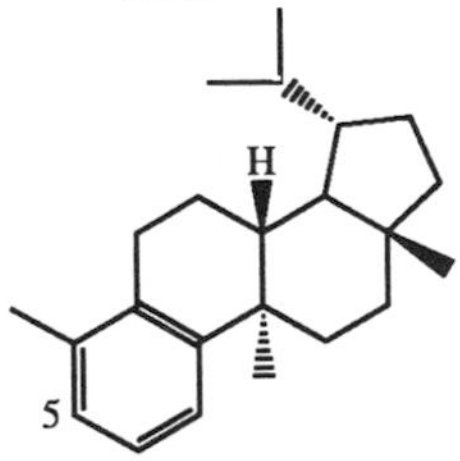

(130)

578

Lupeolactone (**131**) is a rare triterpenoid β-lactone from *Antidesma pentandrum* (H. Kikuchi *et al.*, *Chem. Lett.*, 1983, 603).

(131)

6. The Oleanane Group

Pfaffic acid (**132**) is an unusual rearranged oleanane from *Pfaffia paniculata* (T. Takemoto *et al.*, *Tetrahedron Lett.*, 1983, 24, 1057). It also occurs as glycosides in *P.* species (Y. Shiobara *et al.*, *Phytochemistry*, 1992, 37, 1737). Its structure was determined by X-ray analysis. The avenacins, for example avenacin A-1 (**133**) are resistance factors to the 'take-all' disease and occur in the roots of oats (*Avena sativa*) (M.J. Begley *et al.*, *J. Chem. Soc., Perkin Trans. 1*, 1986, 1905; L. Crombie, W.M. Crombie and D.A. Whiting, *J. Chem. Soc., Perkin Trans. 1*, 1986, 1917). Other unusual ring E arrangements are found in steganogenin (**134**) from *Steganotaenia araliacea* (A.G. González *et al.*, *Tetrahedron*, 1992, 48, 769) and radermasinin (**135**), a cytotoxic constituent of *Radermachia sinica* (G.K. Rice *et al.*, *J. Chem. Soc., Chem. Commun.*, 1986, 1397). The β-lactone papyriogenin G (**136**) has been isolated from the leaves of *Tetrapanax papyriferum* (M. Asada *et al.*, *J. Chem. Soc., Perkin Trans. 1*, 1980, 325).

The structures of soyasapogenols A (**137**), B and E have been revised on the basis of an X-ray analysis (I. Kitagawa *et al.*, *Chem. Pharm. Bull.*, 1982, 30, 2294: R.L. Baxter, K.R. Price, and G.R. Fenwick, *J. Nat. Prod.*, 1990, 53, 298). Germanidiol from *Rhododendron macrocephalum* has been shown to be olean-18-ene-2α,3α-diol and not the 2β,3β-diol initially suggested (H. Ageta and T. Ageta, *Chem. Pharm. Bull.*, 1984, 32, 369). Consideration of the ¹H nmr shifts of 2,3-diols and 2,3,23- and 2,3,24-triols in the oleanane and ursane series has led to the structural revision of several structures (H. Kojima and H. Ogura, *Phytochemistry*, 1989, 28,1703). Backbone rearrangement of 3β,4β-epoxyfriedelin affords germanicol (olean-18-en-3β-ol) or dendropanoxide (**138**) depending on the conditions used (T. Tori *et al.*, *Tetrahedron Lett.*, 1975, 2283; *Bull. Chem. Soc. Jpn.*, 1977, 50, 338). Total syntheses of alnusenone (**139**) and friedelin have been published (R.E. Ireland and D.M. Walba, *Tetrahedron Lett.*, 1976, 1071; T. Kametani *et al.*, *J. Am. Chem. Soc.*, 1978, 100, 554).

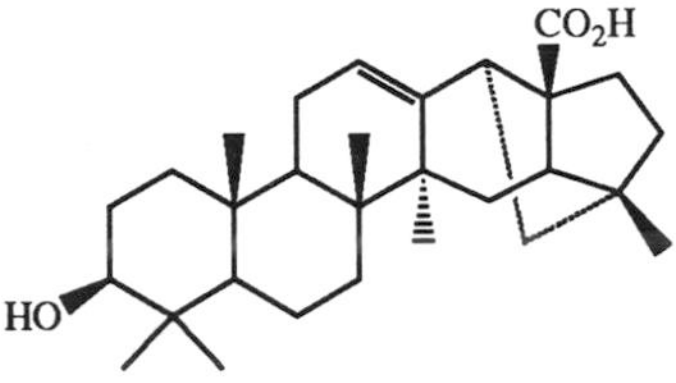

(132)

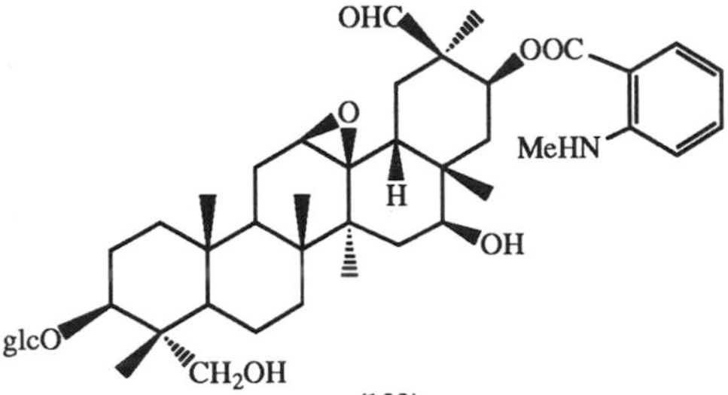

(133)

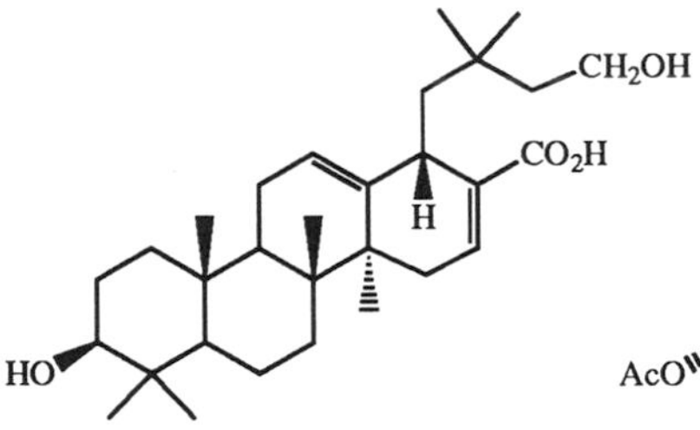

(134)

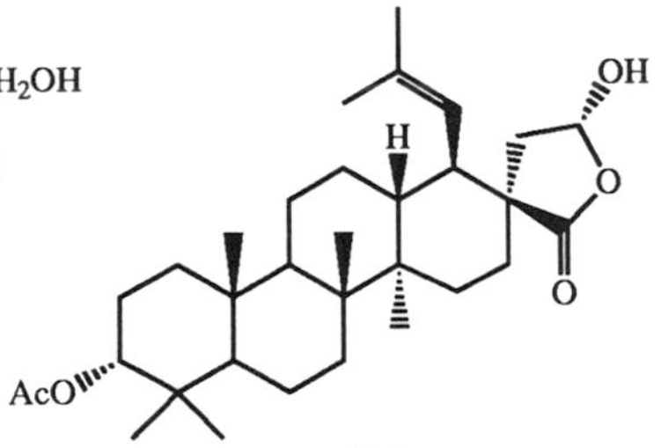

(135)

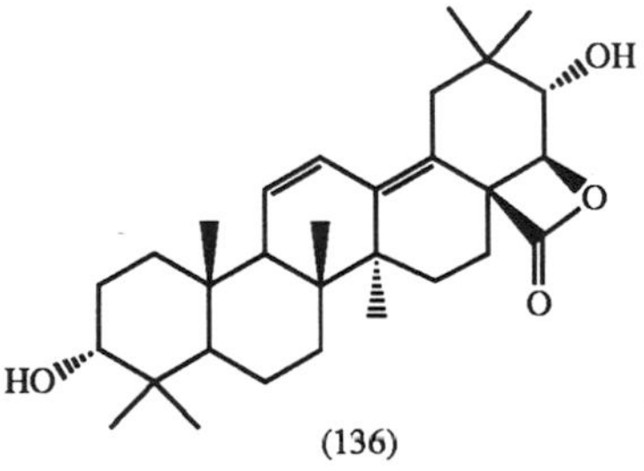

(136)

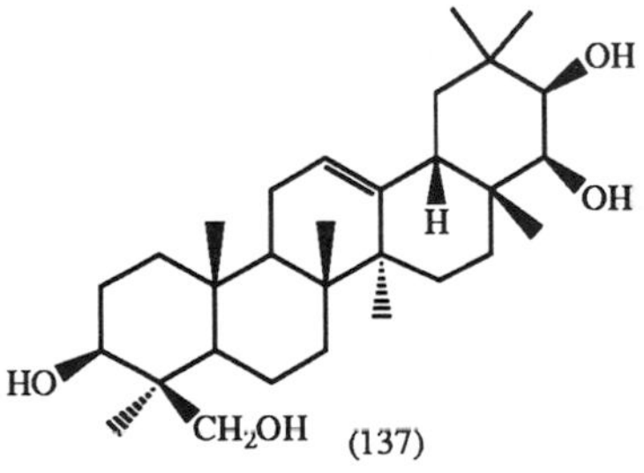

(137)

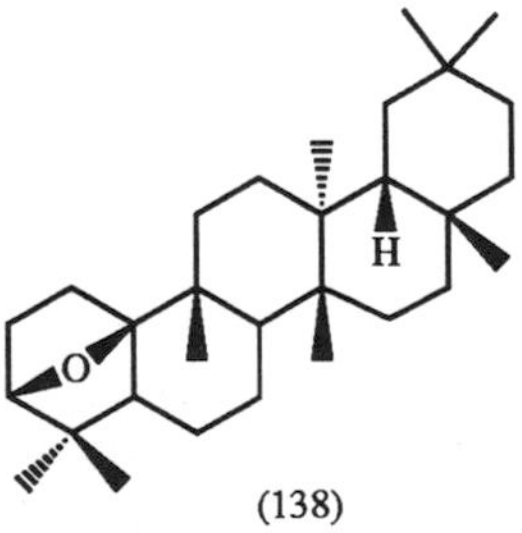

(138)

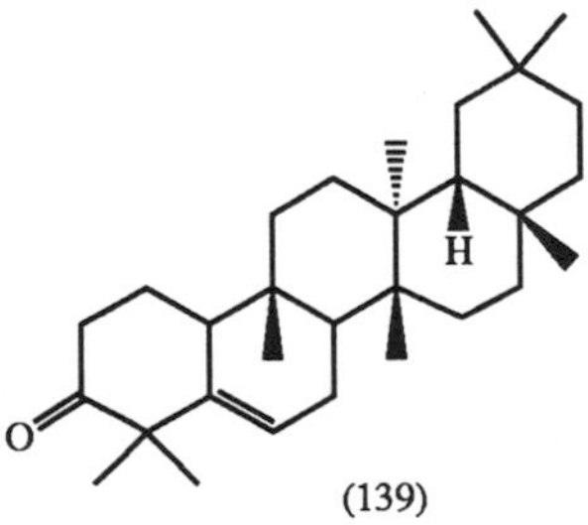

(139)

580

As with other groups of triterpenoids, oleanane derivatives, often degraded and aromatised, are found in geological samples. 18α-Oleanane (D.T. Fowell, B.G. Melsom and G.W. Smith, *Acta Crystallogr. Sect. B*, 1978, 34, 2264) and 24,28-dinor-18α-oleanane (J.M. Trendel *et al.*, *Tetrahedron Lett.*, 1991, 32, 2959) are constituents of Nigerian and Egyptian petroleum respectively. Several chrysene derivatives, for example the tetrahydrochrysene (**140**), have been identified from sediments and other geological sources (C. Spyckerelle *et al.*, J. Chem. Res., (S), 1977, 330; 332 J.M. Trendal *et al.*, *Tetrahedron*, 1989, 45, 4457). Another sediment constituent is the unusual thiophene derivative (**141**) (P. Adam, J.M. Trendel and P. Albrecht, *Tetrahedron Lett.*, 1991, 47, 4179) which can be prepared by heating β-amyrin acetate with elemental sulfur, followed by removal of the C-3 oxygen substituent.

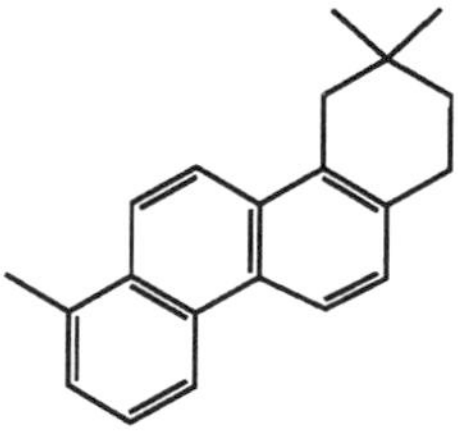

(140)

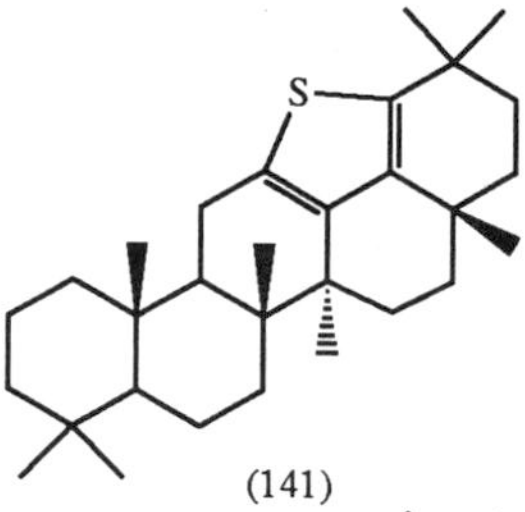

(141)

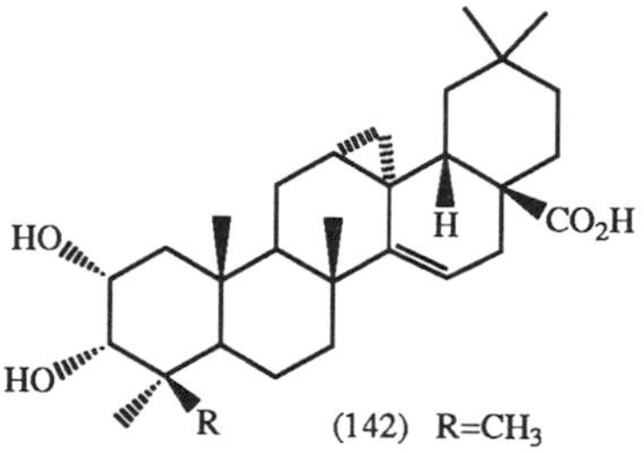

(142) R=CH₃

(143) R=CH₂OH

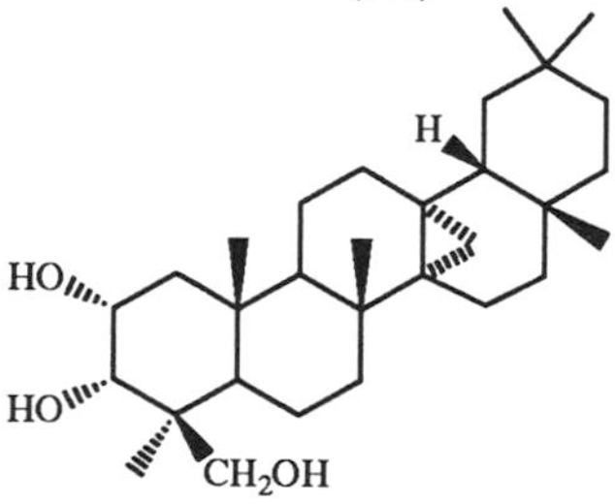

(144)

Several hexacyclic oleanane derivatives have been described. The 12α,27-cyclo compounds przewanoic acid A (**142**) and its 24-nor congener przewanoic acid B ($\Delta^{4(23)}$) occur in *Salvia przewalskii* (N. Wang, M. Niwa and H.-W. Luo, *Phytochemistry*, 1988, 27, 299) while (**143**) and the corresponding 13α,27-cyclo compound (**144**) are found in *Prunella vulgaris* (H. Kojima *et al.*, *Phytochemistry*, 1988, 27, 2971). The structure of 4β,24-cyclofriedelan-3-one (**145**) from *Euphorbia neriifolia* was established by X-ray analysis (J.D. Connolly *et al.*, *Acta Crystallogr., Sect. C*, 1986, 42, 1352).

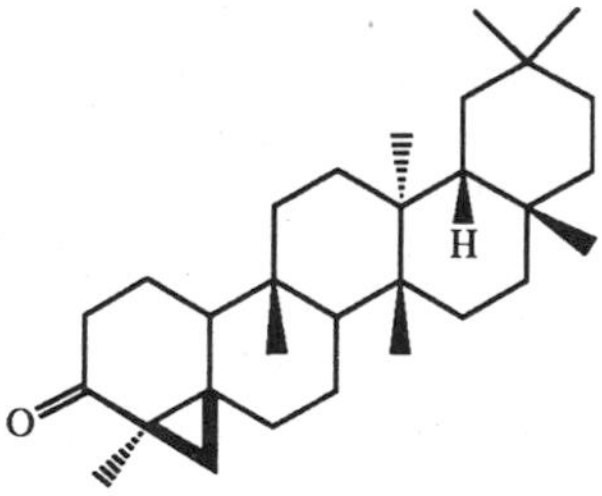

(145)

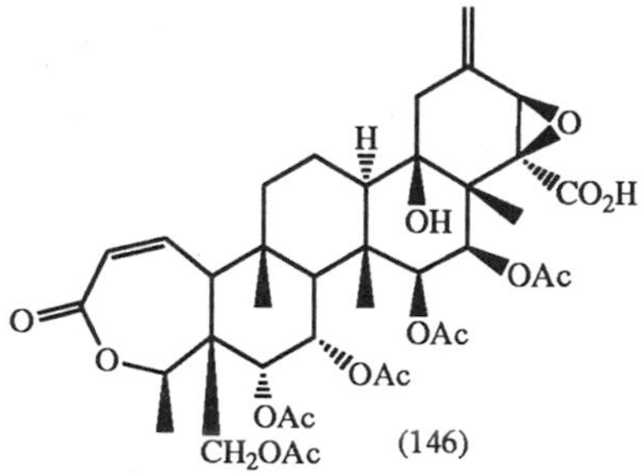

(146)

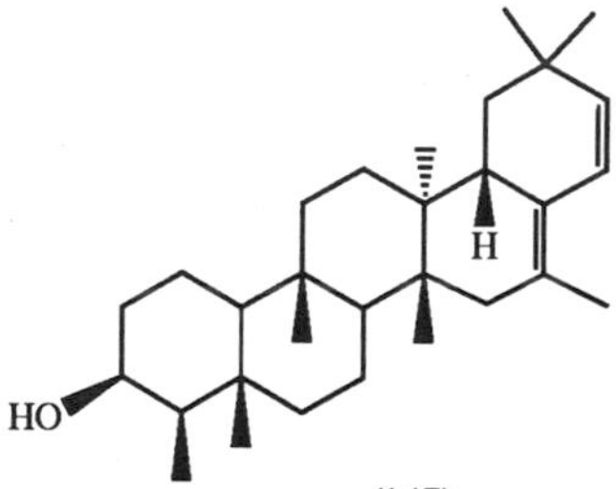

(147)

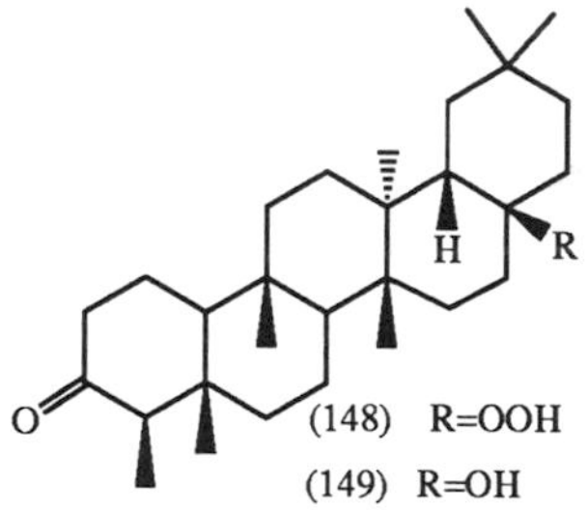

(148) R=OOH
(149) R=OH

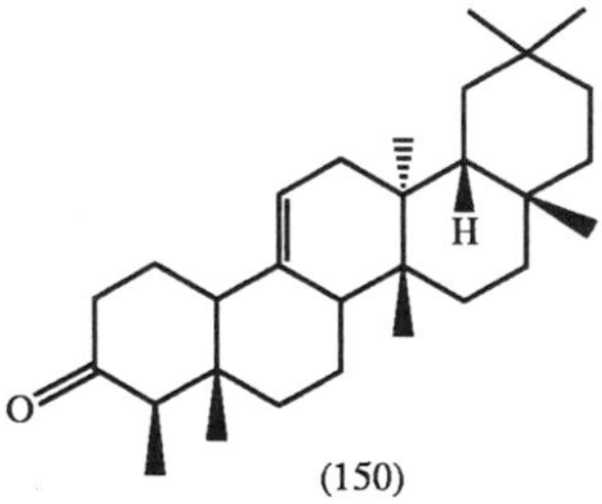

(150)

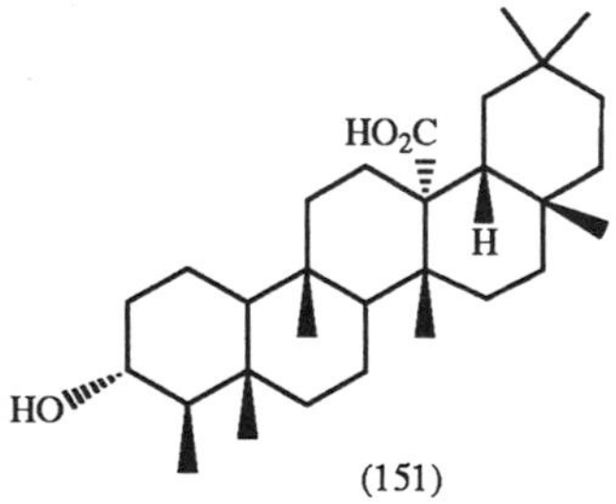

(151)

One of the most interesting friedelane derivatives to appear recently is the highly oxygenated compound (**146**) from *Lophanthera lactescens* (H. dos S. Abrea *et al.*, *Phytochemistry*, 1990, 29, 2257). Its rearranged ring E presumably arises by a cleavage and recyclisation process. Pachysandienol A (**147**) (T. Kikuchi *et al.*, *Chem. Pharm. Bull.*, 1981, 29, 2531) is unusual in having the C-28 methyl group attached to C-16. Several compounds of this type have been reported from *Pachysandra terminalis* (M.I.M. Wazeer, *Magn. Reson. Chem.*, 1985, 23, 250). Maytensifolin A (**148**) is a 28-nor-28-hydroperoxide of friedelin from *Maytenus diversifolia* (K.-H. Lee, H. Nozaki and A.T. McPhail, *Tetrahedron Lett.*, 1984, 25, 707). The corresponding 28-hydroxyfriedelan-3-one, elaeodendrol (**149**), has been

582

found in *Elaeodendron glaucum* (A.S.R. Anjaneyulu and M.N. Rao, *Phytochemistry*, 1980, 19, 1163). Putrone, from *Putranjiva roxburghii*, is 25-norfriedel-9(11)-en-3-one (**150**). This structure was confirmed by partial synthesis (P. Sengupta *et al.*, *J. Chem. Soc., Perkin Trans. 1*, 1979, 60; *Indian J. Chem., Sect. B*, 1989, 28, 21). The following species are rich sources of friedelane derivatives: *Kokoonia zeylanica* (A.A.L. Gunatilaka, N.P.D. Nanayakkara and M.U.S. Sultanbawa, *J. Chem. Soc., Chem. Commun.*, 1979, 434; *Tetrahedron Lett.*, 1979, 1727), *Trichadenia zeylanica* (S.P. Gunasekera and M.U.S. Sultanbawa, *J. Chem. Soc., Perkin Trans. 1*, 1977, 483) and *Hydnocarpus octandra* (S.P. Gunasekera and M.U.S. Sultanbawa, *J. Chem. Soc., Perkin Trans. 1*, 1977, 418). A recent X-ray analysis of trichaodenic acid B (**151**) reveals that the *Trichadenia* compounds are oxygenated on C-27 and not C-26 as originally proposed (R. Tanaka, S. Matsunaga and T. Ishida, *Tetrahedron Lett.*, 1988, 29, 4751).

(152)

(153) R=H
(154) R=OH

The group of friedelane quinone-methide derivatives has grown steadily over the years since the early work on pristimerin (**152**) and tingenins A (**153**) and B (**154**) (K. Nakanishi *et al.*, *J. Am. Chem. Soc.*, 1973, 95, 6473; F. Delle Monache *et al.*, *An. Quim.*, 1974, 70, 1040). Ring A aromatic derivatives, exemplified by zeylasterone (**155**) from *K. zeylanica*, also occur (G.M.K.B. Gunaherath *et al.*, *Tetrahedron Lett.*, 1980, 21, 4749; Y. Tezuka *et al.*, *J. Chem. Res.*, 1989, (S) 268; (M) 1901). Zeylasterone has been prepared from pristimerin (G.M.K.B. Gunaherath and A.A.L. Gunatilaka, *J. Chem. Soc., Perkin Trans. 1*, 1983, 2845). Extension of conjugation and migration of the 26-methyl group are apparent in netzahualcoyone (**156**) from *Orthosphenia mexicana* (A.G. González *et al.*, *Tetrahedron Lett.*, 1982, 24, 3033), the structure of which was solved by X-ray analysis. A series of related compounds has been reported from *O. mexicana, Maytenus horrida* and *Cassine balae* (A.G. González *et al.*, *J. Chem. Res.*, 1988, (S) 20; (M) 0273; H.C. Fernando *et al.*, *Tetrahedron*, 1989, 45, 5867). Celastanhydride (**157**) is a ring A cleaved derivative, found in several species (C.B. Gamlath *et al.*, *Tetrahedron Lett.*, 1988, 29, 109; *Phytochemistry*, 1990, 29, 3189). Recently bisfriedelanes have been found in *Maytenus umbalata* and *M. ilicifolia* (H. Itokawa *et al.*,, *Tetrahedron Lett.*, 1990, 31, 6881). Umbellatin α (**158**) is a representative example (A.G. González *et al.*, *Tetrahedron*, 1992, 48, 769).

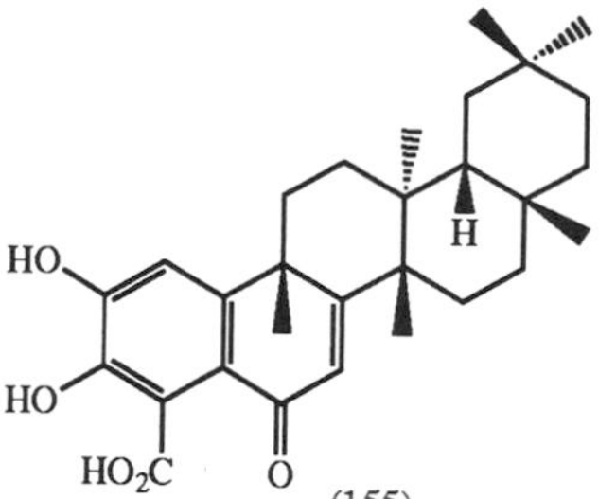

(155)

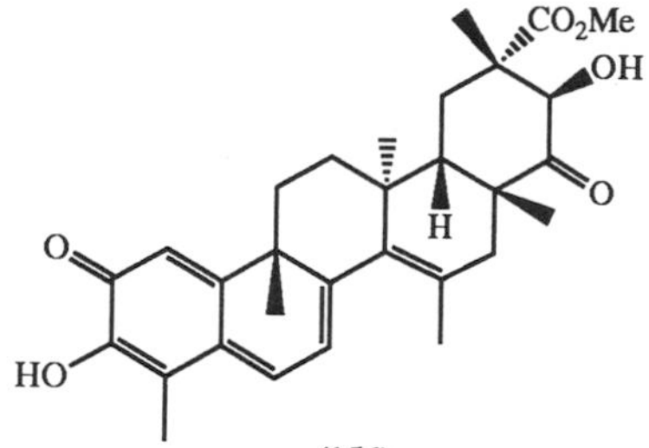

(156)

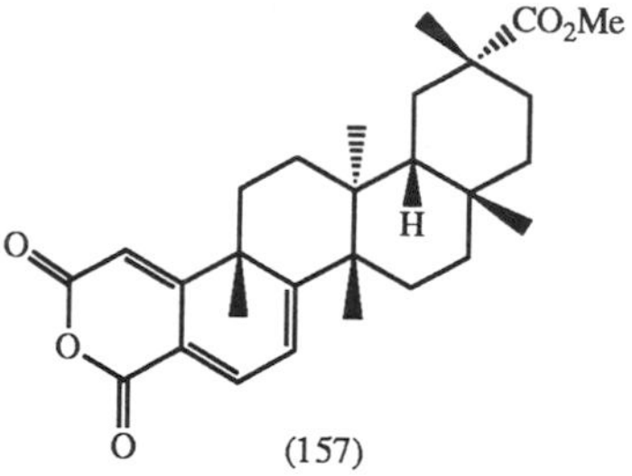

(157)

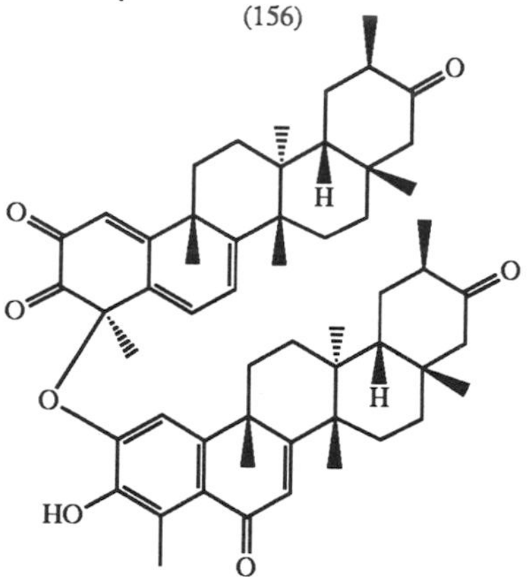

(158)

7. The Ursane Group

A study of the biosynthesis of ursolic acid in cell cultures of *Perilea putrescens*, using [3,5-$^{13}C_2$]- and [2-^{13}C, 4-$^2H_2$]-mevalonolactone, has provided experimental proof for the generally accepted mechanism of formation of rings D and E (Y. Tomita, M. Arata and Y. Ikeshiro, *J. Chem. Soc., Chem. Commun.*, 1985, 1087).

The degraded norursane (159) occurs in various sediments and geological samples together with the partially aromatised derivatives (160), (161) and (162) (J.M. Trendal *et al.*, *Tetrahedron*, 1989, 45, 4457).

A recent example of the many 19α-hydroxyursanes which have appeared is the glucosyl ester (163) from *Bencomia candata* and *Dendriopterium menendezii* (G. Reher and M. Buděšínský, *Phytochemistry*, 1992, 31, 3909). Several ring-A modified ursolic acid derivatives, including (164), have been reported from the fruit of *Pseudopanax arboreum* (B.F. Bowden, R.C. Cambie and J.C. Parnell, *Aust. J. Chem.*, 1975, 28, 785). The structure of marsformoxide A (165) from *Marsdenis formosana* has been confirmed by partial synthesis from α-amyrin (K. Ito and J. Lai, *Chem. Pharm. Bull.*, 1978, 26, 1908).

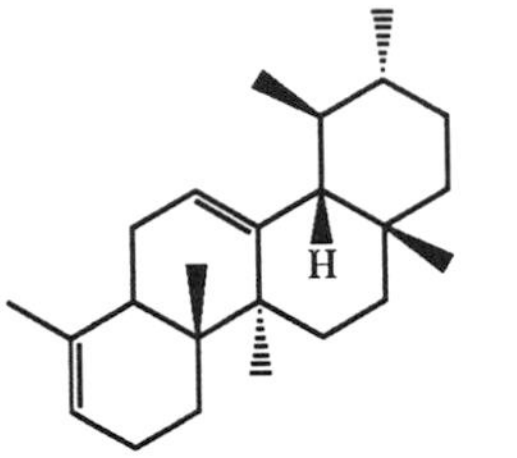

584

(159)

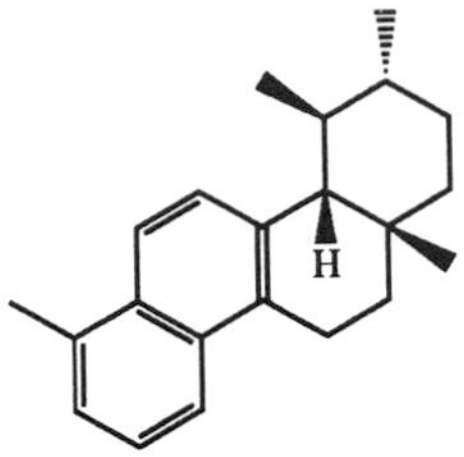

(160)

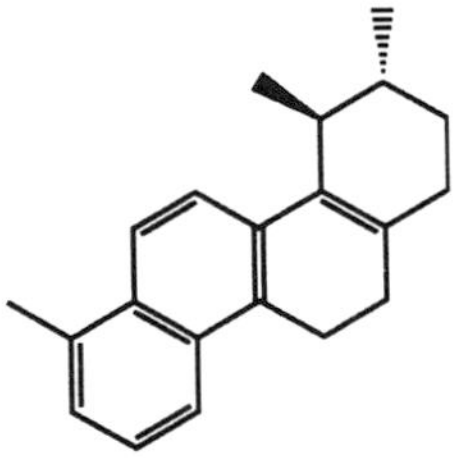

(161)

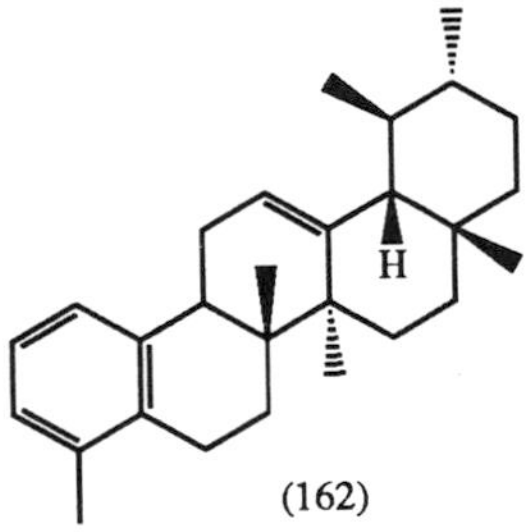

(162)

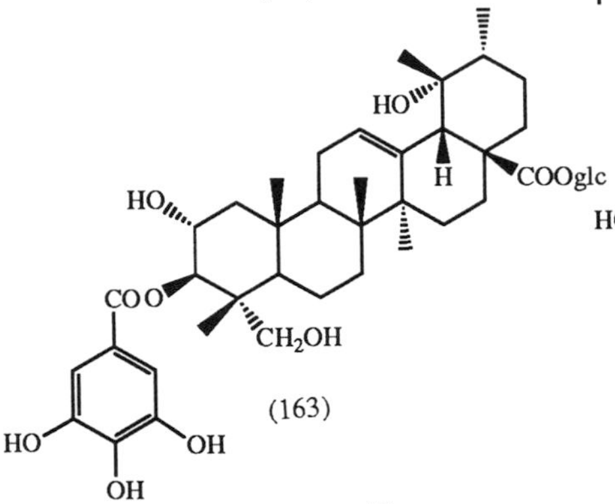

(163)

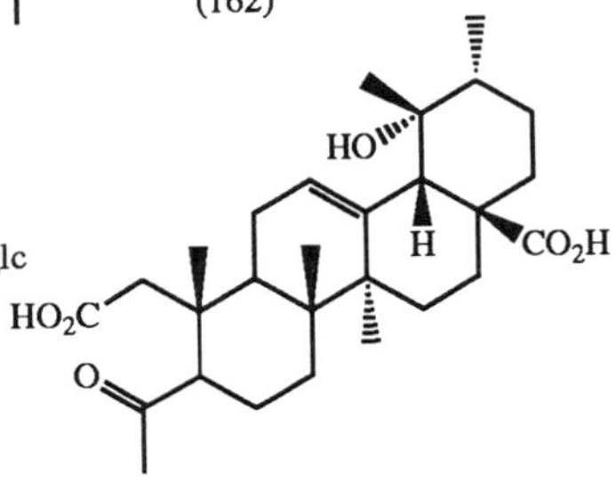

(164)

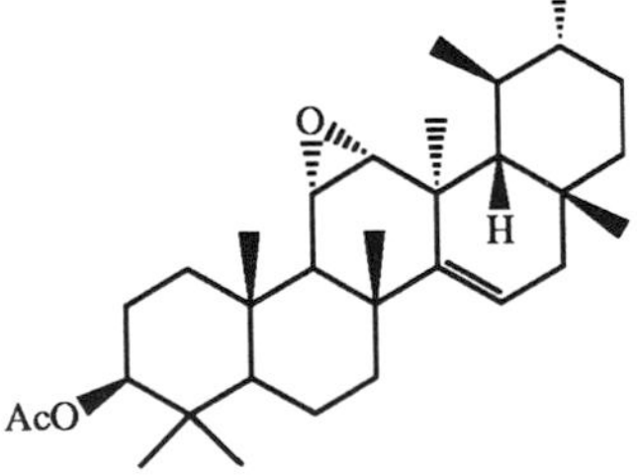

(165)

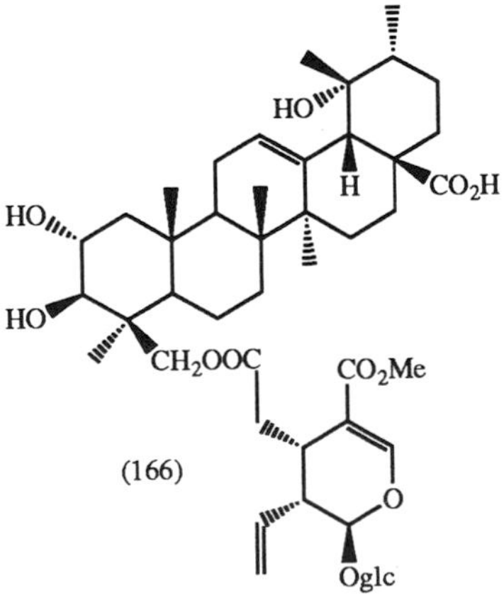

(166)

Desfontainic acid (**166**), from *Desfontainia spinosa*, has the unusual combination of a triterpenoid and an iridoid (P.J. Houghton and L.M. Lian, *Phytochemistry*, 1986, <u>25</u>, 1907).

The genin of ilexosides A - D from *Ilex crenata* is α-ilexanolic acid (**167**) with a novel 18,19-cleaved skeleton (T. Takano, K. Yoshikawa and S. Arihara, *Tetrahedron Lett.*, 1991, <u>32</u>, 3535). Hyptadienic acid (**168**) from *Hyptis suaveolens* (K.V.R. Rao, L.J.M. Rao and N.S.P. Rao, *Phytochemistry*, 1990, <u>29</u>, 1326) and *Coleus forskohlii* (R. Roy *et al.*, *Tetrahedron Lett*, 1990, <u>31</u>, 3467) has a ring A contracted skeleton (*cf* lupanes). A different type of cyclopentanic ring A is found in musancropic acid A (**169**) from *Musanga cecropioides*. The ring contraction presumably proceeds by a benzilic acid rearrangement of an intermediate 2,3-diketone (D. Lontsi *et al.*, *Phytochemistry*, 1991, **30**, 2361).

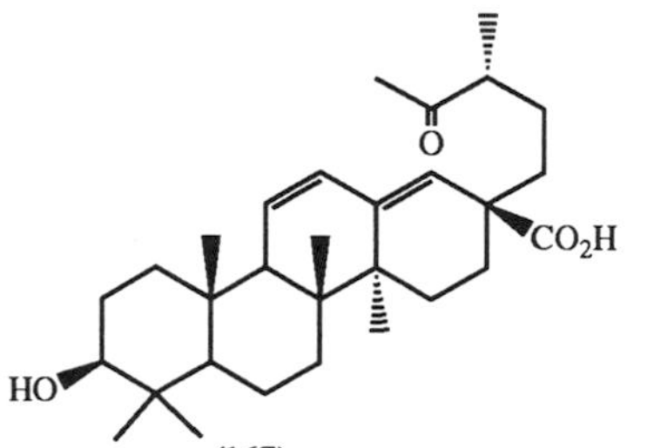

(167)

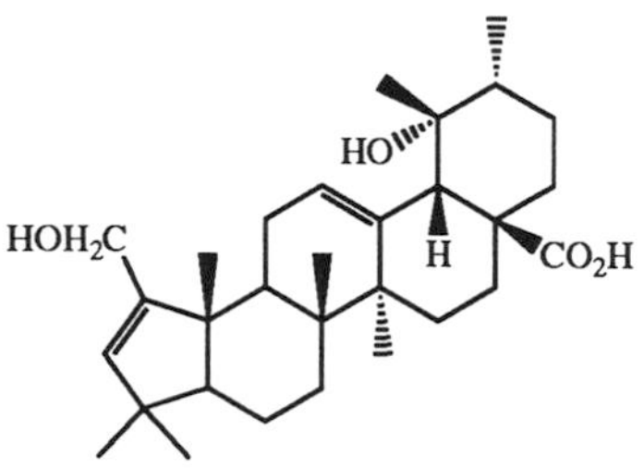

(168)

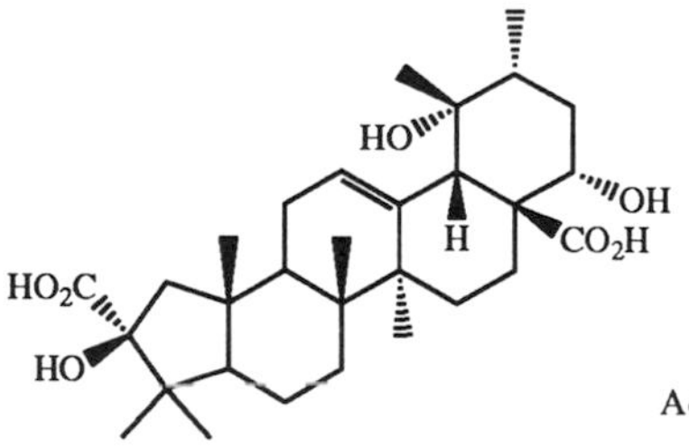

(169)

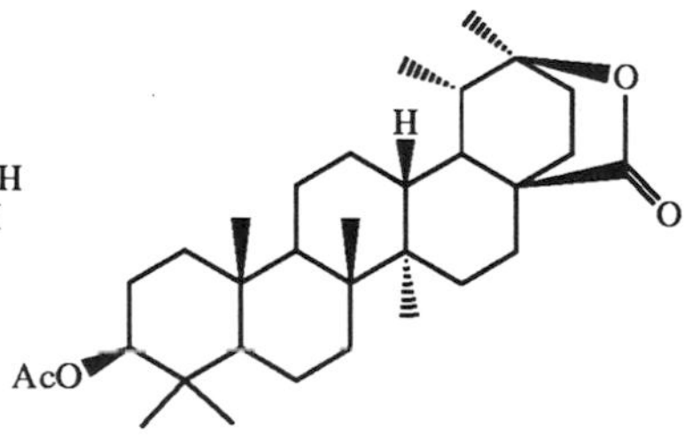

(170)

X-Ray analysis revealed the taraxastane structure (**170**) for randiflorin from *Opilia celtidifolia* (D. Druet *et al.*, *Can. J. Chem.*, 1987, <u>65</u>, 851). Careyagenolide from *Careya arboria* and nahagenin from *Fagonia indica* are the 2α,3β-diol (M.C. Das and S.B. Mahato, *Phytochemistry*, 1982, <u>21</u>, 2069) and 3β,23-diol (Atta-ur-Rahman *et al.*, *Heterocycles*, 1982, <u>19</u>, 217) of (**170**) respectively.

8. The Hopane Group

There is a continuing interest in this group because of the presence of extended and degraded hopane derivatives in bacteria, sediments and petroleum. Since the structure of bacteriohopanetetrol (171) was proposed (H. J. Forster *et al.*, *Biochem. J.*, 1973, 135, 133; M. Rohmer and G. Ourisson, *Tetrahedron Lett.*, 1976, 3633) many variations on this theme have appeared. Aminobacteriohopanetriol (172) has been correlated with bacteriohopanetetrol revealing that both have the same absolute configuration (S. Neunlist and M. Rohmer, *J. Chem. Soc., Chem. Commun.*. 1988, 830; P. Bissert and M. Rohmer, *J. Org. Chem.*, 1989, 54, 2958). The side chain of bacteriohopanetetrol is derived from ribose (G. Flesch and M. Rohmer, *J. Chem. Soc., Chem. Commun.*, 1988, 868; M. Rohmer, B. Sutter and H. Sahm, *J. Chem. Soc., Chem. Commun.*, 1989, 1471). *Acetobacter aceti* and *A. pasteureinus* are rich sources of bacteriohopanes, many of which, for example (173), contain a 3β-methyl group and/or Δ^6 and/or Δ^{11} unsaturation (B. Peiseler and M. Rohmer, *J. Chem Res. (S)* 1992, 298). The aminoglycoside (174) is a metabolite of a *Synechocystis* species (P. Simonin, U. Jürgens and M. Rohmer, *Tetrahedron Lett.*, 1992, 33, 3629).

Methylation of the normal hopane skeleton can occur at C-17 (175) and both C-3 and C-17 (176) (D.L. Howard and D.J. Chapman, *J. Chem. Soc., Chem. Commun.*, 1981, 468), C-2β (177) from the methanol-oxidizing bacterium *Corynebacterium* sp. XG (A. Babadjamiam *et al.*, *J. Chem. Soc., Chem. Commun.*, 1984, 1657) and C-2α (178) from *Methylobacterium organophilum* (P. Stampf *et al.*, *Tetrahedron*, 1991, 47, 7081).

(171) R=H ; R'=OH
(172) R=H ; R'=NH₂
(173) R=CH₃ ; R'=OH ; Δ⁶

(175) R=βOH ; R'=CH₃
(176) R=CH₃ ; R'=CH₂OH

(174)

(177) R=βCH₃
(178) R=αCH₃

A series of hydrocarbons C_{27}-C_{35} based on the 17αH-hopane skeleton has been identified in geological samples. These compounds, which occur with the corresponding C_{29}-C_{31} moretanes such as (179), may arise by degradation of a bacteriohopanetetrol precursor (A. Van Dorsselaer, P. Albrecht and G. Ourisson, *Bull. Soc. Chim. France,* 1977, 165). The structures of a C_{28} hopane (180) from Monterey shale (G.W. Smith, *Acta Crystallogr.,* 1979, 35B, 2173) and C_{27} (181) and C_{29} (182) hopanes from Nigerian crude oil (G.W. Smith, *Acta Crystallogr.,* 1975, 31B, 522; 526) have been established by X-ray analyses.

The unusual aromatic compounds, the benzohopanes (183), occur widely in sediments and petroleum (G. Hussler *et al., Tetrahedron Lett.,* 1984, 25, 1179) and the thiophene (184) is found in immature sediments (J. Valisolalao *et al., J. Chem. Soc., Chem. Commun.,* 1984, 1657). Other aromatic compounds from sediments include (185) (A.C. Greiner, C. Spyckerelle and P. Albrecht, *Tetrahedron,* 1976, 32, 257) and (186) (A.C. Greiner *et al., J. Chem. Res (S),* 1977, 334) and several arborinanes such as (187) (V. Hauke *et al., Tetrahedron,* 1992, 48, 3915).

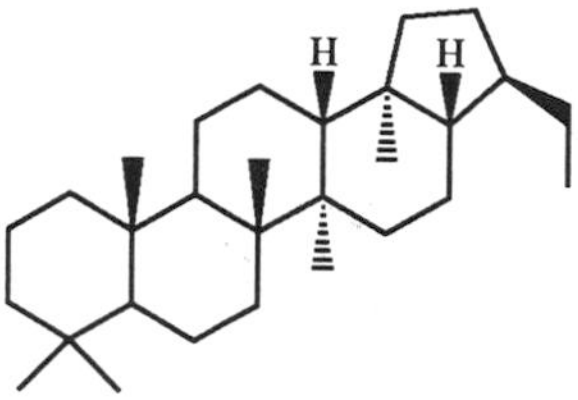

(179)

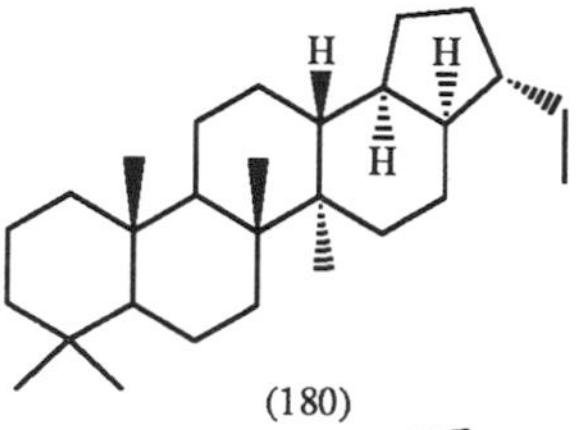

(180)

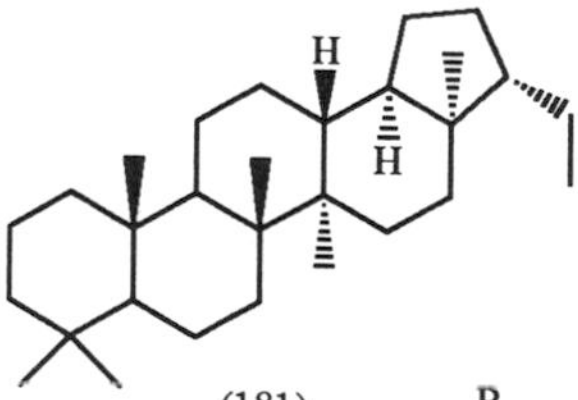

(181)

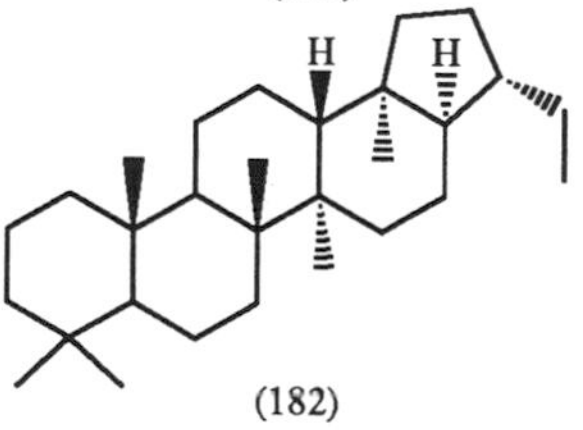

(182)

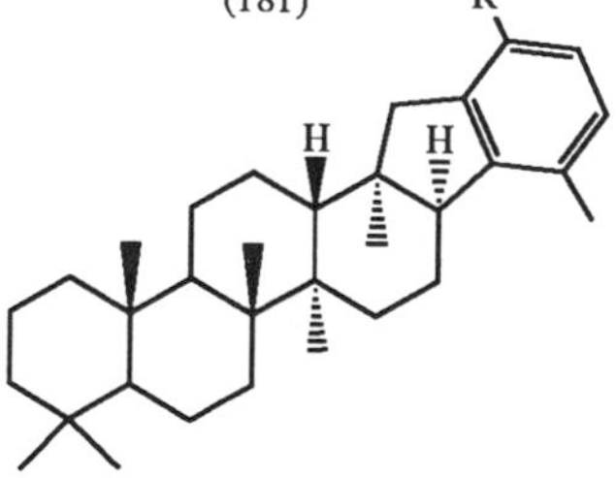

(183) R=H,Me,Et,Pr

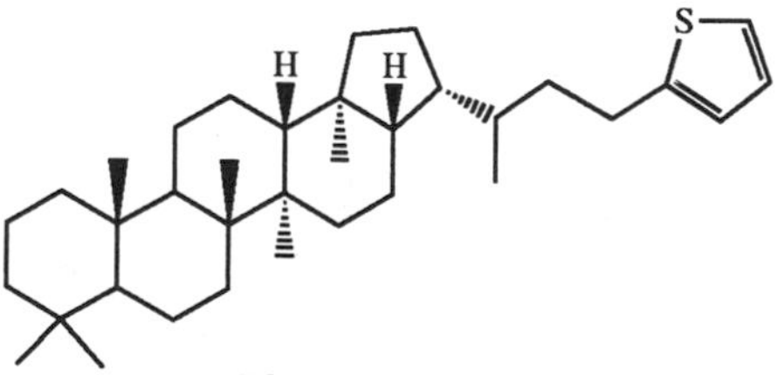

(184)

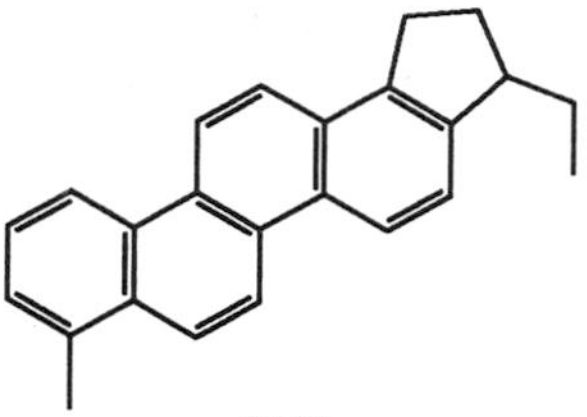

(185)

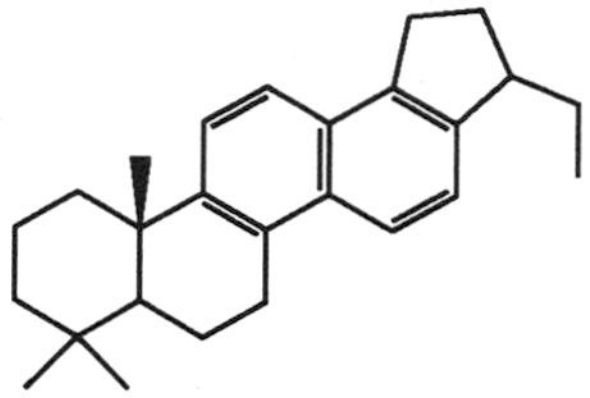

(186)

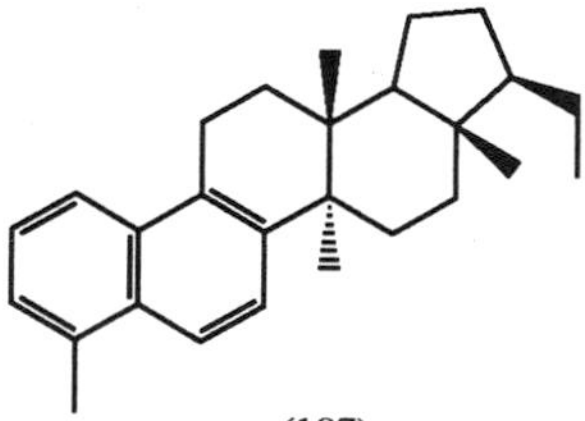

(187)

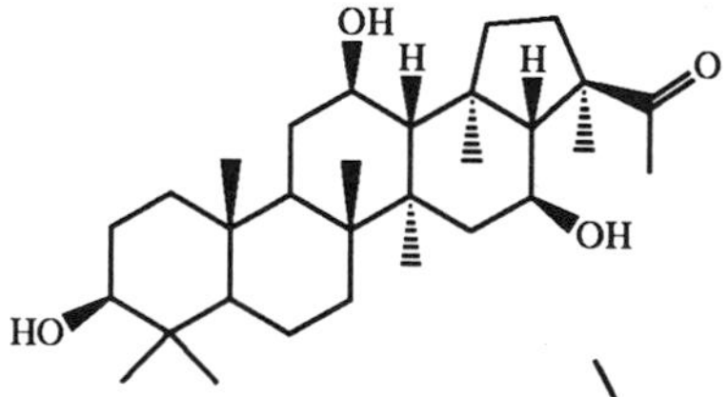

(188)

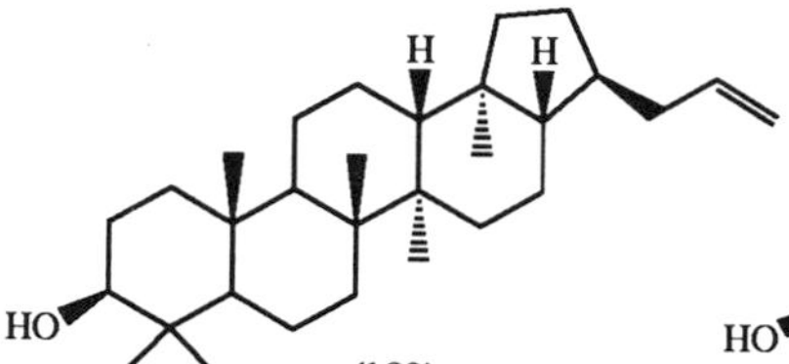

(189)

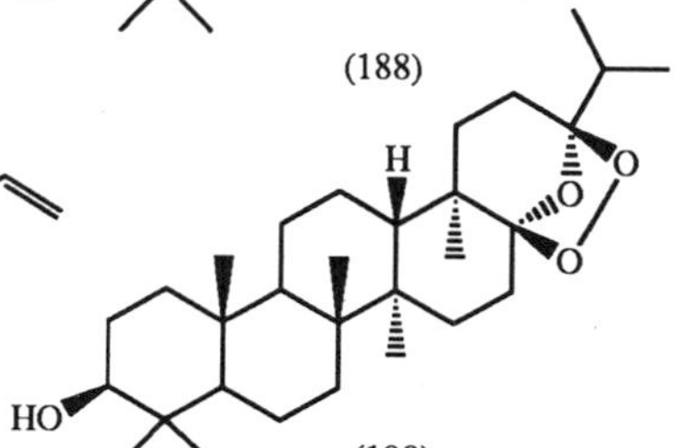

(190)

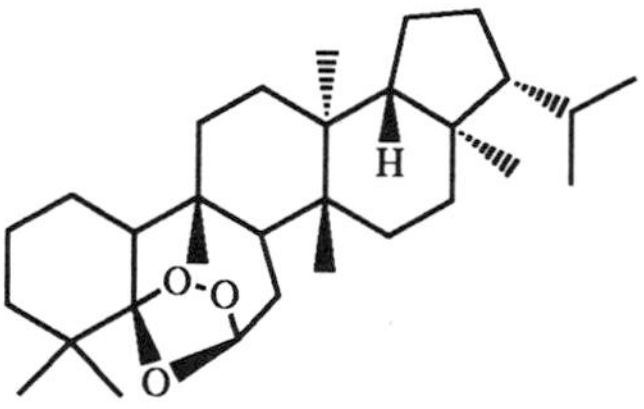

(191)

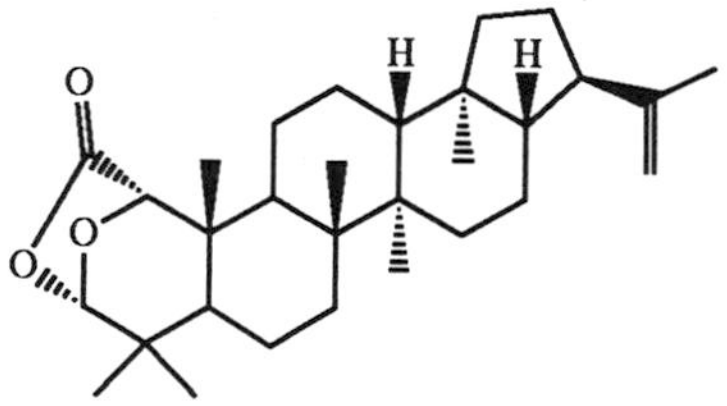

(192)

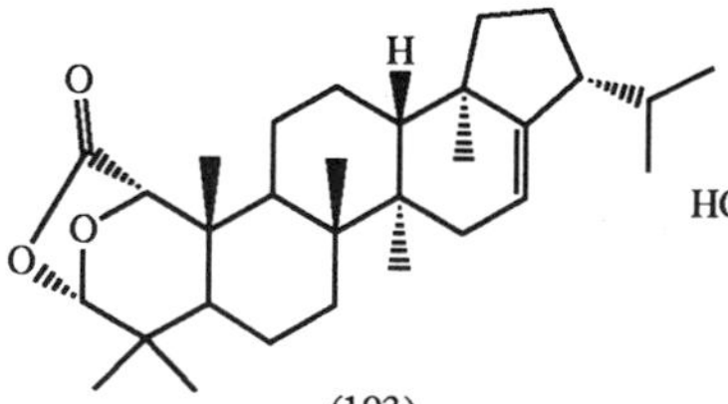

(193)

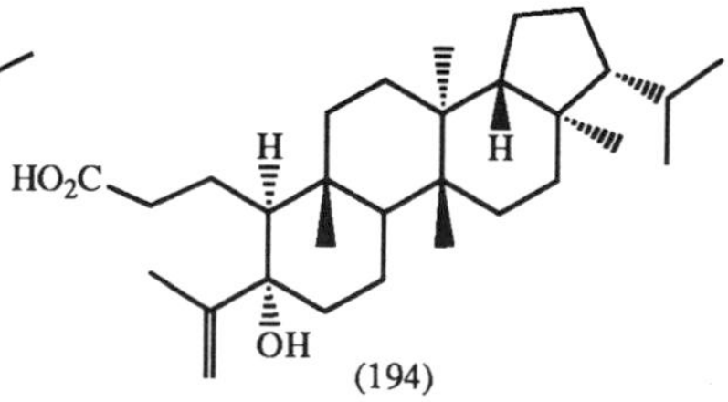

(194)

Other interesting hopanes include spergulagenin A **(188)**, from *Mollugo spergula* (I. Kitagawa *et al., Chem. Pharm. Bull.*, 1975, 23, 355), the structure of which was established by X-ray analysis), leptadenol **(189)** a rearranged 3β-hydroxyhopane from *Leptadenia pyrotechnica* (F. Noor *et al., Phytochemistry*, 1993, 32, 211) and gilvanol **(190)**, a naturally occurring ozonide of hop-17(21)-en-3β-ol, from *Quercus gilva* (H. Itokawa *et al., Chem. Pharm. Bull.*, 1978, 26, 331). The ozonide **(191)** of adian-5-ene has been isolated from *Adiantum monochlamys* (H. Ageta *et al., Tetrahedron Lett.*, 1978, 899). The structure of thysanolactone **(192)**, a novel ring-A secomoretane from *Thysanospermum diffusum*, was established by X-ray analysis (N. Aimi *et al., Tetrahedron*, 1981, 37, 983). A partial synthesis of thysanolactone **(192)** from hydroxyhopanone (N. Aimi, K. Kawada and S. Sakai, *Chem. Pharm. Bull.*, 1983, 31, 3765) and the ¹³C nmr assignments (N. Aimi *et al., Heterocycles*, 1985, 23, 1863) have been reported. Similar ring-A cleaved compounds such as swertialactone D **(193)** have been found in *Swertia petiolata* (S. Bhan *et al., Phytochemistry*, 1987, 26, 3363).

Several 3,4-secoadiananes have been isolated from *Euphorbia supina* (R. Tanaka *et al., Aust. J. Chem.*, 1989, 30, 1661). The structures of espinenediol A **(194)** and espinenoxide **(195)** were confirmed by X-ray analysis. Neospirosupinanetrione **(196)**, also from *E. supina*, may be the product of a pinacol rearrangement of an 8β,9β-dihydroxyfernane derivative (R. Tanaka and S. Matsunaga, *Phytochemistry*, 1991, 30, 293). Antibiotic WF 11605 **(197)**, from fungus strain F 11605, is a cleaved fernane glycoside (E. Tsujii *et al., J. Antibiot.* 1992, 45, 698; 704).

(195)   (196)

(197)   (198) R=O

(199) R=H,βOH

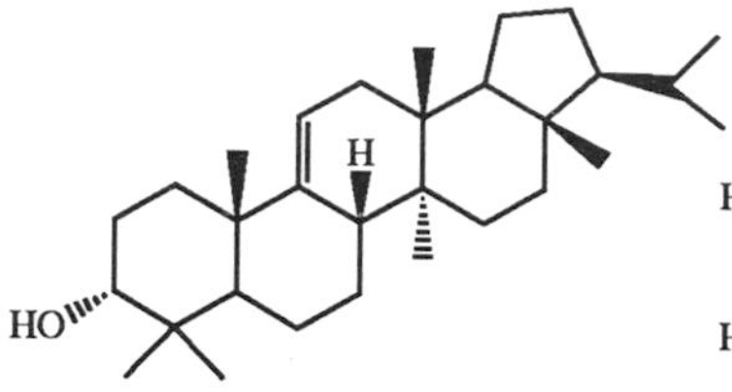

590

(200)

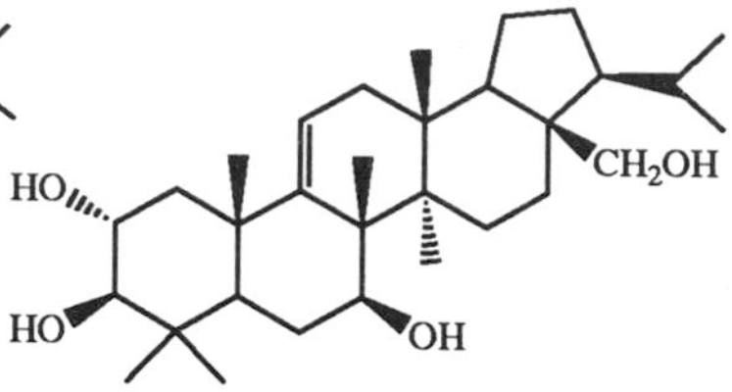

(201)

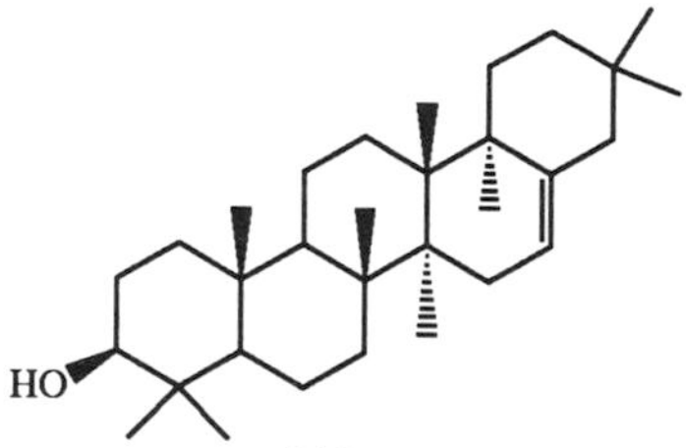

(202)

The carbon framework of boehmerone (198) and boehmerol (199) from the bark of *Boehmeria excelsa* is clearly related to that of arborinol (200). (H.C. Fernando *et al.*, *Tetrahedron*, 1989, 45, 5867). These structures are supported by detailed nmr analyses (Y. Tezuka *et al.*, *Phytochemistry*, 1993, 32, 1531). The rubiarborinols, for example rubiarborinol F (201), are arborinane derivatives from *Rubia* species (H. Itokawa, Y.-F. Qiao and K. Takeya, *Chem. Pharm. Bull.*, 1990, 38, 1435). Chiratenol (202) from *Swertia chirata* has a novel rearranged hopane skeleton (A.K. Chakravarty *et al.*, *Tetrahedron Lett.*, 1990, 31, 7649).

9. The Stictane Group

Triterpenoids with a stictane carbon skeleton, e.g. (203) (A.L. Wilkins, *Phytochemistry*, 1977, 16, 2031), occur in *Sticta*, *Pseudocyphellaria* and other species of lichen (W.J. Chin *et al.*, *J. Chem. Soc.*, *Perkin Trans. 1*, 1973, 1437; E.M. Goh, A.L. Wilkins, *J. Chem. Soc.*, *Perkin Trans. 1*, 1978, 1560). The skeleton can be derived by cyclisation of squalene in a chair-boat-chair-chair-boat conformation to give the classical ion (204) which is also the precursor of arborinol (200).

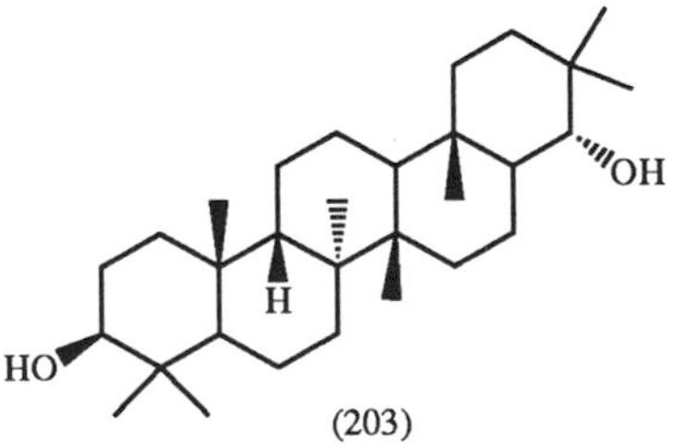

(203)

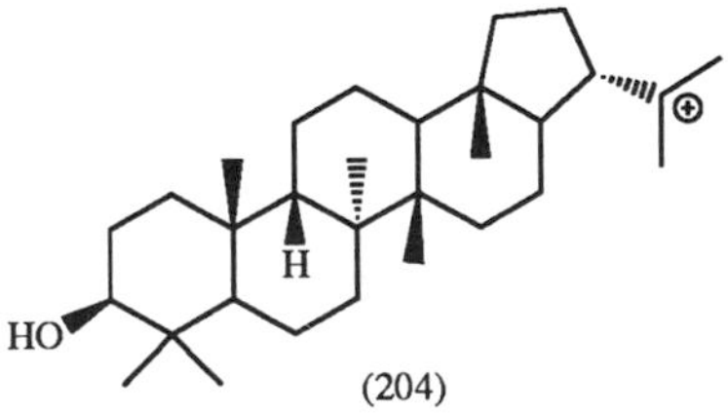

(204)

10. Miscellaneous

Achilleol A (**205**) has a new monocyclic carbon skeleton (A.F. Barrero, E.J. Alvarez-Manzaneda R. and R. Alvarez-Manzaneda R., *Tetrahedron Lett.*, 1989, 30, 3351) and occurs in *Achillea odorata* with the tricyclic achilleol B (**206**) (A.F. Barrero *et al.*, *Tetrahedron*, 1990, 46, 8161). Another achillane (**207**) has been found in *Polypoides formosana* (Y. Arai *et al.*, *Tetrahedron Lett.*, 1992, 33, 1325). The diol (**208**) from Pistachia resin has a bicyclic polypodane skeleton (R.B. Boar *et al.*, *J. Am. Chem. Soc.*, 1984, 106, 2476). Several examples of compounds with this skeleton have been reported (K. Shiojima *et al.*, *Tetrahedron Lett.*, 1983, 24, 5733; G.J. Bennett *et al.*, *Phytochemistry*, 1993, 32, 1245; Y. Arai *et al.*, *Tetrahedron Lett.*, 1992, 33, 1325). A total synthesis of (+)-ambrein (**209**) has been achieved (K. Mori and H. Tamura, *Liebigs Ann. Chem.* 1990, 361). A series of triterpenoid galactosides, for example pouoside A (**210**), with a novel carotenoid-type skelton has been isolated from an *Asteropus* species of marine sponge (M.B. Ksebati, F.J. Schmitz and S.P. Gunaskera, *J. Org. Chem.*, 1988, 53, 3917).

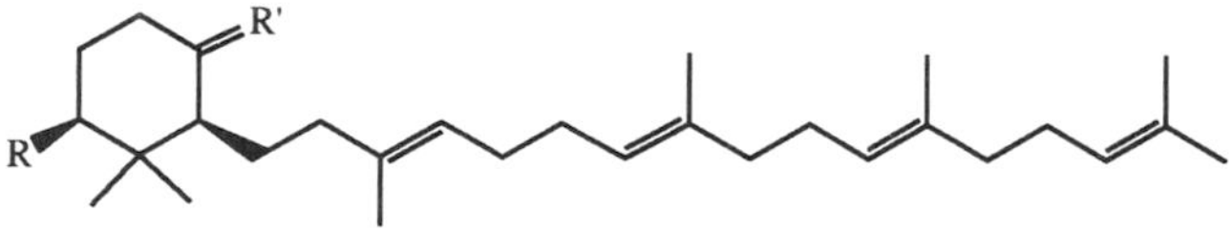

(**205**)   R=OH ; R'=CH$_2$

(**207**)   R=H ; R'=CH$_3$,αOH

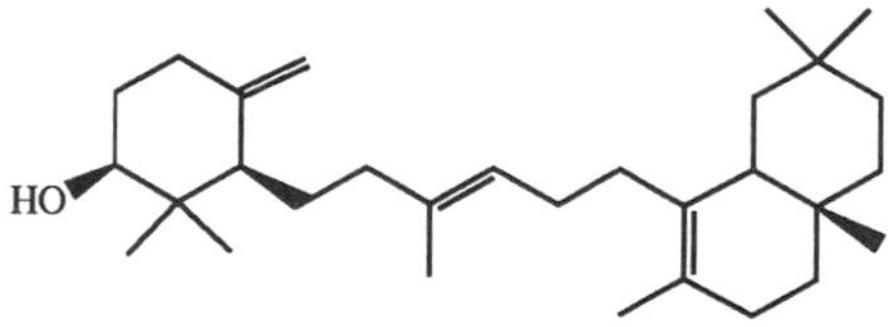

(**206**)

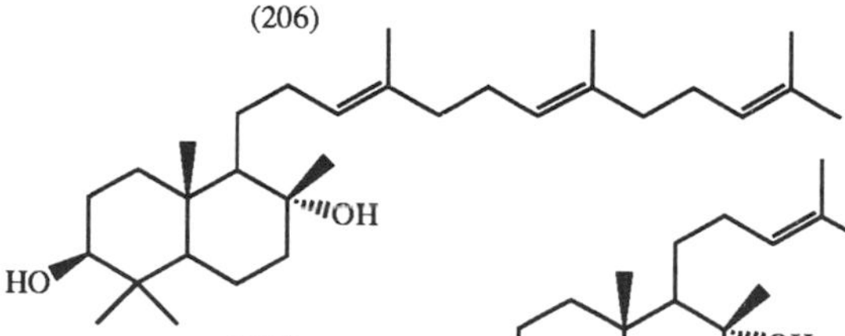

(**208**)

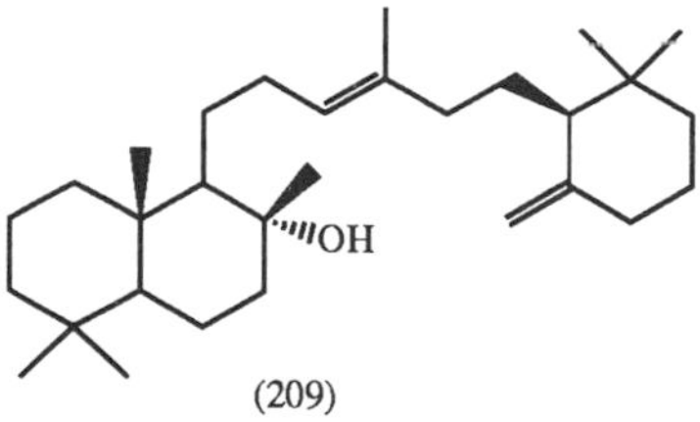

(**209**)

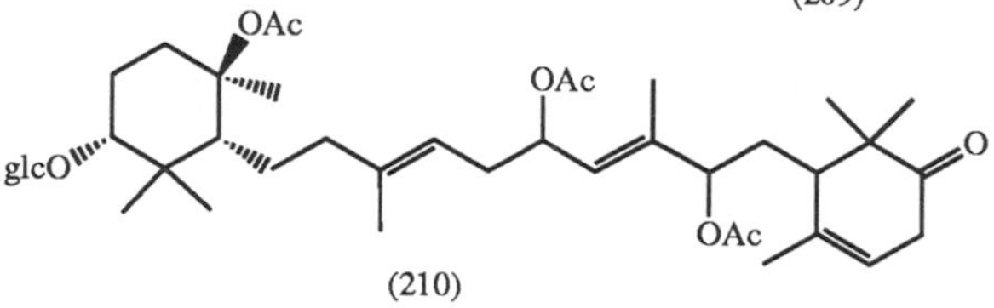

(**210**)

The structure of malabaricol (**211**) has been confirmed by X-ray analysis (W.F. Paton *et al., Tetrahedron Lett.*, 1979, 4153; *Cryst. Struct. Commun.*, 1979, 8, 481). The sponge *Jaspis stelifera* is a rich source of malabaricane derivatives (B.N. Ravi, R.J. Wells and K.D. Croft, *J. Org. Chem.*, 1981, 46, 1998; B.N. Ravi and R.J. Wells, *Aust. J. Chem.*, 1982, 35, 39). An X-ray analysis of the yellow pigment (**212**) from a sponge of the genus *Stellata* reveals it to have the 8αMe,9βH-isomalabaricane stereochemistry (T. McCabe *et al., Tetrahedron Lett.*, 1982, 23, 3307). Two migrated malabaricanes (**213**) and (**214**) have been found in *Polypodioides niponica* (Y. Arai, M. Hirohara and H. Ageta, *Tetrahedron Lett.*, 1989, 30, 7209). Limatulone, a potent defensive substance of the limpet *Collisella limatula*, has the unusual structure (**215**) (K.F. Albizati, J.R. Pawlik and D.J. Faulkner, *J. Org. Chem.*, 1985, 50, 3428). A similar but symmetrical structure has been assigned to naurol A (**216**) which occurs in a sponge together with the bisepimer diol, naurol B (**217**) (F.S. de Guzman and F.J. Schmitz, *J. Org. Chem.*, 1991, 56, 55).

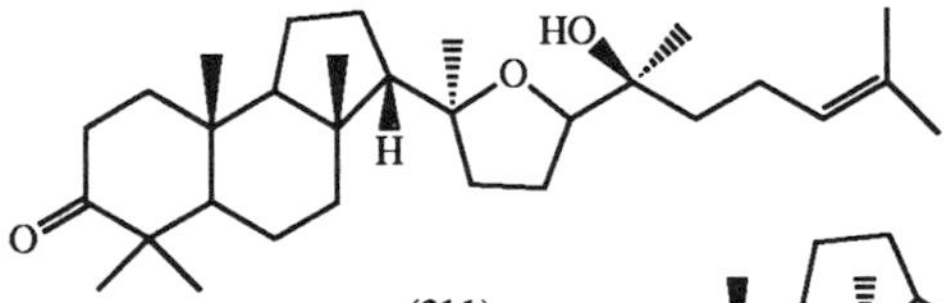

(211)

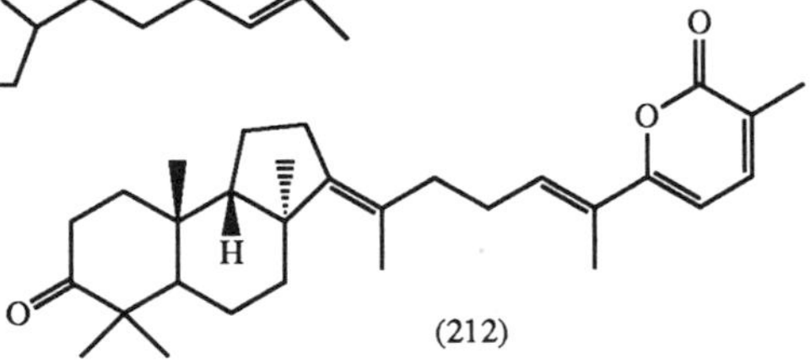

(212)

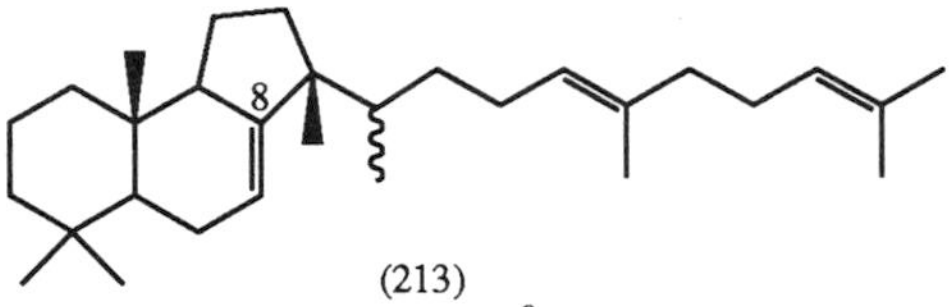

(213)
(214) Δ⁸

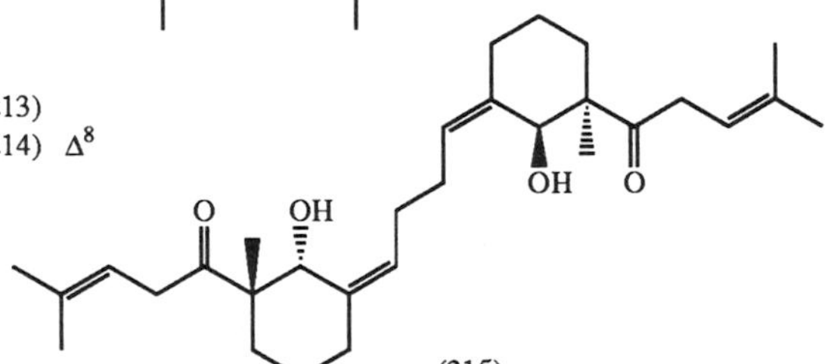

(215)

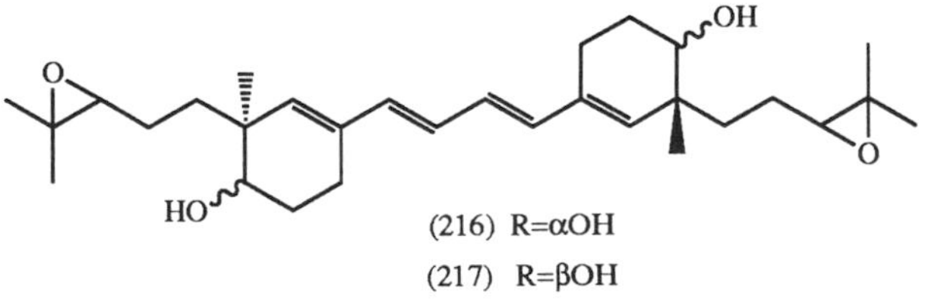

(216) R=αOH
(217) R=βOH

A fascinating group of modified triterpenoids, the iridals, is found in the in the roots and rhizomes of *Iris* species (L. Jaenicke and F.-J. Marner, *Prog. Chem. Org. Nat. Prod.*, 1986, 50, 1; *Pure Appl. Chem.*, 1990, 62, 1365). The iridals are the precursors of the irones, much valued perfumery compounds. Representative types are γ-irigermanal **(218)** from *Iris germanica*, the structure of which was confirmed by X-ray analysis (F.-J. Marner *et al.*, *J. Org. Chem.*, 1982, 40, 2531) and belamcandal **(219)** from *Belamcanda chinensis* and *I. japonica* (F. Abe, R.-F. Chen and T. Yamauchi, *Phytochemistry*, 1991, 30, 3379). The absolute configuration of the cycloiridals has been established (F.-J. Marner and L. Jaeniche, *Helv. Chim. Acta*, 1989, 72, 287).

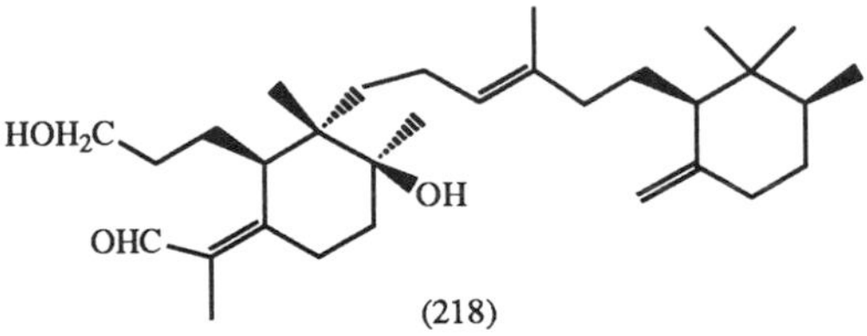

(218)

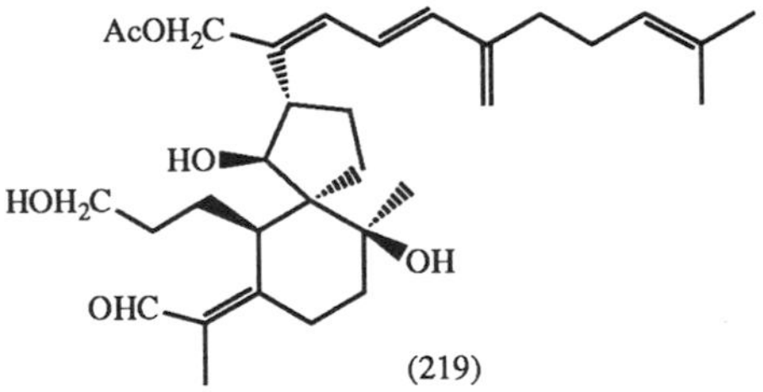

(219)

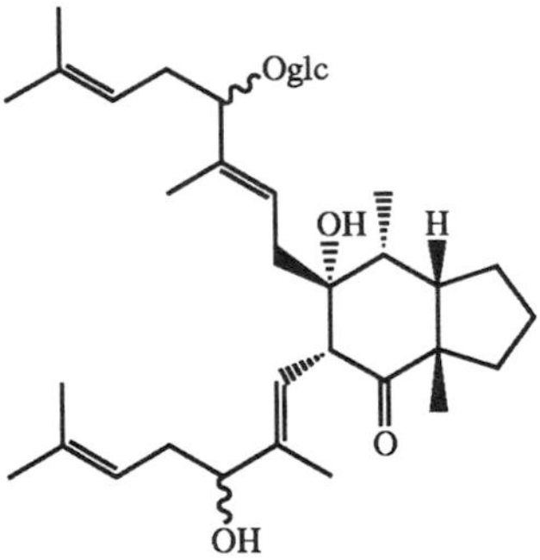

(220)

The sponge *Xestospongia vanilla* contains several unusual triterpenoid glycosides. Xestovanin **(220)** can be regarded as the parent system which can undergo a retro-aldol to give secoxestovanin A **(221)** (P.T. Northcote and R.J. Anderson, *J. Am. Chem. Soc.*, 1989, 111, 6276). A subsequent intramolecular aldol condensation converts secoxestovanin A **(221)** into isoxestovanin **(222)** (S.A. Morris, P.T. Northcote and R.J. Anderson, *Can. J. Chem.*, 1991, 69, 1352).

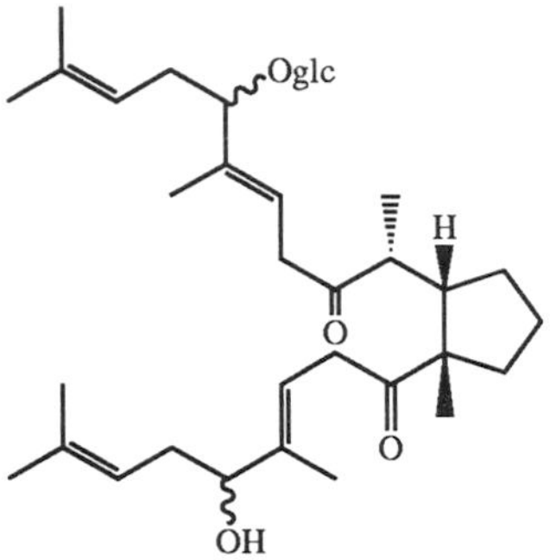

(221)

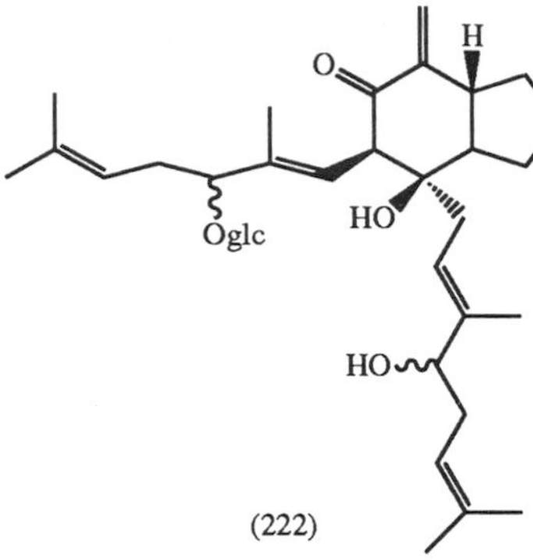

(222)

The sponge *Siphonochalina siphonella* is a rich source of intriguing triterpenoids. The most abundant group has a sipholane carbon skeleton (P. Sengupta *et al.*, *Indian. J. Chem., Sect. B*, 1983, 22, 882). The structure of sipholenol A (223) was established by X-ray analysis and its absolute configuration has recently been determined (I. Ohtani *et al.*, *J. Org. Chem.*, 1991, 56, 1296). A different folding of the acyclic precursor leads to raspacionin (224) from another sponge *Raspaciona aculeata* the structure of which was also determined by X-ray analysis (G. Cimino *et al.*, *Tetrahedron Lett.*, 1990, 31, 6565.). Raspacionin A (225), which co-occurs with raspacionin, has a partially migrated skeleton (G. Cimino *et al.*, *Tetrahedron*, 1992, 48, 9013).

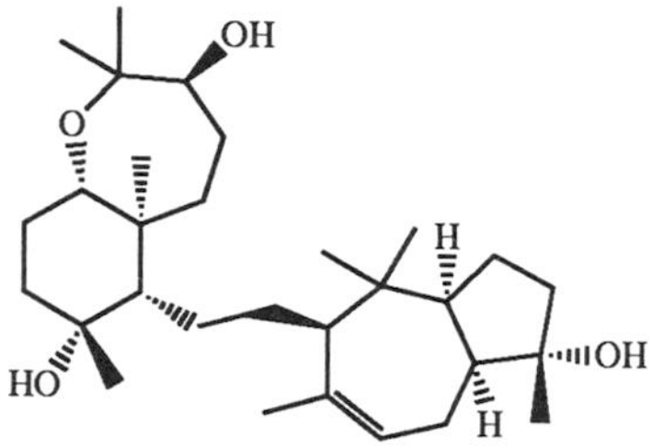

(223)

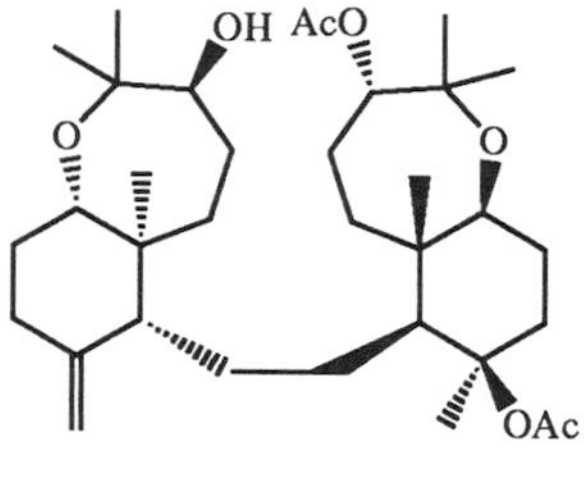

(224)

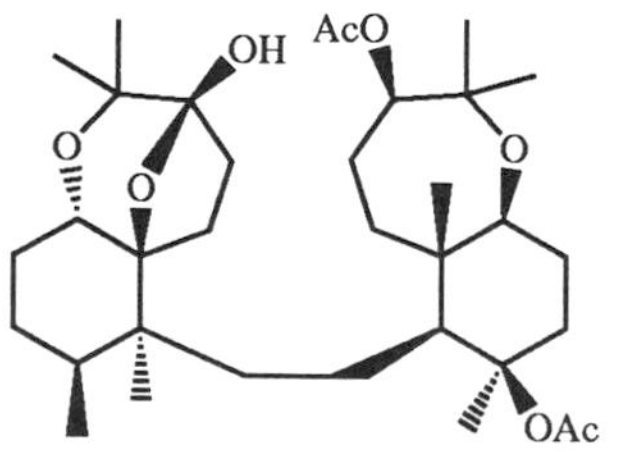

(225)

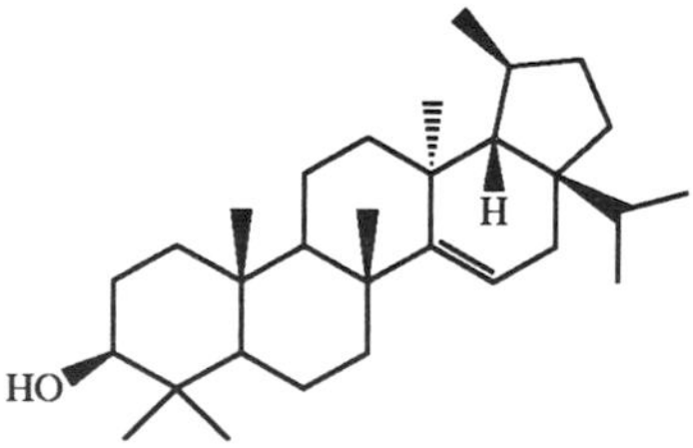

(226)

D-Friedimadeir-14-en-3β-ol   (**226**) and D:C-friedomadeir-7-en-3β-ol   (**227**), from *Euphorbia mellifera*, possess the novel madeirane carbon skeleton (M.-J.U. Ferreira *et al.*, *Helv. Chim. Acta*, 1991, 74, 1329). The biosynthesis of madeirane (as the C-18 cation) can be envisaged as proceeding via the spirodammarane C-24 cation (**228**) as shown. Subsequent backbone rearrangement leads to the friedomadeirane derivatives (**227**) and (**228**). Both structures were confirmed by X-ray analysis.

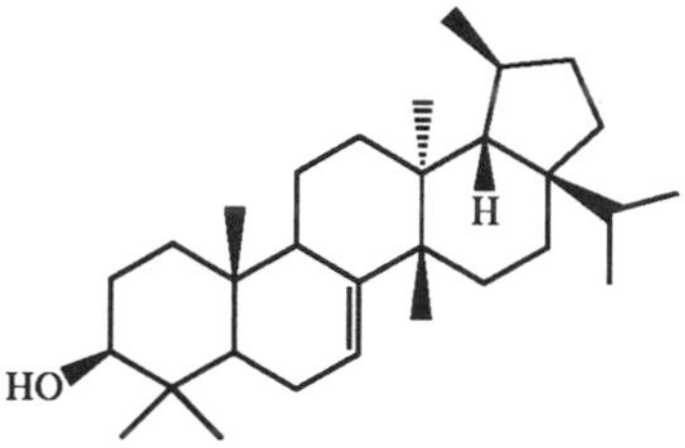

(227) 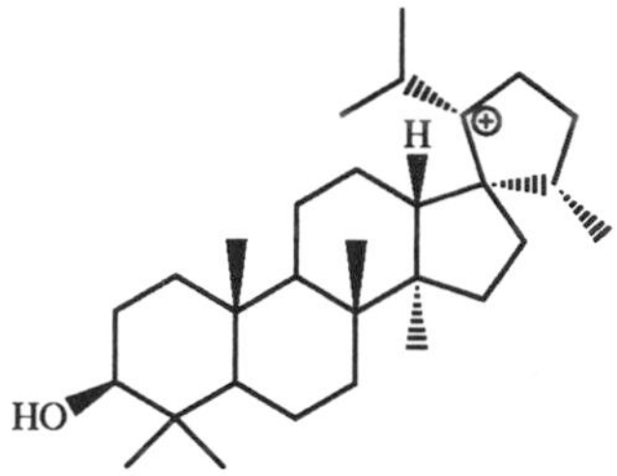   (228)

Several migrated gammacerane triterpenoids have recently been described. Swertanone (**229**) (A.K. Chakravarty, B. Das *et al.*, *J. Chem. Soc., Chem. Commun.*, 1989, 438) and derivatives have been found in *Swertia chirata* (A.K. Chakravarty *et al.*, *Indian J. Chem.*, 1992, 31B, 70). The carbon skeleton of swertanone has been named swertane. The roots of *Picris hieracioides* contain 3β-acetoxyswert-9(11)-ene (pichrenyl acetate) and the corresponding Δ^8-isomer (isopichierenyl acetate) (K. Shijima *et al.*, *Tetrahedron Lett.*, 1989, 30, 6873) together with their probable biogenetic precursors, the Δ^{16}-gammacerane derivatives (**230**), (**231**) and (**232**) (K. Shijima *et al.*, *Tetrahedron Lett.*, 1989, 30, 4977). Acid treatment of the first of these afforded, as expected, isopichierenyl acetate. The details of the biosynthesis of tetrahymanol have been reviewed (E. Caspi, *Acc. Chem. Res.*, 1980, 13, 97; P. Bouvier *et al.*, *Eur. J. Biochem.*, 1980, 112, 549).

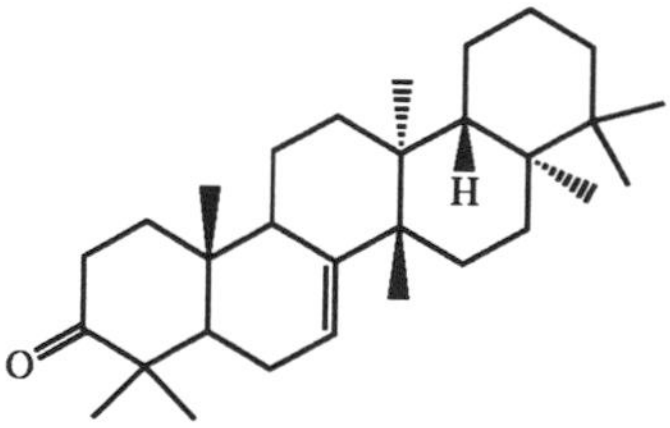

(229)

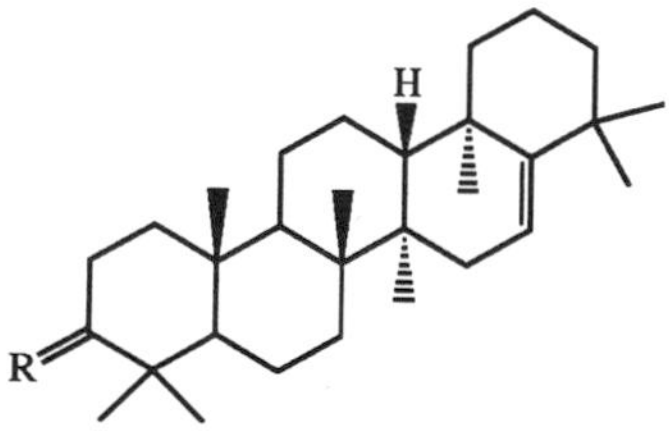

(230) R=H,βOAc
(231) R=H,βOH
(232) R=H,αOH

14β,26-Epoxyserratane-3β,21α-diol **(233)**, from *Primula rosea,* is a recent addition to the serratane series (R. Kapoor *et al., Planta Med.,* 1985, 334). The full details of some of the early chemistry of this series have been published (Y. Tsuda *et al., Chem. Pharm. Bull.,* 1974, <u>22</u>, 2383). The lansiosides are onocerane glycosides, based on lansiolic acid **(234),** which co-occur in the peel of *Lansium domesticum* (M. Nishizawa *et al., J. Org. Chem.,* 1983, <u>48</u>, 4462).

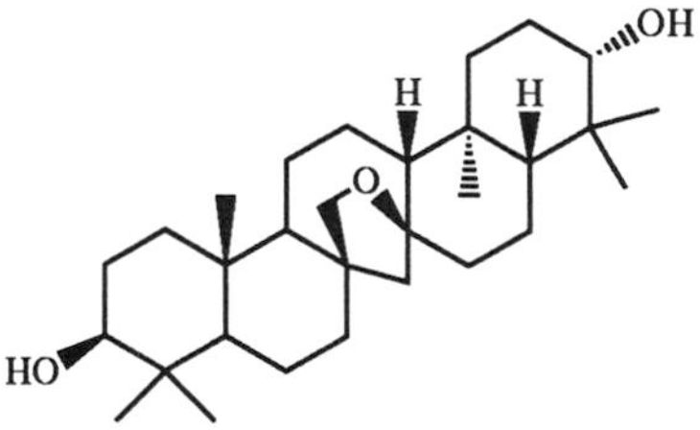

(233) 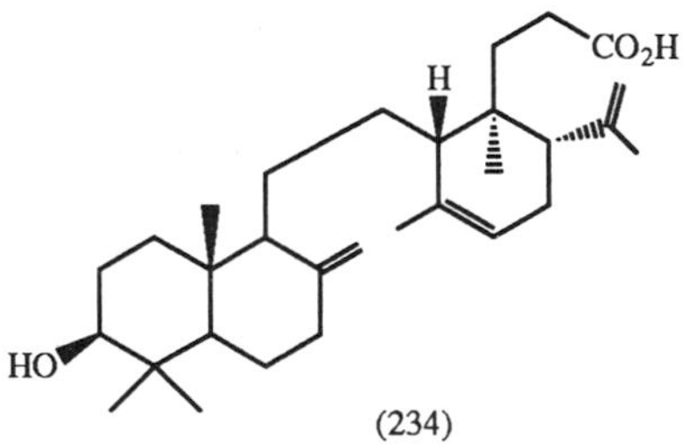   (234)

Guide to the Index

This index is constructed in a similar manner to the volume indexes of the first edition of the Chemistry of Carbon Compounds. However, to make the index easier to use, more descriptive entries have been made for the commonly occurring individual, and groups of chemicals.

The indexes cover primarily the chemical compounds mentioned in the text, and also include reactions and techniques, where named, and some sources of chemical compounds such as plant and animal species, oils, etc.

Chemical compounds have been indexed alphabetically under the names used by authors, editing being restricted to ensuring uniformity of entries under the same heading. In view of the alternative nomenclature that can often be used, a limited amount of cross-referencing has been done where it is considered to be helpful, but attention is particularly drawn to Convention 2 below.

For this and the succeeding volumes, the indexing conventions listed below have been adopted.

1. Alphabetisation

(a) A letter by letter alphabetical sequence is followed for entries, firstly for the main entry, followed by the descriptive entry.

(b) The following prefixes have not been counted for alphabetising:

n-	*o-*	*as-*	*meso-*	*C-*	*E-*
	m-	*sym-*	*cis-*	*O-*	*Z-*
	p-	*gem-*	*trans-*	*N-*	
	vic-			*S-*	
		lin-		*Bz-*	
				Py-	

Some prefixes and numbering have been omitted in the index, where they do not usefully contribute to the reference.

(c) The following prefixes have been alphabetised:

Allo	Epi	Neo
Anti	Hetero	Nor
Bis	Homo	Pseudo
Cyclo	Iso	

2. Cross references

In view of the many alternative trivial and systematic names for chemical compounds, the indexes should be searched under any alternative names which may be indicated in the main body of the text. Only a limited amount of cross-referencing has been carried out, where it is considered that it would be helpful to the user.

3. Derivatives

Simple derivatives are not normally indexed if they follow in the same short section of the text.

4. Collective and plural entries

In place of "– derivatives" the plural entry has normally been used. Plural entries have occasionally been used where compounds of the same name but differing numbering appear in the same section of the text.

5. Main entries

The main entry of the more common individual compounds is indicated by heavy type. Multiple entries, such as headings and sub-headings over several pages are shown by "–", e.g., 67–74, 137–139, etc.

Index

600

622